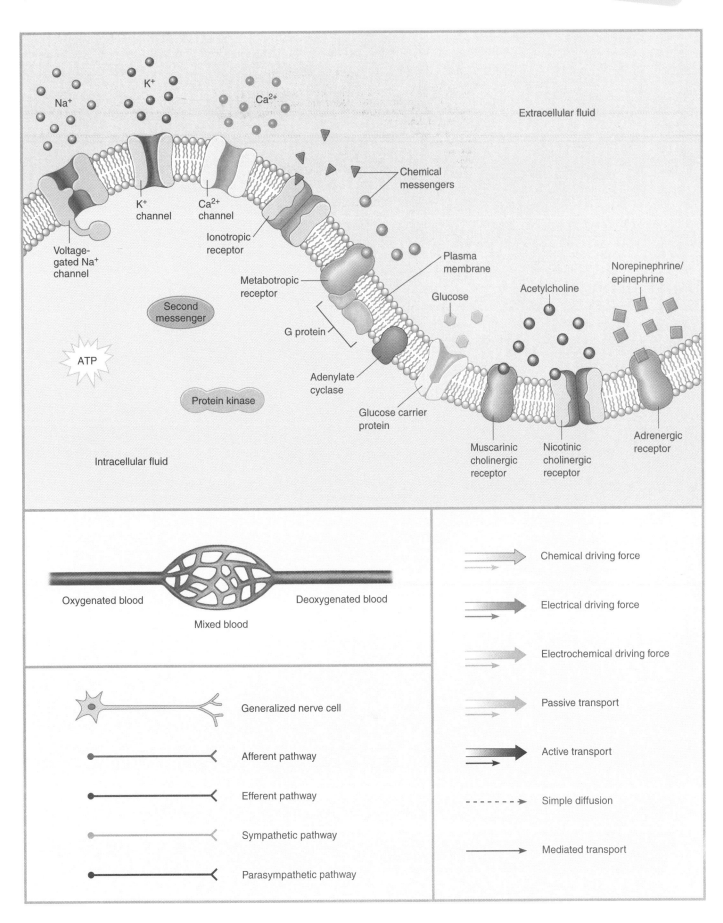

Extracellular fluid

Na⁺ K⁺ Ca²⁺

Chemical messengers

K⁺ channel

Ca²⁺ channel

Voltage-gated Na⁺ channel

Ionotropic receptor

Metabotropic receptor

Second messenger

G protein

ATP

Adenylate cyclase

Protein kinase

Glucose carrier protein

Plasma membrane

Glucose

Acetylcholine

Norepinephrine/epinephrine

Muscarinic cholinergic receptor

Nicotinic cholinergic receptor

Adrenergic receptor

Intracellular fluid

Oxygenated blood

Deoxygenated blood

Mixed blood

Generalized nerve cell

Afferent pathway

Efferent pathway

Sympathetic pathway

Parasympathetic pathway

Chemical driving force

Electrical driving force

Electrochemical driving force

Passive transport

Active transport

Simple diffusion

Mediated transport

PRINCIPLES OF
HUMAN PHYSIOLOGY

SECOND EDITION

WILLIAM J. GERMANN
University of Dallas

CINDY L. STANFIELD
University of South Alabama

CONTRIBUTORS

Mary Jane Niles
University of San Francisco
Chapter 23

Joseph G. Cannon
Medical College of Georgia
Chapter 24

PEARSON

Benjamin
Cummings

San Francisco   Boston   New York
Cape Town   Hong Kong   London   Madrid   Mexico City
Montreal   Munich   Paris   Singapore   Sydney   Tokyo   Toronto

Publisher *Daryl Fox*
Executive Editor *Serina Beauparlant*
Development Manager *Claire Alexander*
Sr. Project Editor *Barbara Yien*
Art Development Editor *Laura Southworth*
Managing Editor *Wendy Earl*
Production Editor *Michele Mangelli*
Art and Photo Coordinator *Bettina Borer*
Text Designer *tani hasegawa*
Copyeditor *Anita Wagner*
Photo Researcher *Kathleen Olson*
Supplements Project Editor *Ryan Shaw*
Supplements Production Editor *Leslie Austin*
Cover Image *Tomo Narashima*
Cover Design *Lillian Carr*
Manufacturing Buyer *Stacey Weinberger*
Executive Marketing Manager *Lauren Harp*
Compositor *GTS Graphics*

Photography and illustration credits appear on page 780.

Library of Congress Cataloging-in-Publication Data
Germann, William J.
    Principles of human physiology / William J. Germann, Cindy L. Stanfield;
contributors
    Joseph G. Cannon, Mary Jane Niles.—2nd ed.
        p. cm.
    Includes index.
    ISBN 0-8053-5691-6
    1. Human physiology. I. Stanfield, Cindy L. II. Title.

QP34.5.G465 2005
612–dc22   2003064674

ISBN 0-8053-5691-6
1 2 3 4 5 6 7 8 9 10—QWV—08 07 06 05 04
www.aw-bc.com

About the Authors

William J. Germann earned a Ph.D. in physiology at the University of Michigan. One of Germann's teachers at Michigan was Arthur Vander, whose commitment to education made a lasting impression and ultimately inspired him to pursue a career in undergraduate teaching. Germann currently teaches human physiology and a number of other biology courses at the University of Dallas in Irving, Texas. He believes that a teacher's job is not just to communicate information, but to inspire students to want to learn. As chair of the Biology Department, Germann spends much of his time advising premedical students and others interested in pursuing careers in healthcare, research, or teaching. Germann's research interest is in the field of electrophysiology and ion channel regulation in epithelial cells. He is a member of the Biophysical Society, the American Association of University Professors, the Human Anatomy and Physiology Society, and the Texas Association of Advisors for Health Professions.

Cindy L. Stanfield earned both a B.S. and a Ph.D in physiology at the University of California at Davis. She was exposed to and became fascinated by neurophysiology research as an undergraduate student. As a graduate student, Stanfield taught several physiology laboratory courses, and developed an interest in teaching. She currently teaches three undergraduate human physiology courses at the University of South Alabama, and continues with her neurophysiology research on sensory modulation. One of Stanfield's goals as a teacher is to expose students not only to the existing knowledge base, but also to the discovery of new knowledge through research experiences in the hope that students will be excited about science. Stanfield is a member of the Society for Neuroscience, the International Association for the Study of Pain, the American Pain Society, Sigma Xi, and the Golden Key Honor Society. She lives in Mobile, Alabama, with her husband, Jim, and their cats and dog.

Preface

In setting out to write the second edition of this text, we kept in mind our primary goal from the first edition: to create a text that makes it as easy as possible for students to learn human physiology while also providing a solid, comprehensive overview of the field. It remains our belief that a physiology textbook should emphasize the understanding of concepts over mere memorization of facts, as well as provide tools for students with varied levels of preparation in biology, chemistry, physics, and related sciences to aid them in their individual studies. In this new edition, we have made a number of enhancements with these goals in mind, while retaining the book's proven hallmarks: a clear and precise writing style; an exceptionally well developed art program; and pedagogical features designed to stimulate student interest, help them think about physiological processes in an integrated way, and provide reinforcement of the most important concepts.

New to the Second Edition

We have listened to the feedback provided by dozens of reviewers and implemented a number of organizational changes to the text, summarized as follows:

- Chemical messengers and endocrine glands are now treated in two discrete chapters (Chapter 5, Chemical Messengers; Chapter 6, The Endocrine System: Endocrine Glands and Hormone Action). In addition, the chapter on regulation of energy metabolism and growth (formerly Chapter 20) has now been moved up to Chapter 7 so that all of the endocrine topics are consolidated in the table of contents.

- The chapters on the cardiovascular system (now Chapters 14–16) have been reorganized so that Chapter 14 covers cardiac function, Chapter 15 covers vasculature, and Chapter 16 covers blood.

We have also added a number of new elements to this edition of the text. Every chapter now begins with an engaging **chapter-opening vignette** designed to draw the student into the reading. We have added a number of **new *Discovery* and *Clinical Connections* boxes** on topics such as jet lag, creatine supplements, and osteoporosis. **New photos and/or illustrations** have been added to the vast majority of the boxes. A new feature called ***Chemistry Review*** is designed to aid students who lack a strong background in chemistry so that they can better understand related topics in physiology. We have created **30 new figures** to further clarify physiological topics and concepts. The end-of-chapter pedagogy has been enhanced to provide **more answer options for the multiple-choice questions** and a new section consisting of **critical thinking questions.** For a closer look at these enhancements and at this text's other features, please turn to p. ix.

In addition, each chapter was thoroughly revised to make the text even more useful in the classroom. The following is a summary of substantive changes to each chapter:

Chapter 1: Introduction to Physiology

- New topic: endocrine and exocrine glands
- New table: the ten organ systems
- New Discovery box: Scientific Method and Physiology

Chapter 2: The Cell: Structure and Function

- Section on transport in membrane-bound vesicles moved to Chapter 4

Chapter 3: Cell Metabolism

- New Toolbox: Ligand-Protein Interactions
- New Discovery box: Can You Lose Weight While Sitting Still?
- New Clinical Connections box: Oxidative Stress and Disease
- Moved Clinical Connections box on lactose intolerance to Chapter 21
- Separated the conversion of pyruvate to acetyl CoA (linking step in first edition) from the Krebs cycle to

emphasize that acetyl CoA can come from many sources

Chapter 4: Cell Membrane Transport

- New Toolbox: Energy of Solutions
- Topics have been reorganized for better understanding
- Description of ion channels moved to its own sub-heading under passive transport

Chapter 5: Chemical Messengers

- Section on endocrine system has been moved to Chapter 6
- New Exercise Link on epinephrine

Chapter 6: The Endocrine System: Endocrine Glands and Hormone Actions

- New Clinical Connections box: Pituitary Adenomas
- New Discovery box: Circadian Rhythms and Jet Lag

Chapter 7: The Endocrine System: Regulation of Energy Metabolism and Growth

- Now follows Chapter 6 on endocrine glands
- Deleted review of absorption across gastrointestinal tract

Chapter 8: Nerve Cells and Electrical Signaling

- New Toolbox: Electrical Circuits in Biology
- New figure showing dissipation of current as it spreads through neuron
- New figure to illustrate refractory period
- Added enteric nervous system to general description of nervous system
- Simplified section on summation to be more general instead of synapse-specific

Chapter 9: Synaptic Transmission and Neural Integration

- Enhanced discussion of electrical synapses

Chapter 10: The Nervous System: Central Nervous System

- New figure of glial cells
- Enhanced description of glial cells, including their possible role in neurodegenerative diseases
- Updated descriptions of communication disorders and brain lateralization
- Enhanced description of sleep

Chapter 11: The Nervous System: Sensory Systems

- New Clinical Connections box: Phantom Limb Pain
- New figure demonstrating the visual blind spot

Chapter 12: The Nervous System: Autonomic and Motor Systems

- Clarified presence of afferent fibers in autonomic nerves

Chapter 13: Muscle Physiology

- New Discovery box: Creatine Supplements
- New table comparing skeletal, smooth, and cardiac muscle
- New figure showing energy metabolism pathways in skeletal muscle
- New figure showing the role of DHP and ryanodine receptors in excitation-contraction coupling
- Crossbridge cycle renumbered to start with binding of myosin to actin in presence of calcium
- Description of summation and tetanus enhanced
- Section on physics of skeletal muscle contraction moved into a Toolbox

Chapter 14: The Cardiovascular System: Cardiac Function

- The chapters on the cardiovascular system (now Chapters 14–16) have been reorganized so that Chapter 14 covers cardiac function, Chapter 15 covers vasculature, and Chapter 16 covers blood.
- New Discovery box: Coronary Arteries
- New tables on ion channels involved in pacemaker potentials and ventricular contractile cell action potentials
- Moved description of blood flow through cardiovascular system to beginning of chapter
- Added description of portal circulations to section on parallel flow
- Enhanced description of excitation-contraction coupling in cardiac muscle, including a new figure
- Moved all sections on autonomic control of the heart to the section on regulation of cardiac output at the end of the chapter
- Added description of cellular effects of autonomic nervous system on cardiac contractile and autorhythmic cells

Chapter 15: The Cardiovascular System: Blood Vessels, Blood Flow, and Blood Pressure

- New Discovery box: Measuring Arterial Blood Pressure
- Physical laws of blood flow and blood pressure moved to beginning of chapter
- Topics on one type of blood vessel grouped together
- Added description of metarterioles and precapillary sphincters in regulating blood flow through capillary beds
- Introduced concept of cardiovascular control center in medulla oblongata

Chapter 16: The Cardiovascular System: Blood

- New chapter on blood
- New Discovery box: Leeches and Bloodletting
- Clinical Connections box on anemia moved here from respiration chapter
- Enhanced description of red blood cell life cycle, including new art
- Added description of hematopoiesis, including new art
- Expanded table on components of blood

Chapter 17: The Respiratory System: Pulmonary Ventilation

- Renamed chapter to better suit contents

Chapter 18: The Respiratory System: Gas Exchange and Regulation of Breathing

- New Clinical Connections box: The Bends

Chapter 19: The Urinary System: Renal Function

- Moved description of sodium reabsorption to Chapter 20
- Figures on excretion and clearance revised to simplify and match text calculations

Chapter 20: The Urinary System: Fluid and Electrolyte Balance

- New Clinical Connections box: Osteoporosis

Chapter 21: The Gastrointestinal System

- Enhanced description of vitamin, water, and mineral absorption
- Enhanced description of gastrointestinal motility

Chapter 22: The Reproductive System

- New table comparing Sertoli cells to granulosa cells
- New table on relationships between stage of ovum development and stage of follicle development
- Updated Discovery box: Birth Control Methods
- Added information on screening for prostate cancer

Chapter 23: The Immune System

- Added illustration of gene therapy methods to Discovery box: Gene Therapy for Severe Combined Immunodeficiency Disease

Chapter 24: The Whole Body: Integrated Physiological Response to Exercise

- Revised terminology as needed to match rest of text

The following pages provide a visual walk-through of this text's features and supplements.

Stimulating Student Interest

Instructors across the country have confirmed what we have observed in our classrooms: Students are more motivated to learn when they understand how a topic relates to real life. To capitalize on this interest, we discuss topics of clinical and/or everyday relevance within the text where appropriate, as well as in **new chapter-opening vignettes (Figure 1), Clinical Connections** boxes **(Figure 2)**, and **Discovery** boxes **(Figure 3)**. In addition, photos and/or illustrations have been added to the majority of the boxed features to lend added visual interest.

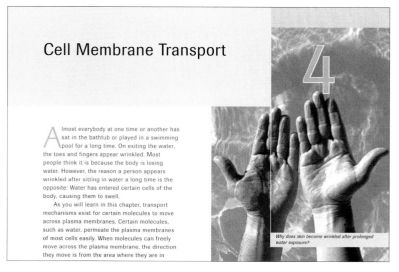

Cell Membrane Transport

4

Almost everybody at one time or another has sat in the bathtub or played in a swimming pool for a long time. On exiting the water, the toes and fingers appear wrinkled. Most people think it is because the body is losing water. However, the reason a person appears wrinkled after sitting in water a long time is the opposite: Water has entered certain cells of the body, causing them to swell.

As you will learn in this chapter, transport mechanisms exist for certain molecules to move across plasma membranes. Certain molecules, such as water, permeate the plasma membranes of most cells easily. When molecules can freely move across the plasma membrane, the direction they move is from the area where they are in

Why does skin become wrinkled after prolonged water exposure?

Figure 1 New chapter-opening vignettes are designed to engage students' interest with eye-catching photos and examples of how physiological topics relate to real life.

Figure 2 Clinical Connections boxes focus on clinical topics of interest. Topics new to this edition include osteoporosis, "the bends," phantom limb pain, and oxidative stress.

CLINICAL CONNECTIONS

OSTEOPOROSIS

Osteoporosis includes a group of disorders that result in a decrease in bone mass, which leads to bone fragility. Some forms of osteoporosis are localized to a specific bone. For example, immobilization of an extremity for a long period results in *disuse osteoporosis*. Most forms of osteoporosis, however, are more widespread and affect bones throughout the body. These widespread forms of osteoporosis can be *primary osteoporosis*—the osteoporosis is the disorder—or they can be *secondary osteoporosis*, in which the osteoporosis is secondary to some other [dis]eases can result in [de]f osteoporosis. For [ving] endocrine dis-[osteoporosis: hyper]perthyroidism, diabetes [leg]ally. Several gastroin-[may] also cause osteo-[malnutrition, malab-][t] or D deficiency, or [.] Prepubertal dietary [es]pecially deleterious. [increase the likelihood][of] osteoporosis include a [tion, obesity, and][al] activity. [develops with age, as][n changes in the rates]

of bone formation and resorption. Bone formation is carried out by cells called *osteoblasts*, whereas bone resorption is carried out by cells called *osteoclasts*. Bone formation exceeds bone resorption until the early 30s, when bone reaches its maximum density. After the 30s, bone density decreases about 0.7% per year. Certain bones degenerate faster, including the vertebrae, pelvis, and femur.

Certain hormones may be important in the development of osteoporosis. Estrogen has received considerable attention in this regard, since postmenopausal women are more subject to the development of osteoporosis. A decrease in estrogen causes osteoblasts to increase production of a paracrine called *interleukin-6*. Interleukin-6 stimulates osteoclast activity, causing bone resorption. Estrogen may also stimulate osteoblast activity, so a decrease in estrogen would result in a decrease in osteoblast activity and thus less bone formation. More recently, testosterone levels have been linked to the development of osteoporosis in men. Testosterone levels tend to be low in males with osteoporosis. However, whether the low testosterone levels contribute to the development of osteoporosis is still unknown.

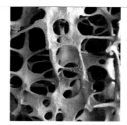

Scanning electron micrograph of osteoporosis from an 89 year old female.

Diagnosis of osteoporosis is difficult during the early stages because bone degeneration must reach a level observable by X rays. Often, diagnosis is not made until after a bone fracture has resulted from the weakened state.

Treatment of osteoporosis often includes hormonal or dietary supplements. Estrogen supplements are commonly given to postmenopausal women to decrease the rate of bone resorption. However, the estrogen cannot reverse any damage that has already occurred. Calcitonin supplements also appear to help prevent bone resorption. Calcium and fluoride supplements are also used.

DISCOVERY

CIRCADIAN RHYTHMS AND JET LAG

The invention of air travel had an unexpected consequence: rapid travel from one time zone to another results in a body imbalance commonly called "jet lag." Jet lag occurs because we have an internal clock that sets a *circadian rhythm* for many body functions. A circadian rhythm follows a cycle that is approximately 24 hours (*circa* means "almost" and *dies* means "day"). If a person is deprived of light, then the circadian rhythm is approximately 25 hours; this is the innate rhythm, or the rhythm that exists in the body independent of external influences. The actual circadian rhythm is 24 hours because the presence of the normal light-dark variation that occurs in 24-hour cycles causes the circadian rhythm to reset each day. In fact, approximately 60–70% of totally blind people have a 25-hour cycle, indicating that light input to the eyes has a strong influence on circadian rhythms. That a significant number of blind people have a 24-hour cycle suggests that the detectors of light for vision and for the circadian rhythm are distinct.

Because the light-dark cycle influences our circadian rhythm, alterations in

light, such as occurs when one travels from one time zone to another, can change the rhythm. Jet lag is a common term used to describe the physiological changes occurring in a traveler who has crossed time zones. Jet lag lasts for several days, and the duration increases as the number of time zones crossed increases. Because the circadian rhythm drives many physiological processes, including sleep, alterations in the rhythm can be disturbing to the individual.

Several interventions help travelers adjust to jet lag. Because a change in the

light cycle is what causes the lag, artificial bright light can be used to adapt more rapidly to a new time zone. Artificial light at night benefits those traveling westward, whereas artificial light in the morning benefits those traveling eastward. Medications may also help some travelers. The use of melatonin for jet lag has grown considerably over the last several years, although scientific evidence of its effectiveness is not conclusive. Melatonin is a hormone of the pineal gland that is hypothesized to function in establishing or maintaining the circadian rhythm.

Figure 3 Discovery boxes reveal relationships between concepts presented in different chapters, and discuss interesting topics and emerging discoveries in more detail. Topics new to this edition include jet lag, creatine supplements, weight loss and metabolism, and the medicinal use of leeches.

Beautiful, Pedagogically Sound Art Program

Instructors tell us that the illustrations and photos in a book can greatly impact students' chances of understanding and retaining material. We carefully thought through each figure in this text for maximum pedagogical effectiveness. Beautifully drawn **anatomy diagrams** **(Figure 4)** succinctly illustrate the structures that students need to know to understand the physiology. We added **30 new illustrations** to this edition **(Figure 5)**. Consistent colors and icons help students intuitively understand what they are seeing from figure to figure.

Figure 4 Appealing, accurate anatomy diagrams appear where students may need review of structures.

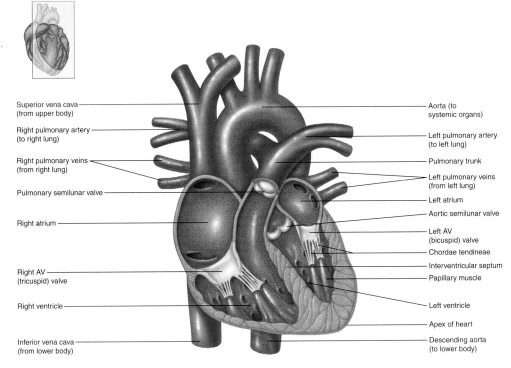

Superior vena cava (from upper body)
Right pulmonary artery (to right lung)
Right pulmonary veins (from right lung)
Pulmonary semilunar valve
Right atrium
Right AV (tricuspid) valve
Right ventricle
Inferior vena cava (from lower body)

Aorta (to systemic organs)
Left pulmonary artery (to left lung)
Pulmonary trunk
Left pulmonary veins (from left lung)
Left atrium
Aortic semilunar valve
Left AV (bicuspid) valve
Chordae tendineae
Interventricular septum
Papillary muscle
Left ventricle
Apex of heart
Descending aorta (to lower body)

Figure 5 30 new illustrations have been added to this edition, including this figure from Chapter 14, to further clarify and explain concepts in the text.

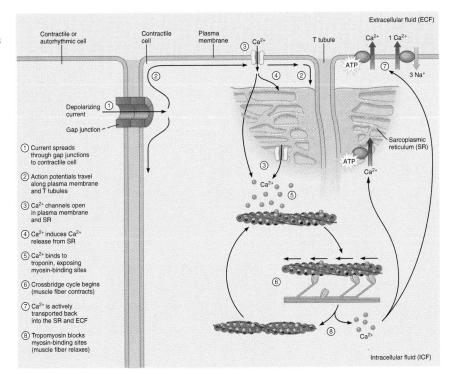

Extracellular fluid (ECF)

Contractile or autorhythmic cell
Contractile cell
Plasma membrane
Ca^{2+}
T tubule
Ca^{2+} 1 Ca^{2+}
ATP
3 Na^+

Depolarizing current
Gap junction
Sarcoplasmic reticulum (SR)
ATP
Ca^{2+}

① Current spreads through gap junctions to contractile cell
② Action potentials travel along plasma membrane and T tubules
③ Ca^{2+} channels open in plasma membrane and SR
④ Ca^{2+} induces Ca^{2+} release from SR
⑤ Ca^{2+} binds to troponin, exposing myosin-binding sites
⑥ Crossbridge cycle begins (muscle fiber contracts)
⑦ Ca^{2+} is actively transported back into the SR and ECF
⑧ Tropomyosin blocks myosin-binding sites (muscle fiber relaxes)

Ca^{2+}
Ca^{2+}

Intracellular fluid (ICF)

Numerous **flowcharts (Figure 6)** visually describe processes, neatly summarizing cause-and-effect relationships. **Layered, step-by-step PowerPoint flowcharts (Figure 6A)** are also available to instructors, providing a powerful way of presenting physiological pathways in the classroom.

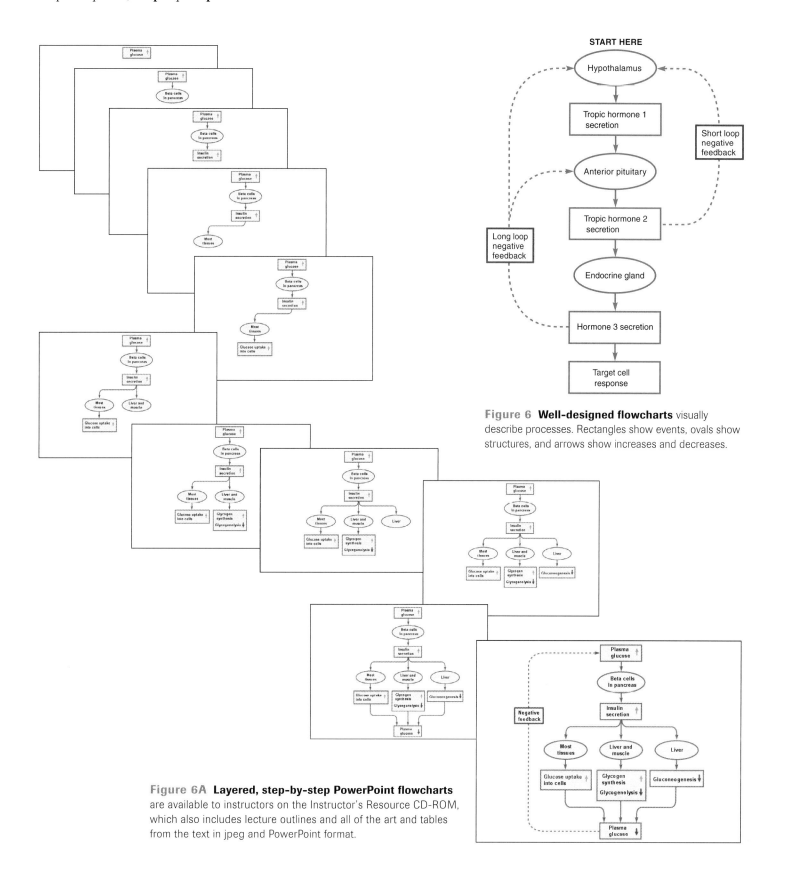

Figure 6 **Well-designed flowcharts** visually describe processes. Rectangles show events, ovals show structures, and arrows show increases and decreases.

Figure 6A **Layered, step-by-step PowerPoint flowcharts** are available to instructors on the Instructor's Resource CD-ROM, which also includes lecture outlines and all of the art and tables from the text in jpeg and PowerPoint format.

Toolboxes and Chemistry Review

New Chemistry Review boxes provide a quick resource for students to understand the chemistry associated with a particular physiological conept **(Figure 7)**. In addition, **Toolboxes (Figure 8)** use real examples to provide quantitative analysis of chemical or physical properties that apply to body function. The chemistry in a Toolbox is at a higher level than that in the Chemistry Review boxes and is generally quantitative in nature. This presentation allows students to read about the physiology first, and then apply the relevant math, chemistry, or physics necessary to round out their understanding.

Figure 7 Chemistry Review boxes provide quick refreshers on chemistry topics needed to understand physiological concepts.

CHEMISTRY REVIEW

SOLUTIONS AND CONCENTRATIONS

A mixture of molecules dissolved in a liquid is known as a *solution*. The dissolved substance, which is usually a solid or gas in its pure form, is known as the *solute*, whereas the liquid is referred to as the *solvent*. Solute molecules are said to be *dissolved* when they are completely separate from one another and are surrounded by solvent molecules.

The *concentration* of a solution is a measure of the quantity of solute contained in a unit volume of solution. Solute quantity is most commonly expressed in terms of moles (mol), and volume is most often given in liters (L). If 1 mole of solute is present in 1 liter of solution, the concentration is said to be 1 *molar* (1 M = 1 mol/L). If 1/1000 of a mole (that is, 1 *millimole*) of solute is dissolved in 1 liter of solution, the concentration is 0.001 molar or 1 *millimolar* (1 mM = 1×10^{-3} mol/L). For very

dilute solutions, concentrations may be expressed in *micromoles* per liter (μmol/L = 1×10^{-6} mol/L) or *nanomoles* per liter (nmol/L = 1×10^{-9} mol/L). Concentrations of specific substances are often indicated using brackets. For example, [Na$^+$] represents sodium ion concentration.

Occasionally, concentrations are expressed in terms of the *mass* of solute contained in a unit volume of solution—grams per liter (g/L) or micrograms per liter (μg/L = 1×10^{-6} g/L), for instance. Often in physiology, the unit of volume is a *deciliter* (dL = 0.1 L), which is equivalent to 100 milliliters (100 mL). The concentration of a solution containing 1 gram of solute per deciliter is frequently expressed as 1 *percent* (1%) because physiological solutions are mostly water, 100 mL of which weighs 100 grams.

Figure 8 Toolboxes provide quantitative analyses or more in-depth analyses of chemical or physical principles as they relate to physiology.

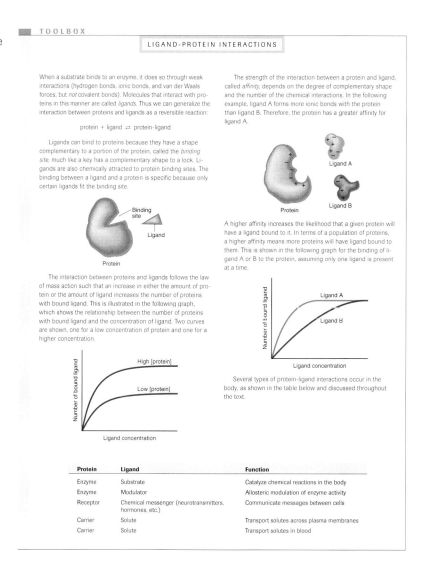

TOOLBOX

LIGAND-PROTEIN INTERACTIONS

When a substrate binds to an enzyme, it does so through weak interactions (hydrogen bonds, ionic bonds, and van der Waals forces, but *not* covalent bonds). Molecules that interact with proteins in this manner are called *ligands*. Thus we can generalize the interaction between proteins and ligands as a reversible reaction:

protein + ligand ⇌ protein-ligand

Ligands can bind to proteins because they have a shape complementary to a portion of the protein, called the *binding site*, much like a key has a complementary shape to a lock. Ligands are also chemically attracted to protein binding sites. The binding between a ligand and a protein is specific because only certain ligands fit the binding site.

The interaction between proteins and ligands follows the law of mass action such that an increase in either the amount of protein or the amount of ligand increases the number of proteins with bound ligand. This is illustrated in the following graph, which shows the relationship between the number of proteins with bound ligand and the concentration of ligand. Two curves are shown, one for a low concentration of protein and one for a higher concentration.

The strength of the interaction between a protein and ligand, called *affinity*, depends on the degree of complementary shape and the number of the chemical interactions. In the following example, ligand A forms more ionic bonds with the protein than ligand B. Therefore, the protein has a greater affinity for ligand A.

A higher affinity increases the likelihood that a given protein will have a ligand bound to it. In terms of a population of proteins, a higher affinity means more proteins will have ligand bound to them. This is shown in the following graph for the binding of ligand A or B to the protein, assuming only one ligand is present at a time.

Several types of protein-ligand interactions occur in the body, as shown in the table below and discussed throughout the text.

Protein	Ligand	Function
Enzyme	Substrate	Catalyze chemical reactions in the body
Enzyme	Modulator	Allosteric modulation of enzyme activity
Receptor	Chemical messenger (neurotransmitters, hormones, etc.)	Communicate messages between cells
Carrier	Solute	Transport solutes across plasma membranes
Carrier	Solute	Transport solutes in blood

Integrating Subject Material

One of the challenges and rewards of studying physiology is understanding how concepts tie together. To help students understand the overall function of individual systems, we summarize topics in **orientation charts (Figure 9)**. These charts show how the topics presented in a chapter, or series of chapters, fit together. They clarify how organ systems work as a whole to perform a function, such as delivering oxygen to and removing carbon dioxide from tissues. Where applicable, **partial orientation charts (Figure 10)** summarize chunks of information in the chapters. The large orientation charts pull together all of the pieces from the smaller charts.

Single organ systems do not function alone; therefore, the first paragraph of each chapter describes the relevance of material learned in previous chapters to the chapter at

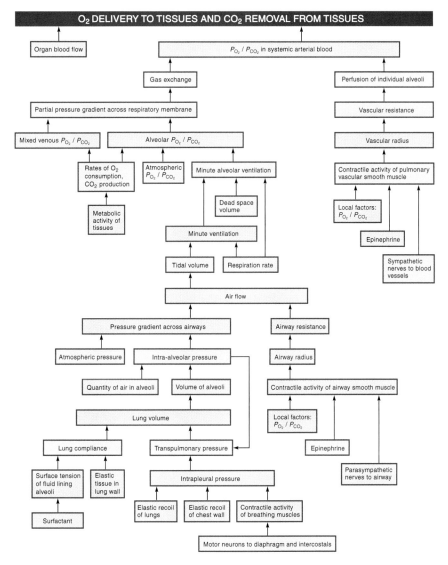

Figure 9 Orientation charts tie concepts within systems together.

Figure 10 Partial orientation charts tie manageable chunks of information together, which are then pulled into a larger orientation chart.

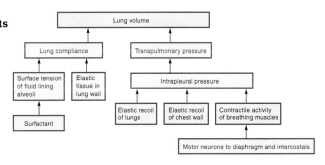

hand. In addition, at the end of each system chapter, or series of chapters, a **Systems Integration** chart **(Figure 11)** shows how the organ system just discussed influences the others.

Finally, to present physiology as an integrated whole, the text concludes with a chapter on the body's physiological response to exercise. Exercise affects virtually all aspects of body function and is a topic of interest for many students, making it an ideal "capstone" for the entire book.

Chapter 1 concludes with a story that follows two fictional characters, Bill and Jane, as they run a marathon race from beginning to end. In each chapter, **Exercise Links (Figure 12)** discuss the relevance of specific topics to exercise in the context of the marathon story. Thus, the marathon story serves as a thread that wends its way through the entire book, guiding the students toward the story's culmination in the final chapter.

Figure 11 Systems Integration charts at the end of each set of system chapters reinforce how each body system connects to the others.

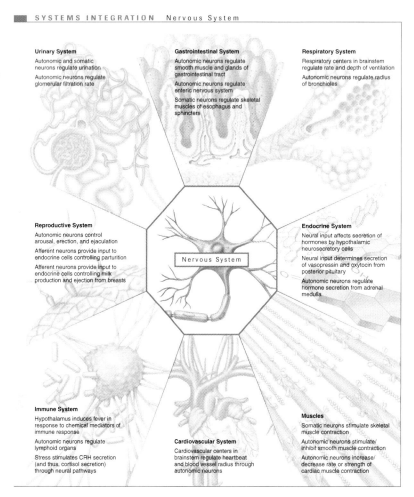

Figure 12 Exercise Links tie each chapter to a marathon story introduced in Chapter 1 and discussed in detail in Chapter 24.

Exercise Link

As Bill and Jane ran the marathon, they began to sweat. Sweat is a fluid, composed of water and solutes (especially sodium and chloride), that is derived from blood plasma, but it is less concentrated than plasma. As a result, Bill and Jane's plasma became more concentrated, causing water to move out of their cells, making them shrink about 2%. This shrinkage during exercise is not usually a serious issue, but cellular volume changes that occurred after Bill drank a large amount of pure water following the marathon caused his severe headache. Based on what you know about osmotic pressure and its influence on cell volume, can you determine what might have caused Bill's headache?

Study Tools Reinforce Key Concepts Learned

To learn effectively, students need to know they are on the right track as they are studying. Five study tools help students test their knowledge and go back to review material when necessary: Before You Begin, Quick Tests, Chapter Summaries, End-of-Chapter Exercises, and Figure Questions. Each chapter opens with **Before You Begin (Figure 13)**, a short list of topics discussed in previous chapters that students should know in order to get the most out of the chapter at hand. **Quick Tests (Figure 14)**, short sets of review questions at the ends of chapter sections, let students check their understanding of material before tackling new material. Each chapter closes with a detailed but concise **Chapter Summary (Figure**

15) and a complete bank of **End-of-Chapter Exercises (Figure 16)**, which include multiple-choice, objective, critical thinking, and essay questions. Multiple choice and objective questions help students to test their knowledge. Essay questions require students to synthesize material or to explain involved concepts. Critical thinking questions require students to apply their knowledge. **Figure Questions (Figure 17** on page xvi) appear with selected figures throughout the text. Long a hallmark of Benjamin Cummings texts, these questions ask students to take their studying one step further. Answers are provided at the bottom of the page.

Figure 13 Before You Begin boxes give references to topics in previous chapters that students need to understand fully in order to get the most out of the current chapter.

> ▨ *Before You Begin*
>
> *Make sure you have mastered the following topics:*
>
> **1.** *Biomolecules, p. 24*
>
> **2.** *Mitochondria, p. 40*
>
> **3.** *Membrane structure, p. 36*

Figure 14 Quick Tests provide strategic checkpoints in the text to help students test their comprehension of the material and reread the previous section if necessary before moving forward. Answers to the Quick Tests are available online at www.physiologyplace.com.

> **Quick Test 3.3**
>
> **1.** An increase in which of the following would speed up the net rate of a reaction proceeding in the forward direction? temperature, reactant concentrations, product concentrations, the height of the activation energy barrier (choose all that apply)
>
> **2.** Define the following terms: *enzyme, substrate, substrate specificity, active site, affinity, cofactor, coenzyme, percent saturation, regulatory site, modulator.*
>
> **3.** How is the rate of an enzymatic reaction affected by the catalytic rate of an enzyme? By its affinity for the substrate?
>
> **4.** What is the primary distinction between allosteric regulation and covalent regulation? Between feedback inhibition and feedforward activation?

Figure 15 Chapter Summaries pull together the main points of each chapter. IP (InterActive Physiology®) references point students to relevant pages in this award-winning software, where students can find lively animation and interactive quizzing to help them review topics further.

▨ *CHAPTER SUMMARY*

Skeletal Muscle Structure, p. 376

Most skeletal muscles are connected to bones by tendons and contain numerous elongated cells (muscle fibers) that generate contractile force using energy from ATP hydrolysis. Within muscle fibers are myofibrils that contain the contractile machinery. The sarcoplasmic reticulum surrounds the myofibrils, stores calcium ions, and is closely associated with transverse (T) tubules, which penetrate into the cell interior from the sarcolemma. Skeletal and cardiac muscle are striated, reflecting the orderly arrangement of thick and thin filaments in the myofibrils, which are made up of fundamental force-generating units (sarcomeres) joined end to end. Thick and thin filaments contain the contractile proteins myosin and actin, respectively. The heads (crossbridges) of myosin molecules are responsible for generating the motion that drives contraction and possess two important sites: an actin-binding site and an ATPase site. Two regulatory proteins (troponin and tropomyosin) present

on the thin filaments serve to initiate and terminate contractions.

▢ Muscular, Anatomy Review: Skeletal Muscle Tissue, pp. 4–11

▢ Muscular, Sliding Filament Theory, pp. 3–15

The Mechanism of Force Generation in Muscle, p. 380

When a muscle contracts, thick and thin filaments slide past one another. This sliding is driven by the crossbridge cycle, in which the motion of crossbridges is coupled to their cyclic binding and unbinding to actin molecules in adjacent thin filaments. In skeletal muscle, each fiber receives input from one motor neuron, which branches and innervates more than one fiber. An action potential in a motor neuron triggers the release of acetylcholine, which binds to receptors in the muscle fiber's motor end plate. The result is an electrical signal (end-plate potential) that triggers an action potential in the sarcolemma. This is followed by propagation of the action potential through the T tubules, release of Ca²⁺ from the sarcoplasmic reticulum, binding of Ca²⁺ to troponin,

movement of tropomyosin away from actin's myosin-binding sites, and initiation of the crossbridge cycle.

The Mechanics of Skeletal Muscle Contraction, p. 388

A motor neuron plus the muscle fibers it innervates constitutes a unit. When a motor neuron fir an action potential, all fibers in motor unit contract together. T chanical response of a motor u a single action potential is a twi which is reproducible in size. T can be isometric, in which case muscle generates force but doe shorten, or isotonic, in which c muscle shortens. The force gen by an entire muscle is determin by both the force generated by vidual fibers (which depends o frequency of stimulation, fiber ter, and changes in fiber length the number of fibers that are ac Stimulation at high frequencies summation of twitches, such th the force eventually reaches a p (tetanus).

Figure 16 End-of-Chapter Exercises allow students to self-test on the material covered in the chapter.

▨ *EXERCISES*

Multiple-Choice Questions

1. When a muscle cell is relaxed and intracellular ATP levels are normal, a crossbridge will remain in which of the following states?
   a) bound to actin and in the low-energy form
   b) bound to actin and in the high-energy form
   c) in the high-energy form, with ADP and Pᵢ bound to it
   d) in the high-energy form, with ATP bound to it
   e) in the low-energy form with nothing bound to it

2. During a muscle contraction, which of the following does *not* change length?
   a) the distance between Z lines
   b) the width of I bands
   c) the width of A bands
   d) none of the above

3. Which of the following would tend to *reduce* the concentration of lactic acid that accumulates in a muscle cell as a result of contractile activity?
   a) increasing the concentration of glycolytic enzymes
   b) decreasing the oxygen supply to the cell
   c) increasing the diameter of the cell
   d) increasing the number of mitochondria in the cell
   e) all of the above

4. Which of the following statements is a valid generalization regarding the properties of smooth muscle?
   a) Neurotransmitters can either excite or inhibit smooth muscle contraction, but any given neurotransmitter is always excitatory or inhibitory, regardless of where the muscle is located.
   b) A given smooth muscle cell can respond to more than one type of neurotransmitter.
   c) Smooth muscle cells are generally unresponsive to neurotransmitters of all types.
   d) Smooth muscle cells can respond to neural input from the somatic or autonomic nervous systems.
   e) none of the above

5. Which of the following is *not* a determinant of whole muscle tension?
   a) the number of muscle fibers contracting
   b) the tension produced by each contracting fiber
   c) the proportion of each motor unit that is contracting at any given time
   d) the extent of fatigue
   e) the frequency of action potentials in the motor neurons

6. In an isotonic contraction,
   a) muscle length shortens.
   b) muscle tension exceeds the force of the load.
   c) the load is moved.

   d) a and c
   e) all of the above

7. Which of the following is true for the excitation-contraction coupling of *all* muscle types (skeletal, cardiac, and smooth)?
   a) An action potential causes calcium levels in the cytosol to increase.
   b) Calcium binds to troponin.
   c) Thick and thin filaments slide past each other.
   d) a and c
   e) all of the above

8. During contraction of a skeletal muscle fiber,
   a) the thick filaments contract.
   b) the thin filaments contract.
   c) the A band becomes shorter.
   d) the I band becomes shorter.
   e) all of the above

9. Which of the following statements concerning the characteristics of different types of muscle fibers is *false*?
   a) The higher the myosin ATPase activity, the faster the speed of contraction.
   b) Muscles that have high glycolytic capacity and large glycogen stores are more resistant to fatigue.
   c) Oxidative types of muscle fibers contain myoglobin.
   d) Oxidative fibers have a richer blood supply.
   e) Larger diameter fibers can produce greater tension.

Media

Benjamin Cummings has long been a leader in technology with fun, innovative software programs that truly teach conceptual material. We developed the text with this in mind. You'll find that the media offerings described here will help students take their learning a step beyond what is in the book. Animations and tutorials bring the concepts in the book to life and help students with different learning styles. The following illustrations show, for example, how **glomerular filtration** is covered in multiple ways: in a figure in the text **(Figure 17)**, in a case study at *The Physiology Place* **(Figure 18)**, in a tutorial in *InterActive Physiology®* **(Figure 19)**, and in an experiment on *PhysioEx™* **(Figure 20)**. Quizzes allow students to test their knowledge as they work through the material online or on CD-ROM.

Figure 17 A figure from the discussion on glomerular filtration illustrates the concept in the text. Note the figure question, which challenges students to take their studying one step further.

Glomerular filtration. (a) Glomerular filtration pressure is the result of four Starling forces: (1) hydrostatic pressure in the glomerular capillaries (P_{GC}), (2) hydrostatic pressure in Bowman's capsule (P_{BC}), (3) oncotic pressure in the glomerular capillaries (π_{GC}), and (4) oncotic pressure in Bowman's capsule (π_{BC}). The net filtration pressure is 16 mm Hg. (b) The filtration fraction is the proportion of renal plasma that is filtered into Bowman's capsule. The normal filtration fraction is 20%.

If proteins leaked out of the glomerular capillaries (which would decrease (π_{GC}) and increase (π_{BC})), what would happen to the glomerular filtration pressure and to the glomerular filtration rate?

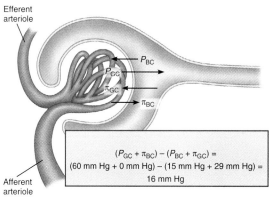

$$(P_{GC} + \pi_{BC}) - (P_{BC} + \pi_{GC}) =$$
$$(60 \text{ mm Hg} + 0 \text{ mm Hg}) - (15 \text{ mm Hg} + 29 \text{ mm Hg}) =$$
$$16 \text{ mm Hg}$$

(a) Glomerular filtration pressure

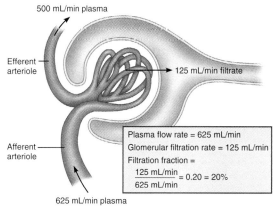

Plasma flow rate = 625 mL/min
Glomerular filtration rate = 125 mL/min
Filtration fraction =
$$\frac{125 \text{ mL/min}}{625 \text{ mL/min}} = 0.20 = 20\%$$

(b) Filtration fraction

▮ *Both would increase.*

Figure 18 With challenging activities, flowchart exercises, and case studies, **The Physiology Place** is the perfect site for online learning. In this case, students apply what they have learned about renal physiology to the real world. Visit www.physiologyplace.com for a demo. **A subscription to The Physiology Place comes free with the purchase of a new copy of** *Principles of Human Physiology.*

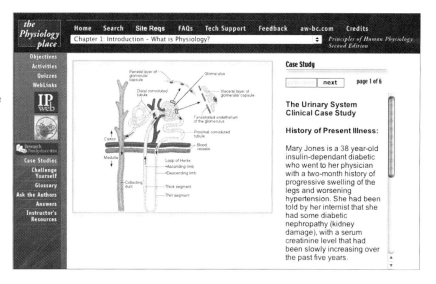

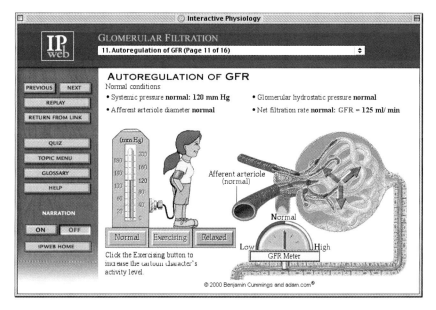

Figure 19 InterActive Physiology® has already helped thousands of students understand and visualize complex physiological processes and thus succeed in this challenging course. Animations, tutorials, and quizzing help students master difficult concepts. IP brings the topic of glomerular filtration alive with animations that clarify processes. Here the animation shows autoregulation of GFR during different states of activity. References to IP are found in the book's chapter summaries. **Every new copy of** *Principles of Human Physiology* **includes the complete IP CD-ROM.**

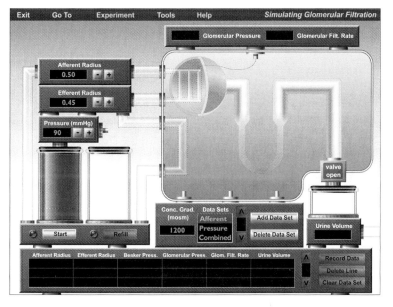

Figure 20 PhysioEx™ offers an alternative lab environment in which to master difficult concepts. Computerized simulations allow students to repeat lab experiments as needed to achieve a sound understanding of physiological processes. Here, students simulate the process of glomerular filtration.

Supplements

For Instructors

Instructor's Resource CD-ROM 0-8053-5417-4
This powerful cross-platform CD-ROM includes:

- PowerPoint lecture outlines for each chapter of *Principles of Human Physiology*, 2nd edition
- Step-by-step "layered" flowcharts in PowerPoint slide format (see pg. xi)
- All illustrations from the text in both jpeg and PowerPoint format, with and without labels
- All photos from the text in both jpeg and PowerPoint format, with and without labels

Instructor's Guide to Text and Media 0-8053-5416-6
The Instructor's Guide to Text and Media includes a detailed chapter-by-chapter correlation guide for integrating powerful media supplements such as Interactive Physiology and PhysioEx into the text; thumbnail snapshots of all the art available on the Instructor's Resource CD-ROM; chapter outlines; chapter cross-references; and detailed answers to end-of-chapter material in the text.

Transparency Acetates 0-8053-5418-2
Contains full-color acetates of every illustration and table in *Principles of Human Physiology*, 2nd edition (650 in all)

Test Bank
Printed test bank: 0-8053-5436-0; Cross-platform computerized test bank: 0-8053-5435-2

The test bank contains approximately 3500 multiple-choice, true/false, matching, fill-in-the-blank, and essay questions. The easy-to-use, cross-platform computerized test bank allows instructors to view and edit questions electronically, create multiple tests, and print tests in a variety of formats.

The Physiology Place www.physiologyplace.com
For instructors, this website contains downloadable supplements; quizzes, multimedia activities, and case studies which may be used as student assignments; and access to Research Navigator, a powerful online research tool that links to three exclusive databases of credible and reliable source material including the EBSCO Academic Journal and Abstract Database, New York Times Search by Subject Archive, and "Best of the Web" Link Library.

Course Compass, Blackboard, WebCT Course Management Systems
Preloaded content specific to *Principles of Human Physiology*, 2nd edition, is available in CourseCompass, Blackboard, and WebCT formats, including testing and assessment, activities, and weblinks. To learn more about these course management systems, please visit: http://suppscentral.aw.com/browse.php

For Students

Study Guide 0-8053-5751-3
The Study Guide provides a variety of questions designed to help students review their understanding of physiology, including multiple-choice, fill-in-the-blank, completion, matching, short-answer, true/false, "concept map," and art labeling questions.

The Physiology Place www.physiologyplace.com
For students, this website contains an unsurpassed wealth of study resources, including multiple-choice, fill-in-the-blank, and true/false questions for each chapter of the book; case studies; multimedia activities; "Challenge Yourself" physiology topics for further exploration; flash-cards to review key terms from the text; an "Ask the Author" tutoring feature; and a link to Research Navigator, a powerful online research tool that grants students access to exclusive online databases such as the New York Times science archives.

InterActive Physiology **CD-ROM/website**
This award-winning tutorial series now comes packaged **FREE** with every new copy of the text. InterActive Physiology uses detailed, engaging animations and quizzes to help students advance beyond memorization to a genuine understanding of the most difficult topics in physiology. InterActive Physiology is referenced in chapter summaries throughout the text, and is also available in web format via The Physiology Place website (see p. xvii).

PhysioEx 5.0 Laboratory Simulations (online version)
PhysioEx Version 5.0 consists of thirteen modules containing 36 lab simulations that may be used to supplement or substitute for wet labs. This easy-to-use software allows students to repeat labs as often as they like, perform experiments without harming live animals, and conduct experiments that may be difficult to perform in a wet lab environment due to time, cost, or safety concerns. Students also have the flexibility to change the parameters of an experiment and observe how outcomes are affected. Access to the online version of PhysioEx 5.0 is available **FREE** with every new copy of the text, via The Physiology Place website (see p. xvii).

Acknowledgments

Writing a textbook has been a challenging yet rewarding experience. Revision to the second edition went smoothly, thanks to the hard work of a number of individuals who worked together as a team.

Special thanks go to Mary Jane Niles (University of San Francisco) and Joseph G. Cannon (Medical College of Georgia). Mary Jane wrote Chapter 23, The Immune System, and masterfully made this complex subject easily intelligible and a joy to read. Joe wrote the marathon story at the end of Chapter 1, the Exercise Link boxes that appear in each chapter, and Chapter 24, The Whole Body: Integrated Physiological Responses to Exercise. The innovative and entertaining marathon story and links provide a thread that runs through the entire book, and integrate many concepts into a coherent whole. We think Mary Jane and Joe's work have improved the book immeasurably.

We also thank the outstanding publishing professionals at Benjamin Cummings. We are grateful to Daryl Fox, publisher, and Serina Beauparlant, executive editor, for their support of this book's vision. We thank our project editor, Barbara Yien, whose leadership put together a cohesive team and kept the project moving smoothly. We thank Claire Alexander, project editor of the first edition and development manager on this one, for lending guidance when needed.

Many thanks, too, go to all those involved in developing the art. Laura Southworth, art development editor, has been with us since the first edition and deserves more credit than one can express. Not only did Laura take our ideas (and poor excuses for sketches) and create the beautiful, concise illustrations that you will find in the text, she also helped develop the art to improve its teaching effectiveness. Her experience in publishing was instructional to us all; her experience in life kept us smiling through even the roughest of times. We also thank Dr. Patsy Itaya, who diligently reviewed the anatomy art under the tightest of schedules; her kindness made her a pleasure to work with.

Wendy Earl and Michele Mangelli headed up a stellar production team, helping us create a beautiful new design (the work of the talented tani hasegawa), keeping the team to a timely production schedule, and keeping track of countless details to make sure every last piece of the book came together. It was also a pleasure to work with Anita Wagner, copyeditor, who scrutinized the work and provided critical feedback that enhanced the quality of the writing. We are grateful to the artists at Imagineering and Precision Graphics for their fine work rendering the illustrations in this text. We also thank Kathleen Olson, who spent many hours searching for just the right photos; and Bettina Borer, art coordinator, who expertly tracked all of the book's figures.

This edition's striking cover was hand-painted by Tomo Narashima and designed by Lillian Carr, both of whom we thank very much. In addition, Ryan Shaw and Leslie Austin deserve much credit for their fine work managing the supplements package. We are grateful to Ziki Dekel, Steve Wright, Rhatia Carr, Michelle Levine, and Barbara Yien for their work on the excellent website and media accompanying this text. And we thank Lauren Harp, executive marketing manager, for forwarding us the comments and suggestions she hears from instructors and students using our text.

Throughout the text preparation, a number of individuals reviewed the manuscript and provided invaluable feedback. We thank all the reviewers (listed on the following pages) for their thoroughness. Every comment was considered, and most were incorporated into the text. You made a difference.

Cindy Stanfield thanks her husband Jim, whose continued support and patience made the project possible for her. She would also like to thank her friends and colleagues, including Vicki Barrett, Daniel and Laura Sellers, Paula Medveal, Camille Hegg, and Gail Chryslee. Special thanks to her mom, brothers, and sisters for their support. She also thanks her students for challenging her to continue to learn and for providing feedback on the effectiveness of the book as a teaching aid.

William Germann extends gratitude to his colleagues in the Department of Biology at the University of Dallas, who supported him over the years and who also took up some of the slack while he was writing this book: Marcy Brown-Marsden, Frank Doe, David Pope, and Cathy Shupe. Most of all, he thanks Joe Cannon, whom he has known since the days they were both graduate students in the Department of Physiology at the University of Michigan, for superlative contributions to the book.

Nothing he can say could adequately express his gratitude and admiration for someone who has turned out to be such a great physiologist.

In closing, we hope that students will find this text truly helps in studying physiology, and that it reflects our enthusiasm for the subject and our dedication to helping students learn. We welcome your suggestions and comments for future editions.

William J Germann   Cindy Lou Stanfield

William Germann and Cindy Stanfield

Applied Sciences
Benjamin Cummings
1301 Sansome St.
San Francisco, CA 94111

REVIEWERS

Second Edition

Michelle Cleary
Florida International University

David Cochrane
Tufts University

Doris Daniels
Kansas City Kansas Community College

John Davis
Alma College

John Dobson
University of Florida

Jonathan Fowles
Acadia University

Nicholas R. Geist
Sonoma State University

Katja Hoehn
Mt. Royal College

Jim Hunter
Texas A&M University

Rosemary Knapp
University of Oklahoma

Clancy Leahy
Lynchburg College

William Lutterschmidt
Sam Houston State University

Vikki McCleary
University of North Dakota

Stacia Moffett
Washington State University

Stephen Price
Virginia Commonwealth University

Mary Anne Rokitka
University of Buffalo

Beth Scalione-Sewell
Valparaiso University

Robert Vick
Elon College

Paul White
Monash University

David Wilson
Miami University

First Edition

Thomas Adams
Michigan State University

Brian Adrian
Grand Valley State University

David Anderson
Pennsylvania State University

Patricia D. Ashby
Eastern New Mexico University

Albert Baccari
Montclair Community College

Ralph Barclay
Wayne State University

Debra Barnes
Contra Costa College

Janis Beaird
University of West Alabama

Cynthia Beck
George Mason University

Dale Benos
University of Alabama, Birmingham

Joseph Berger
Springfield College

David G. Bernard
University of Texas, Arlington

Ron Beumer
Armstrong Atlantic State University

Eric Bittman
University of Massachusetts, Amherst

Sunny Boyd
University of Notre Dame

Mark Bracken
Utah Valley State College

Edward Brandt
Shenandoah University

Virginia Brooks
Oregon Health Sciences University

William Brothers
San Diego Mesa College

Nishi Bryska
University of North Carolina, Charlotte

Michael Canute
Hawaii Pacific University

John Capeheart
University of Houston

Andrew N. Clancy
Georgia State University

William Cliff
Niagara University

David Cochrane
Tufts University

Frank Corotto
North Georgia College and State University

Charles Costa
Eastern Illinois University

Joseph Crivello
University of Connecticut

Leon Cuervo
Florida International University

Dan Deaver
Advanced Inhalation Research

Russell DiFiore
Pasadena City College

Leon Dorosz
San Jose State University

Gary Dudley
University of Georgia

Jean Pierre Dujardin
Ohio State University

John Dziak
Community College of Allegheny County

Steven Eiger
Montana State University

Yasir El-Sharif
City University of New York, College of Staten Island

Cory Etchberger
Longview Community College

Kenneth Etzel
University of Pittsburgh

Bridget Falkenstein
Sierra College

Richard Falvo
Southern Illinois University

Ralph E. Ferges
Palomar College

Philip J. Fernandez
Grand Canyon University

Milton Fingerman
Tulane University

Kathleen A. Fitzpatrick
Merrimack College

George Fortes
University of California, San Diego

Frank Frisch
Chapman University

Irja Galvan
Western Oregon University

Greg Garman
Centralia College

Nicholas R. Geist
Sonoma State University

Debabrata Ghosh
Texas Southern University

Alice Gibb
Northern Arizona University

Roger Gilchrist
University of Alabama, Birmingham

Bruce Gladden
Auburn University

Jack Goldberg
University of California, Davis

Sheldon R. Gordon
Oakland University

Todd Gordon
Kansas City Kansas Community College

Jean Grassman
City University of New York, Brooklyn College

Tamara Greco
Eastern Michigan University

Michael Griffin
Angelo State University

Charles J. Grossman
Xavier University

Michael Guinan
University of California, Davis

Douglas C. Gula
Miami University

David L. Hammerman
Long Island University

Dennis C. Haney
Furman University

John P. Harley
Eastern Kentucky University

Janet L. Haynes
Long Island University

Laura Hebert
Angelina College

James Herman
Texas A & M University

Tina Hines
University of Pittsburgh

James Hoffman
Diablo Valley College

Charles W. Holliday
Lafayette College

Michael J. Holmes
University of Pittsburgh

Sandra Hsu
University of San Francisco

Mark Hubley
Washington College

Kerry Hull
Bishops University

Christine Iltis
Salt Lake Community College

William Jackson
Western Michigan University

Paul Jarrell
Pasadena City College

Najma Javed
Ball State University

J. Kelly Johnson
University of Kansas

Penny Knoblich
Minnesota State University, Mankato

Nuran Kumbaraci
Steven's Technical Institute

David Kurjiaka
Ohio University

David C. Lennartz
Mount St. Mary's College

John Lepri
University of North Carolina, Greensboro

James Long
Boise State University

John Lovell
Kent State University

Jennifer Lundmark
California State University, Sacramento

Ronald Lynch
University of Arizona

Duncan MacDougall
McMaster University

Jennifer Marcinkiewicz
Kent State University

Christel Marschall
Lansing Community College

Theresa Martin
College of San Mateo

Todd McBride
California State University, Bakersfield

Jaqueline S. McLaughlin
Pennsylvania State University, Berks/Lehigh Valley

Katherine Mechlin
Wright State University

Robert Meckler
City College of San Francisco

Essie Meisami
University of Illinois, Urbana-Champaign

Jon S. Miller
Northern Illinois University

John W. Mills
Clarkson University

Jeanne Mitchell
Truman State University

Stacia Moffett
Washington State University

Ann Motekaitis
California State University, Sacramento

David P. Muehleisen
University of Utah

Donald Mulcare
University of Massachusetts, Dartmouth

Patricia Munn
Longview Community College

Ruth K. Nash
College of Marin

Linda R. Nichols
Santa Fe Community College

Gemma Niermann
University of California, Berkeley

Colleen Nolan
St. Mary's University

Margaret Nordlie
University of Mary

Howard J. Normile
Wayne State University

Brian J. Norris
California State University, San Marcos

Nathan Norris
West Valley College

Bruce O'Gara
Humboldt State University

Angela Ottava
Michigan State University

T. Lon Owen
Northern Arizona University

M.J. Parkes
University of Birmingham, UK

Shawn Pearcy
Wayne State College

John Perry
Oklahoma City Community College

Dickson Phiri
Mesa College

John Pigage
University of Colorado

Steven Price
Virginia Commonwealth University

Richard Puzdrowski
University of Houston, Clear Lake

David Quadagno
Florida State University

Ralph Reiner
College of the Redwoods

David Rivers
Loyola College

Mary Anne Rokitka
State University of New York, Buffalo

Barry Rothman
San Francisco State University

Jane Rudolph
Salt Lake Community College

Elizabeth M. Rust
University of Michigan

Allen Sanborn
Barry University

Brian Schmaefsky
Kingwood College

Lynne Schneider
University of Vermont

Thomas Secrest
Austin Community College

Arnold Sillman
University of California, Davis

C. Wayne Simpson
University of Missouri, Kansas City

James W. Small
Rollins College

Davis Smith
San Antonio College

George Snow
Mount St. Mary's College

Michael Stamper
College of West Virginia

Mary Lynne Stephanou
Santa Monica College

Philip Stephens
Villanova University

James Stockand
University of Texas, San Antonio

Kevin Strang
University of Wisconsin

James A. Strauss
Pennsylvania State University

Alan F. Sved
University of Pittsburgh

Karen Swearingen
Mills College

Steven Swoap
Williams College

Sherry Tamone
Sonoma State University

Deborah Taylor
Kansas City Kansas Community College

Carl Thurman
University of Northern Iowa

H. Ti Tien
Michigan State University

Paola S. Timiras
University of California, Berkeley

Tom Tomasi
Southwest Missouri State University

Mark Tomita
City University of New York, Brooklyn College

Joseph Tupper
Syracuse University

Christine K. Wade
University of Wyoming

David R. Wade
Southern Illinois University

Tracy Wagner
Washburn University

Bruce Wainman
McMaster University

Tim Wakefield
John Brown University

Benjamin Walcott
State University of New York, Stony Brook

Stephen Warburton
New Mexico State University

David Washington
University of Central Florida

James Watrous
St. Joseph's University

Ralph E. Werner
The Richard Stockton College of New Jersey

Eric P. Widmaier
Boston University

Nancy Williams
Pennsylvania State University, Fayette

F. Ray Wilson II
Baylor University

Darla Wise
Concord College

Nicholas J. D. Wright
Eastern New Mexico University

Marianna J. Zamlauski-Tucker
Ball State University

FOCUS GROUPS

Faculty

Allen Bedford
Bryn Athyn College

Mary Bober
Santa Monica College

Denise Cavener
Santa Monica College

Theresa Chen
College of Notre Dame

Richard Connett
Monroe Community College

Paul Emerick
Monroe Community College

Ana Escadon
L.A. Harbor College

Michael Guinan
University of California, Davis

David Hammerman
Long Island University

Robert Hyde
San Jose State University

William Jackson
Western Michigan University

Vasily Kolchenko
Pace University

David Lennartz
Mount St. Mary's College

Eileen Lynd-Balta
St. John Fisher College

Theresa Martin
College of San Mateo

Robert Meckler
City College of San Francisco

James Murphy
Monroe Community College

Ruth Nash
College of Marin

Nathan Norris
West Valley College

Dougald Scott
Cabrillo College

Rachel Simons
Monroe Community College

Gary Skuse
Rochester Institute of Technology

Michael Sneary
San Jose State University

Edmund Tong
Wheaton College

Carlene Tonini
College of San Mateo

Mary Jo Witz
Monroe Community College

Student

Bryn Athyn College

Francis Darkwah
Nicola Echols
Ruth Homber
Shondra Perry
Shada Sullivan

College of San Mateo

Deborah Bilafer
Nidia Medina
Robin Garoutte
Colleen Mahoney
Manolo Cacella
Angela Stratton
Maribeth King
Nina Breiz
Jim Tupas
Noel Segali
Helen Murphy
Steven Fulton

Texas A&M University

Mallika Anand
Ben Alexander
Katie Jackson
Nick Campbell
Andrew Svetlick

CLASS TESTERS

Magdalene Abraham
Mississippi Delta Community College

Ateegh Al-Arabi
Johnson County Community College

William Andresen
William Rainey Harper College

Gwen Bachman
University of Nebraska

Floyd Banks
Chicago State University

Cynthia Beck
George Mason University

Charles Biles
East Central University

Edward Bilsky
University of Northern Colorado

William Boyko
Sinclaire Community College

Mark Bracken
Utah Valley State College

Andrew Browe
Indiana University of Pennsylvania

Elaine S. Brubacher
University of Redlands

Maureen Burton
University of South Dakota

Michael Canute
Hawaii Pacific University

John Capeheart
University of Houston Downtown

Chun-fan Chen
Florida International University

Chris Christopher
Santa Rosa Junior College

Craig Clifford
Northeastern State University

Jonathan Day
California State University, Chico

Nick Despo
Thiel College

Kathryn Dickson
California State University, Fullerton

Christopher Dunbar
Brooklyn College

Cory Etchberger
Longview Community College

Robert Farrell
Pennsylvania State University

Michael Ferrari
University of Arkansas

Charles Foster
State University of New York, Potsdam

Rhonda Gamble
Mineral Area College

Jorge Golowasch
Rutgers University

Tamara Greco
Eastern Michigan University

Joseph Gregorek
Gannon University

Michael Griffin
Angelo State University

John Harley
Eastern Kentucky University

Elizabeth Hayes
University of Wisconsin, Fond du Lac

Lisa Hays
Evergreen Valley College

James Herman
Texas A & M University

Stephan Hourdez
Pennsylvania State University

Najma Javed
Ball State University

Clyde E. Johnson
Southern University at Baton Rouge

Robert David Jones
Adelphi University

Henry Kayongo-Male
South Dakota State University

Peter King
Francis Marion University

Gopal Krishna
Moberly Area Community College

Jill Kruper
Murray State University

Nuran M. Kumbaraci
Stevens Institute of Technology

Gavin Lawson
Bridgewater College

Curtis Lee
Dallas Baptist University

John Lepri
University of North Carolina, Greensboro

David Mallory
Marshall University

Theresa Martin
College of San Mateo

Bill Mathena
Kaskaskia College

Vikki McCleary
University of North Dakota School of Medicine

Kerry McDonald
University of Missouri

Jackie McLaughlin
Pennsylvania State University, Berks-Lehigh Valley College

Robert Meckler
City College of San Francisco

Robb Moats
University of Florida, Jacksonville

Suzanne Moshier
University of Nebraska, Omaha

Donald J. Mulcare
University of Massachusetts, Dartmouth

Greg Mullen
University of Northern Colorado

Patricia Munn
Longview Community College

Mahalashmi Nagarajan
Wilberforce University

Colleen Nolan
St. Mary's University Texas

Robert Okazaki
Weber State University

Gibson Oriji
William Paterson University

Marie O'Rourke
College of St. Mary

Sujata Patel
University of Chicago

Shawn Pearcy
Wayne State College

Dave Pendergast
State University of New York, Buffalo—South Campus

Judy Penn
Shoreline Community College

Lisa Perrino
University of Maryland

Mary Perry
Allan Hancock College

Larry A. Reichard
Maple Woods Community College

Tricia A. Reichert
Colby Community College

Carmen Rexach-Zellhoefer
Merced College

Barbara Roller
Florida International University

Jane Rudolph
Salt Lake Community College

Arleen Sawitzke
Salt Lake Community College

Brian Shmaefsky
Kingwood College

Linda Smith
Stockton State College

John R. Steele
Ivy Tech State College

Mary Lynne Stephanou
Santa Monica College

Cathryn Stevens
Louisiana State College

Bonnie Tarricone
Ivy Tech State College

Phillip Taylor
Glenville State College

Carl Thurman
University of Northern Iowa

James Toepfer
Youngstown State University

Edmund Tong
Wheaton College

William Trotter
Des Moines Area Community College

Dana Vaughan
University of Wisconsin, Oshkosh

Tracy Wagner
Washburn University

Curt Walker
Dixie College

Wade Warren
Louisiana College

David Washington
University of Central Florida

John Water
Pennsylvania State University

Cheryl Watson
Central Connecticut State University

Flora Watson
South Carolina State University, Stanislaus

Allison Wilson
Benedict University

Hing-Sing Yu
University of Texas

Henry H. Ziller
Southeastern Louisiana State University

ART BOARD MEMBERS

Andrew Clancy
Georgia State University

Cathy Stevens
Louisiana State University

David Wilson
Miami University

Balwant Khatra
California State University, Long Beach

Peter Bannerman
Children's Hospital of Philadelphia

Robert Mitchell
Pennsylvania State University

Stephen Dodd
University of Florida

Thomas Adams
Michigan State University

John McReynolds
University of Michigan

Judith Gibber
Columbia University

Robert Moats
University of North Florida

Scott Hadley
Grand Valley State University

Scott Moye-Rowley
University of Iowa

Brief Contents

Contents

8 Nerve Cells and Electrical Signaling 209

9 Synaptic Transmission and Neural Integration 241

10 The Nervous System: Central Nervous System 262

11 The Nervous System: Sensory Systems 301

The Nervous System: Autonomic and Motor Systems 353

Muscle Physiology 375

The Cardiovascular System: Cardiac Function 413

The Cardiovascular System: Blood Vessels, Blood Flow, and Blood Pressure 451

16 The Cardiovascular System: Blood   497

17 The Respiratory System: Pulmonary Ventilation   516

18 The Respiratory System: Gas Exchange and Regulation of Breathing   542

23   The Immune System   725

24   The Whole Body: Integrated Physiological Responses to Exercise   756

List of Boxes

Physiology, the science of body function, may be a child of human reason, but it was undoubtedly conceived in passion—our natural desire to understand ourselves and the universe we inhabit. For some people this desire persists beyond mere curiosity and becomes the driving force for a lifetime of study. It is because of the efforts of such people that the science of physiology exists today.

Like many other fields of study, physiology has a simple definition but is almost impossible to describe succinctly. Simply put, physiology attempts to explain the workings of the body using established principles from physics and chemistry. This description may seem clear enough, but the same could also be said of other biological sciences such as biochemistry and cell biology. It is difficult to say what makes physiology distinctive because it, like many other fields of scientific study, lacks distinct boundaries. To explain how the body works, physiologists use tools from many different fields, including biochemistry, cell biology, genetics, physics, and even engineering. In physiology, there is something for everyone. If you enjoy thinking about things on a cellular or molecular level, for example, you will find topics such as membrane transport or nerve cell signaling especially interesting. If you like thinking on a larger, organ-level or "whole-body" scale, the study of cardiovascular or respiratory function might be your cup of tea. Most physiologists would say that the best thing about studying physiology is that you get to explore *all* these things.

Because nearly everyone is curious about how the human body works, there is a good chance that studying physiology will be one of your most satisfying academic experiences. The study of physiology not only offers you the satisfaction of knowing something about the workings of the body, but it also may provide you a deep, perhaps even profound, understanding of it. You will also come to realize that physiology, like the other sciences, is not just a collection of well-worn facts but rather a work in progress. You will recognize that there are significant gaps in our understanding of how the body works, and you will also see that much of our current understanding will be subject to change as new discoveries are made.

Regardless of your background or current affinities, your study of physiology will broaden your scientific interests and will widen the scope of your outlook. You will begin to see the "big picture," understanding body function not as a collection of unrelated phenomena but as a connected whole. You might even discover something else—that physiology is beautiful. Most of us who have decided to make it our life's work think so.

Organization of the Body

If you have ever spent time examining a detailed anatomical chart or model of the human body, you have seen that it is an exceedingly complex and intricate structure. Despite the complexity of its structure, however, there is an underlying simplicity to the function of the human body.

To a physiologist, perhaps the most interesting thing about the body is that its operation can be explained in terms of a relatively small set of easily understood principles. For this reason, a physiologist's approach to studying the body is to strip away all unnecessary details so that the essentials—that is, the unifying themes and principles—can be seen more clearly. To get an idea of what this means, consider **Figure 1.1**, which shows four kinds of cells found in the brain. Although these cells are all clearly different, you can see that they can be classified into a relatively small number of categories on the basis of similarities in their morphology (shape). This is fortunate, because the brain contains billions of such cells and it is unlikely that any two are exactly alike! When you consider the function of these cells, however, the similarities among them extend even further, allowing them to be grouped into just one category: All these cells (and many others throughout the body) are specialized to transmit information in the form of electrical signals from one body location to another. Because of these similarities, all these cells are classified as *neurons* (or *nerve cells*).

If the body's underlying simplicity is one of physiology's major themes, another is the degree of interaction among its various parts. Although each of the body's **cells**, the smallest living units, is independently capable of carrying out its own basic life processes, the various types of cells are specialized to perform different functions important to the operation of the body as a whole, and for this reason all the cells ultimately depend on each other for their survival. Similarly, the body's organs are also specialized to perform certain tasks vital to the operation of other organs. You know, for example, that your cells need oxygen to live and that oxygen is delivered to your cells by the bloodstream, but consider some of the many things that must occur in order to ensure that

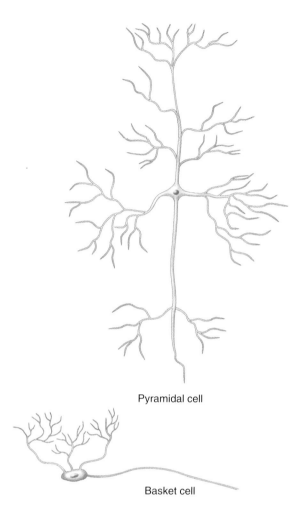

Pyramidal cell

Basket cell

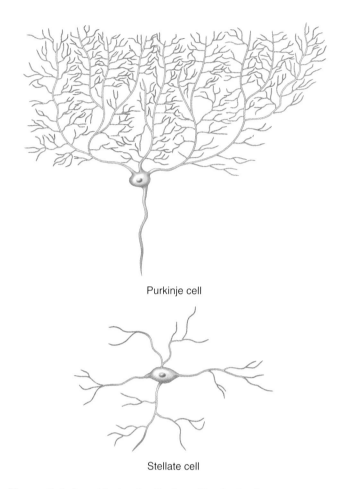

Purkinje cell

Stellate cell

Figure 1.1 **Four kinds of cells found in the brain.**

oxygen delivery is sufficient to meet the cells' needs. Oxygen is carried in the bloodstream by cells called *erythrocytes,* which are manufactured by *bone marrow,* a tissue found inside certain bones. To ensure that adequate numbers of erythrocytes are present in the blood, the synthesis of these cells is regulated by a hormone called *erythropoietin,* which is secreted by the kidneys. To ensure adequate blood flow to the body's tissues, the heart must pump a sufficient volume of blood every minute, and for this reason the rate and force of its contractions are regulated by the nervous system. To ensure that the blood carries enough oxygen, the lungs must take in sufficient quantities of air, which requires the control of breathing muscles (such as the diaphragm) by the nervous system. Finally, to provide the energy necessary to drive these and other processes, the gastrointestinal system breaks ingested food down into smaller molecules, which are absorbed into the bloodstream and distributed to cells throughout the body.

This example shows that proper body function requires not only that each part be able to carry out its own particular function, but also that the parts be able to work together in a coordinated manner. To help you better understand how the body's parts work together,

the remainder of the chapter outlines broad principles pertaining to body function in general; the functions of specific organs and organ systems are the topics of later chapters.

Cells, Tissues, Organs, and Organ Systems

The human body is a remarkable structure consisting of cells arranged in an orderly fashion. Cells are grouped together to form tissues, which are grouped together to form organs. Organs work together as organ systems. We now describe each of these hierarchical components.

Cells

Although over 200 distinguishable kinds of cells are present in the body, there are only four major types of cells: (1) *neurons,* (2) *muscle cells,* (3) *epithelial cells,* and (4) *connective tissue cells.* Representative cells belonging to each of these cell types are shown in **Figure 1.2**. These classifications are very broad and are based primarily on functional differences. There are other, more rigorous ways to classify cells based on anatomical distinctions and embryological origins.

respond to that stimulus. During an LH surge, for example, the LH concentration rises rapidly to a peak and then begins to fall because the surge triggers ovulation, which temporarily inhibits the ovaries' ability to secrete estrogens. The resulting fall in plasma estrogen levels removes the stimulus that caused LH secretion to rise in the first place, thereby allowing LH levels to fall.

Homeostasis in Action: Thermoregulation

When nighttime comes to the desert, the snakes, lizards, and insects that were active in the noonday sun begin to sink into a state of relative torpor; many are barely able to move at all. This change occurs because the falling temperature causes these animals' bodies to cool, which slows down biochemical reactions and other metabolic processes. By contrast, humans and other mammals are less affected by changes in the ambient temperature because they have the ability to maintain their body temperature within a fairly narrow range (a process called **thermoregulation**). Animals with this ability are said to be *homeothermic*, whereas those lacking this ability are *poikilothermic*.

All animals (and other living things) produce heat as a by-product of metabolism, but homeothermic animals are able to control body temperature by regulating the rates at which heat is produced and lost from the body. By increasing the rate of heat production and/or decreasing the rate of heat loss, body temperature is made to rise; by decreasing heat production and/or increasing heat loss, body temperature is made to fall.

However, there are limits to what these thermoregulatory mechanisms can accomplish. Prolonged exposure to very cold temperatures can cause body temperature to fall below the set point, a condition known as **hypothermia**. This happens, for example, to those unfortunate enough to be trapped in mountain snowstorms or swept from a boat into icy waters. Such misfortunes can quickly lead to stupor, loss of consciousness, multiple organ failure, and ultimately death. In contrast, overexertion or exposure to very high environmental temperatures can cause the body temperature to rise above the set point, a condition called **hyperthermia**. If severe enough it can lead to loss of consciousness, convulsions, respiratory failure, and death. Adverse effects begin to appear when body temperature approaches 41°C (106°F); a temperature of 43°C (109°F) or higher is usually fatal.

Mechanisms of Heat Transfer Between the Body and the External Environment

Under most conditions, the body loses heat to the environment because the surrounding temperature is normally lower than body temperature. When the rate of heat loss equals the rate of heat generation, body temperature does not change. Generally speaking, heat loss occurs via three different mechanisms: (1) *radiation,* (2) *conduction,* and (3) *evaporation.*

In **radiation**, thermal energy is transferred from the body to the environment in the form of electromagnetic waves. It is a general law of physics that all objects emit and absorb these waves, though to varying degrees. When an object is warmer than its surroundings, it emits more energy than it absorbs and loses heat. By contrast, if an object is cooler than its surroundings, it absorbs more energy than it emits and gains heat. If you have ever felt heat from a fire when you were standing several feet away, it is because you were absorbing radiated heat energy.

Conduction is the transfer of thermal energy between objects that are in direct contact with each other. As in radiation, heat is always transferred from the warmer object to the cooler object. When you touch cold metal or cold water, for example, you feel colder because thermal energy is transferred directly from your skin to the surrounding medium.

In **evaporation**, heat is lost from an object through the evaporation of water from its surface. When water evaporates from your body, the process that converts it from liquid form to gaseous form absorbs thermal energy. Some evaporation occurs through the skin, the lining of the lungs, and other moist surfaces such as the lining of the mouth, and is known as *insensible water loss* because it occurs on a continual basis without your being aware of it. Your body also loses water through the evaporation of *sweat,* a salt-containing solution secreted by numerous small *sweat glands* in the skin. Unlike insensible water loss, which is unavoidable, sweat production is regulated according to the body's needs. When increased heat loss is desirable, sweat production is increased. As a result, more water evaporates from the skin surface, which carries away more thermal energy from the body.

When the environmental temperature is warmer than body temperature, radiation and conduction actually transfer heat *into* the body. Because this transfer only adds to the heat generated by the body itself, how can the body regulate its temperature under these conditions? The answer is that the body relies solely on *evaporation* to carry heat away and therefore increases the production of sweat. Sweating cools the body under these conditions because water continues to evaporate even when it is cooler than its surroundings (assuming that the humidity of the surroundings is not too high).

Convection, the transfer of heat from one place to another by moving gas or liquid, contributes to heat loss on a windy day. Under still conditions, the air that is closer to your skin warms up due to the absorption of heat from the body's surface. This warmer air forms a kind of "blanket" around you that slows down the rate of heat loss by conduction. Because the air in this protective layer contains moisture that has evaporated from the skin, it also tends to be moister than the surrounding air.

The presence of this moisture near the skin reduces the rate of evaporative heat loss. When the surrounding air is moving, as occurs on a windy day, the thickness of this protective "blanket" of air is reduced, and conductive and evaporative heat loss both increase.

The Components of the Body's Thermoregulatory System

The body's thermoregulatory efforts are coordinated by a number of different centers in the brain, the most important of which is in the *hypothalamus,* a region located in the base of the brain just above the pituitary gland. Input to these brain centers comes from thermoreceptors located within the brain itself, the spinal cord, and other internal organs. These receptors, known as *central thermoreceptors,* are important because they monitor the temperature deep within the body (known as the **core temperature**). Other thermoreceptors, called *peripheral thermoreceptors,* are located in the skin, whose temperature is usually well below core temperature and is more variable. These receptors, which are of less importance in thermoregulation, enable us to sense the temperature of the environment. This ability is indirectly important in temperature regulation because it allows us to compensate for changes in environmental temperature by making behavioral adjustments, such as dressing appropriately or avoiding extreme temperatures altogether.

Output from brain thermoregulatory centers is transmitted by neurons to various effectors that vary the rate of heat production or loss. The effectors include *sweat glands,* which control evaporative heat loss by increasing or decreasing sweat secretion; *blood vessels in the skin,* which control conductive and radiative heat loss by increasing or decreasing blood flow to the skin; and *skeletal muscles,* which control heat production through *shivering*—bursts of rapid, involuntary contractions that generate heat as a metabolic by-product. When the core temperature changes, brain thermoregulatory centers attempt to compensate by sending appropriate commands to these effectors. As long as the outside temperature is within a narrow range called the *thermoneutral zone* (25–30°C), alteration of skin blood flow is usually sufficient to adjust the core temperature to normal. Outside this range, the other effector responses are called into play.

The Thermoregulatory Response to a Drop in Temperature: A Negative Feedback System in Operation

Suppose the temperature of the body's surroundings drops from 30°C to 15°C (**Figure 1.9**a). As the body's rate of heat loss increases, the core temperature gradually drops below the normal set point, from 37°C to an eventual minimum of 36°C (Figure 1.9b). As this is occurring, information concerning the body's temperature is relayed by central thermoreceptors to the brain's thermoregulatory centers, which detect the error signal and orchestrate the following responses (Figure 1.9c):

1. *Decreased sweat production:* If the sweat glands are active initially, sweat production decreases or even stops altogether, which reduces evaporative heat loss.

2. *Decreased blood flow to skin:* Certain blood vessels in the skin constrict in response to neural commands, leading to a decrease in blood flow to the skin. Because the blood normally carries heat from the warmer deep regions of the body to the cooler outer regions, this reduction in flow causes skin temperature to decrease. As a result, rates of radiative and conductive heat loss decrease.

3. *Stimulation of shivering:* Skeletal muscles begin to shiver, which generates heat. This heat adds to the heat generated by other metabolic processes, and the body's overall rate of heat production increases.

These responses are part of a negative feedback system because the reduction in heat loss coupled with the increase in heat production raises core temperature, which counters the initial decrease in temperature. If these compensatory changes are able to restore the balance between heat production and loss, body temperature eventually returns to near its normal value and stays there (see Figure 1.9b). But no negative feedback system is perfect, and thermoregulatory mechanisms cannot hold body temperature absolutely constant. Instead, the temperature undergoes small variations and fluctuates above and below the set point; such fluctuations are normal and occur in all physiological variables.

In contrast to the events depicted in Figure 1.9, a rise in the surrounding temperature will cause a rise in core temperature, which triggers the opposite responses—increased sweat production, increased blood flow to the skin, and a decrease or complete cessation of shivering. The resulting increase in heat loss and decrease in heat production should act to reduce the core temperature, returning it to near normal.

Fever, an elevation of body temperature that frequently occurs during illness, does *not* result from a failure of the thermoregulatory system. Rather, fever results from a resetting of the set point that causes the system to raise the body's temperature in a controlled fashion. Fever is most often triggered in response to bacterial or viral infection, which stimulates certain blood cells (*white blood cells*) to proliferate and secrete various chemical substances. One or more of these substances act on thermoregulatory centers in the brain to increase the set point and therefore act as *pyrogens* (fever-inducers). Because the resulting temperature elevation stimulates an increase in the rate of many immune responses against invading microorganisms, fever is actually considered beneficial because it enhances the body's ability to defend itself.

Table 1.2 ▏ Physical Characteristics of the Two Marathon Runners

	Bill	Jane
Age (years)	33	31
Height (cm)	177	166
Weight (kg)	72	57
Percent fat	16	22
Aerobic capacity (mL of O_2/kg/min)	55	51

Different types of exercise can elicit different patterns of physiological response. We will look at the physiological responses to running a marathon because it imposes large stresses on healthy individuals. People participating in a marathon can increase their energy expenditure (and hence their oxygen consumption and heat production) by 15-fold or more. In addition, they must maintain this increased metabolism for relatively long periods of time—anywhere between two and six hours, depending on their talents and training status. The runners either finish the 26.2-mile race, or they are forced to drop out along the way. The factors that determine how fast (or if) they finish will be described here.

Although we must wait until the final chapter for our full treatment of exercise physiology, we don't have to wait that long to begin our exploration of the subject. In fact, we can begin shortly by following the experiences of our two runners, who as we join them are beginning their preparations to run their marathon. As you read through the following story, you may not understand all the terms or statements in the narrative, but do not worry about that now. When you study later chapters in this textbook and learn about various physiological mechanisms, we will use features called *Exercise Links* to point out how particular mechanisms play a role in the experiences of our intrepid runners. At the end of the book you will be asked to read the story again. By then, all the terminology will be familiar, and you will have a more complete understanding of the events.

Before we consider their race experiences we turn to some relevant physical characteristics for Bill and Jane, displayed in **Table 1.2**. (Male and female runners were chosen as the subjects of this story to illustrate some physiological differences between the sexes that affect exercise performance.) Bill is taller and heavier because he has more muscle and a larger skeleton, but Jane has a higher percentage of body fat. Bill has a higher *aerobic capacity;* that is, he has a greater ability to use oxygen to generate energy when exercising at his maximal effort. (This measure is one indicator of physical fitness.) Aerobic capacity is measured as the volume of oxygen (mL of O_2) that is consumed per minute. Because a larger person usually has more cells that are using oxygen, regardless of physical fitness, the rate of oxygen consumption is adjusted to the person's body weight (per kilogram)

when calculating aerobic capacity. Based simply on aerobic capacity, one might predict that Bill can run slightly faster than Jane for a prolonged period of time. Over shorter race distances ("10K" races, which are 10 kilometers or 6.2 miles) this is in fact true: Bill's best time is 43 minutes, and Jane's best is about a minute slower. But as we will see, aerobic capacity is only one of many factors that determine performance in a marathon.

Many integrated physiological processes will play out during the marathon, and several decisions by the two runners will have profound influences on these processes. For those of you who have participated in endurance events, some of these situations will be familiar. Now, let's examine Bill and Jane's marathon experiences.

Bill and Jane's Marathon

Our narrative is divided into three time periods. It begins not as the runners leave the starting line, but six months earlier, as Bill and Jane begin to train for the marathon.

Training

If Bill and Jane had decided impulsively to run a marathon without preparation, the chances that they might finish would be limited. Exercise causes depletion of resources stored in the body (water, energy, and nutrients) and a certain amount of "wear and tear" on the muscles, tendons, and joints. During recovery from exercise, physiological processes not only restore the resources and repair the strained tissues but also overshoot the original (untrained) condition by increasing the storage capacities and strengthening the tissues (adapted condition). Thus training comprises repetitive cycles of strain and depletion followed by a recovery process of overcompensation resulting in an upward spiral of increasing physical and functional capacity (**Figure 1.10**).

Without worrying about the actual physiological mechanisms yet, several aspects of training are apparent in Figure 1.10. First, each exercise session causes some debilitation; thus a person in training must be careful not to "overdo it" in any single session. Second, a recovery period is essential for the positive adaptations to occur; thus proper rest intervals are as important as the exercise sessions themselves. A third issue is not obvious in the figure: As an individual reaches a new level of adaptation, the exercise stimulus needed to keep the spiral rising increases. The increased stimulus can be accomplished by increasing either the intensity or the duration of the exercise (or a little of both). After many months of training, Bill and Jane have developed the capacity and strength to complete a marathon with relatively little debilitation. As listed in **Table 1.3**, some organs and tissues actually grow larger in response to training, but many of the changes are the result of improved function rather than increased size.

Bill and Jane (as do many marathon trainees) reduce their training intensity in the week immediately preceding the race. With their goal so close at hand, they want to minimize the "debilitation" of training (and also want to avoid accidental injury after training so long).

Prelude to the Race

In the 24–48 hours just before the race, Bill and Jane "carbo-load" (eat lots of starchy and sweet foods such as bread, pasta, and fruits). These types of food are the source of the most rapidly used energy source—carbohydrate—and the runners are "topping off their fuel tanks" in preparation for the long journey ahead of them. Some runners, including Bill, drink plenty of coffee on the morning of the race. The caffeine in coffee is a stimulant, and Bill knew enough physiology to understand that caffeine also promotes use of carbohydrates as an energy source. Jane, however, knew an additional important fact—that caffeine adversely affects the body's ability to maintain proper water volumes (blood, for example, is mostly water). Jane drank less coffee. These and other decisions will have important consequences in each runner's marathon experience.

The Race

The race starts at 9 A.M. under clear skies and comfortable (68°F) temperatures. It is a popular race that draws several thousand runners. The "elite" runners are placed in a reserved area immediately behind the starting line; everyone else is asked to begin farther back, according to their expected finishing time. Bill and Jane are in the middle of the pack. When the gun goes off, only those up front are actually able to start running. Farther back,

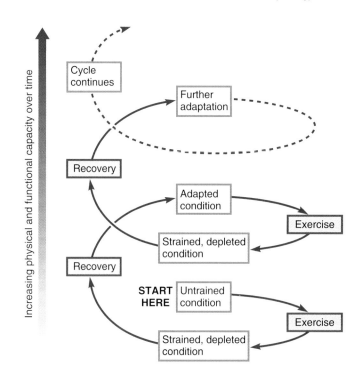

Figure 1.10 The upward spiral of exercise training. Small increments of strain and debilitation resulting from each exercise session are followed by overcompensation during recovery, which leads to an adapted condition with higher physical and functional capacity than the previous condition.

Table 1.3 ▌ Anatomical and Functional Changes Resulting from Training

Organ, tissue, or structure	Anatomical changes	Functional changes
Heart	Size increases	Resting heart rate decreases; resting stroke volume increases, which increases pumping efficiency
Blood	Volume of blood and red cell mass increase	Oxygen-carrying capacity of blood increases
Skeletal muscle	Muscle mass, capillary density, and glycogen stores increase	Ability to develop useful power increases
Lungs	N/A	Ability of airways to widen in response to neural and hormonal influences increases, which increases ease of breathing during exercise
Bone	Density increases	Bone strength increases, which increases the body's ability to withstand the physical stresses of exercise
Endocrine and autonomic nervous systems	N/A	Sensitivity of tissues to insulin and pituitary hormones, and to autonomic input increases, which enhances mobilization of energy stores during exercise
Various cells	N/A	Synthesis of stress proteins inside cells increases
Brain	N/A	Learning leads to improved technique, which decreases exertion; coping mechanisms decrease perceived exertion
Sweat glands	N/A	Production of more copious and more dilute sweat, which enhances the body's ability to cool itself

CHEMISTRY REVIEW

POLAR MOLECULES AND HYDROGEN BONDS

When two atoms are covalently bonded, they share electrons. However, this sharing may or may not be equal. Certain atoms, such as oxygen (O), nitrogen (N), or sulfur (S), hold onto electrons tightly. Therefore, when bonded to other atoms, they have a tendency to pull electrons away from the other atoms. As a result, each oxygen, nitrogen, or sulfur atom acquires a partial negative charge because it has slightly more than its "fair share" of electrons; that is, it has more electrons than are necessary to balance the positive charge of the protons in its nucleus. In contrast, the atom to which the oxygen, nitrogen, or sulfur is bonded is left with less than its "fair share" of electrons, which gives it a partial positive charge. This can be illustrated as follows for molecules in which oxygen, nitrogen, or sulfur is bound to hydrogen, with the R representing the remainder of the molecule and the δ representing a partial charge:

$$\overset{\delta^-\;\;\delta^+}{R:O:H} \qquad \overset{\delta^-}{R:N:H} \qquad \overset{\delta^-\;\;\delta^+}{R:S:H}$$
$$\underset{\overset{|}{H}\;\delta^+}{}$$

Bonds characterized by such unequal electron sharing are known as *polar bonds*. In contrast, in *nonpolar bonds* electrons are shared more or less equally, such that the atoms remain uncharged. Carbon to carbon (C—C) and carbon to hydrogen bonds (C—H) are common examples of nonpolar bonds found in the molecules of the body.

The presence of electrical charges within a polar molecule produces electrical forces. Direction of electrical forces follow a simple rule: *Opposite charges attract; like charges repel.* The positive region of one polar molecule is electrically attracted to the negative region of another polar molecule (or in some cases, the positive and negative regions can be within the same large polar molecule). This electrical attraction holds the two polar molecules together, forming a *hydrogen bond*. The polar bond is called a hydrogen bond because in a polar molecule it is usually a hydrogen that contains the partial positive charge. The most common polar molecule is water. Hydrogen bonds between polar water molecules can be represented as follows:

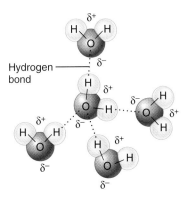

Even though hydrogen bonds are weak bonds and can easily be broken, they are important in determining the structure of large biomolecules such as proteins and nucleic acids, and in establishing the properties of substances in water. How molecules behave in water is critical to cell function. Polar molecules are electrically attracted to the polar water molecules, and therefore they dissolve in water. Polar molecules are called hydrophilic ("water-loving") because of their ability to dissolve in water. Nonpolar molecules are called hydrophobic ("water-fearing") because they do not dissolve in water.

Although water is the most common solvent in the body, membranes are composed of lipids, and how a substance behaves in lipids is also important to physiology. Nonpolar molecules dissolve in lipids and can permeate the phospholipid bilayer of membranes, and thus are called lipophilic ("lipid-loving"). Polar molecules cannot dissolve in lipids nor permeate the phospholipid bilayer of membranes, and thus are called lipophobic ("lipid-fearing"). The table below summarizes the nature of molecules held together by covalent bonds. Notice that hydrophilic molecules are lipophobic, and hydrophobic molecules are lipophilic.

Electron sharing in covalent bond	Class of covalent bond	Property in water	Property in lipid
Equal sharing	Nonpolar	Hydrophobic	Lipophilic
Unequal sharing	Polar	Hydrophilic	Lipophobic

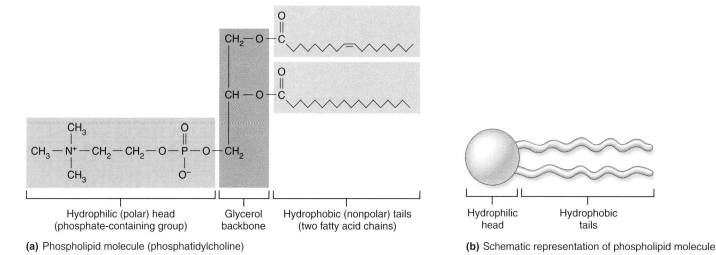

(a) Phospholipid molecule (phosphatidylcholine)

(b) Schematic representation of phospholipid molecule

Figure 2.3 Phospholipids. (a) A phospholipid, which consists of a glycerol backbone linked to two fatty acid chains and a phosphate-containing group. The phospholipid shown here is *phosphatidylcholine.* (b) The standard way to schematically depict a phospholipid, emphasizing the hydrophilic head and hydrophobic tails of the molecule.

Phospholipids

Phospholipids are lipids that contain a phosphate group. They are similar in structure to triglycerides in that a glycerol forms the backbone. However, instead of three fatty acids, a phospholipid contains two fatty acids and has a phosphate group attached to the third carbon of glycerol (**Figure 2.3**). The two fatty acids form the *tail* region of the phospholipid, which is nonpolar because of their long chains of carbon atoms. The phosphate group is generally attached to another chemical group, and together they form the *head* region of the phospholipid, which is polar. Therefore, phospholipids have both a polar region and a nonpolar region, making them amphipathic molecules.

The amphipathic property of phospholipids gives them unique behaviors in an aqueous or watery environment. The polar regions can dissolve in water, but the nonpolar regions cannot. Therefore, when phospholipids are placed in an aqueous environment, the polar regions face the water, and the nonpolar regions face each other. Phospholipids form two physiologically important structures when placed in an aqueous environment: phospholipid bilayers and micelles (**Figure 2.4**). In a phospholipid bilayer (Figure 2.4a), which is the core structure of cell membranes, the phospholipids are arranged in two parallel layers: The tails of parallel phospholipids face inward toward each other, and the heads face the outside, where they come into contact with the aqueous environment. A micelle (Figure 2.4b) is a spherical structure composed of a single layer of phospholipids; it functions in the transport of nonpolar molecules in an aqueous environment. The heads of the phospholipids face outward, where they come into contact with the aqueous environment; the tails face inward, forming a hydrophobic interior.

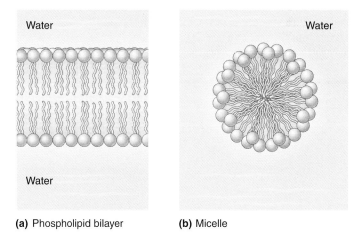

(a) Phospholipid bilayer   (b) Micelle

Figure 2.4 Structures formed by phospholipids in an aqueous environment. (a) A lipid bilayer, which conists of two sheets of phospholipids aligned such that the polar heads face the aqueous environment and the nonpolar tails face each other. (b) A micelle, a sphere formed by a single layer of phospholipids aligned such that the polar heads face the aqueous environment and the nonpolar tails face each other.

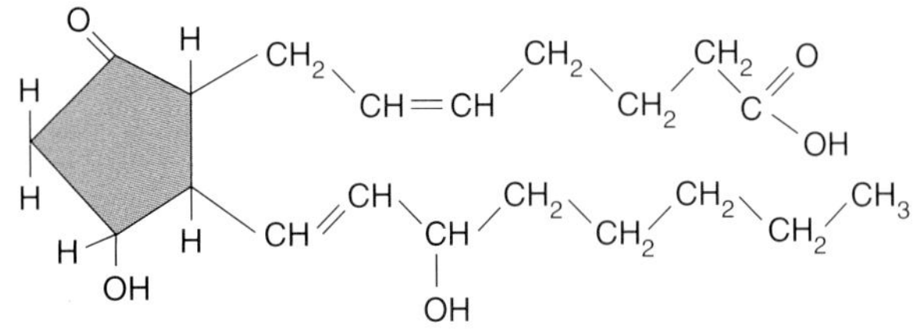

Figure 2.5 Eicosanoids. In these modified 20-carbon fatty acids, the carbons in the middle of the carbon chain form a ring that causes the molecule to fold upon itself.

Eicosanoids

Eicosanoids are modified fatty acids that function in intercellular communication. An eicosanoid is derived from a 20-carbon fatty acid and contains a five-carbon ring in the middle (**Figure 2.5**). This ring causes the molecule to fold upon itself, such that two chains of carbon atoms extend parallel to each other away from the ring. Eicosanoids include *prostaglandins, thromboxanes,* and *leukotrienes.*

Steroids

Steroids have a unique chemical structure consisting of three six-carbon rings and one five-carbon ring (**Figure 2.6**a). The most common steroid is cholesterol (Figure 2.6b). Due to the polar hydroxyl group on one end of the mostly nonpolar cholesterol molecule, it is a slightly amphipathic molecule. Cholesterol is an important component of the *plasma membrane,* the membrane surrounding cells. In addition, cholesterol is the precursor to all other steroids. Many steroids function as hormones, including *testosterone* (Figure 2.6c), *estradiol, cortisol,* and *calcitriol* (also called *1,25-dihydroxyvitamin D₃*).

Amino Acids and Proteins

Proteins are polymers of **amino acids,** relatively small biomolecules that contain a central carbon, an amino group, a carboxyl group, a hydrogen, and an R or residual group (**Figure 2.7**a). There are 20 different R groups, and thus 20 different amino acids. The three amino acids shown in Figure 2.7b—alanine, tyrosine, and glutamate—have R groups with different chemical properties: nonpolar, polar, and charged, respectively. Although we will be discussing amino acids as components of proteins, they have other functions as well, including intercellular communication.

Polymers of amino acids are formed by joining two amino acids together by a *peptide bond;* these polymers are therefore called **polypeptides.** A peptide bond forms between the amino group of one amino acid and the carboxyl group of another amino acid by a *condensation reaction,* which is a reaction that releases water as two small molecules are joined together (**Figure 2.8**). Polypeptides vary in length from just two to several hundred amino acids; the name given to them differs based on the length or function. **Peptides** are short chains of amino acids, usually less than 50. Proteins consist of chains that generally are longer than 50 amino acids and often are hundreds of amino acids long.

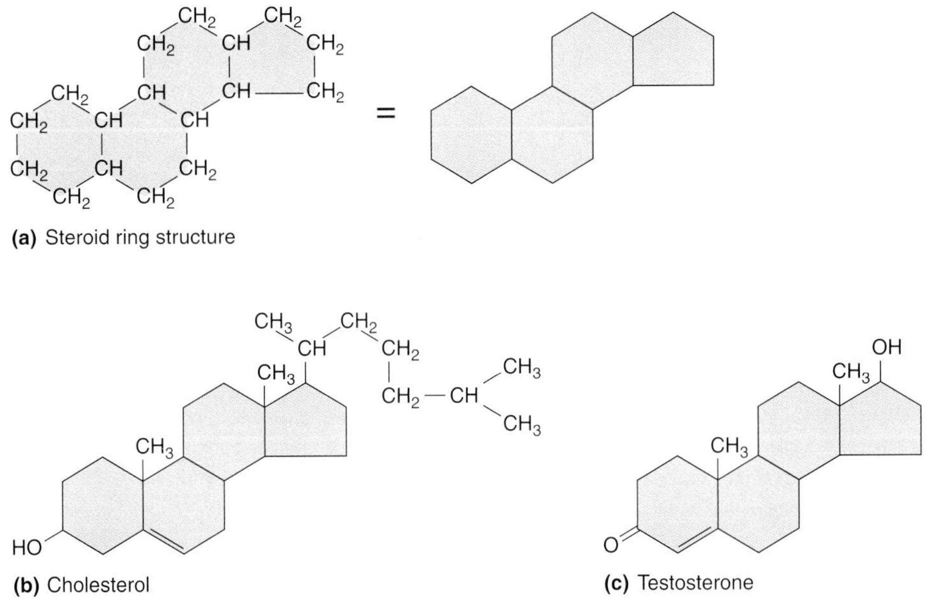

(a) Steroid ring structure

(b) Cholesterol

(c) Testosterone

Figure 2.6 Steroids. (a) The basic structure of all steroids, which consists of three six-carbon rings and one five-carbon ring. (b) Cholesterol, the most common steroid and the precursor for all other steroids in the body. (c) Testosterone, an example of a steroid hormone.

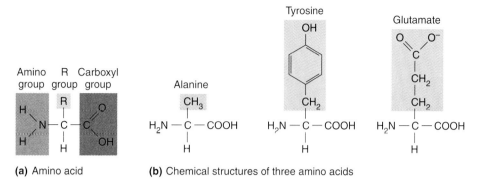

Figure 2.7 Amino acids. (a) The basic structure of an amino acid. The central carbon is bonded to an amino group, a carboxyl group, a hydrogen, and an R group. (b) Structures of three of the 20 amino acids. The different R group in each of these amino acids gives them different chemical properties: Alanine is nonpolar, tyrosine is polar, and glutamate is charged.

The function of a protein is highly dependent on its three-dimensional structure or *conformation*. Protein structure can be described at as many as four different levels: *primary, secondary, tertiary,* and *quaternary structure* (**Figure 2.9**). The levels of protein structure can be likened to a telephone receiver's cord: Primary structure is the cord stretched out straight, secondary structure is the coiled cord, tertiary structure is the loops and bends in the coiled cord, and quaternary structure is the existence of two cords wrapped together. Now let's relate this analogy to the details of protein structure.

Primary protein structure simply refers to the sequence of amino acids, which is determined by peptide bonds within the peptide chain (see Figure 2.9a). Secondary protein structure is the folding pattern produced by hydrogen bonds between the hydrogen atom in the amino group of one amino acid and the oxygen atom in the carboxyl group of another amino acid in the same polypeptide. These hydrogen bonds fold the polypeptide into various shapes, such as α-helixes and β-pleated sheets (see Figure 2.9b), which contribute to the three-dimensional structure of proteins.

Tertiary protein structure is the folding produced by interactions between the R groups of different amino acids in the same polypeptide (see Figure 2.9c). Recall that the R groups have different chemical properties. The types of chemical interactions that can occur depend on the amino acids involved and include the following: (1) *hydrogen bonds,* (2) *ionic bonds,* (3) *van der Waals forces,* and (4) *covalent bonds.* Hydrogen bonds can form between polar R groups. Ionic bonds can form between ionized or charged R groups (**Chemistry Review: Ions and Ionic Bonds**, p. 37). Van der Waals forces are electrical attractions between the electrons of one atom and the protons of another atom. Covalent bonds form between the R groups of two cysteines. The R group of cysteine is a sulfhydryl group, and the resulting covalent bond is called a *disulfide bridge.*

Quaternary protein structure exists only in proteins containing more than one polypeptide chain. An example is the protein hemoglobin (see Figure 2.9d), which functions in the transport of oxygen in the blood. Hemoglobin is a single protein consisting of four separate polypeptide chains. Hemoglobin, like all proteins, can function properly only when in the correct conformation.

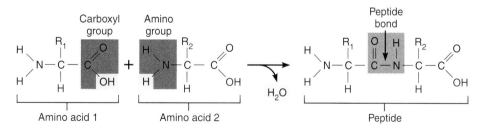

Figure 2.8 Formation of a peptide bond by a condensation reaction. The peptide bond is formed between two amino acids, and water is released.

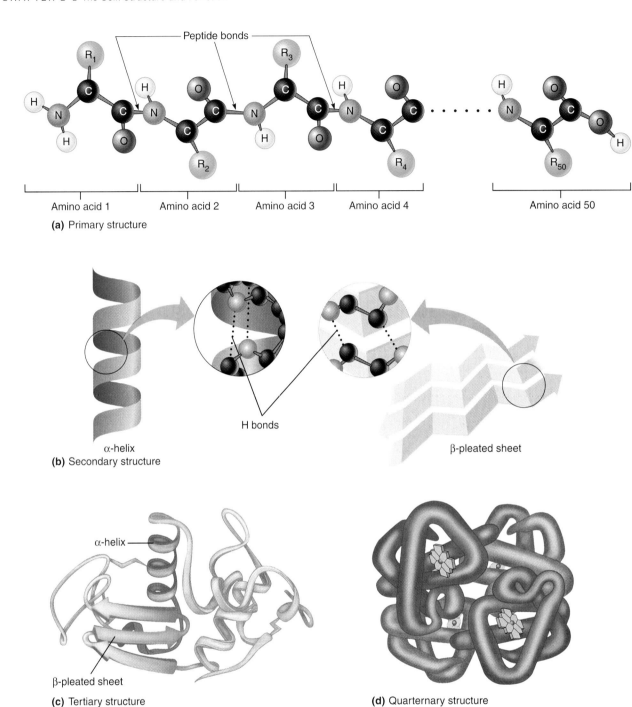

(a) Primary structure

(b) Secondary structure

α-helix

β-pleated sheet

H bonds

(c) Tertiary structure

α-helix

β-pleated sheet

(d) Quarternary structure

Figure 2.9 Levels of protein structure. (a) Primary protein structure, which is the sequence of amino acids. (b) Secondary protein structure, which is caused by hydrogen bonding between the amino hydrogen of one amino acid and the carboxyl oxygen of another amino acid. Common secondary structures include α-helixes and β-pleated sheets. (c) Tertiary protein structure, which is the folding pattern produced by interactions between the R groups of amino acids. The protein shown here is the enzyme *lysozyme.* (d) Quaternary protein structure, which is the arrangement of more than one polypeptide chain in a single protein. Shown here is *hemoglobin,* which consists of four polypeptide chains.

The three-dimensional conformation of proteins can be classified as either *fibrous* or *globular* (**Figure 2.10**). *Fibrous proteins* are generally extended, elongated strands that function in structure or contraction. Examples of fibrous proteins include *collagen,* a protein found in tendons and bone, and *tropomyosin,* a protein found in muscle cells. *Globular proteins* are coiled, folded, irregular, and bulky. Among their many functions, they act as *chemical messengers* for intercellular communication, as *receptors* that bind chemical messengers, as *carrier proteins* that transport substances in blood or across membranes, and as *enzymes* that catalyze chemical reactions in the body. Examples of globular proteins include the oxygen binder *myoglobin,* the chemical messenger *growth hormone,* and a membrane transport protein called the Na^+/K^+ *pump.* Some proteins have both globular and fibrous components; one example is the muscle protein *myosin,* which has a globular head and fibrous tail. All the proteins mentioned here are discussed further in detailed descriptions of organ systems later in the book.

Some proteins have other classes of organic molecules attached to them. For example, *glycoproteins* have carbohydrates attached to the polypeptide chains, whereas *lipoproteins* have lipids attached to them. Glycoproteins are important components of the plasma membrane that surrounds cells; they also contribute to cell recognition, which is the ability of the immune system to recognize cells that are part of the body. Lipoproteins play an important role in the transport of lipids in blood.

Nucleotides and Nucleic Acids

Nucleotides function in the transfer of energy within cells, and they form the genetic material of the cells. The basic structure of a nucleotide is shown in **Figure 2.11**a. **Nucleotides** contain a five-carbon carbohydrate, a nitrogenous base, and one or more phosphate groups. The carbohydrates found in nucleotides are ribose and deoxyribose. The nitrogenous bases in nucleotides include two classes: (1) *pyrimidines,* which contain a single carbon ring and include *cytosine, thymine,* and *uracil,* and (2) *purines,* which contain a double carbon ring and include *adenine* and *guanine.* A nucleotide that has one phosphate is called a *nucleotide monophosphate* (Figure 2.11b); similarly, a nucleotide with two phosphates is a *nucleotide diphosphate,* and a nucleotide with three phosphates is a *nucleotide triphosphate.*

Several nucleotides function in the exchange of cellular energy. For example, the energy contained in the phosphate bonds of *adenosine triphosphate (ATP)* can be used to drive cell processes when ATP is hydrolyzed to ADP. *Nicotinamide adenine dinucleotide (NAD)* and *flavin adenine dinucleotide (FAD)* transfer energy in the form of electrons.

Some nucleotides form a ring due to covalent bonding between an oxygen of the phosphate group, and a

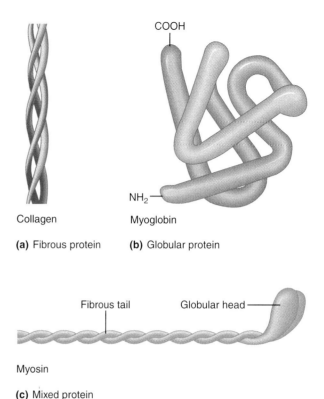

(a) Fibrous protein   **(b)** Globular protein

Collagen   Myoglobin

COOH

NH_2

Fibrous tail   Globular head

Myosin

(c) Mixed protein

Figure 2.10 Three-dimensional structures of proteins. (a) A fibrous protein, collagen, which consists of thin strands. (b) A globular protein, myoglobin. (c) A mixed protein, myosin, with a fibrous tail region and a globular head region.

What types of chemical forces cause the myoglobin molecule to fold into its three-dimensional structure?

carbon of the carbohydrate group, as shown in **Figure 2.12**. These nucleotides, called *cyclic nucleotides,* include chemical messengers (molecules that function as signals for the cell to do something) inside cells, such as *cyclic AMP (cAMP)* and *cyclic GMP (cGMP).*

Polymers of nucleotides include the **nucleic acids** that function in the storage and expression of genetic information: **deoxyribonucleic acid (DNA)** and **ribonucleic acid (RNA)** (**Figure 2.13**). DNA molecules are found in a cell's nucleus, where they store the genetic information. RNA molecules are found in both a cell's nucleus and its cytoplasm, and they are necessary for the expression of genetic information. Both DNA and RNA are polymers of nucleotides, with the phosphate group of one nucleotide linked to the carbohydrate of another, forming a chain with the bases sticking out to the sides. However, there are important differences in their structures, which we discuss next.

DNA (see Figure 2.13a) consists of two strands of nucleotides coiled together into a double helix. The

Hydrogen bonds, ionic bonds, and van der Waals forces

Figure 2.11 Nucleotides. (a) The basic structure of a nucleotide. The carbohydrate in a nucleotide is either deoxyribose or ribose. The five possible bases include the pyrimidines, which contain a single ring, and the purines, which contain two rings. (b) A schematic representation of nucleotides. Nucleotide monophosphates contain a single phosphate group, nucleotide diphosphates contain two phosphate groups, and nucleotide triphosphates contain three phosphate groups.

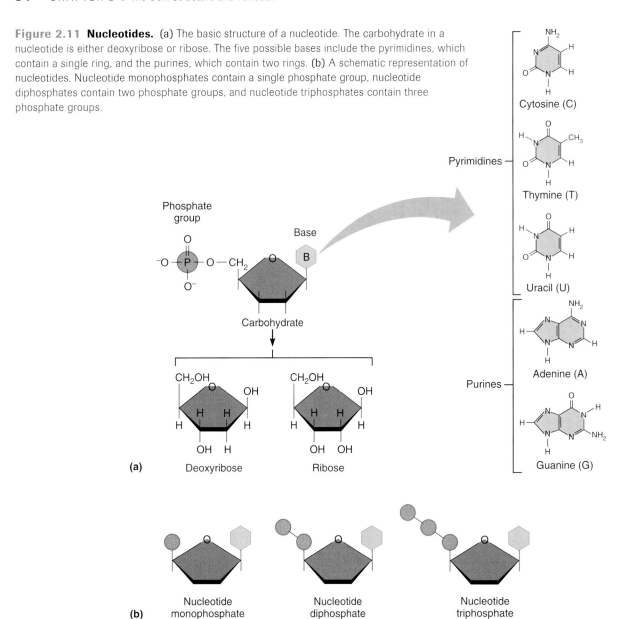

carbohydrate in DNA is deoxyribose. The bases in DNA include adenine (A), guanine (G), cytosine (C), and thymine (T). The ends of the strands are labeled as 3′ or 5′, with 3′ corresponding to the carbohydrate end and 5′

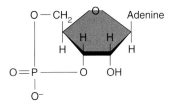

Cyclic AMP (cAMP)

Figure 2.12 The cyclic nucleotide cyclic AMP (cAMP), a common messenger molecule.

corresponding to the phosphate end. The two strands of DNA are held together by hydrogen bonding between bases according to the **law of complementary base pairing**, which states that *whenever two strands of nucleic acids are held together by hydrogen bonds, G in one strand is always paired with C in the opposite strand, and A is always paired with T in DNA (or with U in RNA).* Cytosine and guanine form three hydrogen bonds between them, and adenine and thymine form two hydrogen bonds between them (see Figure 2.13b). Because of this pairing, the two strands of DNA are complementary to each other.

RNA (see Figure 2.13c) consists of a single strand of nucleotides with a 3′ end and a 5′ end. The carbohydrate in RNA is ribose. The bases in RNA include adenine, guanine, cytosine, and uracil. The bases of RNA also follow the law of complementary base pairing; guanine and cytosine form hydrogen bonds between them, as in DNA, but adenine forms hydrogen bonds with uracil instead of

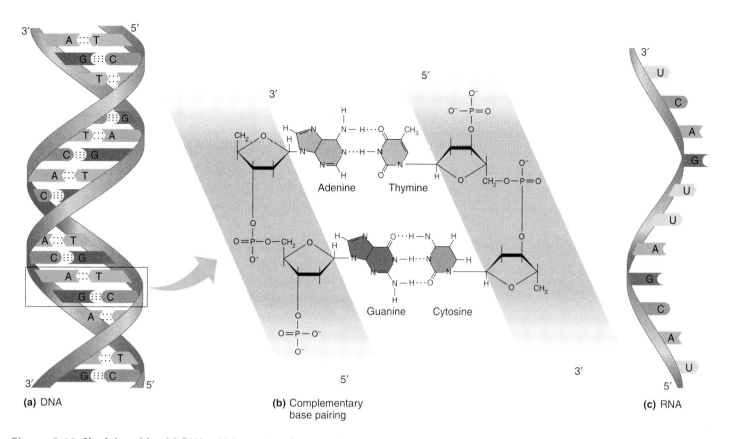

(a) DNA

(b) Complementary base pairing

(c) RNA

Figure 2.13 Nucleic acids. (a) DNA, which consists of two strands of nucleotides held together by hydrogen bonding according to the law of complementary base pairing (b) Complementary base pairing, with adenine (A) and thymine (T) forming two hydrogen bonds between them, and cytosine (C) and guanine (G) forming three hydrogen bonds between them. (c) RNA, which consists of a single strand of nucleotides.

What type of chemical bond links the phosphates and carbohydrates together to form a chain?

thymine. Even though RNA is single stranded, complementary base pairing is necessary to synthesize RNA from DNA (the RNA will be complementary to the DNA, except that thymine will be replaced with uracil), and to enable folding of a single RNA molecule upon itself.

> **Quick Test 2.1**
>
> 1. Are carbohydrates generally polar or nonpolar? What about triglycerides?
>
> 2. Name the two specialized structures phospholipids can form in an aqueous environment. What physiological functions are associated with these structures?
>
> 3. Name the subunits that make up the following polymers: glycogen, proteins, and nucleic acids. What are the general functions of each of these polymers?
>
> 4. Cholesterol is what type of biomolecule?

Cell Structure

When put together in a certain way, the biomolecules form the basic units of life: cells. The human body contains more than 100 trillion cells that work together to maintain homeostasis. Remarkably, all of these cells derive from a single fertilized egg. During development, the cells differentiate into over 200 different types of cells, each with its own specialized function. Although cells can be specialized both anatomically and functionally, they generally have the same basic components. In this section we describe the basic components of a "typical" cell (**Figure 2.14**).

Each cell is bound by a **plasma membrane**, which separates the cell from the extracellular fluid. Inside the cell are two main components: the **nucleus**, which is a membrane-bound structure that contains the genetic information for the cell, and the **cytoplasm**, which

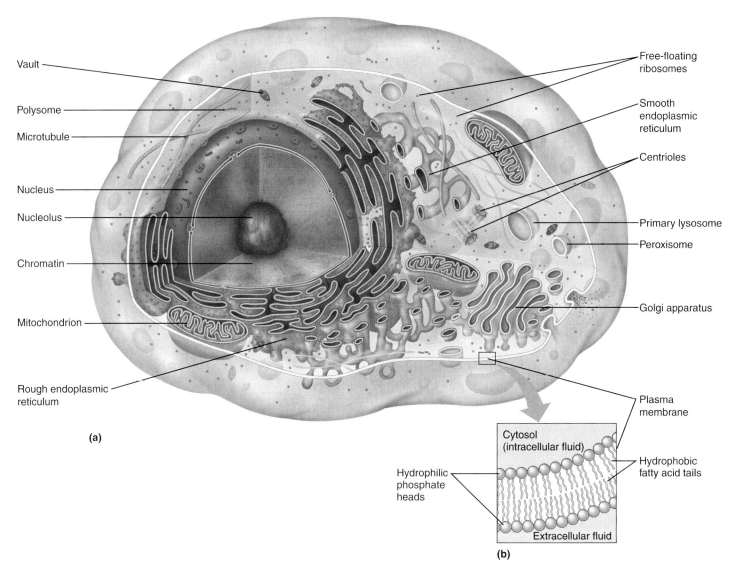

(a)

(b)

Vault

Polysome

Microtubule

Nucleus

Nucleolus

Chromatin

Mitochondrion

Rough endoplasmic reticulum

Free-floating ribosomes

Smooth endoplasmic reticulum

Centrioles

Primary lysosome

Peroxisome

Golgi apparatus

Plasma membrane

Cytosol (intracellular fluid)

Hydrophilic phosphate heads

Hydrophobic fatty acid tails

Extracellular fluid

Figure 2.14 Structures of a typical cell. (a) Three-dimensional view of the cell. The plasma membrane separates the cell from the extracellular fluid. The nucleus contains the genetic information. The cytoplasm contains the organelles and the cytoskeleton, both of which are surrounded by a fluid called cytosol. Organelles include mitochondria, endoplasmic reticulum, Golgi apparatus, lysosomes, peroxisomes, ribosomes, and vaults. (b) Enlargement of a portion of the plasma membrane showing the lipid bilayer.

includes everything inside the cell except the nucleus. The cytoplasm itself consists of two main components: the *cytosol,* or intracellular fluid, and the *organelles.* The **cytosol** is a gel-like fluid. The **organelles**, which are structures made up of a variety of biomolecules, carry out specific functions in the cell, much as the organs carry out specific functions in the body. There are *membranous organelles,* which are separated from the cytosol by one or more membranes, and *nonmembranous organelles,* which have no such boundary with the cytosol.

This section describes the anatomy of a cell; general functions of the various components of a cell are discussed in a subsequent section.

Structure of the Plasma Membrane

The cell contains several types of membrane that function as barriers between compartments. This section focuses on the structure of the plasma membrane, which separates the cell from its external environment. The structures of the *nuclear envelope,* which separates the nucleus

IONS AND IONIC BONDS

Some atoms have a tendency to gain or lose electrons completely, in the process acquiring an excess or deficit of electrons. Particles such as this have net negative or positive charges, respectively, and are known as *ions.* Negatively charged ions are known as *anions;* positively charged ions are known as *cations.* When anions and cations are present in solids, they tend to form crystals in which the cations and anions are closely associated. A familiar example is *sodium chloride (NaCl)* or table salt, which contains sodium ions (Na^+) and chloride ions (Cl^-). Sodium ions are formed when sodium atoms lose an electron, producing an ion with 11 protons and 10 electrons. Chloride ions are formed when chloride atoms gain an electron producing an ion with 17 protons and 18 electrons. This process occurs as follows:

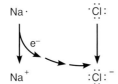

In a crystal of NaCl, the cations (Na^+) and anions (Cl^-) are held together by electrical forces of attraction due to their opposite charges. These forces are sometimes called *ionic bonds.* When ionic solids are dissolved in water, ionic bonds are disrupted by electrical attractions between the ions and polar water molecule, leaving cations and anions free to dissociate into separate particles. For sodium chloride this process can be illustrated as shown at right.

Solutions containing dissolved ions are described as *electrolytic* because they are good conductors of electricity, and ionic substances are referred to as *electrolytes.* Body fluids are electrolytic and contain a number of small ions (known as *inorganic ions*), including sodium, potassium (K^+), calcium (Ca^{2+}), hydrogen (H^+), magnesium (Mg^{2+}), chloride, sulfate (SO_4^{2-}), and bicarbonate (HCO_3^-).

Ionized chemical groups can also be found on certain types of biomolecules. Molecules containing significant numbers of polar bonds or ionized groups are described as *polar* or *hydrophilic;* those possessing mostly nonpolar bonds and lacking ionized groups are described as *nonpolar* or *hydrophobic.*

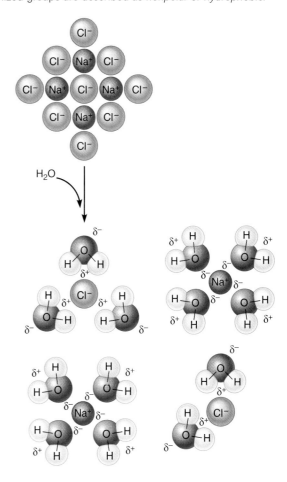

from the cytoplasm, and those of the membranes around organelles, which separate the internal compartment(s) of the organelles from the cytosol, are similar to those of the plasma membrane.

The structure of the plasma membrane, described as a *fluid mosaic,* consists of phospholipids, cholesterol, proteins, and carbohydrates (**Figure 2.15**). Each of these components is described next.

Phospholipid Bilayer

The phospholipids are arranged in a bilayer such that the hydrophilic heads face the aqueous environments inside the cell (the cytosol) and outside the cell (the extracellu-

lar fluid); the hydrophobic tails face each other (see Figure 2.15). This bilayer forms the basic structure of the membrane. The membrane is considered *fluid* because the phospholipids and the other molecules in the membrane are not linked together by chemical bonds and can therefore move about laterally, and even occasionally move from one side of the bilayer to the other. Cholesterol molecules are found within the lipid bilayer, where they interfere with hydrophobic interactions between phospholipid tails, which could cause crystallization of the bilayer and decrease the fluidity.

The phospholipid bilayer is a barrier to the movement of large polar molecules. Although water molecules are polar, they can generally cross the lipid bilayer because of

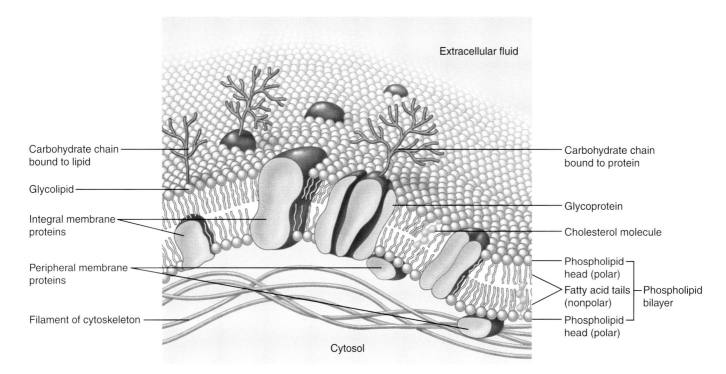

Figure 2.15 Plasma membrane. The flexible plasma membrane is primarily a double layer of phospholipids, with the hydrophobic tails of the phospholipid buried within the bilayer, and the hydrophilic phosphate heads facing the cytosol and extracellular fluid. Integral membrane proteins are scattered throughout the bilayer. Peripheral membrane proteins are associated with the integral membrane proteins, primarily on the side facing the cytosol, where they function as part of the cytoskeleton. Cholesterol molecules are scattered throughout the lipid bilayer. Carbohydrates bound to membrane lipids or membrane proteins face the extracellular fluid. These lipids and proteins are called glycolipids and glycoproteins, respectively.

their small size. Cholesterol decreases the ability of water to cross the lipid bilayer.

Membrane Proteins

The plasma membrane is described as a *mosaic* because of the presence of proteins that are dispersed in the bilayer, like islands in a sea of phospholipids. There are two main classes of membrane proteins: *integral membrane proteins* and *peripheral membrane proteins* (see Figure 2.15). **Integral membrane proteins** are embedded within the lipid bilayer, and therefore they can be dissociated from the membrane only by physically disrupting the bilayer. Integral membrane proteins are amphipathic molecules that are in contact with both the lipid bilayer and the aqueous environment. The polar surface(s) of the protein face the aqueous environment, which could be the cytosol, extracellular fluid, or both. The nonpolar areas are embedded within the lipid bilayer.

Some integral membrane proteins are called **transmembrane proteins**, because they span the lipid bilayer, with surfaces exposed to both the cytosol and extracellular fluid. Often, a transmembrane protein crosses the membrane at several places. Transmembrane proteins include channels that allow ions to *permeate* (or

cross) the membrane, and carrier proteins that transport molecules from one side of the membrane to the other. Other integral membrane proteins are located on only one side of the membrane. Some of the proteins facing the cytosol function as enzymes that catalyze chemical reactions in the cytosol or as a special class of proteins called G proteins (described later). Some of the proteins facing the extracellular fluid function as enzymes that catalyze reactions in the extracellular fluid or as receptor molecules that bind chemical messengers from other cells.

Peripheral membrane proteins are loosely bound to the membrane by associations with integral membrane proteins or phospholipids. Peripheral membrane proteins can be dissociated from the membrane and still leave the membrane intact. Most peripheral membrane proteins are located on the cytosolic surface of the plasma membrane and often function as part of a group of proteins that make up the *cytoskeleton.*

Membrane Carbohydrates

Also associated with the plasma membrane are carbohydrates, which are covalently bound to the membrane lipids or to proteins to form glycolipids or glycoproteins,

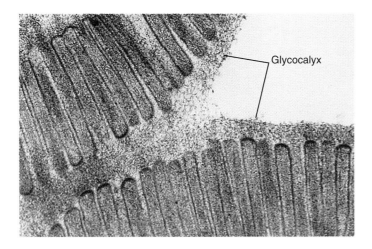

Figure 2.16 **Electron micrograph of the plasma membrane showing the glycocalyx.**

respectively (see Figure 2.15). Plasma membrane carbohydrates are located primarily on the extracellular surface, where they have two main functions: They form the *glycocalyx*, a protective layer that also functions in holding cells together (**Figure 2.16**), and they function in cell recognition; that is, they label the cell as a part of the body or as a distinct type of cell. Recognition is an important aspect of both the immune response and tissue growth.

Structure of the Nucleus

Most cells have a single nucleus that contains the genetic material of the cell, DNA. The DNA exists as thin threads called **chromatin**, except during cell division. The nucleus generally appears as a prominent spherical structure in the cell (**Figure 2.17**). Surrounding the nucleus is the **nuclear envelope**, which consists of two membranes. These membranes fuse intermittently, leaving gaps called **nuclear pores** that allow selective movement of molecules between the nucleus and the cytoplasm. Within the nucleus is a structure called the *nucleolus,* which is the site of synthesis of a type of RNA called *ribosomal RNA (rRNA).* The nucleus functions in the transmission and expression of genetic information. Encoded in the DNA is the genetic information for RNA and protein synthesis. The very important roles of DNA and RNA in protein synthesis are described later in this chapter.

Contents of the Cytosol

The cytosol is more than simply a fluid that bathes the organelles; it is a site of important chemical reactions and also a site for storage of molecules. Several enzymes that catalyze specific chemical reactions are located in the cytosol. Energy in the form of triglycerides or glycogen is stored in masses called **inclusions**. Molecules that are to

be released from the cell, or **secreted**, are stored in membrane-bound sacs called **secretory vesicles**. In addition, the molecular composition of the cytosol is critical to the function of cells. For example, the ionic composition of the cytosol plays a crucial role in the activities of nerve and muscle cells.

Structure of Membranous Organelles

The membranes that surround certain organelles provide a barrier between the cytosol and the interior of the organelle, which creates compartments within the

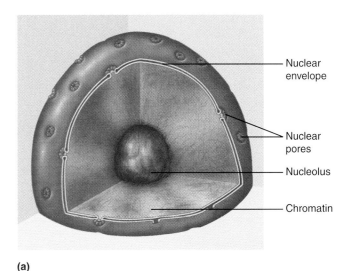

(a)

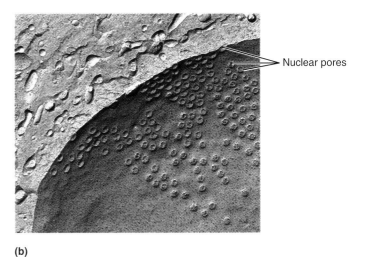

(b)

Figure 2.17 **Nucleus.** (a) The nucleus is separated from the cytoplasm by the nuclear envelope. Pores in the nuclear envelope allow the movement of specific substances between the nucleus and cytoplasm. Located in the nucleus is the DNA that contains the genetic information of each individual. The DNA exists as thin threads called chromatin. Also within the nucleus is the nucleolus, the site of rRNA synthesis. (b) Electron micrograph of a freeze-fractured nuclear envelope showing nuclear pores.

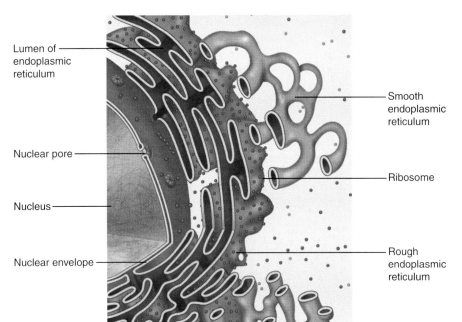

Lumen of endoplasmic reticulum

Nuclear pore

Nucleus

Nuclear envelope

Smooth endoplasmic reticulum

Ribosome

Rough endoplasmic reticulum

Figure 2.18 Endoplasmic reticulum. The endoplasmic reticulum's membrane is continuous with the outer membrane of the nuclear envelope. The rough endoplasmic reticulum is closest to the nucleus and consists of flattened sacs with ribosomes attached to its external surface. The membrane of the rough endoplasmic reticulum is continuous with that of the smooth endoplasmic reticulum, which is tubular and lacks ribosomes.

cytoplasm. In some cases, two membranes surround an organelle, creating compartments within the organelle. The structure of each membranous organelle is described next.

Endoplasmic Reticulum

The **endoplasmic reticulum** consists of an elaborate network of membrane enclosing an interior compartment called the lumen (**Figure 2.18**). There are two types of endoplasmic reticulum that differ in appearance and function: *rough endoplasmic reticulum* and *smooth endoplasmic reticulum.* The rough endoplasmic reticulum gets its name from its granular or "rough" appearance under magnification, due to the presence of **ribosomes**, which are complexes of rRNA and proteins that function in protein synthesis. In addition, rough endoplasmic reticulum looks like flattened sacs. Smooth endoplasmic reticulum, by contrast, consists of tubules and does not have ribosomes attached to it, giving it a "smooth" appearance. The membrane of the rough endoplasmic reticulum is continuous on one side with the outer membrane of the nuclear envelope, and on the other side with the smooth endoplasmic reticulum (see Figure 2.18).

The endoplasmic reticulum is important in the synthesis of several types of biomolecules. The rough endoplasmic reticulum is associated with the synthesis of proteins that will be secreted from the cell or incorporated into the plasma membrane, or are destined for another organelle. The smooth endoplasmic reticulum is the site of the synthesis of lipids, including triglycerides and steroids, and is a site for storage of calcium ions. In addition, the smooth endoplasmic reticulum is specialized in certain cells. In liver cells, for example, the smooth endoplasmic reticulum contains detoxification enzymes that break down toxic substances in the blood.

Golgi Apparatus

The **Golgi apparatus** consists of membrane-bound flattened sacs called *cisternae* (**Figure 2.19**). The Golgi apparatus is closely associated with the endoplasmic reticulum on one side, called the *cis face,* although the membranes of the Golgi apparatus and endoplasmic reticulum are separate. The other side of the Golgi apparatus faces the plasma membrane and is called the *trans face.* The Golgi apparatus processes molecules synthesized in the endoplasmic reticulum, and prepares them for transport to their final location. The Golgi apparatus packages molecules into vesicles and directs the vesicles to the appropriate location. Some vesicles transport substances to intracellular sites, whereas secretory vesicles transport substances out of the cell.

Mitochondria

Mitochondria are bound by two membranes (**Figure 2.20**). The outer mitochondrial membrane separates the mitochondrion from the cytosol, whereas the inner mitochondrial membrane divides each mitochondrion into two compartments: the *intermembrane space,* the area between the two membranes, and the **mitochondrial matrix**, the innermost compartment. The inner mitochondrial membrane is a functional compartment in itself because it houses a series of proteins and other molecules called the *electron transport chain.* To increase the area for the electron transport system, the inner membrane is folded into tubules called *cristae.* Mitochondria are often called the "powerhouse" of the cell because most of the cell's usable energy form, ATP, is produced in these organelles.

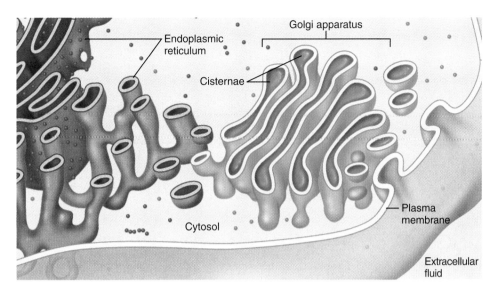

Figure 2.19 Golgi apparatus. The Golgi apparatus consists of stacks of flattened sacs called cisternae. One side of the Golgi apparatus faces the endoplasmic reticulum; the other side faces the plasma membrane.

The number of mitochondria per cell varies considerably among different cell types based on the particular cell's energy needs. For example, because muscle cells demand large amounts of energy they contain numerous mitochondria. By contrast, red blood cells, which function primarily in the transport of blood gases, contain no mitochondria.

Lysosomes

Lysosomes are small spherical organelles surrounded by a single membrane (**Figure 2.21**). Lysosomes contain enzymes that degrade intracellular debris and extracellular debris that has been taken into the cell. For example, after lysosomes fuse with old organelles that are no longer functioning, the enzymes inside the lysosomes break down the organelles and usable components are recycled and waste products are eliminated from the cell. In the case of extracellular debris, cells can engulf extracellular particles by a process called *endocytosis,* in which the particles are enclosed within a vesicle and brought inside the cell. Lysosomes then fuse with the vesicle, enabling their enzymes to degrade the particles. The process of endocytosis is described in Chapter 4.

Peroxisomes

Peroxisomes are spherical organelles that are slightly smaller than lysosomes and are surrounded by a single membrane. They function in the degradation of molecules such as amino acids, fatty acids, and toxic foreign matter. In the process of this degradation, they often produce hydrogen peroxide (H_2O_2), which in itself is toxic. One of the most notable characteristics of peroxisomes

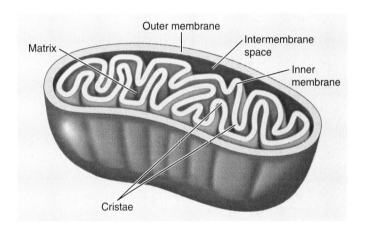

(a)

(b)

Figure 2.20 Mitochondria. (a) Drawing of a mitochondrion. The outer mitochondrial membrane separates the mitochondrion from the cytosol, whereas the inner mitochondrial membrane separates the mitochondrion into an inner matrix and the intermembrane space. The inner membrane has numerous folds called cristae. (b) Electron micrograph of a mitochondrion.

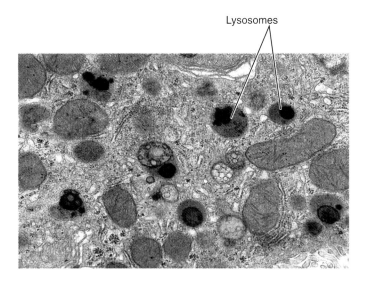

Lysosomes

Figure 2.21 **Electron micrograph of lysosomes.**

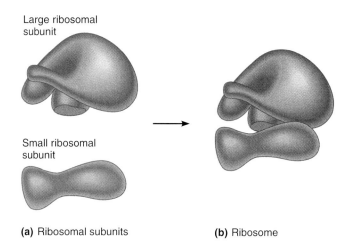

Large ribosomal subunit

Small ribosomal subunit

(a) Ribosomal subunits

(b) Ribosome

Figure 2.22 **Ribosomes.** In order for protein synthesis to begin, (a) the two free subunits in the cytosol must come together to form (b) a ribosome.

is the presence of the enzyme *catalase,* which catalyzes the breakdown of hydrogen peroxide to form water and oxygen:

$$2\ H_2O_2 \rightarrow 2\ H_2O + O_2$$

This reaction prevents the buildup of toxic hydrogen peroxide.

Structure of Nonmembranous Organelles

Even though nonmembranous organelles have no barrier separating them from the cytosol, they are still considered organelles because they consist of biomolecules organized into structures that perform specific functions within the cell.

Ribosomes

Ribosomes are dense granules composed of rRNA and proteins that function in protein synthesis. Each ribosome consists of a small subunit and a large subunit that are separate and located in the cytosol when the ribosome is not active (**Figure 2.22**a). During the process of protein synthesis (discussed later in this chapter), the two subunits come together in the cytosol to form a functional ribosome (Figure 2.22b). After protein synthesis is initiated, some ribosomes remain free in the cytosol, whereas others become attached to the rough endoplasmic reticulum. Proteins synthesized in association with free ribosomes can remain in the cytosol or enter a mitochondrion, the nucleus, or a peroxisome. Proteins synthesized in association with the rough endoplasmic reticulum will cross or enter a membrane, such as the plasma membrane or the membrane of an organelle.

Vaults

Vaults are the most recently discovered organelle, and their function is not fully understood. Vaults are barrel-shaped structures, approximately three times larger than ribosomes, that consist of two identical subunits. Vaults may be involved in the transport of molecules, such as messenger RNA (mRNA), between the nucleus and the cytoplasm. Although little is known about them, vaults are receiving a lot of research attention because they may be involved in the development of resistance to chemotherapy in the treatment of cancer (**Discovery: Vaults and Chemotherapy**).

Centrioles

Centrioles are short cylindrical structures consisting of bundles of protein filaments. Each cell has two centrioles oriented perpendicular to each other (see Figure 2.14). They function in directing the development of a structure called the *mitotic spindle* during cell division.

Cytoskeleton

The **cytoskeleton** is a flexible lattice of fibrous proteins, also called filaments, that gives the cell structure and support, much as the skeleton provides support for your body (**Figure 2.23**). The cytoskeleton is not rigid, nor does it have a fixed structure. The fibrous proteins that form it can disassemble and reassemble as necessary. Functions of the cytoskeleton include mechanical support and structure, intracellular transport of materials, suspension of organelles, formation of adhesions with other cells, contraction, and movement of certain cells. Several types of filaments form the cytoskeleton, including *microfilaments, intermediate filaments,* and *microtubules.* The classification of filaments is based on their diameters.

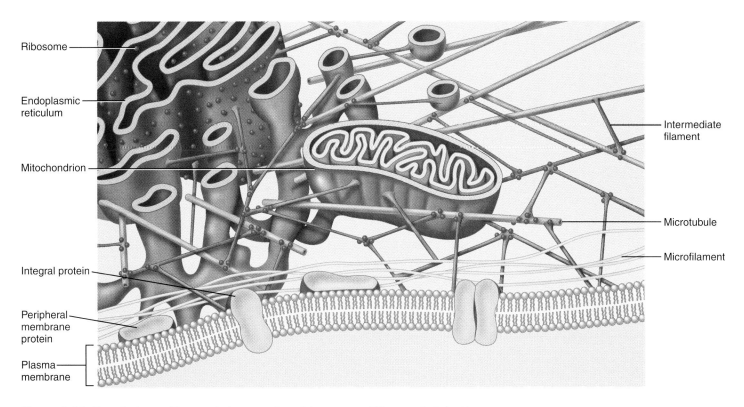

Ribosome

Endoplasmic reticulum

Mitochondrion

Integral protein

Peripheral membrane protein

Plasma membrane

Intermediate filament

Microtubule

Microfilament

Figure 2.23 Cytoskeleton. The cytoskeleton consists of three types of filaments that give the cell structural support and enable some degree of motility or contractility. Organelles are suspended in the cytosol by the filaments.

DISCOVERY

VAULTS AND CHEMOTHERAPY

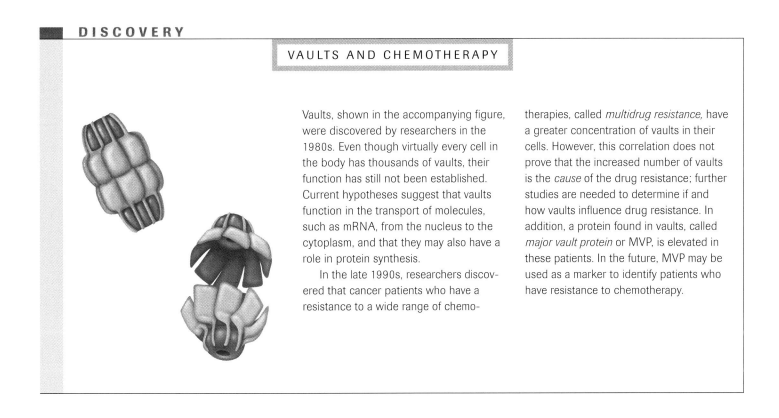

Vaults, shown in the accompanying figure, were discovered by researchers in the 1980s. Even though virtually every cell in the body has thousands of vaults, their function has still not been established. Current hypotheses suggest that vaults function in the transport of molecules, such as mRNA, from the nucleus to the cytoplasm, and that they may also have a role in protein synthesis.

In the late 1990s, researchers discovered that cancer patients who have a resistance to a wide range of chemo-

therapies, called *multidrug resistance,* have a greater concentration of vaults in their cells. However, this correlation does not prove that the increased number of vaults is the *cause* of the drug resistance; further studies are needed to determine if and how vaults influence drug resistance. In addition, a protein found in vaults, called *major vault protein* or MVP, is elevated in these patients. In the future, MVP may be used as a marker to identify patients who have resistance to chemotherapy.

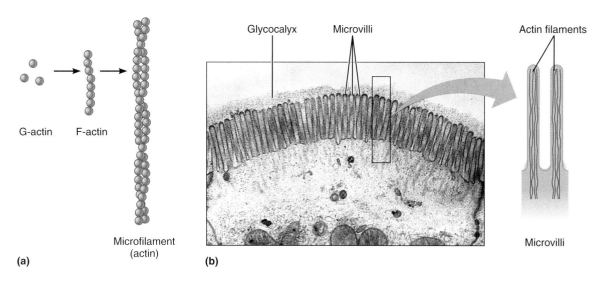

Figure 2.24 Microfilaments. (a) An actin filament, the most common microfilament. Actin filaments consist of two helically arranged strands of F-actin, a fibrous protein formed of chains of the globular protein G-actin. (b) Electron micrograph of microvilli on the surface of an epithelial cell. The enlarged drawing shows a schematic of the arrangement of microfilaments that maintain the fingerlike microvilli.

Microfilaments have the smallest diameter among the fibrous proteins. One microfilament, called *actin* (**Figure 2.24**a), has several functions in cells, including muscle contraction, "amoeboid-like" movement of cells, and separation of the cytoplasm during cell division. Other actin microfilaments provide structural support for special cell projections called **microvilli** (Figure 2.24b), which are often found in epithelial cells that are specialized for the exchange of molecules.

Intermediate filaments have a diameter between that of microfilaments and microtubules. Intermediate filaments tend to be stronger and more stable than microfilaments. Intermediate filaments called *keratin* are found in skin and hair cells. *Myosin* is an intermediate filament

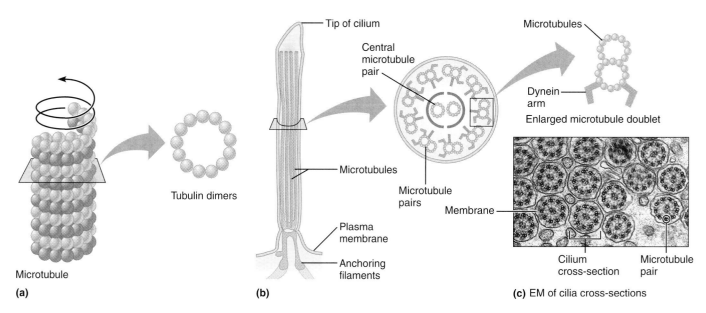

Figure 2.25 Microtubules. (a) A microtubule, the largest of the cytoskeletal elements. Each is composed of a long hollow tubule of a globular protein called tubulin organized in dimers. (b) The arrangement of microtubules in cilia and flagella. Nine pairs of microtubules are arranged around a tenth central pair. The cross section shows the pairs of microtubules with their dynein arms. (c) Electron micrograph of cross sections through cilia, showing the ten pairs of microtubules.

found in muscle cells, where it works with actin to produce contraction.

Microtubules, the filaments with the largest diameter, are composed of long hollow tubes of a spherical protein called *tubulin* (**Figure 2.25**a). Microtubules provide strength to the cytoskeleton. They also form *spindle fibers,* which aid in the distribution of chromosomes during cell division, and assist with directional movement of vesicles and large biomolecules within the cytosol. In addition, microtubules are components of two motile structures: *cilia* and *flagella* (Figure 2.25b).

Cilia and flagella are hairlike protrusions from the cell surface that can move in a wavelike manner. Movement of cilia propels particles along hollow conduits in the body. In the respiratory tract, for example, cilia move mucus containing trapped inhaled particles out of the lungs and up toward the mouth, where it can be swallowed. A cilium consists of ten pairs of microtubules, with nine of the pairs surrounding a central pair. The paired microtubules are connected by proteins called *dynein arms* that help to generate the force needed for the microtubules to slide past each other; the sliding movement leads to the wavelike motion of a cilium. Flagella are similar to cilia in basic structure, but they are longer and are found in only one cell type in humans, the sperm. Flagella move sperm through the female reproductive tract toward the ovum.

Table 2.2 summarizes the different components of the cell and their major functions.

Table 2.2 ∎ Summary of Cell Structures and Functions

Cell part	Structure	Function
Plasma membrane	Lipid bilayer with scattered proteins and cholesterol molecules	Maintains boundary of cell and integrity of cell structure; embedded proteins serve multiple functions
Nucleus	Surrounded by double-layered nuclear envelope	Houses the DNA, which dictates cellular function and protein synthesis
Nucleolus	Dark oval structure inside the nucleus	Synthesis of ribosomal RNA
Cytosol	Gel-like fluid	Cell metabolism, storage
Membranous organelles		
Rough endoplasmic reticulum	Continuous with the nuclear envelope; flattened sacs dotted with ribosomes	Protein synthesis and post-translational processing
Smooth endoplasmic reticulum	Continuous with rough endoplasmic reticulum; tubular structure without ribosomes	Lipid synthesis and post-translational processing of proteins; transport of molecules from endoplasmic reticulum to Golgi apparatus
Golgi apparatus	Series of flattened sacs near the endoplasmic reticulum	Post-translational processing; packaging and sorting of proteins
Mitochondria	Oval-shaped, with an outer membrane and an inner membrane with folds called cristae that project into the matrix	ATP synthesis
Lysosomes	Granular, saclike; scattered throughout the cytoplasm	Breakdown of cellular and extracellular debris
Peroxisomes	Similar in appearance to lysosomes, but smaller	Breakdown of toxic substances
Nonmembranous organelles		
Vaults	Small, barrel-shaped	Unknown; possibly transport of molecules between nucleus and cytoplasm
Ribosomes	Granular organelles composed of proteins and rRNA; located in cytosol or on surface of rough endoplasmic reticulum	Protein synthesis
Centrioles	Two cylindrical bundles of protein filaments that are perpendicular to each other	Direction of mitotic spindle development during cell division
Cytoskeleton	Composed of protein filaments, including microfilaments, intermediate filaments, and microtubules	Structural support of cell; cell movement and contraction

(**Figure 2.26**). Because of this barrier, solutes generally must cross the epithelial cell layer to go from one side of an epithelium to the other *(transepithelial transport),* rather than going around the cells *(paracellular movement).*

Epithelial cells often line hollow structures, such as the organs of the gastrointestinal tract or the tubules in the kidneys. The internal compartment of a hollow organ, the *lumen,* is separated from the cytosol of each epithelial cell by the *apical membrane.* The membrane facing the extracellular fluid is called the *basolateral membrane.* Epithelial cells lining the organs often have tight junctions that restrict or regulate the movement of molecules from the lumen of the organ into the blood, or from the blood into the lumen. Because of these tight junctions, the composition of the contents of the lumen can differ from that of the blood.

Cell-to-Cell Adhesions

In many tissues cells are held together by special membrane proteins called *cell adhesion molecules.* Other proteins function in more specialized cell adhesion, in which the purpose is more than merely holding the cells together. There are three main types of special junctions: *tight junctions, desmosomes,* and *gap junctions.*

Tight Junctions

Tight junctions are commonly found in epithelial tissue that is specialized for molecular transport. In **tight junctions**, integral membrane proteins called *occludins* fuse adjacent cells together to form a nearly *impermeable* barrier to the movement of substances between cells

Desmosomes

Desmosomes are found in tissues subject to mechanical stress, such as those in the heart, uterus, and skin. A **desmosome** is a filamentous junction between two adjacent cells that provides strength so that the cells do not tear apart when the tissue is subject to stress. **Figure 2.27**a shows the structure of desmosomes in the heart, where they are closely associated with another type of adhesion called *gap junctions* (described next). At the site of each desmosome is a *plaque* formed by glycoproteins clustered inside each cell. Extending from this plaque are both intracellular protein filaments and protein filaments called *cadherins* that cross the plasma membrane into the extracellular space. The cadherins are linked to intracellular filaments at the plaque.

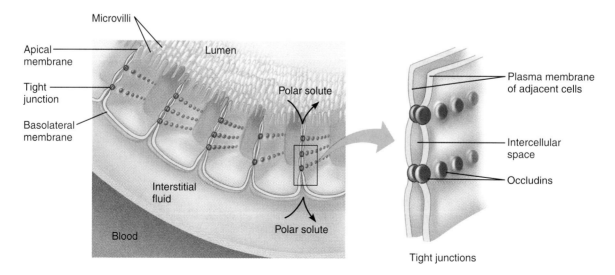

Figure 2.26 Tight junctions. Tight junctions are commonly found in epithelial tissue, where they limit the movement of molecules through the interstitial space between epithelial cells. Proteins called occludins link two adjacent cells together such that they form nearly impermeable adhesions. Polar molecules that cannot cross lipid bilayers also generally cannot move between cells connected by tight junctions.

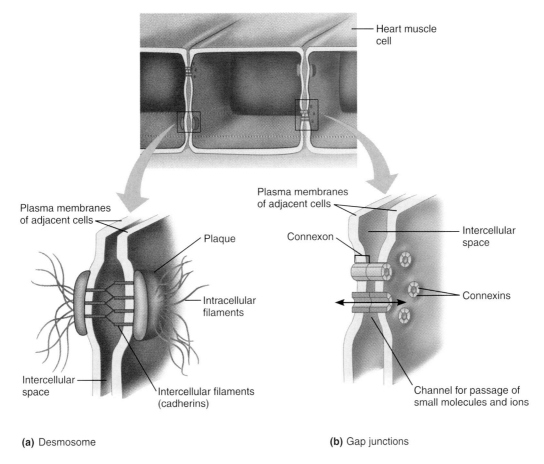

Figure 2.27 Desmosomes and gap junctions, as shown in heart muscle.
(a) Desmosomes are strong junctions between cells in tissues subjected to stress. Protein filaments extend from a plaque located near the plasma membrane where the two cells join together. Some of these filaments are located inside the cells (intracellular filaments), whereas other filaments called cadherins (intercellular filaments) extend into the intercellular space, linking the two cells together. (b) Gap junctions are areas of membrane containing proteins called connexons that form channels linking the cytoplasm of two adjacent cells. These channels allow ions and small molecules to pass between adjacent cells.

Gap Junctions

Gap junctions, although found in a variety of tissue types, are most notable for their presence in smooth muscle and in the muscle of the heart, where the gap junctions allow the muscle to contract as a unit. **Gap junctions** are areas where two adjacent cells are connected by membrane proteins called *connexons* (Figure 2.27b). Each connexon is composed of six smaller membrane proteins called *connexins*. The connexons in the adjacent cells bind to each other, forming small channels that enable ions and small molecules to move between the two cells.

Gap junctions provide direct electrical and metabolic coupling of adjacent cells. Movement of ions between cells electrically couples the cells, such that if the electrical properties of one cell change, ions move from this cell to the adjacent cell through the gap junction, producing the same type of electrical change. This process allows the cells of the heart, for example, to act in a coordinated fashion so that they can contract in unison. In places such as bone, gap junctions allow nutrients to travel from cells close to the bloodstream to cells deep within the tissue. Gap junctions also allow the passage of some chemical messengers such as the molecule cAMP. In this manner, a signal initiated in one cell can be transmitted to adjacent cells.

General Cell Functions

Our cells have specialized functions that contribute to homeostasis. For example, muscle cells contract to generate force, and nerve cells use electrical impulses and chemical messengers for communication. Despite these special functions, most cells of the body carry out several general functions, which we discuss next. The specialized functions of cells are described in subsequent chapters.

Metabolism

All cells of the body carry out a variety of chemical reactions. **Metabolism** refers to all the chemical reactions that occur in the body. Metabolism can be divided into two classes of chemical reactions: (1) **anabolism**, the synthesis of large molecules from smaller molecules, which requires an input of energy, and (2) **catabolism**, the breakdown of large molecules to smaller molecules, which releases energy. Every second, thousands of anabolic and catabolic reactions are occurring in virtually every cell of the body. These reactions are highly regulated by the presence of enzymes. You will learn more about cell metabolism and enzymes in Chapter 3.

Exercise Link

The rates of all these anabolic and cata-  bolic reactions are constantly changing (adapting) to environmental and behavioral circumstances. The energy required for exercise comes from the catabolism of glucose and triglycerides. In extreme exercise, such as running a marathon, even protein molecules are broken down for energy. When individuals are resting after a meal, anabolic processes are prominent: New proteins are synthesized from dietary amino acids, glycogen is synthesized from dietary glucose, and triglycerides are formed from dietary fatty acids and stored in fat cells.

Cellular Transport

We have seen that the lipid bilayer is a barrier to hydrophilic molecules. However, cells require hydrophilic molecules to carry out their functions, and cells also release hydrophilic molecules into the extracellular environment. The process by which molecules move into and out of cells is called *membrane transport*.

Nonpolar molecules can permeate the plasma membrane and therefore will simply move across the membrane by diffusion. The smallest of the polar molecules, such as water, can also permeate most plasma membranes. However, larger polar molecules cannot permeate the lipid bilayer; some of these molecules can cross plasma membranes only with the assistance of transmembrane proteins or vesicles. For example, some molecules such as glucose and amino acids are transported across the membrane by *carrier proteins*. Other molecules, such as ions, can move through *ion channels*. Proteins and other *macromolecules* are too large to be transported across the plasma membrane by carrier proteins or to move through channels. Instead, movement of these macromolecules involves packaging the molecules into vesicles. Details regarding cellular transport are described in Chapter 4.

Intercellular Communication

Communication is another function that all cells perform. A single cell in the body cannot exist in isolation. Cells must communicate with each other to maintain an environment in which they can survive. Communication can be direct, as occurs through gap junctions. Often, however, communication involves the release from one cell of a chemical messenger that acts on a second cell. The two communicating cells can be adjacent to each other, or they can be in different areas of the body. In general, messenger molecules are released from one cell and reach other cells, called *target cells*, which contain proteins that function as *receptors* for that specific messenger. The binding of the messenger molecule to its specific receptor triggers a response in the target cell.

In one example of such intercellular communication, a chemical messenger, antidiuretic hormone (ADH), is released into the blood by exocytosis from an endocrine gland in the brain. When ADH reaches its target cells—certain kidney cells—the hormone binds with its receptors. This binding stimulates the insertion of water pores into the kidney cells, which increases water conservation by removing some of the water from urine being formed. The different types of chemical messengers and the mechanisms by which they produce a response in target cells are the topics of Chapter 5.

Quick Test 2.3

1. Briefly describe the three types of cell adhesions, and give an example of where each is found.

2. Define the following terms: *metabolism, anabolism,* and *catabolism.* Which requires an input of energy, anabolism or catabolism?

3. Describe the role of transmembrane proteins in the transport of molecules across the plasma membrane.

Protein Synthesis

By now you should realize that proteins are very important biomolecules in our cells. Contained within the DNA are the codes for the synthesis of all the proteins in our cells. This code is called the **genetic code**, and it is universal among all animal species. Even though DNA is located in the nucleus, protein synthesis takes place in the cytoplasm. Therefore, the code must be transmitted to another molecule, called **messenger RNA (mRNA)**, that can move into the cytoplasm. The general steps of protein synthesis are as follows:

1. DNA is *transcribed* according to the genetic code to form a complementary mRNA in the nucleus.

2. mRNA moves from the nucleus to the cytoplasm.

3. mRNA is *translated* by ribosomes to form the correct amino acid sequence of the protein in the cytoplasm.

The Role of the Genetic Code

Recall that DNA consists of two strands of nucleotides held together by hydrogen bonding between complementary base pairs. The sequence of the four bases in DNA codes for the sequence of amino acids in proteins. The section of DNA that codes for a particular protein or proteins is called a **gene** (**Figure 2.28**a). Only one DNA strand, the *sense* strand, contains the actual code.

Twenty different amino acids are found in proteins. The sequence of the DNA bases must provide enough information to specify each amino acid in a polypeptide chain. A sequence of three bases, called a **triplet**, codes for a single amino acid (Figure 2.28b). Each base can be thought of as a letter of an alphabet that is used in words that are all three letters long and indicate a particular amino acid. With four bases that can take up three possible positions in a triplet, there are $4^3 = 64$ possible words in the genetic code. During transcription, which is described shortly, this code is passed on to mRNA, whose base sequence is complementary to the sense strand of DNA. Therefore, a DNA triplet is complementary to a three-base sequence in mRNA, called a **codon**, that codes for the same amino acid (Figure 2.28c).

Note that there are 64 possible codons, but only 20 amino acids. Thus, even though each codon codes for only one amino acid, some amino acids are coded for by more than one codon. For example, CCC always codes for the amino acid proline, but CCG also codes for proline. One codon, called an *initiator codon,* is found in every mRNA. This initiator codon, AUG, also codes for the amino acid methionine. There are three *termination codons* that indicate that the end of the protein code has occurred. Because these termination codons do not code for any amino acid, only 61 of the 64 codons actually code for amino acids.

Transcription

Transcription is the process in which RNA is synthesized using information contained in the DNA, and it occurs in the nucleus. Three types of RNA can be transcribed from DNA: (1) mRNA, (2) rRNA, and (3) *tRNA (transfer RNA).* Each type of RNA is involved in protein synthesis. This section describes the process of transcribing RNA from DNA, using mRNA as the example.

Figure 2.29 illustrates the process of transcription. The first step is the uncoiling of DNA and its separation into two strands at the site where transcription is occurring. One of the strands of DNA contains the gene and functions as a template for the synthesis of mRNA. The section of DNA with the gene is identified by the **promoter sequence**, a specific base sequence to which the enzyme *RNA polymerase* can bind. When RNA polymerase binds to the promoter sequence, it initiates the separation of DNA into two strands. This allows free ri-

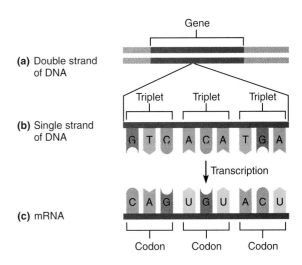

Figure 2.28 The genetic code. (a) Genes, which are portions of DNA that code for a particular protein(s). (b) Triplets, the three-base sequences that code for an amino acid sequence. (c) Transcribed mRNA codons, which are complementary to the code in DNA triplets.

bonucleotides in the triphosphate form (ATP, UTP, GTP, and CTP) to align with the DNA template according to the law of complementary base pairing. (Recall that the base uracil in RNA replaces thymine in DNA.) RNA polymerase also catalyzes the formation of bonds between the aligned ribonucleotides, making a polynucleotide that will eventually be the single-stranded mRNA.

The mRNA undergoes *post-transcriptional processing* in the nucleus before the final mRNA product moves through the nuclear pores and enters the cytoplasm (**Figure 2.30**). Within a gene are regions of excess bases, called *introns,* that do not code for the amino acid sequence of proteins. Although these introns are transcribed in the initial mRNA molecule, they must be removed before mRNA leaves the nucleus. After introns are cut out of the initial mRNA molecule, the remaining coding segments, called *exons,* are joined together. Other aspects of post-transcriptional processing include addition of a chemical group called a *CAP* to the 5′ end, which is necessary for initiating translation, and addition of several adenine nucleotides called a poly A tail to the 3′ end, which protects the mRNA from degradation in the cytoplasm. After processing is complete, the mRNA enters the cytoplasm, and translation of mRNA can occur.

Translation

Translation is the process in which polypeptides are synthesized using mRNA codons as a template for the assembly of the correct amino acids along the sequence. Another type of RNA molecule, tRNA, carries the appropriate amino acid to the ribosome based on interactions with the mRNA codon. Due to complementary base

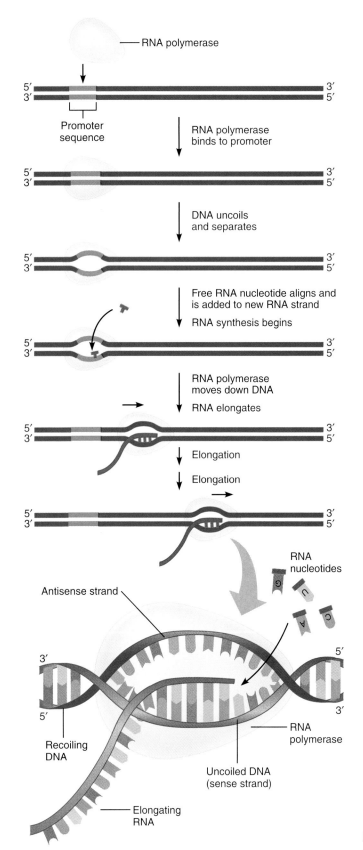

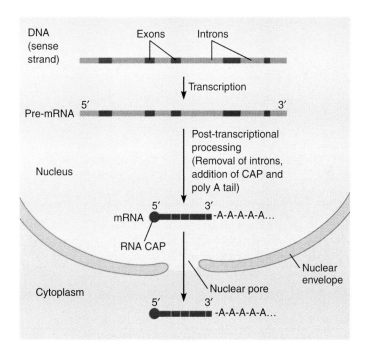

Figure 2.30 Post-transcriptional processing. Following transcription, mRNA must undergo further processing in the nucleus. Transcribed introns must be removed, and the remaining exons spliced together. A chemical structure called a CAP is added to the 5′ end, and a poly A tail (several nucleotides with the adenine base) is added to the 3′ end. After processing is complete, the mRNA molecule can move from the nucleus to the cytoplasm.

pairing within its single strand of nucleotides, tRNA is shaped somewhat like a cloverleaf (**Figure 2.31**). The 3′ end of the tRNA molecule contains a binding site for a specific amino acid. Another region of the tRNA contains a base sequence, called the *anticodon,* that is complementary to the mRNA codon. It is the correct interaction between the codon on the mRNA and the anticodon on the tRNA that ensures that the correct amino acid is added to the elongating polypeptide chain.

Translation occurs in the cytoplasm in association with ribosomes. Recall that ribosomes can be found free in the cytosol or attached to the rough endoplasmic reticulum. The ribosome aligns the mRNA with the tRNA carrying the appropriately coded amino acids, and contains the enzymes that catalyze the formation of peptide bonds between the amino acids, thus forming a polypeptide. Ribosomes can hold two tRNA molecules at the same time in two distinct ribosomal regions identified as the *P site* and the *A site.*

Figure 2.29 Transcription. RNA polymerase binds to the promoter sequence of a DNA molecule, causing the DNA to uncoil. The two strands of DNA separate, exposing the bases for complementary base pairing with free nucleotides. The free nucleotides align, and RNA polymerase catalyzes the formation of a bond between them, generating an elongating strand of mRNA. The enlargement shows that the process of elongation occurs according to the law of complementary base pairing.

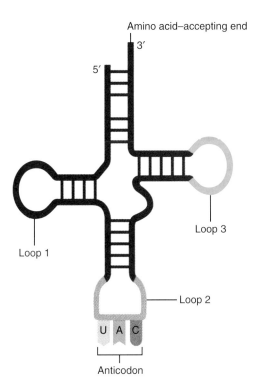

Figure 2.31 **Transfer RNA.** Hydrogen bonding between complementary base pairs is responsible for the cloverleaf structure of transfer RNA. The tRNA binds a specific amino acid on one end, and it has a base sequence called an anticodon on the other end that recognizes the complementary mRNA codon.

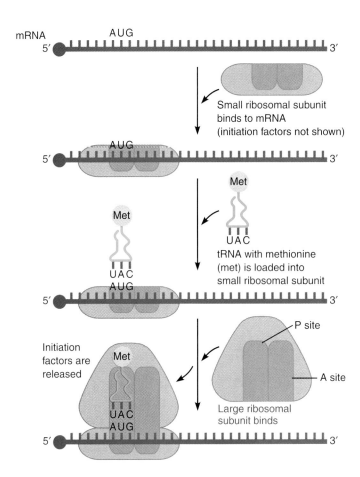

Figure 2.32 **Initiation of translation.** In the cytosol, initiation factors bind to the CAP region of a mRNA molecule and to the small ribosomal subunit, triggering the binding of the small ribosomal subunit to the mRNA. A tRNA with the anticodon complementary to the initiation codon of mRNA binds to the mRNA by the law of complementary base pairing. The large ribosomal subunit binds such that the first tRNA is located in the P site of the ribosome, and the initiation factors are released.

Now that we have some knowledge of the components needed for translation, we turn to the process itself. The first step of translation is *initiation* (**Figure 2.32**). Ribosomes not actively involved in translation are found free in the cytosol as their individual subunits. To start translation, certain *initiation factors* bind to the CAP that was added to the 5′ end of the mRNA molecule during post-transcriptional processing; other initiation factors form a complex with the small ribosomal subunit and a *charged tRNA* (tRNA bound to its amino acid) bearing the anticodon complementary to the initiation codon (AUG). Binding of the initiation factors triggers the binding of the small ribosomal subunit to the mRNA, such that the first initiation codon (AUG) found within the mRNA is aligned correctly for the start of translation. The large ribosomal subunit then binds, causing dissociation of the initiation factors and alignment of the first tRNA in the P site of the ribosome. A second charged tRNA molecule with the appropriate anticodon then enters the A site of the ribosome, and translation begins.

During the process of translation (**Figure 2.33**), an enzyme in the ribosome catalyzes the formation of a peptide bond between the two amino acids. The first amino acid, which is always methionine, is released from the tRNA molecule, and the *free tRNA* (tRNA without its amino acid) leaves the P site on the ribosome. The ribosome then moves down the mRNA three bases (one

codon), placing the second tRNA in the P site on the ribosome. The next charged tRNA enters the A site, bringing in the third amino acid to be added to the polypeptide. This process continues until a termination codon is reached on the mRNA. At that point, the polypeptide is released, and the ribosome and mRNA dissociate. More than one ribosome translates a given mRNA molecule at the same time, forming what is called a *polyribosome*.

Destination of Proteins

Proteins synthesized in a cell can be destined for the nucleus, cytosol, plasma membrane, or one of the organelles, or they can be secreted from the cell into the extracellular fluid. Unless the final destination of a protein is to remain in the cytosol, the first functioning sequence of amino acids that is translated in a polypeptide chain is the **leader sequence**. The leader sequence is synthesized during translation by ribosomes free in the

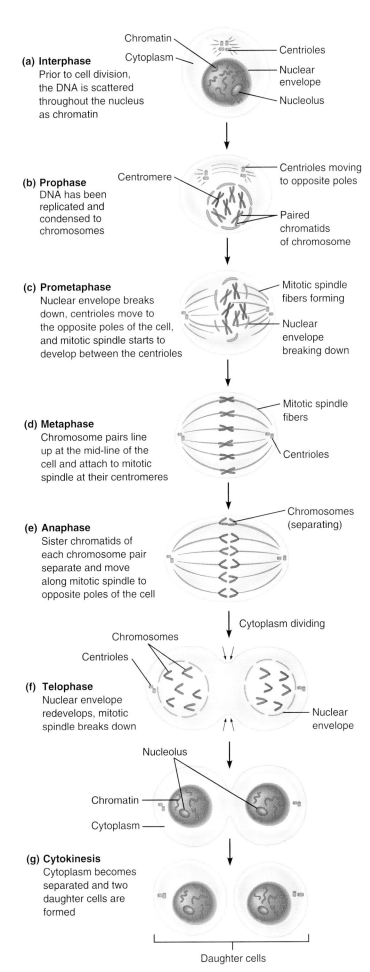

(a) Interphase
Prior to cell division, the DNA is scattered throughout the nucleus as chromatin

Chromatin
Cytoplasm
Centrioles
Nuclear envelope
Nucleolus

(b) Prophase
DNA has been replicated and condensed to chromosomes

Centromere
Centrioles moving to opposite poles
Paired chromatids of chromosome

(c) Prometaphase
Nuclear envelope breaks down, centrioles move to the opposite poles of the cell, and mitotic spindle starts to develop between the centrioles

Mitotic spindle fibers forming
Nuclear envelope breaking down

(d) Metaphase
Chromosome pairs line up at the mid-line of the cell and attach to mitotic spindle at their centromeres

Mitotic spindle fibers
Centrioles

(e) Anaphase
Sister chromatids of each chromosome pair separate and move along mitotic spindle to opposite poles of the cell

Chromosomes (separating)

Cytoplasm dividing

(f) Telophase
Nuclear envelope redevelops, mitotic spindle breaks down

Chromosomes
Centrioles
Nuclear envelope

Nucleolus

Chromatin
Cytoplasm

(g) Cytokinesis
Cytoplasm becomes separated and two daughter cells are formed

Daughter cells

centromere. Microtubules in the cell's cytoskeleton disassemble into their tubulin components, which will be used subsequently to form the mitotic spindle. The centriole pairs start moving to opposite poles of the cell, and the mitotic spindle develops between them.

2. Prometaphase: The nuclear envelope breaks down, the nucleolus is no longer visible, the centrioles are at opposite poles of the cell, and the chromosomes become linked at their centromeres to the spindle fibers.

3. Metaphase: The chromosomes are aligned in a plane at the middle of the cell.

4. Anaphase: The chromatid pairs separate, and the chromosomes move along the mitotic spindle toward opposite poles of the cell.

5. Telophase: New nuclear envelopes develop on the two sides of the cell, the chromosomes begin to decondense to chromatin, and the mitotic spindle begins breaking down.

Cytokinesis is the division of the cytoplasm. The cytoplasm divides by cleavage, which starts during the latter phases of mitosis (anaphase or telophase). The cleavage is accomplished by contraction of microfilaments (primarily actin) that form a ring around the midline of the cell, and the mitotic spindle disassembles. The two resulting daughter cells then enter interphase at G0.

Quick Test 2.5

1. Define *replication.* How does replication differ from transcription?

2. Name the four phases of interphase. During which phase does DNA replication occur?

3. Name the five stages of mitosis. During which stage does the nuclear envelope start to disintegrate?

Figure 2.40 Events in cell division. A hypothetical cell with only 6 chromosomes (instead of 46) is shown. (a) The appearance of the cell in interphase, which precedes cell division. (b–f) The major events in the five phases of mitosis. (g) The major events in cytokinesis.

What type of filaments make up the mitotic spindle fibers?

Microtubules

CLINICAL CONNECTIONS

CANCER

Each year in the United States over 1 million individuals are diagnosed with cancer, and approximately 500,000 people die every year from this disease. Even though cancer can arise in nearly every type of body tissue and leads to illness in diverse ways, at least one theme is common to cancer: Cancerous cells divide uncontrollably and eventually displace the body's normally functioning cells and consume energy vital to normal body cells.

Cell growth and division is controlled by genes called *proto-oncogenes*. A mutation in a proto-oncogene can produce an *oncogene* that leads to abnormal cell growth and division. Mutations may occur spontaneously, or they may be induced by substances called *mutagens*. One example of a mutagen is DDT, which was used in pesticides in the 1960s. In some cases, including cancer, the mutated cells divide rapidly, passing the mutation on to their daughter cells, which then also divide rapidly. To rid the body of cancer, these mutated cells must be eliminated.

Most therapeutic strategies, including radiation therapy and chemotherapy, are directed at different stages of actively dividing cells. Because cancer cells divide rapidly, these therapies harm cancer cells more than they harm the normal, infrequently dividing cells of the body. Unfortunately, normal body cells that divide frequently are also affected by these therapeutic strategies. Thus because hair cells undergo rapid cell division, hair loss is one of the consequences of cancer therapy.

The role of genetics and of environmental agents (such as tobacco) in the initiation of abnormal cell growth and division is currently the topic of intense research. Researchers are learning to manipulate the body's immune system, hoping to halt the growth of cancerous cells by "tricking" the patient's own immunity into destroying the cancer. Sometimes a patient's immune system will not recognize cancer cells as foreign because the cancer cells are not different enough from normal cells to stimulate an immune

reaction. At other times the immune system may recognize cancer cells, but its response is not strong enough to destroy the cancer. Various kinds of immunotherapies have been designed to help the immune system recognize cancer cells as distinctly foreign, and to strengthen the attack in the hope that malignant cells will be destroyed.

Color-enhanced scanning electron micrograph of cancerous tumor in lung tissue.

CHAPTER SUMMARY

Biomolecules, p. 24

The four basic types of biomolecules are carbohydrates, lipids, proteins, and nucleotides. Carbohydrates are polar molecules and include monosaccharides, disaccharides, and polysaccharides. Lipids are generally nonpolar molecules and include triglycerides, phospholipids, eicosanoids, and steroids. The phospholipids are amphiphathic molecules. Proteins are polymers of amino acids. Nucleic acids are polymers of nucleotides and include DNA and RNA.

Cell Structure, p. 35

The plasma membrane separates the cell from the extracellular fluid. The plasma membrane is composed of a phospholipid bilayer and proteins and

cholesterol embedded within the bilayer. Carbohydrates are found on the outer surface of the membrane as glycolipids or glycoproteins.

Inside the cell are the nucleus and cytoplasm. The nucleus contains the genetic material of the cell, DNA, and thus functions in the transmission and expression of genetic information. The cytoplasm consists of cytosol and organelles. The cytosol is a gel-like fluid that contains enzymes, glycogen and triglyceride stores, and secretory vesicles. The organelles include the rough and smooth endoplasmic reticulum, which function in protein and lipid synthesis, respectively; the Golgi apparatus, which processes molecules and prepares them for transport; mitochondria,

which are the site of most ATP synthesis in the cells; lysosomes and peroxisomes, which degrade waste products; and ribosomes, which are the sites of protein synthesis.

The cytoskeleton gives the cells their shape, provides structural support, transports materials within the cell, suspends organelles, forms adhesions with other cells to form tissues, and causes contraction or movement in certain cells. The cytoskeleton consists of protein filaments including microfilaments, intermediate filaments, and microtubules.

Cell-to-Cell Adhesions, p. 46

Three specialized types of cell adhesions are commonly found between cells. Tight junctions are commonly found in epithelial tissue, where they form impermeable barriers to the movement of substances between cells. Desmosomes are strong junctions that provide resistance to mechanical stress. Gap junctions provide channels between two adjacent cells that enable ions and small molecules to move from the cytosol of one cell to the cytosol of the other cell.

General Cell Functions, p. 47

Almost all cells of the body carry out several general functions, including metabolism, cellular transport of molecules across membranes, and intercellular communication.

Protein Synthesis, p. 48

Genes in the DNA contain the codes for protein synthesis. This code is transcribed to mRNA in the nucleus. The mRNA then moves to the cytoplasm where it is translated by ribosomes to synthesize proteins.

The leader sequence of a protein determines its fate. A protein will be synthesized either in the cytosol or in association with the endoplasmic reticulum. For those proteins synthesized in the cytosol, the leader sequence determines whether the protein will remain in the cytosol or enter a mitochondrion, a peroxisome, or the nucleus. For those proteins synthesized in association with the

endoplasmic reticulum, the protein will eventually be packaged into vesicles by the Golgi apparatus and directed to the appropriate site in the cell or secreted from the cell.

Cell Division, p. 56

The cell cycle includes interphase and cell division. Interphase is the period between cell divisions; mitosis and cytokinesis make up the process of cell division. During the S phase of interphase, the DNA is replicated so that each daughter cell will receive an exact DNA copy. The five phases of mitosis are prophase, prometaphase, metaphase, anaphase, and telophase. Mitosis is followed by cytokinesis.

■ EXERCISES

Multiple-Choice Questions

1. Which of the following biomolecules is *not* a polymer?
   a) polysaccharide
   b) phospholipid
   c) protein
   d) nucleic acid

2. A fatty acid that has two double bonds between carbon atoms is called
   a) a saturated fatty acid.
   b) a desaturated fatty acid.
   c) a monounsaturated fatty acid.
   d) a polyunsaturated fatty acid.
   e) an eicosanoid.

3. Which of the following molecules is *not* a component of a phospholipid?
   a) cholesterol
   b) glycerol
   c) phosphate
   d) fatty acid

4. Hydrogen bonding between the amino hydrogen of one amino acid and the carboxyl oxygen of another amino acid in a protein is responsible for
   a) primary protein structure.
   b) secondary protein structure.
   c) tertiary protein structure.
   d) quaternary protein structure.

5. Which of the following nucleic acids is *not* a pyrimidine?
   a) cytosine
   b) thymine
   c) uracil
   d) adenine

6. Glycogen and lipids are stored
   a) as inclusions in the cytosol.
   b) in storage vesicles.
   c) in secretory vesicles.
   d) in lysosomes.
   e) in the nucleolus.

7. Which organelle produces most of a cell's ATP?
   a) nucleus
   b) peroxisome
   c) Golgi apparatus
   d) mitochondrion
   e) smooth endoplasmic reticulum

8. Which cell-to-cell adhesion allows quick transmission of electrical signals between adjacent cells?
   a) desmosome
   b) gap junction
   c) tight junction
   d) all of the above

9. To initiate transcription of DNA, the enzyme RNA polymerase binds to the
   a) initiator codon.
   b) termination codon.
   c) promoter sequence.
   d) centromere.
   e) leader sequence.

10. The base sequence on tRNA that recognizes the mRNA codon by the law of complementary base pairing is called the
    a) tRNA codon.
    b) anticodon.
    c) amino codon.
    d) leader codon.
    e) initiator codon.

11. For proteins to be synthesized in association with the rough endoplasmic reticulum, the leader sequence must bind to a _____ on the membrane of the endoplasmic reticulum.
    a) transport vesicle
    b) secretory vesicle
    c) coated vesicle
    d) signal recognition protein
    e) promoter

12. During what phase of mitosis is the mitotic spindle developing?
    a) prophase
    b) metaphase
    c) anaphase
    d) telophase
    e) prometaphase

13. During what phase of mitosis do the chromosome pairs move to opposite poles of the cell?
    a) prophase
    b) metaphase
    c) anaphase
    d) telophase
    e) prometaphase

Objective Questions

1. Monosaccharides are (polar/nonpolar/amphipathic) molecules.

2. Triglycerides are (polar/nonpolar/amphipathic) molecules.

3. The precursor molecule for all steroids is _____.

4. According to the law of complementary base pairing, in RNA, adenine base pairs with _____.

5. Transmembrane proteins are examples of (integral/peripheral) membrane proteins.

6. The membrane of the smooth endoplasmic reticulum is continuous with the membrane of the Golgi apparatus. (true/false)

7. Microfilaments provide structural support for hairlike projections of the plasma membrane called _____.

8. The proteins that form gap junctions between two adjacent cells are called _____.

9. A section of DNA that codes for a specific protein is called a _____.

10. The mRNA codon that is transcribed from the DNA triplet ATC is _____.

11. More than one codon may code for a single amino acid. (true/false)

12. The first section of a polypeptide to be translated is called the _____, and it is important in directing the translated protein to its final destination.

13. Replication of DNA occurs during (interphase/mitosis).

14. Division of the cytoplasm into two daughter cells during cell division is called _____.

Essay Questions

1. Name the four major classes of biomolecules. Describe the major function of each class.

2. Define the four levels of protein structure, and explain the chemical interactions responsible for each level.

3. Describe the structure of the plasma membrane. Name at least one function for each of the major components.

4. Certain cells in the adrenal gland synthesize and secrete lots of steroid hormones. What organelle would be abundant in these cells, and why?

5. Name the three special types of cell adhesions and list the special functions of each.

6. Define each of the following terms: *transcription, translation,* and *replication.* Where does each of these processes occur in a cell?

7. Describe the anatomical and functional relationship between the rough endoplasmic reticulum, smooth endoplasmic reticulum, and Golgi apparatus in protein synthesis.

Critical Thinking

1. The fluid portion of blood, plasma, is mostly water. The plasma membrane consists of mostly phospholipids. Based on what you know about the biomolecules, predict which classes of biomolecules can dissolve in plasma and which can cross the plasma membrane.

2. Name some factors that cause cancer and potential targets of these factors in cells. How can radiation both cause cancer and be used to treat it?

3. What are some of the differences you would expect to find between muscle cells, fat cells, saliva-secreting cells (saliva contains mostly water and proteins), and skin cells?

4. Explain how a mutation of a single base (called a *point mutation*) in a gene may or may not produce a disease. Can you name a disease that is caused by a point mutation in the protein *hemoglobin*? What is the role of mutations in evolution? Are all mutations deleterious?

Find the answers to these exercises, and additional study tools, at the
Physiology Place (www.physiologyplace.com).

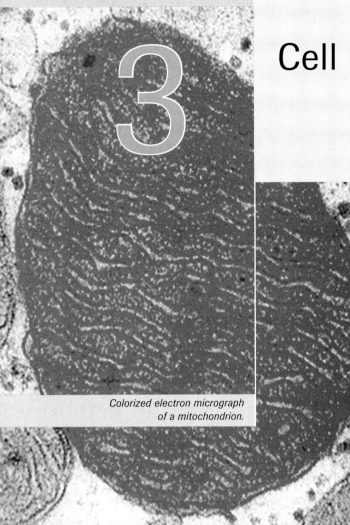

Colorized electron micrograph of a mitochondrion.

Cell Metabolism

magine a factory that manufactures hundreds of chemical compounds using only raw materials that can be found in any grocery store. This factory operates 24 hours a day and can speed up or slow down production on a moment's notice according to the demand for its products. The factory generates only environmentally safe waste products and can repair itself when the need arises. Sounds like a utopian dream of the future, right? Well, it actually happens every day right now: You carry billions of chemical factories that operate just like this in your own body—your cells. Each cell is its own "factory," but the cells cannot function alone. They rely on one another for the raw materials, waste removal, and energy generation. In this chapter, however, we focus on the reactions that take place within a single cell.

Before You Begin

Make sure you have mastered the following topics:

I n Chapter 1, homeostasis was described as the fundamental theme of physiology. Homeostasis requires coordination of the organ systems in maintaining the internal environment of the body in a state compatible for life. To carry out their functions, cells that make up the organ systems must work efficiently, which requires many different chemical reactions and energy. The sum total of all chemical reactions that occur in cells is called **metabolism**. In this chapter we focus on the types of reactions that provide the energy necessary to perform cellular work. First this energy must be extracted from biomolecules ingested in food; then it must be processed before it is either used for work or set aside in storage for future use. The set of reactions involved in energy storage and use (otherwise known as *energy exchange*) is called **energy metabolism**.

Types of Metabolic Reactions

Any chemical reaction involves the transformation of materials that enter the reaction, called *reactants*, into different materials, called *products*. In the reaction

$$A + B \rightarrow C + D$$

the reactants A and B are transformed into the products C and D. By convention, reactions are usually written with reactants on the left and products on the right. A reaction is said to "go forward" when it proceeds from left to right—that is, when reactants are being converted to products. Under certain conditions, some reactions may go in the opposite direction; in that case a reaction is said to go "in reverse."

A proper understanding of chemical reactions requires us to realize that even though we say a reaction goes forward or in reverse, it actually is *bidirectional,* meaning that it goes in both directions at the same time; that is, when reactant molecules are being converted to products, some product molecules are also being converted to reactants. To indicate this, reactions are often drawn with double arrows, as follows:

$$A + B \rightleftharpoons C + D$$

When we say that a reaction goes forward or in reverse, we are referring to the direction of the overall (net) reaction. Suppose, for example, that over a certain time interval some product molecules are converted to reactants, but an even greater number of reactant molecules are converted to products. In this case, the net reaction goes forward because more reactant molecules are converted to products than vice versa. Unless noted otherwise, we use the terms *forward* or *reverse* to refer to the direction of the *net* reaction.

Metabolic reactions are broadly classified as being either *catabolic* or *anabolic,* depending on whether the product molecules are larger or smaller in size than the reactant molecules. A reaction is **catabolic** if it involves the breakdown of larger molecules into smaller ones. Examples of catabolic reactions are the breakdown of proteins into amino acids and the breakdown of glycogen into individual glucose molecules, both of which occur when food is being digested. A reaction is **anabolic** if it involves the production of larger molecules from smaller reactants. Examples are the synthesis of proteins from amino acids and the synthesis of glycogen from glucose, both of which occur in cells following the absorption of ingested nutrients. The breakdown and synthesis of larger molecules are sometimes referred to as *catabolism* and *anabolism,* respectively.

In most cases, metabolic reactions are linked together in a series of steps such that the products of one reaction serve as the reactants in the next. Such a series of reactions is referred to as a **metabolic pathway**, which can be written either as

$$A + X \rightarrow B \rightarrow C \rightarrow D + Y$$

or as

$$A \underset{X}{\rightarrow} B \rightarrow C \overset{Y}{\rightarrow} D$$

This particular pathway consists of a series of three reactions. In the first step, the starting material A reacts with X to form B, which is then transformed into C in the second reaction. C then enters the third reaction, which yields D and Y. The final products, D and Y, are referred to as *end-products* of the pathway. Substances in the middle of the pathway (B and C in this example) are called *intermediates.*

Later in this chapter we take a detailed look at a group of metabolic pathways involved in *glucose oxidation,* an important reaction because it supplies much of a cell's energy needs. Although each of the individual reactions that make up these pathways is unique, many have features in common and can be classified into one of a few general categories. In the following sections we examine three general types of metabolic reactions, which occur not only in glucose oxidation but in all metabolic processes: (1) *hydrolysis and condensation* reactions, (2) *phosphorylation and dephosphorylation* reactions, and (3) *oxidation-reduction* reactions.

Hydrolysis and Condensation Reactions

When you eat a meal containing protein, your gastrointestinal tract breaks down the relatively large protein molecules into amino acids, which are considerably smaller and more easily absorbed into the bloodstream. This reaction, like most other catabolic reactions that occur in the body, is an example of *hydrolysis.* In **hydrolysis**, which means "splitting with water," water reacts with molecules, causing breakage of the bonds that link a molecule together. In this case, the bonds that are broken are the peptide bonds that link amino acids together to form proteins. The general form of a hydrolysis reaction is as follows:

$$A—B + H_2O \rightarrow A—OH + H—B$$

In this reaction, the molecule A—B is broken down into its constituent parts, A and B, as a result of the breaking of the bond that originally linked them together. In this process, a water molecule (H_2O) splits into two parts, a hydroxyl group (—OH) and a hydrogen (—H), one of which combines with A while the other combines with B. As a result, the bond between A and B is broken and replaced with new bonds.

The reverse of hydrolysis, called *condensation,* involves the joining together of two or more smaller molecules to form a larger one, as when amino acids are joined together to form proteins. In this process, water is generated as a product:

$$A—OH + H—B \rightarrow A—B + H_2O$$

Phosphorylation and Dephosphorylation Reactions

In many metabolic reactions, particularly those involved in energy exchange, a phosphate group is either added to a molecule or removed from it. Addition of a phosphate group (abbreviated as P) is known as **phosphorylation** and can be written in an abbreviated fashion as follows:

$$A + P_i \rightarrow A—P$$

Here, the free phosphate group is abbreviated as P_i, for inorganic phosphate, which under physiological conditions exists mostly in the ionized forms HPO_4^{2-} or $H_2PO_4^{-}$. The bond that is formed in this reaction (that is, the bond linking A to P) is known as a *phosphate bond.* Removal of a phosphate group, known as *dephosphorylation,* can be written as follows:

$$A—P \rightarrow A + P_i$$

Oxidation-Reduction Reactions

The removal of electrons from any molecule is called **oxidation**; the addition of electrons to a molecule is called **reduction**. Thus an *oxidation-reduction reaction* is any reaction in which electrons are removed from one reactant molecule and transferred to another. Once electrons have been removed from a molecule, that molecule is said to be *oxidized;* a molecule that has gained electrons (or that has not yet had its electrons removed) is said to be *reduced.*

Depending on the context, the term *oxidation* can be narrowly defined as the reaction of any molecule with oxygen. One example is the oxidation of glucose ($C_6H_{12}O_6$), which occurs in cells as follows:

$$C_6H_{12}O_6 + 6\ O_2 \rightarrow 6\ CO_2 + 6\ H_2O$$

In this reaction, O_2 is the molecular form of oxygen, the form in which oxygen is present in the atmosphere, and CO_2 is carbon dioxide. This narrow definition of oxidation is consistent with the broader definition (removal of electrons) because an oxygen atom holds onto electrons very tightly and tends to pull them away from other atoms to which it is bonded, even though it doesn't remove these electrons completely.

The term *oxidation* is also used in reference to the removal of hydrogen atoms from a molecule, as shown in the following reaction:

$$HA—BH \rightarrow A{=}B + 2\ H$$

In this example, removal of two hydrogen atoms causes the single bond linking A and B to be replaced with a double bond. Even though hydrogen atoms (not electrons) are removed, the reaction is still considered oxidation because these atoms each carry an electron. Therefore, removal of these atoms involves the removal of electrons that originally belonged to the reactant molecule. (As a reflection of this, hydrogen atoms are sometimes referred to as *reducing equivalents*.) Addition of hydrogens to a molecule, the reverse of the previous reaction, is called *reduction* and is shown as follows:

$$A{=}B + 2 \text{ H} \rightarrow HA{-}BH$$

Some oxidation or reduction reactions involve the removal or addition of actual electrons, which causes changes in a molecule's electrical charge. As we will see, this occurs in *oxidative phosphorylation,* a process occurring in mitochondria that is crucial to energy metabolism. At certain points in this process, electrons (e^-) are removed from pairs of hydrogen atoms, causing these atoms to become positively charged hydrogen ions (H^+), an example of oxidation:

$$H_2 \rightarrow 2 \text{ H}^+ + 2 \text{ e}^-$$

At other points in oxidative phosphorylation the reverse occurs; electrons combine with hydrogen ions to form uncharged hydrogen atoms, an example of reduction:

$$2 \text{ H}^+ + 2 \text{ e}^- \rightarrow H_2$$

Quick Test 3.1

1. What is the primary distinction between an anabolic reaction and a catabolic reaction?

2. Define the following terms: *reactant, product, metabolic pathway, intermediate, end-product.*

3. Describe in general terms what occurs in hydrolysis, phosphorylation, and oxidation reactions.

4. For each of the reaction types in question 3, give the term that applies to the opposite reaction.

Metabolic Reactions and Energy

The metabolic reactions that occur in the body perform many functions. One of the most important functions is enabling cells to transform raw materials from the environment into the great variety of substances from which the body is made. Metabolic reactions also provide us with **energy**, broadly defined as the capacity to perform work. The body most obviously performs work during movement. When you push a grocery cart, for example, you exert a certain amount of force against the cart and move it a certain distance. The amount of work you perform is the product of the force exerted and the distance moved (work = force × distance). However, the body performs

work in other ways, too—even while at rest—and it must do so on a continual basis just to stay alive. The heart performs work when it pumps blood, the kidneys perform work when they form urine, and cells perform work when they multiply to repair damaged tissues. All this work is performed using energy derived from metabolic reactions. In this section we explore the nature of the relationship between chemical reactions and energy.

Energy-Releasing and Energy-Requiring Reactions

Although certain metabolic reactions release energy that can be used by cells to perform work, other metabolic reactions consume or require energy. As a general rule, reactions that release energy are catabolic. You know, for example, that you obtain energy from the food you eat, but before nutrients can be absorbed, the food must be digested. In this process, large molecules such as proteins and starch are broken down into smaller molecules such as amino acids and glucose. These nutrients are then absorbed into the bloodstream and taken up by cells throughout your body. To obtain energy from these nutrients, cells must break them down into still smaller molecules. Cells can also use amino acids, glucose, and other small nutrient molecules for other purposes, such as the synthesis of glycogen or cellular proteins, but these reactions are anabolic and do not release energy. Instead, energy must be put into them if they are to occur, which is true of anabolic reactions in general.

Energy Changes in Reactions

It is an inescapable truth of nature that chemical reactions are always accompanied by either the release of energy or the input of energy. In other words, all reactions involve energy exchange of one sort or another. This exchange occurs because molecules possess energy; different types of molecules possess different amounts of energy. When a reaction releases energy, it does so because the reactant molecules possess more energy than the product molecules; in this case, reactants enter the reaction with a certain amount of energy, and products come out of the reaction with less energy. The *first law of thermodynamics* states that *the energy in a closed system is constant;* that is, energy can be neither created nor destroyed but instead can only change form. Therefore, if the products come out of the reaction with less energy than the reactants had when they went in, then the "extra" energy must be released in one form or another during the course of the reaction:

$$\text{reactants} \rightarrow \text{products} + \text{energy}$$

The released energy might be heat or take some other form, and under certain conditions it can be used to perform work. A familiar example of an energy-releasing reaction is the combustion of gasoline. When gasoline is

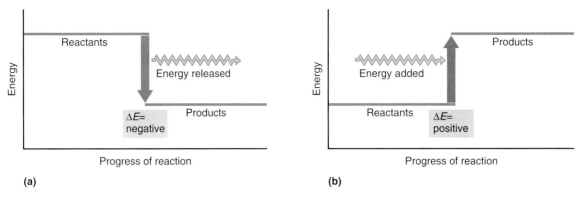

Figure 3.1 Energy changes in reactions. Wavy arrows indicate the transfer of energy in the reaction. (a) When the energy of the reactants is higher than that of the products, the energy change is negative, and energy is released in the reaction. (b) When the energy of the products is higher than that of the reactants, the energy change is positive, and energy is added to the reaction.

burned in the open air, it generates energy—heat and light—in the form of flames. When burned in the internal combustion engine of a car, some of this energy is used to perform the work of moving the car.

The energy change of any reaction (expressed as ΔE) is the difference between the energy of the products ($E_{products}$) and the energy of the reactants ($E_{reactants}$): $\Delta E = E_{products} - E_{reactants}$. (The symbol Δ or delta is frequently used to denote a change in some quantity.) Because the products have less energy than the reactants in an energy-releasing reaction, the energy change ΔE in such a reaction is negative. This is illustrated in **Figure 3.1**a, where distance along the y-axis represents the amount of energy present in reactant or product molecules, and distance along the x-axis represents the progress of the reaction—that is, the transformation of reactants into products.

Figure 3.1b depicts the energy change that occurs in an energy-requiring reaction, which we can write as follows:

$$reactants + energy \rightarrow products$$

In this case, the products possess more energy than the reactants, so that the energy change ΔE is positive. Because energy cannot be created, the "extra" energy acquired by the products must come from some source other than the reactant molecules themselves. Perhaps the most familiar example of an energy-requiring process is the boiling of water. To make water boil, energy must be put into it (by heating it on a stove, for instance).

The energy change ΔE is typically expressed in units of energy or heat, such as calories (cal) or kilocalories (kcal), but it may be given in joules (J) or kilojoules (kJ). A **calorie** is the amount of energy that must be put into 1 gram (or 1 mL) of water to raise its temperature by 1 degree centigrade (°C) under a standard set of conditions. One calorie is equivalent to 4.18 joules, and a kilocalorie or kilojoule is 1000 calories or joules, respectively. The

amount of energy released or consumed in a reaction depends not only on the nature of the reactants and products involved, but also on the quantity of reactants consumed or products produced. As this quantity increases, the size of the energy change increases in direct proportion. For this reason, ΔE is commonly expressed in units of kcal/mole or some equivalent.

Kinetic Energy Versus Potential Energy

When we say that reactions release or require energy, or that reactant or product molecules possess a certain amount of energy, what do we mean? To answer this question, we first must realize that there are two basic types of energy: kinetic energy and potential energy. **Kinetic energy**, sometimes called *energy of motion,* is the type of energy possessed by moving objects. The kinetic energy of an object depends on the object's mass and how fast it is moving; kinetic energy increases when either of these factors increases. Atoms and molecules possess kinetic energy because they undergo *thermal motion,* a random movement or vibration that occurs at any temperature above absolute zero. This kinetic energy increases in direct proportion to the molecules' temperature and is often referred to as *thermal energy* or *heat.* In contrast, **potential energy** is not associated with motion but instead is the type of energy that can be stored and eventually converted to kinetic energy. (In other words, potential energy has the *potential* to become kinetic energy.) When we lift an object or push it up a hill, for example, we expend energy and perform work. At the same time, some of this energy is stored as potential energy because the potential energy of the object increases as it rises higher and higher above the ground or gains elevation. This potential energy can be converted to kinetic energy by allowing the object to fall to the ground or roll down the hill (**Figure 3.2**).

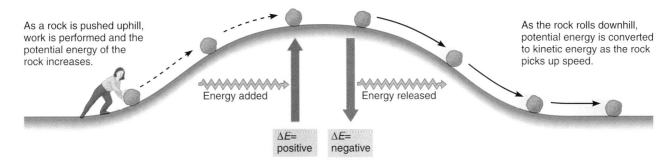

Figure 3.2 Potential energy and kinetic energy.

Chemical energy, the energy stored in the chemical bonds holding molecules together, is a form of potential energy. When we are talking about the energy change in a reaction, we are talking about change in *potential* energy. In energy diagrams such as Figure 3.1, it is assumed that reactant and product molecules are at the same temperature (that is, have the same kinetic energy), so any difference in energy between reactants and products must reflect differences in potential energy. In the next section we see that the energy change in a reaction not only determines whether energy is released or required, but also the direction in which the reaction proceeds.

How the Energy Change of a Reaction Determines Its Direction

We know from experience that chemical reactions have a natural tendency to run in particular directions. For instance, once you light a piece of paper, it burns until it is reduced to ashes (provided there is sufficient oxygen), but you never see the combustion of paper stop and go in reverse, such that ashes are reconstituted into paper. The reason for this is that the burning of paper is an energy-releasing reaction, and *energy-releasing reactions always proceed spontaneously in the forward direction.* (To say that a process occurs spontaneously means that it occurs on its own—that is, without any energy being put into it.) To reconstitute ashes into paper would require the paper-burning reaction to go in reverse, which would require energy rather than release it. This does not happen because *energy-requiring reactions do not go forward spontaneously; they go forward only when energy is put into them.* (In fact, energy-requiring reactions will spontaneously go in reverse if energy is not put into them.)

These rules reflect another fundamental truth of nature: Systems have a natural tendency to go from states of higher potential energy to states of lower potential energy. We know, for example, that a rock will roll downhill spontaneously but not uphill (not unless it has momentum to begin with), because the potential energy of the rock decreases when it rolls downhill; to roll uphill, its potential energy would have to increase. Likewise, an energy-releasing reaction (reactants → products + energy) goes forward spontaneously because the potential energy decreases as high-energy reactants are transformed into lower-energy products. In contrast, an energy-requiring reaction (reactants + energy → products) does not go forward spontaneously because the potential energy increases as low-energy reactants are converted to higher-energy products. (Strictly speaking, it is the *free energy* change of a reaction, not the *potential energy* change, that determines its direction, but free energy is a kind of chemical potential energy.)

Given the energy-releasing nature of catabolic reactions and the energy-requiring nature of anabolic reactions, we can surmise that catabolic reactions should occur spontaneously in cells, whereas anabolic reactions should occur only when energy is put into them. Anabolic reactions are able to occur in cells because certain cellular mechanisms couple these reactions with catabolic reactions. Such coupling harnesses the energy that is released from catabolic reactions and uses it to drive anabolic reactions. In the following example, the reaction A → B releases energy. This energy is used to drive the reaction C → D.

$$A \rightarrow B + energy$$

$$C + energy \rightarrow D$$

A reaction is said to be in **equilibrium** when there is no net reaction direction; that is, when the reactant is converted to product at the same rate product is converted to reactant. Equilibrium occurs when the reactant and product have the same energy, such that the energy change, ΔE, is zero.

The Law of Mass Action

The energy of reactant or product molecules is determined by many different factors, including the numbers and types of chemical bonds that are present, the nature of the solvent in which the molecules are dissolved, and the molecules' *concentrations*. In general, as the concentration of molecules increases, the energy of those molecules increases (**Chemistry Review: Solutions and Concentrations**, p. 69). Because of this connection between concentration and energy, a given reaction can be made

An energy-releasing reaction goes forward spontaneously...

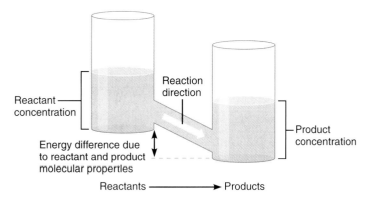

(a)

...but can be forced to go in reverse if the concentration of products is increased.

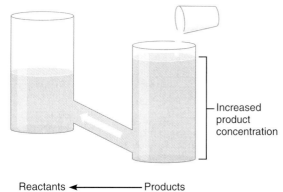

An energy-requiring reaction goes in reverse spontaneously...

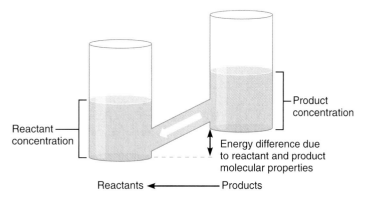

(b)

...but can be forced to go forward if the concentration of reactants is increased.

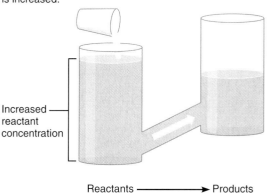

Figure 3.3 The law of mass action. In this model, the volume of water in the reservoirs represents the concentrations of products and reactants. **(a)** Energy-releasing reactions can be forced to go in reverse when product concentrations are increased relative to reactant concentrations. **(b)** Energy-requiring reactions can be forced to go forward when reactant concentrations are increased relative to product concentrations.

to go in either direction—forward or in reverse—by raising or lowering the concentrations of reactants relative to products. This phenomenon is commonly referred to as the **law of mass action**: *An increase in the concentration of reactants relative to products tends to push a reaction forward, and an increase in the concentration of products relative to reactants tends to push a reaction in reverse.*

The law of mass action is illustrated in **Figure 3.3** using a model consisting of two water reservoirs connected by a tube, with the left-hand reservoir representing "reactants" and the right-hand reservoir representing "products." The difference between the reservoirs' heights indicates differences in energy between reactants and products based on their molecular properties. The volume of the water in the reservoirs represents reactant and product concentrations. The

direction of water flow in the tube represents the direction of the reaction.

In Figure 3.3a, the left-hand reservoir is above the right-hand reservoir, indicating a reaction that normally releases energy. When water volume is the same in the two reservoirs, water flows from left to right, indicating that the reaction goes forward spontaneously, from reactants to products. However, by adding water to the right-hand reservoir (that is, by increasing the concentration of products), we can make water flow in the opposite direction. Thus, the reaction is forced to go in reverse. In Figure 3.3b, the right-hand reservoir is above the left-hand reservoir, representing a reaction that normally requires energy. When water volume is the same in the two reservoirs, water flows from right to left, indicating that the reaction goes in reverse spontaneously. When

CHEMISTRY REVIEW

SOLUTIONS AND CONCENTRATIONS

A mixture of molecules dissolved in a liquid is known as a *solution.* The dissolved substance, which is usually a solid or gas in its pure form, is known as the *solute,* whereas the liquid is referred to as the *solvent.* Solute molecules are said to be *dissolved* when they are completely separate from one another and are surrounded by solvent molecules.

The *concentration* of a solution is a measure of the quantity of solute contained in a unit volume of solution. Solute quantity is most commonly expressed in terms of moles (mol), and volume is most often given in liters (L). If 1 mole of solute is present in 1 liter of solution, the concentration is said to be 1 *molar* (1 M = 1 mol/L). If 1/1000 of a mole (that is, 1 *millimole*) of solute is dissolved in 1 liter of solution, the concentration is 0.001 molar or 1 *millimolar* (1 mM = 1×10^{-3} mol/L). For very

dilute solutions, concentrations may be expressed in *micromoles* per liter (μmol/L = 1×10^{-6} mol/L) or *nanomoles* per liter (nmol/L = 1×10^{-9} mol/L). Concentrations of specific substances are often indicated using brackets. For example, [Na^+] represents sodium ion concentration.

Occasionally, concentrations are expressed in terms of the *mass* of solute contained in a unit volume of solution—grams per liter (g/L) or micrograms per liter (μg/L = 1×10^{-6} g/L), for instance. Often in physiology, the unit of volume is a *deciliter* (dL = 0.1 L), which is equivalent to 100 milliliters (100 mL). The concentration of a solution containing 1 gram of solute per deciliter is frequently expressed as 1 *percent* (1%) because physiological solutions are mostly water, 100 mL of which weighs 100 grams.

---

enough water is added to the left-hand reservoir, however, the flow goes from left to right. In this case, the reaction is forced to go forward by increasing the concentration of reactants.

Activation Energy

When reactant molecules enter into a reaction, their conversion into product molecules does not occur abruptly. Instead, the reacting molecules go into a high-energy intermediate form called a *transition state,* which then breaks down into the products. The reverse reaction (the conversion of products to reactants) also must go through this transition state. Thus, the transformations that occur in a reaction are continuous and gradual, not sudden, as shown in **Figure 3.4**a. The "hump" in the middle of the curve, known as an *activation energy barrier,* is due to the fact that the potential energy of the transition state is greater than that of either the reactants or the products.

For example, we previously described the burning of paper as an energy-releasing process. However, paper does not spontaneously combust. Instead energy, such as a flame or heat provided through a magnifying glass, must get the process started by overcoming the activation energy barrier. For reactants to become products or vice versa, molecules must have sufficient potential energy to surmount the activation energy barrier, and they must acquire some "extra" energy, called **activation energy**, which is the difference between the energy of the transition state and the energy of either the reactants or products. The activation energy is indicated in Figure 3.4b as the vertical distance between the initial or final energies

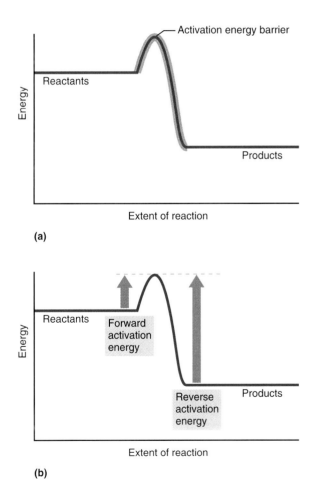

(a)

(b)

Figure 3.4 The activation energy barrier. The reactions depicted are energy-releasing reactions. (a) An energy diagram similar to that in Figure 3.1a, but with the activation energy barrier included. (b) An energy diagram in which the forward and reverse activation energies are indicated by the vertical arrows.

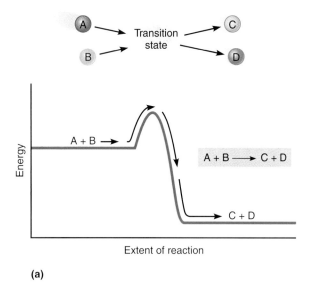

(a)

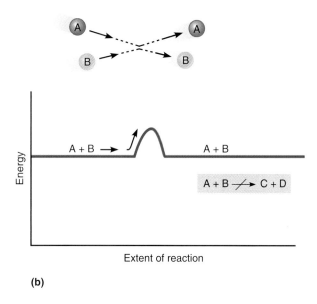

(b)

Figure 3.5 Effect of the activation energy barrier.
(a) Collision of two reactant molecules, A and B, with enough energy to enter the transition state. The energy of the collision surmounts the activation energy barrier, resulting in the formation of products C and D. (b) Collision of A and B with insufficient energy to enter the transition state. The activation energy barrier is not surmounted, so no reaction occurs.

> As the activation energy barrier is lowered, does the energy change in the reaction increase, decrease, or stay the same?

and the peak of the curve. Note that for the reaction shown in the figure, which is an energy-releasing reaction, the activation energy for the forward reaction is less than the activation energy for the reverse reaction. For an energy-requiring reaction, the reverse would be true.

Stays the same

If molecules must acquire extra energy to surmount the activation energy barrier and react, where does this "extra" energy come from? Molecules acquire this energy by colliding with one another, which happens all the time because they are in constant thermal motion. Consider, for example, the reaction A + B → C + D. For molecules of A and B to react, they first must collide. When this happens, some of the molecules' thermal kinetic energy is converted to potential energy. If this gain in potential energy is equal to or greater than the activation energy, then the two molecules will enter the transition state and be converted to the products C and D (**Figure 3.5**a).

The significance of the activation energy barrier is that it limits how fast a reaction can go, a topic that is explored in detail in the next section. Here it is important to note that not every collision between reactant molecules produces a reaction, because some collisions do not generate enough potential energy to surmount the activation energy barrier. The more quickly two molecules about to collide are moving, the more potential energy they gain in the collision. If the energy gained is less than the activation energy, then the colliding molecules will not enter the transition state and will not react (Figure 3.5b); they will simply emerge from the collision unaltered.

> ## Quick Test 3.2
>
> 1. Are energy-releasing reactions usually catabolic or anabolic?
> 2. In an energy-releasing reaction, which has more energy—the reactants or the products?
> 3. What factor determines the direction of a reaction? Are reactions that go forward spontaneously energy-releasing or energy-requiring?
> 4. Give a brief description of the law of mass action.
> 5. What is activation energy? How does it affect a reaction?

Reaction Rates

The *rate* of a chemical reaction is a measure of how fast it consumes reactants and generates products; it is usually expressed as a change in concentration per unit time (moles/liter-second, or some equivalent). For example, as the reaction A + B → C + D proceeds, the concentrations of A and B diminish while the concentrations of C and D increase. The rate of this reaction can be expressed as the change in the concentrations of C or D per unit time.

The rate of a metabolic reaction is of great physiological significance because proper body function demands that reactions proceed at a rate that matches the body's needs at the moment. Rates higher or lower than required ultimately lead to severe impairment of cellular function. Perhaps the most vivid demonstration of this danger occurs in *hypothermia,* when body temperature falls below normal (see Chapter 1). Any decline in body temperature causes metabolic reactions to slow down, and a decline in core temperature of only a few degrees can cause a person to become weak and disoriented, and possibly even to lose consciousness. A further drop in temperature can precipitate cardiac arrest (stoppage of the heartbeat) and death.

Later in this section we will see that metabolic reactions are accelerated or *catalyzed* by special molecules called *enzymes,* and that the rates of metabolic reactions are normally regulated through changes in enzyme activity. First, however, we begin our exploration of reaction rates with a look at the factors that affect the rates of chemical reactions in general.

Factors Affecting the Rates of Chemical Reactions

We know from experience that certain chemical reactions progress more quickly than others. For example, one oxidation reaction—the combustion of gasoline—can occur with explosive speed, whereas the rusting of iron (another oxidation reaction) occurs so slowly that it may take weeks to see the effects. The rate of a reaction is determined by a variety of factors, including (1) reactant and product concentrations, (2) temperature, and (3) the height of the reaction's activation energy barrier.

Reactant and Product Concentrations

When we speak of the rate of a reaction, we are usually referring to its *net rate:* the difference between the rate of the forward reaction (reactants → products) and the rate of the reverse reaction (products → reactants). According to the law of mass action, any increase in the concentration of reactants will bring about an increase in the forward rate without affecting the reverse rate. Likewise, any increase in the concentration of products will cause the reverse rate to increase without affecting the forward rate. Therefore, an increase in the concentration of reactants relative to products will increase the *net rate* in the forward direction. Conversely, an increase in the concentration of products relative to reactants will decrease the net forward rate, and can even make the reaction go in reverse if the change in concentration is large enough. The effect of reactant and product concentrations on reaction rates is due to the fact that concentrations affect the frequency of collisions between molecules: As the concentration of molecules increases, the number of collisions occurring in a given time period also increases.

Temperature

In general, the rate of a reaction increases with increasing temperature and decreases as the temperature falls. We refrigerate food, for example, because cooling slows down the rate of decomposition reactions, which helps the food "keep" longer.

Temperature influences reaction rates because it affects the frequency and energy of molecular collisions. As temperature increases, the average kinetic energy of molecules increases, which increases the energy of collisions. The result is an increase in the proportion of collisions having enough energy to surmount the activation energy barrier, which increases the rate of a reaction.

The Height of the Activation Energy Barrier

The height of the activation energy barrier differs among reactions. All else being equal, the rate of a reaction increases as the height of the barrier decreases. The reason for this is that at any given temperature, only a fraction of the collisions between molecules has sufficient energy to surmount the barrier and produce a reaction. As the height of the barrier decreases, the proportion of collisions having the requisite energy increases—not because the collisions themselves are any more energetic, but because the minimum energy requirement for a "successful" collision has been reduced. (This situation is analogous to a high-jump competition: If the bar is lowered, the number of contestants making a successful jump will increase.)

Note that as the height of the activation energy barrier decreases, rates of both the reverse reaction and forward reaction increase because the activation energy of both reactions becomes lower. However, for reasons beyond the scope of this book, the net rate of the reaction still increases despite the increase in the rate of the reverse reaction.

The Role of Enzymes in Chemical Reactions

If you were to set up a typical metabolic reaction by mixing reactants together in a beaker, it would proceed very slowly. In fact, most metabolic reactions would proceed too slowly to be compatible with life. These reactions are able to proceed at considerably faster rates in cells because of the presence of **enzymes**, biomolecules (almost always proteins) specialized to act as *catalysts,* a general term for substances that increase the rates of chemical reactions. In this section we examine how enzymes accomplish this task, and how the regulation of enzyme activity controls reaction rates according to the body's needs.

Mechanisms of Enzyme Action

Cells contain a wide variety of enzymes, each specialized to catalyze a particular reaction or group of related reactions. To catalyze a reaction, an enzyme molecule must first bind to a reactant molecule, which in the context of enzyme-catalyzed reactions is called a **substrate**. The enzyme then acts on the substrate to generate a product molecule, which is subsequently released. Thus, the

ACIDS, BASES, AND pH

Certain substances release hydrogen ions or *protons* (H^+) when dissolved in water and are known as *acids*. A familiar example is *hydrochloric acid* (HCl), which dissociates into hydrogen and chloride ions, as follows:

$$HCl \rightarrow H^+ + Cl^-$$

HCl is an example of a *strong acid*, an acid that dissociates completely. Certain other acids are *weak acids*, meaning that they dissociate incompletely. For example, *carboxyl groups* (—COOH), which are found on amino acids and other biomolecules, act as weak acids and dissociate as follows:

$$R—COOH \rightleftharpoons R—COO^- + H^+$$

Here, the double arrow signifies that protons not only can dissociate from the anion (R—COO$^-$) but can also combine with it.

Substances that combine with protons are called *bases* and are classified as strong or weak depending on whether they combine completely or incompletely. Examples of weak bases

are *amino groups* (—NH$_2$), which are found on amino acids and other compounds and react with protons as follows:

$$R—NH_2 + H^+ \rightleftharpoons R—NH_3^+$$

The *acidity* of a solution is determined by its hydrogen ion concentration, which can be expressed in terms of molarity or denoted by a number called the *pH*, which is defined according to the following expression:

$$pH = \log(1/[H^+]) = -\log[H^+]$$

where [H^+] is the hydrogen ion concentration. Note that as [H^+] increases, the pH decreases. Note also that because the pH scale is logarithmic, a change of one pH unit represents a tenfold change in the hydrogen ion concentration. The hydrogen ion concentration in pure water is 10^{-7} M, giving a pH of 7. Solutions with a pH of 7 are said to be *neutral*. A solution whose pH is below 7 is said to be *acidic*; if the pH is above 7, the solution is *basic*.

Other Factors Affecting Enzyme Activity The rates of enzymatic reactions can be affected by still other conditions, such as the temperature of the reaction mixture or its pH (**Chemistry Review: Acids, Bases, and pH**). Changes in temperature or pH can alter the shape of an enzyme molecule, which is so crucial to its function. In general, enzyme activities decline if the temperature or pH becomes significantly higher or lower than normal. Within the body, changes in temperature are rarely significant because body temperature is regulated to stay constant. pH is also regulated to stay relatively constant in intracellular and extracellular fluid (within the range 7.2 to 7.4), but it varies appreciably in certain locations within the gastrointestinal tract. The lumen of the stomach, for instance, contains fluid that is highly acidic (the pH can be as low as 2); in the intestine, the luminal fluid is more basic, with a pH over 8. Significantly, the activity of pepsin, an enzyme that works in the stomach, is highest when the pH is in the acidic range, whereas that of *trypsin*, a similar protein-digesting enzyme that works in the intestine, is highest when the pH is in the basic range.

Regulation of Enzyme Activity

We have just seen how enzymes catalyze reactions, and how an enzyme's ability to catalyze a reaction is influenced by various factors. Now we look at how the body regulates the activity of enzymes to adjust the rates of reactions.

As conditions in the body change, the rates of metabolic reactions are continually adjusted to meet the body's needs. This adjustment is accomplished by means of mechanisms that regulate the activities of certain enzymes. In some cases, cellular enzyme concentrations are adjusted to speed up or slow down reactions. More enzyme molecules may be synthesized, for example, which raises the enzyme concentration and increases reaction rates. Alternatively, enzymes may be broken down or inactivated by other means, which effectively lowers the enzyme concentration and decreases reaction rates. More often, reaction rates are adjusted through changes in the activities of existing enzyme molecules, which is considerably faster than altering enzyme concentration. Two common mechanisms for altering the activity of existing enzymes are *allosteric* and *covalent regulation*.

Allosteric Regulation Certain enzyme molecules possess a binding site called the **regulatory site** that is specific for molecules known generally as **modulators**. (The regulatory site is distinct from the active site.) A modulator induces a change in an enzyme's conformation that alters the shape of the active site, causing a change in the enzyme's activity by altering its catalytic rate, its affinity for substrate, or both. This type of regulatory mechanism is known as **allosteric regulation** (**Figure 3.9**).

In allosteric regulation, the binding of the modulator to the regulatory site is reversible, so the modulator can readily leave the regulatory site after binding to it. Normally, an equilibrium exists between the enzyme and

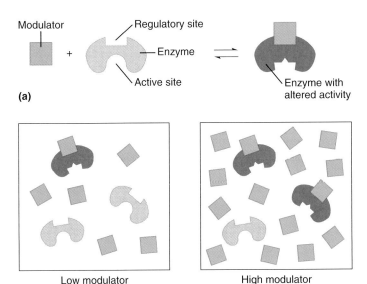

(a)

Modulator — Regulatory site — Enzyme — Active site

Enzyme with altered activity

(b) Low modulator concentration — High modulator concentration

Figure 3.9 Allosteric regulation of enzyme activity.
(a) Reversible binding of a modulator molecule to a regulatory site on an enzyme. When bound, the modulator alters the active site and thus the activity of the enzyme. (b) The effect of modulator concentration on enzyme activity. When the modulator is present at low concentration, relatively few enzyme molecules are bound. At higher modulator concentrations, more enzyme molecules are bound, increasing the modulator's influence on overall enzyme activity.

Why does the number of bound enzyme molecules increase as the modulator concentration increases?

the modulator. When the modulator concentration increases, the equilibrium shifts according to the law of mass action, with more enzymes becoming bound by the modulator (see Figure 3.9b). When the modulator concentration decreases, the equilibrium shifts in the opposite direction, so fewer enzymes are bound by the modulator.

Binding of a modulator to an enzyme can bring about either an increase or a decrease in the enzyme's activity, depending on the enzyme in question. A modulator that increases the activity of an enzyme is called an *activator;*

a modulator that decreases the activity of an enzyme is called an *inhibitor.* Some enzyme molecules have more than one regulatory site and can respond to more than one type of modulator. Sometimes these different modulators affect the enzyme in different ways—some are inhibitory while others are stimulatory.

Covalent Regulation Another type of enzyme regulation is **covalent regulation**, in which changes in an enzyme's activity are brought about by the covalent bonding of a specific chemical group to a site on the enzyme molecule (**Figure 3.10**). In this process, formation of the covalent bond between the chemical group and the enzyme protein is catalyzed by another enzyme (enzyme A in the figure). Because covalent bonds are relatively strong, the chemical group remains attached to the enzyme unless these bonds are broken, which requires the action of other enzymes (enzyme B in the figure).

A common form of covalent regulation involves the phosphorylation and dephosphorylation of enzyme molecules. Phosphorylation of a target enzyme molecule (E) is catalyzed by a type of enzyme called a **protein kinase**, as follows:

$$E + P_i \xrightarrow{\text{protein kinase}} E\text{—}P$$

Dephosphorylation is catalyzed by an enzyme called a **phosphatase** and can be written as follows:

$$E\text{—}P \xrightarrow{\text{phosphatase}} E + P_i$$

These kinases and phosphatases are usually specific for particular target enzymes and are themselves the targets of other allosteric or covalent regulatory mechanisms.

Feedback Inhibition Through allosteric or covalent regulation, the activity of an enzyme may be increased or decreased, but the reactions in a metabolic pathway are generally catalyzed by different enzymes, with only certain enzymes being regulated. Through control of certain regulated enzymes, cells can exert control over entire pathways.

Certain enzymes in metabolic pathways tend to be targets of regulation more than others. For example,

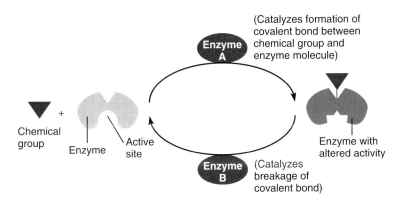

Chemical group + Enzyme — Active site

Enzyme A (Catalyzes formation of covalent bond between chemical group and enzyme molecule)

Enzyme with altered activity

Enzyme B (Catalyzes breakage of covalent bond)

Figure 3.10 Covalent regulation of enzyme activity.
Covalent bonding of a chemical group to an enzyme molecule alters the enzyme's activity. Formation of these covalent bonds is catalyzed by another enzyme (A); removal of the chemical group requires the action of yet another enzyme (B), which breaks the covalent bonds.

Because of the law of mass action

within a metabolic pathway the rates of the individual reactions vary. The rate of the entire pathway can be only as fast as the rate of the slowest reaction within the pathway, which is therefore called the *rate-limiting step*. The enzyme that catalyzes the slowest reaction is thus called the *rate-limiting enzyme*, and is frequently subject to a regulatory process called *feedback inhibition*.

In **feedback inhibition**, an intermediate product of a metabolic pathway allosterically inhibits an enzyme that catalyzes an earlier reaction in the same pathway. Consider for example a metabolic pathway consisting of three steps catalyzed by three different enzymes (E_1, E_2, and E_3):

$$A \xrightarrow{E_1} B \xrightarrow{E_2} C \xrightarrow{E_3} D$$
$$(-)$$

Here, substance C acts as an allosteric inhibitor of E_2, the enzyme catalyzing the second step, as indicated by the arrow with the negative sign.

Normally, feedback inhibition regulates enzyme activities so as to hold the rates of metabolic reactions steady. Under these conditions (referred to as a *steady state*), all three steps in the pathway just described proceed at the same rate, so that C is consumed in the third step as fast as it is generated in the second, which keeps [C] steady. If the rate of the second step increased suddenly for some reason, [C] would rise, which would then suppress the activity of E_2 and counteract the change in concentration.

Feedback inhibition can also speed up or slow down metabolic reactions according to changes in the body's needs. Suppose, for instance, that in the previous example the third step speeds up as an appropriate response to an increased cellular demand for the final product, D. The speed-up of the third step will cause C to be consumed at a greater rate, which should cause [C] to begin falling. This fall in [C] will reduce the inhibitory action of C on E_2, thereby allowing the second step to speed up.

Often, the allosteric inhibitor in feedback inhibition is the end-product of the metabolic pathway, and this type of regulation is called *end-product inhibition*. For example, in the pathway below, the end-product of the pathway, D, inhibits E_1.

$$A \xrightarrow{E_1} B \xrightarrow{E_2} C \xrightarrow{E_3} D$$
$$(-)$$

The enzyme inhibited is often the rate-limiting enzyme or the first enzyme in a metabolic pathway. Some metabolic pathways, however, have branch points. In these pathways, the first enzymes at branch points are often regulated to determine the relative rates of the branched reactions. Consider the following metabolic pathway:

$$A \longrightarrow B \xrightarrow{E_1} C$$
$$\xrightarrow{E_2} D$$

Intermediate B can be converted to C or D, depending on the relative activity of enzymes 1 and 2.

Feedforward Activation Another, less common, mechanism of enzyme regulation is **feedforward activation**, which involves the activation of an enzyme by an intermediate appearing upstream in a metabolic pathway, as follows:

$$A \xrightarrow{E_1} B \xrightarrow{E_2} C \xrightarrow{E_3} D$$
$$(+)$$

In this example, E_3, which catalyzes the third step, is activated by substance B, which is produced in the first step. As with end-product inhibition, feedforward activation helps to keep reaction rates steady under normal conditions, but it also allows reactions to speed up or slow down when conditions change.

> **Quick Test 3.3**
>
> 1. An increase in which of the following would speed up the net rate of a reaction proceeding in the forward direction? temperature, reactant concentrations, product concentrations, the height of the activation energy barrier (choose all that apply)
>
> 2. Define the following terms: *enzyme, substrate, substrate specificity, active site, affinity, cofactor, coenzyme, percent saturation, regulatory site, modulator.*
>
> 3. How is the rate of an enzymatic reaction affected by the catalytic rate of an enzyme? By its affinity for the substrate?
>
> 4. What is the primary distinction between allosteric regulation and covalent regulation? Between feedback inhibition and feedforward activation?

Glucose Oxidation: The Central Reaction of Energy Metabolism

We have seen that energy is released in certain chemical reactions and consumed in others, and that released energy can be used to perform cellular work. We have also learned about how enzymes catalyze these reactions, so that they can occur quickly enough to meet the body's needs. In this section we bring all this information together to understand how a cell utilizes the breakdown of nutrient molecules to produce and store enough energy to maintain all bodily processes. In this discussion we focus on glucose oxidation, the reaction that virtually all cells rely on to provide their energy needs.

Have you ever wondered why we breathe? The simple answer is that breathing enables us to obtain oxygen from the air when we inhale and get rid of carbon dioxide when we exhale. This exchange of gases is crucial to our survival because we derive most of our energy from the reaction of oxygen with glucose and other nutrient molecules that serve as fuels. When oxygen reacts with these fuels, energy is released and carbon dioxide is generated as a waste product. This is apparent in the reaction of oxygen with glucose ($C_6H_{12}O_6$), which proceeds as follows:

$$C_6H_{12}O_6 + 6\ O_2 \rightarrow 6\ CO_2 + 6\ H_2O + energy$$

Here we can see that complete oxidation of one molecule of glucose requires six oxygen molecules and generates six molecules each of carbon dioxide and water. We also see that this reaction releases energy, which is why this reaction is of central importance to cells. The energy released amounts to 686 kcal for every mole of glucose that enters the reaction ($\Delta E\ -686$ kcal/mole).

ATP: The Medium of Energy Exchange

When energy is released in a reaction, it must be "captured" in certain forms before it can be used to do work; otherwise it is simply dissipated (released into the environment) as heat. When coal burns in a furnace, for example, it releases energy in the form of heat. If this energy is not captured somehow, it just dissipates into the atmosphere, without any work being performed. A power plant, in contrast, captures energy released from the burning of coal by using the heat to generate steam under pressure. This steam is then used to drive an electrical generator, and in doing so it performs work. Thus, the steam acts as a temporary energy store or "middleman" that facilitates the transfer of energy from the burning coal to the electric generator.

Cells harness the energy released in a reaction such as glucose oxidation to synthesize a compound called **adenosine triphosphate (ATP)**, which serves as a temporary energy store (**Figure 3.11**). ATP is synthesized from a nucleotide called *adenosine diphosphate (ADP)* and inorganic phosphate (P_i) according to the following reaction:

$$ADP + P_i + energy \rightarrow ATP$$

ATP synthesis is an example of a condensation reaction because water is generated as a product. (This water is often omitted from the reaction notation for simplicity's sake.) The quantity of energy required to make one mole of ATP under normal cellular conditions is about 7 kcal ($\Delta E = +7$ kcal/mole).

The formation of ATP is a *phosphorylation* reaction because it involves the addition of a phosphate group (P) to another compound, ADP. ATP synthesis occurs through two basic processes: *substrate-level phosphorylation* and *oxidative phosphorylation*. In **substrate-level phosphorylation**, a phosphate group is transferred from a metabolic intermediate (X) to ADP to form ATP:

$$X\text{-}P + ADP \rightarrow X + ATP$$

For example, in muscle cells, *creatine phosphate* can donate its phosphate group to ADP to form ATP and creatine. In

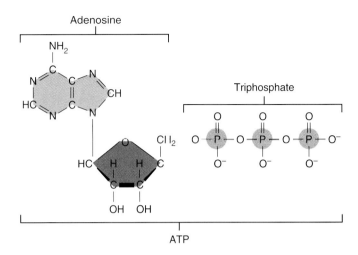

Figure 3.11 ATP.

oxidative phosphorylation, ADP binds with a free inorganic phosphate (P_i) to form ATP:

$$ADP + P_i \rightarrow ATP$$

Oxidative phosphorylation requires the electron transport system and oxygen. In subsequent sections of this chapter, you will learn specific examples of substrate-level phosphorylation and oxidative phosphorylation. We now look at the breakdown of ATP to release energy.

We noted that steam produced in a power plant is a temporary energy store because it is eventually used for generating electricity; during this process it loses the energy that it gained from the burning coal. Likewise, ATP is also a temporary energy store because it is eventually broken down to ADP and Pi, in the process losing the energy that was put into it when it was made:

$$ATP \rightarrow ADP + P_i + energy$$

This reaction is often referred to as *ATP hydrolysis* because water is a reactant, even though it is usually omitted from the reaction notation. (Because ATP hydrolysis releases energy and involves the splitting of a single bond—the bond between ATP and one of the attached phosphate groups—that bond is commonly termed a *high-energy phosphate bond*.)

When cells need energy to perform work or to run energy-requiring reactions, they obtain it by hydrolyzing previously synthesized ATP. Since glucose oxidation and other energy-releasing reactions supply the energy for making ATP, these reactions are the ultimate sources of cellular energy.

ATP in Action

Cells are able to synthesize ATP using energy derived from glucose oxidation because the reaction serving as the energy source (glucose oxidation) has a negative energy change and therefore occurs spontaneously. In

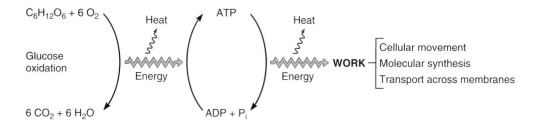

Figure 3.12 The coupling of reactions in energy transfer in cells. Reactions such as glucose oxidation release energy that is used to synthesize ATP. The subsequent breakdown of ATP releases energy that is used to perform various types of cellular work.

As ATP is broken down to ADP and P_i, does the molecules' potential energy increase or decrease?

contrast, the synthesis of ATP has a positive energy change and does not occur spontaneously. When cells use glucose oxidation (or any other energy-releasing reaction) to make ATP, some of the energy released in this reaction is used to drive the energy-requiring process of ATP synthesis. This is accomplished by mechanisms that link or *couple* the energy-releasing reaction to the energy-requiring reaction, so that they must occur together (**Figure 3.12**). When ATP is subsequently broken down, that energy-releasing reaction is coupled to mechanisms that harness the energy to perform work.

We have seen that the oxidation of one mole of glucose releases 686 kcal of energy. Because only 7 kcal are required to synthesize one mole of ATP, the oxidation of one mole of glucose releases enough energy to make 98 moles of ATP. When the actual amount of ATP is measured, however, it turns out that the quantity of ATP synthesized is only about one-third this amount—about 38 moles of ATP per mole of glucose. Why so few? To understand we must consider the energy change for the entire process.

When a mole of glucose is oxidized and 38 moles of ATP are produced, the entire process can be written in the form of a single reaction, as follows:

$$C_6H_{12}O_6 + 6\ O_2 + 38\ ADP + 38\ P_i \rightarrow$$
$$6\ CO_2 + 6\ H_2O + 38\ ATP$$

The energy change for this reaction is the sum of the energy change for glucose oxidation (-686 kcal) and the energy change for the synthesis of 38 moles of ATP (38 moles $\times$ 7 kcal/mole = 266 kcal). Therefore glucose oxidation releases more energy than is used in ATP synthesis, giving a net energy change that is negative ($\Delta E = -686$ kcal + 266 kcal = -420 kcal). Because the net energy change is negative, the reaction can proceed in the forward direction, yielding ATP as a product.

This example illustrates the general principle that when an energy-releasing reaction is used to drive an energy-requiring process, the released energy cannot be used 100% efficiently; some of the released energy

always goes to waste! Because cells normally use about 266 kcal of energy for ATP synthesis for each 686 kcal of energy released from the oxidation of one mole of glucose, only about 40% of the released energy is put to use (266 kcal/686 kcal = 0.388 ≈ 0.40). As for the remaining 60% of the released energy, experience tells us at least some of this appears as heat (see Figure 3.12). In fact, "body heat" is actually heat generated as a by-product of metabolic reactions, including glucose oxidation.

> ### Quick Test 3.4
>
> 1. Is the oxidation of glucose catabolic or anabolic? Does it release energy or require it?
>
> 2. Where does the energy for ATP synthesis come from? When ATP is broken down, what is the released energy used for?
>
> 3. When energy from glucose oxidation is used to make ATP, only a certain fraction of the released energy is used for this purpose. What happens to the rest of the released energy?

Stages of Glucose Oxidation: Glycolysis, the Krebs Cycle, and Oxidative Phosphorylation

We have seen that the oxidation of glucose releases energy, and that cells utilize some of this released energy to synthesize ATP. Exactly how cells perform this feat is the topic of this section.

In a cell, the oxidation of glucose is not carried out in a single reaction; instead, it occurs in three distinct stages, which are carried out in different parts of the cell. The stages are (1) *glycolysis,* which takes place in the cytosol; (2) the *Krebs cycle* (also called the *citric acid cycle, tricarboxylic acid cycle,* or *TCA cycle*), which occurs in the mitochondrial matrix; and (3) *oxidative phosphorylation,* which occurs within the inner mitochondrial membrane.

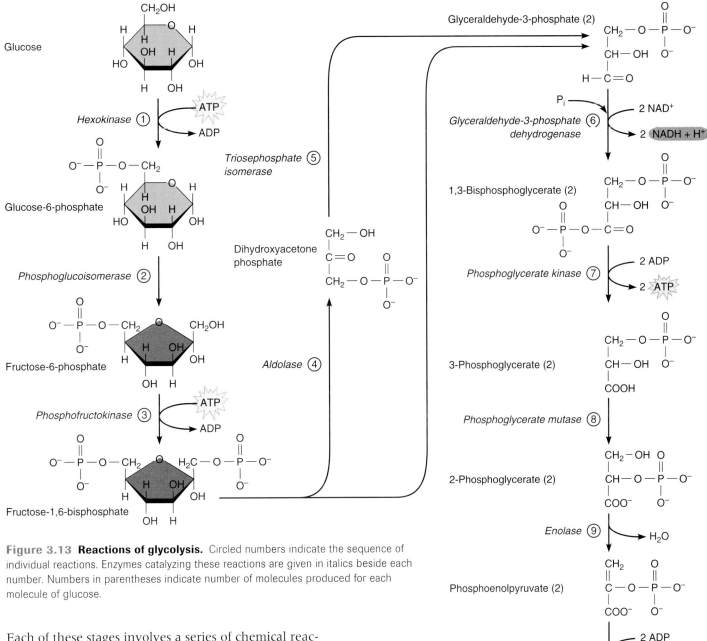

Figure 3.13 Reactions of glycolysis. Circled numbers indicate the sequence of individual reactions. Enzymes catalyzing these reactions are given in italics beside each number. Numbers in parentheses indicate number of molecules produced for each molecule of glucose.

Each of these stages involves a series of chemical reactions that are summarized next, beginning with a look at glycolysis.

Glycolysis

Glycolysis, which means "splitting of sugar," is a metabolic pathway comprising ten reactions, each catalyzed by a different cytosolic enzyme (**Figure 3.13**). In glycolysis, each glucose molecule (which contains six carbons) is broken down into two molecules of *pyruvate* (the ionized form of *pyruvic acid*) containing three carbons apiece. These molecules of pyruvate are normally broken down further in the subsequent stages of glucose oxidation.

The major results of glycolysis are shown in **Figure 3.14** and can be summarized as follows:

1. By the end of glycolysis, each glucose molecule has been split into two molecules of pyruvate.

2. During this process two ATPs are consumed (one each in reactions 1 and 3), but four more are produced by substrate-level phosphorylation (two in reaction 7 and two in reaction 10). This gives a net synthesis of two molecules of ATP for each molecule of glucose consumed.

3. Two molecules of NAD^+ are reduced in step 6, yielding two molecules of NADH for every molecule of glucose.

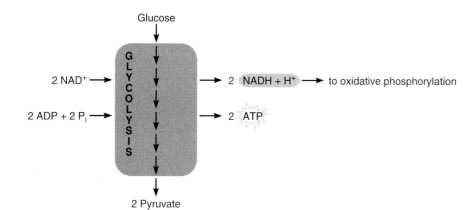

(a)

Figure 3.14 Summary of glycolysis. (a) A diagrammatic summary of the major reactants and products of the glycolysis pathway. Numbers indicate quantities produced or consumed for each molecule of glucose. (b) Net changes occurring in glycolysis. Numbers with minus signs indicate quantities of reactants consumed; those with plus signs indicate quantities of products produced. (c) The overall reaction of glycolysis.

	Glucose	Pyruvate	O_2	CO_2	H_2O	NAD^+	NADH	H^+	FAD	$FADH_2$	ADP	P_i	ATP
(b) Net changes:	−1	+2	0	0	0	−2	+2	+2	0	0	−2	−2	+2

(c) Overall reaction: Glucose + 2 NAD^+ + 2 ADP + 2 P_i ⟶ 2 Pyruvate + 2 NADH + 2 H^+ + 2 ATP

When all reactants and products are accounted for, the overall reaction of glycolysis is as follows:

$$glucose + 2\ NAD^+ + 2\ ADP + 2\ P_i \rightarrow 2\ pyruvate + 2\ NADH + 2\ H^+ + 2\ ATP$$

(The water produced in step 9 is ignored for the sake of simplicity.) Note that during glycolysis, no oxygen is consumed and no carbon dioxide is produced. Although oxygen is indeed consumed in glucose oxidation and carbon dioxide is produced, these events do not occur in glycolysis, but instead in the later stages of glucose oxidation.

Glycolysis is useful to cells because it produces some ATP, but its primary importance is that it sets the stage for subsequent events that yield even more ATP. The NADH that is produced in glycolysis will eventually give up its electrons, thereby releasing energy that will be used to synthesize more ATP by oxidative phosphorylation. Also, the pyruvate that is generated will eventually be catabolized in the next stage of glucose oxidation, the Krebs cycle.

As we see later in the chapter, pyruvate can proceed through further stages of glucose oxidation only when oxygen is readily available inside the cell. Should the availability of oxygen become limited, pyruvate is instead converted to a compound called *lactic acid* (or *lactate*), which is not broken down further. As we continue our discussion of glucose metabolism, assume that oxygen is readily available unless otherwise noted.

The Krebs Cycle

Glycolysis resembles most metabolic pathways in that it is a sequence of steps with distinct starting and ending

points. In contrast, the **Krebs cycle**, the next stage of glucose oxidation, has no starting or ending points because it is circular or *cyclic,* as its name implies.

Glucose oxidation does not proceed beyond glycolysis until pyruvate, which is produced in the cytosol, enters the mitochondrial matrix. Although the Krebs cycle is the next major stage of glucose oxidation, pyruvate does not enter the Krebs cycle directly. Instead, it undergoes a reaction that converts it to acetyl CoA, which then enters the cycle (**Figure 3.15**). (As we will see, acetyl CoA is

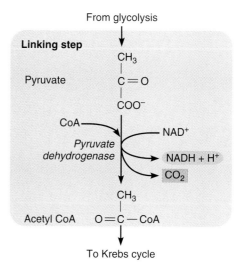

Figure 3.15 Linking step between glycolysis and the Krebs cycle. The end-product of glycolysis, pyruvate, is converted to the initial substrate of the Krebs cycle, acetyl CoA, in the mitochondrial matrix.

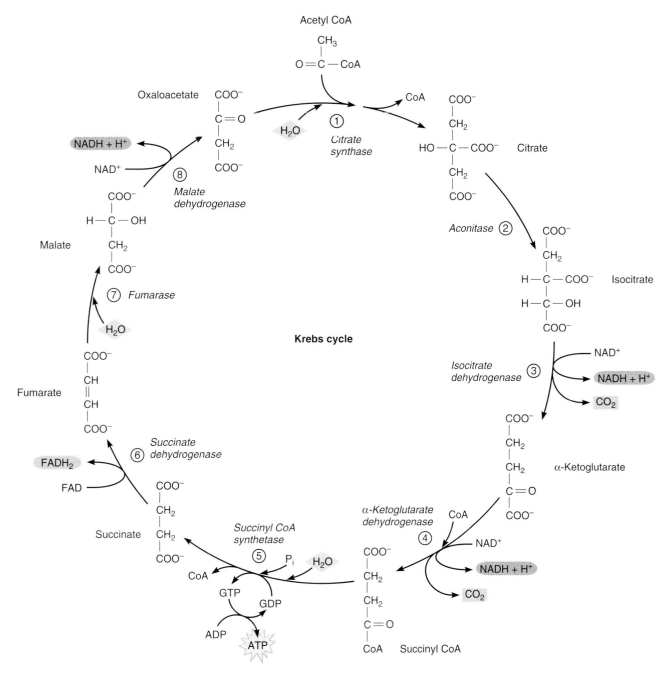

Figure 3.16 **Reactions of the Krebs cycle.** Circled numbers indicate the sequence of individual reactions. Enzymes catalyzing these reactions are given in italics beside each number.

Step 8 of the Krebs cycle is an example of what type of reaction?

also generated in other reactions.) We refer to this reaction as the *linking step* because it links glycolysis to the Krebs cycle. Because glycolysis generates two pyruvate molecules for every molecule of glucose, one glucose molecule ultimately yields two molecules of acetyl CoA in the linking step. Each acetyl CoA molecule participates in one complete "turn" of the Krebs cycle.

Reactions of the Krebs cycle are given in detail in **Figure 3.16**. This diagram begins with acetyl CoA, which

can come from several sources including the linking step. Major results of the Krebs cycle (plus the linking step) are shown in **Figure 3.17** and can be summarized as follows:

1.   By the end of each turn of the Krebs cycle a total of three carbon dioxide molecules have been generated as end-products (one each in the linking step and in

An oxidation-reduction reaction

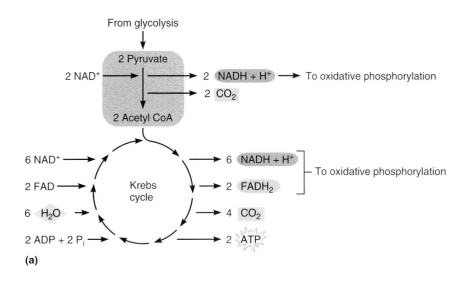

Figure 3.17 Summary of the Krebs cycle. (a) A diagrammatic summary of the major reactants and products of the Krebs cycle, including the linking step. Numbers indicate quantities produced or consumed for each molecule of glucose that enters glycolysis. (b) Net changes occurring in the Krebs cycle and linking step. Plus and minus signs indicate products and reactants, respectively. (c) The overall reaction of the Krebs cycle and the linking step.

	Glucose	Pyruvate	O_2	CO_2	H_2O	NAD^+	NADH	H^+	FAD	$FADH_2$	ADP	P_i	ATP
(b) Net changes:	0	−2	0	+6	−6	−8	+8	+8	−2	+2	−2	−2	+2

(c) Overall reaction: $2 \text{ Pyruvate} + 8 \text{ NAD}^+ + 2 \text{ FAD} + 2 \text{ ADP} + 2 \text{ P}_i + 6 \text{ H}_2\text{O} \longrightarrow 6 \text{ CO}_2 + 8 \text{ NADH} + 8 \text{ H}^+ + 2 \text{ FADH}_2 + 2 \text{ ATP}$

steps 3 and 4), the first time that any carbon dioxide has appeared in glucose oxidation. Because two pyruvate molecules are generated for every molecule of glucose that enters glycolysis, the Krebs cycle must "turn" twice for each glucose molecule, producing two carbon dioxides during each turn plus two generated in the linking step. Thus for each molecule of glucose that enters glycolysis, the Krebs cycle and the linking step generate a total of six carbon dioxides.

2. Only one ATP is generated directly during the Krebs cycle by substrate-level phosphorylation (in step 5), which translates to two ATP per glucose molecule.

3. In the course of the linking step and the subsequent single turn of the Krebs cycle, a total of five reduced coenzymes—4 NADH and 1 FADH$_2$—are produced (in the linking steps and in steps 3, 4, 6, and 8), which is equivalent to ten reduced coenzymes per glucose (8 NADH and 2 FADH$_2$).

The overall reaction for the linking step plus one turn of the Krebs cycle is as follows:

$$\text{pyruvate} + 4 \text{ NAD}^+ + \text{FAD} + \text{ADP} + \text{P}_i + 3 \text{ H}_2\text{O} \rightarrow$$
$$3 \text{ CO}_2 + 4 \text{ NADH} + 4 \text{ H}^+ + \text{FADH}_2 + \text{ATP}$$

which translates into the following for each molecule of glucose that enters glycolysis:

$$2 \text{ pyruvate} + 8 \text{ NAD}^+ + 2 \text{ FAD} + 2 \text{ ADP} + 2 \text{ P}_i +$$
$$6 \text{ H}_2\text{O} \rightarrow 6 \text{ CO}_2 + 8 \text{ NADH} + 8 \text{ H}^+ + 2 \text{ FADH}_2 + 2 \text{ ATP}$$

When combined with the two ATPs per glucose that are netted in glycolysis, the additional two ATPs that are generated in the Krebs cycle makes a running total of four

molecules of ATP for every molecule of glucose. As we will see, however, almost all the ATP that is formed is generated during the final stage of glucose oxidation, oxidative phosphorylation. The Krebs cycle and linking step play an important role in this process because these pathways supply 10 of the 12 reduced coenzyme molecules that eventually go to oxidative phosphorylation to give up their electrons and release energy for making ATP.

The Krebs cycle and linking step produce 100% of the six carbon dioxide molecules that result from the complete oxidation of glucose. Note that absolutely no oxygen has been consumed so far, meaning that 100% of the expected oxygen consumption (six molecules of oxygen per molecule of glucose) must occur in oxidative phosphorylation, the final stage of glucose oxidation, if it is to occur at all. As we will see next, oxygen plays a vital role in oxidative phosphorylation, because oxygen is the ultimate acceptor of all the electrons that are given up by NADH or FADH$_2$. Without it, the electrons would have nowhere to go, and oxidative phosphorylation would stop.

Oxidative Phosphorylation

Most ATP that is made in cells is made by oxidative phosphorylation, which involves two simultaneous processes: (1) the transport in the inner mitochondrial membrane of hydrogen atoms or electrons through a series of compounds, known as the *electron transport chain*, which releases energy, and (2) the harnessing of this energy to make ATP, which is carried out by a mechanism called *chemiosmotic coupling*.

In oxidative phosphorylation, the reduced coenzymes (NADH and FADH$_2$) generated in glycolysis, the linking step, and the Krebs cycle serve as the energy source for making ATP. NADH and FADH$_2$ release electrons to the electron transport chain, and as these electrons travel through the chain, energy is released. Much of that energy is captured and harnessed to drive the synthesis of ATP, which is catalyzed by an enzyme called **ATP synthase**.

The Electron Transport Chain

The **electron transport chain** comprises a set of diverse compounds located in the inner mitochondrial membrane. Most of these compounds are proteins specialized to function as electron carriers; that is, they bind electrons reversibly and thus have the ability to pick up electrons and subsequently give them up. Among these electron carriers are a number of compounds called *cytochromes,* which possess special iron-containing chemical groups known as *hemes.* (Heme groups are also found in hemoglobin, as its name suggests.) Also present are various *iron-sulfur proteins,* which contain iron atoms bound to sulfur. Most of these proteins are aggregated into large complexes that are firmly embedded in the inner mitochondrial membrane. A few electron carriers are individual molecules that move freely in the lipid bilayer and shuttle back and forth, carrying electrons from one complex to another. One, called *coenzyme Q,* is not a protein at all, but instead a small molecule composed mainly of hydrocarbon.

Electrons are carried to the electron transport chain by reduced coenzymes (NADH and FADH$_2$) generated in glycolysis, the linking step, and the Krebs cycle. These reduced coenzymes then release their electrons to certain components of the chain that function as electron acceptors. These electron acceptors then donate their electrons to other electron acceptors, which pass them on to still other acceptors, and so on. Each time electrons move from one component to the next, they lose some energy, and it is this energy that ultimately is used in making ATP.

As electrons pass through the chain, they go from one electron acceptor to the next in the specific sequence shown in **Figure 3.18**. NADH donates its pair of electrons to the first electron acceptor in the chain, called *flavin mononucleotide (FMN).* In the process, NADH becomes oxidized to NAD$^+$ (which is then free to pick up more electrons in glycolysis or the Krebs cycle), and FMN becomes reduced. FMN then passes its electrons to the next component of the chain, an iron-sulfur protein, and becomes oxidized and ready to pick up another pair of electrons from NADH.

Although electrons leave NADH in the form of hydrogens, most components of the electron transport chain do not carry actual hydrogen atoms, only free electrons. Consequently, at some point electrons are stripped from the original hydrogen atoms, which are left as hydrogen ions (H$^+$) in solution, and the electrons continue to move through the chain. However, certain components of the chain *do* carry hydrogen atoms, and when the electrons reach these components, they are "reconstituted" into hydrogen atoms by combining with hydrogen ions from solution. In essence, then, we can think of the electron transport chain as something that carries either hydrogen atoms or an equivalent combination of hydrogen ions and electrons. The release of electrons (denoted by the symbol e$^-$) to the electron transport chain by NADH is represented at the top of Figure 3.18 as follows:

$$NADH + H^+ \rightarrow NAD^+ + 2\,H^+ + 2\,e^-$$

Note that the release of electrons by FADH$_2$ is written in Figure 3.18 in a similar fashion.

After moving from FMN to the iron-sulfur protein, electrons are passed on to coenzyme Q, which passes them on to a cytochrome called cytochrome *b,* and so forth. When eventually the electrons reach the final component of the chain, cytochrome *a*$_3$, they recombine with hydrogen ions to form hydrogens, which react immediately with oxygen to form water:

$$2\,e^- + 2\,H^+ + \tfrac{1}{2}\,O_2 \rightarrow H_2O$$

(The single oxygen atom is written as ½ O$_2$ because oxygen enters the reaction in its molecular form, O$_2$.) By combining this reaction with the previous one, we see that the net result of electron transport is the same as if NADH had simply given up its electrons directly to oxygen to form water, which is an energy-releasing reaction:

$$NADH + H^+ + \tfrac{1}{2}\,O_2 \rightarrow NAD^+ + H_2O + energy$$

If this is the net result, then what use is the electron transport chain? The electron transport provides a way of harnessing this energy so that it can be used to do something useful—namely, to make ATP. (We will see how this is accomplished shortly.) The precise quantity of ATP made depends on conditions in the cell; two or three ATPs are synthesized for every pair of electrons released by NADH.

FADH$_2$ also donates electrons to the electron transport chain, but it does not donate its electrons to FMN for the following reasons: FAD is permanently bound to its partner enzyme succinate dehydrogenase (see step 6 of the Krebs cycle, Figure 3.16), which unlike other Krebs cycle enzymes is embedded in the inner mitochondrial membrane. For this reason, FADH$_2$ is not free to move about, as is NADH. Because of its location, FADH$_2$ is physically removed from FMN and therefore cannot donate electrons to it; instead, it donates its electrons to coenzyme Q (see Figure 3.18). These electrons, like those that come from NADH, eventually combine with oxygen to form water, but because they enter the chain at a point downstream, they release less energy as they move through the chain. As a consequence, only one or two

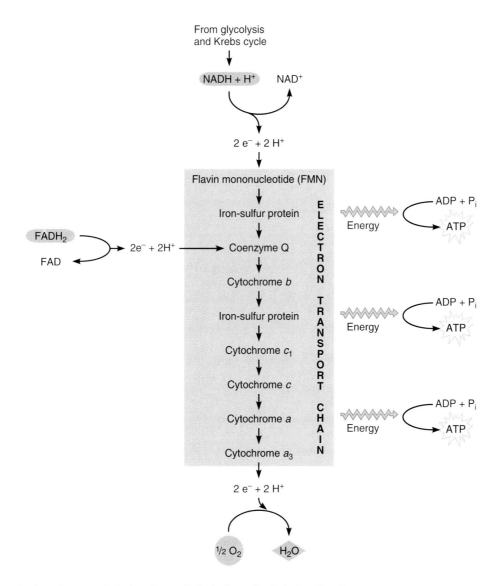

Figure 3.18 The electron transport chain. Arrows indicate the path of electron flow from initial electron donors (NADH or $FADH_2$) to oxygen, the final electron acceptor. Wavy arrows at right indicate that the movement of electrons through the chain releases energy, which is harnessed to drive ATP synthesis by oxidative phosphorylation.

ATPs are produced for each pair of electrons released by $FADH_2$, compared to the two or three ATPs produced when electrons are released by NADH.

We know that energy is released when electrons move through the electron transport chain, and that this energy is used to make ATP, but what is the mechanism that couples energy release to ATP synthesis? The answer is chemiosmotic coupling, which we discuss next.

Chemiosmotic Coupling

The process that couples electron transport to ATP synthesis, known as **chemiosmotic coupling**, first uses energy released in the electron transport chain to transport hydrogen ions across the inner mitochondrial membrane against their *concentration gradient;* then it utilizes the energy stored in this gradient to make ATP (**Figure 3.19**).

The inner mitochondrial membrane contains four distinct complexes containing components of the electron transport chain (see Figure 3.19). Three of these complexes not only function as electron carriers but also are able to transport hydrogen ions (H^+) across the membrane. As electrons pass through these complexes (designated by Roman numerals I, III, and IV), the complexes utilize the released energy to transport hydrogen ions from the mitochondrial matrix into the intermembrane space. This movement of hydrogen ions creates a difference in hydrogen ion concentration (a concentration gradient) across the membrane such that the concentration outside is higher than in the matrix (see Figure 3.19). This concentration gradient represents a store of potential energy. (A difference in electrical potential across the membrane also adds to this stored energy.)

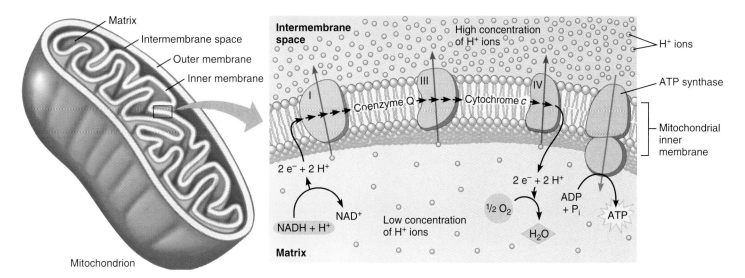

Figure 3.19 Chemiosmotic coupling. The oxidation of NADH yields a pair of electrons that pass through three large inner mitochondrial membrane complexes (I, III, and IV) that contain components of the electron transport chain. (Electrons are carried from complex I to complex III by coenzyme Q, and from complex III to complex IV by cytochrome c.) As these electrons move through the complexes, energy is released, some of which is used to transport hydrogen ions (H^+) from the mitochondrial matrix to the intermembrane space. Hydrogen ions flow in the opposite direction (down their concentration gradient) through the enzyme ATP synthase, in the process releasing energy that is used to synthesize ATP. (For simplicity, the release of electrons by $FADH_2$ is omitted from the diagram.)

The enzyme ATP synthase, which catalyzes the formation of ATP, resides in the inner mitochondrial membrane along with components of the electron transport chain and utilizes energy stored in the hydrogen ion gradient to perform its function. Like the complexes just described, it is also able to transport hydrogen ions across the membrane. In this case, however, the hydrogen ions move down their concentration gradient, not up it (see Figure 3.19). This flow of hydrogen ions releases energy, which ATP synthase harnesses to make ATP. Sometimes, hydrogen ions leak across the membrane independent of the ATP synthase (**Discovery: Can You Lose Weight While Sitting Still?**, p. 90).

Summary of Oxidative Phosphorylation

We have seen that in oxidative phosphorylation, the final stage of glucose metabolism, the following events occur:

1. NADH and $FADH_2$, which are generated in the earlier stages of glucose metabolism, release their electrons to the electron transport chain. These electrons flow through the chain and then emerge from it to combine with oxygen, forming water as an end-product.

2. The movement of electrons through the chain releases energy, which is utilized to transport hydrogen ions across the inner mitochondrial membrane. This transport creates a concentration gradient of

hydrogen ions across the membrane that stores some of the energy released during electron transport.

3. This stored energy is released when hydrogen ions flow through ATP synthase, which uses the energy to make ATP. At maximum, three ATPs are made for every pair of electrons released from NADH, and two ATPs are made for every pair of electrons released from $FADH_2$.

Given that 10 molecules of NADH and 2 molecules of $FADH_2$ are generated for every molecule of glucose that is oxidized, we expect a maximum of 34 molecules of ATP to be made during oxidative phosphorylation [(10 NADH $\times$ 3 ATP/NADH) + (2 $FADH_2$ $\times$ 2 ATP/$FADH_2$) = 34 ATP]. Because these 12 molecules of reduced coenzymes carry a total of 12 pairs of hydrogens, we also expect 12 molecules of water to be generated, which requires that 6 molecules of oxygen be consumed. Therefore, the overall reaction for oxidative phosphorylation is as follows (**Figure 3.20**):

$$10\ NADH + 10\ H^+ + 2\ FADH_2 + 34\ ADP + 34\ P_i +$$
$$6\ O_2 \rightarrow 10\ NAD^+ + 2\ FAD + 12\ H_2O + 34\ ATP$$

Note that this accounts for 100% of the oxygen consumed in the oxidation of one molecule of glucose. Perhaps surprisingly, the reaction also shows that more water molecules are produced—12 instead of 6. However, when the 6 water molecules consumed in the Krebs cycle are taken

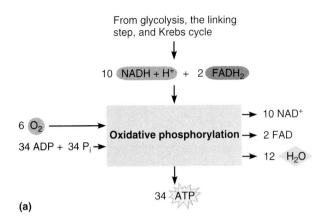

From glycolysis, the linking step, and Krebs cycle

10 NADH + H⁺ + 2 FADH₂

6 O₂ →

34 ADP + 34 Pᵢ →

Oxidative phosphorylation → 10 NAD⁺

→ 2 FAD

→ 12 H₂O

34 ATP

(a)

Figure 3.20 Summary of oxidative phosphorylation. (a) A diagrammatic summary of the major reactants and products of oxidative phosphorylation. Numbers indicate quantities produced or consumed for each molecule of glucose that enters glycolysis. (b) Net changes occurring in oxidative phosphorylation. Plus and minus signs indicate products and reactants, respectively. (c) The overall reaction for oxidative phosphorylation.

	Glucose	Pyruvate	O_2	CO_2	H_2O	NAD^+	NADH	H^+	FAD	$FADH_2$	ADP	P_i	ATP
(b) Net changes:	0	0	−6	0	+12	+10	−10	−10	+2	−2	−34	−34	+34

(c) Overall reaction: 10 NADH + 10 H⁺ + 2 FADH₂ + 34 ADP + 34 Pᵢ + 6 O₂ ⟶ 10 NAD⁺ + 2 FAD + 12 H₂O + 34 ATP

into account, the net result is 6 molecules of water produced, which is in accord with our expectations.

When compared to the quantity of ATP generated by substrate-level phosphorylation in glycolysis and the Krebs cycle, clearly the output from oxidative phosphorylation is responsible for most of the ATP that cells produce. Given that substrate-level phosphorylation and oxidative phosphorylation generate 4 and 34 molecules of ATP per glucose, respectively, the latter accounts for nearly 90% of the total of 38 molecules. (Although oxygen is essential to produce most of the energy used by our cells, its high reactivity contributes to several disease states. See **Clinical Concepts: Oxidative Stress**, p. 97.)

Summary of Glucose Oxidation

Now that we have examined the details of all three stages of glucose metabolism, we can put it all together to see the "big picture." To help us accomplish this, the whole of glucose metabolism is summarized in **Figure 3.21**, a compilation of Figures 3.14, 3.17, and 3.20. Note that when all the reactants and products in all three stages are tallied up in Figure 3.21b, the result is that for every molecule of glucose consumed, six molecules of oxygen are also consumed. In addition, six molecules each of carbon dioxide and water are produced. These figures are in exact agreement with the generalized reaction for glucose oxidation ($C_6H_{12}O_6 + 6 O_2 \rightarrow 6 CO_2 + 6 H_2O$). Because 38 molecules of ATP are generated, the overall reaction for the three stages of glucose metabolism is as follows:

glucose + 6 O_2 + 38 ADP + 38 P_i →
6 CO_2 + 6 H_2O + 38 ATP

Neither NAD nor FAD appears anywhere in this reaction because these coenzymes are reduced in glycolysis and the Krebs cycle but then oxidized in oxidative phosphorylation and returned to their original forms.

Glucose is not the only sugar that can be broken down by cells and used for energy. Cells can also metabolize other monosaccharides, such as fructose, a component of table sugar; galactose, a component of milk sugar; and another monosaccharide called *mannose*. To utilize these sugars, cells first convert them to certain intermediates of glycolysis, such as glucose-6-phosphate and fructose-6-phosphate. These intermediates then are simply "fed into" the glycolysis pathway at the appropriate steps and oxidized in the usual manner.

Conversion of Pyruvate to Lactic Acid

Because oxygen is the ultimate electron acceptor in oxidative phosphorylation, it must be supplied to tissues on a continual basis by the blood if glucose oxidation is to go to completion. The rate at which oxygen must be supplied depends on how fast the tissues are consuming it; in other words, it depends on the tissues' *metabolic demand*. If oxygen can be delivered to tissues at a rate high enough to match demand, oxygen will always be available to accept electrons, and glucose oxidation will proceed to completion. In the event that oxygen delivery is no longer able to keep up with demand (due to a reduction in blood flow to a particular organ or tissue, for instance), oxygen concentration in the affected tissue will fall to very low levels, perhaps even approaching zero.

Under these conditions, fewer oxygen molecules will be available to accept electrons as they reach the end of

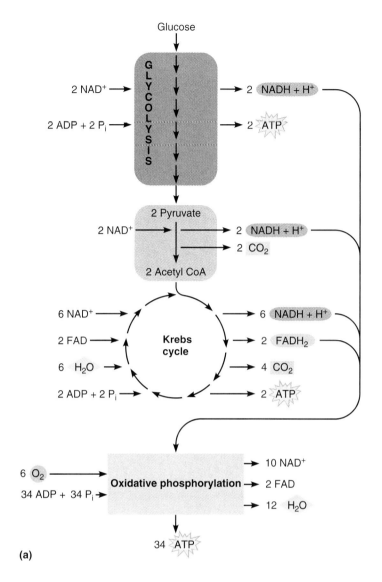

(a)

Figure 3.21 Summary of glucose oxidation. (a) A compilation of Figures 3.14a, 3.17a, and 3.20a, showing all three stages of glucose oxidation and their interconnections. (b) Net changes occurring in glucose oxidation. The first three rows indicate changes occurring during the individual stages; the last row indicates overall changes in the three stages combined. (c) The overall reaction of glucose oxidation. A maximum of 38 ATP molecules can be produced for every molecule of glucose that is oxidized.

	Glucose	Pyruvate	O_2	CO_2	H_2O	NAD^+	NADH	H^+	FAD	$FADH_2$	ADP	P_i	ATP
Glycolysis:	−1	+2	0	0	0	−2	+2	+2	0	0	−2	−2	+2
Krebs cycle:	0	−2	0	+6	−6	−8	+8	+8	−2	+2	−2	−2	+2
Oxidative phosphorylation:	0	0	−6	0	+12	+10	−10	−10	+2	−2	−34	−34	+34
Total:	−1	0	−6	+6	+6	0	0	0	0	0	−38	−38	+38

(b) Net changes

(c) Overall reaction: Glucose + 6 O_2 + 38 ADP + 38 P_i ⟶ 6 CO_2 + 6 H_2O + 38 ATP

the electron transport chain, so it becomes difficult to "off-load" them from the chain's electron carriers. As a result, the flow of electrons through the chain slows, as does the rate of ATP synthesis by oxidative phosphorylation. Under these conditions, an electron "traffic jam" develops in the electron transport chain, which eventually decreases the number of FMN molecules available to accept electrons from NADH. When this happens, coenzyme molecules become "trapped" in their reduced forms, and levels of the coenzyme NAD^+ decrease.

Any drop in NAD^+ levels in a cell is a potential threat to all ATP production because a supply of NAD^+ is necessary for proper operation of both the glycolysis pathway and the Krebs cycle. Of particular interest here is step 6 of glycolysis, which occurs before any of the steps in which ATP is produced by substrate-level phosphorylation. This means that if oxygen becomes depleted and the supply of NAD^+ is decreased, not only does oxidative phosphorylation slow, but substrate-level phosphorylation slows as well. Thus, a reduction in oxygen availability has the

DISCOVERY

CAN YOU LOSE WEIGHT WHILE SITTING STILL?

Is it possible to lose weight by simply sitting still? The idea may not be as far-fetched as it seems. In addition to low-calorie diets, exercise, or the use of calorie-burning supplements, Americans may one day be able to lose weight by manipulating the electron transport system to increase the body's metabolic rate.

The movement of electrons down the electron transport chain is coupled to ATP synthesis by chemiosmotic coupling. If this coupling was perfect, then three ATP would be produced for each NADH + H$^+$ that provides electrons and protons (hydrogen ions) to the electron transport chain, and two ATP would be produced for each FADH$_2$ that provides the electrons and protons. However, in actuality, some of the protons that are transported into the mitochondrial intermembrane

space leak back across the inner mitochondrial membrane into the matrix independently of the ATP synthase channel. These "leaked" protons do not contribute to ATP synthesis and are said to be "uncoupled" from oxidative phosphorylation. A basal leak exists all the time, and *uncoupling proteins* can increase the amount of proton leak, as shown in the figure. The greater the uncoupling, the less ATP produced per oxygen consumed by the body.

Uncoupled mitochondria are found in nearly every tissue, but uncoupling proteins are found particularly in adipose tissue, liver, and skeletal muscle. There are two types of adipose tissue: *White adipose tissue* is present in mammals, including adult humans. White adipose tissue stores energy in the form of triglycerides. *Brown*

adipose tissue is present in most mammals, but only in infants and young children in humans. Brown adipose tissue has considerably more mitochondria than white adipose tissue, and the uncoupling of these mitochondria generates heat through a process called *nonshivering thermogenesis*. Nonshivering thermogenesis is critical in newborns, animals exposed to the cold, and hibernating animals.

What does all of this have to do with weight loss? Although the physiological role of uncoupling proteins in humans is unknown, they are of interest to weight loss researchers because uncoupling increases the body's metabolic rate. An increased metabolic rate enables the body to "burn" more energy and potentially lose weight—even while sitting still.

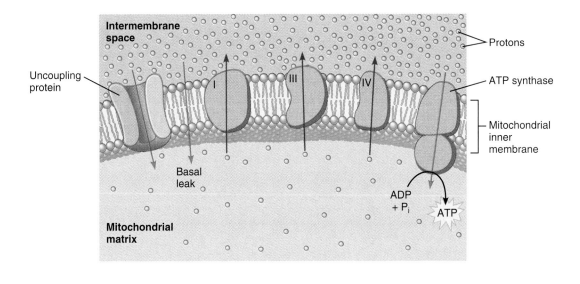

potential to slow (or perhaps even stop) all ATP-producing steps, which would be disastrous for a cell, for its supply of ATP would eventually become exhausted.

Fortunately, this disaster can be averted. Most cells contain an enzyme called *lactate dehydrogenase (LDH)*, which can convert pyruvate (the end-product of glucose breakdown in glycolysis) to *lactate* according to the following reaction:

$$\text{pyruvate} + \text{NADH} + \text{H}^+ \xrightarrow{\text{LDH}} \text{lactate} + \text{NAD}^+$$

The significance of this reaction is as follows: In the event that NADH cannot unload its electrons to the electron transport chain due to a limitation in the availability of oxygen, this reaction provides an alternate pathway that allows electrons to be unloaded in a reaction that generates free NAD$^+$, which can then pick up electrons in step 6 of glycolysis. Therefore, when the electron transport chain is not available to accept electrons, NADH and NAD$^+$ can still shuttle back and forth between the conversion of pyruvate to lactate and step 6 of glycolysis

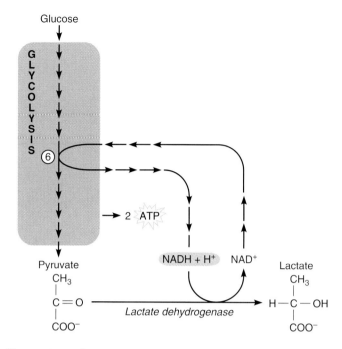

Figure 3.22 Conversion of pyruvate to lactate. NADH, which is generated in step 6 of glycolysis, can be converted back to NAD$^+$ in a reaction that generates lactate from pyruvate. This reaction ensures a steady supply of NAD$^+$, which allows for the continued operation of step 6 and the continued production of ATP even under conditions of reduced oxygen availability.

(**Figure 3.22**). This reaction ensures a steady supply of NAD$^+$, which enables step 6 to run even when oxygen availability is limited. More importantly, the fact that step 6 can run means that the remaining steps in glycolysis can also run, including steps 7 and 10, which synthesize ATP.

The ability to synthesize ATP in the manner just described allows muscles and other tissues to continue working even when the availability of oxygen is low, but it is important to realize that this does not mean the body's tissues can live indefinitely in the absence of oxygen. For one thing, only two ATP molecules are produced in glycolysis for every molecule of glucose consumed, which is only about 5% of the ATP that is normally produced as a result of complete glucose oxidation. Unless a cell's demand for ATP is very low, it is unlikely that it will be able to supply enough ATP through glycolysis alone. (Interestingly, red blood cells, which transport oxygen, must obtain all their ATP from glycolysis because they lack mitochondria.) Furthermore, because the conversion of pyruvate to lactate is a "dead end" reaction, lactate tends to accumulate in cells when produced at a rapid rate, which can cause acidification of the intracellular fluid. Lactate can also leak into extracellular fluid and "spill over" into the bloodstream, causing acidification of the blood. Acidification of tissues will eventually begin to interfere with proper cellular function and therefore cannot continue indefinitely.

Because lactate is potentially harmful to cells, they must get rid of it. When oxygen availability returns to normal, pyruvate begins to proceed to the Krebs cycle as it usually does, and the concentration of pyruvate in cells decreases. As a result of mass action, the lactate dehydrogenase reaction begins to run in reverse, so that lactate is converted back to pyruvate and NAD$^+$ is converted to NADH:

$$\text{pyruvate} + \text{NADH} + \text{H}^+ \xrightarrow{LDH} \text{lactate} + \text{NAD}^+$$

The NADH that is generated in this reaction proceeds to oxidative phosphorylation, and the pyruvate proceeds to the linking step and the Krebs cycle. Thus, the conversion of pyruvate to lactate represents but a momentary "side trip" from the normal pathway of glucose oxidation, a diversion that ends when lactate is converted back to pyruvate.

Quick Test 3.5

1. Name the three stages of glucose oxidation in the order in which they occur. In which stage(s) is oxygen consumed? In which stage(s) is carbon dioxide produced? Which stage normally produces the most ATP?

2. Distinguish between substrate-level phosphorylation and oxidative phosphorylation. In which stage(s) does substrate-level phosphorylation occur? In which stage(s) does oxidative phosphorylation occur?

3. What conditions are likely to cause an increase in the rate of lactate production by cells? Why?

Energy Storage and Use: Metabolism of Carbohydrates, Fats, and Proteins

So far, our examination of energy metabolism has been restricted to one energy source—glucose. Nevertheless, the metabolism of fats and proteins (as well as that of carbohydrates other than glucose) also makes a significant contribution to our energy needs. Although a detailed description of fat and protein metabolism is beyond the scope of this book, a general understanding of the connection between these metabolic pathways and those we have just studied is crucial. Eventually we will need this knowledge to understand the role of the endocrine system in regulating energy metabolism, a topic covered in Chapter 7.

The body is able to draw upon fats and proteins as alternate sources of energy whenever glucose is scarce and needs to be conserved. When glucose supplies are limited, the body can break down fats and proteins into smaller molecules, a catabolic process that releases energy. When the supply of glucose is plentiful, the body can run these reactions in reverse to synthesize fats, proteins, and a

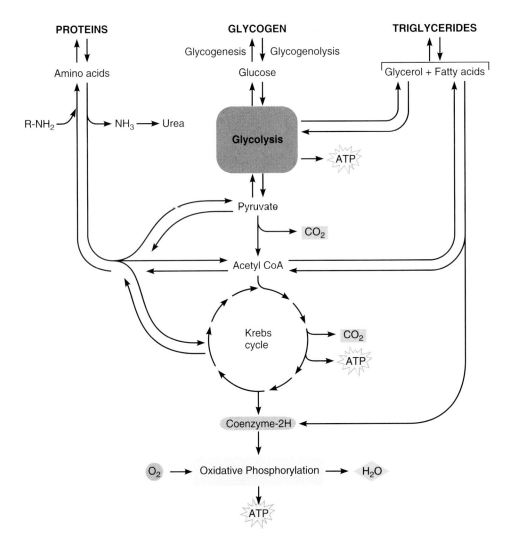

Figure 3.23 Metabolic pathways involved in protein, glycogen, and fat metabolism.
The three stages of glucose metabolism are shown in the center. One-way arrows indicate irreversible reactions; double arrows indicate reversible reactions. Arrows leading into or out of this central pathway indicate points of entry or exit for the breakdown products of protein or triglyceride metabolism.

large glucose storage molecule called *glycogen*. In this manner the body stores energy for future use.

Figure 3.23 shows the relationships among pathways for glucose, fat, and protein metabolism. The three stages of glucose oxidation—glycolysis, the Krebs cycle, and oxidative phosphorylation—are the centerpiece of the figure. Note that the depiction of these three stages is essentially the same as that shown in Figure 3.21, except that only the major products of each stage are shown. In addition, NADH and $FADH_2$ are simply given the abbreviation *coenzyme-2H*. (For simplicity, reduced coenzymes generated in glycolysis or in the linking step are not shown.) Double arrows indicate reactions that can go in both directions; single arrows denote reactions that essentially go one way only. These reactions are described as being *irreversible*, whereas the two-way reactions are said to be reversible.

One-way reactions are not "irreversible" in the absolute sense, but only in the sense that they require large quantities of energy in order to run in reverse. In many cases, however, one-way reactions do run in reverse under certain circumstances. When this occurs, a reaction does not follow exactly the same course while running in reverse as it does when running in the forward direction, and it is catalyzed by a different set of enzymes. For this reason, the reverse reaction is sometimes referred to as a *bypass reaction* because it follows a "route" that bypasses the enzymes that would normally catalyze the reaction when it is running in the forward direction.

In contrast, reversible reactions have smaller energy changes and can be made to go in reverse without large amounts of energy. These reactions readily reverse direction in response to changes in reactant or product concentrations in accordance with the law of mass action.

Furthermore, these reactions are catalyzed by the same enzymes, whether they are going forward or in reverse.

Whenever a metabolic pathway consists of an irreversible reaction in combination with a bypass reaction, the reactions are generally regulated so that they do not run simultaneously, which would be counterproductive. When the irreversible (forward) reaction is running, enzymes catalyzing the bypass reaction are turned off; when the bypass reaction is running, enzymes catalyzing the forward reaction are turned off. In this manner, cells can force the reaction to go in the direction that best suits the body's needs. Thus, when cells need to break down fats for energy, for example, they can turn on the enzymes that perform this function. When they need to synthesize fats to store energy, they can turn these enzymes off and turn on other enzymes that catalyze the necessary bypass reactions. As we will see in Chapter 7, these metabolic adjustments are largely coordinated by hormonal signals that depend on our eating and fasting patterns.

Glycogen Metabolism

Glycogen is a branched-chain molecule found in animal cells; it is similar to starch, a carbohydrate found only in plants. Glycogen is composed of glucose molecules joined together to form a polymer (see Chapter 2). When glucose is in abundant supply, as after a meal, most of the glucose molecules are not metabolized immediately; they are stored as glycogen. This process, which is sometimes called **glycogenesis**, is represented in Figure 3.23 by the arrow running up from glucose to glycogen.

Tissues differ in their capacity to synthesize and store glycogen; the liver and skeletal muscle are particularly adept at this process. During fasting or when glucose is being used up quickly, the glucose supply is replenished by the breakdown of glycogen into individual glucose molecules, a process known as **glycogenolysis**. The fate of glucose released during glycogenolysis depends on the location where it occurs. When glycogen is broken down in skeletal muscle cells, for example, the glucose is used by those cells. In contrast, when glycogen is broken down in liver cells, the liver releases most of the glucose into the bloodstream for uptake and utilization by other tissues.

Exercise Link

Recall that Bill and Jane "carbo-loaded"—ate lots of starchy and sugary foods—prior to running the marathon. This practice is common among endurance athletes because it increases the glycogen content of skeletal muscles. Obviously, these muscles will have an especially high need for glucose during the race.

Gluconeogenesis: Formation of New Glucose

Although fats and proteins can be used by most tissues as energy substitutes for glucose, an adequate supply of glucose must be maintained in the bloodstream at all times because the nervous system, particularly brain tissue, has a lower capacity than other tissues for switching to alternate energy sources. The nervous system requires an uninterrupted glucose supply; otherwise, loss of consciousness and possibly death may result. Nervous tissue is never entirely free of this glucose requirement, although a limited capacity for utilizing other energy sources does exist.

Under normal conditions, the body's glycogen reserves are sufficient to provide energy for a few hours. If a person fasts for a longer period of time, glycogen stores become depleted, which could potentially cause a dangerous fall in blood glucose concentrations. Fortunately, new glucose molecules can be synthesized from noncarbohydrate precursors via a process called **gluconeogenesis** (**Figure 3.24**), which is carried out mainly by the liver.

Glucose can be made via gluconeogenesis from three sources: (1) glycerol, which is produced by breaking down triglycerides; (2) lactate; and (3) amino acids, which can be produced by the breakdown of proteins. In gluconeogenesis, glycerol is first converted to glycerol phosphate, an intermediate of the glycolysis pathway, and lactate is first converted to pyruvate, the normal end-product of glycolysis (see Figure 3.24). These molecules then proceed through the glycolysis pathway *in reverse,* such that glucose molecules are generated.

Certain amino acids can be converted to glucose after first being converted to pyruvate, which then enters the glycolysis pathway in reverse. Certain other amino acids, however, are converted to glucose via a more indirect route, which is indicated by the dashed line in Figure 3.24. These amino acids are first converted into a Krebs cycle intermediate (oxaloacetate), which is then converted into a glycolytic intermediate (phosphoenolpyruvate); this molecule then participates in the glycolysis pathway in the reverse direction. Note, however, that some amino acids cannot be converted to glucose; even though these amino acids can be converted to acetyl CoA, there is no pathway permitting the synthesis of glucose from acetyl CoA. (Acetyl CoA can be converted to oxaloacetate in the Krebs cycle, but in the process an equal amount of oxaloacetate is consumed due to the cyclic nature of this series of reactions. Consequently, acetyl CoA cannot be converted to glucose via the dashed line pathway shown in Figure 3.24.) Fatty acids, which like glycerol are generated from triglyceride breakdown, cannot be converted to glucose either because they too are converted to acetyl CoA.

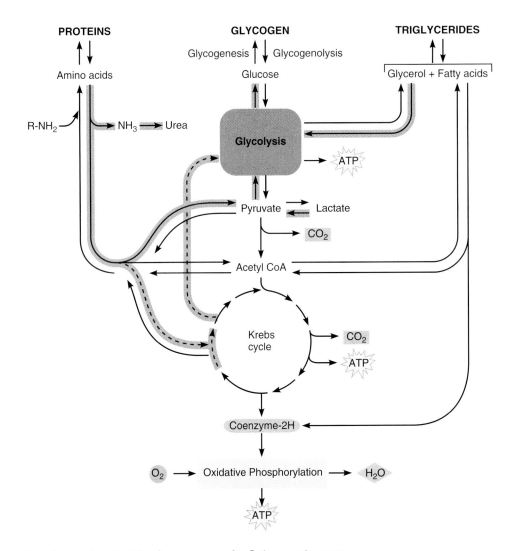

Figure 3.24 Metabolic pathways involved in gluconeogenesis. Pathways relevant to gluconeogenesis are highlighted.

Fat Metabolism

Like glycogen, fats (lipids) can be broken down and used for energy when other energy supplies are running low. Fats can also be synthesized to store energy when they are in abundant supply in the diet. *Adipose tissue,* which contains fat cells, is the primary storage depot for fats.

Lipids are stored predominantly in the form of *triglycerides,* which consist of three fatty acids joined to a glycerol backbone. The first stage of fat breakdown **(lipolysis)** is the separation of fatty acids from the glycerol molecule (**Figure 3.25**). Glycerol then enters the glycolysis pathway (as dihydroxyacetone phosphate) and proceeds from there to the Krebs cycle and oxidative phosphorylation. Fatty acids are converted to acetyl CoA, which also enters the Krebs cycle and oxidative phosphorylation. Because fatty acids obtained in the diet are usually 12–18 carbons in length, several molecules of acetyl CoA can be made from a single fatty acid molecule. (Recall that glucose yields only two acetyl CoAs.) Thus,

many more ATPs can be made from a molecule of fatty acid than from a glucose molecule. The catabolism of fatty acids also generates a large number of reduced coenzyme molecules, which go to the electron transport chain and eventually yield even more ATPs by oxidative phosphorylation. Because a gram of fat yields more energy than a gram of glucose or other carbohydrates, fats are known as high-calorie foods.

Note that when fats are broken down for energy, compounds called *ketones* are generated as a by-product (see Figure 3.25). These compounds are synthesized from acetyl CoA in a set of reversible reactions and are synthesized at higher rates when the concentration of acetyl CoA rises. Because many acetyl CoA molecules can be generated from a single fatty acid, the breakdown of fats tends to generate large quantities of ketones, which is important because ketones can be utilized by the nervous system as a partial alternative to glucose. This helps conserve the body's supply of glucose if energy intake becomes limited such that the body must turn to stored fats for energy.

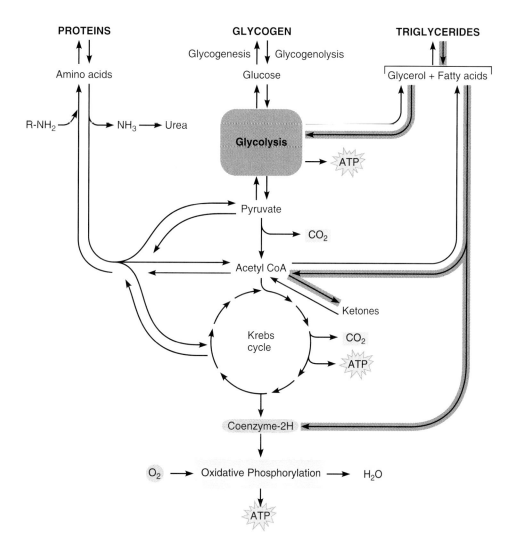

PROTEINS   **GLYCOGEN**   **TRIGLYCERIDES**

Figure 3.25 Metabolic pathways involved in lipolysis, the breakdown of fats (triglycerides) for energy. Relevant pathways are highlighted. Note that fatty acid breakdown leads to the production of ketones, which are generated from acetyl CoA through a series of reversible reactions.

Note as well that the pathways from glycerol to glycolysis, and from fatty acids to acetyl CoA, are double-arrowed in Figure 3.25, which means that it is possible to synthesize fats from other nutrients, a process called **lipogenesis**. This explains why people can become obese from eating too much of nonlipid foods such as carbohydrates or proteins.

Protein Metabolism

In the metabolic breakdown of proteins for energy (**Figure 3.26**), proteins are first broken down to amino acids, a process known as **proteolysis**. The amino acids are then *deaminated*; that is, an amino group ($-NH_2$) is removed. This deamination process produces ammonia (NH_3), which is toxic. Fortunately, ammonia is converted to another compound that is carried by the bloodstream to the liver, where it is converted to *urea*, which is

relatively innocuous and is eventually eliminated by the kidneys.

> ### *Exercise Link*
>
>
>
> In the later stages of the marathon, Bill and Jane's bodies needed to rely more heavily on deaminated proteins as sources of energy because their bodies' glucose stores were almost completely consumed. As a result, their sweat (which is produced from plasma) began to smell of ammonia.

After the amino acid is deaminated, the remainder of the molecule is either converted to pyruvate or acetyl CoA (which then goes into the Krebs cycle), or it is converted to an intermediate that can enter the Krebs cycle directly. Note in Figure 3.26 that the pathways from

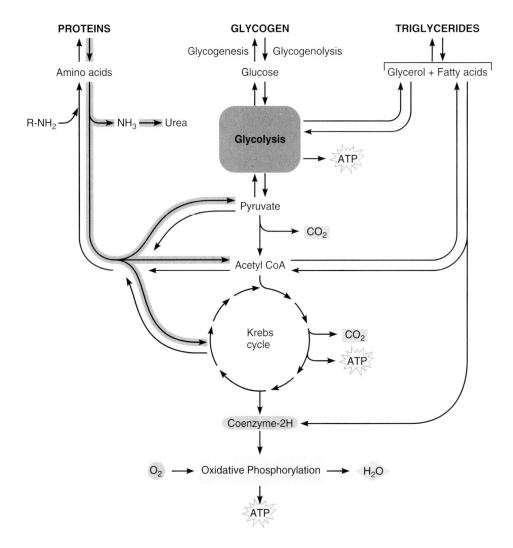

Figure 3.26 Metabolic pathways involved in the breakdown of proteins for energy.
Relevant pathways are highlighted. Note that the breakdown of amino acids generates ammonia (NH_3), which is converted to urea in the liver.

amino acids to pyruvate, acetyl CoA, and the Krebs cycle are double-arrowed, meaning that nonprotein nutrient molecules such as carbohydrates or fats can serve as raw materials for the synthesis of amino acids, which can then be used to make proteins. (As the figure shows, this requires the addition of an amino group, which is obtained from another molecule, indicated as R—NH_2.) Consequently, certain amino acids can be present in cells even when those amino acids are not present in the diet. Other amino acids, however, cannot be synthesized in the body and must be obtained from the diet; these amino acids are referred to as *essential* amino acids. An **essential nutrient** is any biomolecule necessary for proper body function that cannot be synthesized in cells and therefore must be obtained from dietary sources.

Quick Test 3.6

1. Define the following terms: *glycogen, glycogenesis, glycogenolysis, lipogenesis, lipolysis, proteolysis.*

2. What is gluconeogenesis, and why is it important? What three substances can serve as raw materials for this process?

3. What happens to the ammonia that is produced when amino acids are broken down for energy?

4. What is the distinguishing characteristic of an essential nutrient?

OXIDATIVE STRESS

We all know that oxygen is good for us. Our need for oxygen exceeds any other of the body's demands. But the metabolism of oxygen also results in the production of *free radicals,* which are highly reactive molecular species that can damage the body's cells. Thus, high levels of oxygen consumption can prove to be too much of a good thing.

Atomic oxygen has 6 electrons in its outer shell: 4 electrons exist as two pairs and 2 electrons are unpaired. To fill the outer shell, oxygen needs 2 more electrons, which it can obtain by forming covalent bonds with another oxygen to form molecular oxygen, O_2, as shown:

$$\ddot{O}::\ddot{O}$$

At rest, our bodies consume approximately 250 mL of oxygen per minute; 98% of the consumption occurs in the electron transport chain. During the electron transport chain, oxygen is reduced to water according to the following reaction:

$$O_2 + 4\,e^- + 4\,H^+ \rightarrow 2\,H_2O$$

The electron transport system is not infallible, and sometimes an electron "escapes" the chain and binds to molecular oxygen to form *superoxide anion,* $O_2{}^-$. Studies suggest that up to 2% of the electrons that enter the electron transport system escape and form superoxide anion. Superoxide anion can also be produced by some of our immune cells through an *NADPH-oxidase* system.

Superoxide anion is a *free radical,* a molecular species with one or more un-paired electrons, or *free* electrons. Because electrons are more stable when paired, free radicals are highly reactive. Free radicals can gain stability in cells by reacting with another molecule and gaining an electron. However, this leaves the other molecule as a free radical starting a chain reaction. The chain can be broken by two free radicals reacting with each other such that the unpaired electrons become paired or by the action of *antioxidant systems.*

To counter the effects of free radicals and other oxidizing agents, collectively referred to as *oxidants,* cells have antioxidant systems. Antioxidants include enzymes that break down oxidants and molecules that react with free radicals, changing their reactive nature. Common antioxidant enzymes include *superoxide dismutase,* which converts superoxide anion to hydrogen peroxide; and *catalase* and *glutathione peroxidase,* which convert hydrogen peroxide to water and oxygen. Antioxidant molecules include *d-alpha tocopherol* (vitamin E) and *ascorbic acid* (vitamin C). In a healthy cell, the production of oxidants is balanced by the action of antioxidants.

Oxidants have normal functions in cells, but they can also be toxic if produced in excess. For example, free radicals are produced during *inflammation,* a series of reactions in response to an infection whereby free radicals help destroy bacteria. Free radicals and other oxidants also function as intracellular signals during cell stress to induce protective mechanisms within the cell. However, if the production of oxidants exceeds the antioxidant capacity of a cell, then the oxidants may react with critical biomolecules in the cell (including DNA, proteins, and phospholipids) and cause cell damage, a condition known as *oxidative stress.*

Most cells of the body generate oxidants and are subject to oxidative stress, thereby opening the doors for a variety of disease states. Because oxidants are produced as part of inflammation, oxidative stress is thought to contribute to inflammatory diseases such as arthritis. Oxidants produce some of the damage associated with *ischemia* (inadequate blood flow), and are therefore associated with ischemic diseases such as heart disease and stroke. When a cell is severely damaged, oxidants may contribute to *apoptosis,* or programmed cell death as a defense mechanism. Therefore, oxidative stress is associated with degenerative diseases, such as Alzheimer's disease and insulin-dependent diabetes mellitus.

With all the potential damage that oxidants can produce, scientific research has focused much attention on antioxidants as potential preventatives to disease. For example, dietary supplements of vitamins C and E are hypothesized to help prevent heart disease, as is consumption of products containing these vitamins or other antioxidants, including certain fruits and vegetables, and red wine. The research on these vitamins is not yet conclusive, but some evidence exists that these antioxidant vitamins can slow down diseases, and maybe even aging.

▩ CHAPTER SUMMARY

Types of Metabolic Reactions, p. 63

The sum total of chemical reactions occurring in the body is metabolism; reactions of energy metabolism are specifically involved in energy exchange. Catabolic reactions generate smaller products from larger reactants; anabolic reactions generate larger products from smaller reactants. Three common types of metabolic reactions are: (a) hydrolysis and condensation reactions, (b) phosphorylation and dephosphorylation reactions, and (c) oxidation-reduction reactions.

Metabolic Reactions and Energy, p. 65

Metabolic reactions enable cells to transform raw materials from the environment into structural and functional components and also provide cells with energy. The energy change particular direction. Thus, we refer to a concentration gradient as a **chemical driving force**, the direction of which is always down the concentration gradient (**Figure 4.1**). As we will see, the rate at which a substance is transported varies with the size of the concentration gradient and generally increases as the size of the gradient increases. Therefore, we can say that the magnitude of the chemical driving force increases as ΔC increases. When more than one substance is present, as is the case with real cells, more than one concentration gradient exists. Any chemical driving force that might be acting on a given substance depends only on the concentration gradient of *that particular substance.*

Note that a chemical driving force is fundamentally different from the more familiar types of forces because it does not really act on molecules in the same way that the force of gravity pulls on a rock or a magnetic force pulls on a piece of iron. When molecules move across a membrane down a concentration gradient, they do so simply because there are more molecules on one side of the membrane than on the other. *Individual* molecules are not pushed down the gradient but in fact are equally likely to move in either direction. This issue is explored later in this chapter.

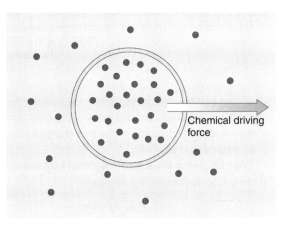

(a)

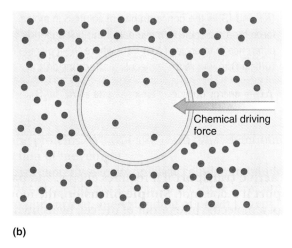

(b)

Figure 4.1 Chemical driving forces. (a) When the concentration of molecules (dots) is higher inside a cell than in extracellular fluid, the direction of the chemical driving force (arrow) is outward. (b) When the concentration of molecules is higher in extracellular fluid, the direction of the chemical driving force is inward. In both cases, molecules will move passively in the direction of the driving force, or down the concentration gradient.

> ### Quick Test 4.1
>
> 1. What is the difference between passive transport and active transport? Between simple diffusion and mediated transport? Is simple diffusion active or passive?
>
> 2. Why is a concentration gradient referred to as a driving force?
>
> 3. In what direction (into or out of a cell) is the chemical driving force for amino acids? (See Table 4.1.)

Electrical Driving Forces

In general, molecules moving passively across membranes can be affected by factors other than the chemical driving force. This is particularly true of ions, which are influenced by *electrical driving forces* in addition to chemical driving forces. Electrical driving forces arise due to the **membrane potential**, a difference in *electrical potential* or *voltage* that exists across the membranes of most cells. In the next section we see that the existence of a membrane potential reflects an unequal distribution of positively charged ions and negatively charged ions across the plasma membrane. (We will not worry about what causes this unequal charge distribution until Chapter 8, when we explore the origin of membrane potentials in neurons.) Once we have a firm understanding of what a membrane potential is, we then consider its influence on the forces that affect ion transport.

The Membrane Potential The fluids in the body contain a wide variety of solutes, and many of these are ions,

which possess electrical charge. Some ions are *cations,* which have a positive charge; others are *anions,* which have a negative charge. Ions are also present in salt solutions, such as seawater, but one normally cannot detect the presence of the ions' electrical charges because the number of positive and negative charges are equal. Such a solution is said to be electrically *neutral* because the positive and negative charges cancel one another, giving a net (total) electrical charge of zero. Likewise, the total electrical charge of your body is zero because the number of cations in your body equals the number of anions.

In intracellular or extracellular fluid, cations and anions are present in unequal numbers; consequently, these fluids are not electrically neutral. Intracellular fluid contains a slight excess of anions over cations, giving it a net negative charge. Extracellular fluid contains a slight excess of cations over anions, giving it a net positive charge. Because positive and negative charges are distributed unequally between the inside and outside of a cell, a *separation of charge* is said to exist across the membrane (**Figure 4.2**). The excess negative and positive charges of intracellular and extracellular fluid tend to be clustered close to the membrane because the excess negative charges on one side of the membrane are attracted to the excess positive charges on the other side.

A cell's membrane potential reflects this separation of charge and is given in units of electrical potential—in *millivolts* (mV), which are 1/1000 of a volt. The magnitude of the membrane potential (number of millivolts) depends on the degree of charge separation; the greater the difference in charge between the two sides of a membrane, the larger the membrane potential. By convention, the *sign* of the membrane potential (positive or negative) is taken to be the sign of the net charge *inside* the cell relative to outside. Since the inside of a cell is typically more negatively charged than the outside, the membrane potential is usually negative. (Under certain conditions, however, the membrane potential can be positive, as we will see in Chapter 8.) For most cells, the membrane potential, which is given the symbol V_m, is around negative 70 millivolts ($V_m = -70$ mV).

How the Membrane Potential Creates an Electrical Driving Force That Acts on Ions As an ion crosses a membrane in the presence of a membrane potential, its electrical charge is attracted by the net electrical charge existing in the fluid on one side and repelled by that existing on the other side. Assuming that no other force (for example, a chemical driving force) is acting on the ion, it experiences a driving force and moves spontaneously from the side that repels it toward the side that attracts it. In this case, the force that is causing it to move is an **electrical driving force**, meaning that it is simply a consequence of the attractive and repulsive forces that inevitably arise between electrically charged objects. Note that this electrical driving force adds to or subtracts from any other driving forces that might also be present, such as a chemical force due to a concentration gradient.

Factors Affecting the Direction and Magnitude of the Electrical Driving Force When an ion is crossing a membrane, the *direction* of the electrical driving force that is acting on it depends on only two things: the sign of the membrane potential and the sign of the ion's *valence* or charge. Determining the direction of the force is easy; just remember that like charges repel and opposite charges attract. If the membrane potential is negative, the electrical force acting on a positively charged ion draws it into the cell because the ion is repelled by the positive charges outside and attracted by the negative charges inside (**Figure 4.3**). If the ion is negatively charged, a negative membrane potential exerts an electrical driving force that pushes it out of the cell because

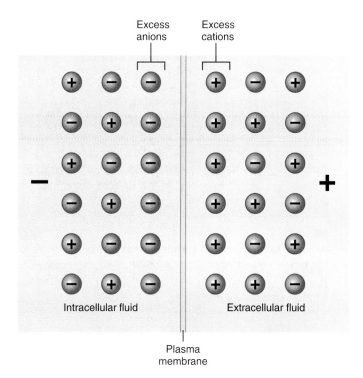

Figure 4.2 Separation of charge across a cell membrane. Under normal conditions, the fluid inside a cell has a slight excess of anions (negative charges), and the fluid outside the cell has a slight excess of cations (positive charges). These excess charges are clustered in the region near the membrane. Net charges inside and outside the cell are indicated by (+) and (−) signs on either side of the membrane.

the ion is repelled by the negative charges inside and attracted by the positive charges outside. Because electrical charges do not act on uncharged molecules such as glucose, substances lacking charges are not affected by the membrane potential.

The *magnitude* of the electrical driving force on an ion depends on the size of the membrane potential and the quantity of charge carried by the ion, and it increases as either of these factors gets larger (**Figure 4.4**). A larger negative membrane potential, for instance, means a greater number of negative charges inside and positive charges outside, which increases the attractive and repulsive forces acting on an ion. If an ion carries more charge, the attractive and repulsive forces are also increased, which makes the electrical driving force stronger.

Electrochemical Driving Forces

To determine whether ions are being transported passively or actively, a physiologist must identify all the driving forces that might be acting on them. In general, when ions are transported across membranes, two driving forces are influential: (1) a chemical force reflecting the ions' tendency to move down their concentration gradient (from higher to lower concentration), and (2) an electrical force reflecting the ions' tendency to be

In the second condition (Figure 4.6b), the concentration on side 1 is twice what it originally was, and twice that of side 2. Under these conditions, the one-way flux from side 1 to side 2 has *doubled* (because the number of molecules near the membrane on that side has doubled). Now the one-way flux from side 1 to side 2 is larger than the one-way flux in the opposite direction. As a result, there is net movement of molecules from side 1 to side 2. The rate of this movement is the *net flux,* the difference between the two unidirectional fluxes. Whenever physiologists say that molecules are being transported passively or actively in a certain direction, they are always referring to the direction of the net flux.

As we will see later in this chapter, transport rates are influenced by many variables, some of which affect certain transport mechanisms but not others.

Passive Transport

In passive transport, molecules move across the membrane down their chemical or electrochemical gradients. No energy is required. The types of passive transport include *simple diffusion, facilitated diffusion,* and *diffusion through ion channels.*

Simple Diffusion: Passive Transport Through the Lipid Bilayer

Simple diffusion is the least complicated of all transport mechanisms, but in a way this makes it all the more interesting. Although we have already learned about the forces for simple diffusion, here we learn about what makes it work. We will also learn about some of the factors that affect the rate of diffusion and the permeability of a membrane to substances moving by simple diffusion.

The Basis for Simple Diffusion

We use the term **simple diffusion** to describe the passive transport of molecules through a biological membrane's lipid bilayer, but in fact the mechanism of simple diffusion is not strictly biological. When someone opens a bottle of perfume nearby, for example, you will eventually smell the fragrance even if there are no air currents to carry it to your nose. This occurs because certain molecules from the perfume travel from the bottle to your nose by simple diffusion through the air. The molecules travel away from the bottle because they are more concentrated near the bottle and less concentrated farther away, and because these molecules, like any others, move spontaneously down their concentration gradient due to the fact that they are in a state of constant thermal motion.

The movement of molecules from one location to another simply as a result of their own thermal motion is called **diffusion**, a general term that applies to molecules moving through any medium. Thermal motion is often called *random* thermal motion because individual

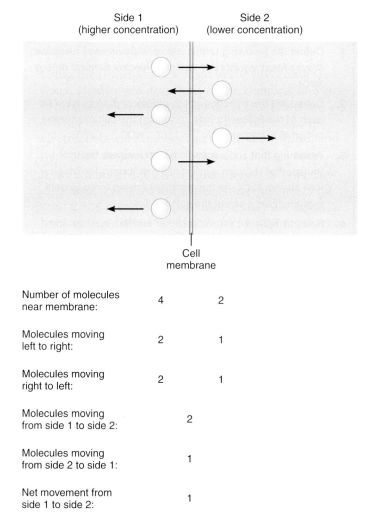

	Side 1 (higher concentration)	Side 2 (lower concentration)
Number of molecules near membrane:	4	2
Molecules moving left to right:	2	1
Molecules moving right to left:	2	1
Molecules moving from side 1 to side 2:	2	
Molecules moving from side 2 to side 1:	1	
Net movement from side 1 to side 2:	1	

Figure 4.7 **A simplified model showing how random thermal molecular motion gives rise to movement down a concentration gradient.** Initially, four molecules are present on the left side of a membrane, and two are present on the right, indicating that the concentration is higher on the left side. Because movement due to thermal motion is random, half of the molecules move to the left, and half move to the right. Considering only those molecules that cross the membrane, the general direction of movement is from left to right, or down the concentration gradient.

molecules move helter-skelter in various directions due to collisions with other molecules. As a consequence, molecules do not go in any particular direction for long before changing course, and when this change occurs the direction is unpredictable.

But if thermal motion is random, then how is it that diffusing molecules always move *down* their concentration gradient, which is clearly not a random event? The answer lies in the distinction between *individual molecules,* which move randomly, and a *population of molecules,* which always moves down its concentration gradient (**Figure 4.7**). Because the motion of each of the molecules in Figure 4.7 is random, each is equally likely to move left or right (and thus to cross the membrane); the net flux from

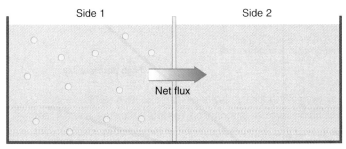

(a)
Side 1   Side 2
Concentration: 1 M   0 M (Pure water)

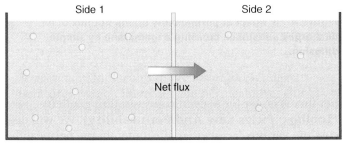

(b)
Side 1   Side 2
Concentration: .75 M   0.25 M

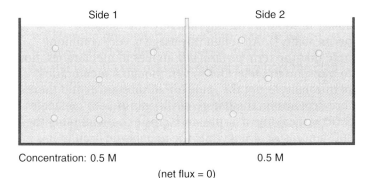

(c)
Side 1   Side 2
Concentration: 0.5 M   0.5 M
(net flux = 0)

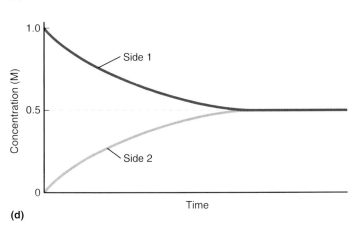

(d)

left to right in the figure is simply the result of the greater number of molecules near the left side of the membrane.

Factors Affecting Rates of Simple Diffusion

When a substance is transported passively across a membrane by simple diffusion, the rate at which it is transported depends on three factors: the magnitude of the driving force, the membrane surface area, and the **permeability** of the membrane, a measure of the ease with which molecules are able to move through it. Each of these factors is discussed next.

The Magnitude of the Driving Force When a driving force acts on molecules crossing a membrane, it influences not only the direction in which they move but also the rate at which they are transported. In most cases the net flux increases as the magnitude of the driving force increases, but not always.

In simple diffusion, the rate of transport is directly related to the size of the driving force. Consider a situation in which a membrane separates two solutions in a chamber (**Figure 4.8**). Initially (Figure 4.8a), side 1 contains solute at a concentration of 1 M, while side 2 contains an equal volume of pure water. Due to the presence of this concentration gradient, a net flux of solute molecules from side 1 to side 2 occurs. As a result, the concentration on side 1 falls while the concentration on side 2 rises. In Figure 4.8b, the concentration on side 1 has fallen to 0.75 M while that on side 2 has risen to 0.25 M. Because the size of the concentration gradient has decreased, the net flux has also decreased, as indicated by the thinner arrow. These changes continue until the two concentrations become equal, at which point the net flux equals zero (Figure 4.8c). The graph in Figure 4.8d shows that the concentrations change less rapidly as they become more nearly equal; that is, the net flux diminishes as the size of the concentration gradient decreases. Once the concentrations become equal, they no longer change, because the net flux is zero. Under these conditions, molecules still move back and forth across the membrane, but at equal rates in both directions.

In simple diffusion, the net flux of a substance is directly proportional to the size of the concentration gradient, or the electrochemical gradient if the substance is

Figure 4.8 Changes in net flux of molecules across a membrane as the concentration gradient changes. (a) A large concentration gradient creates the net flux indicated by the arrow. (b) Due to the net flux from side 1 to side 2, the concentration on side 1 has fallen to 0.75 M, while that on side 2 has risen to 0.25 M. The net flux has also decreased. (c) Concentrations on both sides of the membrane have become equal, and the net flux is now zero. (d) Changes in concentration on the two sides over time.

As time passes, does the one-way flux from side 2 to side 1 increase, decrease, or stay the same?

Increases

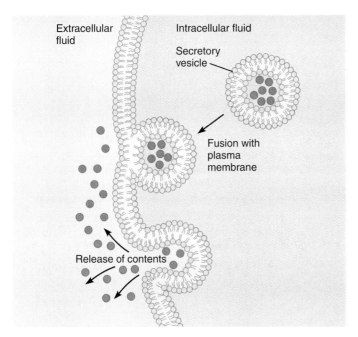

Exocytosis

Figure 4.22 Exocytosis. An intracellular vesicle fuses with the plasma membrane, and the contents of the vesicle are secreted from the cell.

Figure 4.21b). Pinocytosis is a nonspecific process, and the contents of the resulting vesicle are extracellular fluid containing dissolved solutes.

Receptor-mediated endocytosis is similar to pinocytosis in that the plasma membrane indents to form the vesicle (see Figure 4.21c). Unlike pinocytosis, however, receptor-mediated endocytosis is specific. Proteins in the plasma membrane function as *receptors* that recognize and bind specific particles in the extracellular fluid. Binding of particles to receptors concentrates the particles to areas where endocytosis will occur. The area of plasma membrane that forms the vesicle is coated with proteins (called clathrin) on its cytosolic surface. The membrane indents in this area, forming what is called a **coated pit**. The coated pit becomes a coated vesicle containing the receptors and the particles bound to them. The protein coat rapidly leaves the vesicle, and the clathrin molecules are recycled. The now uncoated vesicle fuses with a lysosome, forming an *endolysosome*. The enzymes in the lysosome will degrade the particles brought into the cell. The receptors are often recycled by exocytosis, which is described next.

Transport of Molecules Out of Cells by Exocytosis

Exocytosis is basically endocytosis in reverse: A vesicle inside the cell fuses with the plasma membrane and releases its contents into the extracellular fluid (**Figure 4.22**). Exocytosis has three functions: (1) to add compo-

nents to the plasma membrane, (2) to recycle receptors removed from the plasma membrane by endocytosis, and (3) to secrete specific substances out of the cell and into the extracellular fluid.

The first two functions are related in that both add components to the plasma membrane. During exocytosis, whatever components are present in the vesicle membrane will be added to the plasma membrane. A cell can add certain proteins, phospholipids, or carbohydrates to the plasma membrane, or it can replace the membrane that is lost during endocytosis. In fact, endocytosis and exocytosis must be balanced in a cell; otherwise, the size of the plasma membrane will change.

The third function of exocytosis, the secretion of materials, serves a number of functions. Certain white blood cells secrete antibodies to fight infections. Most cells—in particular nerve cells and endocrine cells—secrete chemical messengers that communicate with other cells. Cells lining certain hollow ducts or passageways, such as the gastrointestinal tract or respiratory airways, secrete a sticky fluid called mucus, which acts as a protective coating. You will learn about many other examples of secretion in the chapters to come.

> ### Quick Test 4.7
>
> 1. Name the three types of endocytosis. Which type(s) transports *specific* molecules across the plasma membrane?
> 2. What are the three functions of exocytosis?
> 3. Define the following terms: *phagosome, endolysosome,* and *coated pit.*

Epithelial Transport: Movement of Molecules Across Two Membranes

Up to this point we have been focusing on cell membrane function as it pertains to the transport of materials into or out of cells. But in many epithelial tissues, cell membranes function to transport materials *across* cells. This occurs, for instance, when the intestine delivers nutrients to the bloodstream, or when sweat glands produce sweat.

Recall from Chapter 1 that one function of epithelial tissues is to form barriers between the body's internal environment and the external environment, and also between different fluid compartments within the body. In addition to acting as barriers, certain epithelia (such as those lining the stomach, intestine, and secretory glands) are also able to transport materials into the internal environment from outside (a process called **absorption**) or from the internal environment to the outside **(secretion)**.

In order for an epithelium to absorb or secrete materials, the cells that make up the epithelial tissue must transport substances inward across the membrane on one side

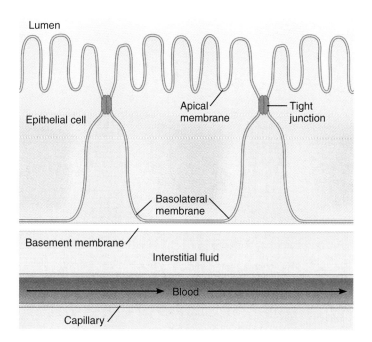

Lumen

Epithelial cell

Apical membrane

Tight junction

Basolateral membrane

Basement membrane

Interstitial fluid

Blood

Capillary

Figure 4.23 **Diagram of the general structure of an epithelium.**

of the cell and outward across the membrane on the opposite side of the cell. For this to occur, the membranes on either side must possess different transport systems. Because the membranes on the two sides are distinctly different (in both structure and function), epithelial cells are said to be *polarized.* In this section we see how the polarity of epithelial cells gives them the ability to absorb or secrete materials. We start first with an overview of epithelial structure.

Epithelial Structure

In an epithelial cell layer of the type that is specialized for absorption or secretion (**Figure 4.23**), one side of an epithelial cell faces the lumen of a body cavity. The membrane on this side is called the *lumen-facing* membrane or **apical membrane**. The membrane on the opposite side faces the internal environment and is in contact with interstitial fluid, which exchanges materials with the blood. This *blood-facing* membrane or **basolateral membrane** rests on a **basement membrane** consisting of noncellular material that is relatively permeable to most substances. The basement membrane anchors the basolateral membrane and provides physical support for the epithelial layer.

An important feature of epithelial tissues is that adjacent cells are joined by tight junctions that limit the passage of material through the spaces between cells called *paracellular* spaces. (Movement of molecules between the cells is called *paracellular transport.*) These junctions permit the fluids on either side of the cell layer to differ in composition. Tight junctions are important in maintaining homeostasis because even though the composition of fluid in the lumen of an organ may vary widely (consider

the variety of foodstuffs that enter the lumen of the stomach), the composition of interstitial fluid should not because it is part of the body's highly regulated internal environment. The "tightness" of these junctions varies from location to location; *tight epithelia* have junctions with extremely low permeabilities, whereas *leaky epithelia* have junctions that are more permeable.

Epithelial Solute Transport

The mechanisms whereby epithelial cells transport solute molecules across an epithelial layer are illustrated in **Figure 4.24**, which depicts epithelial cells of the kinds that absorb nutrients in the intestines or that function in body fluid regulation in kidney tubules. Note first that a comparison of the transport systems in the apical and basolateral membranes of both cells reveals the previously mentioned polarity: Only the basolateral membranes of these cells have Na^+/K^+ pumps, which transport Na^+ out of the cell and K^+ into the cell. In addition, only the basolateral membranes possess K^+ channels, which allow K^+ to leak out of the cells down its electrochemical gradient. This pump-leak system maintains a nearly constant concentration of K^+ inside the cells, a concentration higher than that outside the cells. The Na^+/K^+ pump also maintains a low Na^+ concentration inside the cells, which creates an inwardly directed Na^+ gradient. Note as well that in both cells the apical membrane possesses transport systems that the basolateral membrane lacks.

In the cell in Figure 4.24a, which absorbs Na^+, Na^+ leaks into the cell through Na^+ channels, which are present in the apical membrane only. This leak is counteracted by the active transport of Na^+ out of the cell across the basolateral membrane. Sodium ions enter the cell on one side and exit it on the other side—they are transported completely across the cell from the lumen to the interstitial fluid.

Note additionally that although Na^+ flows passively across the apical membrane, Na^+ transport across the cell as a whole is *active,* because entry of Na^+ into the cell across the apical membrane depends on the presence of an inwardly directed electrochemical gradient, which in turn depends on the ability of the Na^+/K^+ pump to remove Na^+ from the cell. If the pump were to stop working, the concentration of Na^+ inside the cell would rise; eventually, the Na^+ gradient across the apical membrane would disappear, bringing Na^+ entry to a halt. Because Na^+ transport across the cell is active, Na^+ can be absorbed from lumen to interstitial fluid up an electrochemical gradient.

In the cell in Figure 4.24b, which absorbs both Na^+ and glucose, Na^+ again leaks passively inward across the apical membrane, but it enters the cell by way of a sodium-linked cotransport system, which couples the inward movement of Na^+ with the inward movement of glucose. The Na^+ electrochemical gradient provides the energy that drives the secondary active transport of

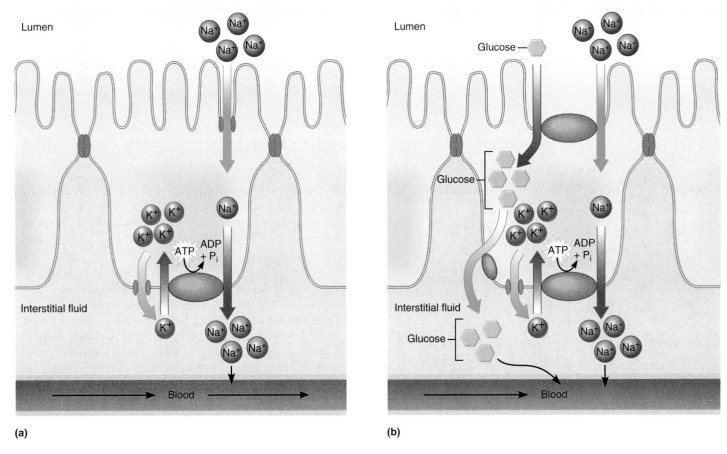

(a)

(b)

Figure 4.24 Mechanisms of epithelial solute transport. (a) Absorption of sodium ions. (b) Absorption of glucose and sodium ions.

If glucose is being transported across the basolateral membrane by a carrier (see Figure 4.24b), is the glucose concentration in interstitial fluid higher or lower than that inside the cell?

glucose up its concentration gradient into the cell. Glucose then passively exits the cell by moving down its concentration gradient via a carrier protein in the basolateral membrane. Although glucose exit across the basolateral membrane is passive, glucose transport across the epithelium as a whole is active because it depends on the presence of the inwardly directed Na^+ electrochemical gradient, which in turn depends on the action of the basolateral Na^+/K^+ pump. Glucose can therefore be transported from the lumen to interstitial fluid up a concentration gradient.

Epithelial Water Transport

Whenever epithelial cells secrete or absorb fluid, water transport occurs by osmosis. This happens, for instance, when you sweat, when the pancreas secretes pancreatic juice, or when you replenish your body fluids by drinking a glass of water.

Epithelia absorb or secrete water by first using the active transport of solutes to create a difference in osmotic pressure—an osmotic pressure gradient—between the

solutions on either side of the cell layer. Water then flows across the epithelium passively by osmosis. Because water flow occurs in response to transport of solutes, water transport is said to be *secondary to* solute transport.

In epithelial water transport (**Figure 4.25**), an epithelium creates an osmotic pressure gradient (or simply an *osmotic gradient*) to absorb water. First, epithelial cells actively transport solute molecules (dots) across the basolateral membrane into the interstitial fluid (Figure 4.25a). As a result, the solute concentration is slightly higher outside the cells than inside—particularly in the paracellular spaces, where the diffusion of solute molecules is hampered by narrowness of the passages. (This is indicated in the figure by the higher density of dots between the cells.) Because this solute transport creates a difference in osmotic pressure between the solutions on either side of the epithelium, water then flows across the cell layer by osmosis (Figure 4.25b). Note that the direction of water flow is toward interstitial fluid, or up the osmotic pressure

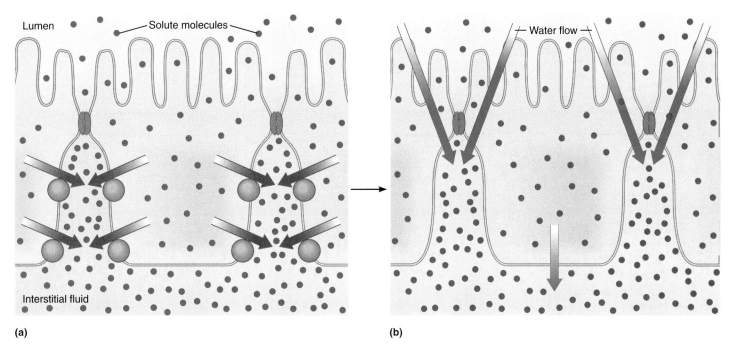

Lumen • Solute molecules

Interstitial fluid

(a)

Water flow

(b)

Figure 4.25 Epithelial water transport in an epithelium that absorbs water and solutes. (a) Pumps (circles) in the basolateral membrane actively transport solute molecules (dots) into interstitial fluid, raising the solute concentration in interstitial fluid, particularly in the paracellular spaces. (b) Active solute transport creates a gradient of osmotic pressure across the epithelium that drives passive water flow from the lumen to interstitial fluid (large arrows).

In this situation, does solute transport tend to increase or decrease the water concentration in interstitial fluid?

gradient; for epithelia that secrete fluids, the same idea pertains, except that solute and water transport go in the opposite direction—away from interstitial fluid.

In certain pathological conditions, epithelial water transport is either excessive or insufficient. In *cholera*, for example, bacterial toxins stimulate certain epithelial cells in the small intestine to oversecrete solutes and water, resulting in massive fluid loss through diarrhea. In *cystic fibrosis*, the opposite problem occurs: The epithelium lining the respiratory airways does not transport enough solute to attain adequate fluid production. As a result the lungs become clogged with mucus, which hampers breathing and can lead to serious respiratory infection (**Clinical Connections: Cystic Fibrosis**, p. 130).

Transcytosis

Macromolecules cross epithelial cells by a process called **transcytosis** (**Figure 4.26**), which involves both endocytosis and exocytosis. During transcytosis, a large

molecule is taken into the cell by endocytosis, but the endocytotic vesicle does not fuse with a lysosome. Instead, the vesicle travels to the opposite side of the cell and fuses with the plasma membrane to release its content by exocytosis.

Quick Test 4.8

1. Define the following terms: *absorption, secretion, apical membrane, basolateral membrane, basement membrane.*

2. Why are epithelial cells said to be polarized? How is polarity important to the function of certain epithelia?

3. What does it mean to say that epithelial water transport is secondary to solute transport?

4. In epithelial water transport, is water flow passive or active?

Decrease

CYSTIC FIBROSIS

Of all lethal hereditary diseases, cystic fibrosis is the most common among Caucasians, affecting about 1 in every 2500 individuals. This condition is notorious for its effects on the respiratory tract, although it affects other systems as well. In afflicted individuals, respiratory passages become clogged with thick, viscous mucus that is difficult to dislodge, even with vigorous coughing. Breathing is difficult, and victims run the risk of choking to death on their own secretions unless strenuous effort is made to clear the lungs several times a day. Victims frequently die in their late 20s of pneumonia, as the mucus-clogged airways offer a fertile environment for bacteria.

In all individuals, mucus is secreted by certain cells in the epithelium that lines respiratory passages. Under normal conditions, the epithelium also secretes a watery fluid that dilutes the mucus, making it thinner and easier to clear from airways. In cystic fibrosis the secretion of the watery fluid is impaired. The undiluted mucus is thicker and quite difficult to clear from the respiratory passages.

In the respiratory epithelium, as in all epithelia that transport fluid, water transport is secondary to solute transport. To drive water secretion, cells of the respiratory epithelium actively transport chloride (Cl^-) from interstitial fluid to the airway lumen, creating a negative electrical potential in the lumen that drives the passive flow of sodium (Na^+) in the same direction (figure a). Movement of Na^+ and Cl^- raises the osmotic pressure of fluid bathing the luminal side of the epithelium. As a consequence, water moves passively down the osmotic gradient from interstitial fluid to lumen.

Scientists recently discovered that cystic fibrosis is caused by a defect in a type of chloride channel protein found in the apical membranes of epithelial cells in the respiratory tract and elsewhere. The defect directly impedes Cl^- transport, which indirectly interferes with transport of Na^+ (figure b). As a result, the epithelium cannot create the osmotic gradient necessary for water secretion.

It has long been known that cystic fibrosis is caused by a single abnormal gene, and recent biochemical and electrophysiological studies have shown that this gene codes for a portion of the chloride channel protein. Efforts are under way to develop a therapy that will permit normal genes to be inserted into the DNA of diseased epithelial cells. If this can be accomplished, these cells will then manufacture normal channel proteins, which should allow epithelia to secrete fluid in the normal manner.

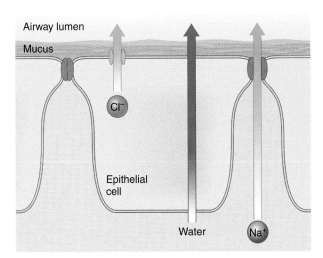

(a) Normal solute and water transport

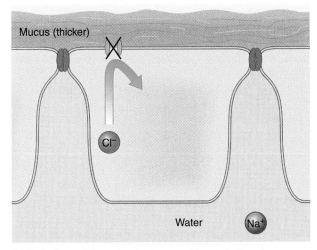

(b) Defective solute and water transport in cystic fibrosis

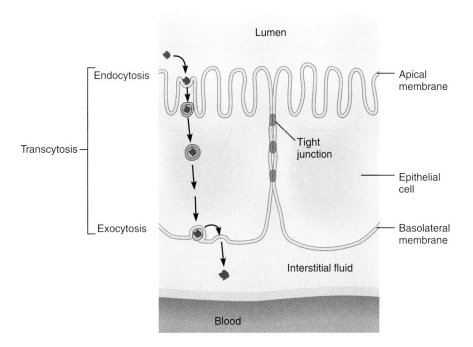

Figure 4.26 Transcytosis. Macromolecules can be transported across epithelial cells by entering the cell on one side via endocytosis and exiting the cell on the other side via exocytosis.

CHAPTER SUMMARY

Factors Affecting the Direction of Transport, p. 102

Molecules that cross cell membranes move either by simple diffusion through the lipid bilayer or by mediated transport, which involves specialized transport proteins. Active transport requires energy and is carried out by proteins called pumps. Passive transport does not require energy and includes simple diffusion and some forms of mediated transport. Transported molecules are generally influenced by three types of driving forces: (a) chemical driving forces, which are due to the presence of concentration gradients; (b) electrical driving forces, which reflect the influence of a cell's membrane potential on the movement of ions; and (c) electrochemical driving forces, a combination of the chemical and electrical driving forces that represents the net force acting on molecules. Passively transported molecules move in the direction of the electrochemical force or down their electrochemical gradient. Actively transported molecules go against the

electrochemical force or up their electrochemical gradient.

IP Nervous 1, The Membrane Potential, pp. 1–17

IP Fluids and Electrolytes, Introduction to Body Fluids, pp. 12–15

Rate of Transport, p. 109

The rate at which a substance moves across a membrane is measured as a flux. The flow of molecules in one direction or the other is the unidirectional flux; the difference between the unidirectional fluxes is the net flux.

Passive Transport p. 110

Diffusion is movement of molecules from one location to another as a result of their random thermal motion. In simple diffusion, membrane permeability is determined by the following factors: (a) the lipid solubility of the diffusing substance, (b) the size and shape of diffusing molecules, (c) temperature, and (d) membrane thickness. Facilitated diffusion involves carrier proteins, which bind molecules on one side of a membrane

and transport them to the other side by means of a conformational change. Ion channels are pores that extend from one side of the membrane to the other.

For a passively transported substance, the rate of transport depends on three factors: (a) the magnitude of the driving force, (b) membrane surface area, and (c) the permeability of the membrane to that particular substance. In facilitated diffusion and ion channels, membrane permeability is determined by two factors: (a) the transport rate of individual carriers or channels, and (b) the number of carriers or channels in the membrane.

Active Transport, p. 116

There are two basic forms of active transport: primary active transport, which uses ATP or some other chemical energy source, and secondary active transport, which uses the electrochemical gradient of one substance as a source of energy to drive the active transport of another substance. For an actively transported substance, the rate of transport is determined by two factors: (a) the rate at which individual pump proteins transport the substance, and (b) the number of pumps in the membrane.

   IP Nervous 1, The Membrane Potential, pp. 13–15

   IP Urinary, Early Filtrate Processing, pp. 5–8

Osmosis: Passive Transport of Water Across Membranes, p. 120

Water flow across membranes affects cell volumes and is involved in the secretion and absorption of fluids. Water movement (osmosis) is always passive and driven by a concentration gradient of water. Because the water concentration of a solution decreases as its solute concentration increases, a difference in total solute concentration (osmolarity) across a membrane implies the existence of a water concentration gradient. Because the osmotic pressure of a solution increases with increasing solute concentration, the flow of water down its concentration gradient is equivalent to flow up an osmotic pressure gradient. The volume of a cell is determined by the tonicity of the solution surrounding it, which depends on the solute concentration and the permeability of the membrane to the solutes that are present.

   IP Urinary, Early Filtrate Processing, pp. 5–8

Transport of Material Within Membrane-Bound Compartments, p. 125

Certain macromolecules can be moved into and out of cells through formation of vesicles in the processes of endocytosis and exocytosis, respectively. There are three forms of endocytosis: phagocytosis, pinocytosis, and receptor-mediated endocytosis. In all three processes, the plasma membrane forms a ring surrounding extracellular fluid and the edges of the membrane fuse together, forming a vesicle inside the cell. Phagocytosis and receptor-mediated endocytosis are specific in that a certain molecule or other matter is brought into the cell. During exocytosis, a substance in an intracellular vesicle is released from the cell when the vesicle fuses with the plasma membrane.

Epithelial Transport: Movement of Molecules Across Two Membranes, p. 126

Certain epithelia are specialized to transport materials into or out of the body's internal environment, termed absorption and secretion, respectively. Epithelial cells can do this because the membranes on either side differ structurally and functionally, a phenomenon known as polarity. Epithelia form barriers between body fluid compartments or between the internal and external environments. The basolateral membrane, which rests on a noncellular basement membrane, faces the internal environment; the opposite membrane is the apical membrane.

   IP Urinary, Early Filtrate Processing, pp. 5–8

▮ *E X E R C I S E S*

Multiple-Choice Questions

1. For a substance crossing a cell membrane, the chemical driving force
   a) depends only on the concentration gradient, regardless of whether or not the substance is an ion.
   b) depends only on the concentration gradient if the substance is uncharged, but also depends on the electrical force if the substance is an ion.
   c) is the total driving force on the substance, even if it is an ion.
   d) is the force that pushes molecules across the membrane, but only if the substance is actively transported.
   e) always favors movement of a molecule into the cell.

2. Which of the following is located in greater concentration inside cells compared to outside?
   a) Potassium ions
   b) Sodium ions
   c) Proteins
   d) Potassium and sodium ions are both located in greater concentration inside cells.
   e) Potassium ions and proteins are both located in greater concentration inside cells.

3. One example of primary active transport is the
   a) transport of Ca^{2+} up an electrochemical gradient by a protein that hydrolyzes ATP.
   b) transport of Ca^{2+} up an electrochemical gradient by a protein that couples Ca^{2+} flow to the flow of Na^+ down an electrochemical gradient.
   c) movement of Ca^{2+} down an electrochemical gradient through channels.
   d) transport of glucose molecules down a concentration gradient by carriers.
   e) transport of glucose up a concentration gradient by a protein that couples glucose flow to the flow of Na^+ down an electrochemical gradient.

4. If a certain anion is located in greater concentration inside the cell and a negative membrane potential exists, then which of the following statements is true?
   a) The electrical force on the anion is to move into the cell.
   b) The chemical force on the anion is to move into the cell.
   c) The equilibrium potential for the anion is a positive value.
   d) a and c
   e) all of the above

5. Given that the potassium equilibrium potential is −94 mV and the sodium equilibrium potential is +60 mV, which of the following statements is true for forces acting on sodium and potassium when a cell is at −70 mV?
   a) The electrochemical gradient for Na^+ is to move into the cell.
   b) The electrochemical gradient for K^+ is to move into the cell.
   c) a and b
   d) neither a nor b

6. When the membrane potential is equal to the equilibrium potential of Na^+ ($E_{Na} = +60$ mV)
   a) Na^+ moves into a cell down its electrochemical gradient.
   b) Na^+ moves out of a cell down its electrochemical gradient.
   c) The net flux of Na^+ is zero because it is at equilibrium.

7. The osmotic pressure of a solution depends on
   a) the concentrations of all solute particles contained in it.
   b) the concentrations of all permeant solute particles contained in it.
   c) the concentrations of all impermeant solute particles contained in it.
   d) pressure exerted on the solution by the atmosphere.
   e) the volume of water in which the solute particles are dissolved.

8. Assuming that only impermeant solutes are present, which of the following will occur when a cell is placed in a solution whose osmolarity is 200 mOsm?
   a) Water will move into the cell.
   b) Water will move out of the cell.
   c) Water will not cross the cell membrane.

9. A solution is hypotonic if
   a) the concentration of all solutes contained in it is less than 300 mOsm.
   b) the concentration of all permeant solutes contained in it is less than 300 mOsm.
   c) the concentration of all impermeant solutes contained in it is less than 300 mOsm.
   d) its osmolarity is less than 300 mOsm.

10. Movement of Na^+ in sodium-linked glucose transport, in sodium-proton exchange, and via the sodium-potassium pump are all examples of
    a) active transport.
    b) passive transport.
    c) mediated transport.
    d) simple diffusion.

11. Which of the following molecules would be most likely to cross the lipid bilayer by simple diffusion?
    a) a small polar molecule
    b) a large polar molecule
    c) a small nonpolar molecule
    d) a large nonpolar molecule

12. Assuming that a substance is uncharged and is transported across a membrane by carriers, the net flux of that substance will tend to increase as
    a) the membrane surface area decreases.
    b) the magnitude of the concentration gradient decreases.
    c) the membrane potential becomes more positive.
    d) the number of carriers in the membrane increases.

13. What do pumps and carriers have in common?
    a) They both transport molecules up electrochemical gradients.
    b) They both transport molecules down electrochemical gradients.
    c) They both transport lipid-soluble substances preferentially.
    d) They both utilize ATP to transport molecules.
    e) They both are specific for certain molecules.

14. A leukocyte, a type of white blood cell, fights bacterial infections by sending out projections of its plasma membrane that surround an invading bacterium. The membrane then fuses together, entrapping the bacterium in a vesicle inside the cell. This is an example of
    a) exocytosis.
    b) transcytosis.
    c) receptor-mediated endocytosis.
    d) pinocytosis.
    e) phagocytosis.

15. Which of the following transport mechanisms functions to bring a specific extracellular substance into the cell?
    a) receptor-mediated endocytosis
    b) pinocytosis
    c) phagocytosis
    d) a and c
    e) a, b, and c

Objective Questions

1. Substances that cross cell membranes by simple diffusion are mostly (hydrophilic/hydrophobic).

2. A channel carries out (active/passive) solute transport.

3. In simple diffusion, an uncharged solute always flows from a region of higher concentration to a region of lower concentration. (true/false)

4. In facilitated diffusion, passive flow of an uncharged solute always goes from higher to lower concentration. (true/false)

5. Passive ion flow always goes from higher to lower concentration, regardless of the membrane potential. (true/false)

6. When an uncharged solute such as glucose flows from a region of lower concentration to a region of higher concentration, its energy (increases/decreases).

7. If the membrane potential is not equal to the equilibrium potential of an ion, that ion will not be at equilibrium. (true/false)

8. A concentration gradient is also referred to as a(n) _____ driving force.

9. When a membrane potential is positive, there is an excess of cations over anions inside the cell. (true/false)

10. Transport of water from the bloodstream to the lumen of the intestine is an example of (secretion/absorption).

11. A cell will shrink if it is placed in a hypertonic solution. (true/false)

12. When water diffuses across a membrane, it normally flows from a region of higher osmotic pressure to a region of lower osmotic pressure. (true/false)

13. The osmotic pressure of a solution depends on the concentrations of all solute particles present. (true/false)

14. Junctions connecting adjacent epithelial cells are _____ junctions.

15. When a substance moves by simple diffusion down a concentration gradient, individual molecules all move in the same direction. (true/false)

Essay Questions

1. To determine whether a substance is being transported actively or passively requires knowledge of only two factors: the direction of the electrochemical gradient and the direction of the net flux. Explain.

2. Describe the various factors that determine membrane permeability in simple diffusion.

3. Explain the mechanism of glucose absorption by intestinal epithelial cells. Include a discussion of the significance of cellular polarity.

4. Compare the diffusion of molecules through the lipid bilayer to the diffusion of ions through channels.

Compare simple diffusion to facilitated diffusion.

5. A cell is placed in a saltwater solution of 0.2 mM NaCl. Neither Na^+ nor Cl^- can permeate the plasma membrane. What happens to the cell?

Critical Thinking

1. Assuming that $E_{Na} = +60$ mV, $E_{Cl} = -90$ mV, and $V_m = -70$ mV, find the direction of the electrochemical driving forces acting on Na^+ and Cl^- ions. In which direction will the ions move if they are transported passively? If they are transported actively? (Extra challenge: From the sign of E_{Cl}, determine the direction of the concentration gradient of Cl^- ions.)

2. Repeat question 1 for the case in which $V_m = -100$ mV.

3. Design an epithelium that moves potassium from the lumen to the interstitial fluid with a minimum number of proteins. Include all *required* proteins to accomplish the task.

4. Ion channels can be regulated to open or close, changing the permeability of the membrane to a specific ion. Assume that a cell at a membrane potential of -70 mV has few open sodium channels. Knowing that the equilibrium potential for sodium is $+60$ mV, predict what would happen to the membrane potential if many sodium channels suddenly went from a closed state to an open state. Explain.

Find the answers to all of these exercises, and additional study tools,
at the Physiology Place (www.physiologyplace.com).

Chemical Messengers

Polarized light micrograph of epinephrine crystals.

A community is a group of people living in the same area or having a common interest, such as the city you live in. Can you imagine how the city would function without a means of communication between people making up the community? Even under normal circumstances, communication is necessary for people to get food, proper waste disposal services, and transportation from one area of the city to another. Now think what could happen if a disaster such as a tornado or earthquake cut off all means of communication. Chaos would ensue.

Our bodies can be thought of as a community of cells. Just like the people living in a city, our body cells must communicate with one another to maintain order. When communication is disrupted, a break in homeostasis almost inevitably follows. For example, when a person gets bitten by a black widow spider, a toxin causes certain neurons to communicate excessively to skeletal muscle cells. As a result, muscle cramping occurs, especially in the abdomen. Respiratory muscles can also be affected. As another example, insulin-dependent diabetes mellitus is caused by an insufficient amount of a chemical messenger, insulin. Insulin increases cells' uptake of glucose, and thus too little insulin decreases the amount of glucose available to the cells and increases blood glucose levels. Thus either too much or too little of communication in the body can cause chaos.

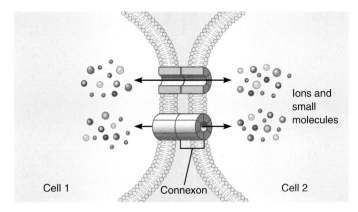

(a) Direct communication through gap junctions

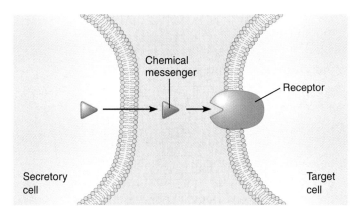

(b) Communication via chemical messengers

Figure 5.1 Types of intercellular communication.
(a) Direct communication through gap junctions. Gap junctions are composed of membrane protein structures called connexons that link the cytosols of two adjacent cells, allowing ions and small molecules to move between cells. (b) Communication via chemical messengers. After a secretory cell releases a messenger into the extracellular fluid, the messenger binds to receptors on target cells, triggering a response in the target cell.

In previous chapters we studied cells and their basic functions. We learned in Chapters 2 and 3 that most cells of the body have the machinery to sustain life. In Chapter 4 we learned how molecules move in and out of cells, a critical process because none of our cells exist in isolation and they must exchange materials with the extracellular fluid to survive. Our bodies are made up of approximately 100 trillion (10^{14}) cells and each depends on the others for survival, as they must work together to maintain homeostasis. To do so, cells must be able to communicate with each other to carry out coordinated activities, such as the maintenance of body temperature, as described in Chapter 1, or the beating of the heart. Communication occurs not only between neighboring cells but also between cells at different locations within the body. This chapter describes the general mechanisms by which cells communicate over both short and long distances.

Mechanisms of Intercellular Communication

Virtually all body functions require communication between cells. Seeing an apple requires communication between the cells in the eyes and the cells in the brain.

Fighting an infection requires communication among several cell types in the blood. Growing and developing from an infant to an adult requires communication among cells throughout the body. There are hundreds of other possible examples. Remarkably, however, all the cells in the body use only a few mechanisms to communicate with one another. In relatively few instances cells are physically linked by gap junctions; in most instances cells communicate through chemical messengers.

Direct Communication Through Gap Junctions

Recall from Chapter 2 that gap junctions link adjacent cells and are formed by plasma membrane proteins, called *connexins,* that form structures called *connexons* (**Figure 5.1**a). These connexons form channels that allow ions and small molecules to pass directly from one cell to another. The movement of ions through gap junctions electrically couples the cells, such that electrical signals in one cell are directly transmitted to the neighboring cells. For example, gap junctions found in heart muscle and the smooth muscle of other internal organs, such as the intestines and blood vessels, cause the muscle cells to contract as a unit; that is, the cells contract at the same time. The movement of small molecules through gap junctions metabolically couples the cells, such that one cell can provide necessary nutrients to other cells. For example, gap junctions allow nutrients to reach certain bone cells distant from the bloodstream. Gap junctions are also found in some glands and between some neurons in the brain and retina, where they function in communication.

Indirect Communication Through Chemical Messengers

Most often, cells communicate via chemical messengers (Figure 5.1b), which are all *ligands,* molecules that bind to proteins reversibly (see Toolbox: Ligand-Protein Interactions, p. 73). The body has hundreds of chemical messengers with a multitude of functions.

Communication through chemical messengers occurs when one cell releases a chemical into the interstitial fluid, usually by a process called secretion, and another cell, called the *target cell,* responds to the chemical messenger. The target cell, in essence, is the cell at which a message is "aimed." A target cell responds to the chemical messenger because it has certain proteins, called **receptors**, that specifically recognize and bind the messenger.

The binding of messengers to receptors produces a response in the target cell through a variety of mechanisms referred to as **signal transduction**. Generally speaking, the strength of the target cell response increases as the number of bound receptors increases. The number of bound receptors depends on both the concentration of messenger in the interstitial fluid and the concentration of receptors on the target cell.

Chemical Messengers

Chemical messengers can be classified on the bases of their function and chemical structure. First we consider the functional classes of chemical messengers.

Functional Classification of Chemical Messengers

Although there are hundreds of chemical messengers, most can be classified into six main categories:
(1) paracrines, (2) autocrines, (3) neurotransmitters,
(4) hormones, (5) neurohormones, and (6) cytokines

(**Figure 5.2**). When released into the interstitial fluid, each of these messenger categories transmits a signal by binding to receptors on a target cell, as described next.

Paracrines are chemicals that communicate with neighboring cells. The target cell must be close enough that once the paracrine is secreted into the extracellular fluid, it can reach the target cell by simple diffusion (Figure 5.2a). An example of a paracrine messenger is *histamine,* a chemical that is important in allergic reactions and inflammation and is secreted by *mast cells* scattered throughout the body (**Discovery: Antihistamines**, p. 139). During allergic reactions, histamine is responsible for the runny nose and red, watery eyes. In response to bacterial infections and various forms of tissue damage, the release of histamine by mast cells is part of a complex response called *inflammation,* which is characterized in part by redness and swelling. In inflammation, histamine increases blood flow to affected tissues (producing redness) and causes fluid to leak out of the blood vessels and into the tissue (producing swelling).

Autocrines are similar to paracrines, except that autocrines act on the same cell that secreted them (Figure 5.2b). Thus, the secretory cell is also the target cell. Often an autocrine also functions as a paracrine or other messenger type and regulates its own secretion.

Neurotransmitters are chemicals released into interstitial fluid from nervous system cells called *neurons.* Neurotransmitters are released from a specialized portion of the neuron called the *axon terminal* (Figure 5.2c), which is very close to the target cell. Because the juncture between the two cells is called a *synapse,* communication by neurotransmitters is often called *synaptic signaling.* The cell that releases the neurotransmitter is called the **presynaptic neuron**, whereas the target cell (which can be another neuron or a gland or muscle cell) is called the **postsynaptic cell**. Upon release from the presynaptic neuron, the neurotransmitter quickly diffuses the short distance from the axon terminal and binds to receptors on the postsynaptic cell, triggering a response. Communication between a neuron and its target cell(s) is very specific because it is directed only to cells with which it has a synapse. An example of a neurotransmitter is *acetylcholine,* which is released by the neurons that trigger contraction of skeletal muscles.

Hormones are chemicals released from *endocrine glands* (or occasionally other types of tissue) into the interstitial fluid, where they can then diffuse into the blood (Figure 5.2d). The hormone then travels in the blood to its target cells, which can be distant from the site of hormone release. The bloodstream distributes a hormone to virtually all cells of the body, but only cells possessing receptors specific for the hormone are able to respond and thus serve as target cells. An example of a hormone is *insulin,* which is secreted by the pancreas and acts on target cells throughout the body to regulate energy metabolism.

Table 5.1 ▮ Functional Classification of Chemical Messengers

Class	Secretory cell type	Distance to target cell	Mode of transport to target cell	Chemical classification of messenger
Paracrine	(Several)	Short	Diffusion	Amines, peptides/proteins, eicosanoids
Autocrine	(Several)	Short	Diffusion	Amines, peptides/proteins
Neurotransmitter	Neuron	Short*	Diffusion	Amino acids, amines, peptides/proteins
Hormone	Endocrine	Long	Blood	Amines, steroids, peptides/proteins
Neurohormone	Neuron	Long	Blood	Amines, peptides/proteins
Cytokine	(Several)	Short†	Diffusion†	Peptides/proteins

*Even though neurotransmitters diffuse only a short distance to the postsynaptic cell, some neurons are involved in long-distance communication because the neuron that releases the neurotransmitter is often long (up to 1 meter).
†Some cytokines are transported in blood to distant targets.

help to defend the body against pathogens, including bacteria and viruses. Cytokines usually function like paracrines in that following release into the extracellular fluid they diffuse a short distance to their target cells. However, some cytokines function like autocrines, and others travel via the bloodstream to more distant target cells, thus functioning more like hormones. Examples of cytokines are *interleukins* and *interferons,* groups of small proteins released from white blood cells as part of the immune response.

Some characteristics of the functional classes of chemical messengers are summarized in **Table 5.1**.

Chemical Classification of Messengers

A messenger's chemical structure determines its mechanisms of synthesis, release, transport, and signal transduction. The most important chemical characteristic is whether the messenger can dissolve in water or cross the lipid bilayer in the plasma membrane of cells. **Lipophilic** (hydrophobic) molecules are lipid-soluble and therefore readily cross the plasma membrane, but they do not dissolve in water. Hydrophilic (or **lipophobic**) molecules are water-soluble and therefore do not cross the plasma membrane.

In the following sections we discuss the five major classes of chemical messengers: (1) amino acids, (2) amines, (3) peptides/proteins, (4) steroids, and (5) eicosanoids (**Table 5.2**). Other chemical messengers, such as acetylcholine and nitric oxide, do not fit into any of these classes and are discussed in later chapters.

Amino Acid Messengers

Four amino acids are classified as chemical messengers because they function as neurotransmitters in the brain and spinal cord: *glutamate, aspartate, glycine,* and *gamma-aminobutyric acid (GABA).* Glutamate, aspartate, and glycine are among the 20 amino acids (alpha amino acids) that are used in protein synthesis, whereas GABA belongs to a different class of amino acids (gamma amino acids). Amino acids are lipophobic; therefore, they dissolve in water but do not cross plasma membranes. Because amino acid messengers function only as neurotransmitters, they are described in detail in Chapter 9.

Amine Messengers

Amines, which are chemical messengers derived from amino acids, are so named because they all possess an amine group ($-NH_2$). The amines include a group of compounds called **catecholamines**, which contain a *catechol* group (a six-carbon ring) and are derived from the amino acid tyrosine. Catecholamines include *dopamine, norepinephrine,* and *epinephrine.* Dopamine and norepinephrine are primarily neurotransmitters, whereas epinephrine is primarily a hormone. Other amines include the neurotransmitter *serotonin,* which is derived from tryptophan; the *thyroid hormones,* which are derived from tyrosine; and the paracrine *histamine,* which is derived from histidine. Most of the amines are lipophobic; therefore, they dissolve in water and do not cross plasma membranes. The thyroid hormones are an exception: They are lipophilic and thus do not dissolve in water but readily cross plasma membranes.

Peptide/Protein Messengers

Most chemical messengers, including many neurotransmitters and hormones and all cytokines, are polypeptides, chains of amino acids linked together by peptide bonds. These messengers are classified as peptides or proteins based on their size, which varies considerably, from just two amino acids to over a hundred amino acids. The term *peptide* generally refers to chains of fewer than 50 amino acids, whereas proteins are longer chains of amino acids. Polypeptides are lipophobic; therefore, they dissolve in water but cannot cross plasma membranes.

Table 5.2 ▐ Chemical Classification of Messengers

Class	Chemical property	Location of receptors on target cell	Functional classification
Amino acids	Lipophobic	Plasma membrane	Neurotransmitters
Amines*	Lipophobic	Plasma membrane	Paracrines, autocrines, neurotransmitters, hormones, neurohormones
Peptides/proteins	Lipophobic	Plasma membrane	Paracrines, autocrines, neurotransmitters, hormones, neurohormones, cytokines
Steroids	Lipophilic	Cytosol†	Hormones
Eicosanoids	Lipophilic	Cytosol	Paracrines

*One exception is the thyroid hormones, which, although amines, are lipophilic and have receptors in the nucleus of target cells.
†A few steroid hormones have receptors on the plasma membrane.

Steroid Messengers

Steroids are a class of compounds derived from *cholesterol.* All the body's steroid messengers function as hormones. Recall from Chapter 2 that cholesterol is a lipid with a distinctive four-ring structure. Because steroids are derived from cholesterol, which is lipophilic, they too are lipophilic and readily cross plasma membranes and are insoluble in water.

Eicosanoid Messengers

Eicosanoids include a variety of paracrines that are produced by almost every cell in the body. Most eicosanoids are derivatives of arachidonic acid, a 20-carbon fatty acid that is found in various plasma membrane phospholipids. Because eicosanoids are lipids, they readily cross the plasma membrane and are insoluble in water. Eicosanoids include the following families of chemically related compounds: prostaglandins, leukotrienes, and thromboxanes.

> **Quick Test 5.1**
>
> 1. Name the six functional classes of messengers. Which messengers are transported in the blood to their target cells? Which are secreted by neurons?
>
> 2. Name the five chemical classes of messengers. Which are lipophobic and which are lipophilic?
>
> 3. Which chemical class of messengers is derived from cholesterol? To which functional class do these messengers belong?

Synthesis and Release of Chemical Messengers

The general synthetic pathways and mechanisms of release for chemical messengers are similar within a chemical class. In this section we examine the synthesis and release of each class of messenger.

Amino Acids

Although amino acids can be obtained from the diet, the four amino acids that function as neurotransmitters must be synthesized within the neuron that will secrete them.

Glutamate and aspartate are synthesized from glucose through a three-step series of reactions. First, glucose is catabolized to pyruvic acid by glycolysis; pyruvic acid is then converted to acetyl CoA, which then enters the Krebs cycle; and finally the amine groups are added to certain Krebs cycle intermediates to form glutamate or aspartate. Glycine is synthesized from a glycolytic intermediate, 3-phosphoglycerate, in a series of four reactions. GABA is synthesized from glutamate in a single reaction catalyzed by the enzyme *glutamic acid decarboxylase.*

Following their synthesis in the cytosol, amino acid neurotransmitters are transported into vesicles where they are stored until they are released by exocytosis.

Amines

All amines are derived from amino acids, and all except thyroid hormones are synthesized in the cytosol by a series of enzyme-catalyzed reactions. (The synthesis and release of thyroid hormones are described in Chapter 7.) Which amine is produced depends on the enzymes present in a given cell.

Figure 5.3 shows the pathway for synthesis of the catecholamines, which are derived from the amino acid tyrosine. Note that in this pathway, dopamine is the precursor for norepinephrine, which in turn serves as the precursor for epinephrine. Because dopamine is a precursor for the

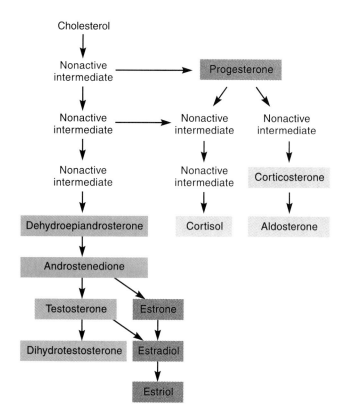

Figure 5.5 Synthetic pathway for steroids. Each arrow indicates an enzyme-catalyzed reaction. Green boxes indicate hormones produced in the adrenal cortex; blue boxes indicate male sex hormones; orange boxes indicate female sex hormones.

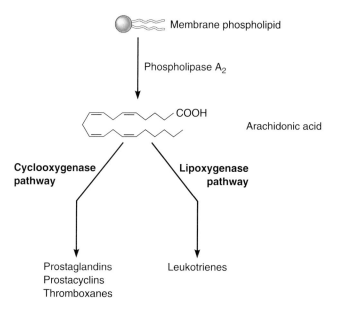

Figure 5.6 Eicosanoid synthesis. Phospholipase A_2 catalyzes the conversion of a membrane phospholipid to arachidonic acid, the precursor for all eicosanoids. Arachidonic acid is converted into eicosanoids via two pathways: The cyclooxygenase-dependent pathway leads to the production of prostaglandins, prostacyclins, and thromboxanes, whereas the lipoxygenase-dependent pathway leads to the production of leukotrienes.

active, this enzyme catalyzes the release of arachidonic acid from membrane phospholipids. Once arachidonic acid is released from the membrane, the final product depends on the complement of enzymes present in the particular cell.

To become an eicosanoid, a molecule of arachidonic acid first reacts with one of two enzymes: either *cyclooxygenase* or *lipoxygenase*. Cyclooxygenase is the first enzyme in a series of reactions, called the *cyclooxygenase pathway,* that leads to the synthesis of *prostacyclins, prostaglandins,* or *thromboxanes.* Prostacyclins and thromboxanes are important in blood clotting; prostaglandins are involved in several systems, including the inflammatory response (described in Chapter 23). Lipoxygenase is the first enzyme of a series of reactions, called the *lipoxygenase pathway,* that leads to the synthesis of *leukotrienes,* which also contribute to the inflammatory response.

Because of the eicosanoids' role in inflammation, many anti-inflammatory drugs, such as aspirin, act by targeting enzymes involved in eicosanoid synthesis. By inhibiting the activity of the enzyme cyclooxygenase, aspirin decreases not only inflammation but also blood clotting. It is this latter effect of aspirin that has led physicians to prescribe low doses of aspirin to patients

prone to a heart attack or stroke, which can be caused by blood clots in the coronary and cerebral arteries, respectively.

Transport of Messengers

Once released, a messenger must first reach and then bind to receptors on the target cell for the signal to be transmitted. In many instances, the messenger is released from a cell that is near the target cell, such that the messenger reaches the receptor by simple diffusion. This is true of paracrines, autocrines, most cytokines, and neurotransmitters. Typically these messengers are quickly degraded in the interstitial fluid and become inactive, minimizing the spread of their signaling. However, hormones and neurohormones (and a few cytokines) are transported in the blood and thus have access to most cells in the body.

Messengers can be transported in the blood either in dissolved form or bound to *carrier proteins.* To be transported in dissolved form, the messenger must be a hydrophilic messenger (**Figure 5.7**a). Peptides and amines (except thyroid hormones) are transported in this manner. Because steroids and the thyroid hormones are hydrophobic and thus do not dissolve well in blood, these hormones are largely transported bound to carrier proteins (Figure 5.7b). Although most of the catecholamines that function as neurohormones are hydrophilic and thus

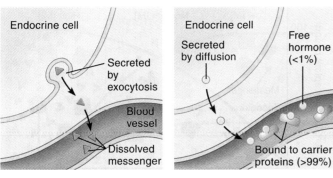

(a) Hydrophilic messenger   **(b)** Hydrophobic messenger

Figure 5.7 Transport of messengers in blood.
(a) Hydrophilic messengers are secreted by exocytosis, enter the bloodstream, and dissolve in the plasma. (b) Hydrophobic messengers are secreted by simple diffusion and then enter the bloodstream. Most of the messenger molecules are transported bound to carrier proteins. Only the small amount of free hormone in the plasma is immediately available for binding with target cell receptors.

are transported in dissolved form, some are bound to carrier proteins. Some carrier proteins are specific for a particular hormone; one example is *corticosteroid-binding globulin,* which transports the steroid hormone *cortisol.* Other carrier proteins—for example, *albumin*—are not specific and can transport many different hormones.

Even though hydrophobic hormones are transported primarily in bound form, a certain fraction of the hormone molecules dissolve in plasma (generally <1%). For each such hormone, an equilibrium develops in the bloodstream between the amount of hormone that is bound to a carrier protein (Pr) in the form of a complex (H-Pr) and the amount of free hormone (H) that is dissolved in the plasma:

$$H\text{-}Pr \rightleftharpoons H + Pr$$

Only free hormone is available to bind to receptors on target cells. However, once the hormone binds, it is removed from the blood and the equilibrium between bound and free hormone shifts to the right, causing more hormone to be released from the carrier proteins. Likewise, the secretion of hormones into the blood causes the equilibrium to be shifted to the left, such that more hormone binds to carrier proteins.

Once in the bloodstream, hormones will ultimately be degraded. How long a hormone persists in blood is measured in terms of **half-life**, the time it takes for half of the hormone in the blood to be degraded. Bloodborne messengers are generally degraded by the liver and excreted by the kidneys. Hormones that are present in dissolved form have relatively short half-lives, usually minutes. However, hormones that are bound to carrier proteins are protected from degradation and have longer half-lives, generally hours.

Quick Test 5.2

1. Name the three catecholamines. What amino acid is the precursor for catecholamines? Which catecholamine generally functions as a hormone?
2. Phospholipase A_2 causes the release of what fatty acid from membrane phospholipids? What chemical class of messengers is produced from this fatty acid?
3. What chemical class(es) of hormones is (are) transported in blood bound to carrier proteins? Does such binding generally *increase* or *decrease* the half-life of a hormone?

Signal Transduction Mechanisms

Chemical messengers transmit their signals by binding to target cell receptors located either on the plasma membrane, in the cytosol, or in the nucleus. The location of the receptor depends on whether the messenger is lipophilic or lipophobic. In either case, binding of messenger to receptor either changes the activity of proteins (for example, enzymes) already present in the cell or stimulates the synthesis of new proteins. This section describes properties of receptors and the different signal transduction mechanisms that are set into motion by them.

Properties of Receptors

Receptors show *specificity* for the messenger; that is, they generally bind only one messenger or a class of messengers. Observe that in **Figure 5.8**, messenger 1 can bind to receptor A but not to receptor B or C. Therefore, only cell A is a target cell for messenger 1. Likewise, cell C is the

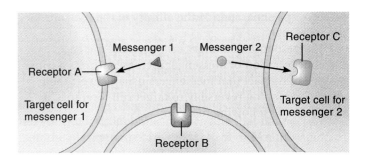

Figure 5.8 Receptor specificity. Receptor A is specific for messenger 1, receptor C is specific for messenger 2, and neither messenger can bind to receptor B. Note that receptors can be located either on the plasma membrane (receptors A and B) or inside the cell (receptor C).

If messenger 2 is to bind receptor C (located inside the target cell), it must cross the plasma membrane of the target cell. What chemical property must messenger 2 possess to enable entry into the cell?

Messenger 2 must be lipophilic (or hydrophobic).

Table 5.6 | Characteristics of the Nervous and Endocrine Systems

Characteristic	Nervous system	Endocrine system
Secretory cell	Neuron	Endocrine cell
Target cell	Neuron, muscle, or gland	Most cell types in body
Messenger	Neurotransmitter	Hormone
Pathway for communication	Across synapse	Via bloodstream
Basis of specificity	Receptors on postsynaptic target cell	Receptors on target cells throughout body
Time to onset of effect	Immediate	Delayed
Duration of effect	Brief	Long

speaking, hormones are secreted into the interstitial fluid and then diffuse into blood, but we often speak of the secretion of hormones into blood for the sake of simplicity. Hormones generally communicate by altering protein synthesis or activating G proteins, processes that are considerably slower than the electrical and chemical signals used by the nervous system. The relative slow-ness of the endocrine system and its ability to broadcast signals over wide areas are important in coordinating metabolic activities among organ systems. The endocrine system is covered in detail in Chapters 6 and 7. Various aspects of the nervous and endocrine systems are compared in **Table 5.6**.

CHAPTER SUMMARY

Mechanisms of Intercellular Communication, p. 136

Virtually all body processes require that cells are able to communicate with each other, which can occur in two basic ways: (1) via gap junctions, which allow electrical signals and small molecules to move directly from one cell to adjacent cells, and (2) via the secretion of chemical messengers, which allow signals to be transmitted from one cell to others that may be at distant locations. Chemical messengers produce responses in target cells by binding to specific receptors.

> **IP** Cardiovascular, Anatomy Review: The Heart, p. 7
> **IP** Nervous II, Anatomy Review, pp. 6–7
> **IP** Nervous II, Synaptic Transmission, p. 4

Chemical Messengers, p. 137

Chemical messengers fall into six major functional categories: (1) paracrines, (2) autocrines, (3) neurotransmitters, (4) hormones, (5) neurohormones, and (6) cytokines. Many messengers exert their effects only on cells that are close to the cells that secrete them, but hormones and neurohormones can act at distant sites because they are carried in the bloodstream to their target cells.

On the basis of their chemical structure, messengers are divided into five major classes: (1) amino acids, (2) amines (amino acid derivatives), (3) peptides, (4) steroids (cholesterol derivatives), and (5) eicosanoids (derivatives of arachidonic acid). Steroids, eicosanoids, and some amines (the thyroid hormones) are lipophilic and pass through cell membranes easily; other messengers are hydrophilic (lipophobic) and do not.

> **IP** Nervous I, Ion Channels, pp. 4,7
> **IP** Nervous II, Synaptic Transmission, pp. 1–6, 9, 12
> **IP** Nervous II, Anatomy Review, pp. 7–8

Signal Transduction Mechanisms, p. 145

The magnitude of a target cell's response to a chemical messenger generally increases with increases in the number of bound receptors, which depends on the affinity of the receptors for the messenger, the messenger's concentration, and the number of receptors present. When exposed for long periods to messenger concentrations that are very low or high, target cells can alter the number of receptors, bringing about a change in their responsiveness to the messenger. A decrease in the number of receptors is called down-regulation; an increase is called up-regulation.

Lipophilic messengers bind to receptors in the cytosol or nucleus of target cells, and the resulting complex binds to DNA to regulate gene transcription and protein synthesis. Hydrophilic messengers bind to cell surface receptors, which are of three types: (1) channel-linked receptors, which affect the opening and closing of fast ligand-gated channels, (2) enzyme-linked receptors, which catalyze reactions inside cells, and (3) G protein–linked receptors, which activate specific membrane proteins called G proteins. Activated G proteins can themselves activate (or inhibit) a variety of intracellular proteins, including enzymes or channels. Many of these enzymes catalyze the formation of second messengers inside the cells. Among the substances known to act as second messengers are cyclic AMP (cAMP), cyclic GMP (cGMP), inositol triphosphate (IP$_3$), and calcium ions (which often work by binding to calmodulin, forming a complex that activates protein kinases).

> IP Nervous II, Synaptic Transmission: pp. 4,13
>
> IP Nervous I, Ion Channels, pp. 4, 7

Long-Distance Communication via the Nervous and Endocrine Systems, p. 157

In the nervous system, neurons send signals to specific groups of target cells, to which they are connected by synapses. Responses triggered by neural signals are generally fast and brief. The endocrine system broadcasts signals to target cells throughout the body, exerting effects that are generally slow and long lasting. Hormones are secreted by specialized endocrine cells, which are usually found in endocrine glands.

> IP Nervous II, Orientation, p. 1
>
> IP Nervous II, Synaptic Transmission, pp. 1–6
>
> IP Urinary. Anatomy Review, p. 4

▧ EXERCISES

Multiple-Choice Questions

1. Arachidonic acid is the raw material for the synthesis of
   a) amines.
   b) thyroid hormones.
   c) eicosanoids.
   d) steroids.
   e) peptides.

2. Epinephrine is a(n)
   a) amino acid.
   b) steroid.
   c) eicosanoid.
   d) adrenocorticoid.
   e) catecholamine.

3. Most chemical messengers fall into which of the following chemical classes?
   a) amines
   b) amino acids
   c) peptides/proteins
   d) eicosanoids
   e) steroids

4. All amino acid chemical messengers function as
   a) paracrines.
   b) autocrines.
   c) cytokines.
   d) neurotransmitters.
   e) hormones.

5. All steroid chemical messengers function as
   a) paracrines.
   b) autocrines.
   c) cytokines.
   d) neurotransmitters.
   e) hormones.

6. Which of the following is likely to cause a rise in intracellular cAMP levels?
   a) stimulation of phosphodiesterase activity
   b) activation of an inhibitory G protein targeting adenylate cyclase
   c) binding of chemical messengers to enzyme-linked receptors
   d) binding of chemical messengers to receptor-operated channels
   e) stimulation of adenylate cyclase activity

7. Which of the following messenger classes bind to intracellular receptors?
   a) catecholamines only
   b) peptides only
   c) steroids only
   d) both catecholamines and steroids
   e) both peptides and steroids

8. The response of a target cell to a messenger depends on which of the following?
   a) concentration of messenger
   b) concentration of receptors on target cell
   c) affinity of the receptor for the messenger
   d) both a and c are true
   e) all of the above are true

9. G proteins are involved whenever
   a) binding of messenger molecules to cell surface receptors triggers a target cell response.
   b) binding of ligand molecules to cell surface receptors triggers activation or inhibition of enzymes.
   c) binding of ligand molecules to cell surface receptors triggers synthesis of second messengers.
   d) binding of ligand molecules to cell surface receptors triggers a change in membrane permeability to ions.
   e) all of the above are true

10. What enzyme catalyzes the synthesis of diacylglycerol?
    a) adenylate cyclase
    b) tyrosine kinase
    c) phospholipase C
    d) protein kinase
    e) phosphoprotein phosphatase

Objective Questions

1. Cells that secrete a messenger are called _____.

2. A (paracrine/autocrine) agent acts on the same cell that secretes it.

3. Endocrine glands release (neurotransmitter/hormone) into the _____ , where it travels to the target cell.

4. An active G protein releases the _____ subunit, which interacts with another membrane protein, altering its activity.

5. The enzyme that catalyzes conversion of ATP to cAMP is called _____.

6. Following activation of the phosphatidylinositol system, (IP$_3$/DAG) liberates calcium from intracellular stores.

7. (Lipophilic/Lipophobic) messengers exert their effect on target cells by activating or inactivating specific genes.

8. Examples of locally acting chemical messengers are (steroids/eicosanoids).

9. Cytosolic calcium often exerts its effect by binding to cytosolic (protein kinase C/calmodulin).

10. Lipophobic messengers are secreted by (exocytosis/diffusion across the cell membrane).

11. Amino acids are (cytokines/neurotransmitters).

12. The (nervous system/endocrine system) is a more rapid means of communication.

Essay Questions

1. Describe the different types of chemical signals (paracrine, autocrine, neurotransmitter, hormone, neurohormone, and cytokine).

▦ Name the primary and secondary endocrine glands, and the hormones associated with each.

▦ Describe the links of the hypothalamus with the anterior and posterior pituitary lobes.

▦ Describe the role of tropic hormones in regulating the release of other hormones. Include feedback loops in your description.

▦ Describe the types of interactions between hormones acting on the same target cell, including additive, synergistic, and permissive interactions.

▦ *Before You Begin*

Make sure you have mastered the following topics:

1. *Epithelial cells, p. 4*

2. *Hormones, p. 137*

3. *Signal transduction mechanisms, p. 145*

In Chapter 5, we learned how cells communicate with one another. For long-distance communication, we have the slow endocrine system that releases hormones and the fast nervous system that releases neurotransmitters. In this chapter and Chapter 7, we focus on the endocrine system. Chapters 8–12 focus on the nervous system.

The organs of the endocrine system consist of endocrine glands, which we learned in Chapter 1 are derived from epithelial tissue. Endocrine glands are found in many organs of the body (**Figure 6.1**). There are two types of endocrine organs: **primary endocrine organs**, whose primary function is the secretion of hormones, and **secondary endocrine organs**, for which the secretion of hormones is secondary to some other function. Some primary endocrine organs are located within the brain, including the hypothalamus, pituitary gland, and pineal gland; other primary endocrine organs are located outside the nervous system, including the thyroid gland, parathyroid glands, thymus, adrenal glands, pancreas, and gonads (testes in the male and ovaries in the female). The placenta also functions as an endocrine gland in pregnant females. Secondary endocrine glands include organs such as the heart, liver, stomach, small intestine, kidney, and skin.

Primary Endocrine Organs

In this section, we examine the functions of the various primary endocrine organs, beginning with the hypothalamus and pituitary gland.

Hypothalamus and Pituitary Gland

Together, the hypothalamus and pituitary gland (**Figure 6.2**) function to regulate virtually every body system. The **hypothalamus** is a part of the brain with many functions in addition to its role as an endocrine gland. It is considered a primary endocrine gland because it secretes several hormones, most of which affect the **pituitary gland** (also called the *hypophysis*), a pea-sized structure that is connected to the hypothalamus by a thin stalk of tissue called the *infundibulum*. The pituitary gland is divided into two structurally and functionally distinct sections called the **anterior lobe** (or *adenohypophysis*) and the **posterior lobe** (or *neurohypophysis*). As we will see shortly, the connections between the hypothalamus and the two lobes of the pituitary gland differ and are critical to the function of both endocrine organs.

Neural Connection Between Hypothalamus and Posterior Pituitary

The posterior lobe of the pituitary gland contains neural tissue consisting of the endings of neurons originating in the hypothalamus (**Figure 6.3**). These neural endings in the posterior pituitary gland secrete two peptide hormones: *antidiuretic hormone* (*ADH;* also called *vasopressin*) and *oxytocin*. These hormones are generally synthesized in neurons originating in different regions of the hypothalamus; ADH is synthesized in the *paraventricular nucleus,* and oxytocin is synthesized in the *supraoptic nucleus.* Following synthesis, the peptides are packaged into secretory vesicles, which are transported to the neural endings in the posterior pituitary. The hormone is released by exocytosis when these neurons receive a signal, usually from other neurons. The hormones are released into the blood, as are other hormones. Because these hormones are secreted by neurons instead of endocrine glands, they are also called *neurohormones.* Antidiuretic hormone regulates water reabsorption by the kidneys and is discussed in Chapter 20; oxytocin stimulates uterine contractions and milk letdown in the breasts and is discussed in Chapter 22.

Blood Connection Between Hypothalamus and Anterior Pituitary

The anterior lobe of the pituitary gland and the cells of the hypothalamus that control it secrete primarily **tropic hormones** (also called *trophic hormones*), which are hormones that regulate the secretion of other hormones. A tropic hormone can be a *stimulating hormone,* which increases the secretion of another hormone, or an *inhibiting hormone,* which decreases the secretion of another hormone. The general signaling pathway is as follows: The hypothalamus releases a tropic hormone that effects the release of another tropic hormone from the anterior pituitary; this tropic hormone effects the release of a third hormone from another endocrine gland and this third hormone exerts effects on target cells throughout the body.

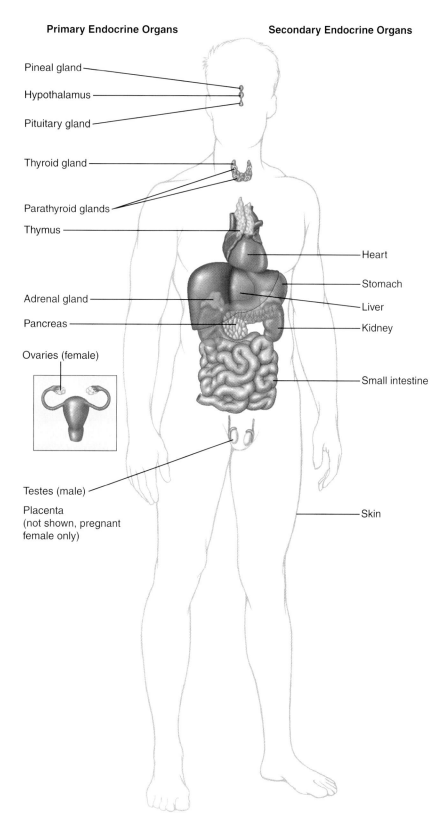

Figure 6.1 **Endocrine organs.**

CIRCADIAN RHYTHMS AND JET LAG

The invention of air travel had an unexpected consequence: rapid travel from one time zone to another results in a body imbalance commonly called "jet lag." Jet lag occurs because we have an internal clock that sets a *circadian rhythm* for many body functions. A circadian rhythm follows a cycle that is approximately 24 hours (*circa* means "almost" and *dies* means "day"). If a person is deprived of light, then the circadian rhythm is approximately 25 hours; this is the innate rhythm, or the rhythm that exists in the body independent of external influences. The actual circadian rhythm is 24 hours because the presence of the normal light-dark variation that occurs in 24-hour cycles causes the circadian rhythm to reset each day. In fact, approximately 60–70% of totally blind people have a 25-hour cycle, indicating that light input to the eyes has a strong influence on circadian rhythms. That a significant number of blind people have a 24-hour cycle suggests that the detectors of light for vision and for the circadian rhythm are distinct.

Because the light-dark cycle influences our circadian rhythm, alterations in light, such as occurs when one travels from one time zone to another, can change the rhythm. Jet lag is a common term used to describe the physiological changes occurring in a traveler who has crossed time zones. Jet lag lasts for several days, and the duration increases as the number of time zones crossed increases. Because the circadian rhythm drives many physiological processes, including sleep, alterations in the rhythm can be disturbing to the individual.

Several interventions help travelers adjust to jet lag. Because a change in the light cycle is what causes the lag, artificial bright light can be used to adapt more rapidly to a new time zone. Artificial light at night benefits those traveling westward, whereas artificial light in the morning benefits those traveling eastward. Medications may also help some travelers. The use of melatonin for jet lag has grown considerably over the last several years, although scientific evidence of its effectiveness is not conclusive. Melatonin is a hormone of the pineal gland that is hypothesized to function in establishing or maintaining the circadian rhythm.

insulin-dependent diabetes mellitus, a disease caused by insufficient secretion of insulin from the beta cells of the pancreas. With too little insulin in the blood, cells cannot use glucose adequately for energy. In cases of hypersecretion or hyposecretion, the disease process can be primary (acting directly in the endocrine gland) or secondary (involving a problem with the tropic hormone).

In a primary secretion disorder, the abnormality originates in the endocrine gland that secretes the hormone. For example, in primary hypersecretion of thyroid hormones, the thyroid gland secretes too much of the thyroid hormones (**Figure 6.13**a). One possible cause of hypersecretion is tumors of the endocrine cells (**Clinical Connections: Pituitary Adenomas**, p. 176). In primary hypersecretion, the blood levels of tropic hormones tend to be lower than normal due to the increased negative feedback from the hormone regulated by the tropic hormones. In the case of thyroid hormones, levels of TRH and TSH in the blood are low because the excess thyroid hormones inhibit their release through negative feedback. In primary hyposecretion of thyroid hormones, the opposite pattern of hormone levels occurs: Thyroid hormone levels are decreased, but TRH and TSH levels increase due to reduced negative feedback.

In a secondary secretion disorder, the abnormality originates in the endocrine cells of either the anterior pituitary or the hypothalamus, which secrete the tropic hormone. In a secondary hypersecretion of thyroid hormones, for example, blood levels of thyroid hormones increase due to either an excess of TSH secretion by an abnormal anterior pituitary (Figure 6.13b) or an excess of TRH secretion by an abnormal hypothalamus. If excess TSH is secreted from the anterior pituitary, thyroid hormone levels in the blood increase, but levels of TRH fall due to increased negative feedback from the thyroid hormones. If excess TRH is secreted from the hypothalamus, then blood levels of both TSH and thyroid hormones also increase.

Hormone Interactions

As we saw earlier, almost all cells are exposed to hormones, so the target cells for a chemical messenger must have receptors that are specific for that messenger. Because

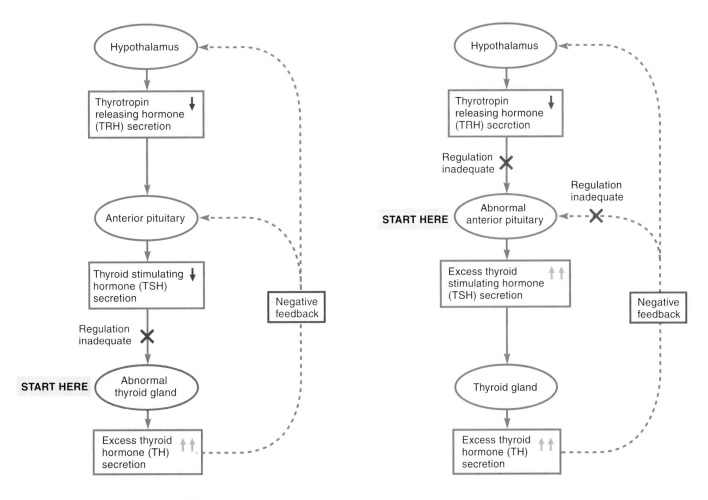

(a) Primary hypersecretion of thyroid hormones

(b) Secondary hypersecretion of thyroid hormones

Figure 6.13 Abnormal secretion of thyroid hormones. (a) Primary hypersecretion of thyroid hormones. An abnormality of the thyroid gland causes it to secrete too much of the thyroid hormones. Excess thyroid hormones in the blood induce strong negative feedback, decreasing TRH and TSH release into the blood. (b) An example of secondary hypersecretion of thyroid hormones. An abnormality of the anterior pituitary causes it to secrete too much TSH. Excess TSH levels in the blood stimulate the thyroid gland to secrete excess thyroid hormones. Negative feedback to the hypothalamus increases, so TRH levels in the blood decrease.

a single hormone may have receptors on different types of cells, that hormone can produce more than one effect in the body. For example, there are ADH receptors both on certain epithelial cells in the kidneys, where ADH increases water reabsorption, and on smooth muscle cells of certain blood vessels, where ADH causes the muscle cells to contract, decreasing the diameter of the blood vessels.

As is evident in Table 6.1 on pages 170–171, often more than one hormone affects a given body function. For example, blood calcium levels are regulated by calcitonin, parathyroid hormone, and calcitriol. Similarly, blood glucose levels are regulated by insulin, glucagon, epinephrine, cortisol, and growth hormone. In some cases the effects of the hormones oppose each other, a process called **antagonism**. For example, parathyroid hormone increases blood calcium levels, whereas calcitonin decreases blood calcium levels. Likewise, glucagon increases blood glucose levels, whereas insulin decreases blood glucose levels.

In other cases, the hormones produce the same direction of effect, generally by different means. When two or more hormones produce the same type of response in the body, the effect can be either **additive**, in which case the net effect equals the sum of the individual effects, or **synergistic**, in which case the net effect is greater than the sum of the individual effects. Assume, for example, that hormones A and B produce equal effects when present at a concentration of 1 nanogram per deciliter (ng/dL) of blood. If both hormones were present at 1 ng/dL blood, then the response in an additive interaction would be the sum of that of the individual hormones, but the response in a synergistic interaction would be greater than the sum of the individual hormone responses.

In some cases the presence of one hormone is needed for another hormone to exert its actions, a process called **permissiveness**. One example involves epinephrine,

PITUITARY ADENOMAS

Tumors are abnormal growths that can occur on most tissues of the body. Some tumors are *benign* or noncancerous. Other tumors are *malignant* or cancerous because unlike benign tumors, they invade and destroy the tissue where they originate and they have the capacity to spread to other tissues, or *metastasize*. Tumors are named based on the cell types and whether the tumor is benign or malignant. Benign tumors of glandular epithelial cells are called *adenomas.*

Because the pituitary gland secretes many different hormones that regulate a variety of body functions, pituitary tumors are potentially dangerous even if they are noncancerous. Pituitary tumors account for approximately 10% of all brain tumors, but 90% of pituitary tumors are adenomas that affect the anterior pituitary (recall that the posterior pituitary is primarily neural tissue, not epithelial).

Pituitary adenomas are classified based on size and functionality. Pituitary adenomas less than 1 cm in diameter are called *microadenomas,* whereas those greater than 1 cm in diameter are called *macroadenomas.* [Figure (a) shows an MRI horizontal section of a normal pituitary gland, and figure (b) shows a macroadenoma.] Pituitary adenomas that increase the secretion of pituitary hormones are called *functioning adenomas* because the tumor cells actively secrete a certain hormone. *Nonfunctioning adenomas* consist of inactive tumor cells. They tend to either decrease the secretion of certain hormones by compressing normal endocrine cells or cause other neurological disturbances by compressing neurons near the tumor. A common neurological symptom is blurred vision because the optic nerves are located below and very near the pituitary gland.

Functioning pituitary adenomas are usually detected early because of the symptoms produced by the hormonal imbalance. Therefore, most functioning pitu-

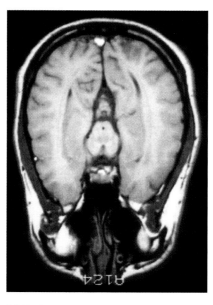

(a) Horizontal MRI of a brain with a normal pituitary gland.

itary adenomas are microadenomas. Approximately 50% of functioning adenomas affect prolactin-secreting cells, resulting in inappropriate breast discharge *(galactorrhea),* irregular menstrual periods *(amenorrhea),* and sometimes sexual dysfunction. Growth hormone–secreting cells are affected in 30% of functioning pituitary adenomas, resulting in excessive height *(gigantism)* in children or excessive thickening of bones and enlargement of soft tissues *(acromegaly)* in adults. ACTH-secreting cells are affected in 20% of functioning pituitary adenomas, resulting in *Cushing's disease.* Due to an increase in cortisol secretion, people with Cushing's disease tend to have elevated blood glucose levels, which can lead to diabetes mellitus. Other symptoms include obesity of the trunk but not extremities, purple striations on the abdomen (stretch marks) due to a collagen deficit in the skin, and a rounded face due to accumulation of fluids. Other hormone-secreting cells of the anterior pituitary are sometimes affected, but much less frequently.

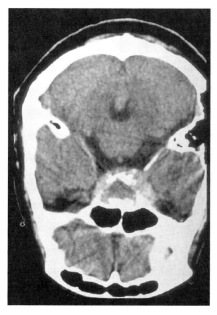

(b) Horizontal MRI of a brain with a macroadenoma.

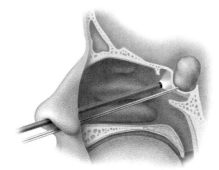

(c) Surgical removal of a pituitary adenoma by the trans-sphenoidal approach.

Treatment of pituitary adenomas is similar to treatment of other tumors and includes pharmacological treatments (usually effective for hypersecretion of prolactin and growth hormone, but not ACTH), radiation therapy, or surgical removal of the tumor. Because of the location of the pituitary gland, surgery usually involves a *trans-sphenoidal approach,* whereby the surgeon goes through the nose or under the upper lip to access the skull at the base, as shown in figure (c).

which by binding to β adrenergic receptors on smooth muscle cells of bronchioles (small airways of the lungs) causes these airways to increase in diameter (dilate). Thyroid hormones are essential for the synthesis of the β adrenergic receptors in these target cells. Therefore, even though thyroid hormones by themselves have no direct effect on the diameter of bronchioles, epinephrine cannot trigger dilation of the bronchioles in the absence of thyroid hormones, because there are no receptors to which epinephrine can bind.

Quick Test 6.3

1. Define the terms *primary hypersecretion, secondary hypersecretion, primary hyposecretion,* and *secondary hyposecretion.*

2. Define the terms *antagonism, additive effect, synergistic effect,* and *permissiveness* as they relate to hormone interactions.

Exercise Link

As Bill and Jane ran the marathon, glucagon and epinephrine concentrations in their blood increased while insulin levels decreased. All these hormonal changes worked together to release stored glucose into the blood for use as a fuel by their exercising muscles. This process kept blood glucose concentrations fairly constant, at least for the first 20 miles or so. The fatigue Bill and Jane felt late in the race was due, in part, to falling blood glucose levels. The delivery of glucose to tissues that need it requires adequate blood perfusion through those tissues, which is closely related to the total amount of water within the body. As Bill and Jane lost water (and salt) through sweating, their antidiuretic hormone concentrations increased (which reduced water loss through the kidneys). Likewise, their aldosterone concentrations increased (which reduced salt loss through the kidneys).

CHAPTER SUMMARY

Primary Endocrine Organs, p. 162

Primary endocrine organs include the pituitary gland (which is divided into anterior and posterior lobes), pineal gland, thyroid gland, parathyroid glands, thymus, pancreas, and gonads. Secretion by the anterior pituitary is regulated by tropic hormones secreted by neurosecretory cells in the hypothalamus. Secretion of these and other hormones is regulated by negative feedback.

> **IP** Endocrine, Endocrine System Review, pp. 1–6

Secondary Endocrine Organs, p. 170

A secondary endocrine organ secretes a hormone in addition to carrying out another primary function. Secondary endocrine organs include the heart, liver, and kidneys.

Hormone Actions at the Target Cell, p. 171

The magnitude of a target cell's response to a hormone varies with the hormone's concentration in the plasma, which depends on the rate of hormone secretion; the amount of hormone that is bound to carrier proteins in the blood; and the rate at which the hormone is metabolized. The liver degrades most hormones, and the breakdown products are eventually excreted in the urine.

> **IP** Endocrine, Actions of Hormones on Target Cells, pp. 1–9

Abnormal Secretion of Hormones, p. 173

Abnormal hormone secretion includes hyposecretion, or too little hormone, and hypersecretion, or too much hormone. The disorder can be primary or secondary. A primary secretion disorder exists when the endocrine organ is abnormal and secretes the wrong amount of hormone. A secondary secretion disorder exists when the hypothalamus or anterior pituitary gland is abnormal and secretes the wrong amount of tropic hormone, which thereby triggers inappropriate secretion of hormone from the target endocrine gland.

Hormone Interactions, p. 174

A single hormone may regulate more than one body function, and a given function may be regulated by two or more hormones, which may exert effects that are additive, antagonistic, synergistic, or permissive.

EXERCISES

Multiple-Choice Questions

1. Epinephrine is released from what area of the adrenal gland?
   a) zona reticularis
   b) zona fasciculata
   c) zona glomerulosa
   d) medulla

2. Which of the following is an accurate statement regarding regulation of pituitary hormone secretion by the hypothalamus?
   a) All pituitary hormones are regulated by tropic hormones from the hypothalamus.
   b) All anterior pituitary hormones are regulated by a releasing hormone and a release inhibiting hormone from the hypothalamus.
   c) All posterior pituitary hormones are regulated by a releasing hormone from the hypothalamus.

d) All anterior pituitary hormones are tropic hormones.

e) none of the above

3. Most hypothalamic and pituitary hormones are
   a) amino acids.
   b) peptides/proteins.
   c) steroids.
   d) eicosanoids.
   e) catecholamines.

4. Gonadotropin releasing hormone stimulates release of which of the following from the anterior pituitary?
   a) sex hormones
   b) follicle stimulating hormone
   c) luteinizing hormone
   d) both follicle stimulating hormone and luteinizing hormone

5. Which of the following adrenal hormones is secreted by chromaffin cells?
   a) cortisol
   b) aldosterone
   c) epinephrine
   d) androgens

6. In primary hyposecretion of thyroid hormones,
   a) levels of thyroid hormones in the blood decrease.
   b) levels of TRH in the blood decrease.
   c) levels of TSH in the blood decrease.
   d) all of the above are true
   e) none of the above is true

7. Which of the following organs secretes glucagon?
   a) liver
   b) anterior pituitary

c) posterior pituitary
d) adrenal gland
e) pancreas

8. Which of the following is an example of permissiveness?
   a) Glucagon increases blood glucose levels, and insulin decreases blood glucose levels.
   b) Glucagon, epinephrine, and cortisol all increase blood glucose levels.
   c) Estrogen stimulates synthesis of progesterone receptors in the endometrium.
   d) all of the above
   e) none of the above

Objective Questions

1. (Endocrine/Exocrine) glands secrete hormones.

2. Neural input to the hypothalamus is involved in regulating secretion of hormones by both lobes of the pituitary. (true/false)

3. Epinephrine is secreted by the adrenal (medulla/cortex).

4. Thyroid hormones are classified as (amines/steroids).

5. Lipophobic messengers are secreted by (exocytosis/diffusion across the cell membrane).

6. Calcitonin is secreted by the (thyroid gland/parathyroid gland).

Essay Questions

1. Describe the role of the hypothalamus in the regulation of anterior pituitary hormone secretion.

2. Describe the various factors that affect the concentration of a hormone in plasma.

3. Describe the anatomy of the adrenal gland. What are its two main subdivisions? What are the minor subdivisions of its outer region? Which hormones are secreted from the different subdivisions?

Critical Thinking

1. Antagonistic hormones have opposing actions on the target tissue. Why might this be advantageous? What would you predict about factors releasing these hormones?

2. When a substance is first discovered and is *thought* to be a hormone, it is referred to as a *factor*. What evidence would you predict a scientist would need to change the name from factor to hormone?

3. Calcitriol is a hormone that increases calcium absorption by the digestive tract and calcium reabsorption by the kidneys, thereby increasing blood calcium levels. Based on what you know about controls of hormone release, what effect does calcium have on the release of calcitriol?

Find the answers to these exercises, and additional study tools, at the Physiology Place (www.physiologyplace.com).

The Endocrine System: Regulation of Energy Metabolism and Growth

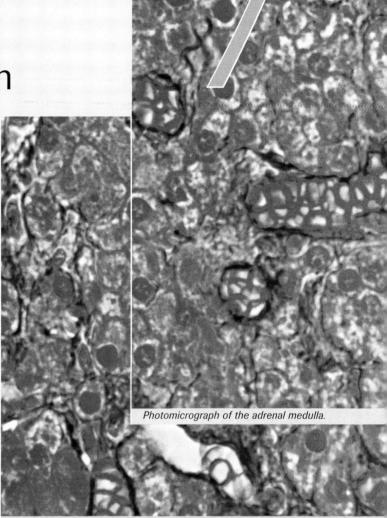

Photomicrograph of the adrenal medulla.

H ave you ever been stuck in classes from 10:00 until 1:00? If so, you probably had difficulty concentrating the entire time because you started feeling hungry before that last class let out. At 1:30, however, you replenished your energy reserves by consuming a late lunch. That allowed you to study until 5:00 before the next wave of hunger set in.

Energy balance differs from other body balances because energy input is intermittent, through meals. This creates a challenge for our organ systems, but the challenge is met by the actions of primarily two hormones, insulin and glucagon. You will learn in this chapter how insulin and glucagon regulate energy metabolism to maintain adequate glucose for our nerve cells to function. You will also learn how other hormones affect energy metabolism as the body grows or copes with stress.

OBJECTIVES

- Compare the metabolic pathways operating during energy mobilization to those operating during energy utilization.

- Explain the concepts of negative energy balance and positive energy balance.

- Describe the hormonal control of metabolism during absorptive and postabsorptive states.

- Explain how growth hormone regulates growth.

- Describe the synthesis and secretion of thyroid hormones. Distinguish between direct and permissive actions of thyroid hormones.

- Describe the effects of glucocorticoids on whole-body metabolism. Compare the physiological effects of glucocorticoids to their pharmacological effects.

- Describe the stress response.

Before You Begin

Make sure you have mastered the following topics:

1. *Enzymes, p. 71*
2. *Metabolic pathways, p. 64*
3. *Glycolysis, p. 81*
4. *Krebs cycle, p. 82*
5. *Electron transport chain, p. 85*
6. *Pancreas, p. 169*
7. *Thyroid gland, p. 167*
8. *Hypothalamus-anterior pituitary tropic hormones, p. 165*
9. *Adrenal gland, p. 168*

In Chapter 3, we explored cellular metabolism, including some specific metabolic pathways and the role of enzymes in regulating metabolic pathways. In this chapter we discuss the coordinated regulation of metabolic pathways in different organs to maintain adequate energy supplies for all the body's cells. Whole-body metabolism is primarily regulated by the hormones, whose basic properties we discussed in Chapter 6. In this chapter, we focus on hormonal control of energy balance.

As we study these hormones and their actions, we will find that their effects are varied and often overlap. To aid in the understanding of these hormones, the first section of this chapter presents general principles relating to energy metabolism and energy balance. A comparison of energy metabolism during and between meals follows. The chapter then describes the hormones regulating blood glucose level, especially insulin and glucagon. Then it discusses the hormones that regulate growth and the primary hormones that regulate whole-body metabolism—thyroid hormones. Finally, the glucocorticoids and their role in adapting to stress are described.

An Overview of Whole-Body Metabolism

The next time you find yourself sitting down to eat a meal, ask yourself this question: Why am I eating? If you do this on several occasions, you will find that there are many possible reasons. Perhaps you smelled something cooking that stimulated your appetite, or you saw an advertisement for a food you like. Perhaps you decided to join some friends who were on their way to dinner, or maybe you just looked at your watch and decided it was mealtime. The number of possibilities is large. In fact, we have so many different motivations for eating that it is easy to forget that the ultimate reason—obtaining nutrition—is a biological necessity, because food is our sole source of energy and the raw materials from which our bodies are made.

Although our need for food is driven by biological necessity, our eating patterns are not constant and are influenced by other factors. (You have probably skipped meals to study for exams, or overeaten during the holidays.) Normally this is no cause for concern because the body has ways of maintaining the steady supply of energy that cells need despite changes in the pattern of food intake. Between meals the body converts energy stores (including large carbohydrates, proteins, and lipids) into smaller molecules that cells can use for energy. When you eat, the body replenishes these stores by converting nutrients into energy storage molecules.

The way the body stores and utilizes energy—*energy metabolism*—is influenced not only by eating patterns but also by such factors as growth, stress, and metabolic rate. In all cases, whether the body stores or utilizes energy is controlled primarily by endocrine signals. Two critical concepts drive the control of energy metabolism:

1. Because food intake is intermittent, the body must store nutrients during periods of intake and then break down these stores during periods between meals.

2. Because the brain depends on glucose as its primary energy source, blood glucose levels must be maintained at all times, even between meals.

To fully appreciate the control of metabolism, we need to review some key concepts of cellular metabolism described in Chapter 3 and relate them to the whole body.

Anabolism

An interesting aspect of metabolism is that the same small biomolecules that provide energy are also used to synthesize larger biomolecules. A good example is acetyl CoA, which can be catabolized in the Krebs cycle for energy and is also a substrate for triglyceride and cholesterol synthesis. Thus because carbohydrates, lipids, and proteins can all be catabolized to acetyl CoA, they can all eventually be converted to lipids. Many other metabolic intermediates of glycolysis and the Krebs cycle can be used to synthesize larger biomolecules. For example, some metabolic intermediates can be converted to amino acids and used for protein synthesis, whereas other intermediates can be used in the synthesis of phospholipids.

Regulation of Metabolic Pathways

If anabolic and catabolic pathways have several of the same intermediates, what governs the direction of metabolism? Most important in determining which metabolic pathways are in operation is the number and activity of the enzymes involved in the pathways. The activity of enzymes can be regulated by changing their concentration through synthesis or degradation, or by changing the activity of individual enzyme molecules through allosteric or covalent regulation. Hormones that regulate metabolic pathways do so by regulating the activity of enzymes in one or more of these ways.

Metabolic pathways are also controlled by compartmentation. In Chapter 3 we examined an example of cellular compartmentation: whereas glycolysis occurs in the cytosol, the Krebs cycle occurs in the mitochondrial matrix. Compartmentation also occurs on the tissue level, because some enzymes are found in the cells of certain tissues only. In addition, hormones differentially affect tissues based on the types of receptors on the cells in that tissue. Tissues or organs that have special metabolic activities include the brain, skeletal muscle, the liver, and adipose tissue. We will explore how these tissues and organs affect whole-body metabolism shortly; first we consider how the body handles different classes of biomolecules, from absorption, to cellular uptake, to utilization by the cell.

Energy Intake, Utilization, and Storage

When we eat, digestion in the gastrointestinal (GI) tract breaks down the large molecules in food into smaller molecules, which are then absorbed into the bloodstream. Of our three main nutrient classes, carbohydrates are transported in the blood as glucose, proteins are transported as amino acids, and lipids are transported in lipoproteins. The blood flow distributes these nutrients to tissues throughout the body, where they are eventually take up by cells. Inside cells these molecules undergo three possible fates:

1. Biomolecules can be broken down to smaller molecules, in the process releasing energy that can be used for driving various cellular processes such as muscular contraction, transport, secretion, or anabolism.

2. Biomolecules can be used to synthesize other molecules needed by cells and tissues for normal function, growth, and repair.

3. Biomolecules in excess of those required for energy and synthesis of essential molecules are converted to energy storage molecules that provide energy during periods between meals. The two primary energy storage molecules are glycogen and triglyceride (fat).

The ultimate fate of consumed molecules depends on their chemical nature and the body's needs at the time of consumption.

Next we examine the uptake of nutrient molecules into cells and the utilization or storage of these nutrients inside the cells.

Uptake, Utilization, and Storage of Energy in Carbohydrates

Although carbohydrates are consumed in a variety of forms, monosaccharides, especially glucose, are the forms found in the bloodstream. **Figure 7.1**a illustrates the fate of glucose in the blood. Molecules of glucose are transported into cells throughout the body by *glucose transporters* ①. Inside cells, glucose can be oxidized for energy ②, which generates carbon dioxide as a waste product; provide substrates for other metabolic reactions ③; or be converted to glycogen for storage ④. If glucose levels in the cell decrease, glycogen can be broken down to glucose by glycogenolysis ⑤.

Although this series of steps accurately describes what happens to glucose *in the body as a whole,* it may or may not describe what happens in individual cells. Most cells, for example, can oxidize glucose but have a limited ability to synthesize and store glycogen.

Uptake, Utilization, and Storage of Energy in Proteins

As depicted in Figure 7.1b, amino acids rather than whole proteins are transported in the bloodstream. Following uptake into cells ①, amino acids are used for the synthesis of proteins ② or catabolized for energy by proteolysis ③. The proteins function as amino acid stores that can subsequently be broken down to amino acids ④, which can then be catabolized for energy or released into the bloodstream for use by other cells. Catabolism of cellular proteins to generate energy occurs only during periods of starvation, and it generates ammonia (NH_3) and carbon dioxide as waste products. The ammonia is converted by the liver to *urea,* which is eventually eliminated in the urine.

Uptake, Utilization, and Storage of Energy in Fats

When the body uses dietary carbohydrates and proteins, they are taken into cells in the form of smaller components (glucose or amino acids), which can either be catabolized for energy or assembled into larger molecules. The same process occurs for fats, although the process is a little more complicated.

Figure 7.1c illustrates the body's handling of triglycerides, the predominant form in which fats are present in

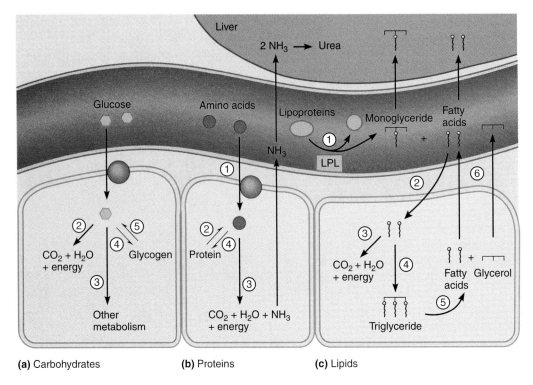

Figure 7.1 Uptake, absorption, and subsequent handling of biomolecules within cells.
Numbered steps are described in the text for the handling of (a) carbohydrates, (b) proteins, and
(c) lipids. LPL is lipoprotein lipase.

▮ *What three metabolic pathways are necessary for the handling of glucose shown in step 2 in part (a)?*

the diet. Triglycerides are transported in the bloodstream in **lipoproteins**, small protein- and lipid-containing particles. An assortment of lipoproteins of varying densities transports lipids to various target cells throughout the body (see Discovery: Lipoproteins and Plasma Cholesterol, p. 668). However, once they reach the target cells, lipids must leave the lipoprotein before entering the cells.

To facilitate entry into cells, triglycerides at the outer surface of lipoproteins are broken down by the enzyme **lipoprotein lipase** ①, which is located on the inside surface of capillaries throughout the body and is particularly dense in capillaries running through adipose tissue (body fat). This enzyme breaks down triglycerides into fatty acids and monoglycerides; the fatty acids are then taken up by nearby cells ② while the monoglycerides remain in the bloodstream and are eventually metabolized in the liver.

Upon entry into cells, fatty acids may be oxidized for energy ③ or combined with glycerol to form new triglycerides ④, which are stored in fat droplets in the cytosol. This storage occurs mainly in **adipocytes**, adipose tissue cells that are specialized for fat storage. (The glycerol used in triglyceride synthesis is not derived from absorbed triglycerides, but instead is synthesized

within adipocytes.) Stored triglycerides can subsequently be broken down to glycerol and fatty acids ⑤, which can be catabolized for energy or released into the bloodstream for use by other cells ⑥. The catabolism of glycerol and fatty acids produces carbon dioxide as a waste product. The breakdown of triglycerides to fatty acids and glycerol, such as occurs in steps 1 or 5, is called *lipolysis*.

■ ■ ■

The body's processing of carbohydrates, proteins, and lipids is summarized in **Table 7.1**. Fatty acids are included among the smaller nutrient molecules (even though they are not absorbed in this form) because they are the form in which fats are made available to most cells. Some small nutrients can be *interconverted*; for example, glucose can be synthesized from amino acids, and fatty acids can be synthesized from glucose or amino acids. These interconversions have a significant role in whole-body metabolism, as we will see shortly.

▮ *Glycolysis, the Krebs cycle, and the electron transport chain*

Table 7.1 | Summary of Carbohydrates, Protein, and Lipid Processing

	Form absorbed across GI tract	Form circulating in blood	Form stored	Storage site	Percent of total energy stored
Carbohydrates	Glucose	Glucose	Glycogen	Liver, skeletal muscle	1%
Proteins	Amino acids, some small peptides	Amino acids	Proteins	Skeletal muscle*	22%
Lipids	Monoglycerides and fatty acids (in chylomicrons)	Free fatty acids, lipoproteins	Triglycerides	Adipose tissue	77%

*Even though proteins are found in all cells of the body, most of the proteins mobilized for energy come from skeletal muscle cells.

Energy Balance

To maintain homeostasis, the human body must be kept in balance. To be in balance in this context means that what comes into the body and what is produced by the body must equal the sum of what is used by the body and what is eliminated by the body, as shown in the following equation:

$$input + production = utilization + output$$

In terms of energy balance, the body does not produce energy (an impossibility according to the laws of thermodynamics) and so this equation becomes

$$energy\,input = energy\,utilization + energy\,output$$

Energy is used by the body to perform work, and energy output exists as heat released. Therefore, the balance equation becomes

$$energy\,input = work\,performed + heat\,released$$

The endocrine system regulates energy balance to ensure that a steady supply of small nutrients is always available to all body cells to meet their energy demands. As cells expend energy, they draw upon stores of nutrients within cells and in the bloodstream to provide more energy. This pool of nutrients must be continually replenished if uninterrupted energy expenditure is to occur. This replenishment can be accomplished in two ways: by absorption of more nutrients into the bloodstream, or by *mobilization* of energy stores—that is, the catabolism of stored macromolecules into small nutrient molecules that are released into the bloodstream. The body mobilizes its energy stores when the rate of energy intake is insufficient to meet its energy needs.

Energy Input

Energy input into the body arrives in the form of absorbed nutrients. When a particular nutrient molecule (such as glucose) is oxidized in the body, a certain quantity of energy is liberated; this quantity represents the *energy content* of the molecule. A person's *energy intake* is the total energy content of all nutrients absorbed.

The energy content of a nutrient is generally determined by burning a known quantity of the substance in an instrument called a *calorimeter,* which measures the total amount of energy released in the form of heat as the substance burns. This amount is usually expressed in calories per gram of the substance burned.

Energy Output

In a calorimeter, all the energy contained in a molecule is converted to heat during oxidation, and thus the energy content of the molecule (input) is equal to the quantity of energy in the heat (output). Although the body also utilizes the oxidation of molecules to liberate energy, the energy released during oxidation or other catabolic reactions takes two forms: heat and work (**Figure 7.2**). Approximately 60% of the energy in consumed nutrients goes to heat production, which is necessary to maintain body temperature. Most of the remaining 40% of energy is used to synthesize ATP, which is used to perform cellular work (a process that releases still more heat).

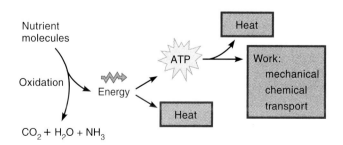

Figure 7.2 The forms of energy produced by the oxidation of nutrient molecules. Some of the energy liberated during oxidation is used to generate ATP, which can perform various types of work within cells; the rest takes the form of heat.

The energy-requiring processes of cells fall into three main categories: *Mechanical work* uses intracellular protein filaments to generate movement, such as occurs in muscle contraction or the beating of cilia lining the respiratory tract. *Chemical work* is used in the formation of bonds during chemical reactions, such as occurs when small molecules are used to synthesize large molecules. *Transport work* utilizes energy to move a molecule from one side of a cell membrane to another, such as occurs in active transport (the Na^+/K^+ pump, for example) or in vesicular transport (exocytosis and endocytosis).

Metabolic Rate

When the body breaks down nutrients, it either releases energy as heat or uses it to perform work. The amount of energy so expended per unit time is the body's **metabolic rate**. A person's metabolic rate is influenced by a number of factors, including muscular activity, age, gender, body surface area, and environmental temperature. When sitting still, for example, the rate of energy expenditure is about 100 kilocalories per hour; when riding a bicycle at a moderate rate, it is over 300 kilocalories per hour.

The **basal metabolic rate (BMR)** is the rate of energy expenditure of a person who is awake, is lying down, is physically and mentally relaxed, and has fasted for at least 12 hours; under these conditions, both metabolic rate and work performed are minimal. The BMR is usually estimated by measuring a person's rate of oxygen consumption, which correlates with the rate at which nutrients are oxidized in the body.

The BMR represents the energy requirement of performing such necessary tasks as pumping blood and transporting ions. The BMR generally increases as body weight increases, because a larger mass of tissue requires greater energy expenditure for its upkeep; therefore, BMR is usually expressed as the rate of energy expenditure per unit body weight. For adults, the BMR averages 20–25 kilocalories per kilogram of body weight per day. Most of this expenditure is due to activity in the nervous system and skeletal muscles, which account for 40% and 20–30% of the BMR, respectively. The BMR varies from tissue to tissue; muscle tissue, for example, has a higher resting metabolic rate than adipose tissue. The BMR (per unit body weight) also varies with age: It is greater in growing children because of the energy expended in the synthesis of new tissue, and it is usually lower in the elderly than in young adults.

Negative and Positive Energy Balance

In today's society, virtually everyone is aware that connections exist among body weight, diet, and exercise. When people eat a lot of food but don't get much exercise, their body weight tends to increase; when they eat less and exercise more, their body weight decreases. These changes in body weight occur when energy input and output are not balanced.

If the body is not in energy balance—that is, if energy input and energy output are not equal—then the difference between energy input and output determines whether the amount of stored energy increases or decreases (is reduced through the catabolism of energy stores). From our balance equation, we can determine the amount of energy that is stored:

$$\text{energy stored} = \text{energy input} - \text{energy output}$$
$$= \text{energy input} - (\text{work performed} + \text{heat released})$$

When a person takes in energy at a rate greater than that at which it is expended as heat and work, the quantity of stored energy increases. This condition, *positive energy balance,* tends to be associated with increases in body weight; a net synthesis of macromolecules from absorbed nutrients occurs. (Later in this chapter we see that most excess nutrients are converted to lipids for storage.) When the rate of energy intake is less than the rate at which energy is expended as heat or work, the quantity of stored energy decreases. This condition, *negative energy balance,* tends to be associated with decreases in body weight. Under these conditions, a net breakdown of macromolecules (including lipid stores) provides energy for body functions.

When someone goes on a diet to lose weight, the idea is to decrease food intake and achieve negative energy balance. The same result can be attained through exercise, which increases work and heat production (energy output). When the increased energy output exceeds energy intake, the energy balance becomes negative.

Although the concept of energy balance is useful for explaining why diet and exercise affect body weight, note that a change in the body's energy content is not necessarily equivalent to a change in mass. If during a given time span 100 grams of glucose are absorbed and 100 grams of triglycerides are oxidized, then body weight does not change; however, the body's energy content decreases because 1 gram of glucose contains less energy than 1 gram of triglycerides.

> ### Quick Test 7.1
>
> 1. Define the terms *metabolic rate* and *basal metabolic rate.*
> 2. What are the storage forms of carbohydrates and lipids?
> 3. Once energy is taken into the body, it is either stored or it appears in what two other forms?

Energy Metabolism During the Absorptive and Postabsorptive States

We have seen that the maintenance of energy balance requires that energy input equals energy output. However, the body is generally not in energy balance at any given

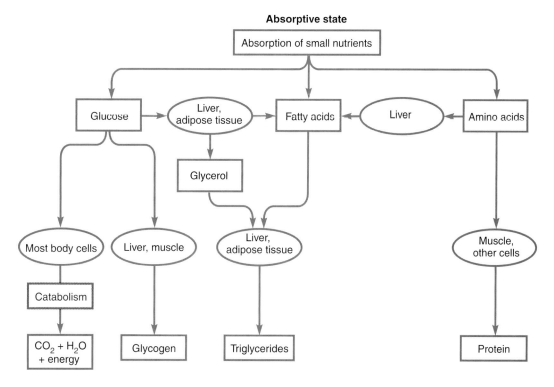

Figure 7.3 **Major metabolic reactions of the absorptive state.**

time, because the rate of energy input is determined by feeding, which is intermittent. For about 3–4 hours after a typical meal, nutrients are absorbed during the **absorptive state**, after which time absorption stops until the next meal. During this time the rate of energy input generally exceeds energy output, putting the body in positive energy balance. Nutrients in the blood are plentiful. Glucose serves as the primary energy source for cells, while fats, amino acids, and excess glucose are taken up by liver, muscle, and fat cells and converted to energy storage molecules. The **postabsorptive state** corresponds to the time between meals when nutrients are not being absorbed; during this time the rate of energy expenditure is greater than the rate of energy intake. Energy stores are mobilized to provide the energy cells need. Whereas glucose serves as the energy source for cells in the central nervous system, other cells in the body utilize other energy sources (such as fatty acids), thereby *sparing* glucose for the central nervous system.

Energy metabolism during the absorptive and postabsorptive states can be summarized by the following rule: *During the absorptive state, energy is stored in macromolecules; during the postabsorptive state, these energy stores are mobilized.* This rule is important because it means that despite our eating different amounts at different times, the body provides a constant supply of nutrients to cells—a supply that cannot be interrupted for even a minute because the body must expend energy continuously just to stay alive. In the following sections we see how the body performs this necessary function.

Metabolism During the Absorptive State

The absorptive state is primarily an anabolic state; that is, the majority of reactions involve synthesis of macromolecules. However, cellular metabolism differs among cell types. Next we look at the typical absorptive state metabolic responses in body cells in general, in skeletal muscle cells, in liver cells, and in adipocytes (**Figure 7.3**).

Body Cells in General

The body's energy needs are supplied primarily by absorbed glucose, which is plentiful after a typical meal. Glucose is taken into the cells and catabolized as the primary fuel. Absorbed fatty acids and amino acids can also be catabolized for energy, particularly if the diet is rich in these nutrients but poor in carbohydrates. Fatty acids undergo oxidation to provide acetyl CoA subunits for the Krebs cycle, and amino acids are converted to keto acids (organic acids with a carbonyl group, $C=O$), many of which are intermediates for the Krebs cycle. Amino acids can also be used to synthesize proteins. Note, however, that proteins are not synthesized as "storage molecules." Instead, most body proteins have important structural and functional roles in cells and are continuously turned over; that is, old proteins are degraded and replaced with new ones. For this reason, the body's protein mass is relatively stable and does not increase simply in response to the absorption of an excess of amino acids. Body cells catabolize proteins for energy only under extreme conditions, because they do so at the expense of losing functioning molecules.

Table 7.2 ▌ Energy Stores (As Percent of Total) in a Healthy 70-kg Male

	Glycogen	Triglycerides	Proteins (mobilizable)
Skeletal muscle	71	<1	98
Liver	24	<1	2
Adipose tissue	5	99	<1
Brain	<1	0	0

Skeletal Muscle Cells

Like body cells in general, skeletal muscle cells take up glucose and amino acids from the blood for their own needs. However, unlike most body cells, skeletal muscle cells can convert glucose to glycogen for storage. Within individual muscle cells, these glycogen stores are limited, but taken together they constitute the majority (approximately 70%) of the body's total stored glycogen.

Liver Cells

The liver converts nutrient molecules to energy stores that can subsequently be mobilized to supply energy to most cells in the body. The liver converts glucose to glycogen or fatty acids, and fatty acids to triglycerides. The glycogen is stored in the liver (which contains approximately 24% of the body's glycogen stores), whereas the triglycerides are transported to adipose tissue for storage. Between liver and skeletal muscle, the body can store only about 500 grams of glycogen, which is enough to meet the body's energy demands for only a few hours. Any absorbed glucose that exceeds the quantity that is needed for energy or that can be stored as glycogen is converted to fatty acids and then to triglycerides.

The liver also takes up amino acids. Although the liver uses some amino acids to synthesize proteins (including plasma proteins), most amino acids are converted to keto acids, many of which are intermediates of glycolysis or the Krebs cycle and can be used for energy. However, most of the keto acids are used to synthesize fatty acids, and therefore ultimately end up as triglycerides.

The triglycerides synthesized in the liver must be transported to adipose tissue, which is achieved by packaging the triglycerides into particles called **very low density lipoproteins (VLDLs)**. (See Discovery: Lipoproteins and Plasma Cholesterol, p. 668.) Briefly, VLDLs transport triglycerides to the cells of the body. The plasma membranes of most cells contain the enzyme lipoprotein lipase, which catabolizes triglycerides at the outer surface of the VLDLs to fatty acids and monoglycerides. The fatty acids then diffuse into cells, where they can be used for energy (by most body cells) or converted back into triglycerides for storage (in adipocytes). Adipocytes have a high concentration of lipoprotein lipase on their plasma

membranes and therefore take up most of the fatty acids transported in VLDLs.

Adipocytes

Adipocytes store energy in the form of triglycerides, or fat. Absorbed triglycerides are transported to adipocytes by chylomicrons, the smallest of the lipoproteins. Lipoprotein lipase catabolizes the triglycerides in chylomicrons in the same manner as just described for VLDLs. Excess absorbed glucose enters the adipocytes and is converted to triglycerides. In addition, triglycerides synthesized in the liver are transported to adipocytes by VLDLs for storage.

Energy Reserves

Whereas the body is limited in its ability to store energy in the form of glycogen or protein, it is practically unlimited in its ability to store energy as fats. Triglyceride synthesis thus represents the final common pathway for all nutrients that are absorbed in excess of the body's needs. In most people, fat accounts for 20–30% of body weight, but in very overweight individuals it can account for as much as 80%. To those who are weight conscious, the body's propensity for storing nutrients as fats is at best an annoyance. However, given that triglycerides contain roughly 9 kilocalories per gram, as opposed to 4 kilocalories per gram for carbohydrates and 5–6 kilocalories per gram for proteins, triglyceride synthesis is clearly the best way to store the most energy in the least weight.

Under normal circumstances, glycogen stores account for 1% or less of the body's total energy reserves and can supply energy needs for only a few hours of quiet activity. Proteins account for about 20–25% of the total energy reserves. Although large amounts of protein can be mobilized for energy without serious consequences, particularly from skeletal muscle, continual use of protein for energy is harmful and potentially fatal because it eventually compromises cellular function. Thus a significant portion of the energy contained in protein stores is, in reality, unavailable for use. Fat stores represent about 75–80% of the total energy reserves and contain enough energy to last for about two months. For this reason, fats are absolutely essential to the body's ability to withstand prolonged periods of fasting. **Table 7.2** tabulates the average energy stores of a 70-kilogram (kg) male.

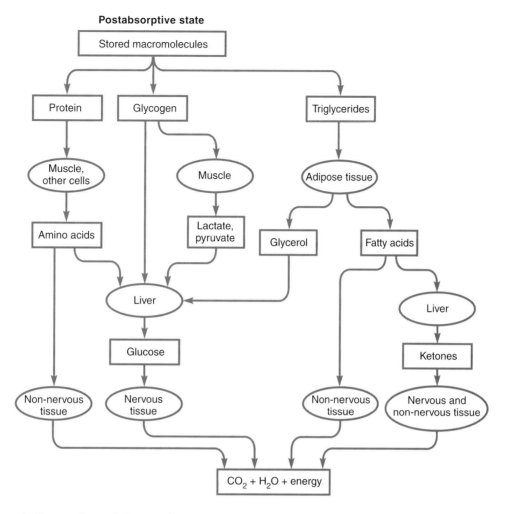

Figure 7.4 **Major metabolic reactions of the postabsorptive state.**

When the liver converts amino acids, lactate, or pyruvate to glucose, what type of reaction is occurring?

Metabolism During the Postabsorptive State

Within a few hours after a typical meal, absorption of nutrients ceases. During this time, the body cannot rely on absorbed nutrients for its energy needs, and so it is forced to catabolize glycogen, proteins, and fats for energy (**Figure 7.4**). Thus the postabsorptive state is primarily a catabolic state. In addition, unlike most body cells, central nervous system cells rely on glucose as their sole energy source (cells of the central nervous system can also obtain energy from ketone bodies during extreme conditions such as starvation). Therefore, a primary function of the postabsorptive state is to maintain plasma glucose levels. Too large a reduction in plasma glucose can result in serious impairment of brain function, loss of consciousness, and even death.

Given the importance of maintaining a steady supply of glucose during the postabsorptive state, a question naturally arises: Because glucose is derived from the breakdown of glycogen, which is in relatively short supply (enough to last only a few hours), how is glucose made available for longer periods? The body synthesizes new

glucose from amino acids, glycerol, and other breakdown products of catabolism, a process known as *gluconeogenesis*. Most tissues turn almost exclusively to other energy sources, primarily fatty acids, thereby conserving glucose for use by the central nervous system; this is called **glucose sparing**.

As in the absorptive state, cellular metabolism during the postabsorptive state differs among the types of cells.

Body Cells in General

Most cells utilize fatty acids for energy instead of glucose, sparing the glucose for the central nervous system.

Skeletal Muscle Cells

In a skeletal muscle cell, any glucose formed from glycogen during glycogenolysis can be used for energy only within that muscle cell. Glycogen is catabolized to glucose-6-P (a glucose molecule with a phosphate group attached to

Table 7.3 | Energy Metabolism During the Absorptive and Postabsorptive States

	Absorptive (anabolic) state	Postabsorptive (catabolic) state
Carbohydrates	In liver and skeletal muscle, glucose is converted to glycogen (glycogenesis) for storage. Liver also converts some glucose to triglycerides, which are transported to adipose tissue by lipoproteins. In adipose tissue, glucose is converted to triglycerides for storage. In body cells, glucose undergoes oxidation for energy.	In liver, glycogen is catabolized to glucose (glycogenolysis) and new glucose is formed from noncarbohydrate precursors (gluconeogenesis); glucose is then transported into the bloodstream. In muscle, glycogen is catabolized to glucose-6-P, which can be used by the muscle cell for energy.
Lipids	Triglycerides are synthesized from ingested fats, proteins, and glucose by adipocytes; triglycerides synthesized in the liver are transported to adipose tissue for storage.	In adipose tissue, triglycerides are broken down to fatty acids and to glycerol, which enters the bloodstream and travels to the liver, where it is converted to glucose (gluconeogenesis). Fatty acids enter the bloodstream and provide the primary energy source for most body cells. In liver, fatty acids are converted to ketones.
Amino acids	In all cell types, amino acids are used for protein synthesis. In liver, amino acids can be converted to keto acids for energy production or triglyceride synthesis. Triglycerides are then transported to adipose tissue by lipoproteins.	In muscle, proteins are catabolized to amino acids, which are transported to the liver for gluconeogenesis.

the sixth carbon; it is an intermediate in glycolysis). The phosphate group cannot be removed from the glucose because skeletal muscle cells lack the enzyme (glucose-6-phosphatase) that catalyzes its removal. For glucose to be transported out of a cell, it must be in its unphosphorylated form. Thus the glucose formed by glycogenolysis in skeletal muscle cells remains in the cell and is catabolized by glycolysis to form pyruvate or lactate. Any lactate produced then travels to the liver for further processing, as described shortly.

Skeletal muscle cells can also catabolize proteins to amino acids, which are then transported into the bloodstream and carried to the liver for further processing.

Liver Cells

The liver is the primary source of plasma glucose during the postabsorptive state. The liver has glycogen stores that can be broken down by glycogenolysis to glucose-6-P, and the liver also has glucose-6-phosphatase, which catalyzes conversion of glucose-6-P to glucose. Glucose can then be transported out of the liver cell and into the bloodstream. Therefore, liver glycogen stores, unlike skeletal muscle glycogen stores, can be mobilized to provide glucose to the blood.

The liver is also the primary site of gluconeogenesis. (Some gluconeogenesis also occurs in the kidneys.) Like the glucose produced by glycogenolysis, the newly synthesized glucose is transported from the liver into the bloodstream for use by other cells in the body.

During the postabsorptive state the liver converts some of the fatty acids entering it to ketone bodies, which are released into the bloodstream and eventually catabolized by most tissues. The production of ketones is important because during prolonged fasting, the central nervous system acquires the ability to use ketones for energy, thereby freeing it from some of its dependence on glucose.

Exercise Link

These metabolic reactions of the postabsorptive state were important for ensuring that Jane and Bill's blood glucose concentrations remained relatively stable through most of their marathon run. A portion of the lactate produced by their exercising muscles was converted by their livers (and kidneys) back into glucose via gluconeogenesis. It has been estimated that perhaps 10% of the glucose released into the blood during moderate-intensity exercise is derived from gluconeogenesis. Liver glycogen was also slowly broken down, providing about 15% of the total glucose consumed during the race. Ketone body production increased as well, but the contribution of these ketones to total energy consumption was small compared to glucose.

Adipocytes

In the postabsorptive state, adipose tissue supplies fatty acids to the bloodstream as energy sources for body cells, thereby sparing glucose for the central nervous system. Adipose tissue does this by catabolizing stored triglycerides into glycerol and free fatty acids. The glycerol is also released into the bloodstream, where it travels to the liver and is catabolized by glycolysis.

■ ■ ■

Table 7.3 summarizes the metabolism of energy molecules during the absorptive and postabsorptive states.

Quick Test 7.2

1. When glucose or amino acids are absorbed in excess of the quantities oxidized or stored as glycogen or proteins, what happens to them?

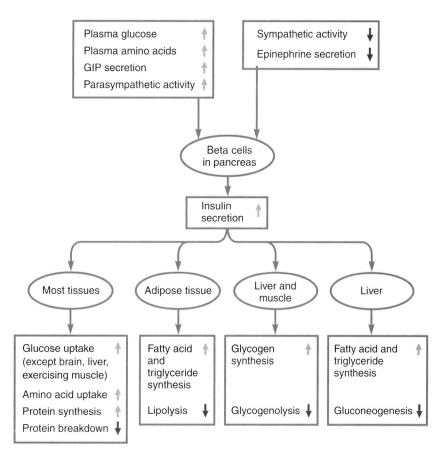

Figure 7.5 Factors affecting the secretion of insulin, and its actions on target tissues.

2. Where is most of the body's glycogen stored? What is the storage site of most of the glycogen that can supply glucose for cells throughout the body?

3. During the postabsorptive state, most tissues use fatty acids instead of glucose as their primary energy source. Why is this important in whole-body metabolism?

4. Name two processes that help the body maintain blood glucose at normal levels during fasting.

Regulation of Absorptive and Postabsorptive Metabolism

The transitions between the absorptive and postabsorptive states are marked by profound alterations in the metabolic activity of tissues throughout the body. In this section we see how these metabolic changes are triggered primarily by endocrine signals involving the pancreatic hormones insulin and glucagon. In addition, epinephrine and sympathetic nervous activity play a role.

The Role of Insulin

The metabolic adjustments that occur as the body switches between the postabsorptive and absorptive states are largely triggered by changes in the plasma concentration of **insulin**, a peptide hormone secreted by beta cells located in the pancreatic islets of Langerhans (see Chapter 6). Even though the actions of insulin and the factors that affect its secretion are numerous, they all have a common thread: *Insulin promotes the synthesis of energy storage molecules and other processes characteristic of the absorptive state* (**Figure 7.5**). In other words, insulin is an anabolic hormone. Accordingly, its secretion is stimulated by signals associated with feeding and the absorption of nutrients into the bloodstream.

Factors Affecting Insulin Secretion

During the absorptive period, insulin secretion by beta cells increases, causing an increase in the plasma concentration of insulin, which promotes many of the metabolic processes characteristic of the absorptive state. During the postabsorptive state, insulin secretion is decreased, causing a decrease in the plasma concentration of insulin, which helps turn off the absorptive processes. This raises a key question: How do beta cells know when to increase or decrease insulin secretion?

Figure 7.5 shows that insulin secretion is influenced by a variety of factors. Particularly important among these is plasma glucose concentration. During the absorptive period, plasma glucose levels increase as glucose is transported into the bloodstream from the gastrointestinal tract. This increase stimulates insulin secretion by a direct effect of glucose on beta cells, which are sensitive to the concentration of glucose in the fluid surrounding them. During the postabsorptive period, plasma glucose levels decrease, causing insulin secretion to fall. Insulin secretion is influenced in a similar fashion by the plasma amino acid concentration: Rising amino acid levels in plasma stimulate insulin secretion, whereas falling amino acid levels decrease insulin secretion.

Hormones and input from the autonomic nervous system also influence insulin secretion. Secretion is stimulated by parasympathetic nervous activity and by glucose-dependent insulinotropic peptide (GIP), a hormone secreted by cells in the wall of the gastrointestinal tract. This is significant because parasympathetic activity and secretion of GIP are both stimulated in response to the presence of food in the gastrointestinal tract. Because feeding occurs prior to the absorption of nutrients, these signals prepare the body for transitions to the absorptive state by triggering insulin secretion in advance. Insulin secretion is inhibited by sympathetic nervous activity and circulating epinephrine.

Actions of Insulin

Through its actions on a variety of target tissues, insulin influences almost every major aspect of energy metabolism (see Figure 7.5). It promotes energy storage by stimulating the synthesis of fatty acids and triglycerides in the liver and adipose tissue, glycogen in liver and skeletal muscle, and proteins in most tissues. At the same time, it opposes the catabolism of energy stores by inhibiting the breakdown of proteins, triglycerides, and glycogen, and by suppressing gluconeogenesis by the liver. In short, insulin promotes reactions of the absorptive state and suppresses reactions of the postabsorptive state.

Along with its metabolic actions, insulin affects the transport of nutrients across the membranes of all body cells except those in the liver and central nervous system. In most tissues, insulin stimulates the uptake of amino acids by cells, which facilitates the hormone's stimulatory effect on protein synthesis. Insulin also stimulates the uptake of glucose in many tissues by increasing the number of glucose transport proteins in cell plasma membranes. The particular protein affected is a type of glucose transporter called GLUT 4. Many cells have pools of GLUT 4 transporters stored in vesicles in the cytoplasm. Insulin can either trigger insertion of these *stored* transporters into the plasma membrane by exocytosis, or stimulate the synthesis of *new* transporters.

The transport of glucose in the central nervous system and the liver is not affected by insulin. This is critical, because when insulin levels are low during the postabsorptive state, glucose uptake by most cells is decreased, sparing the glucose for use by cells of the central nervous system, where glucose transport is not affected by insulin.

Insulin also has little effect on the glucose permeability of cells in exercising muscle because exercise triggers an increase in the number of glucose transporters by a mechanism that operates independently of insulin. In

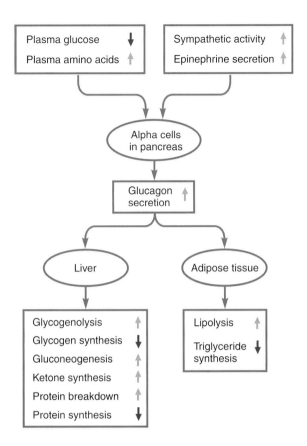

Figure 7.6 Factors affecting the secretion of glucagon, and its actions on target tissues.

resting muscle, however, insulin stimulates increased glucose transport in the normal manner.

In addition to the actions shown in Figure 7.5, insulin also has important growth-promoting effects. Although insulin by itself does not stimulate growth, it must be present in the blood for growth hormone to exert its normal effects, a form of permissiveness discussed in Chapter 6. This need results, at least in part, from insulin's role in promoting protein synthesis, DNA synthesis, and cell division, all of which are essential to tissue growth.

The Role of Glucagon

Insulin's actions to bring about the body's metabolic adaptations to the absorptive and postabsorptive states are reinforced by contrary changes in **glucagon**, a peptide hormone secreted by alpha cells of pancreatic islets of Langerhans. Put another way, insulin and glucagon are antagonists, hormones whose actions oppose each other: Insulin promotes processes of the absorptive state; glucagon promotes processes of the postabsorptive state.

Glucagon secretion decreases during the absorptive state and increases during the postabsorptive state. Because insulin levels are also changing with these states, the metabolic adjustments from one state to the other are orchestrated by contrary changes in plasma levels of insulin and glucagon.

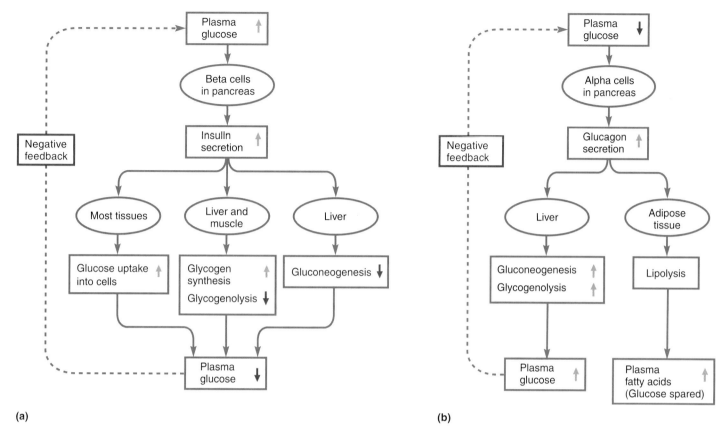

Figure 7.7 Regulation of plasma glucose concentration. (a) Negative feedback control of plasma glucose by insulin. (b) Negative feedback control of plasma glucose by glucagon.

Factors Affecting Glucagon Secretion

Most of the signals that stimulate the secretion of glucagon (**Figure 7.6**) are the same signals that inhibit the secretion of insulin. Decreases in blood glucose both stimulate glucagon secretion and inhibit insulin secretion. Glucagon secretion is also stimulated by sympathetic nervous activity and epinephrine, which have a suppressive effect on insulin secretion. Some studies suggest that glucagon and insulin function as paracrines in the islets of Langerhans, with insulin inhibiting the secretion of glucagon from alpha cells, and glucagon inhibiting the secretion of insulin from beta cells. Because of the opposite controls of these hormones, plasma glucagon levels tend to rise as insulin levels fall, and vice versa.

Actions of Glucagon

Figure 7.6 also shows that the actions of glucagon, though more limited than those of insulin, oppose insulin's actions. In the liver, glucagon promotes glycogenolysis and gluconeogenesis (which increase blood glucose levels), ketone synthesis, and breakdown of proteins, while inhibiting the opposing processes of glycogen and protein synthesis. In adipose tissue, glucagon stimulates lipolysis and suppresses triglyceride synthesis. These actions classify glucagon as a catabolic hormone. The overall effect of glucagon promotes mobilization of

energy stores and synthesis of "new" energy sources (glucose and ketone bodies) that can be used by tissues; all of these actions are characteristic of the postabsorptive state.

Negative Feedback Control of Blood Glucose Levels by Insulin and Glucagon

Plasma glucose levels are normally tightly regulated by the antagonistic actions of insulin and glucagon to maintain stability. (Other hormones also regulate plasma glucose levels.) This stability is important because deviations too far from normal in either direction can have serious adverse effects on health. The normal fasting level of glucose in the blood is 70–110 mg/dL (clinical measures are usually of blood, not plasma, glucose levels). Fasting blood glucose levels greater than 140 mg/dL constitute *hyperglycemia,* which is often indicative of **diabetes mellitus**, a serious and fairly common disease involving defects in insulin production or signaling (**Clinical Connections: Diabetes Mellitus**, p. 192). Fasting blood glucose levels below 60 mg/dL constitute *hypoglycemia,* which has widespread deleterious effects on nervous system function because the nervous system uses glucose almost exclusively as its source of energy.

Insulin and glucagon together control plasma glucose concentration through negative feedback (**Figure 7.7**). An

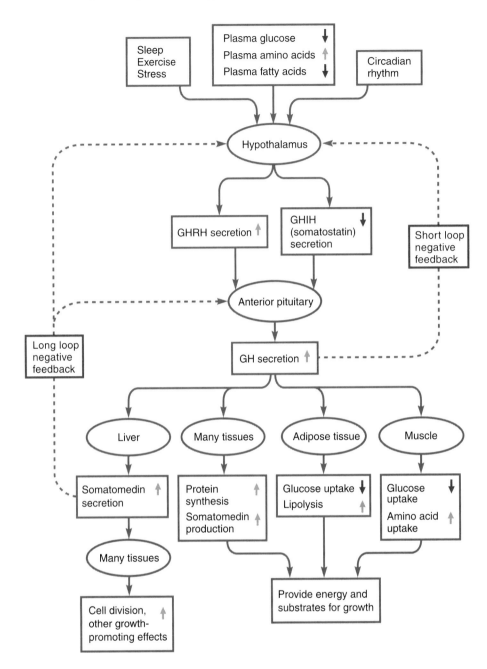

Figure 7.10 **Factors affecting the secretion of growth hormone, and its actions on target tissues.**

Factors Affecting Growth Hormone Secretion

Secretion of growth hormone by the anterior pituitary is regulated by two hypothalamic hormones: **growth hormone releasing hormone (GHRH)**, which stimulates growth hormone secretion, and **growth hormone inhibiting hormone (GHIH)** or **somatostatin**, which inhibits growth hormone secretion (see Figure 7.10). Although the relative importance of these two hormones is unclear, variations in GH secretion are likely triggered primarily by GHRH, with GHIH playing a relatively minor role. Like other anterior pituitary hormones, growth hormone secretion is regulated through negative feedback loops (see Chapter 6). GH limits its own secretion via short loop negative feedback on the hypothalamus. Plasma somatomedins also exert long loop negative feedback controls on the hypothalamus and anterior pituitary to inhibit GHRH and GH secretion, respectively.

GHRH secretion is regulated by neural input of various types to the hypothalamus (see Figure 7.10). Secretion is affected by plasma nutrient concentrations; decreases in plasma glucose or fatty acid levels, or increases in plasma amino acid levels, stimulate GHRH secretion. Because it promotes changes in the opposite direction (by reducing glucose uptake and increasing lipolysis and amino acid uptake), growth hormone acts by negative feedback to limit variations in these nutrient concentrations in plasma. Growth hormone secretion is also stimulated in response to exercise, stress, or sleep. A boost in GH during exercise or stress is useful because it tends to counteract reduced plasma levels of glucose and fatty acids, thereby helping to maintain a steady supply of these much-needed energy sources. The significance of heightened GH secretion during sleep is not understood. Growth hormone secretion is also subject to a circadian rhythm that is mediated by neural input to the hypothalamus. Experiments have shown that GH release exhibits a regular pattern, increasing at night and falling during the day. (Secretion reaches its peak about 1–2 hours after the onset of sleep.)

Average daily plasma levels of growth hormone also vary with a person's age. GH levels generally reach a maximum during puberty and then decline with age. Decreased GH levels are thought to be at least partially responsible for some signs of aging, such as decreased muscle mass and increased body fat.

Bone Growth

Because stimulation of bone growth is an important part of growth hormone's actions, it is appropriate to consider the nature of bone in some detail here. Bone is also an important reserve for calcium, which can be liberated and moved into the bloodstream when plasma calcium levels decrease.

To support the weight of the body and to withstand forces placed on it by contracting muscles, bone must be strong but not brittle. Crystals of calcium phosphate in a form known as *hydroxyapatite* $[Ca_{10}(PO_4)_6(OH)_2]$ give bone a mineral component that helps it withstand compressive forces (that is, "squeezing" or "crushing" forces). **Osteoid**, an organic component that consists of collagen fibers embedded in a gel-like substance, gives bone its ability to withstand tensile or stretching forces, making it less prone to fracture.

Despite its nonliving organic and mineral components, bone is a dynamic living tissue that contains cells. The dynamic nature of bone is evident not only in its ability to grow during childhood but also in the fact that bone can heal following a fracture and adapt its structure in response to forces placed on it. In a person who engages regularly in heavy lifting, for instance, the weight-bearing bones gradually increase in thickness and strength. In a person who is sedentary or bedridden, bone mass diminishes over time. Such restructuring of bone is called *remodeling*, and it is critical to the body's ability to regulate plasma calcium levels.

Central to the remodeling of bone are mobile cells known as osteoblasts and osteoclasts, both of which are found on the outer surfaces and inner cavities of bone tissue (**Figure 7.11**). **Osteoblasts** or "bone makers" are responsible for building up the mass of bone tissue, a

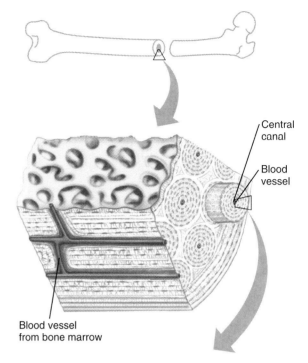

Central canal

Blood vessel

Blood vessel from bone marrow

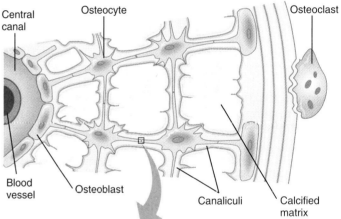

Central canal

Osteocyte

Osteoclast

Blood vessel

Osteoblast

Canaliculi

Calcified matrix

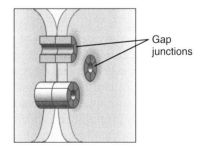

Gap junctions

Figure 7.11 The structure of bone.

Which type of cell breaks down bone?

Osteoclast

CHAPTER SUMMARY

An Overview of Whole-Body Metabolism, p. 180

Whole-body metabolism requires the coordination of cellular metabolic activities. Cells use energy in the form of ATP, which they obtain from the oxidation of small nutrient molecules such as glucose, fatty acids, and amino acids. Cellular metabolism must be coordinated so that nutrients are provided to the appropriate cells when needed.

Energy Intake, Utilization, and Storage, p. 181

Energy is released in cells by the breakdown of nutrients into smaller molecules, as when glucose, amino acids, or fatty acids are oxidized to yield waste products. Energy mobilization is the breakdown of macromolecules into small nutrient molecules that are released into the bloodstream. Energy is stored by converting small nutrient molecules into macromolecules: Glucose is stored as glycogen in skeletal muscle and liver, fatty acids and glycerol are stored as triglycerides in adipose tissue, and amino acids are stored as proteins in all cells, but especially in skeletal muscle cells.

Energy Balance, p. 183

To maintain energy balance, energy input must equal energy output. Energy input comes from ingested nutrients, whereas energy output is the energy expended as heat or work. Positive energy balance occurs when energy input exceeds energy output; negative energy balance occurs when energy output exceeds energy input. The body's metabolic rate is the total amount of energy released per unit time as a result of nutrient oxidation. The metabolic rate at rest is the basal metabolic rate, or BMR.

Energy Metabolism During the Absorptive and Postabsorptive States, p. 184

In the absorptive state, glucose is used by most tissues as the primary fuel. Absorbed nutrients are also converted to glycogen, triglycerides, and proteins. Excess amino acids and glucose are mostly converted to fatty acids and stored as triglycerides. In the postabsorptive state, stored glycogen, triglycerides, and proteins are catabolized for energy. Fatty acids are used by most tissues as the primary fuel. An exception is the nervous system, which relies on a steady supply of glucose for its energy. Utilization of nonglucose fuels conserves glucose for use by the nervous system, a phenomenon called glucose sparing. The liver can also produce more glucose by gluconeogenesis.

Regulation of Absorptive and Postabsorptive Metabolism, p. 189

Metabolic adjustments to the absorptive state are promoted by insulin and include synthesis of energy stores (glycogen, proteins, fatty acids, and triglycerides) and uptake of glucose and amino acids by cells in many tissues. Insulin also suppresses gluconeogenesis and regulates plasma glucose levels via negative feedback control.

Metabolic adjustments to the postabsorptive state are promoted by glucagon and include glycogenolysis, protein breakdown by the liver, lipolysis, gluconeogenesis, and ketone synthesis. Glucagon also helps to regulate blood glucose levels. Postabsorptive metabolic adjustments are also promoted by increased epinephrine secretion and sympathetic nervous activity.

> **IP** Endocrine, Actions of Hormones on Target Cells, pp. 5–6

Hormonal Regulation of Growth, p. 194

Body growth during childhood is promoted by the actions of growth hormone, which is secreted by the anterior pituitary and acts to promote the growth of soft tissues and bones. In adulthood, growth hormone acts to maintain bone mass and lean body mass. Actions promoted by growth hormone include hypertrophy, hyperplasia, protein synthesis, lipolysis, gluconeogenesis, and amino acid uptake by cells. Growth hormone also inhibits glucose uptake by adipose tissue and muscle. Combined metabolic actions work to raise plasma levels of glucose, fatty acids and glycerol, thereby making energy more readily available to growing tissues. Many of growth hormone's actions are mediated by somatomedins synthesized by the liver and other tissues.

Thyroid Hormones, p. 200

Thyroid hormones are normally secreted by the thyroid gland at near-constant rates and increase the metabolic rate in most tissues of the body. At high concentrations, thyroid hormones mobilize energy stores. Thyroid hormones are also necessary for normal growth, development, and maintenance of normal function in many tissues, particularly the nervous system. Thyroid hormones are secreted in two forms, T_3 and T_4. T_4 is the more abundant form and T_3 is the more active one.

> **IP** Endocrine, The Hypothalamic-Pituitary Axis, pp. 6–7

Glucocorticoids, p. 202

Glucocorticoids are released by the adrenal cortex and are important in the body's response to stress. Glucocorticoids are also required for the body's ability to mobilize energy stores during postabsorptive periods.

> **IP** Endocrine, Response to Stress, p. 140

▌ EXERCISES

Multiple-Choice Questions

1. Which of the following is an example of a permissive effect of a hormone?
   a) the effect of thyroid hormones on growth
   b) the effect of insulin on glucose uptake by cells
   c) the effect of sex hormones on the secretion of growth hormone
   d) a and c
   e) all of the above

2. Which of the following is an example of a glucose-sparing effect of cortisol?
   a) inhibition of ACTH release
   b) stimulation of gluconeogenesis by the liver
   c) stimulation of lipolysis
   d) stimulation of glycogen breakdown
   e) stimulation of protein synthesis

3. Which of the following cells of the pancreas secrete insulin?
   a) alpha cells
   b) beta cells
   c) delta cells
   d) exocrine cells
   e) duct cells

4. Stress stimulates secretion of which of the following hormones?
   a) growth hormone
   b) epinephrine
   c) thyroid hormones
   d) ACTH
   e) all of the above

5. Hypoglycemia inhibits secretion of which of the following?
   a) growth hormone
   b) insulin
   c) epinephrine
   d) glucagon
   e) all of the above

6. In the postabsorptive state, the central nervous system uses which of the following as its *primary* source of energy?
   a) fatty acids
   b) amino acids
   c) glucose
   d) glycerol
   e) ketones

7. Which of the following cell types is directly responsible for building new bone material?
   a) osteoblasts
   b) osteoclasts
   c) osteocytes
   d) chondrocytes

8. Which of the following is true of adulthood?
   a) Growth hormone exerts no effects on body tissues.
   b) The secretion of growth hormone ceases altogether.
   c) Growth hormone cannot stimulate increases in the length of long bones.
   d) The structure of bone becomes permanently fixed.
   e) none of the above

9. Which form of thyroid hormone has greater activity at target cells?
   a) T_3
   b) T_4
   c) neither; T_3 and T_4 have equal activity

10. Which of the following hormones is a steroid?
    a) thyroid hormones
    b) insulin
    c) glucagon
    d) growth hormone
    e) cortisol

Objective Questions

1. Energy mobilization is promoted by (insulin/glucagon).

2. Secretion of (insulin/glucagon) is increased during the absorptive period.

3. Insulin and glucagon both help regulate the plasma glucose concentration. (true/false)

4. Breakdown of triglycerides yields fatty acids and _____, which can be used by cells for energy.

5. Conversion of amino acids to fatty acids is more likely to occur in the (absorptive/postabsorptive) state.

6. Conversion of amino acids to glucose is more likely to occur in the (absorptive/postabsorptive) state.

7. An increase in plasma thyroid hormone levels tends to make the body's energy balance more (positive/negative).

8. Energy that is taken into the body is either stored or appears as work or _____.

9. Stress tends to (stimulate/inhibit) GHRH secretion.

10. Many of growth hormone's effects are due to the action on target tissues of other chemical messengers called _____.

11. Closure of the epiphyseal plates is promoted by (growth hormone/sex hormones).

12. Thyroid hormones promote increased responsiveness of target tissues to (sympathetic/parasympathetic) nerve activity.

13. Glucocorticoids promote (increased/decreased) plasma glucose levels.

14. Stimulation of gluconeogenesis by glucagon is an example of a glucose-sparing effect. (true/false)

15. Plasma glucocorticoids have a(n) (stimulatory/inhibitory) effect on the secretion of ACTH.

Essay Questions

1. Describe the regulation of plasma glucose by insulin and glucagon. Include a description of the role of negative feedback.

2. Describe how insulin, glucagon, and the sympathetic nervous system work together to maintain adequate plasma glucose levels during fasting. Why is this important?

3. Describe the various factors that determine the body's energy balance. Be sure to describe what happens to energy that is liberated as a result of fuel oxidation.

4. Describe the similarities between the metabolic actions of thyroid hormones and glucocorticoids.

5. Describe the metabolic actions of growth hormone, and explain how these actions promote growth.

Critical Thinking

1. Some athletes have used anabolic steroids in the hopes of enhancing athletic performance. What type of steroids would they use? Based on what you read in this chapter, what could be some of the deleterious effects of taking such steroids?

2. Many hormone deficiencies are genetic in origin. Explain how a steroid hormone deficiency can be caused by a mutant gene.

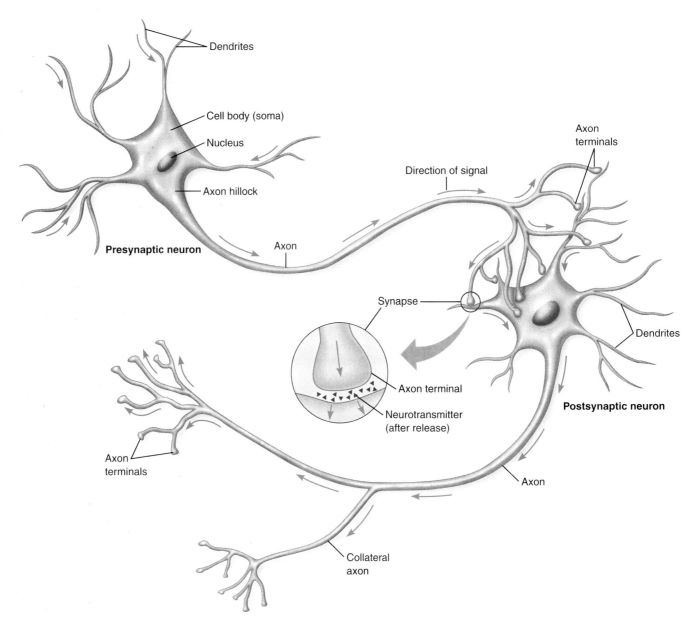

Figure 8.2 Structure of a typical neuron. Two neurons are shown; the upper neuron communicates with the lower neuron, as indicated by the arrows representing information flow. The main parts of a neuron include the cell body (also known as the soma); dendrites, which receive communication from other neurons; and an axon, which is specialized for transmitting electrical impulses. The axon terminals of one neuron release a chemical messenger (neurotransmitter) that communicates with another neuron.

changes the permeability of the plasma membrane for a specific ion, resulting in a change in the electrical properties of the cell or the release of a neurotransmitter. **Leak channels** (or nongated channels), which are found in the plasma membrane throughout a neuron, are always open and are responsible for the resting membrane potential.

Ligand-gated channels open or close in response to the binding of a chemical messenger to a specific receptor in the plasma membrane. In neurons, ligand-gated chan-

nels are most densely located in the dendrites and cell body, areas that receive communication from presynaptic neurons in the form of neurotransmitters.

Voltage-gated channels open or close in response to changes in membrane potential. Voltage-gated sodium channels and voltage-gated potassium channels are located throughout the neuron, but more densely in the axon and in greatest density in the axon hillock. These channels are necessary for the initiation and propagation of action potentials. Voltage-gated calcium channels are

■■ **DISCOVERY**

NEUROGENESIS

For decades neuroscientists believed that the adult human brain could not produce new neurons. Then, in the 1970s, Pasquale Graziadei and colleagues discovered that new olfactory receptor cells (sensory receptor cells for smell) were produced from basal cells in the olfactory epithelium of the nasal cavity. In the 1990s, other researchers have found evidence of *neurogenesis,* the production of new neurons, in the central nervous system.

Neurogenesis in lower animal species has been evident for over 30 years. Elizabeth Gould and her colleagues from Princeton have been studying neurogenesis in primates in recent years. In some of these studies, bromodeoxyuridine (BrdU) was administered to the animals as a marker for cell proliferation. Before cells can divide, they must replicate new DNA. BrdU is incorporated into newly synthesized DNA and therefore marks any DNA formed after its administration. Cells that contain BrdU in their DNA are identified postmortem by histological analysis. Using this technique and others, Gould and her colleagues discovered that neurogenesis occurs in several areas of the brain, including the hippocampus, prefrontal cortex, inferior temporal cortex, and parietal cortex of nonhuman primates.

Fred Gage and his colleagues from the Salk Institute for Biological Studies have studied neurogenesis in the hippocampus in lower animal species, and have recently provided evidence that neurogenesis occurs in the human hippocampus as well. Their studies in humans used BrdU in the same manner as Gould's study in lower primates. The human subjects were cancer patients of Peter Eriksson from Sweden, who was treating the patients with BrdU to measure tumor cell proliferation. These patients agreed to donate their brains to Gage's research group upon their deaths. Postmortem analyses of the

brains showed that BrdU was incorporated into the DNA of neurons in the hippocampus. These findings are the first evidence of neurogenesis in human brains.

The discovery of neurogenesis in human brains is of profound clinical significance. Many neurological diseases, including Parkinson's disease, Alzheimer's disease, and stroke, involve the loss of functioning neurons. If scientists can learn the mysteries of neurogenesis, they may be able to develop techniques to help the central nervous system replace lost neurons.

Fluorescent staining of hippocampal neurons.

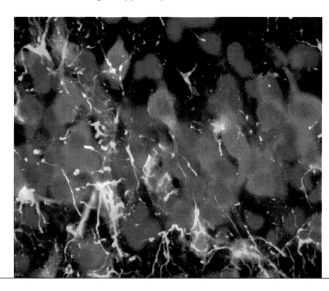

---

located most densely in the axon terminals; these channels open in response to the arrival of an action potential at the axon terminals. When these channels are open, calcium enters the cytosol of the axon terminals and triggers the release of neurotransmitter.

Structural Classification of Neurons

Neurons can be classified structurally according to the number of processes (axons and dendrites) that project from the cell body (**Figure 8.3**). Neurons commonly found in humans include bipolar neurons and multipolar neurons. **Bipolar neurons** are generally sensory neurons with two projections, an axon and a dendrite, coming off the cell body. Typical bipolar neurons (Figure 8.3a)

function in the senses of olfaction (smell) and vision. Most sensory neurons, however, are a subclass of bipolar neurons called *pseudo-unipolar* neurons (Figure 8.3b) because the axon and dendrite projections *appear* as a single process that extends in two directions from the cell body. In actuality, the dendrite is modified to function much like an axon, and therefore is a functional continuation of the axon. The modified dendritic process is called the *peripheral axon* because it originates in the periphery with sensory receptors and functions as an axon in that it transmits action potentials. The axon process is called the *central axon* because it ends in the central nervous system, where it forms synapses with other neurons. **Multipolar neurons**, the most common neurons, have multiple

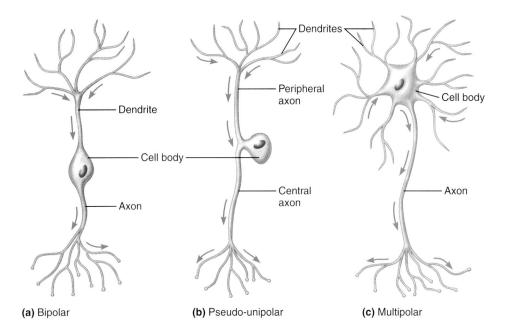

Figure 8.3 Structural classes of neurons. (a) A bipolar neuron. Afferent neurons associated with vision and olfaction are bipolar neurons. (b) A pseudo-unipolar neuron. The vast majority of afferent neurons are pseudo-unipolar. (c) A multipolar neuron. Most neurons are multipolar neurons.

(a) Bipolar   **(b)** Pseudo-unipolar   **(c)** Multipolar

projections from the cell body (Figure 8.3c); one projection is an axon, all the others are dendrites.

Functional Classification of Neurons

There are three functional classes of neurons: efferent neurons, afferent neurons, and interneurons (**Figure 8.4**). **Efferent neurons** transmit information from the central nervous system to effector organs. Recall that efferent neurons include the motor neurons extending to skeletal muscle and neurons of the autonomic nervous system (see Figure 8.1). Notice in Figure 8.4 that the cell body and dendrites of efferent neurons are located in the central nervous system (autonomic postganglionic neurons, described in Chapter 12, are an exception). However, the axon leaves the central nervous system and becomes part of the peripheral nervous system as it travels to the effector organ it innervates.

The function of **afferent neurons** is to transmit either sensory information from *sensory receptors* (which detect information pertaining to the outside environment) or visceral information from *visceral receptors* (which detect information pertaining to conditions in the interior of the body) to the central nervous system for further processing. Most afferent neurons are pseudo-unipolar neurons, with the cell body located outside the central nervous system in a *ganglion* (the general term for a cluster of neural cell bodies located outside the CNS). The endings of the peripheral axon are located in the peripheral organ (sensory organ or visceral organ), where they are either modified into sensory receptors or receive communication from separate sensory receptor cells. The central axon terminates in the central nervous system, where it releases a neurotransmitter to communicate with other neurons.

The third functional class of neurons is **interneurons**, which account for 99% of all neurons in the body. They are located entirely in the central nervous system. Interneurons perform all the functions of the central nervous system, including processing sensory information from afferent neurons, sending out commands to effector organs through efferent neurons, and carrying out complex functions of the brain such as thought, memory, and emotions.

Structural Organization of Neurons in the Nervous System

Neurons are arranged within the nervous system in an orderly fashion, with those having similar functions tending to be grouped together. In addition, neurons are aligned in such a way that cell bodies and dendrites of adjacent cells tend to be grouped together, and axons of adjacent cells tend to be grouped together. In the central nervous system, cell bodies of neurons are often grouped into **nuclei** (singular: *nucleus*), and the axons travel together in bundles called **pathways**, **tracts**, or **commissures**. In the peripheral nervous system, cell bodies of neurons are clustered together in **ganglia** (singular: *ganglion*), and the axons travel together in bundles called **nerves**.

Glial Cells

Glial cells, the second class of cell found in the nervous system, account for 90% of all cells in the nervous system. They do not function directly in signal transmission; instead, they provide structural integrity to the nervous system (*glia* is Latin for "glue"), and they are necessary for neurons to carry out their functions.

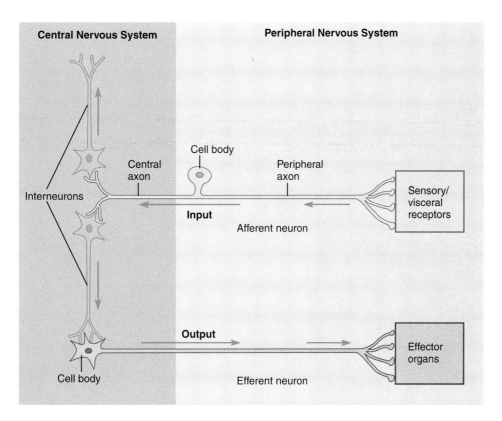

Figure 8.4 Functional classes of neurons. Afferent neurons originate in the periphery with sensory or visceral receptors. The peripheral axons of afferent neurons are part of the peripheral nervous system, but the axon terminals are located in the central nervous system, where they communicate with other neurons. Efferent neurons originate in the central nervous system, where the cell body and dendrites receive synaptic communication from other neurons. Efferent axons, however, are part of the peripheral nervous system and terminate at a synapse with an effector organ. Interneurons lie entirely in the central nervous system and can communicate with afferent neurons, efferent neurons, or other interneurons.

▌ *Name two types of effector organs.*

There are five types of glial cells: *astrocytes, ependymal cells, microglia, oligodendrocytes,* and *Schwann cells.* Of these glial cells, only Schwann cells are located in the peripheral nervous system; the rest are in the central nervous system. The functions of all the glial cells are covered in Chapter 10. Because one of the functions of **oligodendrocytes** and **Schwann cells** (or neurolemmocytes) is crucial to electrical transmission in neurons, we discuss these cells here.

The primary function of oligodendrocytes and Schwann cells is to form an insulating wrap of **myelin** around the axons of neurons. Such insulation enables neurons to transmit action potentials more efficiently and rapidly. **Figure 8.5**a shows the formation of a myelin sheath by a Schwann cell. Myelin consists of concentric layers of the plasma membranes of either oligodendrocytes or Schwann cells. Oligodendrocytes form myelin around axons in the central nervous system; one oligodendrocyte sends out projections providing the myelin

segments for many axons (Figure 8.5b). Schwann cells form myelin around axons in the peripheral nervous system, but each Schwann cell provides myelin for only one axon (Figure 8.5c). Many oligodendrocytes or Schwann cells are needed to provide the myelin for a single axon. Because the lipid bilayer of a plasma membrane has low permeability to ions, the several layers of membrane that make up a myelin sheath substantially reduce leakage of ions across the cell membrane. However, in gaps in the myelin, called **nodes of Ranvier** (or simply nodes), the axonal membrane contains voltage-gated sodium and potassium channels that function in the transmission of action potentials by allowing ion movement across the membrane. We discuss the nature of these electrical signals and how they originate in the next few sections of this chapter.

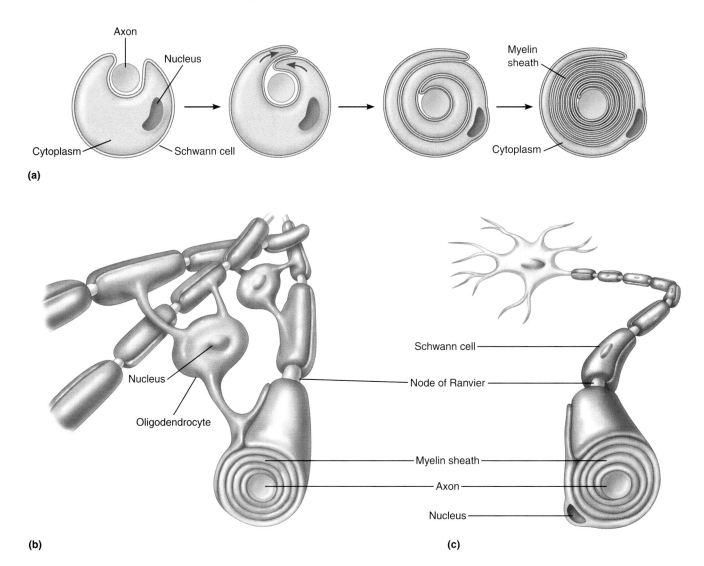

Figure 8.5 Formation and origins of myelin sheaths. (a) Formation of a myelin sheath by a Schwann cell. Myelin, which consists of concentric layers of plasma membrane provided by either a Schwann cell or an oligodendrocyte, forms a layer of insulation around an axon. (b) Arrangement of myelin sheaths formed by oligodendrocytes in the central nervous system. A single oligodendrocyte sends out cytoplasmic processes that form myelin sheaths around several axons. Note the nodes of Ranvier, gaps in the myelin sheaths. (c) Arrangement of myelin sheaths formed by Schwann cells in the peripheral nervous system. A given Schwann cell sheathes only a single axon.

Quick Test 8.1

1. Name the different parts, divisions, and branches of the nervous system and give the basic functions of each.

2. Draw a neuron and label the following structures: cell body, dendrite, axon, and axon terminal. Briefly state the function of these structures, and the type of ion channels (ligand-gated or voltage-gated) that can be found in each.

3. Which glial cell forms myelin in the central nervous system? Which forms myelin in the peripheral nervous system?

Establishment of the Resting Membrane Potential

We learned about the membrane potential in Chapter 4. In this section, we look at the establishment of the baseline, or resting, membrane potential. To further your understanding of electrical forces, see the **Toolbox: Electrical Circuits in Biology**.

Recall from Figure 4.2 that a cell at rest has a potential difference across its membrane such that the inside of the cell is negatively charged relative to the outside. This difference is called the **resting membrane potential** (resting V_m) because the cell is at rest—is not receiving or

TOOLBOX

ELECTRICAL CIRCUITS IN BIOLOGY

Neurons and muscle cells form excitable tissues, meaning that they function through electrical signals in the form of changes in the membrane potential. To better understand these signals, we must first understand the relationship between an electrical potential, current, and resistance.

Electrical forces exist between charged particles. The direction of the electrical force follows a simple rule: Opposite charges attract each other; like charges repel. To separate opposite charges requires energy. When opposite charges are separated they store the energy as an *electrical potential* (or voltage). The strength of the potential depends on the amount of charge separation: the greater the separation of charge, the greater the potential.

Electrical potentials are produced in biological systems by separating oppositely charged ions, which are attracted to each other and will move toward each other if possible. When ions move, they carry their charge with them. The movement of electrical charges is called **current** (I). In biological systems, currents are typically expressed in units of microamps (10^{-6} amperes). The greater the electrical potential, the greater the *force* (voltage) for ion movement, or current. However, force is not the sole determinant of whether ions indeed move.

How easily ions can move depends on the properties of the substance through which they must move. **Resistance** (R) is a measurement of the hindrance to charge movement. The greater

a substance's resistance, the more difficult it will be for ions to move through it, and the weaker the current will be. A neuron's plasma membrane has high resistance to current flow because its permeability to ions is low. The intracellular and extracellular fluids, by contrast, have low resistance to current flow because these fluids are rich in ions.

The inverse of resistance is **conductance** (g):

$$g = 1/R$$

Because the ability of an ion to cross a plasma membrane depends on the permeability of the plasma membrane to that ion, conductance of a particular ion increases as the membrane's permeability to that ion increases.

The relationship between potential difference, current, and resistance is defined by **Ohm's law**:

$$I = E/R$$

where E is the potential difference or voltage. Understanding Ohm's law is crucial to understanding neural physiology because plasma membranes have an electrical potential across them, ions present inside and outside the cell are available to carry charge across the plasma membrane, and resistance to charge movement can be changed by the opening or closing of ion channels.

---

transmitting any signals. The resting membrane potential of neurons is approximately −70 mV. Membrane potentials are *always* described as the potential inside the cell relative to outside. Therefore, the inside of a typical neuron at rest is 70 mV more negative compared to the outside. (Although we are discussing the resting membrane potential of neurons, all cells in the body have a negative membrane potential, ranging from −5 mV to −100 mV.)

Neurons communicate by generating electrical signals in the form of changes in membrane potential. Some of these changes in membrane potential trigger the release of neurotransmitter, which then carries a signal to another cell. **Table 8.1** defines the different types of electrical potentials that are described in this and subsequent chapters. In the following sections we explore (1) what is responsible for the existence of the resting membrane potential and (2) what causes the membrane potential to change.

Determining the Equilibrium Potentials for Potassium and Sodium Ions

The resting membrane potential depends on two critical factors: (1) the concentration gradients of ions (particu-

larly sodium ions and potassium ions) across the plasma membrane and (2) the presence of ion channels in the plasma membrane. Recall from Chapter 4 that the Na^+/K^+ pump creates concentration gradients for sodium and potassium ions by transporting three sodium ions out of the cell and two potassium ions into the cell per ATP hydrolyzed. Sodium ions are more highly concentrated outside the cell, and thus there is a chemical driving force tending to push sodium ions into the cell. Potassium ions are more highly concentrated inside the cell, and thus there is a chemical driving force tending to push potassium ions out of the cell. As we see shortly, the chemical forces for moving sodium and potassium ions across the plasma membrane, and the differences in the permeability of the plasma membrane to these two ions, establish the resting membrane potential.

To understand what causes the resting membrane potential, consider two hypothetical cells that are identical in all respects except for the permeability of their membranes to ions. Cell 1 is permeable to potassium ions only (that is, it has open potassium channels in its plasma membrane), whereas cell 2 is permeable to sodium ions only (that is, it has open sodium channels in its plasma membrane).

Table 8.1 | Types of Electrical Potentials in Biological Systems

Potential	Definition
Potential difference = E	Difference in voltage between two points
Membrane potential = V_m	Difference in voltage across the plasma membrane; always given in terms of voltage inside the cell relative to voltage outside the cell
Resting V_m	Difference in voltage across the plasma membrane when a cell is at rest (not receiving or sending signals)
Graded potential	A relatively small change in membrane potential produced by some type of stimulus that triggers the opening or closing of ion channels; strength of graded potential is relative to strength of stimulus
Synaptic potential	Graded potentials produced in the post-synaptic cell in response to neurotransmitters binding to receptors
Receptor potential	Graded potentials produced in response to a stimulus acting on a sensory receptor
Action potential	A large, rapid change in membrane potential produced by depolarization of an excitable cell's plasma membrane to threshold
Equilibrium potential	The membrane potential that counters the chemical forces acting to move an ion across the membrane, thereby putting the ion at equilibrium

Establishing the Potassium Equilibrium Potential

Figure 8.6 illustrates how the potassium equilibrium potential is achieved in a cell freely permeable to only potassium ions. Sodium ions (Na^+) are at a higher concentration outside the cell and are balanced electrically by the presence of chloride ions (Cl^-) outside the cell. Potassium ions (K^+), by contrast, are at a higher concentration inside the cell and are balanced electrically by the presence of organic anions (A^-, primarily proteins) inside the cell. We assume at first that no potential difference exists across the cell membrane; that is, the membrane potential is 0 mV. Because cell 1 is permeable only to potassium ions, potassium will diffuse down its concentration gradient, or out of the cell (Figure 8.6a). As potassium ions move, they carry their positive charge out of the cell, which leaves the inside of the cell negatively charged relative to the outside. As a consequence, a negative membrane potential develops (Figure 8.6b).

Once the membrane potential has developed, two forces are acting on the potassium ions: a *chemical force* due to the concentration gradient and an *electrical force* due to the membrane potential. (Recall from Chapter 4 that the net force acting on an ion is called the *electrochemical force*, the sum of the electrical and chemical forces.) The direction of the chemical force now present in cell 1 is such that it *pushes* potassium ions out of the cell; the direction of the electrical force is such that it *pulls* potassium ions back into the cell because of the attraction of the positively charged potassium ions for the negative charge inside of the cell.

While the chemical force for potassium ions to move across the membrane remains constant (the number of potassium ions that actually move across the membrane is small relative to the total number of potassium ions in the intracellular and extracellular fluids, less than 0.01%), the electrical force will continue to change as potassium ions move across the membrane until equilibrium is established. Initially the chemical force is greater than the electrical force because the membrane potential is small, and thus potassium ions continue to move out of the cell (see Figure 8.6b). However, the more potassium ions that leave the cell, the greater the membrane potential becomes, and consequently the greater the electrical force to pull potassium back into the cell. Thus, as the membrane potential develops, the rate of potassium movement out of the cell becomes slower and slower as the net (electrochemical) force on potassium decreases. Once the electrical force has become just strong enough to exactly balance the opposing chemical force, no net movement of potassium occurs across the membrane (Figure 8.6c). Under these conditions, potassium is said to be at equilibrium because the electrochemical force is zero. The membrane potential under these conditions is equal to the equilibrium potential for potassium (E_K), which is approximately -94 mV. (The equilibrium potential varies in different neurons based on the concentration gradient for potassium ions.) Recall from Chapter 4 that the equilibrium potential of any ion (E_x) depends only on that ion's charge and the size of its concentration gradient.

Establishing the Sodium Equilibrium Potential

Figure 8.7 illustrates how the sodium equilibrium potential is achieved in a cell freely permeable to only sodium ions. As before, we assume that initially no potential

Cell 1: permeable to potassium only

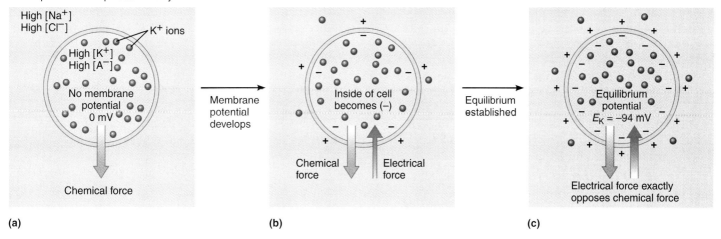

(a)   (b)   (c)

Figure 8.6 Cell 1: Permeable to potassium only. Potassium (K^+) and organic anions (A^-) are located in greater concentration inside the cell. Sodium (Na^+) and chloride (Cl^-) ions are located in greater concentration outside the cell. The width of an arrow is relative to the strength of ion movement in the direction of the arrow. (a) Potassium ions move out of the cell due to a chemical force. (b) As some potassium ions leave the cell, taking with them a positive charge, the inside of the cell becomes negative relative to the outside. This change in charge distribution creates an electrical force to move potassium ions into the cell, opposing the chemical force. (c) Eventually, enough potassium leaves the cell that the electrical force becomes strong enough to oppose further movement of potassium ions out of the cell due to the chemical force, resulting in no net movement of potassium ions. At this membrane potential, potassium is at equilibrium. This potential is thus the potassium equilibrium potential and is approximately -94 mV in neurons.

Cell 2: permeable to sodium only

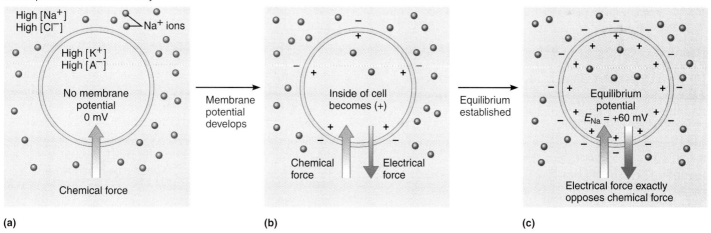

(a)   (b)   (c)

Figure 8.7 Cell 2: Freely permeable to sodium only. Potassium (K^+) and organic anions (A^-) are located in greater concentration inside the cell. Sodium (Na^+) and chloride (Cl^-) ions are located in greater concentration outside the cell. The width of an arrow is relative to the strength of ion movement in the direction of the arrow. (a) Sodium ions move into the cell due to a chemical force. (b) As some sodium ions enter the cell, taking with them a positive charge, the inside of the cell becomes positive relative to the outside. The change in charge distribution creates an electrical force to move sodium ions out of the cell, opposing the chemical force. (c) Eventually, enough sodium enters the cell that the electrical force becomes strong enough to oppose further movement of sodium ions into the cell due to the chemical force, resulting in no net movement of sodium ions. At this membrane potential, sodium is at equilibrium. This potential is thus the sodium equilibrium potential and is approximately $+60$ mV in neurons.

difference exists across the cell membrane. Because sodium can cross the membrane, it diffuses down its concentration gradient into the cell (see Figure 8.7a). As sodium moves, it carries a positive charge into the cell, which makes the inside of the cell positively charged relative to the outside, creating a positive membrane potential. In the presence of this membrane potential, sodium ions are now acted upon by an electrical force in addition to the chemical force. The direction of the chemical force tends to push sodium into the cell, while the electrical force tends to push sodium out of the cell because of the repulsion between the positively charged sodium ions and the net positive charge inside the cell (see Figure 8.7b). Although the chemical force for sodium ions to move across the membrane remains constant, the electrical force will continue to change as sodium ions move across the membrane until equilibrium is established. Sodium continues to flow into the cell, making the membrane potential more positive, until the electrical force becomes just large enough to exactly balance the chemical force (see Figure 8.7c). At this point, sodium comes to equilibrium, with the membrane potential being equal to the equilibrium potential for sodium (E_{Na}), which is approximately +60 mV. (The actual equilibrium potential for sodium varies based on the concentration gradient for sodium ions across a given neuron.)

Because the number of sodium ions that must cross the plasma membrane to cause a potential of +60 mV is very small relative to the concentration of ions in the intracellular and interstitial fluids, the concentration gradient for sodium does not change to any significant degree.

Resting Membrane Potential of Neurons

Now let's consider a more realistic cell, cell 3 (**Figure 8.8**). Cell 3 has the same ion gradients across its cell membrane, but because cell 3 has potassium and sodium channels, the membrane is permeable to both ions. However, the number of open potassium channels far exceeds the number of open sodium channels, and therefore the membrane is approximately 25 times more permeable to potassium than to sodium. Assuming that no membrane potential exists initially, let's consider what happens when potassium and sodium ions are both permeant.

Because potassium and sodium are able to cross the cell membrane, both ions move in the direction of their chemical driving forces: Potassium ions move out of the cell and sodium ions move into the cell (Figure 8.8a). However, the outward movement of potassium at this point exceeds the inward movement of sodium because the permeability of the membrane to potassium is greater than it is to sodium. Under these conditions, a net outward movement of positive charge occurs, which gives rise to a negative membrane potential (Figure 8.8b). As the unequal flows of potassium and sodium continue, the membrane potential becomes more negative, but it does not increase indefinitely, because the negative membrane

potential exerts electrical driving forces on potassium and sodium ions that oppose potassium movement and enhance sodium movement (Figure 8.8c). Therefore, as the membrane potential becomes more negative, outward potassium movement slows down while inward sodium movement speeds up (Figure 8.8d). Eventually, flows of the two ions become equal and opposite, so that there is no net movement of positive charge into or out of the cell (Figure 8.8e). At this point, the membrane potential holds steady at about −70 mV, which is a typical value for the resting membrane potential of a neuron.

In cell 3, then, both sodium and potassium are moving across the membrane, and the movement of each ion tends to bring the membrane potential toward its respective equilibrium potential (as depicted in Figure 8.6 and Figure 8.7). However, neither ion can ever come to equilibrium because the movement of each opposes the other. Given these circumstances, the final resting membrane potential is a weighted sum of the equilibrium potentials of sodium and potassium. The weight given to each equilibrium potential is based on the relative permeabilities. Because the membrane is much more permeable to potassium, the resting membrane potential, −70 mV, is much closer to the potassium equilibrium potential than to the sodium equilibrium potential. The actual resting membrane potential varies among cells, because the types and numbers of ion channels in the plasma membranes of those cells vary, but the resting membrane potential is always negative.

Note that in cell 3 neither sodium nor potassium is at equilibrium, as was the case with cell 1 or cell 2. This is because the membrane potential is not equal to the equilibrium potential of either ion. Therefore, electrochemical forces are acting on both ions, causing sodium to continually *leak into* the cell and potassium to continually *leak out* of the cell. Although these leakages are responsible for creating the membrane potential, they also tend to slowly alter ion concentrations inside the cell, raising the sodium concentration while lowering the potassium concentration. This is a potential problem if left unchecked, because these changes would eventually abolish the concentration gradients of both ions, and the membrane potential would go to zero. However, Na^+/K^+ pumps in the cell membrane avert this problem by actively transporting sodium out of the cell and potassium into the cell using ATP for energy. Normally, sodium is pumped out as fast as it leaks in, and potassium is pumped in as fast as it leaks out. Therefore, the Na^+/K^+ pump not only establishes the concentration gradients but maintains them as well. Furthermore, because the Na^+/K^+ pump is *electrogenic*—that is, it transports a net positive charge out of the cell—it contributes directly to the resting membrane potential, but this effect is minimal and accounts for only a few millivolts of charge separation. Because energy is required to sustain the resting state of cell 3, the cell is not at equilibrium; instead it is in a *steady state*.

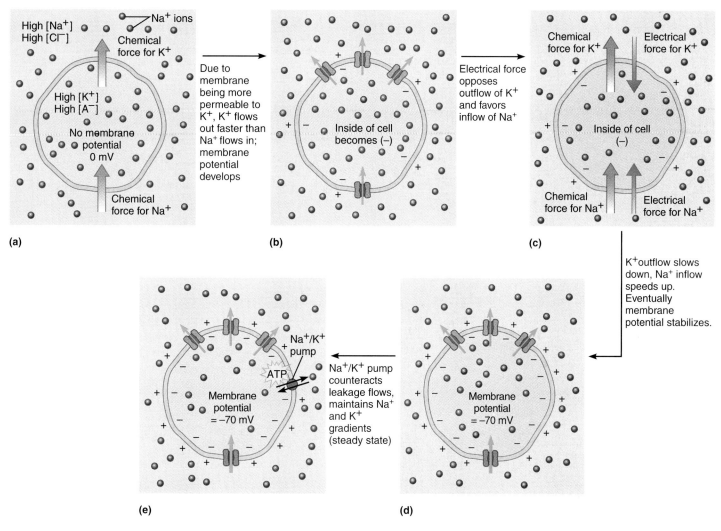

Figure 8.8 **Cell 3: How a typical animal cell achieves a steady state resting membrane potential.** Potassium and organic anions are located in greater concentration inside the cell. Sodium and chloride ions are located in greater concentration outside the cell. The width of an arrow is relative to the strength of ion movement in the direction of the arrow. The cell is permeable to both sodium and potassium ions, but more permeable to potassium. **(a)** Chemical forces act on potassium ions to leave the cell and sodium ions to enter the cell. **(b)** More potassium leaves the cell than sodium enters because of the greater permeability for potassium. With more positive charge leaving the cell, a negative membrane potential develops. **(c)** Electrical forces now act on the ions, drawing both sodium and potassium ions into the cell, creating a stronger electrochemical force for sodium to enter the cell and a weaker electrochemical force for potassium to leave the cell. **(d)** Eventually, a steady state is established whereby the movement of sodium into the cell is balanced by the movement of potassium out of the cell, and no net charge movement occurs. This potential is called the resting membrane potential and is approximately −70 mV in neurons. **(e)** To prevent the sodium and potassium concentration gradients from dissipating, the Na^+/K^+ pump moves sodium out of the cell and potassium into the cell, establishing a steady state at −70 mV.

Neurons at Rest

Although cell 3 is similar to a real neuron, it has several simplifications. First, some cells are permeable to ions other than sodium and potassium, and the flow of these other ions contributes to the resting membrane potential. For example, in some neurons chloride ions make a significant contribution to the membrane potential. Generally speaking, the membrane potential can be affected by two, three, or more ions. Nevertheless, the following rule generally holds true: *A cell's membrane potential is a weighted average of the equilibrium potentials of permeant ions, with each ion's contribution depending on its relative permeability.* (Those ions at concentration magnitudes less than other ions do not contribute significantly to the resting membrane potential.) *As the cell's permeability to a particular ion increases, the membrane potential moves closer to that ion's equilibrium potential.* (This relationship can be explored further in **Toolbox: Resting Membrane Potential and the GHK Equation**, p. 222.)

RESTING MEMBRANE POTENTIAL AND THE GHK EQUATION

In Chapter 4, we saw how the Nernst equation can be used to calculate the equilibrium potential for a specific ion. The Nernst equation cannot be used, however, to calculate the membrane potential because a membrane is permeable to more than one ion, and the membrane's permeability to various ions differs. The membrane potential thus depends on the concentration gradients for all ions across the plasma membrane, and on the permeability of the membrane to those ions. If the permeability for it is zero, an ion will not contribute to the membrane potential.

For situations in which only sodium (Na^+) and potassium (K^+) are permeant, the membrane potential (V_m) can be approximated using the Goldman, Hodgkin, Katz (GHK) equation (named after its developers):

$$V_m = 61 \log \frac{P_{Na}[Na^+]_o + P_K[K^+]_o}{P_{Na}[Na^+]_i + P_K[K^+]_i}$$

where the subscripts o and i indicate concentrations outside and inside the cell, respectively, and P_{Na} and P_K are the membrane's permeabilities to sodium and potassium.

If we divide the numerator and denominator both by P_K, then this equation becomes

$$V_m = 61 \log \frac{(P_{Na}/P_K)[Na^+]_o + [K^+]_o}{(P_{Na}/P_K)[Na^+]_i + [K^+]_i}$$

This form of the equation gives the membrane potential in millivolts.

Using concentrations given in Table 4.1, we can calculate the membrane potential under resting conditions, assuming that P_K is 25 times as large as P_{Na} ($P_{Na}/P_K = 1/25 = 0.04$):

$$V_m = 61 \log \frac{(0.04)(145 \text{ mM}) + 4 \text{ mM}}{(0.04)(15 \text{ mM}) + 140 \text{ mM}}$$
$$= 61 \log(0.0697) = -70.6 \text{ mV}$$

Note that this value is closer to the potassium equilibrium potential (-94 mV) than the sodium equilibrium potential ($+60$ mV), as we would expect. Should there be a change in the membrane's permeability to sodium relative to potassium, the GHK will give you the new membrane potential, so long as the ratio P_{Na}/P_K is known.

Another simplification in Figure 8.8 shows sodium, potassium, and chloride ions localized on only one side of the membrane to illustrate the directions of the ions' concentration gradients. In actual cells, each ion is present both inside and outside the cell, with potassium at a greater concentration in the cytosol and sodium and chloride ions at greater concentrations in the extracellular fluid. Table 4.1 (p. 102) lists the actual ion distribution across a typical cell membrane.

Because neither sodium nor potassium is at equilibrium at the resting membrane potential, a net electrochemical force acts on each ion. The following rule describes electrochemical forces on ions: *The net electrochemical force on an ion tends to move that ion across the membrane in the direction that will move the membrane potential toward that ion's equilibrium potential—that is, bring the ion closer to equilibrium.* Therefore, sodium tends to move into the cell to bring the membrane potential toward +60 mV, and potassium tends to move out of the cell to bring the membrane potential toward −94 mV. The strength of the electrochemical force acting on a specific ion is proportional to the difference between the membrane potential and the equilibrium potential for that ion. Thus, because sodium is 130 mV away from equilibrium (at a resting membrane potential of −70 mV) whereas potassium is only 24 mV away from equilibrium, the electrochemical force moving sodium into the cell is much greater than the electrochemical force moving potassium out of the

cell (indicated by the width of the arrows in Figure 8.8e). The sodium and potassium channels that are responsible for the resting membrane potential are leak channels, which are always open. In addition to these leak channels, neurons also have gated ion channels.

In the next section we see that ion movement across the membrane can be altered by changing the permeability of the membrane for that ion—that is, by opening or closing specific ion channels in the plasma membrane.

Quick Test 8.2

1. Describe the concentration gradients for sodium and potassium across the plasma membrane. What establishes these concentration gradients?

2. If a neuron had equal permeability to sodium and potassium ions, would the resting membrane potential of that cell be more negative or less negative than −70 mV?

3. When a neuron is at rest, what are the directions of the electrochemical forces that drive sodium and potassium movement?

4. If channels that permitted both sodium ions and potassium ions to move through them suddenly opened in the plasma membrane, in which direction would each ion move, into or out of the neuron? Which ion would move more? Why? What change in membrane potential would occur?

Electrical Signaling Through Changes in Membrane Potential

Electrical signals occur in neurons via changes in membrane potential that occur when certain ion channels, called *gated channels,* open or close in response to particular stimuli. When gated ion channels open or close, they change the membrane permeability for that specific ion, thereby affecting the movement of that ion across the plasma membrane. For example, if sodium ion channels open, then sodium movement into the cell increases, driving the membrane potential toward the sodium equilibrium potential.

There are three types of gated ion channels: voltage-gated channels, ligand-gated channels, and mechanically gated channels. Voltage-gated and ligand-gated channels were described previously. Mechanically gated channels open or close in response to a mechanical force on the membrane. These channels are usually found associated with sensory or visceral receptors located at the end of afferent neurons. Gated channels are crucial for normal function of the nervous system; in fact, many toxins exert their poisonous effects by interfering with the normal function of ion channels (**Clinical Connections: Neurotoxins**).

Describing Changes in Membrane Potential

Changes in membrane potential are described based on the direction of change relative to the resting membrane potential, as illustrated in **Figure 8.9**. Because the membrane potential is a *difference* in potential across the membrane, the membrane is *polarized*. The sign of the membrane potential (positive or negative) always refers to the potential inside the cell relative to the potential outside. Because the resting membrane potential is a negative value (approximately -70 mV in neurons), a change to a more negative value is a **hyperpolarization** because the membrane becomes *more polarized*. In contrast, a change to a less negative or to a positive potential is a **depolarization** because the membrane becomes *less polarized*. **Repolarization** occurs when the membrane potential returns to the resting membrane potential following a depolarization.

Neurons communicate via two different types of electrical signals that result from the opening or closing of gated ion channels: (1) *graded potentials*, which are small electrical signals that act over short ranges only because they diminish in size with distance, and (2) *action potentials,* which are large signals capable of traveling long distances without decreasing in size. We examine these potentials in the next two sections.

Graded Potentials

Graded potentials are small changes in membrane potential that occur when ion channels open or close in

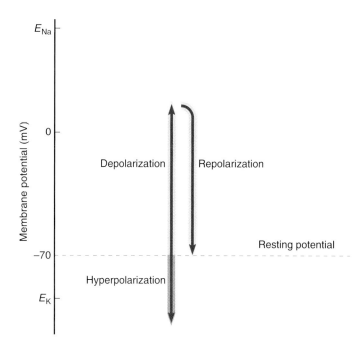

Figure 8.9 Changes in membrane potential. The membrane potential can change with the opening or closing of ion channels. A change in membrane potential to a less negative value is called depolarization. The return of a depolarized membrane to the resting membrane potential is called repolarization. A change in membrane potential to a more negative value is called a hyperpolarization. If gated potassium channels open, then potassium moves out of the cell, bringing membrane potential toward the potassium equilibrium potential (E_K), or hyperpolarizing the cell. If gated sodium channels open, then sodium moves into the cell, bringing membrane potential toward the sodium equilibrium potential (E_{Na}), or depolarizing the cell.

response to a stimulus acting on the cell (**Figure 8.10**). Graded potentials can be produced as a result of neurotransmitter molecules binding to receptors on a dendrite or the cell body of a neuron; graded potentials can also result from a sensory stimulus, such as touch or light, acting on a sensory receptor at the peripheral ending of an afferent neuron. The magnitude of the change in membrane potential varies with—is *graded*—according to the strength of the stimulus: A weak stimulus produces a small change in membrane potential, whereas a stronger stimulus produces a greater change in membrane potential (Figure 8.10a). For example, if the presence of 500 neurotransmitter molecules acting on ligand-gated channels caused a cell to depolarize 5 mV, then 1000 neurotransmitter molecules (a stronger stimulus) will cause the cell to depolarize about 10 mV.

Some graded potentials are depolarizations, whereas others are hyperpolarizations (Figure 8.10b). The direction of change depends on the particular neuron in question, the stimulus applied to it, and the specific ion channels that open or close in response to the stimulus. For example, if one type of neurotransmitter binding to its receptors caused sodium channels to open, then sodium ions would move into the cell and the resulting

NEUROTOXINS

As testimony to their fundamental importance in neural signal transmission, voltage-gated sodium channels are present in distantly related representatives of the animal kingdom, from mammals to mollusks to worms. In further testimony, voltage-gated sodium channels are the targets for some of the most potent neurotoxins known to humanity. A neurotoxin is a toxin that attacks some aspect of nervous system function.

The Chinese proverb "To throw away life, eat blowfish" arose because blowfish (also known as puffer fish) contain the neurotoxin tetrodotoxin (TTX). TTX is very potent, toxic at nanomolar (10^{-9} moles/L) concentrations. Effects of TTX were first described nearly 5000 years ago by the Chinese emperor Shun Nung, who had a dangerous hobby: tasting and cataloguing the effects of various drugs. Apparently, he knew the meaning of "just a taste," because he lived to describe the effects! TTX works by blocking voltage-gated

sodium channels necessary for producing an action potential.

Blowfish is one of the most prized delicacies in the restaurants of Japan. This fish is prized not only for its taste but for the tingling sensation one gets around the lips when eating it. In blowfish, TTX is concentrated in certain organs, including the liver and gonads. Its preparation takes great skill and can only be done by licensed chefs who are skilled at removing the poison-containing organs without crushing them, which can lead to contamination of normally edible parts. The toxin cannot be destroyed by cooking. Lore has it that the most skilled chefs intentionally leave a bit of the poison in, so that diners can enjoy the tingling sensation caused by blockage of nerve signals from the sense receptors on the lips. TTX is also present in some kinds of salamanders, octopus, and goby. Species that carry the toxin have sodium channels that differ slightly, rendering them either less sus-

Puffer fish contain a potent neurotoxin.

ceptible or completely resistant to the toxin.

A closely related toxin that blocks voltage-gated sodium channels is saxitoxin (STX), which is produced by some marine dinoflagellates and by a freshwater cyanobacterium. These microorganisms become a major health threat during certain periods of the year when they multiply rapidly, or "bloom." At high concentrations they may even impart a reddish color to the water, producing the phenomenon known as "red tide." The primary danger to humans during red tide comes from eating shellfish, which are filter feeders that can accumulate the toxin in their bodies. (Shellfish are resistant to the toxin.) Eating even a single contaminated shellfish can be fatal. Like TTX, STX is not destroyed by cooking. Unfortunately, one cannot always tell whether conditions are hazardous by simple observation; seawater can contain dangerous concentrations of dinoflagellates without appearing red. Fortunately, shellfish are routinely monitored by public health officials for the presence of this toxin, and harvesting of shellfish is banned whenever there is cause for concern.

Sodium channels are not the only ion channels affected by neurotoxins. The table in this box lists a variety of neurotoxins, their source, and their effects on ion channels.

Sources and Effects of Selected Neurotoxins

Toxin	Source	Effects
Tetrodotoxin	Blowfish (puffer fish)	Blocks voltage-gated sodium channels
Saxitoxin	Marine dinoflagellates, freshwater cyanobacterium	Blocks voltage-gated sodium channels
Apamin	Honeybee	Blocks potassium channels
Batrachotoxin	Poison arrow frog	Keeps sodium channels from closing
Calciseptin	Black mamba	Blocks calcium channels
Iberiotoxin	Indian red scorpion	Blocks potassium channels
Phoneutriatoxin	Banana spider	Slows closing of sodium channels
Stichodactyla toxin	Sea anemone	Blocks voltage-gated potassium channels

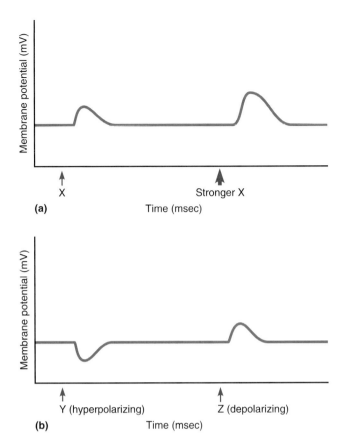

(a)
Time (msec)

(b)
Time (msec)

Figure 8.10 Properties of graded potentials. (a) Effect of stimulus strength on size of graded potential. Graded potentials are small changes in the membrane potential (V_m) in response to a stimulus; a stronger stimulus produces a larger change in V_m. (b) Effect of stimulus type on graded potentials. Depending on the stimulus and the neuron, graded potentials can cause either hyperpolarization or depolarization.

graded potential would be a depolarization. If another type of neurotransmitter binding to its receptors caused potassium channels to open, then potassium ions would move out of the cell and the resulting graded potential would be a hyperpolarization.

The primary significance of graded potentials is that they determine whether or not a cell will generate an action potential. Graded potentials generate action potentials if they depolarize a neuron to a certain level of membrane potential called the **threshold**, a critical value of membrane potential that must be met or exceeded if an action potential is to be generated. Therefore, graded potentials that are depolarizations are described as **excitatory**, because they bring the membrane potential closer to the threshold to generate an action potential. On the other hand, graded potentials that are hyperpolarizations are described as **inhibitory**, because they take the membrane potential away from the threshold to elicit an action potential.

A graded potential can travel away from the site of stimulation for only a short distance because it is

decremental; that is, the change in membrane potential decreases in size as it moves along the membrane away from the site of stimulation. To understand why a graded potential is decremental, let's first look at what happens to water flow through a leaky hose. If the hose is intact, the amount of water exiting the hose at one end is the same as that entering the hose at the faucet end. However, if the hose has leaks (much like a membrane has leaks to ions), then some of the water is lost and the amount of water exiting at the end of the hose is decreased. Now let's look at the flow of current along a membrane.

When a change in potential occurs across a cell membrane at a particular site, this change generates differences in potential within the intracellular and extracellular fluids. Because a separation of charge creates a force for charge to move (current), the graded potential creates charge separation within the intracellular fluid and within the extracellular fluid, which generates currents in these fluids. These currents travel to adjacent areas of the cell membrane, causing voltage changes in these areas. This spread of voltage by passive charge movement is called **electrotonic conduction**. As the graded potential spreads from the site of the stimulation (**Figure 8.11**), the current is spread over a larger area, and some current leaks across the plasma membrane. As a result, the size of the membrane potential change decreases as it moves from the site of initial stimulation.

A single graded potential is generally not of sufficient strength to elicit an action potential. However, if graded potentials overlap in time, then they can sum, both temporally and spatially. In **temporal summation**, stimuli are applied in such rapid succession that the graded potential from one stimulus does not dissipate before the next graded potential occurs. Thus, the effects of the potentials sum. The greater the overlap in time, the greater the summation. In **spatial summation**, the effects of stimuli from different sources occurring close together in time sum. Summation of a hyperpolarizing graded potential and a depolarizing graded potential tend to cancel each other out. **Figure 8.12** graphically depicts the events in temporal and spatial summation.

Quick Test 8.3

1. Define depolarization and hyperpolarization. If the permeability of a membrane to sodium increases, would the membrane depolarize or hyperpolarize? Explain.

2. Name the three different gating mechanisms for ion channels.

3. What properties of graded potentials result in their being "graded"?

4. What is the distinction between an excitatory graded potential and an inhibitory graded potential?

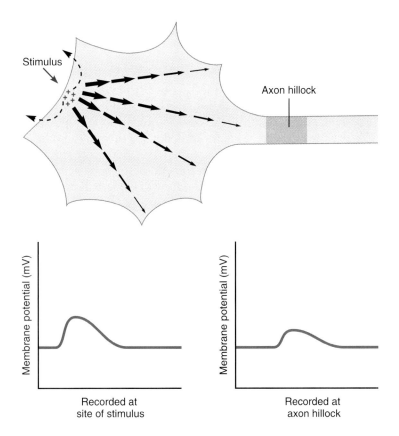

Figure 8.11 Decremental property of graded potentials.
A graded potential dissipates as it moves to adjacent areas of the plasma membrane (solid arrows) and across the plasma membrane (dashed arrows). Therefore, the potential change recorded at the site of stimulus is greater than that recorded distant from the site of stimulus. (Darker arrows indicate a stronger current.)

Action Potentials

In most of the remainder of this chapter we discuss action potentials. Action potentials occur in the membranes of excitable tissue (nerve or muscle) in response to graded potentials that reach threshold. **Table 8.2** compares the properties of graded and action potentials. During an action potential a large, rapid depolarization occurs in which the polarity of the membrane potential actually reverses; that is, the membrane potential becomes positive for a brief time. In fact, the membrane potential changes very quickly (in about 1 msec) from a resting level of approximately -70 mV to $+30$ mV (a change of 100 mV). Once initiated, an action potential, unlike a graded potential, is capable of being propagated long distances along the length of an axon without any decrease in strength.

Ionic Basis of an Action Potential

The generation of an action potential is based on the selective permeability of the plasma membrane and the Na^+ and K^+ electrochemical gradients that exist across

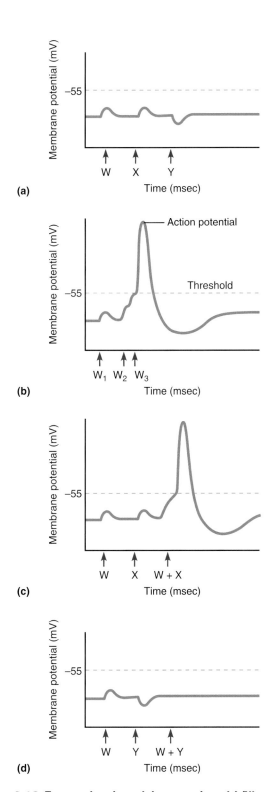

Figure 8.12 Temporal and spatial summation. (a) Effects of stimuli W, X, and Y on membrane potential. Stimuli W and X depolarize the cell; Y hyperpolarizes the cell. (b) Temporal summation of stimulus W resulting in depolarization to above threshold and generation of an action potential. (c) Spatial summation of stimuli W and X resulting in depolarization to above threshold and generation of an action potential. (d) Spatial summation of stimuli W and Y resulting in no change in membrane potential and therefore no action potential.

Table 8.2 │ Comparison of Graded Potentials and Action Potentials

Property	Graded potential	Action potential
Location	Dendrites, cell body, sensory receptors	Axon
Strength	Relatively weak, proportional to strength of stimulus; dissipates with distance from stimulus	100 mV All-or-none
Direction of change in membrane potential	Can be depolarizing or hyperpolarizing depending on stimulus	Depolarizing
Summation	Spatial and temporal	None
Refractory periods	None	Absolute and relative
Channel types involved in producing change in potential	Ligand-gated, mechanically gated	Voltage-gated
Ions involved	Usually Na^+, Cl^-, or K^+	Na^+ and K^+
Duration	Few msec to seconds	1–2 msec (after-hyperpolarization may last 15 msec)

the membrane. Recall that at rest the plasma membrane is approximately 25 times more permeable to potassium ions than to sodium ions because of the presence of many more potassium leak channels than sodium leak channels. In excitable cells, changes in the permeability of the plasma membrane resulting from the opening and closing of gated ion channels can produce action potentials.

An action potential in a neuron consists of three distinct phases (**Figure 8.13**a):

1. Depolarization. The first phase of an action potential is a *rapid depolarization* during which the membrane potential changes from −70 mV (rest) to +30 mV. This depolarization is caused by a sudden and dramatic increase in permeability to sodium (Figure 8.13b) followed by an increase in the movement of sodium ions into the cell, down sodium's electrochemical gradient. With permeability to sodium now greater than permeability to potassium, the membrane potential approaches the sodium equilibrium potential of +60 mV. Although the change in membrane potential produced by sodium movement is large (100 mV), the number of sodium ions that actually cross the membrane to produce this change in potential is relatively small compared to the concentration of sodium ions in the intracellular and extracellular fluids. Therefore, the concentrations of sodium in those fluids do not change appreciably.

2. Repolarization. The second phase of an action potential is a *repolarization* of the membrane potential during which the membrane potential returns from +30 mV back to resting levels (−70 mV). Within 1 msec after the increase in sodium permeability, sodium permeability decreases rapidly, reducing the inflow of

sodium. At approximately the same time, potassium permeability increases. Potassium then moves down its electrochemical gradient out of the cell, repolarizing the membrane potential to bring it back to resting levels.

3. After-hyperpolarization. The third phase of an action potential is termed **after-hyperpolarization**. Potassium permeability remains elevated for a brief time (5–15 msec) after the membrane potential reaches the resting membrane potential, resulting in an after-hyperpolarization. During this time the membrane potential is even more negative than at rest as it approaches the potassium equilibrium potential (−94 mV). As with sodium movement, the movement of potassium ions during repolarization and after-hyperpolarization is small, so the concentrations of potassium inside and outside the cell do not change appreciably.

Next we discuss the ion channels responsible for the changes in permeability to sodium and potassium ions during an action potential.

The Role of Voltage-Gated Ion Channels in Action Potentials

The changes in permeability associated with the phases of an action potential are due to the time-dependent opening and closing of voltage-gated sodium and potassium channels located primarily in the plasma membrane of the axon hillock and axon. (These voltage-gated sodium and potassium channels are also found in the plasma membrane of some muscle cells.) In myelinated axons these channels are at a greater concentration at the nodes

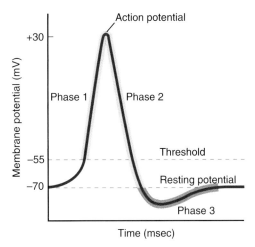

(a) Three phases of an action potential

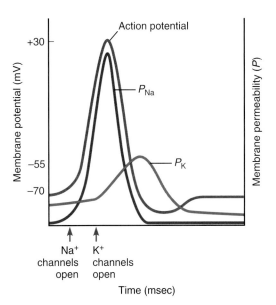

(b) Permeability changes for Na⁺ and K⁺ during an action potential

Figure 8.13 The phases and ionic basis of an action potential. (a) The three distinct phases of an action potential: (1) depolarization, (2) repolarization, and (3) after-hyperpolarization. (b) The permeability changes for sodium ions (P_{Na}) and potassium ions (P_K) that occur during an action potential. The rapid depolarization of phase 1 is caused by a rapid sodium permeability increase that enables sodium to move into the cell. The repolarization of phase 2 is caused by a slower increase in potassium permeability, enabling greater movement of potassium out of the cell compared to resting conditions. The after-hyperpolarization of phase 3 is caused by the continuing movement of potassium out of the cell.

At the beginning of repolarization, permeability to sodium is greater than permeability to potassium. Why does the membrane potential repolarize?

Potassium is further from equilibrium, so potassium movement out of the cell exceeds sodium movement into the cell.

of Ranvier; in unmyelinated axons these channels are evenly distributed along the entire axon. Because the exact mechanisms of gating in the voltage-gated sodium and potassium channels are not known, we use models to describe their function.

The model for explaining the actions of voltage-gated sodium channels involves two types of gates: activation gates and inactivation gates. **Activation gates** are responsible for the opening of sodium channels during the depolarization phase of an action potential, whereas **inactivation gates** are responsible for the closing of sodium channels during the repolarization phase of an action potential. For a sodium channel to be open, both gates must be open. Both types of gates open and close in response to changes in the membrane potential.

Based on the position of these two gates, a sodium channel can exist in three conformations (**Figure 8.14**). At rest the inactivation gate is open, but the activation gate is closed. In this state, the channel is *closed but capable of being opened* by a depolarizing stimulus that causes the activation gate to open. With both gates in their open position, the channel is open and sodium ions move through the channel into the cell; this is what occurs during the depolarization phase of an action potential. However, within approximately 1 msec after the initial stimulus to open the activation gate, the inactivation gate closes. The closing is a delayed response initiated by the same depolarization that caused the activation gate to immediately open. With the inactivation gate closed and the activation gate open, the channel is *closed and incapable of opening* in response to another depolarizing stimulus, because the inactivation gate does not open until the membrane potential returns to near its resting value. Once repolarization has occurred, the inactivation gate opens and the activation gate closes, returning the channel to its resting state.

Exercise Link

Although this chapter focuses on nerve cells, it is important to remember that all cells have a resting membrane potential. Furthermore, action potentials occur in skeletal muscle cells to initiate contraction. During exercise, the repetitive depolarization and repolarization of muscle cells causes a rise in extracellular potassium concentration that is proportional to active muscle mass and to exercise intensity and duration. While running the marathon, Bill and Jane had an increase in plasma potassium concentration of about 10%, which is enough to influence resting membrane potential and possibly contribute to muscle fatigue. Exercise training increases the number of Na⁺/K⁺ pumps in skeletal muscle, an adaptation that helps maintain proper potassium (and sodium) concentrations across the cell membranes during exercise.

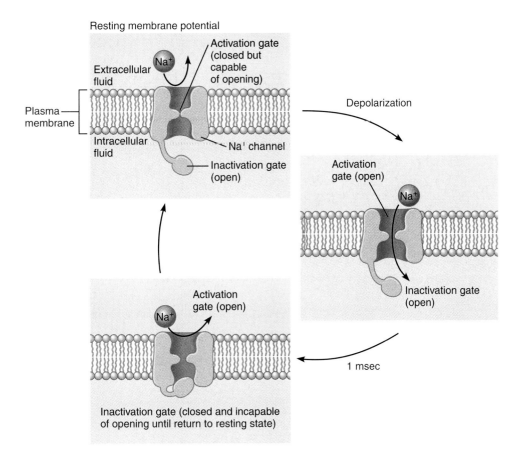

Figure 8.14 A model for the operation of voltage-gated sodium channels. Voltage-gated sodium channels are schematically represented here by two gates. At rest, the sodium inactivation gate is open and the activation gate is closed but can open in response to a depolarizing stimulus. Following a depolarizing stimulus to threshold, both the activation and inactivation gates are open, and sodium can move through the channel. Approximately 1 msec after a depolarizing stimulus, the inactivation gate closes and remains closed until the cell has repolarized to the resting state. Before repolarization the channel cannot open in response to a new depolarizing stimulus.

The opening of sodium activation gates is a **regenerative** mechanism; that is, the opening of some sodium activation gates causes more sodium activation gates to open by *regenerating* the stimulus to open the gates (depolarization). This regenerative mechanism works as follows: At first, depolarization triggers the opening of a few sodium channels, which allows some movement of sodium ions into the cell, which depolarizes the cell further. The increased depolarization causes still more sodium channels to open, leading to a larger inflow of sodium ions and more depolarization, and so on. This positive feedback loop causes the very rapid (less than 1 msec) depolarization phase of the action potential. The positive feedback loop terminates when the sodium inactivation gates close (**Figure 8.15**).

In contrast to the sodium channels just described, the model for voltage-gated potassium channels describes only a single gate that opens more slowly in response to depolarization. At approximately the same time that

sodium inactivation gates are closing (1–2 msec after depolarization to threshold), the potassium channels begin to open. The increased permeability to potassium coupled with the strong electrochemical gradient for potassium to move out of the cell (when the membrane is depolarized, potassium is far from equilibrium) increases the movement of potassium ions out of the cell. This movement of positive charge out of the cell repolarizes the cell. Because the effect of opening potassium channels (repolarization) is an action opposite to the initial stimulus that opened the potassium channels (depolarization), voltage-dependent potassium channels are part of a negative feedback loop during an action potential (see Figure 8.15). Therefore, as the cell repolarizes, the depolarizing stimulus weakens, and potassium channels slowly close.

Table 8.3 summarizes the conditions surrounding voltage-gated sodium and potassium channels at rest and during the three phases of an action potential.

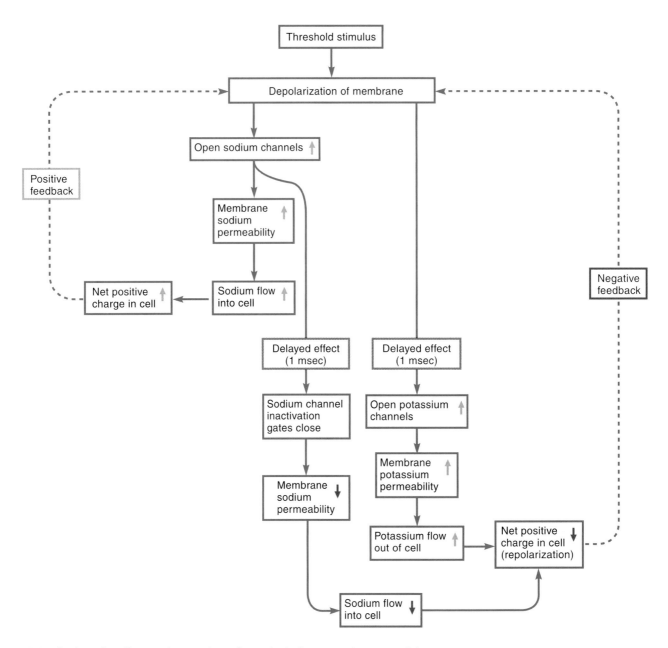

Figure 8.15 Gating of sodium and potassium channels during an action potential.
Sodium channel opening is part of a positive feedback loop that allows for the rapid depolarization of the cell. When the cell is depolarized to threshold, sodium channels open. Opening allows sodium to move into the cell, causing further depolarization and opening more sodium channels. The feedback loop continues until the sodium inactivation gates close, approximately 1 msec after the depolarization to threshold. Potassium channel opening and closing is part of a negative feedback loop. Depolarization stimulates the slow opening of potassium channels. This allows potassium to move out of the cell, repolarizing it. Since repolarization opposes the depolarizing stimulus for opening potassium channels, the potassium channels close.

What stops the positive feedback loop for sodium channels to open?

The threshold for generating an action potential corresponds to the level of depolarization necessary to induce the sodium positive feedback loop. A depolarization that is below threshold (a **subthreshold** stimulus) may open some sodium channels, but not enough of them to produce an inward flow of sodium large enough to overcome the outward flow of potassium through leak channels. **Figure 8.16** illustrates the concept of threshold. Note that a subthreshold stimulus produces no action

The closing of sodium inactivation gates and the opening of potassium channels.

Table 8.3 │ Characteristics of a Neuron at Rest and During the Different Phases of an Action Potential

	Resting	Depolarization	Repolarization	After-Hyperpolarization
Membrane potential	−70 mV	−70 mV to +30 mV	+30 mV to −70 mV	−70 mV to −85 mV
Voltage-gated sodium channel	Closed	Open	Closed	Closed
Activation gate	Closed	Open	Open	Closed
Inactivation gate	Open	Open	Closed	Open
Sodium flow	Low inward, through leak channels	High inward, through voltage-gated channels*	Low inward, through leak channels	Low inward, through leak channels
Voltage-gated potassium channel	Closed	Closed	Open	Closing
Potassium flow	Low outward, through leak channels	Low outward, through leak channels	High outward, through voltage-gated channels*	High outward, through voltage-gated channels, but decreasing*

*Even though at any given time ions move through both voltage-gated channels and leak channels, the conductance through the leak channels is negligible compared to that through voltage-gated channels.

potential, whereas a threshold stimulus elicits an action potential. A stimulus greater than threshold—that is, a **suprathreshold** stimulus—also elicits an action potential; note, however, that the action potential does not increase in size as the strength of a suprathreshold stimulus increases.

Initiation of action potentials follows the **all-or-none principle**: Whether a membrane is depolarized to threshold or above, the amplitude of the resulting action potential is the same; if the membrane is not depolarized to threshold, no action potential occurs.

The level of depolarization reached at the peak of an action potential depends *not* on the strength of the stimulus, but rather on the relative strengths of the electrochemical gradients for sodium and potassium ions and the relative permeabilities of the membrane to these ions. During the depolarization phase, sodium permeability exceeds potassium permeability several hundred-fold and the membrane potential approaches the sodium equilibrium potential of +60 mV.

However, the neuron can never meet or exceed the sodium equilibrium potential for the same reason that the resting membrane potential can never equal or exceed the potassium equilibrium potential: Sodium movement into the cell is countered by potassium movement out of the cell, primarily through potassium leak channels and later through the opened voltage-gated potassium channels. Therefore, the constant peak of an action potential is determined primarily by two factors: (1) the concentration gradients for sodium and potassium ions across the plasma

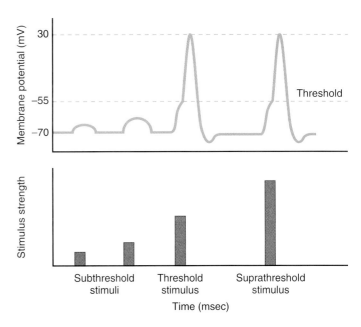

Figure 8.16 The concept of a threshold stimulus. A stimulus must reach a critical level of depolarization—threshold—before an action potential is generated. A stimulus less than threshold (a subthreshold stimulus) cannot generate an action potential. Any stimulus greater than threshold (a suprathreshold stimulus) generates an action potential of the same magnitude and duration as a threshold stimulus.

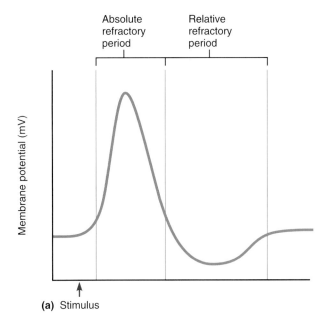

(a) Stimulus

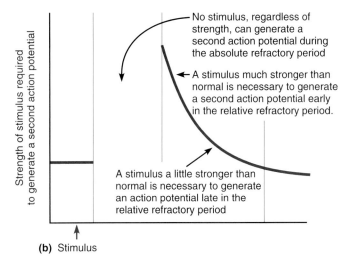

No stimulus, regardless of strength, can generate a second action potential during the absolute refractory period

← A stimulus much stronger than normal is necessary to generate a second action potential early in the relative refractory period.

A stimulus a little stronger than normal is necessary to generate an action potential late in the relative refractory period

(b) Stimulus

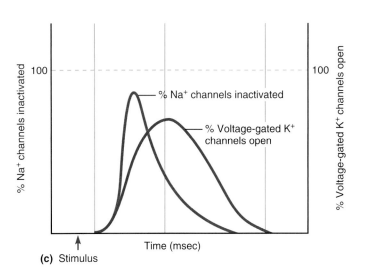

% Na⁺ channels inactivated

% Voltage-gated K⁺ channels open

(c) Stimulus

Figure 8.17 Refractory periods associated with an action potential. The graphs represent (a) membrane potential changes occurring during the action potential; (b) the stimulus strength needed to elicit a second action potential; and (c) the percentages of sodium channels that are inactivated (inactivation gates are closed) and of voltage-gated potassium channels that are open.

membrane, which do not change under normal circumstances, and (2) the number of potassium leak channels in the plasma membrane, which also does not change.

Quick Test 8.4

1. During the depolarization phase of an action potential, is the membrane more permeable to sodium or to potassium? What about during the repolarization phase?

2. Is the action potential produced by a suprathreshold stimulus smaller, larger, or the same size as an action potential produced by a threshold stimulus?

Refractory Periods

During and immediately after an action potential, the membrane is less excitable than it is at rest. This period of reduced excitability is called the **refractory period**. The refractory period can be divided into two phases, the absolute refractory period and the relative refractory period (**Figure 8.17**). The **absolute refractory period** spans all of the depolarization phase plus most of the repolarization phase of an action potential (1–2 msec). During this time, a second action potential cannot be generated in response to a second stimulus, regardless of the strength of that stimulus. There are two reasons for the absolute refractory period: (1) During the rapid depolarization phase of an action potential, the regenerative opening of sodium channels that has been set into motion will proceed to its conclusion and will not be affected by a second stimulus. (2) During the beginning of the repolarization phase, most of the sodium inactivation gates are closed and cannot be opened by a second stimulus. A second action potential cannot be generated until the majority of the sodium channels have returned to their resting state, a situation that occurs near the end of the repolarization phase. At that time the activation gates have closed and the inactivation gates have opened, so the sodium channels are closed but capable of opening in response to depolarization.

The **relative refractory period** occurs immediately after the absolute refractory period and lasts 5–15 msec. During this period, it is possible to generate a second action potential, but only in response to a stimulus stronger than that needed to reach threshold under resting conditions. The relative refractory period is primarily due to the increased permeability to potassium that continues beyond the repolarization phase (during after-hyperpolarization). In addition, some sodium inactivation gates may still be closed, especially early in the relative refractory period. Just how much stronger the stimulus must be to elicit a second action potential is a matter of timing. Early in the relative refractory period it takes a stronger stimulus to generate an action potential than is needed late in the relative refractory period, because more sodium inactivation gates are closed and more potassium channels are open early in the refractory period.

The refractory periods establish several properties of an action potential, including the all-or-none property of action potentials, the frequency with which a single neuron can generate an action potential, and the unidirectional propagation of action potentials along an axon.

Refractory periods contribute to the all-or-none principle of action potentials described previously. Unlike graded potentials, action potentials cannot sum, because the absolute refractory period prevents an overlap of action potentials.

Refractory periods are also important in the coding of information that arrives in the form of action potentials. That action potentials are all-or-none would seem to create a problem: Because graded potentials vary in magnitude based on the strength of a stimulus, the size of a graded potential encodes information about the intensity of the stimulus; but how can action potentials relay information about the intensity of a stimulus (say, the loudness of a sound)? Information pertaining to stimulus intensity is encoded by changes in the *frequency* of action potentials—that is, changes in the number of action potentials that occur in a given period of time. Because graded potentials typically last much longer than action potentials, stronger, longer-lasting graded potentials may generate a burst of action potentials. Depending on the size of the graded potential, these action potentials may occur farther apart or closer together. Thus a loud sound generates action potentials at a higher frequency (more action potentials per unit time) than a soft sound.

Consider **Figure 8.18**a, which compares the effects of a subthreshold stimulus, a threshold stimulus for 10 msec, and a threshold stimulus for 20 msec on the frequency of action potentials generated in a neuron with a relative refractory period of 15 msec. A subthreshold stimulus does not generate an action potential; a stimulus of 10-msec duration that just reaches the threshold generates a single action potential. However, if the threshold stimulus is applied for longer than the relative refractory period, a second action potential can be generated, as seen for the 20-msec threshold stimulus.

Compare Figure 8.18a with Figure 8.18b, where suprathreshold stimuli are applied to the neuron. A suprathreshold stimulus is more likely to generate more than one action potential because it can generate a second action potential during the relative refractory period of the first action potential. To generate a second action potential during the relative refractory period, a stimulus must be strong enough to open enough sodium channels such that sodium inflow overcomes the elevated potassium outflow

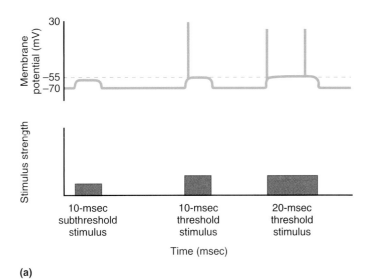

(a)

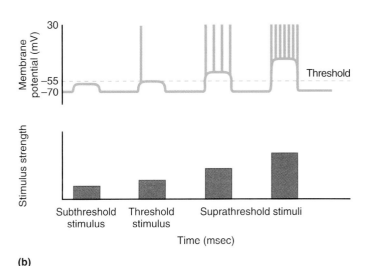

(b)

Figure 8.18 Frequency coding: how action potentials convey intensity of stimuli. The entire refractory period for this neuron is 15 msec. (a) The subthreshold stimulus does not generate an action potential. A 10-msec threshold stimulus generates a single action potential, whereas a 20-msec threshold stimulus (longer than the refractory period) generates a second action potential. (b) A threshold stimulus generates a single action potential, whereas suprathreshold stimuli generate a burst of action potentials. The stronger of the suprathreshold stimuli generates a higher frequency of action potentials.

that occurs during the relative refractory period, and the stimulus may have to overcome some sodium inactivation gates that are still closed. Also, a stronger stimulus can produce a second action potential closer in time to the first action potential. However, because a second stimulus cannot generate a second action potential during the absolute refractory period, the absolute refractory period limits the maximum frequency of action potentials in a neuron to approximately 500–1000 per second.

Figure 8.19 Action potential conduction in unmyelinated axons.
(a) State of the membrane at the resting membrane potential. (b) Initiation of conduction. When an action potential occurs at site A on the membrane, there is separation of charge in the intracellular fluid and extracellular fluid. The charge separation is a force for current to move. The local currents produced by positive ions moving toward the negative regions of the intracellular fluid are shown by arrows. (c) Propagation of conduction. The current depolarizes the adjacent region of membrane (site B) to threshold, eliciting an action potential there. (d) Continuation of propagation. Depolarization of adjacent regions continues until the action potential has been propagated all the way to the axon terminal. Refractory periods prevent action potentials from traveling in the reverse direction.

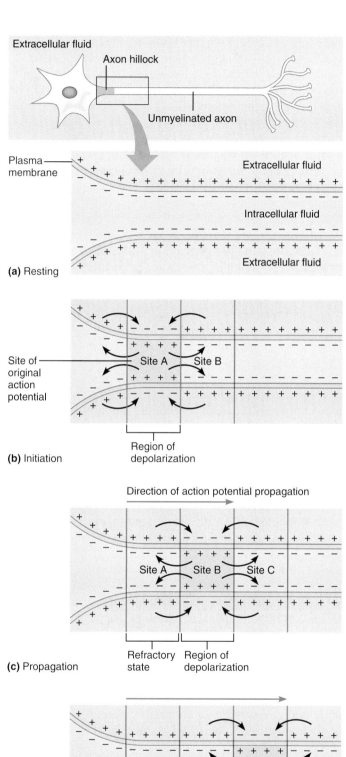

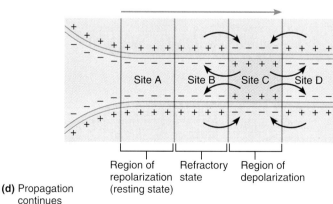

> ## Quick Test 8.5
>
> 1. Can an action potential be generated during the absolute refractory period? During the relative refractory period?
> 2. During the absolute refractory period, are the majority of sodium channels open, closed but capable of opening, or closed and incapable of opening?
> 3. How do action potentials encode the intensity of a stimulus?

Propagation of Action Potentials

Once an action potential is initiated in an axon, it is propagated down the length of the axon from the trigger zone to the axon terminal without decrement. An action potential does not actually travel down the axon; instead, an action potential sets up electrochemical gradients in the extracellular and intracellular fluids. Because the extracellular and intracellular fluids have low resistance to current flow, positive charges move from the area where the membrane has been depolarized to the adjacent area of membrane, depolarizing it as well. Propagation of an action potential down an axon is thus analogous to the sequential falling of a row of dominoes set close together on end: When the first domino falls over, it tips the adjacent one over, and so on all the way to the end of the row of dominoes. The first action potential produced at the trigger zone produces current that causes a second action potential in the adjacent membrane, which produces current that causes a third action potential, and so on until an action potential is produced at the axon terminal. Current flows to adjacent areas of the axon's plasma membrane by electrotonic conduction. The propagation mechanisms differ, however, depending on whether the axon is unmyelinated or myelinated.

Propagation of Action Potentials in Unmyelinated Axons

Recall from our discussion of graded potentials that electrotonic conduction is the passive spread of voltage changes along a neuron, away from the site of origin. Electrotonic conduction is the mechanism by which action potentials are propagated in unmyelinated axons (**Figure 8.19**). When an action potential occurs in an axon, the entire axon does not depolarize at once; instead, depolarization is restricted to just one region of the axon at any one time. Within this region of depolarization, the sign of the membrane potential is reversed, such that the inside of the cell becomes positive and the outside becomes negative. This change in membrane potential in one region creates a difference in potential within both the intracellular fluid and the extracellular fluid because adjacent regions of the membrane are still at rest (that is, the inside is negative and outside is positive). For simplicity, we discuss the changes in the electrical potential and resulting currents occurring inside the cell. However, note that the opposite electrical potentials and current flow are occurring outside the cell.

The positive charge inside the cell at the site where the depolarization phase of an action potential has just occurred is attracted to the adjacent negative charge inside, where the cell is still at rest. Since the intracellular fluid has low resistance to charge movement, the positive charges move to the negatively charged adjacent region, depolarizing it to threshold and causing another action potential. This process continues, with one action potential setting up local currents that cause another action potential in the neighboring region of the axon membrane, until the action potential has been propagated all the way to the axon terminal. Once the first action potential is generated at the axon hillock, the current spread to the adjacent membrane is always of sufficient magnitude to depolarize the neighboring membrane to threshold and elicit an action potential. This is why action potentials, unlike graded potentials, do not get smaller as they travel.

The diameter of an axon determines how quickly current spreads and, therefore, the conduction velocity of action potentials. The larger the diameter, the less resistance to longitudinal current flow (that is, current flow down the axon). This is analogous to a six-lane highway allowing heavy traffic to move faster than a two-lane highway does (there is more space for cars, or ions, to "maneuver"). Therefore, action potentials are propagated more quickly from axon hillock to axon terminal in large-diameter axons.

When a particular region of an axon is depolarized during an action potential, the resulting currents travel both downstream and upstream to adjacent regions of the membrane. What prevents an action potential from traveling upstream is that the portion of the membrane nearer to the axon hillock has just recently experienced an action potential of its own; therefore, it is in an absolute refractory state. The refractory period thus prevents action potentials from traveling backward, ensuring unidirectional propagation of action potentials (see Figure 8.19c and d).

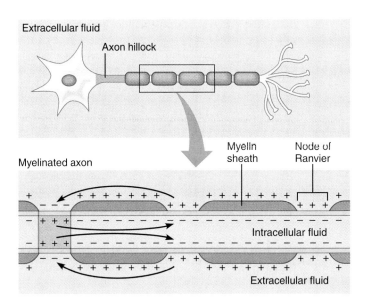

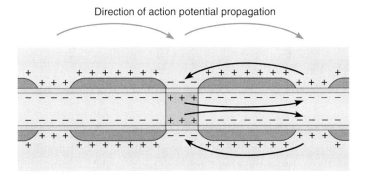

Figure 8.20 Saltatory conduction in myelinated axons.
An action potential in a myelinated axon produces electrical gradients in the intracellular and extracellular fluids that are similar to those observed in unmyelinated axons (see Figure 8.19). However, because very little current flows across the membrane where myelin insulates it, the current must flow all the way to the next node of Ranvier, where it depolarizes this area of the membrane to threshold and initiates an action potential.

Action Potential Propagation in Myelinated Axons

In axons that are sheathed in myelin, action potentials are propagated by a specialized type of electrotonic conduction called **saltatory conduction** (**Figure 8.20**). Myelin provides high resistance to ion movement across the plasma membrane, but longitudinal resistance is low. As noted previously, the nodes of Ranvier are gaps in the myelin where the axon membrane lacks insulation, is exposed to the interstitial fluid, and has the greatest concentration of voltage-gated sodium and potassium channels. In myelinated fibers action potentials are produced at the nodes of Ranvier. The concept is similar to that described for electrotonic conduction, except that action potentials cannot be produced where myelin is present. Therefore, the separation of charge in the

Table 8.4 ▌ Conduction Velocities in Axons of Various Nerve Fiber Types

Fiber type	Myelin present?	Example of function	Fiber diameter (μm)	Conduction velocity (m/sec)
A alpha	Yes	Stimulation of skeletal muscle contraction	12–20	70–120
A beta	Yes	Touch, pressure sensation	5–12	30–70
A gamma	Yes	Stimulation of muscle spindle contractile fibers	3–6	15–30
A delta	Yes, but little	Pain, temperature sensation	2–5	12–30
B	Yes	Visceral afferents, autonomic preganglionics	1–3	3–15
C	No	Pain, temperature sensation, autonomic postganglionics	0.3–1.3	0.7–2.3

intracellular fluid causes current to flow from one node of Ranvier to the next. These currents move rapidly under the myelin sheath but diminish in amplitude because some current leaks across the axon membrane. However, because the distance between nodes of Ranvier is short, the current remains strong enough to depolarize the membrane at the node of Ranvier to threshold and to generate an action potential. In this way, action potentials are generated at each node along the axon until an action potential is generated at the axon terminal. The jumping of action potentials from node to node is the basis of the term *saltatory conduction* (saltatory comes from *saltare,* Latin for "to leap").

We saw in the previous section that conduction velocity is greater in large-diameter axons. Because action potentials can move in large jumps in myelinated axons, conduction velocities in myelinated axons are greater than those in unmyelinated axons. We can think of it this way: Express trains that make few stops (like action potentials in myelinated axons) take less time to reach a destination than local trains that make many stops (like action potentials in unmyelinated axons). The fastest conducting axons, therefore, are both large-diameter and myelinated (**Table 8.4**). The fastest conducting axons are generally found in pathways requiring quick action; thus the axons that control skeletal muscle contractions are myelinated and of the largest diameter, and exhibit the fastest conduction velocities.

Quick Test 8.6

1. What is meant by "all-or-none" in reference to action potentials?

2. In which types of axons does saltatory conduction occur?

3. Explain the effects of fiber diameter and myelination on the conduction velocity of action potentials.

Maintaining Neural Stability

In the beginning of this chapter we saw the importance of ion concentration gradients and equilibrium potentials to the establishment of the resting membrane potential; subsequent sections discussed the role of ion movement into or out of the neuron to produce the resting potential, graded potentials, and action potentials. But this raises questions: If ions move across the cell membrane, would not the gradient dissipate, and the equilibrium potentials change? If the gradients dissipate, what happens to the resting membrane potential?

The key to answering these questions lies in the Na^+/K^+ pumps in the neuron membrane. Early in this chapter we discussed the Na^+/K^+ pump in the context of its importance in developing concentration gradients and sustaining the gradients when the cell is at rest, when sodium and potassium ions are moving in and out of the cell through leak channels. To compensate for the movement of sodium and potassium ions in an active cell (a cell propagating action potentials), the Na^+/K^+ pump transports even more sodium and potassium ions to compensate for the greater movement of these ions through opened voltage-gated channels. Although an action potential is a relatively large change in the membrane potential, we have seen that compared to the total number of ions present in the intracellular and extracellular fluids, very few ions need cross the membrane to produce this change. In fact, even without the Na^+/K^+ pump, no measurable change in the resting membrane potential would occur until a neuron has produced more than a thousand action potentials. Therefore, the action of the Na^+/K^+ pump is sufficient to prevent changes in concentration gradients for sodium and potassium in a cell performing at normal levels of activity.

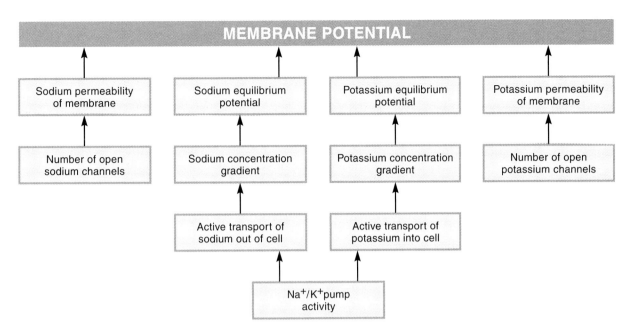

Figure 8.21 **Summary of factors that influence the membrane potential.**

Summary of Factors That Influence Membrane Potential

The main focus of this chapter was the generation of electrical signals within a neuron in the form of graded and action potentials. **Figure 8.21** summarizes the factors that influence the membrane potential. The three primary factors are the Na^+/K^+ pump, the number of open sodium channels, and the number of open potassium channels. The Na^+/K^+ pump establishes the concentration gradients for sodium and potassium ions across the plasma membrane. These concentration gradients, in turn, determine the equilibrium potentials for sodium and potassium ions. The number of open ion channels for a specific ion determines the membrane permeability to that ion. If the membrane is highly permeable to the ion, then the ion moves to drive the membrane potential toward its equilibrium potential. Under resting conditions, the membrane potential is close to the equilibrium potential for potassium because the membrane is highly permeable to potassium. However, we learned that the membrane potential can change when membrane permeability to an ion changes. The changes in membrane potential occur in the form of graded potentials and action potentials. Action potentials in a single neuron transmit information from one end of the neuron (axon hillock) to the other (axon terminal). The role of this action potential propagation in communication between neurons is the focus of the next chapter.

CHAPTER SUMMARY

Overview of the Nervous System, p. 210

The nervous system can be divided into the central nervous system and peripheral nervous system. The central nervous system consists of the brain and spinal cord. The peripheral nervous system includes the afferent and efferent divisions. The afferent division consists of neurons that transmit information from the periphery to the central nervous system, whereas the efferent division consists of neurons that transmit information from the central nervous system to the periphery. The efferent division is divided into two main branches: the somatic nervous system, which communicates to skeletal muscle, and the autonomic nervous system, which communicates to smooth muscle, cardiac muscle, glands, and adipose tissue. The autonomic nervous system is divided into the sympathetic and parasympathetic nervous systems.

Cells of the Nervous System, p. 210

The nervous system contains neurons, which are cells specialized for transmitting electrical impulses, and glial cells, which provide metabolic and structural support to the neurons. Parts of a neuron include the cell body, dendrites, and the axon. The dendrites, and to a lesser extent the cell body, receive information from other neurons at synapses. The axon includes an axon hillock, where electrical impulses (action potentials) are initiated, and an axon terminal. The axon terminal transmits information via neurotransmitters to other neurons at synapses.

Neurons are classified functionally into three classes: Efferent neurons transmit information from the central nervous system to effector organs, afferent neurons transmit information from sensory or visceral organs to the central nervous system, and interneurons communicate within the central nervous system.

Two types of glial cells function in forming myelin around axons: oligodendrocytes in the central nervous system, and Schwann cells in the peripheral nervous system. Myelin enhances the propagation of electrical impulses by providing insulation to the axon.

> **IP** Nervous I, Anatomy Review, pp. 4–6

> **IP** Nervous I, Ion Channels, pp. 3–11

Establishment of the Resting Membrane Potential, p. 216

At rest, cells have a membrane potential across them such that the inside of the cell is negative relative to the outside. This membrane potential exists because of the electrochemical forces for potassium ions to move out of the cell and sodium ions to move into the cell and because the cell membrane is more permeable to potassium ions at rest. Thus the rest-ing membrane potential is close to the potassium equilibrium potential. The membrane potential is maintained by the Na^+/K^+ pump.

> **IP** Nervous I, The Membrane Potential, pp. 1–13

> **IP** Nervous II, Synaptic Potentials and Cellular Integration, pp. 1–10

Electrical Signaling Through Changes in Membrane Potential, p. 223

Changes in the membrane potential can be produced by changing the permeability of the plasma membrane to ions. Graded potentials are small changes in membrane potential in response to a stimulus that opens or closes ion channels. If graded potentials result in a depolarization of the neuron to threshold, an action potential is produced. A single graded potential is usually not of sufficient magnitude to depolarize a neuron to threshold, but graded potentials can be temporally and/or spatially additive.

Action potentials are rapid depolarizations of the plasma membrane that are propagated along axons from the trigger zone to the axon terminal. The rapid depolarization phase of an action potential is caused by the opening of sodium channels and sodium ion movement into the cell. The repolarization phase is caused by the closing of sodium channels and the opening of potassium channels, followed by potassium movement out of the cell. After-hyperpolarization occurs because potassium channels are slow in closing, allowing continued movement of potassium out of the cell for a brief time.

Action potentials are all-or-none phenomena, meaning that their size does not vary with the strength of the stimulus eliciting them. The strength of a stimulus is coded by the frequency of action potentials; stronger stimuli produce more action potentials per unit time. Refractory periods ensure the unidirectional flow of action potentials and limit the frequency of action potentials.

> **IP** Nervous I, The Action Potential, pp 4–17

Maintaining Neural Stability, p. 236

The Na^+/K^+ pump is crucial to the normal function of neurons because it establishes the concentration gradients for sodium and potassium ions, thereby generating the chemical gradients that establish the resting membrane potential. The pump also prevents dissipation of the concentration gradients by returning sodium and potassium ions that have crossed the membrane (either through leak channels at rest or through gated channels during activity) to their original sides.

▦ *EXERCISES*

Multiple-Choice Questions

1. Depolarization of a neuron to threshold stimulates
   a) opening of sodium channels.
   b) delayed closing of sodium channels.
   c) delayed opening of potassium channels.
   d) a and c
   e) all of the above

2. Neurotransmitters are released most commonly from the
   a) cell body.
   b) dendrites.
   c) axon terminals.
   d) axon hillock.

3. If a cation is equally distributed across the cell membrane (that is, its concentration inside the cell equals its concentration outside the cell), then which of the following statements is *false*?
   a) At −70 mV, the chemical force on the ion is zero.
   b) At −70 mV, the electrical force on the ion acts to move it into the cell.
   c) At +30 mV, the chemical force on the ion is zero.
   d) The equilibrium potential for the ion would be zero.
   e) At −70 mV, the electrochemical force on the ion acts to move it out of the cell.

4. The depolarization phase of an action potential is caused by the
   a) opening of potassium channels.
   b) closing of potassium channels.
   c) opening of sodium channels.
   d) closing of sodium channels.

5. During the relative refractory period, a second action potential
   a) cannot be elicited.
   b) can be elicited by a threshold stimulus.
   c) can be elicited by a subthreshold stimulus.
   d) can be elicited by a suprathreshold stimulus.

6. Nerves are found
   a) in the central nervous system.
   b) in the peripheral nervous system.
   c) both a and b
   d) neither a nor b

7. If the membrane potential of a neuron becomes more negative than it was at rest, then the neuron is _____. In this state, the neuron is _____ excitable.
   a) depolarized/more
   b) hyperpolarized/more
   c) depolarized/less
   d) hyperpolarized/less

8. Oubain is a poison that blocks the Na^+/K^+ pump. If this pump is blocked, then the concentration of potassium inside the cell would
   a) increase.
   b) decrease.
   c) not change.

9. If potassium concentrations in the extracellular fluid of the brain increased, activity in the brain would
   a) increase.
   b) decrease.
   c) not change.

10. Which of the following neurons is part of the peripheral nervous system?
    a) motor neuron innervating skeletal muscle
    b) parasympathetic neuron
    c) sympathetic neuron
    d) all of the above

11. Which of the following axons exhibits the greatest conduction velocity?
    a) an unmyelinated axon with diameter 5 μm
    b) a myelinated axon with diameter 5 μm
    c) an unmyelinated axon with diameter 20 μm
    d) a myelinated axon with diameter 20 μm

12. Which of the following best describes the status of sodium channels at the resting membrane potential?
    a) activation gates are open and inactivation gates are closed
    b) activation gates are closed and inactivation gates are open
    c) the activation gates and inactivation gates are closed
    d) the activation gates and the inactivation gates are open

13. Which of the following is not a part of the efferent division of the nervous system?
    a) parasympathetic nervous system
    b) sympathetic nervous system

c) motor neurons
d) sensory receptors

14. Of the following ions, which is (are) located in greater concentration inside the cell?
    a) sodium only
    b) potassium only
    c) chloride only
    d) sodium and potassium
    e) potassium and chloride

15. Which of the following statements about graded potentials is *false*?
    a) The magnitude of a graded potential varies with the strength of the stimulus.
    b) Some graded potentials are hyperpolarizations, others are depolarizations.
    c) Graded potentials are produced at ligand-gated ion channels.
    d) Graded potentials can sum over space and time.
    e) Graded potentials are limited in duration by the refractory period.

Objective Questions

1. What are the subdivisions of the peripheral nervous system?

2. Information from the periphery is brought to the central nervous system by (afferent/efferent) pathways.

3. Which cell type is more abundant in the nervous system, glial cells or neurons?

4. Voltage-gated calcium channels are located at what region(s) of a neuron?

5. (Schwann cells/Oligodendrocytes) form myelin in the peripheral nervous system, and (Schwann cells/oligodendrocytes) form myelin in the central nervous system.

6. Myelin (increases/decreases) conduction velocity in axons.

7. If an anion is located in greater concentration outside the cell compared to inside, would the equilibrium potential for that anion be positive, negative, or zero?

8. Which ion is closer to equilibrium at the resting membrane potential of -70 mV, sodium or potassium?

9. In the peripheral nervous system, cell bodies of afferent neurons are located in _____.

10. The electrochemical force for potassium ions when the membrane po-

tential is at the peak of an action potential is (greater than/less than) the electrochemical force for potassium ions when the membrane potential is at rest.

11. Both sodium and potassium channels have inactivation gates that close shortly after the activation gates open. (true or false)

12. When sodium inactivation gates are closed, a second action potential is impossible. (true or false)

13. In myelinated axons, action potentials are propagated by _____ conduction.

14. The Na^+/K^+ pump causes the repolarization phase of an action potential. (true or false)

15. When a neuron is at the peak of an action potential ($+30$ mV), the direction of the electrical force for potassium ions is (into/out of) the cell.

Essay Questions

1. Draw a typical neuron and label the main structures. Then list the functions of each structure.

2. Compare the chemical and electrical forces acting on sodium and potassium ions when a cell is at rest.

3. List some similarities and some differences between graded potentials and action potentials.

4. Explain the ionic basis of an action potential.

Critical Thinking

1. Predict what the membrane potential would be for a cell with equal permeability to sodium and potassium ions. Assume the concentrations for sodium and potassium ions inside and outside the cell are as given in Table 4.1.

2. Muscle cells, like neurons, are excitable cells in that they can produce action potentials. The resting membrane potential in muscle is approximately -90 mV. Assuming that only sodium and potassium ions establish the membrane potential, explain how muscle cells can have a more polarized potential than neurons.

3. Many local anesthetics (drugs that block sensory input) work by blocking sodium channels. Explain how sensory input would be lost by this type of action.

4. Motor neurons originate in the central nervous system and terminate with a synapse onto skeletal muscle cells in the periphery. Part of the axon, therefore, is in the central nervous system but most of it is in the peripheral nervous system. The axon is myelinated. Which type of glial cell would form the myelin?

5. Predict what would happen to the resting membrane potential and the ability of a neuron to elicit action potentials if the concentration of potassium in the extracellular fluid was decreased to 50% of its normal value. What would happen if the concentration of sodium in the extracellular fluid was decreased to 50% of its normal value?

Find the answers to these exercises, and additional study tools,
at the Physiology Place (www.physiologyplace.com)

Synaptic Transmission and Neural Integration

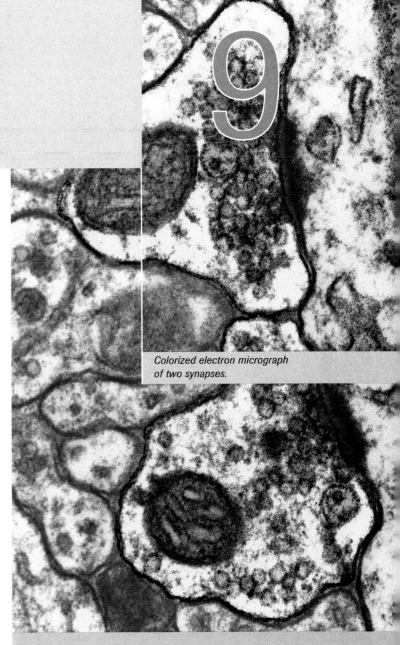

Colorized electron micrograph of two synapses.

Joe is the running back on his college football team. During one game he takes a handoff from the quarterback but is immediately hit low from the side by a defensive lineman. The impact breaks his leg, a serious injury that puts Joe in a lot of pain. He is taken off the field and then to the hospital, where a doctor gives him morphine. Within minutes, his pain starts to subside.

Why does morphine, a product of the opium poppy, relieve Joe's pain? The answer is related to the ways neurons communicate via neurotransmitters. Morphine alleviates pain because it can bind to certain neurons in the central nervous system at receptors for naturally occurring morphinelike messenger molecules called enkephalins and endorphins. As a result of this binding, morphine suppresses the transmission of pain signals.

In this chapter, we discuss how enkephalins and other neurotransmitters communicate across synapses by binding to receptors on the postsynaptic neuron. Anywhere in the body that a chemical messenger acts by binding to a receptor provides a potential target for a drug, as Joe found out with morphine.

▨ Describe the communication across chemical synapses. Explain how neurotransmitters are released and describe their actions after release.

▨ Compare fast and slow responses at synapses.

▨ Describe the process of neural integration and the role of the axon hillock in this process.

▨ Describe the major classes of neurotransmitters, including chemical structure, synthesis, degradation, and signal transduction mechanisms.

■■■ *Before You Begin*

Make sure you have mastered the following topics:

1. *Gap junctions, p. 47*
2. *Exocytosis, p. 126*
3. *Membrane potential, p. 104*
4. *Graded potentials, p. 223*
5. *G proteins, p. 151*
6. *Catecholamine synthesis, p. 141*
7. *Protein synthesis, p. 48*

W e saw in Chapter 8 that neurons generate electrical signals in the form of graded potentials and action potentials that transmit messages from one area of the cell to another. In this chapter, we learn how those electrical signals lead to communication between neurons at synapses.

There are two types of synapses in the nervous system: electrical synapses and chemical synapses. **Electrical synapses** operate by allowing electrical signals to be transmitted from one neuron to another through gap junctions, described in Chapter 2. **Chemical synapses** operate through the release of neurotransmitters that activate signal transduction mechanisms, described in Chapter 5, in the target cell.

Electrical Synapses

Electrical synapses exist between neurons and between neurons and glial cells. At these synapses the plasma membranes of adjacent cells are linked together by gap junctions such that when an electrical signal is generated in one cell, it is directly transferred to the adjacent cell by means of ions flowing through the gap junctions. Second messenger molecules can also move through these junctions.

The existence of electrical synapses in the central nervous system has been known for decades, but the functions of these synapses are just now coming to light. Earlier studies demonstrated the importance of electrical synapses in the development of the nervous system, a topic beyond

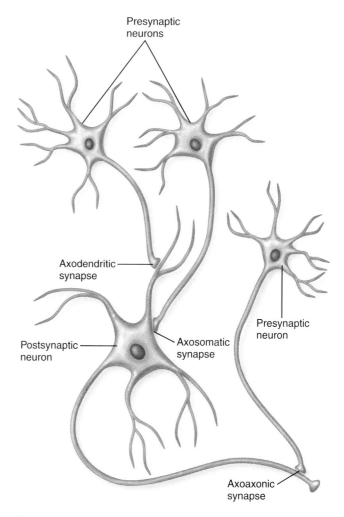

Figure 9.1 Neuron-to-neuron chemical synapses. Synapses can occur at dendrites (axodendritic synapses), at the cell body (axosomatic synapses), or with another axon (axoaxonic synapses).

the scope of this book. More recent studies indicate that these junctions transmit signals in the adult brain.

Electrical synapses allow rapid communication between adjacent neurons that synchronizes the electrical activity in these cells. The communication is often bidirectional, although some gap junctions allow current flow in only one direction. The communication can be excitatory or inhibitory at the same synapse, as either a depolarizing or a hyperpolarizing current can spread through these junctions. Some electrical synapses are always active, whereas others seem to have gating mechanisms that allow current flow only under certain circumstances.

Electrical synapses have been identified in the retina of the eye and certain areas of the cortex, where they are believed to function in transmission of signals. However, there is still much unknown about the function of these synapses. The vast majority of synapses in the nervous system are chemical synapses, which are much better understood in terms of function. We turn now to chemical synapses, the topic of the remainder of the chapter.

Chemical Synapses

Almost all neurons transmit messages to other cells at chemical synapses. In a chemical synapse, one neuron secretes a neurotransmitter into the extracellular fluid in response to an action potential arriving at its axon terminal. The neurotransmitter then binds to receptors on the plasma membrane of a second cell, triggering in that cell an electrical signal that may or may not initiate an action potential, depending on a number of circumstances.

A neuron can form synapses with other cells—the situation that is the topic of this chapter—or with effector cells such as muscle or gland cells. Muscles and glands are generally referred to as *effector organs,* and a synapse between a neuron and an effector cell is called a *neuroeffector junction.* Neuroeffector junctions, examined in more detail in Chapter 12, operate according to the same basic principles presented here for neuron-to-neuron synapses. In a few instances, non-neuronal cells can form synapses with neurons, as occurs, for example, with taste receptor cells in taste buds on the tongue (see Chapter 11).

Functional Anatomy of Chemical Synapses

Figure 9.1 depicts the possible arrangements in typical neuron-to-neuron synapses, close junctions between the axon terminal of one neuron and the plasma membrane of another neuron. The first neuron, which transmits signals to the second, is designated the **presynaptic neuron**; the second neuron, which receives signals from the first, is referred to as the **postsynaptic neuron**. Signaling across a synapse is unidirectional—the presynaptic neuron communicates to the postsynaptic neuron.

Most often the presynaptic neuron's axon terminal forms a synapse with either a dendrite or the cell body of the postsynaptic neuron, in which case the synapses are referred to as axodendritic or axosomatic synapses, respectively. In some cases the presynaptic neuron's axon terminal forms a synapse with the postsynaptic neuron's axon terminal, in which case the synapse is called an axoaxonic synapse.

Figure 9.2 shows a close-up of a synapse and the mechanism of transmitting a signal across the synapse. In all cases, the axon terminal of the presynaptic neuron releases neurotransmitters into the narrow (30–50 nm) space between the two cells, called the **synaptic cleft**. Once released into the synaptic cleft, the neurotransmitters diffuse rapidly across the cleft and bind to receptors on the postsynaptic neuron. The binding of the neurotransmitter to the receptors produces a response in the postsynaptic neuron by signal transduction.

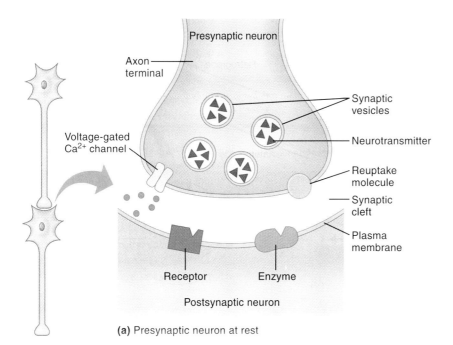

(a) Presynaptic neuron at rest

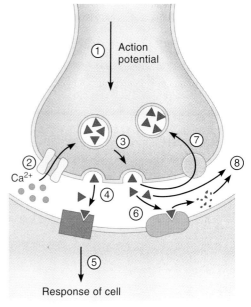

(b) Active presynaptic neuron

Figure 9.2 **Functional anatomy of a synapse.** (a) The presynaptic neuron at rest. Calcium channels are closed, and no neurotransmitter is being released. (b) An active presynaptic neuron. Communication occurs as follows: ① The membrane is depolarized by the arrival of an action potential. ② Calcium channels open. ③ Calcium enters the cell and triggers the release of neurotransmitter by exocytosis. ④ Neurotransmitter diffuses across the synaptic cleft; some of it binds to receptors on the postsynaptic cell membrane. ⑤ A response is produced in the postsynaptic cell. ⑥ Some neurotransmitter is degraded by enzymes. ⑦ Some neurotransmitter is taken up by the presynaptic cell. ⑧ Some neurotransmitter (or products of its degradation) diffuses away from the synaptic cleft.

axon terminal membrane and undergo exocytosis, thereby releasing the neurotransmitters into the synaptic cleft ③.

The amount of neurotransmitter released depends on the frequency of action potentials in the presynaptic neuron. Unless another action potential arrives, neurotransmitter release stops within a few milliseconds because the voltage-gated calcium channels close soon after opening, and because calcium ions are actively transported out of the axon terminal on a continual basis, bringing the cytosolic calcium concentration back to its resting level. However, if a second action potential arrives before neurotransmitter is cleared from the synaptic cleft, then a burst of neurotransmitter release similar to that from the first action potential causes an increase in the amount of neurotransmitter in the synaptic cleft. When a series of action potentials arrives at an axon terminal in a short time, the amount of neurotransmitter released from the presynaptic cell is greater than the amount released in response to a single action potential. Thus the concentration of neurotransmitter in the synaptic cleft increases as the frequency of action potentials increases.

Once in the synaptic cleft, neurotransmitter molecules diffuse away from the axon terminal and toward the postsynaptic neuron, where they bind to receptors (see ④ in Figure 9.2b), thereby inducing a response in the postsynaptic neuron ⑤. The binding of a neurotransmitter molecule to a receptor is a brief and reversible process. If neurotransmitter molecules were to remain indefinitely in the synaptic cleft following their release, they would bind to receptors over and over again, inducing a continual response in the postsynaptic neuron.

Continual binding of neurotransmitter to receptor does not occur because a number of processes quickly clear the neurotransmitter from the cleft, thereby terminating the signal. Some neurotransmitter molecules are degraded by enzymes, which may be located either on the postsynaptic neuron's plasma membrane (see ⑥ in Figure 9.2b), on the presynaptic neuron's plasma membrane, on the plasma membranes of nearby glial cells, or even in the cytoplasm of the presynaptic neuron or glial cells. Other neurotransmitter molecules can be actively transported back into the presynaptic neuron that released them ⑦, a process known as reuptake. Once inside the neuron, these neurotransmitter molecules are usually degraded and the breakdown products recycled to form new neurotransmitter molecules. Still other neurotransmitter molecules in the synaptic cleft simply diffuse out of the cleft ⑧. So long as neurotransmitter molecules are not bound to receptors, they may be subject to any of the fates just described. As a result, neurotransmitter is usually present in the synaptic cleft for only a few milliseconds after its release from the presynaptic neuron.

It takes approximately 0.5–5 msec from the time an action potential arrives at the axon terminal before a response occurs in the postsynaptic cell. This time lag, called the **synaptic delay**, is mostly due to the time required for calcium to trigger the exocytosis of neurotransmitter. Once in the synaptic cleft, diffusion of neurotransmitter to the receptor is so rapid that the time is negligible.

Quick Test 9.1

1. What are the essential differences between a chemical synapse and an electrical synapse?

2. List the steps involved in the transmission of signals at a chemical synapse.

3. What is the meaning of the term *synaptic delay*? What events are occurring during this time?

4. Following its release, a neurotransmitter can be cleared from a synaptic cleft by what three processes?

Signal Transduction Mechanisms at Chemical Synapses

The neurotransmitter released by a presynaptic neuron is a messenger; when it binds to receptors on the postsynaptic neuron, a message is conveyed. Neurotransmitters can induce in a postsynaptic neuron either a fast or a slow response.

The fast response (**Figure 9.3**a) occurs whenever a neurotransmitter binds to a **channel-linked receptor**, also called an **ionotropic receptor**. All channel-linked receptors are ligand-gated channels (see Chapter 8). The binding of the neurotransmitter opens the ion channel, allowing a specific ion or ions to permeate the plasma membrane and change the electrical properties of the postsynaptic neuron. The typical response is a change in the membrane potential, called a **postsynaptic potential (PSP)**, which occurs very rapidly and turns off rapidly (normally within a few milliseconds) because the channel closes as soon as the neurotransmitter leaves the receptor.

Slow responses, on the other hand, are mediated through G protein–linked receptors called **metabotropic receptors**. In the nervous system, G proteins can trigger either the opening or the closing of ion channels, depending on the specific synapse. These G protein–regulated ion channels respond to the binding of neurotransmitter more slowly than the channels that mediate the fast response, with durations ranging from milliseconds to hours, depending on the synapse, but with the same ultimate effect: The opening or closing of ion channels alters the permeability of the postsynaptic neuron to a specific ion or ions, generally resulting in changes in the membrane potential. The G protein can serve as a direct coupling between the receptor and the ion channel (Figure 9.3b), or it can trigger the activation or inhibition of a second messenger system and the second messenger affects the state of the ion channel (Figure 9.3c).

We have seen that, in general, a neurotransmitter exerts its effects on a postsynaptic neuron by triggering the

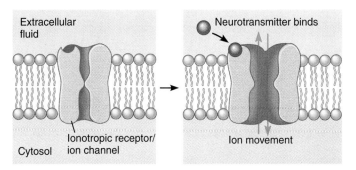

Extracellular fluid

Ionotropic receptor/ion channel

Cytosol

Neurotransmitter binds

Ion movement

(a) Fast response

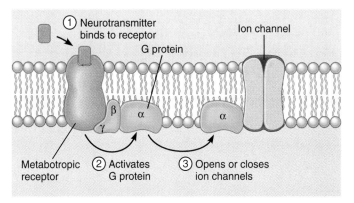

① Neurotransmitter binds to receptor

G protein

Ion channel

Metabotropic receptor

② Activates G protein

③ Opens or closes ion channels

(b) Slow response, direct coupling

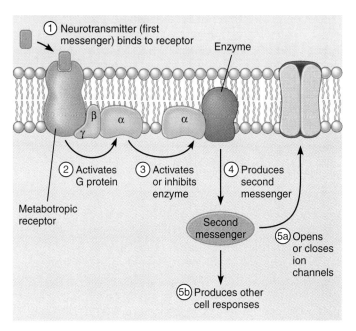

① Neurotransmitter (first messenger) binds to receptor

Enzyme

Metabotropic receptor

② Activates G protein

③ Activates or inhibits enzyme

④ Produces second messenger

Second messenger

⑤a Opens or closes ion channels

⑤b Produces other cell responses

(c) Slow response, second messenger system

Figure 9.3 Signal transduction mechanisms at chemical synapses. (a) Fast responses at a channel-linked receptor. The panel at left shows the closed state of the channel in the absence of neurotransmitter; the panel at right shows the opening of the ion channel by the binding of neurotransmitter to receptor. (b, c) Slow responses, in which the neurotransmitter receptor is coupled to a G protein. In both slow responses, binding of neurotransmitter to receptor activates a G protein. In direct coupling, the G protein opens or closes an ion channel, thereby changing the electrical properties of the cell. When the G protein functions as part of a second messenger system, it either activates or inhibits an enzyme that produces the second messenger, which then either opens or closes an ion channel or produces some other response in the cell.

opening or closing of ion channels, which typically causes a change in membrane potential. Depending on the type of channel that is opened or closed, the resulting change in membrane potential may be a depolarization or a hyperpolarization. Recall that a depolarization is considered an excitation because it brings the membrane potential closer to the threshold to generate an action potential; likewise, a hyperpolarization is considered an inhibition because it takes the membrane potential away from the threshold. At any given synapse, the direction of the response is always the same, and the synapse is classified as being either excitatory or inhibitory.

Excitatory Synapses

An **excitatory synapse** is one that brings the membrane potential of the postsynaptic neuron closer to the threshold for generating an action potential; that is, excitatory synapses depolarize the postsynaptic neuron. This depolarization is called an **excitatory postsynaptic potential (EPSP)** and can occur as either a fast response or a slow response. EPSPs are graded potentials, with the amplitude of depolarization increasing as more neurotransmitter molecules bind to receptors. As with all graded potentials, the depolarization is greatest at the site of origin (usually a dendrite or cell body) and decreases as it moves toward the axon hillock.

Fast EPSPs are generally caused by the binding of neurotransmitter molecules to their receptors on the postsynaptic cell, in the process opening channels that allow small cations (sodium and potassium ions) to move through them (**Figure 9.4**a). Recall that at the resting membrane potential, potassium is much closer to equilibrium than sodium. Therefore, there is a much stronger electrochemical force for sodium ions to move into the cell than for potassium ions to move out of the cell. Thus when channels open that allow both sodium and potassium ions to move through them, more sodium ions move in than potassium ions move out, producing a net depolarization. Fast EPSPs can last from just a few to several hundred milliseconds.

Slow EPSPs can be caused by various mechanisms, but a typical mechanism is the closing of potassium

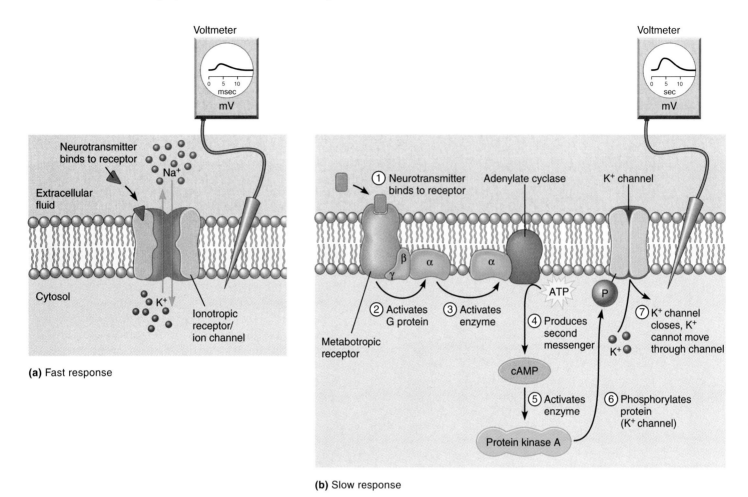

(a) Fast response

(b) Slow response

Figure 9.4　Excitatory synapses.　An electrode inserted into a neuron and connected to a voltmeter can measure electrical activity in the cell. The type of voltmeter depicted in this and subsequent figures is an oscilloscope. **(a)** A fast excitatory synapse. The neurotransmitter opens ion channels, allowing Na^+ to enter the cell and K^+ to leave it. Sodium movement is greater, so the net effect is a depolarization (an EPSP) lasting several milliseconds. **(b)** A slow excitatory synapse. Activation of a G protein by neurotransmitter leads to the series of events depicted. Phosphorylation of the potassium channel in the final step decreases the leakage of potassium out of the cell, producing a depolarization. Note that this response takes seconds to occur.

channels involving the most common second messenger, cAMP (Figure 9.4b). When the neurotransmitter binds to a receptor ① and activates a G protein ②, the G protein then activates the enzyme adenylate cyclase ③, which then catalyzes the reaction converting ATP to cAMP ④. cAMP acts as a second messenger that activates protein kinase A ⑤, which catalyzes the addition of a phosphate group to the potassium channel ⑥. Phosphorylation of the potassium channel closes it ⑦.

Before the potassium channel was phosphorylated, the channel was open, and potassium ions were leaking out of the cell while sodium ions were leaking into the cell. No change in membrane potential was occurring because the sodium and potassium leaks were balanced. When the potassium channel is closed by phosphorylation, fewer potassium ions leak out of the cell. The net effect is a depolarization, because if permeability to potassium decreases while permeability to sodium remains the

same, the membrane potential will move away from the equilibrium potential for potassium and toward the equilibrium potential for sodium (see Chapter 8).

Slow EPSPs take longer to develop and last longer (seconds to hours) than fast EPSPs. In the case of cAMP-dependent closure of potassium channels, the membrane potential will not return to resting conditions until the potassium channels are dephosphorylated, which will not occur until enough cAMP has been degraded by phosphodiesterase to return the cAMP concentration to its resting level.

Inhibitory Synapses

An **inhibitory synapse** is one that either takes the membrane potential of the postsynaptic neuron away from the action potential threshold by hyperpolarizing

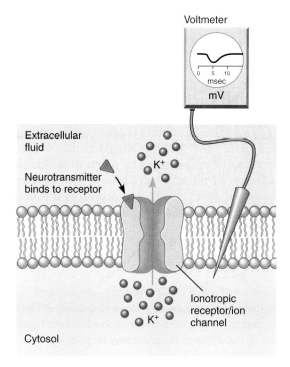

Figure 9.5 An inhibitory synapse involving potassium channels. At this type of inhibitory synapse, potassium channels are opened by the binding of neurotransmitter to the receptor. Potassium flows out of the cell, hyperpolarizing it and producing an IPSP.

the neuron, or stabilizes the membrane potential at the resting value. In either case, activity at the synapse decreases the likelihood that an action potential will be generated in the postsynaptic neuron—which is why the synapse is said to be *inhibitory*. At inhibitory synapses, the binding of a neurotransmitter to its receptors opens channels for either potassium ions or chloride ions.

When a neurotransmitter causes potassium channels to open (**Figure 9.5**), potassium will move out of the cell, hyperpolarizing it. This hyperpolarization is called an **inhibitory postsynaptic potential (IPSP)**. An IPSP, like an EPSP, is a graded potential: When more neurotransmitter molecules bind to receptors, more potassium channels open and the hyperpolarization is greater. IPSPs (like EPSPs and all other graded potentials) decrease in size as they move toward the axon hillock, where the hyperpolarization moves the membrane potential away from threshold, thereby decreasing the likelihood that an action potential will be generated in the postsynaptic neuron.

What happens when a neurotransmitter causes chloride (Cl⁻) channels to open depends on the size and direction of the electrochemical driving force acting to move chloride ions across the postsynaptic neuron membrane.

Because the chloride ion has a negative charge (is an anion), the membrane potential at rest is an electrical force that acts to move chloride out of the cell. However, because chloride ions are found in greater concentration outside the cell than inside, there is a chemical force for moving chloride into the cell. The net force depends on the neuron in question.

Neurons differ in the transport mechanisms that exist for chloride ions. In some neurons, chloride ions are actively transported out of the cell, making the concentration inside much lower than outside the cell. Under these conditions, the equilibrium potential for chloride is negative and larger than the resting membrane potential. In other neurons no active transport of chloride occurs, but there are some ion channels through which chloride can diffuse. Under these conditions, the passive movement of chloride coupled with the lack of active transport allows chloride ions to be at equilibrium.

In a neuron in which chloride ions are actively transported out of the cell, chloride is concentrated in the extracellular fluid (**Figure 9.6**a). Therefore, chloride is not at equilibrium; instead, a net electrochemical driving force for moving chloride into the cell exists. When chloride channels are opened by neurotransmitters binding to receptors, chloride moves into the cell, producing a hyperpolarization, or an IPSP. Thus we see the same type of effect that occurs when potassium channels open.

In a neuron that lacks active transport of chloride either into or out of the cell (Figure 9.6b), chloride simply diffuses through chloride leak channels to distribute itself across the membrane at concentrations that will put it at equilibrium. (Note that the membrane potential determines the concentration gradient for chloride ions when they are not actively transported; this is in contrast to sodium and potassium ions, whose concentration gradients, established by active transport, determine the resting membrane potential.) If chloride is at equilibrium, then the opening of chloride channels will not cause a change in the membrane potential (Figure 9.6b). In this case, perhaps surprisingly, the synapse is still considered to be *inhibitory*. To see why, we must consider how this synapse affects signals that might be generated at other synapses at the same time.

Consider a postsynaptic neuron in which an excitatory synapse and an inhibitory synapse are active at the same time. While the excitatory synapse tends to induce an EPSP (a positive change in membrane potential) in the postsynaptic neuron, the inhibitory synapse also causes chloride channels to open (Figure 9.6c). The result is that the EPSP will be diminished, or may even be absent. The reason for this is that when the membrane becomes depolarized, chloride is no longer at equilibrium. The more positive potential inside the cell pulls the negative chloride ions into the cell, stabilizing the membrane potential and countering the influence of

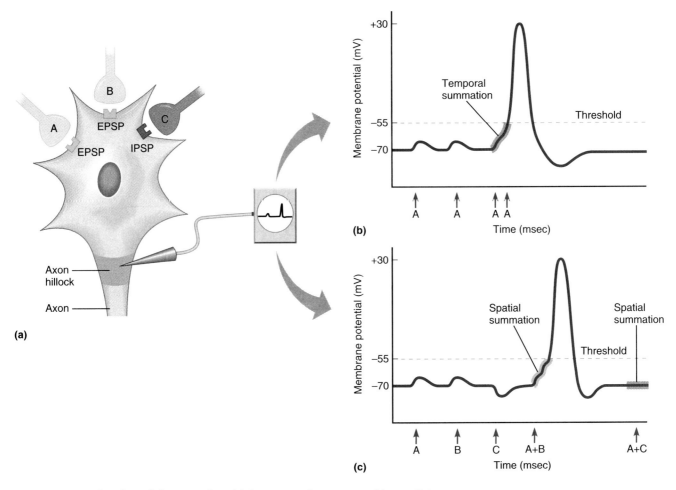

(a)

(b)

(c)

Figure 9.8 Temporal and spatial summation. (a) A postsynaptic neuron receiving excitatory input from neurons A and B, and inhibitory input from neuron C; three different neurotransmitters are involved. If the sum of all synaptic potentials at the axon hillock results in depolarization to threshold, then an action potential is generated. **(b)** Temporal summation occurs when action potentials arriving from a presynaptic axon terminal of neuron A occur close enough together in time such that the EPSPs produced in response to the binding of neurotransmitter overlap and sum. **(c)** Spatial summation occurs when different synapses are simultaneously active. If EPSPs from synapses with neurons A and B occur at the same time, the EPSPs sum and reach threshold, triggering an action potential. Note that if synapses with neurons A and C were simultaneously active, the resulting IPSP and EPSP would tend to cancel each other out, producing little change in membrane potential.

The synapse of neuron A with the postsynaptic cell is closer to the axon hillock than is the synapse of neuron B. If a single action potential in each presynaptic neuron results in an 8-mV depolarization at the axon hillock, which presynaptic neuron had to produce the larger depolarization at the site of the synapse?

causing a greater postsynaptic potential. In contrast, when a presynaptic neuron fires two action potentials farther apart in time, the neurotransmitter released as a result of the first action potential is completely cleared from the synaptic cleft before the next action potential arrives, and therefore the two postsynaptic potentials cannot sum. [Note, however, that when metabotropic (slow response) receptors are involved, a postsynaptic potential may persist even after the neurotransmitter that initiated it has been cleared from the synaptic cleft.]

Spatial Summation

In spatial summation (Figure 9.8c), two or more postsynaptic potentials originating from different synapses are generated at approximately the same time such that they overlap and sum. When neurons A and B have action potentials at different times, each neuron induces an EPSP

Neuron B, because the resulting graded potential would dissipate more by the time it reached the axon hillock

in the postsynaptic neuron, neither of which is large enough to trigger an action potential. When neurons A and B are activated at the same time, however, the resulting EPSPs sum to produce a depolization that *is* large enough to trigger an action potential. Note as well that if neurons A and C were active at the same time, no change in the membrane potential of the postsynaptic neuron would result, because the EPSP that originated from synapse A and the IPSP originating from synapse C would cancel each other out. Spatial summation occurs as postsynaptic potentials originating at different synapses spread to the axon hillock, overlapping along the way.

Figure 9.8 is a highly simplified example of the concept of neural integration. Because the number of synapses with a postsynaptic neuron can reach the hundreds or thousands, and because some synapses are excitatory and others are inhibitory, the number of possible combinations of active synaptic inputs is astronomical. Temporal and spatial summation of postsynaptic potentials occurs continually, and the magnitude of the summed potential at the axon hillock determines if and when action potentials occur.

Frequency Coding

In Chapter 8 we saw that once a depolarizing stimulus exceeds threshold, the degree of depolarization does not affect the size of action potentials, but rather their frequency. We saw that increases in the strength of such suprathreshold stimuli cause the frequency of action potentials to increase, an effect called **frequency coding**. Because the summation of postsynaptic potentials affects the degree of depolarization at the axon hillock, it follows that summation influences the frequency of action potentials in the postsynaptic neuron. When action potentials occur at a higher frequency, more neurotransmitter is released from the neuron. Therefore, a higher frequency of action potentials corresponds to stronger communication to the next neuron in a neural pathway.

The excitatory and inhibitory synapses described above occur primarily at axodendritic and axosomatic synapses. We now describe what happens at axoaxonic synapses.

Presynaptic Modulation

Axoaxonic synapses function as *modulatory synapses,* synapses that regulate the communication across another synapse. Recall that axoaxonic synapses can form between the axon terminal of one neuron and the axon terminal of another neuron (see Figure 9.1). In axoaxonic synapses, neurotransmitter from the presynaptic neuron (also called a *modulating neuron*) does not generate electrical signals in the postsynaptic neuron. Instead, by binding to receptors on the membrane of the axon terminal of the postsynaptic neuron, the neurotransmitter induces a change in the

amount of calcium that enters the axon terminal in response to an action potential, in turn altering the amount of neurotransmitter released from the postsynaptic neuron. In some cases the release of neurotransmitter is enhanced, a phenomenon called **presynaptic facilitation**; in other cases the release of neurotransmitter is decreased, a phenomenon called **presynaptic inhibition**. We consider these phenomena next.

Presynaptic Facilitation

Presynaptic facilitation is illustrated in **Figure 9.9**a. Neurons C and D are presynaptic neurons that form excitatory synapses with postsynaptic neuron X; neuron E is a modulating neuron that forms a synapse on the axon terminal of neuron C. In this example, an action potential in neuron C triggers the release of enough neurotransmitter to produce a 10-mV EPSP in neuron X when E is not active. However, when neuron E is active, a single action potential in neuron C will release enough neurotransmitter to produce a 15-mV EPSP and trigger an action potential. Thus the activity of neuron E increases the amount of neurotransmitter released from neuron C such that neuron C triggers a larger EPSP in neuron X, compared to when neuron E is inactive. Note that activity in neuron E alone has no direct effect on neuron X because neuron E does not form a synapse with neuron X. Note as well that activity in neuron E has no effect on any EPSPs triggered by input from neuron D.

This example illustrates an important characteristic of presynaptic modulation at axoaxonic synapses that differentiates it from axodendritic and axosomatic synapses: Presynaptic modulation affects transmission to the postsynaptic neuron at one specific synapse, thereby altering the ability of that one synapse to excite or inhibit the postsynaptic neuron. In contrast, at axodendritic and axosomatic synapses, presynaptic neurons nonselectively excite or inhibit the postsynaptic neuron by directly producing changes in membrane potential of the postsynaptic neuron in the form of EPSPs and IPSPs.

Presynaptic Inhibition

Figure 9.9b shows an example of presynaptic inhibition. In this example, neurons F and G make direct excitatory synaptic connections with neuron Y; neuron H is a modulating neuron that forms an axoaxonic synapse with neuron F. When either neuron F alone or neuron G alone is active, an EPSP is produced in neuron Y. When neuron H is active by itself, no change in the membrane potential of neuron Y occurs. However, when neurons F and H are active at the same time, the response of neuron Y to input from neuron F is decreased; that is, neuron H suppresses the release of neurotransmitter from neuron F. In contrast, activity in neuron H does not affect the response of neuron Y to input from neuron G.

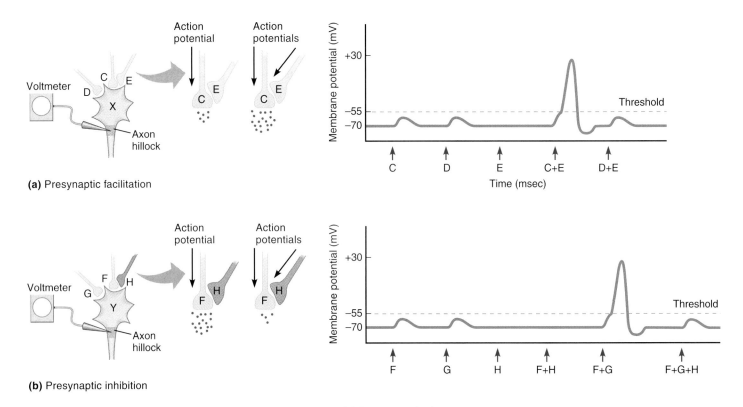

Figure 9.9 Presynaptic modulation at axoaxonic synapses. (a) Presynaptic facilitation. When neurons C and E are both active, neurotransmitter from neuron E enhances the release of neurotransmitter from neuron C, increasing the strength of the resulting EPSP to threshold and generating an action potential in neuron X. **(b)** Presynaptic inhibition. When neurons F and H are both active, neurotransmitter from neuron H decreases the release of neurotransmitter from neuron F, decreasing the strength of the EPSP in neuron Y. When neurons F and G are both active, their EPSPs sum spatially, and neuron Y is depolarized to threshold, generating an action potential. However, when neurons F, G, and H are all activated, the EPSP from neuron F is diminished due to presynaptic inhibition by neuron H, such that the sum of the EPSPs from neurons F and G is subthreshold, and an action potential is not generated.

Note that although Figure 9.9 illustrates presynaptic facilitation and presynaptic inhibition at excitatory synapses, these phenomena can occur at inhibitory synapses as well.

Quick Test 9.3

1. Define the following terms: *spatial summation, temporal summation, presynaptic facilitation,* and *presynaptic inhibition.*

2. During presynaptic inhibition, does the amount of neurotransmitter released increase or decrease?

Neurotransmitters: Structure, Synthesis, and Degradation

Neurotransmitters fall into a variety of chemical classes, including acetylcholine, biogenic amines, amino acids, and neuropeptides. Most are small molecules, with the notable exception of the neuropeptides, which are considerably larger than the others. A few other neurotransmitters, such as *nitric oxide* and ATP, are unlike any others and each is in a class by itself. Selected members of the different classes of neurotransmitters discussed in the following sections are listed in **Table 9.1**.

Acetylcholine

Acetylcholine (ACh) is released from neurons in both the central and peripheral nervous systems. It is the most abundant neurotransmitter in the peripheral nervous system, where it is found in efferent neurons of both the somatic and autonomic branches.

Acetylcholine is synthesized in the cytoplasm of the axon terminals of neurons (**Figure 9.10**). The synthesis of acetylcholine from two substrates, *acetyl CoA* and *choline,* is catalyzed by **choline acetyl transferase (CAT)**, as follows:

$$\text{acetyl CoA} + \text{choline} \xrightarrow{\text{CAT}} \text{acetylcholine} + \text{CoA}$$

Table 9.1 ∎ Selected Members of the Classes of Neurotransmitters

Choline derivative	Biogenic amines	Amino acids	Neuropeptides	Others
Acetylcholine	Catecholamines	Glutamate	TRH	Nitric oxide
	Dopamine	Aspartate	Vasopressin	ATP
	Epinephrine	Glycine	Oxytocin	
	Norepinephrine	GABA	Substance P	
	Serotonin		Endogenous opioids	
	Histamine		Enkephalins	
			Endorphins	

We saw in Chapter 3 that one of these substrates, acetyl CoA, is a two-carbon molecule produced during the catabolism of lipids, carbohydrates, and proteins, and it is the initial substrate for the Krebs cycle. Therefore, it is found in almost all cells of the body, including those neurons that synthesize and release acetylcholine (called *cholinergic* neurons). In contrast, choline cannot be synthesized by neurons (although some is synthesized in the liver); instead, most choline is obtained from the diet and delivered by the bloodstream to cholinergic neurons, where it is taken in via an active transport system.

Once it is synthesized, acetylcholine is stored in synaptic vesicles (see Figure 9.10) until an action potential triggers its release by exocytosis. After release, acetylcholine can bind to receptors on the postsynaptic cell (called **cholinergic** receptors) and/or be degraded by an enzyme called **acetylcholinesterase (AChE)**, which is found on either the presynaptic neuron membrane, the postsynaptic neuron membrane, or both. Acetylcholinesterase catalyzes the breakdown of acetylcholine into acetate (an acetyl CoA molecule without the CoA attached) and choline, as follows:

$$\text{acetylcholine} \xrightarrow{\text{AChE}} \text{acetate} + \text{choline}$$

Neither of the breakdown products of this reaction has any transmitter activity, but the choline that is produced is taken back into the presynaptic cell by an active transport mechanism and can be used to synthesize more acetylcholine. The acetate diffuses away from the synapse and enters the blood.

Cells in the nervous system commonly have different types of receptors for a specific neurotransmitter. Receptors for acetylcholine are of two types: **nicotinic cholinergic receptors** and **muscarinic cholinergic receptors**, so named because the drug nicotine binds to nicotinic receptors but not to muscarinic receptors, whereas the drug muscarine binds to muscarinic receptors but not nicotinic receptors. Even though these receptors share a common messenger—acetylcholine—the effects on the postsynaptic cell are very different (**Figure 9.11**).

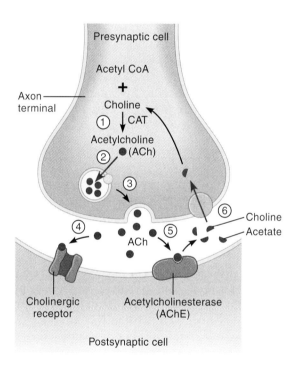

Figure 9.10 Neurotransmitter synthesis, action, and degradation at cholinergic synapses. The sequence of events at a cholinergic synapse. ① The enzyme choline acetyl transferase (CAT) catalyzes the reaction combining acetyl CoA with choline to form ACh (and CoA). ② The ACh is then actively packaged into synaptic vesicles. ③ ACh is released by exocytosis. Once released, ACh may ④ bind to cholinergic receptors on the postsynaptic cell or ⑤ be degraded by the enzyme acetylcholinesterase (AChE) into choline and acetate. ⑥ Choline is actively transported back into the presynaptic neuron, where it can be used to synthesize more acetylcholine.

What triggers the release of acetylcholine by exocytosis?

Calcium ions entering the cell through voltage-gated calcium channels

Figure 9.11 Signal transduction mechanisms at cholinergic receptors. (a) Nicotinic cholinergic receptors are receptor-operated channels that permit both sodium and potassium ions to move through. When acetylcholine binds to these receptors, the channels open. Because more sodium moves in than potassium moves out, the result is an EPSP. (b) Muscarinic cholinergic receptors are coupled to G proteins that can either directly open/close ion channels or activate/inhibit an enzyme that catalyzes the production of a second messenger. The second messengers produced can have a variety of effects on the postsynaptic cell, including opening or closing ion channels. The effects on postsynaptic cells can be either excitatory (EPSPs) or inhibitory (IPSPs).

Which cholinergic receptor type (nicotinic or muscarinic) produces the faster response?

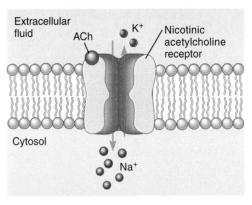

(a) Nicotinic cholinergic receptors

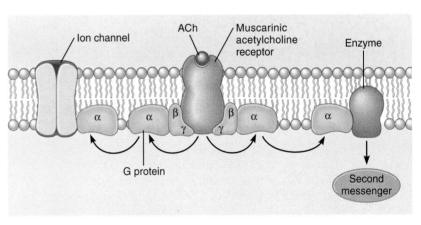

(b) Muscarinic cholinergic receptor

Nicotinic cholinergic receptors are ionotropic, and the binding of acetylcoline to them triggers the opening of channels that allow both sodium and potassium to move through, causing an EPSP in the postsynaptic cell (Figure 9.11a). Nicotinic cholinergic receptors are located in several areas of the peripheral nervous system—for example, in certain autonomic neurons and on skeletal muscle cells, which are the effector cells for somatic motor neurons. Nicotinic cholinergic receptors are also located in some regions of the central nervous system.

Muscarinic cholinergic receptors are metabotropic receptors and operate through the action of a G protein. The effect resulting from the binding of acetylcholine to muscarinic cholinergic receptors depends on the postsynaptic cell in question; these receptors can open or close ion channels or activate enzymes (Figure 9.11b). Muscarinic cholinergic receptors, which are found on some effector organs for the autonomic nervous system, are the dominant cholinergic receptor type found in the central nervous system.

The different effects of acetylcholine at nicotinic and muscarinic cholinergic receptors illustrate an important physiological concept: The action of any chemical messenger ultimately depends not on the nature of the messenger but on the type of receptor to which the messenger binds on the target cell.

Biogenic Amines

The **biogenic amines** are a class of neurotransmitters derived from amino acids. They are called *amines* because they all possess an amine group ($-NH_2$). (Note that they do not include amino acids, which also possess amine groups.) The biogenic amines include the catecholamines, serotonin, and histamine. The catecholamines contain a *catechol group*, a six-carbon ring with two hydroxyl groups, and include the neurotransmitters dopamine, norepinephrine, and epinephrine. (The synthetic pathway for catecholamines is described in Chapter 5.) Like acetylcholine, the biogenic amines are synthesized in the cytosol of the axon terminal and packaged into synaptic vesicles.

Dopamine and norepinephrine are both released primarily by neurons in the central nervous system, but norepinephrine is also released from neurons in the peripheral nervous system. Even though epinephrine is released from some neurons of the central nervous system, it is better known as a hormone that is released from the adrenal medulla in response to commands from the sympathetic nervous system.

The different catecholamines bind to specific receptors. Receptors for epinephrine and norepinephrine are called **adrenergic** receptors, of which there are two main classes, *alpha adrenergic* and *beta adrenergic* receptors. Each of these has subclasses designated by numerical subscripts, with the primary receptor types being alpha₁, alpha₂, beta₁, beta₂, and beta₃. Epinephrine has the highest affinity for beta₂ receptors, whereas norepinephrine has higher affinity for alpha and beta₁ receptors, but both compounds bind to all these receptors. Adrenergic receptors are found in the central nervous system and in the effector organs for the sympathetic branch of the autonomic nervous system. Adrenergic receptors are referred to as being *adrenergic* because alternate terms for epinephrine and norepinephrine are *adrenaline* and *noradrenaline*, respectively. In similar fashion, receptors that bind dopamine are called *dopaminergic*.

Catecholamines generally produce slow responses mediated through G proteins and changes in second messenger systems. In addition, they often function as autocrines, binding to receptors (called **autoreceptors**) on the axon terminal of the cell that released them. These receptors enable a neuron to modulate its own release of neurotransmitter, generally by altering the amount of calcium that enters the cell in response to an action potential. Neurons possess autoreceptors for other neurotransmitters as well.

Following their release, catecholamines can be degraded by two enzymes, **monoamine oxidase (MAO)** and **catechol-*O*-methyltransferase (COMT)**. MAO is located in the synaptic cleft, in the axon terminal of the neurons that release the catecholamines, and in some glial cells. COMT is located in the synaptic cleft. To understand the importance of the enzymes that degrade neurotransmitters, see **Clinical Connections: Treating Depression**, p. 258.

Serotonin and histamine are biogenic amines, but they are not catecholamines. Serotonin is found in the central nervous system, particularly the lower portion of the brain called the *brainstem*. Some functions of serotonin include regulating sleep and emotions. Histamine, better known for its release from non-neuronal cells during allergic and other types of reactions, can also function as a neurotransmitter. Histamine is found in the central nervous system, primarily in an area of the brain called the *hypothalamus*.

Amino Acid Neurotransmitters

The amino acid neurotransmitters are the most abundant class of neurotransmitters in the central nervous system, where they are widely distributed throughout virtually all areas. **Aspartate** and **glutamate** are released at excita-

(a) Amino acid neurotransmitters at excitatory synapses

(b) Amino acid neurotransmitters at inhibitory synapses

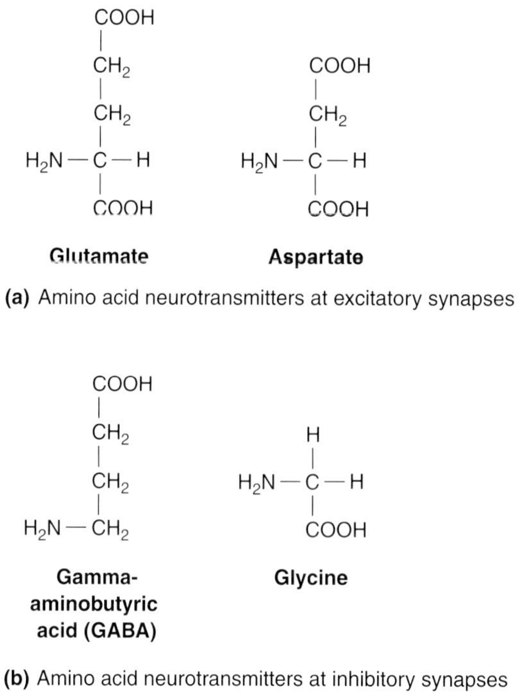

Figure 9.12 The amino acid neurotransmitters. (a) Amino acid neurotransmitters released at excitatory synapses. Note their similar chemical structures. (b) Amino acid neurotransmitters released at inhibitory synapses.

tory synapses, whereas **glycine** and *gamma-aminobutyric acid*, more commonly known as **GABA**, are released at inhibitory synapses (**Figure 9.12**). Because GABA is derived from one of the other amino acid neurotransmitters, glutamate, it has a structure similar to an excitatory amino acid but produces inhibitory effects. The reason for this apparent contradiction is that the receptors for these compounds are very specific; thus even though GABA and glutamate are similar in structure, they bind to completely separate populations of receptors and produce opposite results.

Glutamate is the most commonly released neurotransmitter at excitatory synapses in the central nervous system. Glutamate can bind to three main classes of receptors: α-amino-3-hydroxy-5-methyl-4-isoxazole propionate (AMPA) receptors, *N*-methyl-ᴅ-aspartate (NMDA) receptors, and kainate receptors. When glutamate binds to AMPA or kainate receptors, a fast EPSP is produced by sodium movement into the cell. When glutamate binds to NMDA receptors, calcium channels open, resulting in an influx of calcium. Calcium acts as a second messenger, producing biochemical changes in the postsynaptic neuron.

Neuropeptides

The **neuropeptides** are short chains of amino acids that are synthesized in the same manner as proteins. More

Central Nervous System   **Peripheral Nervous System**

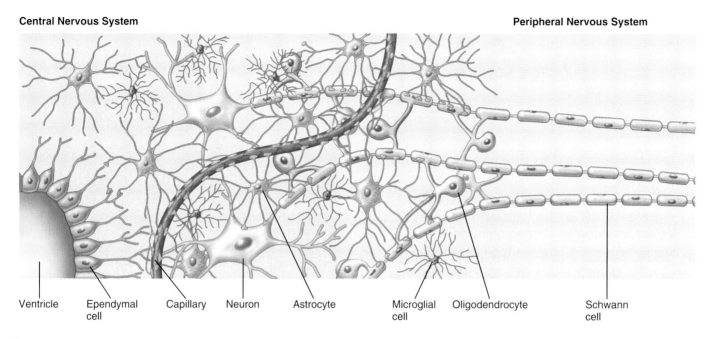

Ventricle   Ependymal   Capillary   Neuron   Astrocyte   Microglial   Oligodendrocyte   Schwann
                   cell   cell   cell

Figure 10.1 Glial cells in the nervous system.

Microglia

Microglia protect the central nervous system from foreign matter, such as bacteria and remnants of dead or injured cells. They carry out this function through phagocytosis and the release of cytokines in a manner similar to certain blood cells. Microglia also protect neurons against oxidative stress.

Glial Cells in Neurodegenerative Diseases

Evidence is growing that glial cells contribute to neurodegenerative diseases such as multiple sclerosis, Alzheimer's disease, and Parkinson's disease. Multiple sclerosis results from the loss of myelin in the central nervous system. Multiple sclerosis is an autoimmune disease, a disease in which the immune system attacks a part of the body, in this case oligodendrocytes. The loss of myelin (and some axons) in the central nervous system slows down or stops communication along certain neural pathways. Symptoms include blurred vision, muscle weakness, and difficulty maintaining balance. Multiple sclerosis is described in more detail in Chapter 23.

Alzheimer's disease is caused by the loss of cholinergic neurons in certain brain areas and replacement of the lost neurons with scar tissue called plaques. During the degeneration of cholinergic neurons, astrocytes and microglia become overly active. These glial cells release inflammatory chemicals that enhance further degeneration of cholinergic neurons. Thus a vicious cycle takes place. Early signs of Alzheimer's disease include loss of memory and confusion. Later signs include motor dysfunction, loss of communication skills, and decrease in cognitive functions. Treatment is limited due to the difficulty of diagnosing the disease until advanced stages. Early stages are treated with acetylcholinesterase inhibitors, but these are not effective in the advanced disease.

Parkinson's disease is a degenerative disease involving the loss of dopaminergic neurons. As in Alzheimer's disease, glial cells are thought to enhance neural degeneration through the production of inflammatory agents.

Glial cells protect neurons in their immediate vicinity. We next look at the physical support of the CNS.

Physical Support of the Central Nervous System

The outermost structures that protect the soft tissues of the central nervous system are the bony skull, or **cranium**, which surrounds the brain, and the bony **vertebral column**, which surrounds the spinal cord (**Figure 10.2**). Even though these bony structures are clearly beneficial in that they act as rigid armor surrounding delicate nervous tissue, their hardness also poses a potential hazard: What would prevent the soft brain from crashing into the hard inner surface of the skull when, for example, you make a sudden stop in a car moving at freeway speeds? Between the bone and the nervous tissue are a series of three membranes called the meninges and a layer of fluid called cerebrospinal fluid, which provide protection against such impact.

The **meninges** are three connective tissue membranes that separate the soft tissue of the CNS from the surrounding bone (see Figure 10.2). The three meningeal membranes are the **dura mater**, the **arachnoid mater**, and the **pia mater**. *Mater* is Latin for mother, indicating the protective nature of the meninges. The dura mater is the outermost layer, closest to the bone. *Dura* is Latin for "hard, durable," and the dura mater is a very tough, fibrous tissue. The arachnoid mater is the middle layer. *Arachnoid* is Greek for "spider," which appropriately describes the arachnoid mater's weblike structure. Normally,

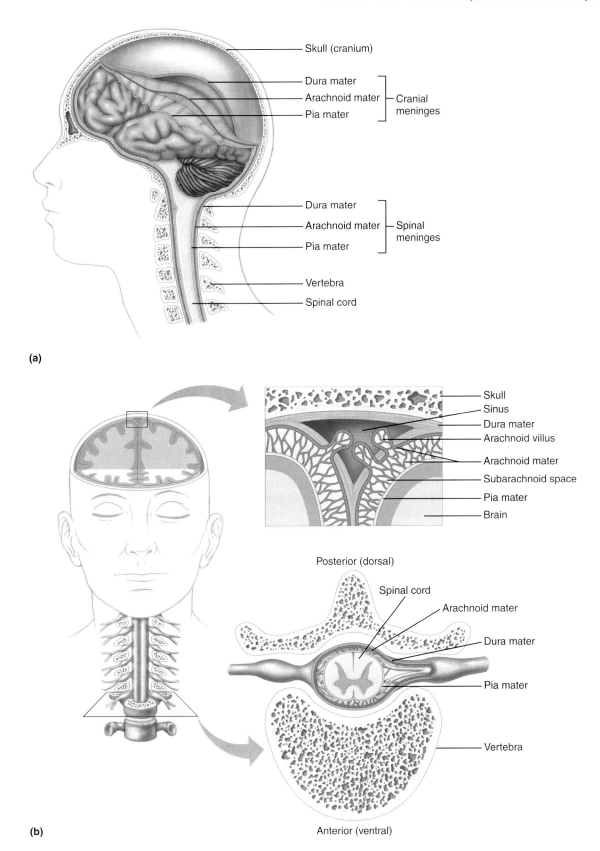

(a)

(b)

Posterior (dorsal)

Anterior (ventral)

Figure 10.2 Protective structures of the CNS. (a,b) Sections of CNS protective structures. Bony outer structures include the cranium and vertebrae. The meninges, located between the bony structures and the soft nervous tissue, are composed of three layers: dura mater, arachnoid mater, and pia mater. The cushioning presence of cerebrospinal fluid within the subarachnoid space provides yet another level of protection.

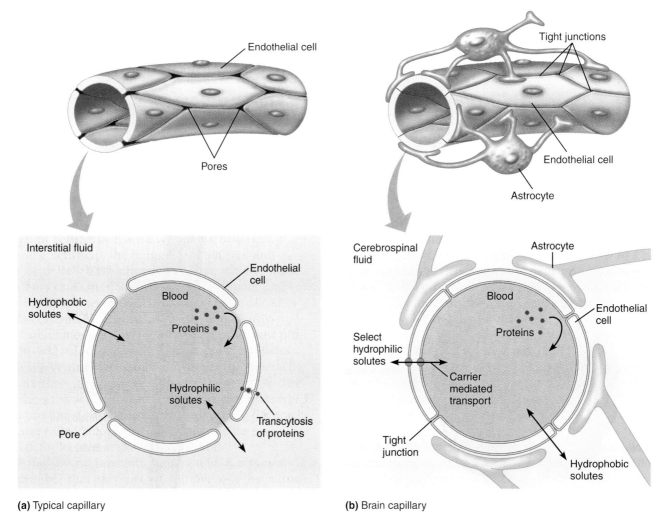

(a) Typical capillary

(b) Brain capillary

Figure 10.4 Blood-brain barrier. (a) Typical capillaries (found in most regions of the body). Whereas exchange of small hydrophilic molecules occurs by simple diffusion between blood and interstitial fluid through pores, proteins are too large to cross through pores; some proteins are transported across capillary walls by transcytosis. (b) Brain capillaries. Because endothelial cells in these capillaries are connected by tight junctions, hydrophilic molecules must be transported across the wall by mediated transport systems. Proteins cannot cross the blood-brain barrier because transcytosis does not occur in brain capillaries. Even though astrocytes are found in close association with brain capillaries, they do not constitute a functional barrier.

capillaries, molecules must cross the endothelial cells themselves. Gases and other hydrophobic molecules penetrate through these cells relatively easily because they are able to move across cell membranes by simple diffusion through the lipid bilayer. As a consequence, these molecules are able to move freely between blood and brain tissue. One example is ethanol (grain alcohol), which depresses CNS function by mechanisms not fully understood.

In contrast, hydrophilic substances such as ions, carbohydrates, and amino acids cannot cross the plasma membranes of any cells by simple diffusion and must therefore rely on mediated transport to cross capillary walls in the CNS. Because mediated transport utilizes

transport proteins, which are selective to certain molecules, the blood-brain barrier is selectively permeable, allowing only certain compounds to move across. Compounds such as glucose and amino acids can penetrate the blood-brain barrier readily by facilitated diffusion because transport proteins for these compounds are present in cell membranes. Other substances transported across the blood-brain barrier include choline (needed for acetylcholine synthesis) and aspirin. Other hydrophilic substances, such as inorganic ions (including H^+) and certain drugs, cannot penetrate this barrier to any great degree because transport proteins for these compounds do not exist in the endothelial cells of CNS capillaries. It is because of the selective

STROKE

Cerebrovascular diseases—those that affect blood vessels in the brain—are the third leading cause of death in the United States today, behind only heart disease and cancer. Cerebrovascular diseases are responsible for impairment of brain function in approximately 2 million Americans.

Stroke (also known as *cerebrovascular accident* or *CVA*) is a cerebrovascular disorder characterized by a sudden decrease or stoppage of blood flow to a part of the brain. Decreased blood flow (ischemia) is dangerous to any tissue, but brain tissue is especially vulnerable, in part because of the high rate of its metabolic reactions, which consume fuel and oxygen quickly. In fact, stoppage of blood flow for no more than three or four minutes may be sufficient to cause death of most brain cells. For this reason, a stroke can kill people within minutes, or leave them with permanent brain damage.

Strokes are classified as either *occlusive* or *hemorrhagic* and may occur in the interior of the brain or on its surface. In occlusive stroke, blood flow through a vessel is blocked, usually as a result of a blood clot. In hemorrhagic stroke, a blood vessel ruptures, causing *hemorrhage* (bleeding). Conditions that increase the likelihood of stroke include *arteriosclerosis* (narrowing of arteries due to buildup of fatty deposits on the inner wall), *aneurysm* (bulging of an artery due to weakening of the wall), and *hypertension* (high blood pressure). Stroke symptoms vary with the location and extent of the affected area, and are sudden in onset, which distinguishes stroke from other neurological disorders. There may be muscle weakness or paralysis, loss of sight (usually in just a part of the visual field), loss of other senses, tingling or other sensory disturbances, disturbances of language or spatial perception,

changes in personality, disorientation, or memory loss.

Effects of a stroke may be confined to the initial symptoms, or they may intensify or become more widespread over the course of a few hours or days. Thus a person who first experiences numbness in a hand may find it spreading through the arm and shoulder. There are a number of possible reasons for this. If the stroke is hemorrhagic, blood could pool in the brain tissue and exert pressure on nearby vessels, thereby occluding blood flow over a wider area. Stroke sometimes induces spasms of smooth muscle in blood vessel walls, which also reduces flow.

Another possible cause of progressively worsening symptoms following a stroke is *glutamate excitotoxicity,* a phenomenon discovered only recently. Recall from Chapter 9 that glutamate is an excitatory neurotransmitter that affects virtually every cell in the CNS. Following a stroke, damaged neurons become chronically depolarized, which causes them to release neurotransmitter continually. As a result,

the neurotransmitter excites nearby cells, which stimulates them to release more glutamate, which excites more cells, and so on. The result of the release of so much glutamate is that it accumulates in the extracellular fluid at high concentrations, such that the wave of excitatory activity cannot be stemmed.

Unfortunately, the story does not end there. Among the receptors activated by glutamate are NMDA receptors in receptor-operated calcium channels. When glutamate binds to NMDA receptors, calcium permeability increases. When glutamate levels are chronically elevated following a stroke, calcium ions diffuse across the plasma membranes of cells at a rate that exceeds the ability of active transport mechanisms to remove calcium from the cytoplasm. Consequently, cytoplasmic calcium concentrations rise to toxic levels, resulting in the activation of proteases and other enzymes whose activities are normally carefully regulated. The result is a derangement of metabolism, and eventually cell death.

Three-dimensional magnetic resonance angiogram showing blood supply to the brain (white) and area of stroke (yellow).

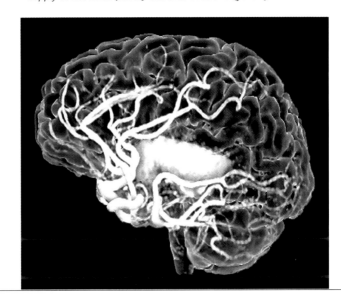

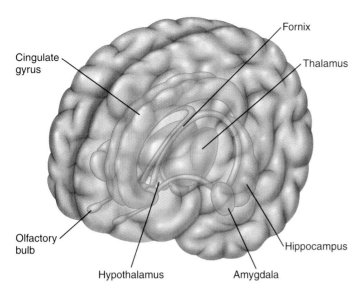

Figure 10.17 The limbic system. The major structures of the limbic system are depicted in this three-dimensional view.

Quick Test 10.3

1. Name the two structures that make up the diencephalon, and list their major functions.

2. What three structures make up the brainstem?

3. Name the four lobes of the cerebral cortex and describe at least one function associated with each.

4. Name five functions of the hypothalamus.

5. For each of the following functions, indicate whether the left or right cerebral hemisphere plays the predominant role: (a) creative activity, (b) logic, (c) motor control on the left side of the body, (d) sensory perception on the left side of the body, (e) language comprehension and expression.

Thus far we have focused on identifying specific areas of the CNS and describing their major functions. Although it is true that different areas specialize in different functions, it is important to realize that most tasks carried out by the CNS require the coordinated action of many different areas working together. For example, we can identify an object as an apple because we can see it, touch it, smell it, taste it, put those sensations together into a coherent "picture" of the object, and compare that picture with our concept of what an apple is, which derives from our remembered experiences with apples. Thus, the seemingly simple task of identifying an apple involves the coordinated activity of several sensory areas, memory storage areas, and association areas. In the next section we examine several CNS functions that involve such coordinated actions, starting with the simplest functions: *reflexes.* We then examine more complex tasks, in-cluding voluntary motor control, language, sleep and consciousness, emotions and motivation, and learning and memory.

Integrated CNS Function: Reflexes

We have seen that neural pathways are a series of neurons connected by synapses that form a line of communication required for a specific task. One example is a visual pathway that starts in the eye with light-sensitive photoreceptor cells and ends in the area of visual cortex responsible for perception of vision. Pathways in the nervous system vary in their complexity. The simplest pathways in the nervous system are *reflex arcs.*

When we react to some stimuli, we often stop to consider the situation before taking action. Sometimes, however, our responses are automatic, involving no conscious intervention on our part, as when we jump in response to a loud sound. Such an automatic, patterned response to a sensory stimulus is called a **reflex**.

Reflexes can be categorized into the following four groups, each of which contains two classes (**Table 10.3**):

1. Reflexes can be either *spinal* or *cranial* according to the level of neural processing involved. In spinal reflexes, the highest level of integration occurs in the spinal cord; cranial reflexes require participation by the brain.

2. Reflexes can be either *somatic* or *autonomic,* depending on which efferent division controls the pathway. Somatic reflexes involve signals sent via somatic neurons to skeletal muscle; autonomic reflexes (also called *visceral reflexes*) involve signals sent via autonomic neurons to smooth muscle, cardiac muscle, or glands.

3. Reflexes can be either *innate* (inborn) or *conditioned* (learned).

4. Reflexes can be either *monosynaptic* or *polysynaptic.* In monosynaptic reflexes, the neural pathway consists of only two neurons and a single synapse; polysynaptic reflexes contain more than two neurons and multiple synapses.

Note that a given reflex can belong in more than one class. The reflex involved when the patellar tendon below the knee is tapped and the leg kicks forward is both a spinal reflex (the brain is not involved) and a somatic reflex (because efferent signals are sent to skeletal muscles via somatic neurons). Similarly, the reflex involved when the pupil constricts as a light is shined in the eye is both a cranial reflex (because it involves the brain) and an autonomic reflex (because efferent signals are sent to the smooth muscle surrounding the pupil via autonomic neurons).

Note as well that these reflexes are innate, as they need not be learned. The classic example of a learned reflex is the *salivation reflex,* first described by Ivan Pavlov, a Russian physiologist who won the Nobel Prize in 1904 for his study on digestion in dogs. In his classic demonstration of the salivation reflex, over time Pavlov rang a bell just before he fed the dogs. The dogs learned to associate the sound of the bell with the arrival of food, so that eventually they began to salivate upon hearing the bell, even if no food was forthcoming. Thus salivation in response to the sound of a bell is a conditioned (learned) reflex.

Neural pathways for reflexes are known as **reflex arcs**, which consist of five components (**Figure 10.18**): (1) a sensory receptor, (2) an afferent neuron, (3) an integration center, (4) an efferent neuron, and (5) an effector organ. To set the reflex into motion, the receptor first detects a stimulus. Information is then transmitted from the receptor to the CNS via the afferent neuron. The CNS, which functions as the integrator, sends signals to the efferent neuron, which transmits signals to the effector organ, stimulating it to produce a specific response.

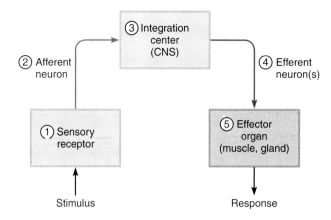

Figure 10.18 Schematic representation of a reflex arc. The five components of a reflex arc are a sensory receptor that detects a stimulus, an afferent neuron that transmits information from the receptor to the CNS, an integration center (which is generally the CNS), an efferent neuron that transmits information from the integration center to the periphery, and an effector organ, which produces a response to the stimulus.

Stretch Reflex

The simplest example of a reflex is the **muscle spindle stretch reflex**, of which the knee-jerk reflex (when the leg kicks in response to a tap just below the knee, as previously described) is an example. **Figure 10.19** depicts this reflex in detail. The stretch reflex is the only known monosynaptic reflex in the human body. The receptor in this reflex is a *muscle spindle,* a specialized structure found in skeletal muscles that responds when muscles are stretched. In the knee-jerk reflex, tapping the patellar tendon below the kneecap stretches the quadriceps muscle in the upper thigh. This stretch excites muscle spindles in that muscle, thereby triggering action potentials that travel in afferent neurons to the spinal cord (integration center). In the spinal cord, the afferent neurons make direct excitatory synaptic connections (synapse 1 in Figure 10.19) with efferent neurons that innervate the quadriceps muscle, the same muscle that contains the muscle

spindles, stimulating the quadriceps to contract and the leg to "kick" forward, or extend.

For extension of the leg to work effectively, the muscles causing extension (quadriceps) should not be opposed by the actions of other muscles that cause the leg to bend or flex (hamstrings). During the stretch reflex, contraction of the hamstrings could slow down or even prevent extension of the leg. This potential problem is avoided because the stretch reflex, once set into motion, not only triggers contraction of the quadriceps but also *inhibition* of the hamstrings. Afferent neurons from muscle spindles have collaterals that synapse with inhibitory interneurons innervating the motor neurons going to the hamstrings (synapse 2 in Figure 10.19). Simultaneous excitation of the quadriceps and inhibition of the hamstrings by muscle spindle afferents causes leg extension during the knee-jerk reflex.

Neural processing does not end here. Collaterals from muscle spindle afferent neurons also ascend to the brain, forming synapses with various interneurons. The brain

Table 10.3 ▌ Classes of Reflexes

Basis of classification	Classes	Example
Level of neural processing	Spinal	Muscle spindle stretch reflex
	Cranial	Pupillary reflex
Efferent division controlling effector	Somatic	Muscle spindle stretch reflex
	Autonomic	Baroreceptor reflex to control blood pressure
Developmental pattern	Innate	Muscle spindle stretch reflex
	Conditioned	Salivation reflex of Pavlov's dogs
Number of synapses in the pathway	Monosynaptic	Muscle spindle stretch reflex
	Polysynaptic	All other reflexes

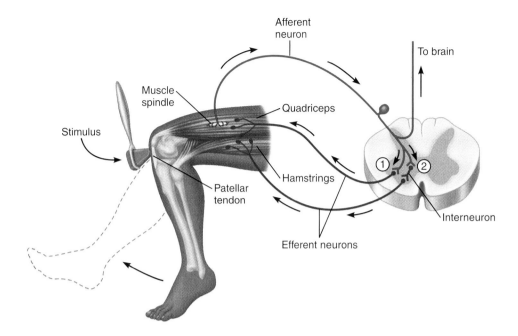

Figure 10.19 The muscle spindle stretch reflex. The knee jerk reflex, an example of the monosynaptic muscle spindle stretch reflex, by which a tap on the patellar tendon causes contraction of the quadriceps muscle. Muscle spindle afferent neurons make two synaptic communications in the spinal cord: ①excitatory synapses with efferent neurons to the quadriceps muscle, and ② synapses with inhibitory interneurons that communicate with efferent neurons to the hamstring muscles in the same leg. The afferent neurons also have collaterals that travel in the white matter of the spinal cord to the brainstem, where they form synapses with interneurons that transmit information about muscle length to various areas of the brain.

uses information from muscle spindles (and other receptors from the muscles, joints, and skin) to monitor the contractile states of the various muscles and the positions of the limbs they control. Because the brain controls skeletal muscle contractions, this information provides feedback that enables the brain to adjust its commands to the muscles, a function that is necessary for the execution of smooth and accurate movements.

Withdrawal and Crossed-Extensor Reflexes

When a portion of the body is subjected to a painful stimulus, it withdraws from the stimulus automatically via a response called the **withdrawal reflex**, as occurs when a person steps on a tack (**Figure 10.20**). Stepping on a tack is perceived as painful because it activates special sensory receptors called *nociceptors*, which respond to intense stimuli that are damaging (or potentially damaging) to tissue. Afferent neurons from nociceptors transmit this information to the spinal cord, where they have excitatory synapses on interneurons. The interneurons then excite efferent neurons (synapse 1 in Figure 10.20) that innervate the skeletal muscles that cause withdrawal of the limb or other body part. In the example in Figure 10.20, the hamstring muscles contract to withdraw the leg.

For the withdrawal reflex to work effectively, the muscles that cause withdrawal should be excited, whereas the muscles that oppose withdrawal (or cause extension) should be inhibited. This is indeed the case because branches of the afferents activate inhibitory interneurons to the efferents innervating the quadriceps (synapse 2 in Figure 10.20). Therefore, the withdrawal reflex simultaneously triggers contraction of the hamstrings and relaxation of the quadriceps, allowing flexion of the leg to proceed unimpeded.

But another problem remains: If you withdraw your leg while it is supporting your weight, you may well lose your balance and fall over. Fortunately, when a painful stimulus triggers the withdrawal reflex, another reflex—the **crossed-extensor reflex**—is initiated simultaneously. The afferent neurons from nociceptors have branches that send signals (via interneurons) to efferent neurons controlling muscles on the opposite leg (synapses 3 and 4 in Figure 10.20). These signals trigger contraction of the extensor muscles and relaxation of the flexor muscles in that leg, so that when the first leg is withdrawn in response to a painful stimulus, the other leg is extended to support the body.

Would the person be aware that he or she stepped on a tack? Of course, because the same afferent neurons that activate both the withdrawal and crossed-extensor

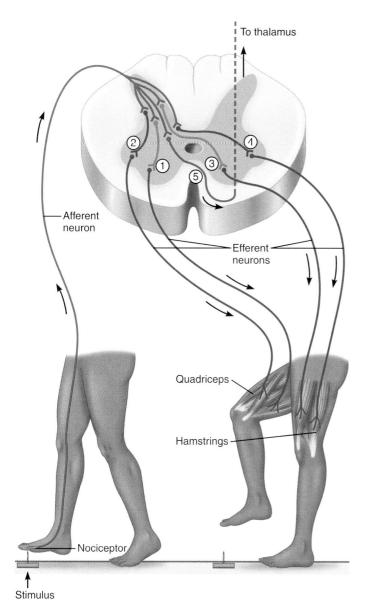

To thalamus

Afferent
neuron

Efferent
neurons

Quadriceps

Hamstrings

Nociceptor

Stimulus

Figure 10.20 Withdrawal and crossed-extensor reflexes.
In response to the activation of a nociceptor, an afferent neuron
synapses on an excitatory interneuron ① and an inhibitory
interneuron ②, ultimately producing contraction of the hamstrings
and relaxation of the quadriceps in the affected leg, and effecting
its withdrawal. Simultaneously, the afferent neuron also synapses
with an excitatory interneuron ③ and an inhibitory interneuron
④, which ultimately produces contraction of the quadriceps and
relaxation of the hamstrings in the other leg. The afferent neuron
also synapses with an interneuron ⑤ that crosses to the opposite
side of the spinal cord and travels to the thalamus to convey
information about the painful stimulus to the brain.

reflexes also communicate with ascending pathways that
send information to the brain that a painful stimulus has
been applied to the body (synapse 5 in Figure 10.20).
Given the existence of the withdrawal reflex, does this
mean that you will withdraw from every painful stimulus
you encounter in life? Probably not, for there are times

when reflexes are detrimental. Suppose you were to pick
up a cup of tea served in your mother's favorite china
and found the cup to be painfully hot. Knowing how
much your mother likes her china, you might attempt to
override the reflex and hold on to the cup, even if you
experience pain by doing so. In this process, the brain activates inhibitory interneurons in the spinal cord that
suppress activity in the neurons that control the withdrawal reflex. Remember, postsynaptic cells (in this case,
the efferent neurons that control the muscles) sum information that comes from different synaptic inputs. If the
inhibitory influence that descends from the brain is
greater than the excitatory influence coming from the
nociceptor afferents, then the withdrawal reflex will not
occur.

Pupillary Light Reflex

The stretch reflex and withdrawal reflex are examples of
somatic spinal reflexes. The pupillary light reflex is an
example of an autonomic cranial reflex in which the
stimulus is light entering one eye and activating photoreceptors. The photoreceptors activate afferent neurons that
transmit signals to areas in the midbrain of the brainstem
that function as the integration center. These midbrain
areas then activate, through a multisynaptic pathway, autonomic efferents that innervate the smooth muscle surrounding the pupils of both eyes. The response is
pupillary constriction (reduction in diameter of the
pupils) in both eyes.

Quick Test 10.4

1. What is the difference between a spinal reflex and a
   cranial reflex? Between a somatic reflex and an
   autonomic reflex? Between an innate reflex and a
   conditioned reflex? Between a monosynaptic reflex and a
   polysynaptic reflex?

2. Is salivation in response to a bell an example of an innate
   reflex or a conditioned reflex? Why? What about salivation
   in response to food entering the mouth?

3. What are the five components of a reflex arc?

4. Briefly describe how a muscle spindle stretch reflex and a
   withdrawal reflex occur.

Integrated CNS Function:
Voluntary Motor Control

Imagine that you are in an operating room, about to undergo surgery for appendicitis. The surgeon explains that
before the procedure begins, you will be given two drugs
intravenously, an anesthetic and a muscle relaxant.
Someone injects a syringe full of liquid into your arm,
and soon you feel your arms and legs getting heavy.

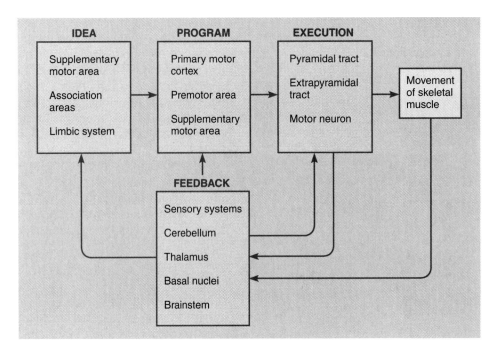

Figure 10.21 Four processes in voluntary movement. Voluntary movement requires four processes to ensure smooth skeletal muscle movement, starting with the idea to move.

Within minutes, you find yourself unable to move at all—not even to breathe. (Fortunately, you have been hooked up to a respirator.) Although you expected to become sleepy, you are wide awake and can see, hear, and feel everything that is going on. To your dismay, you notice that the surgeon has picked up a scalpel. Apparently, someone has made a mistake and forgotten to give you the anesthetic. As the blade approaches your skin, you try desperately to yell, to move your arms, to do anything that would let the staff know you are awake, but you can't. All your skeletal muscles are paralyzed, and you can't even blink an eye!

This horrifying scenario illustrates that in order for our thoughts and intentions to become actions, the nervous system must be able to communicate with muscles and trigger muscle contraction. The purpose of this section is to describe what happens when we control our muscles voluntarily; in other words, what happens when our intentions are translated into actions.

Neural Components for Smooth Voluntary Movements

The successful completion of a voluntary motor task involves four components (**Figure 10.21**):

1. The idea or intention to move must first be formulated. The formulation of intentions, and their translation into plans for movement, is exerted by the cerebral cortex and is the highest level of motor control. Several areas of the cortex are thought to be involved in these processes, including the supplementary motor area (in the frontal lobe), association areas, and the limbic system.

2. A program of motor commands must be developed to ultimately direct the appropriate muscles to contract at the right time to produce the intended motion. Such development involves many areas, including the primary motor cortex, somatosensory areas, the supplementary motor area, and the premotor area (located in the frontal lobe between the primary motor cortex and the supplementary motor area).

3. The program of motor commands must be executed by sending to the muscles commands telling them to contract.

4. As the motor task is being carried out, continual feedback to the CNS enables it to ascertain whether the task is being accomplished according to the intention and the program. Such feedback enables the CNS to make any needed adjustments in the program, and possibly even change the original intention, to ensure smooth and reliable completion of the task. The cerebellum, thalamus, basal nuclei, certain brainstem nuclei, and various cortical association areas provide this feedback and assist with development of motor programming.

The execution of a motor program requires activation of the efferent neurons that innervate skeletal muscle. These efferent neurons originate in the ventral horn of the spinal cord and are called **lower motor neurons**, or simply **motor neurons**. (For those muscles that are

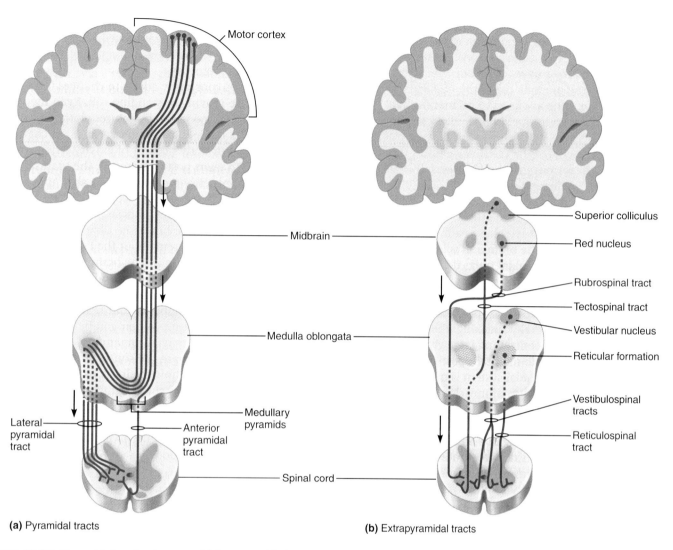

(a) Pyramidal tracts

(b) Extrapyramidal tracts

Figure 10.22 Pyramidal and extrapyramidal tracts. (a) Pyramidal tracts. The lateral pyramidal tract is primarily a crossed pathway, whereas the anterior pyramidal tract is primarily uncrossed. (b) Extrapyramidal tracts. These tracts include all motor tracts except the pyramidal tracts; some cross, some do not, and some are bilateral.

Where do the pyramidal tracts originate?

controlled by cranial nerves, the motor neurons originate in certain cranial nerve nuclei of the brainstem.) Motor neurons are the only neurons that control contraction of skeletal muscle, and the influence of these neurons is always excitatory; that is, signals in a motor neuron always stimulate a muscle cell to contract. Therefore, initiation of a muscle cell contraction requires excitation of the motor neuron innervating that muscle cell. Relaxation of a skeletal muscle cell simply requires that the motor neuron innervating it not be active.

Input to motor neurons originates from many different levels in the nervous system. In the previous section we saw that motor neurons can be excited or inhibited by

input from afferent neurons, as is the case for spinal reflexes. We also saw that activity in motor neurons can be influenced by input descending from the brain, as when one overrides a withdrawal reflex through conscious effort. In this section we are interested primarily in descending input from the brain to motor neurons. In particular, two descending pathways are important in voluntary motor control that provide input to motor neurons: the *pyramidal tracts* (also known as the *corticospinal tracts*) and the *extrapyramidal tracts.*

Cortical Control of Voluntary Movement

The **pyramidal tracts** (**Figure 10.22**a) are direct pathways from the primary motor cortex to the spinal cord. The axons of neurons in these tracts terminate in the

In the primary motor cortex

2. What type of junction between the endothelial cells of brain capillaries produces the blood-brain barrier?

3. Myelinated axons are found in (gray/white) matter.

4. Somatic efferents originate in the (dorsal/ventral) horn of the spinal cord.

5. The major function of the (cerebrum/cerebellum) is to coordinate body movements.

6. What three structures make up the brainstem?

7. Which side of the brain is generally associated with creativity?

8. What are the two main structures of the diencephalon?

9. What is the major sensory relay nucleus to the cortex?

10. The area of the brain most closely associated with fear is the _____.

11. The _____ system is associated with emotions, learning, and memory.

12. The ability to recall information when taking physiology exams is an example of _____ memory.

13. The afferents that activate reflex motor actions are different from those that activate ascending tracts to communicate a sensory input to the brain. (true/false)

Essay Questions

1. Describe the different structures that protect central nervous system tissues.

2. Describe the energy supply to the brain.

3. Identify the four lobes of the cerebral cortex, and name at least one function associated with each.

4. Name the different components involved in accomplishing a voluntary motor task. Identify the structure(s) associated with each component.

5. Name the different stages of sleep. During which sleep stage is a person most likely to roll over?

6. Describe the roles of NMDA and AMPA receptors in long-term potentiation.

Critical Thinking

1. Parkinson's disease is due to a loss of dopaminergic function in the basal nuclei. To treat patients with Parkinson's disease, doctors try to elevate the dopamine levels in the brain. However, dopamine cannot cross the blood-brain barrier. Can you devise methods to elevate brain dopamine levels?

2. A person suffers a stroke that affects Wernicke's area. Describe what it would be like to carry on a conversation with this person.

3. Reflexes can be overridden by voluntary controls. Explain how this can occur, using the information you learned in Chapter 9.

4. Paralysis refers to the inability to voluntarily control movement and can include a loss of excitatory input to motor neurons (flaccid paralysis) or a loss of inhibitory input to motor neurons (spastic paralysis). In addition, some forms of paralysis are coupled with the absence of reflexes. Compare the types of paralysis that would be observed when the following areas of the nervous system are damaged: (1) primary motor cortex, (2) basal nuclei, and (3) motor neuron.

Find the answers to these exercises, and additional study tools, at the Physiology Place (www.physiologyplace.com).

The Nervous System: Sensory Systems

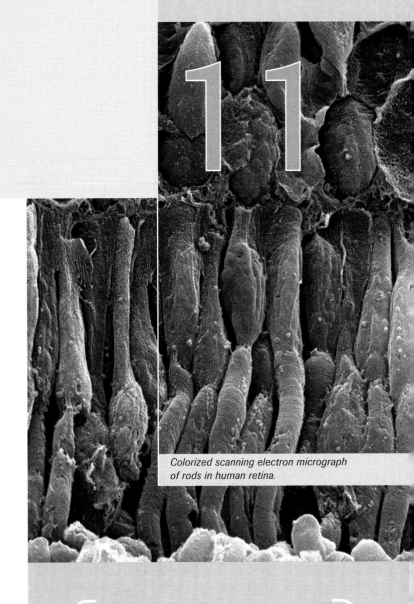

Colorized scanning electron micrograph of rods in human retina.

S it back for a moment and reflect on the world around you. Are you sitting or lying down? If you are indoors, what color are the walls of the room? Are you warm or cold? Are the people near you noisy? Maybe someone has the stereo on so loud that you can feel the vibrations. Your perception of the world around you is achieved via the functioning of your sensory systems. But you cannot perceive everything in your environment. For example, if you have ever had an X ray taken, you know you could neither see nor hear the rays as they passed through your body, and when you put some food in the microwave, you cannot see the waves that heat your food.

Why can we perceive some things in the world and not others? Humans have special sensory receptors that can detect specific forms of energy in the environment, such as light waves, sound waves, or vibrations, but we do not have receptors for X rays or microwaves. Therefore, we cannot perceive these energy forms, even when they are present in our immediate environment. Furthermore, our perceptions of the stimuli that we *can* perceive are not absolute. When you jump into a lake, for example, you immediately perceive the cold water, but over time your body adapts, and within minutes the water does not feel as cold—even though the temperature of the water has not changed.

In Chapter 8, we learned that the peripheral nervous system consists of an afferent and an efferent branch, with the afferent branch sending information about the internal and external environments to the central nervous system and the efferent branch sending information from the central nervous system to effector organs. We then learned in Chapter 10 about some of the structures of the central nervous system involved in our perception of the world around us. This chapter focuses on the sensory systems involved in **perception**, the conscious interpretation of the world based on the sensory systems, memory, and other neural processes. We first describe general properties of sensory systems. Then we discuss those senses that allow us to perceive touch and other stimuli associated with the skin surface, and that also allow us to perceive such things as the position of our limbs and bodies. Finally, we discuss the *special senses*.

General Principles of Sensory Physiology

The afferent division of the peripheral nervous system transmits information from the periphery to the central nervous system. The information is detected by *sensory receptors* that respond to specific types of stimuli. Whereas some of these receptors detect stimuli in the external environment, others, called *visceral receptors*, detect stimuli

that arise within the body. Visceral receptors transmit information to the CNS by a class of afferent neurons known as *visceral afferents.*

Examples of the many types of visceral receptors include *chemoreceptors* in major blood vessels that monitor O_2, CO_2, and H^+ levels in the blood; *baroreceptors* in certain blood vessels that monitor blood pressure; and *mechanoreceptors* in the gastrointestinal tract that monitor the degree of stretch or distension. Although the brain receives information from these receptors and utilizes this information for regulatory purposes, we are not consciously aware of these stimuli. For example, we cannot sense what the pH of our own blood is unless we take a blood sample and analyze it. However, pH receptors in certain arteries and in the brainstem tell the brain whether the pH is normal, information that enables the brain to make regulatory decisions. The specifics of these and other visceral receptors are covered in chapters that discuss particular organ systems.

Exercise Link

Although a person does not consciously sense blood pH or the other individual inputs from visceral receptors, there may be times when we are able to sense the overall message being sent by these receptors. In the later stages of the marathon, Jane and especially Bill experienced two sensations that we call fatigue. One was a reduction in the responsiveness of their leg muscles because metabolic fuels were running low and waste products were building up in ways that hampered normal function. The other sensation is variously called "central" or "psychological" fatigue. Although physiologists do not fully understand this second type of fatigue, it is suspected that visceral receptor inputs, along with influences of body temperature and blood glucose levels, feed back to the motor cortex, reducing the drive to exercise. In addition, these inputs touch on the edge of our consciousness to produce those hard-to-describe feelings we associate with fatigue.

This chapter covers the sensory systems that detect information about the external environment. This information is transmitted to the cerebral cortex, reaches consciousness, and is perceived, making us aware of the world around us. But note that our perception of the world is not always fully accurate. That this is so is demonstrated in **Figure 11.1**.

In Figure 11.1a you perceive something that is not present—a square. Part of your perception is based on previous experiences; in this example, your brain perceives what it expects to see. In Figure 11.1b you perceive curved lines when in actuality the lines are straight, which shows that our perceptions may deceive us. In Figure 11.1c some people perceive a rabbit, others a duck,

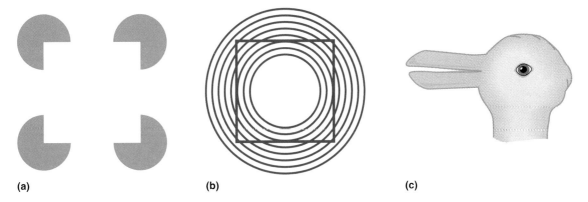

(a)   (b)   (c)

Figure 11.1 **The fallibility of perception.** These optical illusions demonstrate that our perception is not always accurate. (a) Where parts of the four green disks are missing, the brain perceives a square. (b) The effect of the concentric circles is to make straight lines appear curved. (c) This shape can resemble a duck or a rabbit—or both or neither.

others both, and still others neither. Individuals may perceive things differently, based in part on their individual past experiences.

The sensory systems that enable us to perceive the external environment include the somatosensory system and the special sensory systems. The **somatosensory system** is necessary for perception of sensations associated with receptors in the skin **(somesthetic sensations)** and for **proprioception**, the perception of the position of the limbs and the body. Proprioception depends on specific proprioceptors in muscles and joints and on more generalized receptors in the skin. The **special senses** are necessary for senses of vision, hearing, balance and equilibrium, taste, and smell.

Receptor Physiology

Sensory receptors are specialized neuronal structures that detect a specific form of energy in either the internal or external environment. The energy form of a stimulus is called its **modality**. Some examples of modalities include light waves, sound waves, pressure, temperature, and chemicals. The **law of specific nerve energies** states that a given sensory receptor is specific for a particular modality. Thus special cells in the eye called *photoreceptors* detect light waves, but not sound waves. The modality to which a receptor responds best is called the **adequate stimulus**. **Table 11.1** lists the characteristics of several classes of sensory receptors. Modalities

Table 11.1 | Characteristics of Sensory Receptors

Receptor class	Sensation/visceral information	Modality
Photoreceptors	Vision	Photons of light
Chemoreceptors	Taste	Chemicals dissolved in saliva
	Smell	Chemicals dissolved in mucus
	Pain	Chemicals in extracellular fluid
	Blood oxygen	Oxygen dissolved in plasma
	Blood pH	Free hydrogen ions in plasma
Thermoreceptors		
Warm receptors	Warmth	Increase in temperatures between 30°C and 43°C
Cold receptors	Cold	Decrease in temperatures between 35°C and 20°C
Mechanoreceptors		
Baroreceptors	Blood pressure	Stretch of specific blood vessel wall
Osmoreceptors	Osmolarity of extracellular fluid	Swelling (stretch) of receptor cells
Hair cells	Sound	Sound waves
	Balance and equilibrium	Acceleration

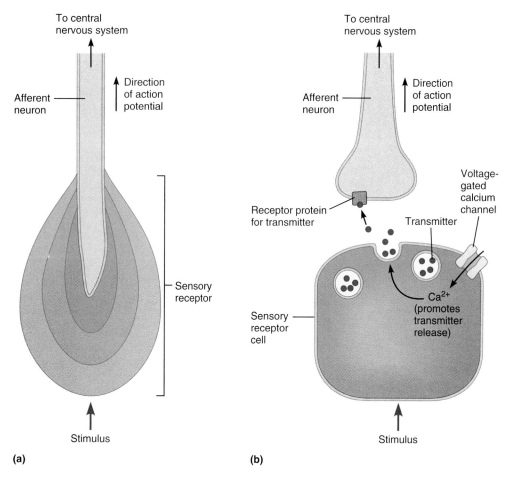

Figure 11.2 Structure and function of sensory receptors. (a) A sensory receptor that is a specialized ending of an afferent neuron. The stimulus acts on the sensory receptor by opening or closing ion channels, thus producing a receptor potential. (b) A sensory receptor that is a separate cell from the afferent neuron. The stimulus changes the membrane potential of the receptor cell, which opens (or closes) a calcium channel, and cytosolic calcium concentration increases (or decreases). Changes in calcium concentration trigger (or inhibit) the release of a chemical transmitter by exocytosis. The transmitter communicates to the afferent neuron by binding to receptors on the afferent ending.

other than the adequate stimulus may activate receptors, but only at relatively high energy levels. Moreover, a sufficiently strong "inadequate" stimulus will induce the perception of the adequate stimulus, as when a blow to the eye (pressure) activates photoreceptors, causing the person to perceive light.

Sensory Transduction

The function of sensory receptors is **transduction**—that is, the conversion of one form of energy into another. In **sensory transduction**, receptors convert the energy of a sensory stimulus into changes in membrane potential called **receptor potentials** or **generator potentials**. Receptor potentials resemble postsynaptic potentials (see Chapter 9) in that they are graded potentials caused by the opening or closing of ion channels. The greater the strength of the stimulus, the greater the change in membrane potential. However, unlike postsynaptic potentials,

which are triggered by the binding of neurotransmitter to receptors, receptor potentials are triggered by sensory stimuli.

Sensory receptors exist in two basic forms, illustrated in **Figure 11.2**. In some cases, a sensory receptor is a specialized structure at the peripheral end of an afferent neuron (Figure 11.2a). When such a receptor is depolarized to threshold, an action potential is generated in the afferent neuron and propagated to the CNS, thereby transmitting information about the stimulus. In other cases, the sensory receptor is a separate cell that communicates via a chemical synapse with an associated afferent neuron (Figure 11.2b). When the sensory receptor is a separate cell, changes in the receptor cell's membrane potential cause the release of a chemical messenger or transmitter. (In some cases a stimulus triggers a change in membrane potential that *inhibits* the release of a transmitter.) The greater the excitatory stimulus, the greater the

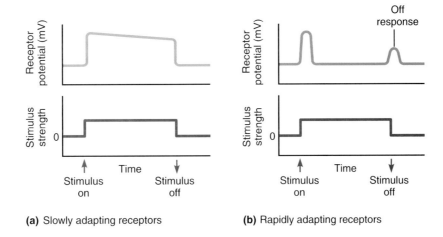

(a) Slowly adapting receptors   **(b)** Rapidly adapting receptors

Figure 11.3 Responses of slowly adapting receptors and rapidly adapting receptors.
(a) Slowly adapting receptors respond with a change in receptor potential that persists for the duration of the stimulus. (b) Rapidly adapting receptors respond with a change in receptor potential at the onset of a stimulus, but then adapt. The "off response" is a second, smaller response that occurs upon termination of a stimulus.

amount of transmitter released. This transmitter then binds to receptors on the afferent neuron and causes changes in the membrane potential of that cell. If the afferent neuron is depolarized to threshold, then an action potential is generated and transmitted by that cell to the CNS. The mechanism of transduction varies for different types of receptors and is covered with the individual sensory systems later in this chapter.

Receptor Adaptation

Some receptors continue to respond to a stimulus for as long as the stimulus is applied. However, most receptors *adapt* to the stimulus; that is, their response declines with the passage of time. **Adaptation** is a decrease over time in the magnitude of the receptor potential in the presence of a constant stimulus. **Slowly adapting** or **tonic receptors** show little adaptation and therefore can function in signaling the intensity of a prolonged stimulus (**Figure 11.3**a). Examples of slowly adapting receptors are muscle stretch receptors, which detect the degree to which muscles are stretched, and proprioceptors, which detect the position of the body in three-dimensional space. For proper movement and balance, it is critical that these receptors continually inform the CNS of the position of muscles and joints. **Rapidly adapting** or **phasic receptors** adapt quickly, and thus function best in detecting changes in stimulus intensity (Figure 11.3b). Rapidly adapting receptors respond at the onset of a stimulus, then adapt. Some rapidly adapting receptors also show a second, smaller response upon the termination of a stimulus, called the "off response." Examples of rapidly adapting receptors are *olfactory receptors,* which detect odors, and *Pacinian corpuscles,* which detect vibration in the skin.

> **Quick Test 11.1**
>
> 1. Define the following terms: *modality, adequate stimulus, transduction, adaptation,* and the *law of specific nerve energies.*
>
> 2. How can a sensory receptor cell that is not part of an afferent neuron cause action potentials in that neuron?
>
> 3. Which type of receptor is better for communicating changes in the intensity of a sensory input, rapidly adapting receptors or slowly adapting receptors?

Sensory Pathways

In Chapter 10, we learned about areas of cerebral cortex specialized for the perception of certain sensory stimuli. Here we focus on the sensory pathways that relay sensory information to the appropriate area of the cortex.

The specific neural pathways that transmit information pertaining to a particular modality are referred to as **labeled lines**, and each sensory modality follows its own labeled line or pathway. Activation of a specific pathway causes perception of the associated modality, regardless of the actual stimulus that activated the pathway. Thus, as we have seen, pressure on the eye may trigger light flashes because the pressure generates signals in the pathway for perception of light. The pathways for different modalities terminate in different sensory areas of the cerebral cortex (**Figure 11.4**). Although the pathways are distinct, we can make some generalizations concerning most pathways, starting with a sensory unit.

Table 11.3 ▎ Sensory Receptors in the Skin

Receptor class	Type	Associated afferent type	Location	Receptive field size	Adaptation	Modality
Mechano-receptors	Free nerve ending	A-delta, C	Superficial, all skin	Small	Slow	Light touch
	Merkel's disk	A-beta	Superficial, all skin	Small	Slow	Pressure
	Pacinian corpuscle	A-beta	Deep, all skin	Large	Rapid	Vibration
	Meissner's corpuscle	A-beta	Superficial, glabrous skin	Small	Rapid	Vibration
	Hair follicle receptor	A-beta	Superficial, hairy skin	Small	Rapid	Bending of hair
	Ruffini's ending	A-beta	Deep, hairy skin	Large	Slow	Pressure
Thermo-receptors	Warm receptors (free nerve endings)	C	Superficial, all skin	Small	Rapid	Increase in skin temperature
	Cold receptors	A-delta	Superficial, all skin	Small	Rapid	Decrease in skin temperature
Nociceptors	Mechanical (free nerve endings)	A-delta	Superficial, all skin	Large	Slow	Intense mechanical stimulus
	Thermal (free nerve endings)	A-delta	Superficial, all skin	Small	Rapid	Intense hot or cold stimulus
	Polymodal (free nerve endings)	C	Superficial, all skin	Large	Slow	Intense mechanical or thermal stimulus; specific chemicals

body's position requires receptors in the muscles, tendons, ligaments, and joints, as well as receptors in the skin. Such receptors are called proprioceptors, one example of which is the muscle spindle, discussed in Chapter 10. Somesthetic sensations of stimuli associated with the surface of the body require **mechanoreceptors** to detect pressure, force, or vibration; **thermoreceptors** to detect temperature; and **nociceptors** to detect tissue-damaging stimuli. Of all the sensory systems, the somatosensory system has the widest variety of receptor types; **Table 11.3** lists and characterizes the various types of receptors in the skin.

Most somatosensory receptors in the skin are specialized structures at nerve endings, which are easily identified under the light microscope. However, a few somatosensory receptor types lack identifiable specialized structures and are therefore called **free nerve endings**. The different structures are designed to respond to particular modalities of stimuli impinging on the skin.

Mechanoreceptors in the Skin

The various types of mechanoreceptors in the skin are depicted in **Figure 11.13**. Some receptors are found in

superficial layers of the skin close to the *epidermis,* or outer layer of the skin. Superficial receptors include *Merkel's disks* and *Meissner's corpuscles.* Other receptors are located deeper, in the *dermis* or inner layer of the skin, including *hair follicle receptors, Pacinian corpuscles,* and *Ruffini's endings.* Meissner's corpuscles are found only in *glabrous* skin (hairless skin), whereas hair follicle receptors are found only in hairy skin. Note that slowly adapting mechanoreceptors respond to pressure (a sustained stimulus), whereas rapidly adapting receptors respond best to vibration (a constantly changing stimulus). Note as well that the sizes of the receptive fields for the various mechanoreceptors vary greatly, and that the smaller receptive fields provide better tactile acuity.

Thermoreceptors in the Skin

Thermoreceptors respond to the temperature of the receptor endings themselves and the surrounding tissue, not the temperature of the surrounding air. There are two types of thermoreceptors: warm receptors and cold receptors. **Warm receptors** are free nerve endings that respond to temperatures between 30°C and 43°C; the frequency of

Table 11.2 | Two-Point Discrimination Thresholds for Selected Areas of the Body

Body region	Two-point discrimination threshold (mm)*
Lips (greatest acuity)	1
Index finger	2
Thumb	3
Palm of hand	10
Big toe	10
Forehead	18
Sole of foot	22
Breast	31
Abdomen	36
Shoulder	38
Back	42
Thigh	46
Upper arm	47
Calf (least acuity)	48

Source: Weinstein and Kenshalo, editors, *The Skin Senses,* copyright 1968. Courtesy of Charles C. Thomas Publisher, Ltd., Springfield, Illinois.
*Smaller distances indicate greater tactile acuity.

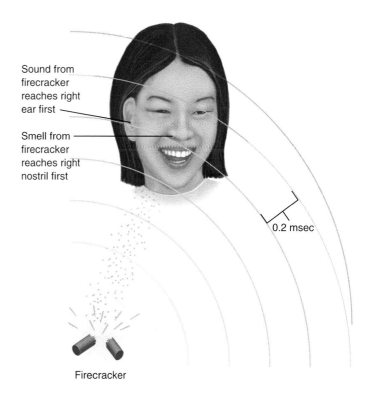

Sound from firecracker reaches right ear first

Smell from firecracker reaches right nostril first

0.2 msec

Firecracker

Figure 11.12 Localization in hearing and olfaction. Localization of a sound or an odor depends on the difference between the time the stimulus reaches the left and right ears or nostrils, respectively.

Tactile acuity varies over different regions of the body. The skin in some areas of the body is innervated by afferent neurons with few branches and, therefore, small receptive fields, whereas other areas are innervated by afferent neurons with extensive branching and large receptive fields. In addition, greater overlap occurs for the smaller receptive fields, which also contributes to greater tactile acuity. **Table 11.2** lists two-point discrimination thresholds in several selected areas of the body. In the lips, the most sensitive areas, points as close together as 1 mm can be distinguished as two distinct points. The fingertips are also quite sensitive. Areas on the back, thigh, and upper arm, by contrast, are not very sensitive. In fact, two points as far apart as 40–50 mm (almost 2 inches!) may be indistinguishable as separate points in these areas.

Localization in some sensory systems is unrelated to coding by receptive fields. Even though a person can determine the direction from which an odor or a sound comes, neither olfactory nor auditory afferent neurons have receptive fields. Instead, these systems code for the quality and intensity of smell, and the pitch and loudness of sound, respectively. Localization in olfaction and in hearing is based on the arrival of stimuli at the two nostrils or at the two ears at slightly different times (**Figure 11.12**). The brain uses the difference in the time of arrival of action potentials at the olfactory or auditory cortex to determine where the stimulus originated.

> ### Quick Test 11.3
>
> 1. What is the difference between population coding and frequency coding? Do these processes code a stimulus's type or its intensity?
> 2. What is the difference between rapidly adapting receptors and slowly adapting receptors?
> 3. Explain the concept of two-point discrimination. Is two-point discrimination better on the palm of the hand or on the thigh? Why? Where is the ability to locate a stimulus better, on the palm of the hand or on the thigh? Why?

The Somatosensory System

The somatosensory system is involved with body sensations such as pressure, temperature, pain, and body position (*soma* = body). We first consider the types of receptors in this system.

Somatosensory Receptors

The somatosensory system responds to a variety of stimuli arising in many areas of the body, and thus it utilizes many receptor types. For example, proprioception of our

Table 11.3 │ Sensory Receptors in the Skin

Receptor class	Type	Associated afferent type	Location	Receptive field size	Adaptation	Modality
Mechano-receptors	Free nerve ending	A-delta, C	Superficial, all skin	Small	Slow	Light touch
	Merkel's disk	A-beta	Superficial, all skin	Small	Slow	Pressure
	Pacinian corpuscle	A-beta	Deep, all skin	Large	Rapid	Vibration
	Meissner's corpuscle	A-beta	Superficial, glabrous skin	Small	Rapid	Vibration
	Hair follicle receptor	A-beta	Superficial, hairy skin	Small	Rapid	Bending of hair
	Ruffini's ending	A-beta	Deep, hairy skin	Large	Slow	Pressure
Thermo-receptors	Warm receptors (free nerve endings)	C	Superficial, all skin	Small	Rapid	Increase in skin temperature
	Cold receptors	A-delta	Superficial, all skin	Small	Rapid	Decrease in skin temperature
Nociceptors	Mechanical (free nerve endings)	A-delta	Superficial, all skin	Large	Slow	Intense mechanical stimulus
	Thermal (free nerve endings)	A-delta	Superficial, all skin	Small	Rapid	Intense hot or cold stimulus
	Polymodal (free nerve endings)	C	Superficial, all skin	Large	Slow	Intense mechanical or thermal stimulus; specific chemicals

body's position requires receptors in the muscles, tendons, ligaments, and joints, as well as receptors in the skin. Such receptors are called proprioceptors, one example of which is the muscle spindle, discussed in Chapter 10. Somesthetic sensations of stimuli associated with the surface of the body require **mechanoreceptors** to detect pressure, force, or vibration; **thermoreceptors** to detect temperature; and **nociceptors** to detect tissue-damaging stimuli. Of all the sensory systems, the somatosensory system has the widest variety of receptor types; **Table 11.3** lists and characterizes the various types of receptors in the skin.

Most somatosensory receptors in the skin are specialized structures at nerve endings, which are easily identified under the light microscope. However, a few somatosensory receptor types lack identifiable specialized structures and are therefore called **free nerve endings**. The different structures are designed to respond to particular modalities of stimuli impinging on the skin.

Mechanoreceptors in the Skin

The various types of mechanoreceptors in the skin are depicted in **Figure 11.13**. Some receptors are found in superficial layers of the skin close to the *epidermis,* or outer layer of the skin. Superficial receptors include *Merkel's disks* and *Meissner's corpuscles.* Other receptors are located deeper, in the *dermis* or inner layer of the skin, including *hair follicle receptors, Pacinian corpuscles,* and *Ruffini's endings.* Meissner's corpuscles are found only in *glabrous* skin (hairless skin), whereas hair follicle receptors are found only in hairy skin. Note that slowly adapting mechanoreceptors respond to pressure (a sustained stimulus), whereas rapidly adapting receptors respond best to vibration (a constantly changing stimulus). Note as well that the sizes of the receptive fields for the various mechanoreceptors vary greatly, and that the smaller receptive fields provide better tactile acuity.

Thermoreceptors in the Skin

Thermoreceptors respond to the temperature of the receptor endings themselves and the surrounding tissue, not the temperature of the surrounding air. There are two types of thermoreceptors: warm receptors and cold receptors. **Warm receptors** are free nerve endings that respond to temperatures between 30°C and 43°C; the frequency of

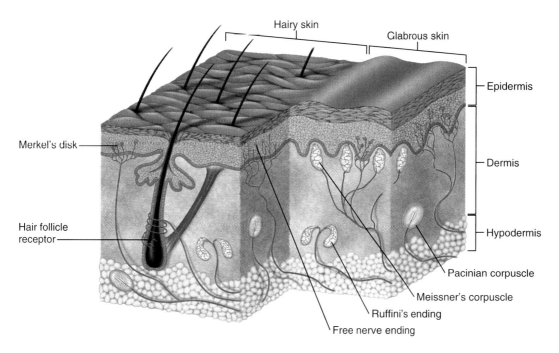

Figure 11.13 Sensory receptors in the skin.

action potentials increases as the temperature increases. **Cold receptors** respond to temperatures between 35°C and 20°C; the frequency of action potentials increases as the temperature falls and the receptors get colder. Cold receptors also respond to temperatures greater than 45°C, a painfully *hot* stimulus, with the frequency of action potentials increasing as temperature increases. The perception of cold at these hot temperatures is referred to as *paradoxical cold.* The structure of cold receptors is not known, although some evidence suggests that they too are free nerve endings.

Our perception of skin temperature depends on properties and location of the thermal sensory units. The sensitivity of thermoreceptors is measured in our ability to discriminate between two temperatures. People can detect changes in skin temperature of only 0.02°C. Thermoreceptors are also rapidly adapting, so they detect changes in temperature especially well. (You have probably sensed this if you ever jumped into a cold swimming pool. After a while, the water did not seem as cold.) Thermal sensory units, like other somesthetic sensors, have receptive fields. Because different sensory units are associated with the perception of warm and cold, and the receptive fields typically do not overlap, some *spots* on the skin are responsive to warm stimuli and other spots are responsive to cold stimuli.

Nociceptors in the Skin

Nociceptors are the sensory receptors responsible for the transduction of *noxious stimuli* that we perceive as pain. Nociceptors are free nerve endings that respond to tissue-damaging (or potentially damaging) stimuli. There are three types of nociceptors: **mechanical nociceptors**,

which respond to intense mechanical stimuli, such as when you stub your toe; **thermal nociceptors**, which respond to intense heat (greater than 44°C), such as when you touch a hot stove; and **polymodal nociceptors**, which respond to a variety of stimuli, including intense mechanical stimuli, intense heat, intense cold, and chemicals released from damaged tissue. Chemicals that are released from damaged tissue and are capable of activating polymodal nociceptors include *histamine, bradykinin,* and *prostaglandins.*

The Somatosensory Cortex

The perception of somatic sensations from all parts of the body begins in the primary somatosensory cortex (although recent studies suggest that some "crude" perception may occur in the thalamus). Recall from Chapter 10 that the somatosensory cortex is topographically oriented; that is, sensory information arising in neighboring areas of the body generally projects to neighboring areas of the cortex. Recall as well that the size of the cortex area devoted to somatic sensations from a specific area of the body is not necessarily proportional to the size of the body region, but rather to the sensitivity of the body region. The lips and fingertips, for example, are very sensitive areas because they have small two-point discrimination thresholds (see Table 11.2), and these body regions also have large areas of the primary somatosensory cortex devoted to them (see Figure 10.15, p. 281).

Recall also that the cerebral cortex has columnar organization. In the primary somatosensory cortex, vertical columns are organized according to sensory modality. For example, in the area of the primary somatosensory cortex

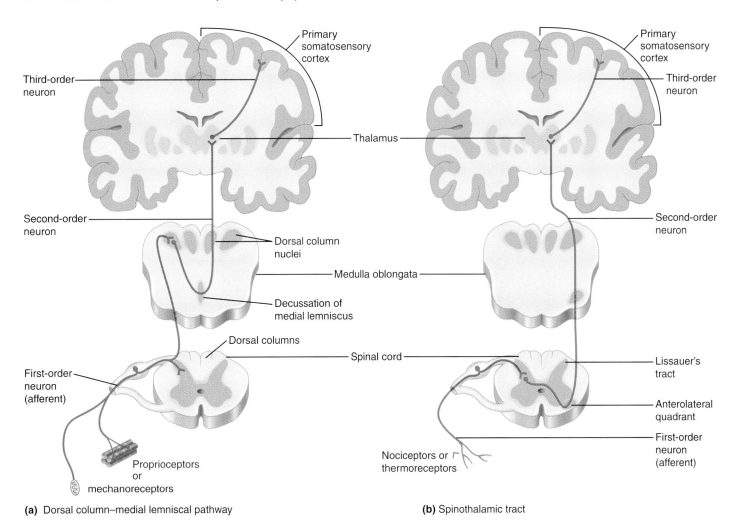

(a) Dorsal column–medial lemniscal pathway

(b) Spinothalamic tract

Figure 11.14 The two somatosensory pathways. (a) Dorsal column–medial lemniscal pathway, which transmits information from mechanoreceptors and proprioceptors to the CNS. (b) Spinothalamic tract, which transmits information from thermoreceptors and nociceptors to the CNS.

If a person received damage to the right dorsal columns, would they lose perception of touch on the right or left side?

for the thumb, one column is associated with pressure to the thumb, another column with vibration to the thumb, another column with cold, and so on. The next section describes the pathways along which information travels from the receptors to the primary somatosensory cortex.

Somatosensory Pathways

Two main pathways transmit information from peripheral somatosensory receptors to the central nervous system: the *dorsal column–medial lemniscal pathway* and the *spinothalamic tract.* These pathways transmit different types of sensory information to the thalamus, and then to the primary somatosensory cortex. In both cases the pathways

enter the spinal cord on one side and cross to the other side before reaching the thalamus. Therefore, somatosensory information from the right side of the body is perceived in the left somatosensory cortex, and vice versa.

The Dorsal Column–Medial Lemniscal Pathway

The **dorsal column–medial lemniscal pathway** transmits information from mechanoreceptors and proprioceptors to the thalamus; it crosses to the other side of the CNS in the medulla oblongata (**Figure 11.14**a). In this pathway, first-order neurons originate in the periphery and enter the dorsal horn of the spinal cord. Collaterals from the main axon may terminate in the spinal cord, communicating with interneurons such as those involved in spinal reflexes. However, the main branch of the axon ascends from the spinal cord to the ipsilateral (same side as the stimulus) brainstem in the *dorsal columns,* which

The right side

are tracts of white matter that run dorsal and medial to the dorsal horn. The first-order neurons terminate in *dorsal column nuclei,* which are located in the medulla, where they form synapses with second-order neurons. The second-order neurons then cross over to the contralateral side of the medulla in a tract called the *medial lemniscus,* and then ascend to the thalamus. In the thalamus the second-order neurons synapse with third-order neurons, which transmit information from the thalamus to the somatosensory cortex.

The Spinothalamic Tract

The spinothalamic tract transmits information from thermoreceptors and nociceptors to the thalamus; it crosses to the other side of the CNS within the spinal cord before it reaches the brain (Figure 11.14b). In this pathway, first-order neurons originate in the periphery at either thermoreceptors or nociceptors and enter the dorsal horn of the spinal cord. Here, first-order neurons may ascend or descend for a short distance (a few spinal segments) along *Lissauer's tract,* but eventually they form synapses with second-order neurons in the dorsal horn. The second-order neurons cross over to the contralateral side of the spinal cord, ascend in the anterolateral quadrant of the spinal cord through the brainstem, and then terminate in the thalamus. In the thalamus, second-order neurons form synapses with third-order neurons that ascend to the somatosensory cortex.

Quick Test 11.4

1. Name the three classes of receptors for somesthetic sensations and the adequate stimulus (or stimuli) for each.
2. Name the two types of thermoreceptors and describe the types of stimuli that excite them.
3. Name the three types of nociceptors and the stimuli that activate them.
4. How do the dorsal column–medial lemniscal and the spinothalamic pathways differ with regard to the types of sensory information they transmit? Where does each pathway cross over to the other side of the CNS?

Pain Perception

Pain, one of the somesthetic sensations, is important because it teaches us to avoid subsequent encounters with potentially damaging stimuli. Pain is also important clinically, because it is a sign that tissue damage may have occurred. Thus the mechanisms of pain perception warrant special consideration.

The Pain Response

The activation of nociceptors leads not only to the perception of pain but also to a variety of other body responses, including one or more of the following: (1) autonomic responses, such as increases in blood pressure and heart rate, increases in blood epinephrine levels, increases in blood glucose, dilation of the pupils of the eye, or sweating; (2) emotional responses such as fear or anxiety; and (3) a reflexive withdrawal from the stimulus. The level of perceived pain varies considerably among individuals based on their past experiences and the circumstances under which the stimulus is applied. Thus a toothache that is barely noticeable during a busy day, for example, may be excruciating when one tries to fall asleep at night. To better understand how pain perception can vary, let's take a closer look at the mechanisms of pain perception.

Each of two types of pain, fast pain and slow pain, is perceived differently and is transmitted by a different class of afferent neurons. **Fast pain** is perceived as a sharp pricking sensation that can be easily localized; it is transmitted by $A\delta$ (A delta) *fibers,* thin, lightly myelinated axons with a conduction velocity of approximately 12–30 m/sec. **Slow pain** is perceived as a poorly localized, dull aching sensation; it is transmitted by *C fibers,* thin, unmyelinated axons with a conduction velocity of approximately 0.2–1.3 m/sec. Think about the last time you stubbed your toe. An initial sharp pain (fast pain), was followed by a prolonged aching pain (slow pain).

The primary afferents, whether $A\delta$ or C fibers, form synapses with second-order neurons in the dorsal horn of the spinal cord. Communication between these first- and second-order neurons involves different neurotransmitters, one of which is *substance P.* Substance P is released from primary afferent neurons and binds to receptors on second-order neurons. The second-order neurons ascend to the thalamus via the spinothalamic tract, the pathway involved in the perception and discrimination of pain. Nociceptive afferents also activate different ascending pathways that are necessary for interpreting the affective components of pain. The affective pathways ascend to the reticular formation of the brainstem, the hypothalamus, and the limbic system.

Visceral Pain

Pain is not limited to the body surface. For example, most people have suffered pain in muscles after overexercising, or experienced a stomachache. The viscera are subject to tissue damage, and nociceptors in the organs detect this damage. But what does a person perceive when the appendix becomes inflamed or when suffering a heart attack? Generally, activation of nociceptors in the viscera produces pain that is called **referred pain** (because it has been "referred" to the body surface). For example, a person having a heart attack generally complains about pain in the left chest, upper arm, and shoulder—not in the heart itself. Referred pain occurs

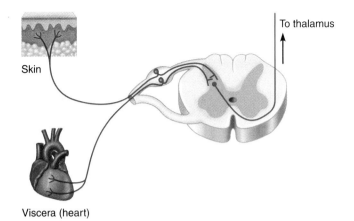

Skin

To thalamus

Viscera (heart)

(a) Mechanism of referred pain

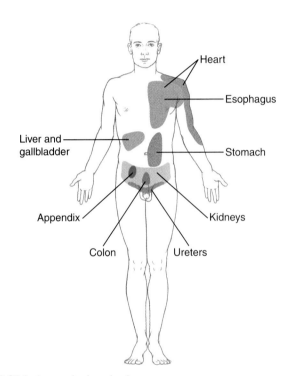

Heart

Esophagus

Liver and gallbladder

Stomach

Appendix

Kidneys

Colon

Ureters

(b) Clinical map of referred pain

Figure 11.15 Mechanisms and sites of referred pain.
(a) Referred pain occurs when visceral and somesthetic afferents converge on the same second-order neurons in the spinal cord. (b) Pain in specific visceral organs is generally referred to the areas of the body surface indicated in this map.

because the second-order neurons that receive input from visceral afferents also receive input from somatic afferents (**Figure 11.15**a). According to one theory, the brain interprets information based on past experiences. Throughout a person's life, these second-order neurons are activated primarily by the somatic afferents, and thus the brain has learned that signals from these neurons are somatic in origin. When a person has a heart attack, the brain interprets the signals as a somatic disturbance because that is what such signals have meant in the past.

Physicians use maps of surface locations of referred pain (Figure 11.15b) to ascertain which internal organ(s) may be causing a patient's pain.

Modulation of Pain Signals

Signals about sensory information can be modulated as they are transmitted along sensory pathways; that is, facilitation or attenuation of signals can result in changes in the final perception of that information. Sensory signals can be modulated wherever there is a synapse in the pathway. The mechanisms involved in the modulation of pain have profound clinical significance, as pain is a common complaint of patients.

The *gate-control theory* of pain modulation states that somatic signals of nonpainful sources can inhibit signals of pain at the spinal level (**Figure 11.16**). Among the various interneurons within the spinal cord are interneurons that inhibit the second-order neurons that transmit pain information. When these interneurons are active, the transmission of pain signals is suppressed, and the perception of pain is lessened. When information about a painful stimulus is being transmitted to the spinal cord by C fibers (Figure 11.16a), the collaterals of these C fibers *inhibit* the activity of the inhibitory interneuron, which allows transmission to proceed to the second-order neuron. But the same inhibitory interneuron is *stimulated* by collaterals from large-diameter myelinated afferents (Aβ fibers) associated with mechanical stimuli such as touch, pressure, and vibration (Figure 11.16b). If a nonpainful mechanical stimulus is applied simultaneously with a painful stimulus, the collaterals from the Aβ fibers stimulate the inhibitory interneuron, thereby decreasing the transmission of pain signals.

The gate-control theory describes why rubbing a painful area relieves the pain. It is also the basis for using TENS, or *transcutaneous electrical nerve stimulation*, to treat pain. In TENS, a small current applied through the skin overlying a nerve activates large-diameter afferents, which relieves the pain.

The gate-control theory describes how afferent signals to the spinal cord can influence the perception of pain. The brain also has the ability to block pain, or produce **analgesia**, through descending pathways that are part of the pain-blocking **endogenous analgesia systems**. Pain can be debilitating, and there are times when perceiving pain is disadvantageous. For example, the chances that seriously wounded soldiers will survive are reduced if they are immobilized by pain while their lives are at stake. Such wounded soldiers can often function without perceiving pain because the endogenous analgesia systems become active and block the pain, allowing the body to cope with a more pressing need—survival.

Many brain areas are involved in the endogenous analgesia systems. One of the best defined pathways is

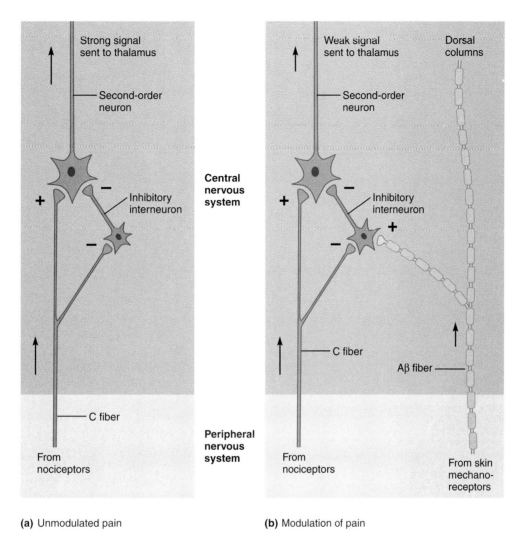

(a) Unmodulated pain   **(b)** Modulation of pain

Figure 11.16 Gate-control theory of pain. (a) In unmodulated pain transmission, collaterals of the nociceptor afferents (C fibers) inhibit inhibitory interneurons, allowing transmission of pain signals to second-order neurons in the dorsal horn of the spinal cord and then to the thalamus. (b) In the modulation of pain transmission, collaterals of large-diameter afferents (Aβ fibers) extending from touch and pressure receptors excite the inhibitory interneuron, thereby decreasing the transmission of pain signals.

illustrated in **Figure 11.17**. Stressful situations can activate an area in the midbrain called the *periaqueductal gray matter.* This area communicates to an area in the medulla called the *nucleus raphe magnus,* and to the *lateral reticular formation,* which extends the length of the brainstem. Neurons from these areas descend to the spinal cord, where they block the communication between nociceptive afferent neurons and second-order neurons, as follows.

Recall that substance P is one neurotransmitter released from nociceptive afferents that communicate with second-order neurons. Inhibitory interneurons in the spinal cord form synapses on the cell body and dendrites of the second-order neurons and also on the axon terminal of the nociceptive afferent neuron. These inhibitory interneurons release the endogenous opiate

neurotransmitter *enkephalin* (discussed in Chapter 9), which binds to opioid receptors on the second-order neuron and induces inhibitory postsynaptic potentials. Enkephalin also binds to opioid receptors on the axon terminal of the nociceptive afferent neuron, which inhibits the release of substance P. Both these actions suppress signal transmission from the afferent neuron to the second-order neuron, thereby decreasing the transmission of pain signals to the brain. These inhibitory interneurons are activated by descending neurons of the nucleus raphe magnus.

More recently, descending modulation has been demonstrated to include pathways that facilitate pain. One example is the *hyperalgesia* that results when a person is ill. Several of the cytokines that induce symptoms of illness (for example, fever and fatigue)

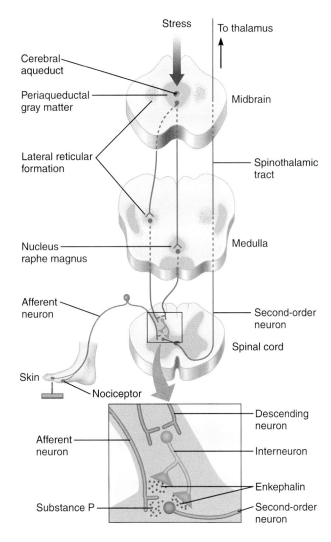

Figure 11.17 Endogenous analgesia systems. When a painful stimulus activates nociceptors, information is transmitted to the CNS via the spinothalamic tract. In the presence of stress, endogenous analgesia systems can block pain transmission at the level of the synapse between the nociceptive afferent neuron and the second-order neuron, as follows: The periaqueductal gray matter in the midbrain communicates to the lateral reticular formation and to the nucleus raphe magnus of the medulla. These regions have neurons that descend to the dorsal horn of the spinal cord and activate inhibitory interneurons that release the neurotransmitter enkephalin, which then blocks communication between the nociceptive afferent and the second-order neuron via two mechanisms: presynaptic inhibition of substance P release from the nociceptive afferent, and production of IPSPs on the second-order neuron.

also act on areas of the brain to enhance the perception of pain.

Sometimes, pain transmitting or modulating systems malfunction, resulting in chronic pain. One example is *phantom limb pain*, the perception of pain in an amputated limb (see **Clinical Connections: Phantom Limb Pain**).

Quick Test 11.5

1. Distinguish between fast pain and slow pain. Which class of afferent neuron is responsible for fast pain, and which is responsible for slow pain?

2. What kind of pain is experienced in the left shoulder by a person having a heart attack?

3. What are endogenous analgesia systems?

Vision

Much of what we learn about the world we learn through seeing. In this section we examine the visual system, which endows us with this important capability. We begin with the anatomy of the eye, the sensory organ of the visual system.

Anatomy of the Eye

The important structures of the eye are shown in **Figure 11.18**. The eye can be divided into three concentric layers. The outermost layer consists of the sclera and cornea. The **sclera**, a tough connective tissue, makes up the "white" of the eye. In the front of the eye, however, the sclera gives way to the **cornea**, a transparent structure that allows light to enter the eye.

The middle layer of the eye consists of the choroid, ciliary body, and iris. The **choroid** is a highly pigmented layer of tissue beneath the sclera. Its pigment absorbs light that reaches the back of the eye so that it is not reflected, which would cause distortion of the visual image. The choroid also contains blood vessels that nourish the inner layer of the eye. The **ciliary body** contains the **ciliary muscles**, which are attached to the lens by strands of connective tissue called **zonular fibers**. The **lens** focuses the light on the retina, where the visual information is transduced. The ciliary muscles change the shape of the lens to focus light waves. The **iris**, which consists of two layers of pigmented smooth muscle, is located in front of the lens. The pigmentation of the iris determines eye color. The **pupil** is a hole in the center of the iris that allows light to enter the posterior part of the eye; it is not a structure. The iris regulates the diameter of the pupil, thereby regulating the amount of light that reaches the back of the eye.

The innermost layer of the eye is the **retina**. The retina consists of neural tissue and contains the **photoreceptors**, cells that detect the light waves. Photoreceptors are of two types, *rods* and *cones,* which detect dim light and bright light, respectively. The retina, therefore, functions in *phototransduction,* the conversion of light energy to electrical energy. Two areas of the retina are worth noting. One is the **fovea**, the central point on the retina, where light from the center of the visual field

PHANTOM LIMB PAIN

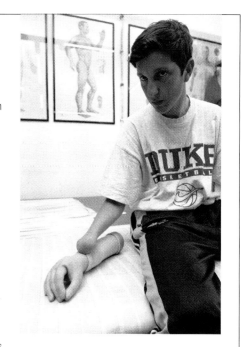

When a person loses an arm or a leg, the damage can be psychologically devastating. To make matters worse, these patients often suffer chronic pain, coming from the limb that is no longer present!

An amputee often perceives other sensations coming from the lost limb (called *phantom sensations*), but the sensation of pain is the most debilitating. More than 50% of amputees perceive some sort of *phantom limb pain*. In some patients, the pain is relatively minor and subsides over time. In other patients, the pain is excruciating and persists almost constantly for the rest of their lives.

How can a person feel something in a missing body part? This question has baffled doctors for centuries. Early theories suggested that nerve endings in the stump can be activated and thereby produce the sensation of pain. However, the pain induced by stimuli to the stump produces *stump pain,* which is better characterized and more easily treated than true phantom limb pain. More recent theories suggest that changes are occurring within the central nervous system whereby second- or third-order pain transmitting neurons that no longer receive input from the missing limb become more sensitive to other types of synaptic input. Because neurons receive converging input from a variety of sources, other synapses become more effective in their ability to excite the second- or third-order neurons. Thus neurons that used to transmit information about painful stimuli in the limb now become excited by other synaptic input unrelated to the limb. The perception is still pain in the limb, however, because the pain pathway normally from the limb is activated.

Treatment of phantom limb pain has been difficult, with only 25% of the patients with pain receiving long-term relief. Traditional chronic pain treatments, such as anti-inflammatory agents and opioids, have not been very successful in reducing phantom limb pain. Other treatments used include surgical manipulations, acupuncture, and transcutaneous electrical nerve stimulation. Scientists continue to study the baffling observation that brings misery to so many patients with the hope of developing an effective pain therapy.

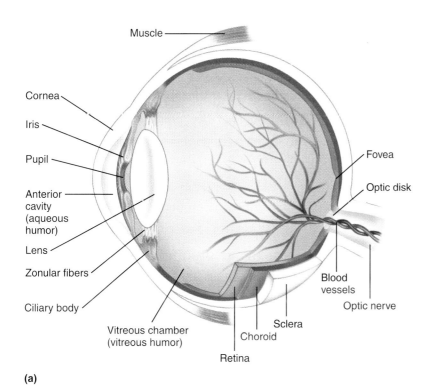

(a)

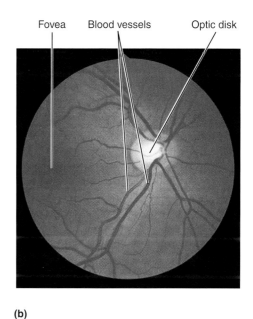

(b)

Figure 11.18 Anatomy of the eye. (a) The major structures of the eye as viewed in a horizontal section. (b) The retina as viewed through an ophthalmoscope.

Figure 11.19 The blind spot. To demonstrate your blind spot, cover your right eye and look at the − sign with your left eye. Move the book close to your left eye while continuing to look at the − sign. When the book is a few inches from your face, the + sign should disappear because the light waves reflecting from it strike the optic disk.

strikes. (The transparent lens focuses light waves on the retina, much as a camera lens focuses light waves on film.) For reasons described later, the fovea is the area of the retina with the greatest visual acuity. The other area of note is the **optic disk**, the portion of the retina where the optic nerve and blood vessels supplying the eye pass through the retina. Because there are no photoreceptors in the optic disk, this area is a **blind spot**, a region where light striking the retina cannot be transduced into neural impulses and thus cannot be perceived.

Figure 11.19 has a design to demonstrate the blind spot. Close your right eye and look at the − sign through the left eye. While moving the book toward you, continue looking at the − sign with your left eye. As the book approaches you, the + sign will disappear from view. That is because the light reflected from the + sign is striking the optic disk and therefore cannot be perceived.

The lens and ciliary body separate the eye into two fluid-filled chambers. In front of the lens and ciliary body is the **anterior cavity** containing a clear, watery fluid called **aqueous humor**, which supplies nutrients to the cornea and lens. The cornea and lens are transparent so that light can pass through them easily. If these structures relied on blood to supply their nutrients, the presence of blood vessels would obstruct the light. Behind the lens and ciliary body is the **vitreous chamber** containing a firmer, jellylike material called **vitreous humor**, which maintains the spherical structure of the eye.

The Nature and Behavior of Light Waves

Light is a form of energy. Specifically, light exists as *electromagnetic waves*. Along with other forms of electromagnetic energy, including radio waves, television waves, X rays, and gamma rays, light is a part of the electromagnetic spectrum (**Figure 11.20**). Visible light includes those electromagnetic waves having wavelengths between about 350 nm and 750 nm; different colors correspond to different wavelengths within this range.

Because light is a wave, it exhibits the usual properties of waves, including reflection and refraction. **Reflection** is a phenomenon in which light waves strike and bounce off a surface. Reflection is important in vision because much of the light we perceive has reflected off the objects we are observing. (We also see emitted light, such as the light coming directly to the eye from the sun or a

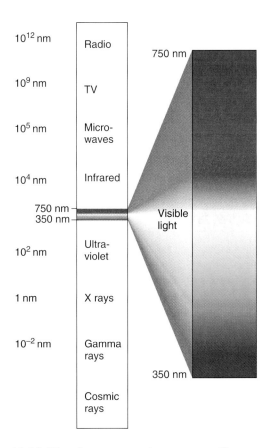

Figure 11.20 The electromagnetic spectrum. The numbers indicate wavelength in nanometers (1 nm = 1×10^{-9} meter). The band of visible light is highlighted.

lightbulb.) Light that is absorbed by objects is not perceived; thus we perceive an object as green because it reflects to the eyes light of the wavelength corresponding to green (approximately 530 nm) while absorbing all other wavelengths. **Refraction** refers to the bending of light waves as they pass through transparent materials of different densities. This property is important in vision because in its path from objects to the photoreceptors in the retina, light must pass through several different materials, including air, the cornea, the lens, and the vitreous and aqueous humors. Because refraction is important in the focusing of light waves on the retina, we consider it here in greater detail.

Try this at home: Place a straw in a clear glass half full of water and let it rest at an angle. So oriented, the straw looks broken at the air-water interface (**Figure 11.21**a). Now hold the straw straight up so that it is perpendicular to the air-water interface. The straw now appears to be in one piece again (Figure 11.21b). Light waves are refracted as they travel from one medium to another if they strike the second medium at an angle other than perpendicular. How much the light waves refract depends on the differences in the densities of the two media and the angle at which the light strikes them. Try holding the straw at different angles.

(a)   **(b)**

Figure 11.21 Refraction of light waves passing through different media. (a) Straw at an angle to the air-water interface. (b) Straw perpendicular to the air-water interface.

Figure 11.22 illustrates the refraction of parallel light waves as they pass through *concave* and *convex* surfaces often used in lenses, such as those in eyeglasses or telescopes. Note that with either type of surface, light waves striking the surface perpendicular to it pass straight through. However, as the light strikes the surfaces at other angles, the concave lens causes the once-parallel light waves to diverge (move farther apart), whereas the convex lens causes the light waves to converge at a single point called the *focal point*. The distance from the long axis of the convex lens to the focal point is called the *focal length*.

Both the cornea and the lens have convex surfaces that function to converge the light waves entering the eye onto the retina, a process that is necessary if visual images are to be in focus. **Figure 11.23** shows how light waves from a viewed object are projected onto the retina. For us to see items in focus, light from a given point in the *visual field*—what we are looking at—must converge to a single point on the retina. Although the cornea has greater refractive power than the lens due to its greater curvature, the refractive power of the lens can be varied as needed to focus light on the retina. For example, to

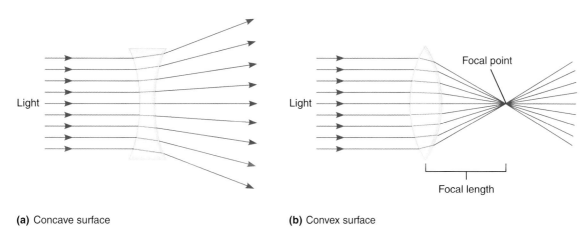

(a) Concave surface   **(b)** Convex surface

Figure 11.22 Refraction of light waves passing through curved surfaces. (a) Concave surfaces cause divergence of light waves. (b) Convex surfaces cause convergence of light waves to a focal point. The distance from the long axis of a convex lens to the focal point is called the focal length.

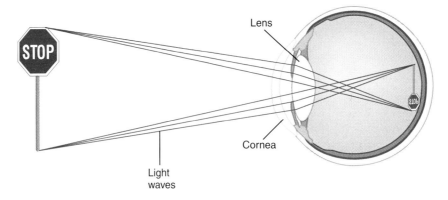

Figure 11.23 Refraction of light waves in the eye. A given point in the visual field comes to focus on a single point in the retina. Refraction of light waves as they pass through the convex cornea and lens of the eye causes the image to be inverted and reversed on the retina.

view very near objects, the lens becomes rounder, increasing its refractive power to focus the image on the retina. The ability of the lens to adjust its refractive power for viewing near objects is a process called **accommodation**, which we explore next.

Accommodation

If an object is to be seen clearly, the light reflected from any given point on the object must converge at a single point on the retina. When viewing something far away, the light waves enter the eye almost parallel to each other (**Figure 11.24**a), so little refractive power is needed to focus the light on the retina. However, light waves from close-up objects diverge as they enter the eye (Figure 11.24b), so the greater refractive power of a rounder lens is necessary to overcome this divergence and focus the light on the retina.

The shape of the lens is controlled by the circularly arranged ciliary muscle through the tension it applies to the zonular fibers, which attach the ciliary muscle to the lens. The more a circular muscle contracts, the smaller the diameter of the circle becomes. For viewing distant objects, the ciliary muscle is relaxed, which increases the diameter of the circle and tightens the zonular fibers, pulling the lens into a flattened shape (**Figure 11.25**a). To achieve accommodation for viewing close-up objects (Figure 11.25b), the ciliary muscle contracts, reducing the diameter of the circle and lessening the tension on the zonular fibers. Due to its inherent elasticity, the lens becomes rounder when the tension on the zonular fibers is reduced. Accommodation is under control of the parasympathetic nervous system, which triggers contraction of the ciliary muscle for near vision. In the absence of parasympathetic activity, the ciliary muscle relaxes.

Clinical Defects in Vision

If light waves are not adequately focused on the retina, then vision is blurred. There are many causes of blurred vision, and different types of corrective lenses to improve vision.

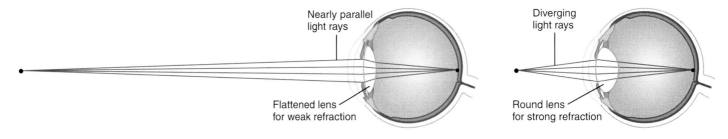

(a) Viewing a distant object

(b) Viewing a near object

Figure 11.24 Focusing light from distant and near sources. (a) Light waves reflected from a distant object approach the lens parallel to one another. A relatively flat (weak) lens is sufficient to converge the light waves on the retina. (b) Light waves reflected from a near object diverge as they approach the lens. A rounder (strong) lens is needed to converge the light waves on the retina.

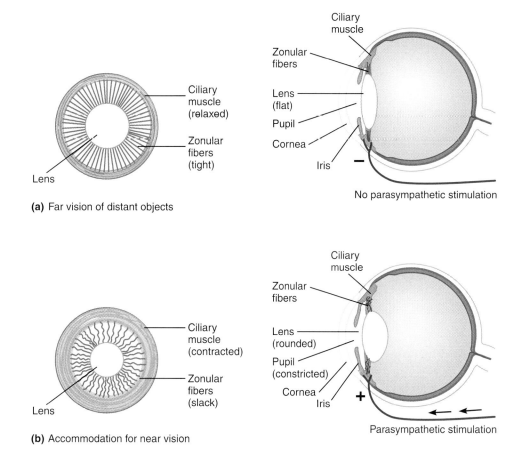

(a) Far vision of distant objects

No parasympathetic stimulation

(b) Accommodation for near vision

Parasympathetic stimulation

Figure 11.25 Mechanism of accommodation. (a) Vision of distant objects. In the absence of parasympathetic stimulation, the ciliary muscle relaxes, putting tension on the zonular fibers. The zonular fibers pull on the lens, flattening the lens. (b) Accommodation for near vision. Under parasympathetic stimulation, the ciliary muscle contracts, reducing the tension on the zonular fibers and enabling the elastic lens to become rounder.

Common visual defects include near-sightedness or **myopia**, and far-sightedness or **hyperopia**. In **emmetropia**, or normal vision (**Figure 11.26**a), a person can see both distant and close-up objects clearly because the eye can focus light from far sources without accommodation, and from near sources with accommodation. In myopia or hyperopia, a mismatch exists between lens or cornea strength and eyeball length. The defect could be related to the lens, the cornea, or the eyeball length.

In myopia (Figure 11.26b), a person can see near objects clearly, but not distant objects because the lens or cornea is too strong for the length of the eyeball and therefore bends light rays too much. In this situation, close-up objects can be focused without accommodation, but light from distant objects is focused in front of the retina, resulting in a blurred image. To correct for myopia, a concave lens is placed in front of the eye. The lens causes light waves to diverge before reaching the eye. Under these conditions, the eye must accommodate to view close-up objects, and distant objects will be in focus without accommodation.

In hyperopia (Figure 11.26c), the lens or cornea is too weak for the length of the eyeball. Therefore, distant objects can be focused only with accommodation, which means that the lens cannot increase accommodation enough to adjust for near vision. Light from close-up objects comes to focus behind the retina, resulting in a blurred image. To correct for hyperopia, a convex lens is placed in front of the eye. The lens causes light waves to converge before reaching the eye. Now the eye can see distant objects without accommodation, which gives the lens enough leeway to enable it to accommodate for near objects.

Many other clinical defects affect the ability to focus light on the retina. In *astigmatism,* irregularities on the surface of the cornea or lens cause erratic bending of light waves. *Presbyopia* is a hardening of the lens that occurs with aging; as the lens hardens, its loss of elasticity decreases its ability to become spherical, making accommodation for near vision more difficult. A *cataract* is an age-related discoloration of the lens that decreases its transparency. In *glaucoma,* an increase in the volume of

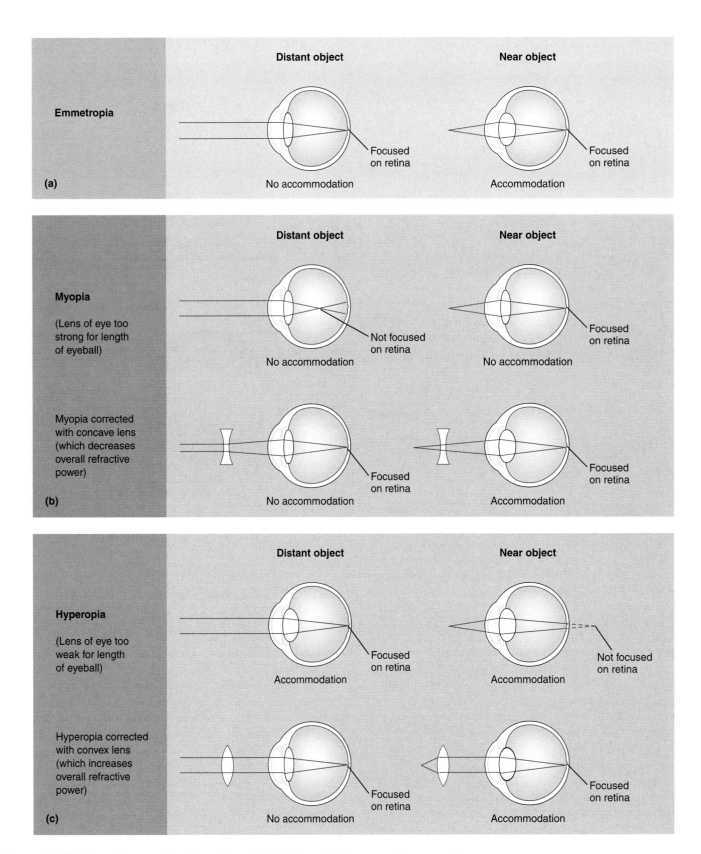

Figure 11.26 Normal, near-sighted, and far-sighted vision. (a) In emmetropia, or normal vision, distant objects are focused on the retina without accommodation, and near objects are focused with accommodation. (b) In myopia, or near-sightedness, the lens (or cornea) is too strong for the length of the eyeball. Near objects are focused without accommodation, and distant objects come into focus in front of the retina even without accommodation. Myopia can be corrected by using a concave lens to produce divergence of light waves before they enter the eye. (c) In hyperopia, or far-sightedness, the lens (or cornea) is too weak for the length of the eyeball. Distant objects are focused with accommodation, and near objects come into focus behind the retina, even with accommodation. Hyperopia can be corrected by using a convex lens to produce convergence of light waves that supplements the convergence produced in the eye.

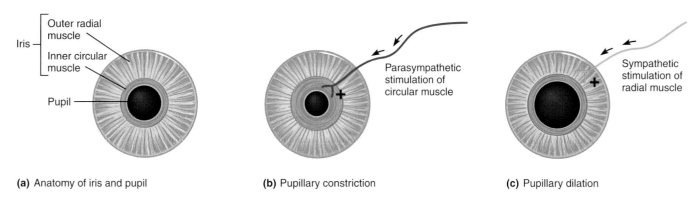

(a) Anatomy of iris and pupil   **(b)** Pupillary constriction   **(c)** Pupillary dilation

Figure 11.27 Regulation of the amount of light entering the eye. (a) The iris, which consists of two layers of smooth muscle—an inner circular layer and an outer radial layer—controls pupil size. The size of the pupil determines the amount of light that enters the eye. (b) Pupillary constriction, which is caused by parasympathetic stimulation of the circular muscle layer of the iris. (c) Pupillary dilation, which is caused by sympathetic stimulation of the radial muscle layer of the iris.

aqueous humor raises pressure in the anterior cavity of the eyeball, which can distort the shape of the cornea and shift the position of the lens. Shifting of the lens can transmit the increased pressure to the vitreous chamber, where it can compress the blood vessels that supply the retina. Excessive pressure can reduce the blood supply significantly, leading to permanent blindness.

Regulating the Amount of Light Entering the Eye

The eyes are capable of regulating the amount of light that enters them by varying the size of the pupils. In bright light, the pupils are small, or *constricted,* so that the photoreceptors do not become "bleached out" by too much light. In dim light, by contrast, the pupils are large or *dilated* to allow more light in, which enhances the ability to see. The size of the pupil is controlled by the iris.

Recall that the iris consists of two layers of smooth muscle around the pupil. These two layers of smooth muscle are an inner **circular muscle** layer, also called the *constrictor muscle,* and an outer **radial muscle** layer, also called the *dilator muscle* (**Figure 11.27**a). The circular muscles form concentric rings around the pupil; when they contract, the diameter of the pupil decreases. Thus contraction of the circular muscles causes *pupillary constriction* (Figure 11.27b). The radial muscles are arranged like spokes in a wheel; when they contract, the diameter of the pupil increases. Thus contraction of the radial muscles causes *pupillary dilation* (Figure 11.27c).

The iris is under control of the autonomic nervous system. Parasympathetic neurons innervate the circular muscles; activity in these neurons causes the circular muscles to contract, producing pupillary constriction.

Sympathetic neurons innervate the radial muscles; activity in these neurons causes the radial muscles to contract, producing pupillary dilation.

The Retina

The retina, which is composed of neural tissue, is the location of photoreceptors, the rods and cones. **Rods** provide the ability to see in black and white during relatively low light conditions, such as the light provided by the moon at night. **Cones** provide us with color vision, but they are active only in relatively bright light, such as the sunlight during the day.

The retina consists of three distinct layers (**Figure 11.28**): (1) an inner layer containing neurons called ganglion cells, (2) a middle layer containing neurons called bipolar cells, and (3) an outer layer containing rods and cones. Also present are *amacrine cells* and *horizontal cells,* neurons that modulate (for example, by lateral inhibition) communication between the cells in the retina.

Note in Figure 11.28 that because the photoreceptors are in the outer layer of the retina, light must pass through the inner and middle layers before striking them. In addition, blood vessels are in the light's path to the photoreceptors. However, to provide light a clear path to the fovea, the bipolar and ganglion cells are laterally displaced, creating a depression in the center of the retina called the **macula lutea** that surrounds the fovea.

The fovea contains cones only; the ratio of rods to cones increases with distance from the fovea, until at the periphery of the retina only rods are present (**Figure 11.29**). This is why in dim light we see objects better if we do not look directly at them, and we see them in black and white only.

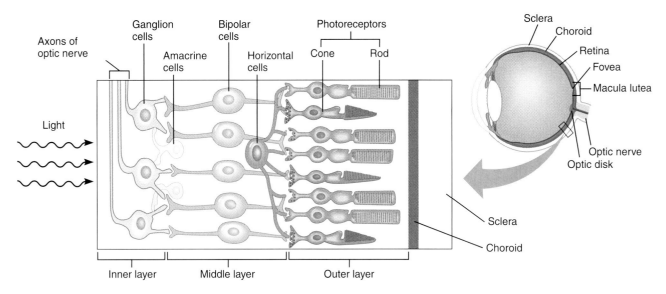

Figure 11.28 Anatomy of the retina. Located on the inner surface of the eye, the retina consists of three layers of neural tissue composed of the various types of cells depicted. Note that light must pass through the inner and middle layers of the retina before striking the photoreceptors in the outer layer. Deep to the retina is the choroid, which absorbs light.

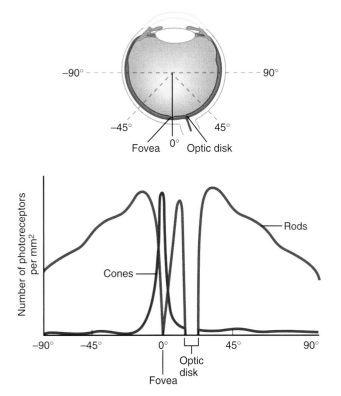

Figure 11.29 Distribution of rods and cones in the retina. The abundance of cones is greatest at the fovea and declines rapidly with distance from it. Rods are absent from the fovea but very abundant near it; they slowly decrease in abundance with distance from the fovea.

Quick Test 11.6

1. What is the effect of contraction of the ciliary muscles on the lens? Does this help vision of near objects or of distant objects?

2. What effects do the two branches of the autonomic nervous system have on the diameter of the pupil?

3. What are the locations of the vitreous humor and the aqueous humor? What is the function of each?

4. What cells are in each of the three layers of the retina?

Phototransduction

Phototransduction, the conversion of light energy into electrical signals, is carried out by the rods and cones. The basic morphology of the two types of photoreceptors is the same; each consists of two major portions referred to as *outer* and *inner segments* (**Figure 11.30**). The outer segment contains invaginations with membranous disks that contain the molecules that absorb light waves, giving the photoreceptors the ability to respond to light. The inner segment contains the cell nucleus and various organelles and ends at the receptor's synaptic terminal, which is analogous to an ordinary neuron's axon terminal and is where a chemical messenger is stored in synaptic vesicles.

The absorption of light is the first step in phototransduction, and the molecule in the photoreceptors that absorbs light is a **photopigment**. Each of the four different types of photoreceptors contains a different photopigment. One type of photopigment is found in rods, and the other three types are found in three types of cones, each of which contains a photopigment that best absorbs light of a particular range of wavelengths and is thus most responsive to certain colors. Each photopigment molecule contains a light-absorbing portion called *retinal* and a protein called an *opsin*. The retinal portion is the same in all photopigments, but the kind of opsin present determines which light wavelengths are absorbed by a given photopigment by altering the electromagnetic energies to which the retinal is sensitive.

The components of photoreceptors involved in phototransduction are shown in **Figure 11.31** using rods as an example. The photopigment of rods, *rhodopsin*, is located in the membrane of the disks. Also within the disk membrane is a G protein called *transducin* and the enzyme *phosphodiesterase*, which catalyzes the degradation of cGMP (which if present in the cytosol, opens sodium channels located in the plasma membrane of the photoreceptor).

Let's examine the process of phototransduction as it occurs in rods. (The same process occurs in the three types of cones, but different photopigments are involved.) We begin by considering the state of the photoreceptor in the dark (**Figure 11.32**a). In the dark, levels of the second messenger cGMP are high inside the outer segment ①, so cGMP opens sodium channels in the plasma membrane of the outer segment ②. Therefore, sodium ions are moving into the cell, and the photoreceptor is *depolarized* ③. This depolarization spreads to the inner segment and ④ opens calcium channels that are also present in the plasma membrane. Calcium enters the

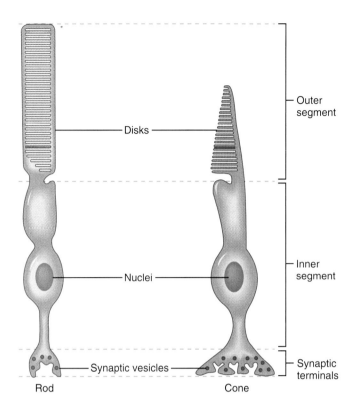

Figure 11.30 Morphology of the photoreceptors. Rods and cones have the same basic structural components: The outer segment consists of disks that contain the photopigment; the inner segment contains the nucleus and most of the organelles. The synaptic terminal contains the synaptic vesicles, which store a chemical transmitter used for communication.

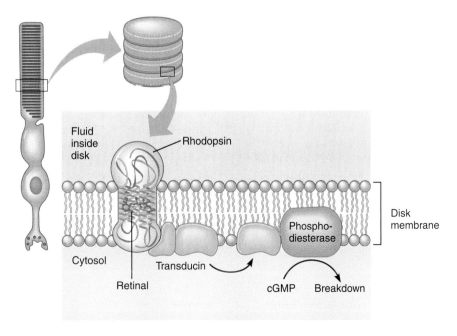

Figure 11.31 Components of rods. The photopigment, rhodopsin, is located in the membrane of the stacked disks located in the rod's outer segment. Rhodopsin is coupled to a G protein called transducin, which activates the enzyme phosphodiesterase, which in turn catalyzes the breakdown of cGMP.

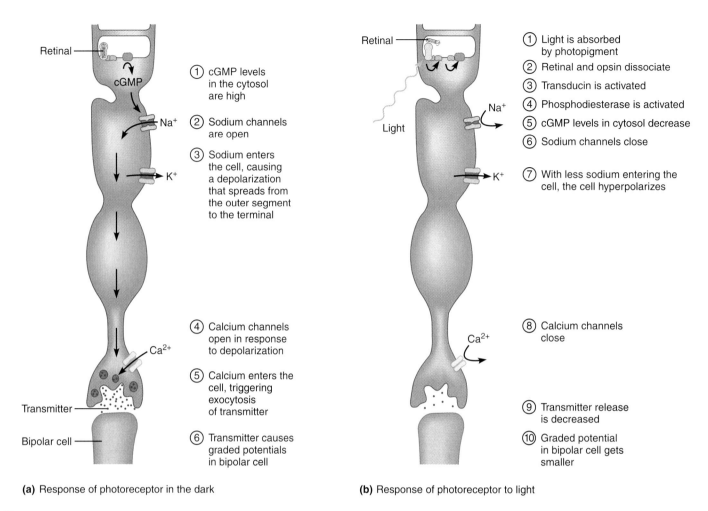

(1) cGMP levels in the cytosol are high

(2) Sodium channels are open

(3) Sodium enters the cell, causing a depolarization that spreads from the outer segment to the terminal

(4) Calcium channels open in response to depolarization

(5) Calcium enters the cell, triggering exocytosis of transmitter

(6) Transmitter causes graded potentials in bipolar cell

(1) Light is absorbed by photopigment

(2) Retinal and opsin dissociate

(3) Transducin is activated

(4) Phosphodiesterase is activated

(5) cGMP levels in cytosol decrease

(6) Sodium channels close

(7) With less sodium entering the cell, the cell hyperpolarizes

(8) Calcium channels close

(9) Transmitter release is decreased

(10) Graded potential in bipolar cell gets smaller

(a) Response of photoreceptor in the dark

(b) Response of photoreceptor to light

Figure 11.32 Phototransduction of light. (a) In the dark, photoreceptors release their chemical transmitter. (b) When light is present, it is absorbed by the photopigment, initiating a sequence of events that decreases release of the transmitter.

cell ⑤, triggering the release of transmitter by exocytosis *in the dark.* The transmitter communicates to bipolar cells ⑥.

When the photoreceptor is exposed to light (Figure 11.32b), light is absorbed by the rhodopsin ①. The retinal component changes its conformation and dissociates from the opsin ②, leaving what is called "bleached opsin." (When opsin is bleached, the photoreceptors become less sensitive to light, a phenomenon called *light adaptation.*) The bleached opsin activates transducin ③, which activates the enzyme phosphodiesterase ④, which then catalyzes the breakdown of cGMP.

With cGMP levels in the outer segment decreased ⑤, sodium channels close ⑥. Potassium leaking out of the cell causes a hyperpolarization that is no longer opposed by sodium movement into the cell ⑦. The hyperpolarization causes closing of calcium channels on the inner segment ⑧. With less calcium entering the cell, release of transmitter decreases ⑨. Therefore, in the light, less transmitter is released from the photoreceptor terminal. Information about the presence of light is relayed,

therefore, by a decrease in signaling to the next cells in the visual pathway, the bipolar cells ⑩.

Rods are very sensitive to light of a wide range of wavelengths, but most sensitive to light in the blue-green range (**Figure 11.33**). Rods are so sensitive to light that they can respond to a single photon (the unit of electromagnetic energy) of light. In bright light, however, rods become saturated (completely bleached); that is, they become as hyperpolarized as possible and thus cannot code for any additional brightness.

The process of phototransduction in cones is similar to that in rods. However, cones are not as sensitive to light as rods and therefore do not function well in dim light. In addition, cones respond best to light within a narrower range of wavelengths than rods. Blue cones are most sensitive to light at a wavelength near 430 nm, green cones are most sensitive to light at 530 nm, and red cones are most sensitive to light at 560 nm (see Figure 11.33). However, because the absorbance spectra of the three types of cones overlap, many colors can be perceived based on the patterns of activation of the different

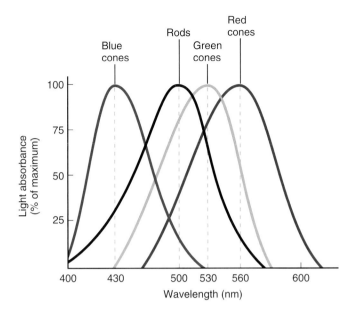

Figure 11.33 Absorbance spectra for the different photoreceptors. Rods can absorb light over the widest range of wavelengths. The absorbance spectra of the three types of cones overlap.

cones. A comparison of the characteristics of rods and cones can be found in **Table 11.4**.

Adaptation to Light and Dark

With small changes in light intensity, the eyes adapt simply through pupillary dilation and constriction, described previously. But with more drastic changes in light intensity, dilation and constriction are not enough. Recall the last time you went to a matinee in a theater, where it was very dark inside. When you walked outside into the bright sunlight, your eyes were overwhelmed by the intensity of the light. Similarly, when a person moves from the sunny outdoors into a dark room, the ability to see is greatly impeded. In either case, however, the eyes adapt in a matter of minutes to enable clear vision. What allows our eyes to adjust to these varying intensities of light?

When exposed to bright light, the rods become "bleached"; that is, most of the rhodopsin has absorbed light, and the opsin is in its active form. As a result, no more light can be absorbed until the rhodopsin has been returned to its original or "unbleached" state. Under these conditions, which correspond to when you first enter a dark room, rods are much less sensitive to light. Unbleaching of rods occurs in dim light, when the opsin returns to its inactive state. The retinal and opsin reassociate, and retinal becomes sensitive to light again. By contrast, unbleached rods are extremely sensitive to light. Therefore, when you have been in the dark for a period of time and then emerge into the daylight, the bright light overwhelms the rods until they become bleached.

Table 11.4 ▎ Characteristics of Rods and Cones

	Rods	Cones
Types of vision	Black and white; night (dim light)	Color; day (bright light)
Sensitivity to light	High	Low
Abundance	100 million per retina	3 million per retina
Visual acuity	Low	High
Site of greatest concentration	Periphery of retina	Fovea
Degree of convergence with bipolar cells	High	Low

Neural Processing in the Retina

The transmitter released from rods and cones communicates light and dark signals to bipolar cells in the retina. Some degree of convergence generally exists between photoreceptors and bipolar cells; that is, more than one photoreceptor communicates to a single bipolar cell. However, the extent of this convergence is greater with rods than with cones. Thus in the fovea and macula, where cones predominate, little convergence occurs; only a few photoreceptors converge onto one bipolar cell. In contrast, in the periphery of the retina, where there are only rods, thousands of rods converge on one bipolar cell. Recall that in the somatosensory system, less convergence results in greater tactile acuity and two-point discrimination. Similarly, less convergence in the visual system provides greater visual acuity, because two separate light sources can be discriminated as distinct sources only if they trigger responses in separate cells along the visual pathway. Greater convergence, on the other hand, provides greater sensitivity to light, because of spatial summation of inputs from several photoreceptors onto one bipolar cell.

Bipolar cells are capable of transmitting graded potentials, but not action potentials. In some synapses between photoreceptor and bipolar cells, the transmitter is excitatory (depolarizes the bipolar cell), whereas in others the transmitter is inhibitory (hyperpolarizes the bipolar cell). Therefore, light excites some bipolar cells and inhibits others. In addition, the synapses between photoreceptors and bipolar cells are subject to lateral modulation (excitation or inhibition) by horizontal cells.

When depolarized, bipolar cells release a transmitter that communicates to ganglion cells. Again, at some synapses the transmitter is inhibitory, whereas at other synapses the transmitter is excitatory. In addition, the synapses between bipolar cells and ganglion cells are subject to lateral modulation by amacrine cells. The ganglion cells are the first neurons in the visual pathway that are

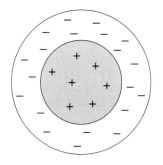

(a) ON-center, OFF-surround ganglion cell

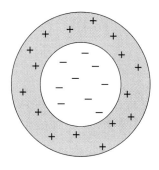

(b) OFF-center, ON-surround ganglion cell

Figure 11.34 Receptive fields of ganglion cells. Ganglion cells respond with both excitation (+) and inhibition (−) to light applied to the visual field. (a) ON-center, OFF-surround ganglion cells respond with excitation to light in the center of their receptive field and with inhibition to light in the surrounding area of their receptive field. (b) OFF-center, ON-surround ganglion cells respond with inhibition to light in the center of their receptive field, and with excitation to light in the surrounding area of their receptive field.

capable of transmitting action potentials. The axons of the ganglion cells make up the **optic nerve** and are the output neurons of the visual pathway.

Based on this description, one would expect the receptive field properties of ganglion cells to be complex. That is indeed the case. The receptive field for a ganglion cell is the area of the visual field in which a light stimulus either increases or decreases the frequency of action potentials in the cell. **Figure 11.34** shows the receptive fields of two types of ganglion cells. The first type of ganglion cell is called an *ON-center, OFF-surround cell* (Figure 11.34a). In these ganglion cells, light in the center of the receptive field excites the cell (turns the cell "on"), whereas light in the surrounding area of the receptive field (the "surround") inhibits the cell (turns the cell "off"). Diffuse light over the entire receptive field produces a small excitation (relative to diffuse dark). The other type of ganglion cell is called an *OFF-center, ON-surround cell* (Figure 11.34b). In these ganglion cells, light in the center of the receptive field inhibits the cell, whereas light in the surround excites the cell. Diffuse light over the entire receptive field produces a small inhibition (relative to diffuse dark).

Receptive fields of ganglion cells become even more complex when color is considered. For example, some ganglion cells are excited by red in the center of the visual field and inhibited by green in the surround; others are excited by blue in the center of the visual field and inhibited by yellow in the surround. (For more on color vision, see **Discovery: Color Vision**, p. 332.)

Neural Pathways for Vision

As previously described, the ganglion cells are the output neurons from the retina because they generate action

potentials that are transmitted to the CNS. The axons of ganglion cells form the optic nerve (cranial nerve II). The two optic nerves exit each eye at the optic disk and combine at the base of the brain just in front of the brainstem to form the **optic chiasm**. In the optic chiasm, half the axons from each eye cross over to the other side of the brain (**Figure 11.35**). Note in Figure 11.35 that input from the left visual field strikes the nasal retina (side closest to the nose) of the left eye and the temporal retina (side closest to the side of the head) of the right eye. Likewise, input from the right visual field strikes the nasal retina of the right eye and the temporal retina of the left eye. Therefore, both eyes receive information from both visual fields.

In the optic chiasm, axons originating from nasal ganglion cells cross to the opposite side, whereas axons originating from temporal ganglion cells stay on the side of origin. The result is that after the optic chiasm, all input from the right visual field travels in axons in the left side of the brain, and all input from the left visual field travels in axons in the right side of the brain. Although the axons are still those of ganglion cells, after the optic chiasm the axons travel in what is called the **optic tract**. The ganglion cells terminate in a nucleus in the thalamus called the **lateral geniculate body**, where they form synapses with neurons that ascend to the primary visual cortex in the occipital lobe. Pathways from the lateral geniculate body to the visual cortex on either side are called the **optic radiations**.

In the somatosensory cortex, the topographic organization of the cortex is such that adjacent areas of the body are usually represented on adjacent areas of cortex. The visual cortex also has topographic organization, but in this case the *visual field* is mapped onto the cortex. Due to the crossing of axons in the optic chiasm, the right visual field is mapped onto the left visual cortex, and the left visual field is mapped onto the right visual cortex.

Parallel Processing in the Visual System

The visual system clearly possesses parallel processing, in which parallel pathways transmit different qualities of a stimulus. For example, information about the color of an observed object is transmitted by some neurons in the visual pathway, whereas other neurons transmit information about shape or movement. The coding stays distinct all the way to the primary visual cortex. However, higher cortical areas integrate the different qualities of a stimulus so that we can perceive, for example, a red fire truck moving quickly away from us.

Depth Perception

Depth perception requires that the brain receives input from both eyes. Figure 11.35 shows that images in most (but not all) areas of the left and right visual fields are

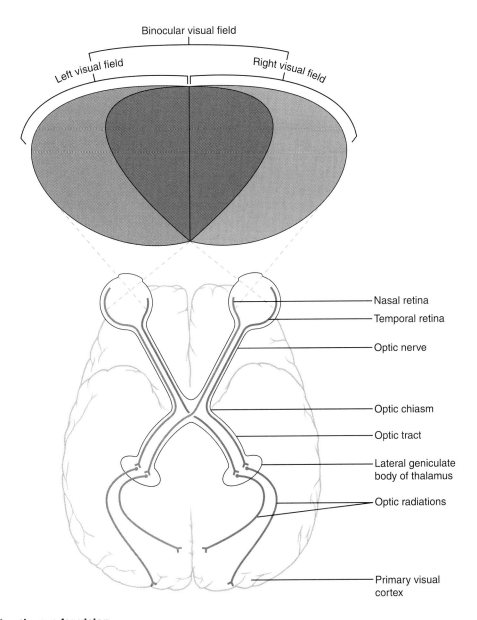

Figure 11.35 Neural pathways for vision.

Damage to the optic radiations on the left side would produce what type of visual deficit?

detected by both eyes. For the portion of the visual fields that is detected by both eyes, called the *binocular visual field,* we are capable of depth perception. Depth perception depends on the fact that the left and right eyes, with their different positions in the head, see images from slightly different angles. The cortex uses these differences to construct a three-dimensional image of the world (that is, one that includes depth) rather than a flat, two-dimensional image. To test your depth perception, try the following: Hold two pencils at arm's length, one in each hand. With one eye open, try to touch the point of one pencil to the point of the other. Repeat with both eyes open.

Loss or impaired perception of the right visual field

Quick Test 11.7

1. Put the following components of the visual pathway in order such that they correctly reflect the path of transmission of visual information: optic tract, ganglion cell, photoreceptor, optic radiation, optic chiasm, bipolar cell, optic nerve, lateral geniculate nucleus, and visual cortex.

2. When are cGMP levels greatest in photoreceptors: when they are exposed to light or when they are exposed to dark? What effect does cGMP have on sodium channels in the outer segment of the photoreceptors?

3. What are the two components of photopigment molecules? Which of these components absorbs light?

COLOR VISION

The ability to perceive colors is based on the presence of three types of cones that respond best to light of different wavelengths. That activation of a single cone cannot distinguish between colors is illustrated in figure (a), which shows the absorbance spectrum for one type of photoreceptor only. This hypothetical photoreceptor absorbs light at 450 nm and at 550 nm equally and thus is unable to distinguish between the two wavelengths. But how do three types of cones allow us to perceive the full spectrum of colors?

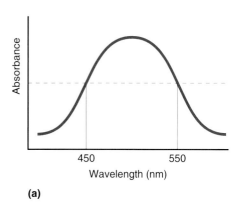

(a)

Each type of cone responds best to a specific wavelength, but each cone also responds over a range of wavelengths, though to different degrees. Whereas a given wavelength of light might elicit a response in more than one type of cone, the different cones generally respond to different degrees. Thus, our brains discern different colors by *comparing* the responses of the different cone types to each wavelength.

However, the relative responses of cones alone does not fully explain color perception. For example, why is there no such color as reddish green or yellowish blue? Properties like these can be explained by the *opponent-process theory,* which states that red and green, or blue and yellow, or black and white are opponent colors such that stimulation of one color inhibits the other. Therefore, we cannot see reddish green because the presence of green inhibits the perception of red. The opponent-process theory

pertains to the level of ganglion cells, where some ganglion cells are excited by red in their visual fields and inhibited by green in the same regions.

The opponent-process theory also explains the concept of afterimages. To observe afterimages, perform the following: Stare at figure (b) for approximately 30 seconds, and then stare at a blank sheet of white paper for approximately 30 seconds. What do you see?

(b)

When you looked at the white paper, you should have seen the opponent colors because of adaptation to the original colors. Recall that most of the light we perceive is light waves reflected off objects. White is observed when the full spectrum of light waves is reflected to the eye and all three cones are activated. To understand your perception, let's concentrate on what happened in the area of the visual field where you were looking at green. While you were looking at the picture, the green cones were activated, or bleached. When vision was shifted to the white paper, all wavelengths of light were reflected to the eye. However, in the visual field that originally detected green, the green cones

were bleached and thus did not respond as strongly to the green wavelengths that were present in the white light. With less inhibition from green, red signals were transmitted to the CNS more strongly, and red was observed.

People who lack a specific type of cone lose their ability to distinguish between certain colors; this is the basis of color blindness. The most common type of color blindness is red-green color blindness, in which red and green colors cannot be distinguished from each other. This type of color blindness is generally caused by a genetic defect in the photo pigments of red or green cones. The genes that code for the red and green photopigments are recessive and are located on the X chromosome; because males have only one X chromosome, they are more likely to inherit this recessive trait. Blue color blindness also exists, but it is rarer and is not linked to the X chromosome.

A common test for color blindness employs what are known as *Ishihara charts.* In these charts, numbers are hidden within a pattern of colored dots. An example is shown in figure (c). A person with normal vision can identify the number, but a person who is color blind is not able to see the number.

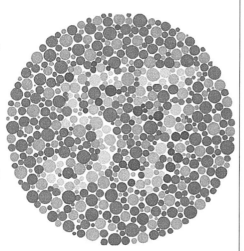

(c)

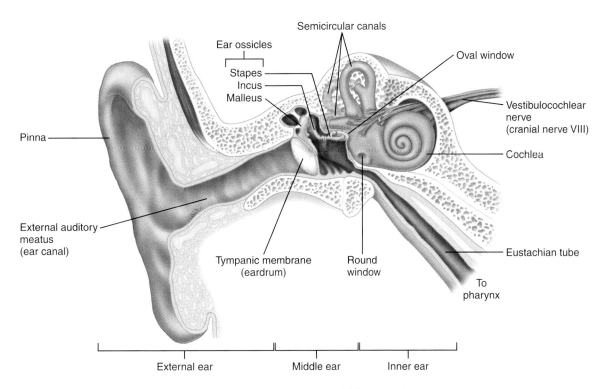

Figure 11.36 Anatomy of the ear. The major structures of the external ear, middle ear, and inner ear are labeled.

We now turn our attention to the sensory systems of the ear: first to the *auditory system,* which is responsible for hearing, and then to the *vestibular system,* which is responsible for balance or equilibrium. Although the stimuli involved vary, both of these sensory systems rely on hair cells to detect the movement of fluid within the cavities of the ear, which causes a receptor potential and release of a transmitter that communicates to an afferent neuron.

The Ear and Hearing

We begin our examination of the auditory system by considering the anatomy of the ear.

Anatomy of the Ear

The ear can be divided into three parts: the *external ear, middle ear,* and *inner ear* (**Figure 11.36**). The external and middle ears are air-filled cavities, whereas the inner ear is fluid filled.

The external ear includes the *pinna* and the **external auditory meatus**, or ear canal. The primary function of the external ear is to gather sound waves and conduct them to the *tympanic membrane* (or eardrum), which separates the external and middle ears.

The function of the middle ear is amplification of sound waves in preparation for the transmission of those waves from air to a fluid environment. Within the middle ear are three **ossicles**, or small bones, called the *malleus, incus,* and *stapes.* The three ossicles extend from the tympanic membrane to a thin membrane called the **oval window**, which is one connection between the middle and inner ears. The **round window** also connects the middle and inner ears.

The **eustachian tube**, which connects the middle ear with the *pharynx,* or throat, helps maintain normal pressure in the middle ear. Pressure changes, which may occur in the middle ear while flying or scuba diving, during ear infections, or even while going up or down in an elevator, can be painful and could cause rupture of the tympanic membrane if they become large enough. Even mild pressure changes can change the ability to hear, because the pressure changes dampen sound vibrations, just as occurs when a tympanist puts a hand on the drum head to stop a note. Opening the eustachian tube allows the pressure in the middle ear to equilibrate with the pressure in the pharynx, which alleviates any pressure difference across the eardrum. Swallowing or yawning can facilitate the opening of the eustachian tube.

The inner ear contains structures associated with both hearing and equilibrium. The **cochlea** is a spiral-shaped structure that contains the receptor cells for hearing. (The structures of the *vestibular apparatus,* including the *semicircular canals,* are discussed in the section on equilibrium.) The nerve that contains the afferents for both hearing and equilibrium, cranial nerve VIII, is also called the **vestibulocochlear nerve**.

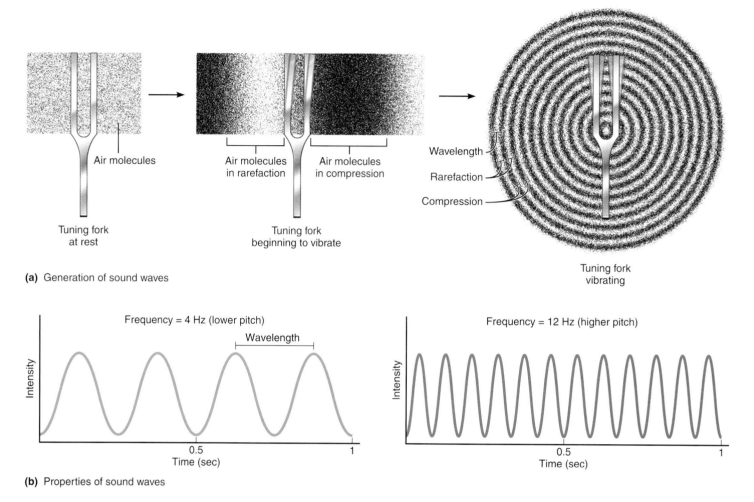

(a) Generation of sound waves

(b) Properties of sound waves

Figure 11.37 The nature of sound waves. (a) Vibrations produce sound waves, which consist of areas of high-density air molecules (compression) separated by areas of low-density air molecules (rarefaction). (b) Sound waves have an intensity (loudness) and a frequency (pitch).

The Nature of Sound Waves

Sound waves are mechanical waves caused by air molecules put into motion. **Figure 11.37**a shows the production of sound waves by a vibrating tuning fork. When the tines of a tuning fork vibrate, they generate waves in the air, similar to those produced when you touch your hand to the surface of a still pond. The waves consist of areas where the air molecules are closer together or *compressed,* and areas where the air molecules are further apart or *rarefied.*

Figure 11.37b illustrates the properties of sound waves: loudness and pitch. The loudness (amplitude) of a sound is proportional to the difference in the densities of air molecules between the areas of compression and the areas of rarefaction. The amplitude of a sound is most conveniently expressed in logarithmic units called *decibels (dB)* (see **Toolbox: Decibels**).

The pitch of the sound is determined by the frequency of sound waves. Low-frequency sound waves produce low-pitch sounds, such as those produced by a tuba. High-frequency sound waves produce high-pitch

sounds, such as the squealing of car brakes. The frequency of sound waves is measured as the number of waves per second, or *hertz (Hz)*. The average person can hear sound waves with frequencies ranging from 20 to 20,000 Hz; the greatest auditory sensitivity occurs in the range between 1000 Hz and 4000 Hz.

The coding of amplitude and pitch of sound must be maintained both while sound waves move from the air-filled middle ear to the fluid-filled inner ear, and during the transduction process in the inner ear. The anatomy of the ear is exquisitely designed to sustain the coding.

Sound Amplification in the Middle Ear

When sound waves enter the ear, they strike the tympanic membrane, causing it to oscillate (vibrate) back and forth. The oscillations occur at the same frequency as that of the sound waves, and with an amplitude proportional to the amplitude of the sound waves. Because the malleus, the first of the small ossicles, is connected to the tympanic membrane (**Figure 11.38**), oscillations of the tympanic

TOOLBOX

DECIBELS

The loudness (amplitude) of sound is based on the difference in density of air molecules in the areas of rarefaction and compression. Because the ear functions over a wide range of amplitudes, loudness is generally expressed in logarithmic units known as decibels (dB).

Measurements in decibels always compare the ratio of two intensities according to the following equation:

$$\text{intensity (dB)} = 20 \log \frac{\text{sound amplitude}}{\text{reference amplitude}}$$

In hearing, the reference amplitude is the threshold for normal hearing, that is, the minimum amplitude that can be perceived by humans, and it is given a value of 1. Thus, with the denominator set at 1, the equation becomes:

$$\text{intensity (dB)} = 20 \log (\text{sound amplitude})$$

At the threshold amplitude, the number of decibels is 0 [20 log (1) = 0]. A sound that is ten times the threshold for hearing is 20 dB [20 log (10) = 20], whereas a sound that is 100,000 times the threshold is 100 dB [20 log (100,000) = 100].

The intensities of various familiar sounds are listed in the table. Sounds approaching 100 dB create the potential for hearing loss; sounds at 130–140 dB approach the pain threshold.

Intensities of Some Familiar Sounds

Sound	Intensity (decibels)
Ticking watch	20
Elevator music	40
Conversational speech	50–60
Alarm clock	80
Live rock band	100
Jackhammer	110
Propeller airplane	120
Jet airplane	130

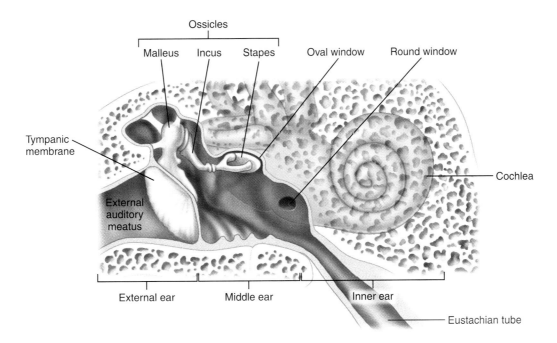

Figure 11.38 Structures that transmit sound waves in the middle ear. Sound waves hit the tympanic membrane, initiating vibrations in the ossicles. The stapes pushes the oval window, and vibrations are passed on to fluid in the cochlea.

membrane cause the malleus to oscillate at the same frequency and with an amplitude reflecting that of the tympanic membrane's vibrations. The three ossicles are arranged in such a manner that they function as a series of levers; movement of the malleus causes a greater movement of the incus, which causes an even greater movement of the stapes. The net effect is amplification of the motion produced by the initial sound waves. The stapes overlies the oval window to the fluid-filled cochlea, such that oscillations of the stapes generate waves in the fluid of the cochlea. However, because it takes a greater pressure to produce waves in fluid than in air, amplification is required.

To generate enough pressure to produce waves in the fluid of the cochlea, the sound waves are amplified as they travel from middle to inner ear by two primary means: First, the ossicles act as a lever system, as described above. Second, the much larger diameter of the tympanic membrane than that of the oval window produces amplification, because a given force acting on a smaller surface produces a greater pressure. As an analogy, consider hammering a nail into wood. The head of a nail has a larger area relative to the point. When you place the point of the nail against a block of wood and hit the head of the nail with a hammer, the force of the blow is transmitted through the nail to the wood. Because the area of the point is so small, however, the point exerts much more pressure on the wood than the hammer exerts on the nail, and therefore the nail penetrates the wood. In essence, the nail amplifies the pressure exerted on it by the hammer.

Signal Transduction for Sound

The cochlea is the organ in which sound transduction occurs. Next we explore the anatomy of the cochlea, and how it relates to the mechanism of sound transduction.

Functional Anatomy of the Cochlea

Understanding the mechanisms of sound transduction requires in-depth knowledge of the anatomy of the cochlea (**Figure 11.39**). From the outside (Figure 11.39a, b), the cochlea looks like a spiral seashell; the point at the end of the spiral is called the **helicotrema**. To view the inside of the cochlea, the spiral in Figure 11.39c, d is partially uncoiled. A cross section of the cochlea is shown in Figure 11.39e. Inside the cochlea are two membranes that separate it into three fluid-filled compartments: The **vestibular membrane** and the **basilar membrane** separate the cochlea into the **scala vestibuli** (or **vestibular duct**), the **scala tympani** (or **tympanic duct**), and the **scala media** (or **cochlear duct**). The vestibular and basilar membranes join at the helicotrema; thus there is an opening between the scala vestibuli and scala tympani at the helicotrema. The fluid in the scala vestibuli and scala tympani is called

perilymph; it differs from the fluid in the scala media, which is called **endolymph**. Perilymph is similar in composition to cerebrospinal fluid, but endolymph is closer in composition to intracellular fluid, with a high concentration of potassium ions and a low concentration of sodium ions.

The cochlea is a closed, fluid-filled structure separated from the middle ear at the oval and round windows. Because fluid is incompressible, to generate waves in the perilymph requires movement within the system without changing volume. When the stapes vibrates in response to sound waves, it causes vibrations of the oval window, producing waves in the perilymph of the scala vestibuli. The waves travel through the scala tympani, where they cause motion of the round window. Thus the movement of the oval window, which would tend to change the volume of the perilymph, is countered by movement of the round window, which dissipates the wave energy. Consider the following analogy: If you fill a syringe with water and cap off the opening, then you cannot move the plunger because the water can neither expand nor compress. However, if you connect two water-filled syringes with a tube, you can push one plunger down because the resulting force pushes the other plunger out. This arrangement allows you to move the plunger (the stapes) back and forth despite the fact that the water remains at a constant volume.

Functional Anatomy of the Organ of Corti

The **organ of Corti**, the sensory organ for sound, is located on top of the basilar membrane (Figure 11.39f). The organ of Corti contains **hair cells**, supporting cells, and an overlying membrane called the *tectorial membrane*. The receptor cells are called hair cells because they end in hairlike projections called **stereocilia**, the tips of which are embedded in the **tectorial membrane**. The anatomy of the organ of Corti is such that sound waves cause mechanical bending of the stereocilia, which causes receptor potentials in the hair cells.

Sound Transduction by Hair Cells

Figure 11.40a shows the conduction of sound waves into the internal ear. Sound waves enter the external ear and cause vibrations of the tympanic membrane, which in turn causes vibrations of the ossicles of the middle ear. Vibrations of the stapes cause movement of the oval window, which sets up waves in the perilymph in the scala vestibuli of the inner ear. The energy from waves in the perilymph causes the vestibular and basilar membranes to move relative to one another (Figure 11.40b), which causes the stereocilia to bend back and forth. Depending on the direction the stereocilia bend, potassium channels in the hair cells either open or close. Because endolymph has a higher concentration of potassium than that inside the hair cell (which is opposite to the conditions in most of the body, where potassium is in greater concentration

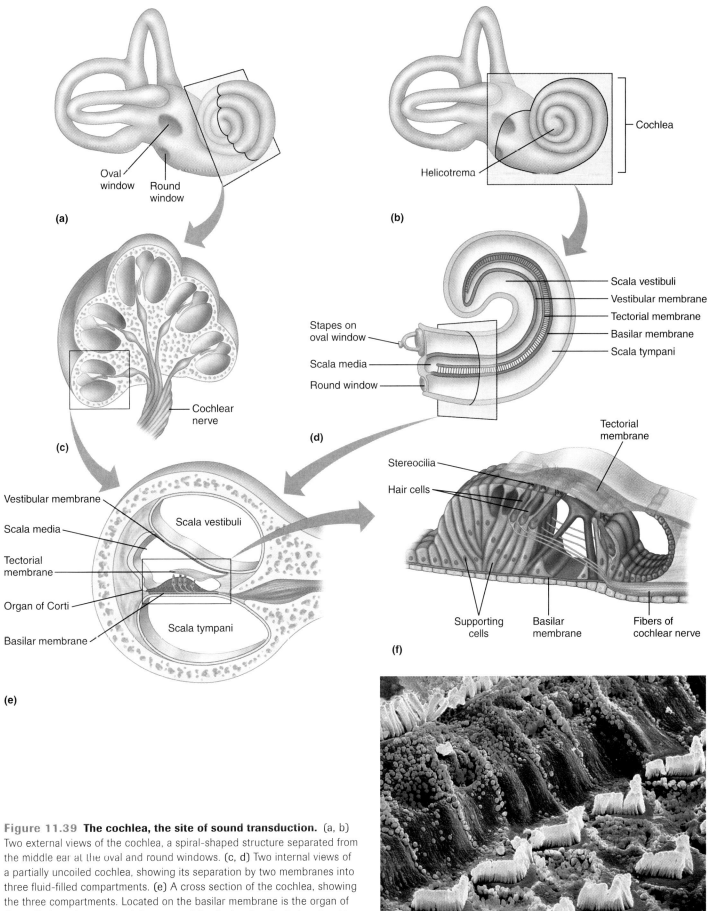

(a)

Oval window Round window

(b)

Helicotrema Cochlea

(c)

Cochlear nerve

(d)

Stapes on oval window Scala vestibuli Vestibular membrane Tectorial membrane Basilar membrane Scala tympani Scala media Round window

(e)

Vestibular membrane Scala media Tectorial membrane Organ of Corti Basilar membrane Scala vestibuli Scala tympani

(f)

Stereocilia Hair cells Tectorial membrane Supporting cells Basilar membrane Fibers of cochlear nerve

(g)

Figure 11.39 The cochlea, the site of sound transduction. (a, b) Two external views of the cochlea, a spiral-shaped structure separated from the middle ear at the oval and round windows. (c, d) Two internal views of a partially uncoiled cochlea, showing its separation by two membranes into three fluid-filled compartments. (e) A cross section of the cochlea, showing the three compartments. Located on the basilar membrane is the organ of Corti. (f) An enlargement of the organ of Corti, showing the hair cells with stereocilia that are embedded in the tectorial membrane. (g) An electron micrograph of hair cells in the cochlea.

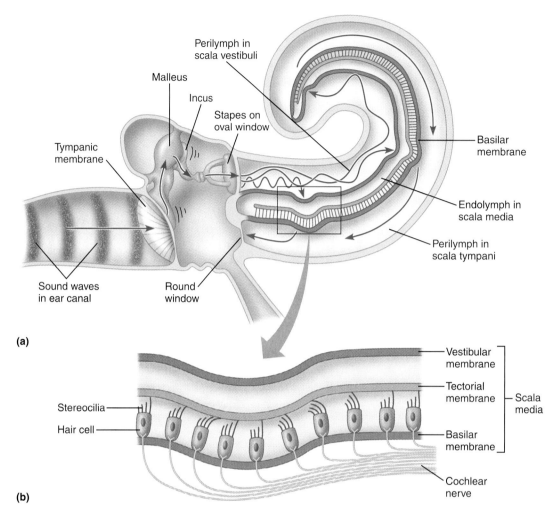

Figure 11.40 Conduction of sound waves in the ear. (a) Sound waves that enter the ear through the pinna and external auditory meatus strike the tympanic membrane, causing it to vibrate. The ossicles vibrate in response to the tympanic membrane and transmit the vibrations to the oval window. The vibrating oval window causes waves in the fluid (perilymph) of the cochlea. (b) Waves in the perilymph cause deflection of the membranes in the cochlea. When the membranes oscillate, the stereocilia of the hair cells bend, causing the opening or the closing of potassium channels.

inside the cell compared to outside), the opening of potassium channels results in potassium diffusing into the hair cell, causing depolarization; by contrast, closing of the potassium channels prevents potassium from diffusing into the hair cell, causing hyperpolarization.

Figure 11.41 shows how the bending of stereocilia causes potassium channels to open or close. Note that the stereocilia are linked together by elastic protein filaments and are of different sizes, such that one side of the hair cell has the tallest stereocilia. When the stereocilia extend straight up (no sound waves present), some tension on the elastic filaments holds the potassium channels in a partially opened state, allowing some potassium to diffuse into the hair cell and partially depolarize it (Figure 11.41a). This partial depolarization of the hair cell at rest causes the opening of some calcium channels in the hair cell, so calcium enters the cell and causes the release of transmitter

by exocytosis. The transmitter communicates to an afferent neuron that is part of the cochlear nerve, resulting in a low frequency of action potentials (see Figure 11.40b).

When the stereocilia are bent in the direction of the tallest stereocilium (Figure 11.41b), then more tension is put on the elastic filaments and the potassium channels are opened farther. More potassium diffuses into the hair cell, producing a more depolarized state. More calcium channels open, resulting in the release of more transmitter, thereby increasing the frequency of action potentials in the afferent neuron.

When the stereocilia are bent away from the tallest stereocilium, the elastic filaments go slack and the potassium channels close (Figure 11.41c). With the potassium channels closed, potassium cannot diffuse into the cell, and the cell becomes hyperpolarized compared to the resting state. Under these conditions,

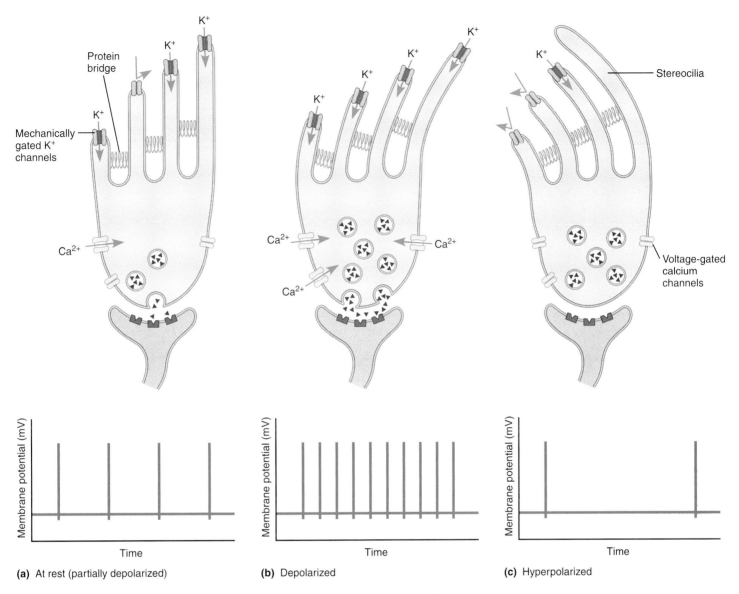

Figure 11.41 The roles of stereocilia in sound transduction by hair cells. (a) The mechanically gated potassium channels in stereocilia are partially opened when the cell is at rest (stereocilia stand upright), and potassium ions enter the cell, producing a small depolarization that is sufficient to release transmitter that communicates to the afferent neuron; the result is a low frequency of action potentials. (b) When the stereocilia bend toward the taller stereocilium, the potassium channels open more, and more potassium ions enter the cell, producing a greater depolarization and a higher frequency of action potentials in the afferent neuron. (c) When the stereocilia bend away from the taller stereocilium, the potassium channels close, and little potassium can enter the cell. Less transmitter is released, and the frequency of action potentials in the afferent neuron decreases.

calcium channels close and less transmitter is released, thereby decreasing the frequency of action potentials in the afferent neuron.

Coding of Sound Intensity and Pitch in the Cochlea

Given that receptor potentials are produced by bending of stereocilia, how are the intensity and pitch of sound coded? Louder sounds cause the stereocilia to bend farther in either direction, causing larger changes in the

number of open potassium channels. This results in larger receptor potentials and larger variations in transmitter release.

The coding of sound frequency is based on the location of hair cells on the basilar membrane. Sound waves of different frequencies cause deflection of the basilar membrane at different regions because the structure of the basilar membrane varies over its length: The basilar membrane is stiff and narrow near the oval and round

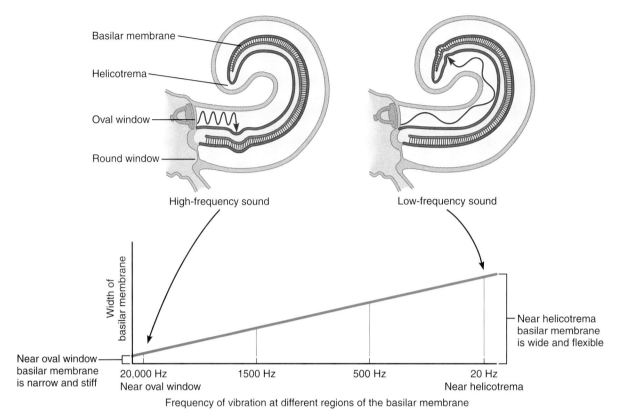

Figure 11.42 Coding for the frequency of sound. The frequency of sound is coded based on the location along the basilar membrane of activated hair cells. Variations in width and flexibility of the basilar membrane along its length dictate that high-frequency sounds cause deflections near the oval window, whereas low-frequency sounds cause deflections near the helicotrema.

windows, and wide and flexible near the helicotrema (**Figure 11.42**). Therefore, high-frequency sound waves associated with high pitch cause greatest deflection of the basilar membrane in the region closer to the oval and round windows, which activates hair cells located in this region. Low-frequency sound waves, by contrast, cause greatest deflection of the basilar membrane in the region closer to the helicotrema, which activates hair cells located in this region of the basilar membrane. The frequency of sound, therefore, is coded by which hair cells are activated most strongly.

Neural Pathways for Sound

The transmitter released from hair cells binds to receptors on afferent neurons of the *cochlear nerve,* part of cranial nerve VIII. The hair cell transmitter depolarizes the afferent neuron; the greater the degree of depolarization, the greater frequency of action potentials in the afferent neuron, which therefore codes for the intensity of the sound. The afferent neurons terminate in the cochlear nuclei in the brainstem, where they indirectly communicate with second-order neurons that travel to a nucleus of the thalamus called the **medial geniculate body**. In the medial geniculate body, the second-order neurons form synapses

with third-order neurons that transmit information to the **auditory cortex** in the temporal lobe. Like other sensory systems, the auditory cortex has topographical organization. Its organization is *tonotopic;* that is, it maps the frequency of sound.

Quick Test 11.8

1. Define the terms *outer ear, middle ear,* and *inner ear,* and describe the major structures contained in each.

2. Explain how the pitch of a sound is coded. How is the loudness of sound coded?

3. When stereocilia of hair cells are bent in a certain direction, what ion moves into the hair cell to depolarize it?

The Ear and Equilibrium

The previous section discussed the anatomy of the ear as it relates to transduction of sound waves. This section presents ear anatomy as it pertains to the ability to detect acceleration of the body and position of the head, information required if balance and equilibrium are to be maintained. Acceleration is a change in an object's velocity, which depends on its speed and direction. A change in speed without a change in direction is called *linear acceleration;* it occurs, for example, in a car that is gaining speed while moving in a straight line. However, acceleration also occurs when the car turns, even if its speed remains constant; this type of acceleration is called *rotational acceleration.*

Anatomy of the Vestibular Apparatus

The vestibular apparatus is located in cavities of the temporal bones called the *bony labyrinth.* Because the vestibular apparatus consists of membrane-bound structures within the bony labyrinth, these structures are also called the *membranous labyrinth* (the cochlea is also part of the membranous labyrinth). The membranous labyrinth is filled with endolymph, whereas the space between the membranous labyrinth and the bony labyrinth contains perilymph; note that these fluids are the same as those found in the cochlea.

Figure 11.43 provides a close-up view of the vestibular apparatus of the inner ear. The **vestibular apparatus** consists of the **semicircular canals** and the **utricle** and the **saccule**. Notice in Figure 11.43 that the three semicircular canals in each ear, which detect rotational acceleration, are oriented in planes that are perpendicular to each other, which enables them to detect rotational movements of the head in three planes. The anterior canal detects rotation of the head up and down, as when nodding "yes." The posterior canal detects rotation of the head up and down to the side, as in moving the ear toward the shoulder. The lateral canal detects rotation of the head from side to side, as when shaking the head to indicate "no." The utricle and saccule detect linear acceleration; the utricle detects acceleration forward and backward, whereas the saccule detects acceleration up or down.

The Semicircular Canals and the Transduction of Rotation

The receptor cells for equilibrium are another type of hair cell located in the **ampulla**, an enlarged area at the base of each semicircular canal (**Figure 11.44**a). Within each ampulla is the **cupula**, a gelatinous area separated from the endolymph by a membrane. Among the support cells at the base of the cupula are hair cells that are similar to those in the cochlea, including stereocilia that project upward into the cupula (Figure 11.44b). However, one of the stereocilia is much larger than the others and is called a **kinocilium**. As in hearing, mechanical bending of the stereocilia causes ion channels to open or close, resulting in a change in membrane potential in the hair cells. Next we examine how rotation can cause the stereocilia to bend.

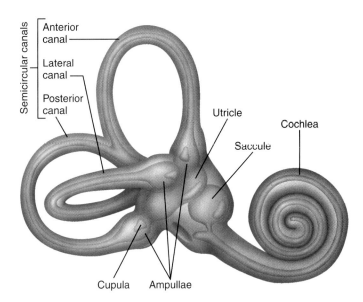

Figure 11.43 Anatomy of the vestibular apparatus in the inner ear. The vestibular apparatus includes the three semicircular canals, the utricle, and the saccule. The hair cells for detecting rotational acceleration are located in the ampullae of the semicircular canals.

When the head is at rest (Figure 11.44c), no force is acting on the cupula, so the stereocilia are upright. In this case the hair cells are partially depolarized, which leads to a low frequency of action potentials in the associated afferent neuron. When the head begins to rotate, the bony labyrinth rotates with it, but the endolymph lags behind the motion of the head, exerting a force on the cupula and causing the stereocilia to bend in the direction opposite that of rotation. When the direction of rotation is such that the stereocilia bend away from the kinocilium (Figure 11.44d), the hair cell is hyperpolarized, and the frequency of action potentials in the afferent neuron declines. When the head rotates such that the stereocilia bend toward the kinocilium (Figure 11.44e), the hair cells are depolarized, and the frequency of action potentials in the afferent neuron increases.

Note that when the head continues to rotate at a constant speed (see Figure 11.44c), the movement of fluid eventually catches up to the movement of the bony labyrinth. In that case, no force is acting on the cupula, so the stereocilia are no longer bent, and the frequency of action potentials returns to that occurring when the head is at rest. Thus the semicircular canals detect only changes in the rate of rotation, not constant rotation.

What happened when as a child you spun around to make yourself dizzy and then suddenly stopped spinning? When the constant rotation suddenly stops, the head and bony labyrinth stop as well, but the endolymph keeps moving for a time. Thus for a while rotation is detected even though you are standing still, but it feels like it is occurring in the direction opposite that of the original rotation.

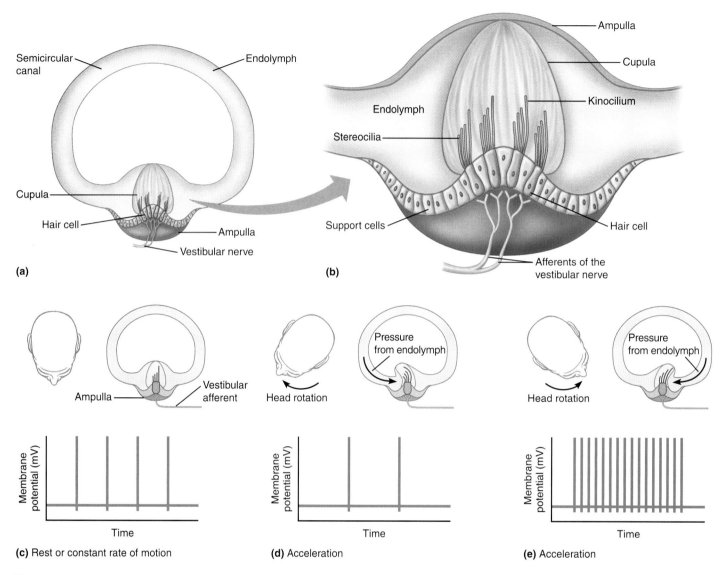

(a)

(b)

(c) Rest or constant rate of motion

(d) Acceleration

(e) Acceleration

Figure 11.44 Functional anatomy of the semicircular canals. (a) Within the ampulla, a swelling located at the base of each semicircular canal, is the cupula, a gelatinous area separated from the endolymph by a membrane. (b) Hair cells, the receptor cells for acceleration, have stereocilia that extend into the gelatinous cupula; the longest stereocilium is the kinocilium. Hair cells communicate to afferent neurons. (c) When the head is still or moving at a constant rate of motion, no net force is acting on the cupula, and the stereocilia are upright. The hair cells are partially depolarized, releasing some chemical transmitter and causing a low frequency of action potentials in the afferent neuron. (d) During acceleration of the head, inertia of the fluid causes it to lag behind the motion of the head. As it does so it pushes against the cupula in the direction opposite that of rotation, causing the stereocilia to bend. When the stereocilia bend away from the kinocilium, the hair cell is hyperpolarized, and the frequency of action potentials in the afferent neuron decreases. (e) During acceleration of the head in the opposite direction, the stereocilia bend toward the kinocilium, the hair cells are depolarized, and the frequency of action potentials in the afferent neuron increases.

Stereocilia are composed of what type of protein filament?

Microtubules

The Utricle and Saccule and the Transduction of Linear Acceleration

The utricle and saccule function in a manner similar to the semicircular canals in that they contain hair cells with stereocilia that bend, but their anatomy is somewhat different. The utricle and saccule, which are bulges located between the semicircular canal and cochlea of the inner ear (see Figure 11.43), have hair cells with stereocilia extending up into a gelatinous material (**Figure 11.45**). Within the upper edges of the gelatinous material are **otoliths**, small calcium carbonate crystals that add mass to the gelatinous material. The hair cells of the utricle are oriented in horizontal rows in the head, with the stereocilia extending up vertically; the hair cells of the saccule are oriented in vertical rows in the head, such that the stereocilia are oriented horizontally. Because of these orientations, the utricle detects forward and backward linear acceleration, whereas the saccule detects up and down linear acceleration.

Figure 11.45b–d illustrates the bending of the stereocilia within the utricle during forward and backward linear acceleration. Although all kinocilia are shown oriented in the same direction (posterior) in this example, some kinocilia are located anteriorly and others posteriorly in the utricle. Hair cells with anterior kinocilia will behave in a manner opposite that described here. When a person is at rest or moving forward or backward at a constant speed (Figure 11.45b), the hair cells are vertical, and the resulting depolarization results in moderately frequent action potentials in the afferent neuron. When the person begins to walk forward (Figure 11.45c), the otoliths cause the gelatinous mass to drag behind, bending the stereocilia toward the kinocilium and causing greater depolarization and more frequent action potentials. Backward linear acceleration (Figure 11.45d) causes bending of stereocilia away from the kinocilium, producing hyperpolarization and less frequent action potentials. When linear motion continues at a constant rate (see Figure 11.45b), the gelatinous material catches up with the movement of the body, and the stereocilia return to the at-rest position.

The utricle also plays a role in detecting the position of the head relative to gravity, a force that can cause acceleration. When the head is upright (Figure 11.45e), the stereocilia in the utricle are vertical, but when the head is tilted forward (Figure 11.45f), gravity pulls the weighted otoliths downward, bending the stereocilia downward and producing hyperpolarization.

The saccule functions in the same way as the utricle, but because of its orientation when the head is held upright it detects up and down linear acceleration, as occurs when one rides in an elevator.

Neural Pathways for Equilibrium

At rest, hair cells of the vestibular apparatus are partly depolarized and release some transmitter that communicates to afferents in the vestibular nerve. Therefore, at rest, a low frequency of action potentials is traveling along the vestibular afferents. When a hair cell becomes depolarized due to bending of the stereocilia toward the kinocilium, more transmitter is released, and the frequency of action potentials in the afferent neuron increases. Similarly, when a hair cell becomes hyperpolarized due to bending of the stereocilia away from the kinocilium, less transmitter is released, and the frequency of action potentials in the afferent neuron decreases.

The vestibular afferents enter the brainstem as part of the vestibular nerve. Most terminate in the *vestibular nuclei*, although some travel directly to the cerebellum to provide immediate feedback for equilibrium and balance, which is important in motor coordination. The vestibular nuclei have projections to different areas of the cortex that allow perception of acceleration, but the pathways are not well understood. The vestibular nuclei are important in feedback control of movement for balance and in controlling eye movements. To carry out these functions, the vestibular nuclei also receive input from different sensory systems, including those involved in vision, somesthetic sensations, and proprioception. This sensory information is used to provide output to motor neurons involved in balance and to match eye movements with body movements. If the eyes and head do not move together, vision will be blurred and motion sickness can occur.

Quick Test 11.9

1. Of the organs of the vestibular apparatus, which detect linear acceleration?

2. Why do the semicircular canals detect rotation when it starts or stops, but not constant rotation?

3. What two nerves combine to form the vestibulocochlear nerve? What types of sensory information are carried by each?

Taste

Although many of us think of chocolate as nectar of the gods, on another level chocolate is simply a bunch of chemicals mixed in just the right way to stimulate chemoreceptors in the mouth in a way that is pleasurable to most of us. This section describes how we detect the flavor of food.

Anatomy of Taste Buds

We can taste food because chemoreceptors in the mouth respond to certain chemicals in food. The chemoreceptors for taste are located in structures called **taste buds**, each of which contains 50–150 receptor cells and numerous

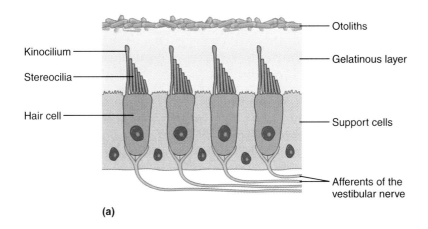

(a)

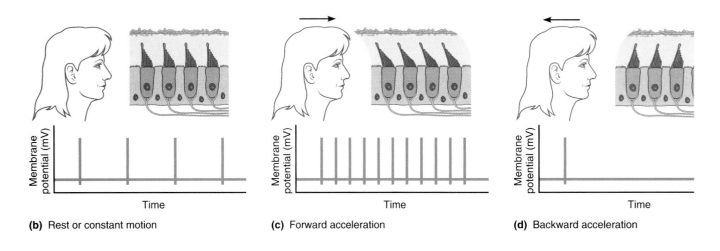

(b) Rest or constant motion

(c) Forward acceleration

(d) Backward acceleration

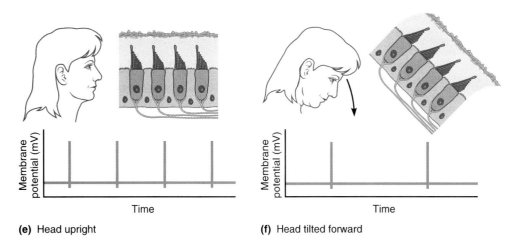

(e) Head upright

(f) Head tilted forward

Figure 11.45 Functional anatomy of the utricle and saccule. (a) The receptor cells for linear acceleration in both the utricle and the saccule are hair cells with stereocilia extending into a gelatinous layer containing otolith crystals. Parts b–f depict events associated with the utricle only. (b) At rest or in constant motion, the stereocilia in the utricle stand erect, and the hair cell is partially depolarized. The afferent neuron has a low frequency of action potentials. (c) During forward acceleration, the otoliths cause a drag on the gelatinous mass of the utricle, causing the stereocilia to bend. In this example, the stereocilia are bending toward the kinocilium. The hair cell is depolarized, and the frequency of action potentials in the afferent neuron is increased. (d) During backward acceleration, the otoliths in the utricle cause a drag on the gelatinous mass in the opposite direction as in part (c). The stereocilia bend away from the kinocilium, causing a hyperpolarization of the hair cell and a decrease in the frequency of action potentials in the afferent neuron. (e, f) Effect of gravity on the utricle. In the example shown, when the head is tilted forward, gravity causes the otoliths to fall, pulling on the gelatinous mass. The stereocilia bend, causing a hyperpolarization and a decrease in the frequency of action potentials in the afferent neuron.

support cells (**Figure 11.46**). At the top of each bud is a pore that allows receptor cells to be exposed to saliva and dissolved food molecules. Each person has over 10,000 taste buds, located primarily on the tongue and the roof of the mouth, but also located in the pharynx.

Taste receptor cells are modified epithelial cells with microvilli that extend into the pore of a taste bud. Located on the plasma membrane of the microvilli are taste receptors that bind selectively to the different chemicals, called **tastants**, that give food its flavor. For a chemical to bind to the taste receptor, the chemical must dissolve in saliva, which is why a person does not taste food as well when the mouth is dry. The interaction of the tastant with the taste receptor changes the membrane potential of the taste receptor cell.

Signal Transduction in Taste

There are four primary tastes—sour, salty, sweet, and bitter—each requiring its own transduction mechanism (**Figure 11.47**). Sour tastes are caused by the presence of hydrogen ions (acid) in the food (Figure 11.47a). These hydrogen ions bind to potassium channels ①, blocking them and preventing potassium from leaving the cell ②. The result is that the taste receptor cell depolarizes, which opens voltage-gated calcium channels and allows calcium to enter the cell ③, where it triggers the release of a transmitter by exocytosis ④.

Salty tastes are caused by the presence of sodium ions in the food (Figure 11.47b). When sodium levels outside the cell are increased by the consumption of salty foods, the inward electrochemical driving force on sodium ions increases, causing an increased flow of sodium into the cell ①, which depolarizes it. Voltage-gated calcium channels then open ②, and calcium triggers the release of a transmitter ③.

Sweet tastes are caused by the presence of organic molecules with structures similar to sucrose (Figure 11.47c). The binding of the organic molecule to a plasma membrane receptor ① activates a G protein called *gustducin* ②, which stimulates the production of cAMP ③. cAMP activates a protein kinase ④ that catalyzes the phosphorylation of potassium channels such that they close ⑤, which decreases the leakage of potassium out of the cell ⑥ and causes a depolarization. Voltage-gated calcium channels then open ⑦, and calcium triggers the release of a transmitter.

Bitter tastes are associated with a wide variety of nitrogen-containing compounds, including some that are toxic; that animals tend to avoid eating things that taste extremely bitter is thus a protective mechanism. There are several different types of bitter substances, and two known mechanisms for transduction of bitter taste (Figure 11.47d). Some bitter foods have molecules, such as quinine, that can block potassium channels ①a and thereby decrease potassium diffusion out of the cell ②a, depolarizing the cell and opening voltage-gated calcium

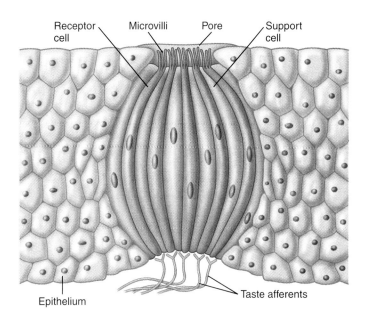

Receptor cell   Microvilli   Pore   Support cell

Epithelium

Taste afferents

Figure 11.46 Taste receptor cells in a taste bud. Taste receptors contain microvilli that extend into a pore at the top of the taste bud, where they are exposed to food particles dissolved in saliva.

channels ③a. Calcium enters the cell and triggers release of a transmitter ④a. Other bitter molecules act as ligands that bind to receptors on the plasma membrane ①b, activating a G protein called transducin ②b. Transducin activates the enzyme phospholipase C ③b, which catalyzes the formation of inositol triphosphate ④b. Inositol triphosphate opens calcium channels on the membrane of organelles ⑤b, causing calcium release into the cytosol. Calcium triggers the release of a transmitter ⑥b.

Researchers have identified a fifth taste that seems to differ from the primary tastes in that it produces no taste sensation of its own, but instead enhances other tastes: *umami,* which is Japanese for "delicious." Chemicals associated with umami seem to be flavor enhancers. Umami is associated with amino acids—most commonly, glutamate—and thus monosodium glutamate (MSG) is often added to food to enhance its flavor. Glutamate binds to receptors on taste receptor cells, opening channels that allow both sodium and potassium to move through. The net effect is a depolarization, which ultimately triggers the release of a transmitter.

Each taste receptor cell has all four transduction mechanisms, and thus each responds to all four primary tastes. However, a receptor cell tends to respond more strongly to one primary taste than to the others. Receptor cells that respond most strongly to a given taste are differentially distributed in the tongue, causing different regions of the tongue to be somewhat more sensitive to certain tastes (**Figure 11.48**a).

Coding of qualities of taste is complex. Several receptor cells communicate with a single afferent neuron. This extensive convergence of information creates complex

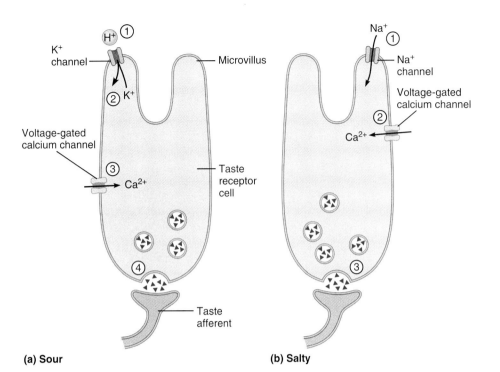

(a) Sour

(b) Salty

Figure 11.47 Taste transduction mechanisms for the four primary tastes. (a) Sour is produced by the presence of hydrogen ions. (b) Salty is produced by the presence of sodium ions. (c) Sweet is produced by molecules that can bind to a receptor on the plasma membrane of receptor cells. (d) Bitter taste can be produced by the two different mechanisms shown here.

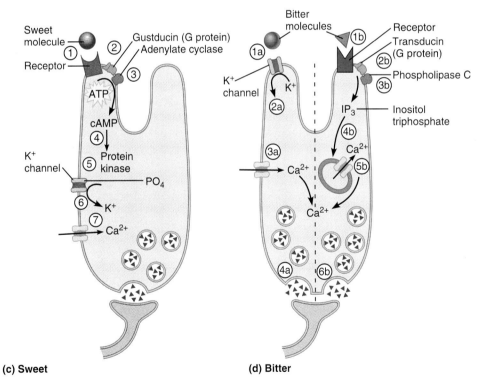

(c) Sweet

(d) Bitter

properties of taste afferent neurons, as shown in Figure 11.48b, which contains a plot of the responses of an afferent neuron that responds most strongly to sweet. Therefore, whereas most sensory systems code the quality of the stimulus based on receptor specificity and a labeled-line pathway, taste does not have that specificity. The perception of taste is complex and seems to depend on the pattern of activity in multiple afferent neurons, and it is further complicated by the fact that it is dependent on the sense of smell.

Neural Pathway for Taste

Taste receptor cells communicate to afferent neurons that travel in three cranial nerves: VII, IX, and X. These taste afferents terminate in the *gustatory nucleus* of the medulla, where they form synapses with second-order neurons. The second-order neurons travel to the contralateral thalamus, where they form synapses with third-order neurons that terminate in the *gustatory cortex,* which is

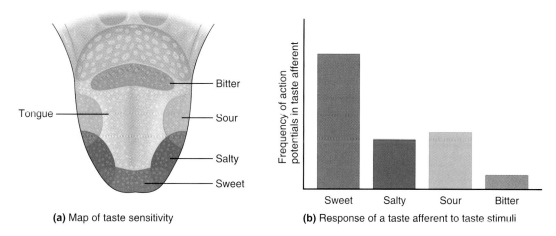

(a) Map of taste sensitivity

(b) Response of a taste afferent to taste stimuli

Figure 11.48 Neural coding of taste. (a) Different areas of the tongue are more sensitive to a particular taste. (b) Responses of a single taste afferent to the four primary tastes. This particular taste afferent responds best to sweet.

near the region of the somatosensory cortex that corresponds to the mouth.

Olfaction

The sensation of *olfaction,* or smell, depends on many of the same mechanisms as the sensation of taste. In olfaction, however, specific chemical substances called **odorants** must dissolve in *mucus* if they are to bind to specific chemoreceptors and be smelled.

Anatomy of the Olfactory System

Within the nasal cavity is the organ for smell, the **olfactory epithelium** (**Figure 11.49**). The epithelium is composed of three main classes of cells: supporting cells,

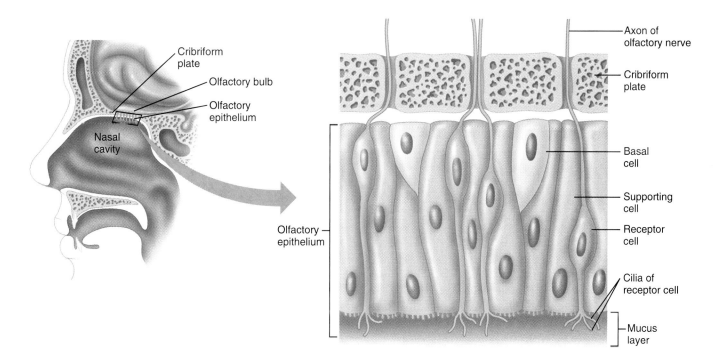

Figure 11.49 The olfactory epithelium. The olfactory epithelium, located in the nasal cavity, contains three types of cells: receptor cells, supporting cells, and basal cells. Cilia of the receptor cells extend into a layer of mucus in the nasal cavity, whereas axons of the receptor cells ascend to the olfactory bulb of the brain through holes in the cribriform plate of the skull.

basal cells, and olfactory receptor cells. The *supporting cells* provide structural support to the epithelium and secrete mucus; the **olfactory receptor cells**, the neurons that respond to olfactory stimuli, are replaced continuously throughout life by development of the **basal cells**. Olfactory receptor cells are the only neurons in the body currently known to be replaced regularly. Axons of olfactory receptor cells enter the CNS through holes in the cribriform plate of the skull and terminate in the olfactory bulb.

The olfactory receptor cells have cilia that project down into the mucus lining the nasal cavity. The cilia contain receptors that bind with specific odorant molecules. Within the mucus are **olfactory binding proteins**, which carry an odorant molecule to the receptor on the cilia, similar to the way that carrier proteins in the blood transport steroid hormones. Binding of the odorant molecule to the receptor initiates signal transduction.

Olfactory Signal Transduction

Although there are probably primary odors, a single basic transduction process seems to be universal in olfactory receptor cells. Binding of an odorant molecule to a membrane receptor activates a G protein called G_{olf}, which in turn activates the enzyme adenylate cyclase, which catalyzes the formation of cAMP. Although cAMP typically functions as a second messenger by activating a protein kinase, in olfactory receptor cells cAMP instead binds to cation channels, opening them and allowing both sodium and calcium ions to enter the cell. The primary effect of sodium and calcium entry is depolarization. However, the entry of calcium into the cell also causes chloride channels to open, allowing chloride to move out of the cell and increasing depolarization of the receptor cell. If the depolarization is great enough, action potentials are generated on the axon of the receptor cell.

Neural Pathway for Olfaction

Unlike other special sense receptors, olfactory receptors are specialized endings of afferent neurons, not separate cells.

The axons of the afferent neurons make up the **olfactory nerve** (cranial nerve I). If a receptor cell is depolarized to threshold, action potentials are produced and transmitted to the axon terminal in the brain. **Figure 11.50** illustrates a portion of the neural pathway for olfaction. The axons terminate in an area of the brain called the olfactory bulb, where they communicate with second-order neurons called *mitral cells*. The synapses between the afferents and second-order neurons occur in clusters called *glomeruli*, with much convergence of input onto the second-order neurons.

The second-order neurons form the *olfactory tract*, which is the neural pathway to an area of the brain called the *olfactory tubercle*. The neural pathways from the olfactory tubercle terminate in two areas of cerebral cortex: the olfactory cortex for the perception and discrimination of smells, and the limbic system for triggering olfactory-driven behaviors such as sexual behavior. The pathway to the olfactory cortex starts with a synapse in the olfactory tubercle with another neuron. It is unknown whether these neurons travel to the thalamus or directly to the olfactory cortex. For decades, olfaction was believed to be the one sensory system whose pathway to the cortex did *not* include a synapse in the thalamus. However, recent studies have identified odor-responsive neurons in the thalamus that might function as a relay to the olfactory cortex. Other second-order neurons do not synapse in the olfactory tubercle but instead travel directly to the limbic system.

Quick Test 11.10

1. Name the four primary tastes.
2. What is the difference between a taste bud and a taste receptor?
3. Name the three types of cells found in the olfactory epithelium, and describe their functions.
4. Give three similarities between the systems responsible for taste and smell.
5. What is unique about the neurons in cranial nerve I?

▩ *CHAPTER SUMMARY*

General Principles of Sensory Physiology, p. 302

Our ability to perceive the world around us depends on the presence of sensory receptors and specific neural pathways to communicate information to the cerebral cortex. Sensory systems must code for different qualities of the stimulus. Stimulus type is coded by the receptor and the pathway activated. Stimulus intensity is coded by frequency coding and population coding. The ability to locate a stimulus depends on the size of receptive fields, the degree of overlap of receptive fields, and lateral inhibition.

The Somatosensory System, p. 311

The somatosensory system enables perception of stimuli associated with the body surface (somesthetic

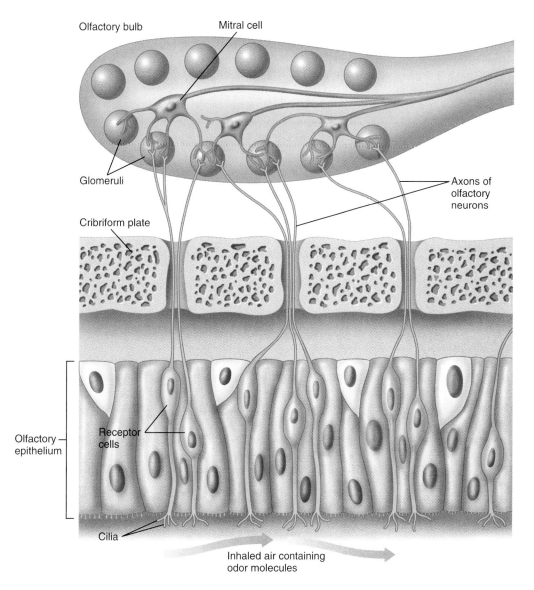

Olfactory bulb

Mitral cell

Glomeruli

Axons of olfactory neurons

Cribriform plate

Olfactory epithelium

Receptor cells

Cilia

Inhaled air containing odor molecules

Figure 11.50 The initial portion of the neural pathway for olfaction. Communication between olfactory afferents and second order neurons called mitral cells occurs in glomeruli of the olfactory bulb.

sensations) or body position (proprioception). Some receptors in the somatosensory system are specialized nerve endings, whereas others are free nerve endings. Information about touch, pressure, vibration, and proprioception is transmitted to the thalamus via the dorsal column–medial lemniscal pathway. Information about pain and temperature is transmitted to the thalamus via the spinothalamic tract. Information from the thalamus is transmitted to the primary somatosensory cortex.

Vision, p. 318

The eyes function in transduction of light energy. To focus light on the retina, the eye can change lens shape and thus its refractive power. Several clinical defects affect the ability of the eyes to focus light, including myopia, hyperopia, astigmatism, presbyopia, cataracts, and glaucoma. The pupils of the eye are capable of constricting or dilating to regulate the amount of light that enters the eye.

Phototransduction occurs in the retina, which consists of neural tissue

forming three layers: photoreceptors, bipolar cells, and ganglion cells. The photoreceptors, rods and cones, contain photopigments that absorb light. In the dark, photoreceptors are depolarized and release a chemical transmitter. In the light, the photopigments absorb the light and dissociate, causing a chain of reactions leading to hyperpolarization of the photoreceptor and decreased release of the transmitter. Photoreceptors communicate to bipolar cells, which communicate to ganglion cells. If the ganglion cell is depolarized to threshold, then an action potential results.

The axons of ganglion cells make up the optic nerve. Information is transmitted from the optic nerve to the optic chiasm, where half of the axons from each eye cross to the opposite side of the CNS such that all input from the right visual field is now on the left side, and all information from the left visual field is on the right side. The axons of ganglion cells after the optic chiasm make up the optic tract. The optic tract terminates in the lateral geniculate body of the thalamus, where the ganglion cell axons communicate with neurons that transmit information to the visual cortex.

The Ear and Hearing, p. 333

The ear contains the receptor cells for two sensory systems, hearing and equilibrium. For hearing, sound waves must enter the external ear, be amplified in the middle ear, and then be transduced into neural impulses in the cochlea of the inner ear. Sound transduction occurs in hair cells of the organ of Corti, located in the cochlea. When sound waves reach the fluid-filled cochlea, they set up waves that cause movement of the basilar membrane, which causes stereocilia to bend. This bending causes the opening or closing of potassium channels, which changes the electrical properties of the hair cells. When a hair cell is depolarized, a chemical transmitter is released and communicates with the afferent neurons in the cochlear nerve. Sound information is transmit-

ted to the medial geniculate body of the thalamus and then to the auditory cortex. In the auditory cortex is a tonotopic map. The loudness of a sound is coded by the degree of bending of the stereocilia, whereas the pitch of sound is coded for by the location of hair cells on the basilar membrane.

The Ear and Equilibrium, p. 340

The vestibular apparatus of the inner ear includes the semicircular canals (for detecting rotation) and the utricle and saccule (for detecting linear acceleration). The vestibular apparatus contains hair cells with stereocilia that bend with acceleration of the head. In the semicircular canals, the hair cells are located in the ampulla. Hair cells in the utricle and saccule have stereocilia that extend into a gelatinous mass containing otoliths. The bending of the stereocilia opens or closes ion channels, affecting the release of a chemical transmitter that communicates with afferent neurons in the vestibular nerve. Vestibular information is transmitted to the vestibular nuclei of the brainstem, which communicate to the thalamus and then the cortex for perception of equilibrium. Vestibular nuclei also communicate to the cerebellum for maintaining balance, and to brainstem nuclei that regulate eye movement.

Taste, p. 343

Both chemical senses—taste (gustation) and smell (olfaction)—depend on the binding of specific chemicals in

food or the air to chemoreceptors on receptor cells. Taste receptor cells are located within taste buds. Molecules that bind to taste receptors are called tastants. Each of the four primary tastes requires a different transduction mechanism. A single taste receptor cell responds to all four primary tastes, but most strongly to only one. Information is first transmitted via cranial nerves to the gustatory nucleus of the medulla, then relayed through the thalamus to the gustatory cortex.

Olfaction, p. 347

Olfactory receptors, located in the olfactory epithelium of the nasal cavity, respond to odorants dissolved in the mucus found there. Also present in the olfactory epithelium are basal cells, which are precursors for receptor cells. To bind to olfactory receptors, an odorant must bind to olfactory-binding proteins in the mucus. Upon binding to receptors, odorants trigger the production of cAMP in the cytosol of the receptor cell, which through a series of steps depolarizes the cell. If depolarized to threshold, an action potential is transmitted along the axon of the receptor cell. Axons of the receptor cells, which together constitute the olfactory nerve, form synapses with mitral cells in glomeruli of the olfactory bulb. Mitral cells transmit information along two pathways, one terminating in the olfactory cortex and the other in the limbic system.

▓ *EXERCISES*

Multiple-Choice Questions

1. The strength of a stimulus is coded by
   a) the size of the receptor potential.
   b) the size of the action potentials.
   c) the frequency of action potentials.
   d) a and c
   e) all of the above

2. The mechanism by which a receptor converts a stimulus into an electrical signal is called
   a) conduction.
   b) convection.
   c) transduction.
   d) modulation.
   e) propagation.

3. In lateral inhibition,
   a) the nervous system produces contrast to emphasize more-important information over less-important information.
   b) afferent neurons with neighboring receptive fields inhibit each other's communication to second-order neurons.

c) the ability to locate the site of a stimulus is enhanced.
d) a and c
e) all of the above

4. Which of the following observations best illustrates the concept of the labeled line?
   a) When a boxer gets punched in the eye, he perceives light.
   b) Rotation of the head stimulates certain receptors in the vestibular system but not those in the visual system.
   c) Information from different photoreceptors converges on a single ganglion cell that projects to the lateral geniculate nucleus.
   d) Hair cells in the cochlea are stimulated by sound vibrations over a wide range of frequencies.

5. Which of the following best illustrates the concept of an adequate stimulus?
   a) When a boxer gets punched in the eye, he perceives light.
   b) Rotation of the head stimulates certain receptors in the vestibular system but not those in the visual system.
   c) Information from different photoreceptors converges on a single ganglion cell that projects to the lateral geniculate nucleus.
   d) Hair cells in the cochlea are stimulated by sound vibrations over a wide range of frequencies.

6. Rubbing a sore area can decrease the sensation of pain by
   a) activating the endogenous analgesia systems.
   b) referring the pain to another area of the body.
   c) activating larger-diameter afferents, which activate an inhibitory interneuron, which inhibits the second-order neurons for pain.
   d) decreasing the number of action potentials in nociceptor afferents.
   e) presynaptic inhibition of substance P release.

7. In the dorsal column–medial lemniscal pathway,
   a) proprioception information is transmitted to the brain.
   b) the first-order neuron communicates to the second-order neuron in the dorsal horn of the spinal cord.
   c) the pathway crosses to the contralateral side in the spinal cord.

d) a and c
e) all of the above

8. Which of the following is the correct name of the pathway from the retina to the optic chiasm?
   a) optic tract
   b) optic radiations
   c) optic nerve
   d) optic disk

9. Which of the following is the correct name of the pathway from the lateral geniculate nucleus of the thalamus to the visual cortex?
   a) optic tract
   b) optic radiations
   c) optic nerve
   d) optic chiasm
   e) optic disk

10. Where would you expect to find the ascending tracts for somatosensory information?
    a) in the white matter of the spinal cord
    b) in a spinal nerve
    c) in the gray matter of the spinal cord
    d) none of the above

11. The ability to perceive different frequencies in sound vibrations is based on the fact that
    a) the stereocilia of any given hair cell respond to one frequency only.
    b) different areas of the basilar membrane resonate at different frequencies, such that sound of a particular frequency causes only a certain region of the membrane to vibrate.
    c) the frequency of action potentials in the cochlear nerve varies in proportion to the frequency of a sound stimulus.

12. The stereocilia for hearing are exposed to
    a) endolymph in the vestibular duct.
    b) perilymph in the vestibular duct.
    c) endolymph in the cochlear duct.
    d) perilymph in the cochlear duct.
    e) endolymph in the tympanic duct.

13. The parasympathetic nervous system causes
    a) contraction of the radial muscle of the iris.
    b) contraction of the ciliary muscle.
    c) pupillary dilation.
    d) a and c
    e) all of the above

Objective Questions

1. The two types of thermoreceptors are _____ and _____.

2. Receptors are most sensitive to energy from the _____ stimulus.

3. A phasic receptor adapts (quickly/slowly) to a constant stimulus.

4. The three types of nociceptors are _____ , _____ , and _____.

5. Information about touch detected on the left side of the body is transmitted to the brain in the dorsal columns on the _____ side of the spinal cord.

6. When a photopigment absorbs light, cGMP levels (increase/decrease).

7. The first neurons that support production of action potentials in the visual pathway are (photoreceptors/bipolar cells/ganglion cells).

8. The pitch of sound vibration reflects its (amplitude/frequency).

9. A hair cell in the cochlea can be excited by sounds of different frequencies. (true/false)

10. The process by which the lens becomes stronger for close-up vision is called _____.

11. Rods and cones differ with regard to the type of (retinal/opsin) they contain.

12. A single ganglion cell will either be excited or inhibited by light applied to its visual field. (true/false)

13. The visual cortex on the left side of the brain receives information from the right eye only. (true/false)

14. Odorant molecules must be dissolved in mucus if they are to bind to olfactory receptors. (true/false)

15. A given taste receptor cell responds to only one of the four primary tastes. (true/false)

Essay Questions

1. Compare the response of rapidly and slowly adapting touch receptors when you place your hand on a vibrating speaker.

2. Explain the concepts of topographic organization of the cerebral cortex. Compare the topographic organization of the somatosensory cortex, the visual cortex, and the auditory cortex.

3. Explain how it is possible for one person's perception to differ from another person's perception.

4. Diagram the general sensory pathway for transmitting information from a receptor to the cortex. Try fitting each sensory system into the general pathway; which sensory systems do not exactly fit this scheme?

5. Describe the sequence of events that occurs when photoreceptors are exposed to light.

6. Make a list of similarities between the olfactory and gustatory systems.

7. Make a list of the different types of sensory receptors in your body that are being activated at this moment.

Critical Thinking

1. A person goes to the hospital complaining of pain in the lower right abdomen. What might be the ailment? What neural pathway is responsible for the sensation of pain?

2. Explain why you are normally unaware of your blind spot.

3. Stare at a bright light for 10 seconds and then stare at a white sheet of paper. What do you observe? Explain.

4. If a person suffers damage to the left side of the thoracic spinal cord (damaging all the white matter on that side), which somatic sensations would be lost, which limbs would be affected, and from which side of the body?

5. Which sensory inputs are generally affected by a severe cold? Explain.

Find the answers to these exercises, and additional study tools, at the Physiology Place (www.physiologyplace.com).

The Nervous System: Autonomic and Motor Systems

12

Photomicrograph of neuromuscular junction.

Imagine it's a Sunday afternoon, and you have just finished a big lunch and are sitting in a recliner watching a movie. Your heart is beating slowly while your gastrointestinal organs are actively digesting and absorbing nutrients. All of a sudden, the fire alarm goes off and you see smoke. Your heart starts pounding. Blood flow is diverted from your gastrointestinal organs to your skeletal muscles, and you flee the house as flames emerge. Shortly after you've reached safety, your body starts to return to its resting state.

How does your body shift so rapidly from a relaxed, resting state to one prepared for emergency action? Your central nervous system controls your muscles and other organs by way of signals sent through the efferent branch of the peripheral nervous system. When your brain detects impending danger, it can quickly communicate to the sympathetic branch of the autonomic nervous system to initiate the fight-or-flight response. In just a matter of seconds your body adjusts to the crisis.

I n Chapter 11, we learned about the afferent branch of the nervous system and the different pathways by which information is transmitted from sensory receptors to the central nervous system for perception. In the efferent branch of the nervous system, commands are transmitted from the central nervous system to effector organs, such as muscles and glands. Recall from Chapter 8 that the efferent nervous system has two branches, the *autonomic* and *somatic* nervous systems, and that the autonomic nervous system is further divided into two subdivisions, the *sympathetic* and *parasympathetic*. The somatic nervous system controls our skeletal muscles, whereas the autonomic nervous system controls glands and muscles of the internal organs. Within the autonomic nervous system, activity in the parasympathetic nervous system is most active during rest, whereas the sympathetic nervous system is most active during excitement or activity. In this chapter, we describe the two branches of the efferent nervous system, starting with the autonomic nervous system.

The Autonomic Nervous System

The autonomic nervous system innervates all effector organs and tissues in the body except the skeletal muscles, including cardiac muscle, the smooth muscle found in blood vessels and various visceral organs (for example, the stomach and the respiratory airways), glands (for ex-

ample, sweat glands, salivary glands, and some endocrine glands), and adipose tissue. It is called "autonomic" because its functions occur at a subconscious level. For example, a person does not consciously decide to increase his or her heart rate; instead, heart rate increases through the operation of subconscious neural mechanisms when the need arises, such as during exercise. For this reason, the autonomic nervous system is also sometimes referred to as the *involuntary nervous system*.

Dual Innervation in the Autonomic Nervous System

Innervation of organs by the two branches of the autonomic nervous system is depicted in **Figure 12.1**; note that *both* branches of the autonomic nervous system innervate most organs, an arrangement called *dual innervation*. Although both branches innervate most organs, their functions are generally opposite in nature, and thus their effects are opposite. This poses no conflict, however, because the two autonomic divisions are typically most active under different conditions. For instance, the parasympathetic nervous system is most active during resting conditions, when it both stimulates the digestive organs (enhancing the digestion and absorption of nutrients) and inhibits the cardiovascular system (decreasing heart rate). In contrast, the sympathetic nervous system is most active during periods of excitation or physical activity, when it coordinates a group of physiological changes known as the **fight-or-flight response** that prepares the body to cope with threatening situations. During the fight-or-flight response, the rate and force of the heart's contractions increase, blood flow shifts from the gastrointestinal organs to skeletal and cardiac muscles, and energy stores are mobilized. These and other changes, all mediated by the sympathetic nervous system, prepare the body for intense physical exertion and otherwise adapt it for a possible response to a threatening situation.

The primary function of the autonomic nervous system is to regulate the function of effector organs in order to maintain homeostasis. At rest, both the sympathetic and parasympathetic branches are active, but the parasympathetic nervous system dominates. When the body gets excited or stressed, the pattern shifts to less parasympathetic activity and more sympathetic activity. The existence of both systems working at cross purposes provides a backup mechanism that permits greater control over the effector organs. Control of heart rate, for example, is exerted by both sympathetic and parasympathetic neurons; to bring about an increase in heart rate, sympathetic activity increases (that is, the frequency of action potentials in sympathetic neurons increases) while parasympathetic activity decreases (that is, the frequency of action potentials in parasympathetic neurons decreases). Additional coverage of the functions of the autonomic nervous system appears in the chapters that discuss the individual organ systems.

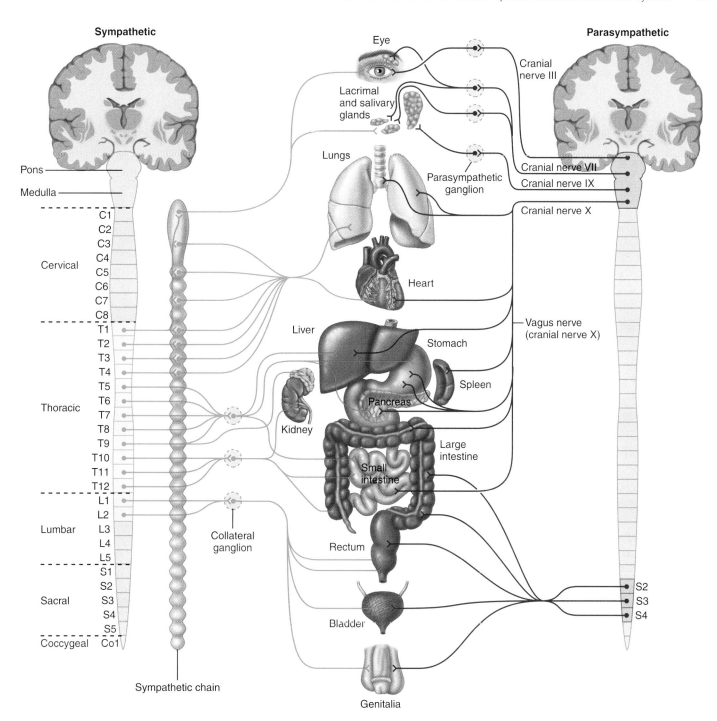

Sympathetic

Parasympathetic

Eye

Cranial nerve III

Lacrimal and salivary glands

Lungs

Parasympathetic ganglion

Cranial nerve VII
Cranial nerve IX
Cranial nerve X

Pons

Medulla

C1
C2
C3
C4
Cervical C5
C6
C7
C8
T1
T2
T3
T4
T5
T6
Thoracic T7
T8
T9
T10
T11
T12
L1
L2
Lumbar L3
L4
L5
S1
S2
Sacral S3
S4
S5
Coccygeal Co1

Heart

Liver

Stomach

Spleen

Pancreas

Kidney

Vagus nerve (cranial nerve X)

Large intestine

Small intestine

Collateral ganglion

Rectum

S2
S3
S4

Sympathetic chain

Bladder

Genitalia

Figure 12.1 Dual innervation in the autonomic nervous system. In this detailed schematic diagram of the autonomic nervous system, parasympathetic pathways are shown in purple, and sympathetic pathways in green. Both branches of the autonomic nervous system generally innervate the same organs. (Only one side of the body is shown for each system; thus there are actually two sympathetic chains, two of each cranial nerve, and so on.)

Anatomy of the Autonomic Nervous System

The autonomic nervous system consists of efferent pathways containing two neurons arranged in series that communicate between the CNS and the effector organ (**Figure 12.2**). The neurons communicate with each other through synapses located in peripheral structures

called **autonomic ganglia**. The neurons that travel from the CNS to the ganglia are called **preganglionic neurons**; the neurons that travel from the ganglia to the effector organs are called **postganglionic neurons**. Within each ganglion, therefore, are the axon terminals of preganglionic neurons and the cell bodies and dendrites of

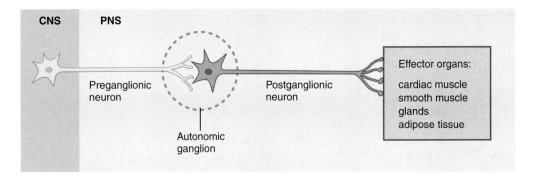

Figure 12.2 Anatomy of autonomic pathways. In autonomic pathways, two neurons transmit information from the CNS to an effector organ. The preganglionic neuron originates in the CNS and travels to a ganglion in the periphery, where it synapses with a postganglionic neuron that innervates one of several types of effector organ.

postganglionic neurons. A single preganglionic neuron generally synapses with several postganglionic neurons. In addition, other neurons located entirely within each ganglion, called *intrinsic neurons,* modulate the flow of information to the target organs. Although the two branches of the autonomic nervous system are similar in this regard, they show characteristic anatomical differences that are discussed in the following two sections.

Anatomy of the Sympathetic Nervous System

Because the preganglionic neurons in the sympathetic nervous system emerge from the thoracic and lumbar portions of the spinal cord (see Figure 12.1), the sympathetic nervous system is sometimes called the *thoracolumbar* division of the autonomic nervous system. The preganglionic neurons originate in a region of gray matter called the **lateral horn** or the *intermediolateral cell column.* Preganglionic and postganglionic sympathetic neurons are anatomically arranged in three patterns (**Figure 12.3**).

In the most common of these arrangements, the preganglionic neurons have short axons that originate in the lateral horn of the spinal cord and exit the spinal cord in the ventral root (Figure 12.3 and **Figure 12.4**). Immediately after the dorsal and ventral roots merge to form a spinal nerve, the axon of the preganglionic neuron leaves the spinal nerve via a branch called a *white ramus* to enter one of several sympathetic ganglia located just outside the spinal cord. Here the preganglionic neuron synapses with several postganglionic neurons whose long axons travel to the effector organ. Most of these postganglionic axons return to the spinal nerve via a branch called the *gray ramus,* and then travel to the effector organ in a spinal nerve.

The various sympathetic ganglia are linked together to form structures that run parallel to the spinal column on either side of it, called the **sympathetic chains** or *sympathetic trunks* (see Figure 12.4). Because a preganglionic neuron that enters a particular ganglion may have

collaterals that travel up or down these chains to make synapses with postganglionic neurons of other ganglia, activation of the sympathetic nervous system produces a widespread action that affects many different target organs at once, as is evident in the fight-or-flight response.

The second anatomical arrangement of sympathetic fibers, and a very important exception to the arrangement just described, is a group of long preganglionic neurons that innervate endocrine tissue—the *adrenal medulla* of the adrenal gland—instead of synapsing on postganglionic neurons (see Figure 12.3). Each of the two adrenal glands, located in fat pads on top of each kidney (**Figure 12.5**a), is divided into an outer cortex and inner medulla. The medulla consists of modified sympathetic postganglionic cells, called *chromaffin cells,*

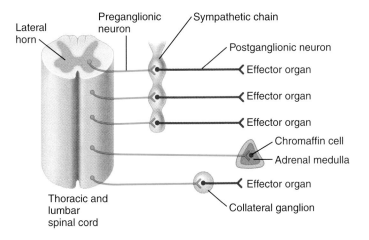

Figure 12.3 Anatomical pathways of preganglionic and postganglionic neurons in the sympathetic nervous system. Most sympathetic preganglionic neurons synapse with postganglionic neurons in ganglia in the sympathetic chain; some sympathetic preganglionic neurons innervate secretory cells of the adrenal medulla; still other sympathetic preganglionic neurons synapse in collateral ganglia that are independent of the sympathetic chain.

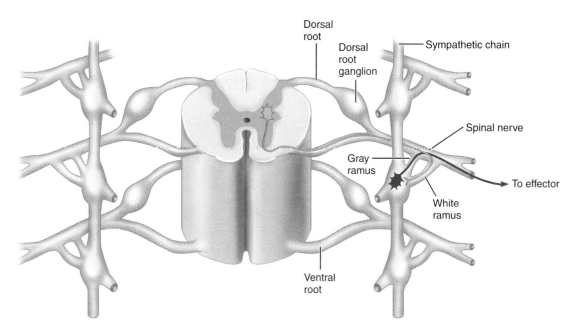

Figure 12.4 The most common pathway of sympathetic fibers. The sympathetic preganglionic neuron exits the spinal cord as part of the ventral root and travels to the spinal nerve. Shortly after entering the spinal nerve the preganglionic neuron exits the nerve and travels via a white ramus to a ganglion in the sympathetic chain, where the neuron synapses with a postganglionic neuron. The postganglionic neuron then returns to the spinal nerve via a gray ramus.

In addition to sympathetic postganglionic axons, what types of axons travel in a spinal nerve?

Axons of afferent neurons and motor neurons

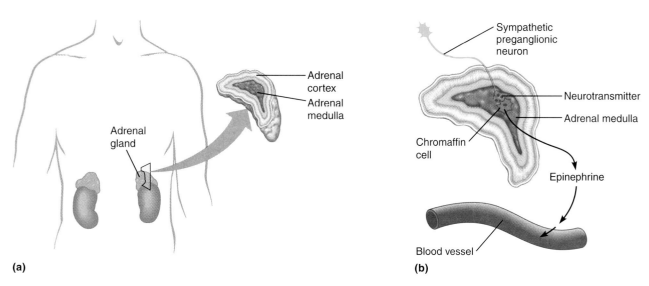

(a)   **(b)**

Figure 12.5 Sympathetic innervation of the adrenal gland. (a) The adrenal glands, located within fat pads on top of each kidney, consist of an outer cortex and inner medulla. (b) Sympathetic preganglionic neurons innervate chromaffin cells of the adrenal medulla, stimulating the release of epinephrine (and a small amount of norepinephrine) into the blood.

What three hormones are secreted by chromaffin cells, and what are the proportions of the total amount secreted?

Epinephrine (80%), norepinephrine (20%), dopamine (less than 1%)

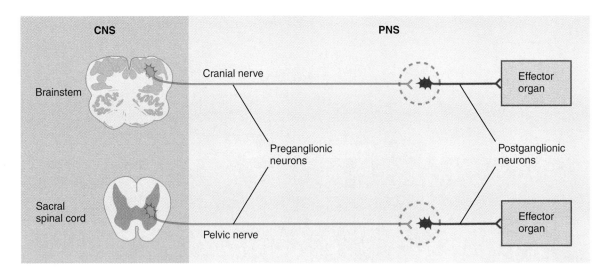

Figure 12.6 Parasympathetic nervous system pathways. Parasympathetic preganglionic neurons originate in either the brainstem or the sacral spinal cord, and their axons are found in cranial and pelvic nerves, respectively. These axons form synapses with postganglionic neurons in ganglia close to or within the effector organs.

that developed into endocrine cells instead of neurons (Figure 12.5b). Upon stimulation by the sympathetic nervous system, the adrenal medulla releases catecholamines. Of the three catecholamines released by the adrenal gland, approximately 80% is epinephrine (also known as adrenaline), 20% is norepinephrine (also known as noradrenaline), and a very small amount is dopamine. Like other endocrine glands, the adrenal medulla releases its products into the bloodstream, and thus these products function as hormones. Because these catecholamines travel throughout the body in the blood, this endocrine component of the sympathetic nervous system contributes to the widespread effects of sympathetic activation.

The third anatomical arrangement of sympathetic fibers includes preganglionic neurons that synapse with postganglionic neurons in structures called **collateral ganglia** situated somewhere between the CNS and the effector organ (see Figure 12.3). An example of a collateral ganglion is the celiac ganglion, which contains the cell bodies of postganglionic neurons innervating some of the digestive organs. In this case, a preganglionic neuron exits the spinal cord in the ventral root and then enters the sympathetic chain via a white ramus, as previously described. However, the axon of the preganglionic neuron continues through this ganglion without forming a synapse, and travels to a collateral ganglion via a sympathetic nerve. Within the collateral ganglion, the preganglionic neuron forms synapses with several postganglionic neurons that travel to target tissues. Because these ganglia are not interconnected, as are those of the sympathetic chain, they provide routes that enable the sympathetic nervous system to target select groups of target tissues and thus exert more discrete effects.

Anatomy of the Parasympathetic Nervous System

The preganglionic neurons of the parasympathetic nervous system originate in either the brainstem or the sacral spinal cord (see Figure 12.1), and thus the parasympathetic nervous system is sometimes called the *craniosacral division* of the autonomic nervous system. Generally speaking, parasympathetic preganglionic neurons are relatively long and terminate in ganglia located near the effector organ (**Figure 12.6**); in the ganglia they form synapses with short postganglionic neurons that travel to the effector organ.

In the cranial portion of the parasympathetic nervous system, axons of the preganglionic neurons originate from cranial nerve nuclei located in the brainstem, and travel with axons in cranial nerves. One important cranial nerve is the **vagus nerve** (cranial nerve X), which originates in the medulla oblongata and innervates much of the viscera, including the lungs, heart, stomach, small intestines, and liver (see Figure 12.1). Other cranial nerves containing parasympathetic preganglionic axons are the *oculomotor nerve* (cranial nerve III), the *facial nerve* (cranial nerve VII), and the *glossopharyngeal nerve* (cranial nerve IX), which innervate structures of the head and neck region.

Unlike sympathetic preganglionic neurons, parasympathetic preganglionic neurons that originate in the spinal cord do not join with the spinal nerve. Instead, they join with other parasympathetic preganglionic neurons to form distinct pelvic nerves (see Figures 12.1 and 12.6).

The Mixed Composition of Autonomic Nerves

Although it was long thought that nerves with fibers of the autonomic nervous system contained only efferent fibers, recent studies have shown that these nerves are generally mixed nerves. In fact, some of these nerves (for example, the vagus) actually contain more afferent fibers than efferent fibers. Only the efferent fibers in these nerves are part of the autonomic nervous system (which by definition is an *efferent* branch). The afferent fibers generally transmit signals from visceral receptors to the CNS and are important in maintaining homeostasis. Visceral receptors and their associated afferents are discussed in later chapters.

Quick Test 12.1

1. Define the following terms: *preganglionic neuron, postganglionic neuron, autonomic ganglia, sympathetic chain,* and *collateral ganglia.*

2. Which autonomic division regulates the release of epinephrine from the adrenal medulla?

3. Which system, sympathetic or parasympathetic, produces the more diffuse response? Why?

Autonomic Neurotransmitters and Receptors

The two neurotransmitters in the peripheral nervous system are acetylcholine and norepinephrine. Neurons that release the more common of the two, acetylcholine, are referred to as **cholinergic**. Acetylcholine is released by preganglionic neurons of both the sympathetic and parasympathetic branches of the autonomic nervous system, and by parasympathetic postganglionic neurons. The sympathetic preganglionic neurons that innervate the chromaffin cells of the adrenal medulla release acetylcholine, just like all other preganglionic neurons. However, in this case, acetylcholine acts on endocrine cells in the adrenal medulla to stimulate the release of epinephrine. Acetylcholine is also the sole neurotransmitter of the somatic branch of the efferent nervous system, a topic that is discussed later in this chapter.

Norepinephrine is released by almost all sympathetic postganglionic neurons. Neurons that release norepinephrine are referred to as **adrenergic**. (Some sympathetic postganglionic neurons release acetylcholine, specifically those that innervate the sweat glands).

Recall from Chapter 9 that the effects of a neurotransmitter depend on the type of postsynaptic receptors activated when the neurotransmitter binds to them. Acetylcholine and norepinephrine can each bind to different classes and subclasses of cholinergic and adrenergic receptors, respectively. We also consider receptors for the hormone epinephrine in this section because it is released from the adrenal medulla during activation of the sympathetic nervous system.

Types of Cholinergic Receptors

The two classes of cholinergic receptors—*nicotinic receptors* and *muscarinic receptors*—are distinguished on the basis of pharmacological studies using two acetylcholine agonists (chemicals that bind to a receptor and produce the biological effect): *nicotine* (found in tobacco) and *muscarine* (a toxin found in certain mushrooms). Nicotinic cholinergic receptors are located on the cell bodies and dendrites of sympathetic and parasympathetic postganglionic neurons (**Figure 12.7**), on chromaffin cells of the adrenal medulla, and on skeletal muscle cells. Muscarinic cholinergic receptors are found on effector organs of the parasympathetic nervous system, such as the heart, smooth muscles controlling the diameter of the pupil of the eye, and smooth muscle in the digestive tract (Figure 12.7).

The signal transduction mechanisms of cholinergic receptors are summarized in **Table 12.1**. Recall from Chapter 9 that nicotinic cholinergic receptors are associated with channels that allow both sodium and potassium ions to move through them. When acetylcholine binds to these receptors, these cation channels open, allowing sodium to diffuse into the cell and potassium to diffuse out of the cell. Because sodium is farther from equilibrium, the flow of sodium into the postsynaptic cell exceeds the flow of potassium out of the cell and leads to depolarization. Thus nicotinic cholinergic receptors are associated with depolarization, or excitation, of the postsynaptic cell.

In contrast, muscarinic cholinergic receptors are coupled to G proteins. The responses triggered by the binding of acetylcholine can be either excitatory or inhibitory, depending on the target cell in question and the nature of the signal transduction pathway. The binding of acetylcholine at muscarinic receptors can induce either changes in second messenger systems or the opening or closing of specific ion channels.

Types of Adrenergic Receptors

There are two major classes of adrenergic receptors located in effector organs of the sympathetic nervous system: **alpha (α) receptors** and **beta (β) receptors**. Each of these is further divided into subclasses: α_1 and α_2, and β_1, β_2, and β_3.

Adrenergic receptors are coupled to G proteins that either activate or inhibit second messenger systems (**Figure 12.8**). Binding of norepinephrine or epinephrine to an α_1 receptor activates a G protein that in turn activates the enzyme phospholipase C, which catalyzes the conversion of phosphatidylinositol biphosphate (PIP_2) to inositol triphosphate (IP_3) and diacylglycerol (DAG) (Figure 12.8a). IP_3 enters the cell and triggers the release of calcium from intracellular stores. Calcium then causes a response in the cell. DAG activates the enzyme protein kinase C, which catalyzes the

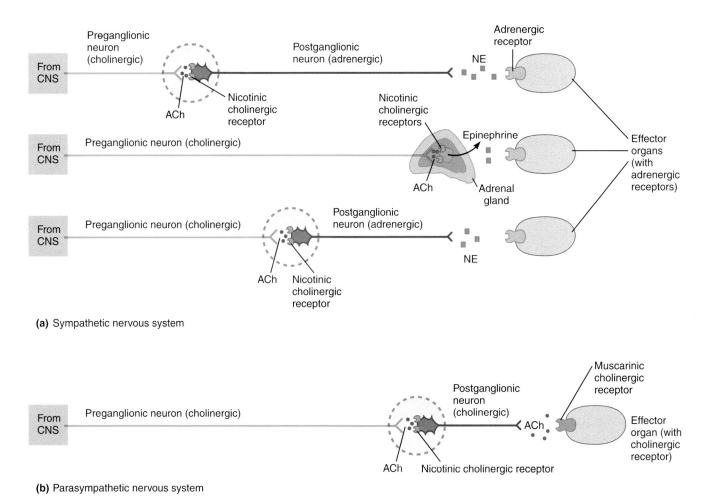

(a) Sympathetic nervous system

(b) Parasympathetic nervous system

Figure 12.7 Neurotransmitters and receptors in the autonomic nervous system.
(a) Neurotransmitters and receptors for the three distinct anatomical pathways of the sympathetic nervous system. In all cases, the preganglionic neuron releases acetylcholine (ACh), which then binds to nicotinic cholinergic receptors on either postganglionic neurons or endocrine cells in the adrenal medulla. The postganglionic neurons release norepinephrine (NE), which binds to adrenergic receptors on the effector organs. **(b)** Neurotransmitters and receptors in the parasympathetic pathway. Acetylcholine is released from both preganglionic and postganglionic neurons; it binds to nicotinic cholinergic receptors on the postganglionic neuron, and to muscarinic cholinergic receptors at the effector organ.

phosphorylation of a protein, causing a response in the cell.

Binding of norepinephrine or epinephrine to an α_2 receptor activates an inhibitory G protein (G_i) that decreases the activity of the enzyme adenylate cyclase, thereby suppressing the synthesis of cAMP (Figure 12.8b). Binding of norepinephrine or epinephrine to a β receptor, on the other hand, activates a stimulatory G protein (G_s) that increases the activity of the enzyme adenylate cyclase, thereby enhancing the synthesis of cAMP (Figure 12.8b).

Whereas acetylcholine is the only endogenous messenger that binds to cholinergic receptors, both norepinephrine and epinephrine bind to adrenergic receptors. However, the subclasses of adrenergic receptors have different affinities for these two messengers (**Table 12.2**).

The α receptors have a greater affinity for norepinephrine than epinephrine and are generally excitatory; that is, they stimulate muscle cells to contract or glands to secrete a product. β_1 and β_3 receptors have approximately equal affinities for norepinephrine and epinephrine, and tend to be excitatory. β_2 receptors have a much greater affinity for epinephrine than norepinephrine, and as a consequence norepinephrine does not normally function as a chemical messenger at these receptors. Activation of β_2 receptors generally produces an inhibitory response.

Autonomic receptors are of great clinical significance. For example, people suffering from asthma or nasal congestion often take β_2 adrenergic agonists, such as epinephrine or ephedrine, to produce dilation of the respiratory airways; α_1 agonists can also be used to treat congestion.

Table 12.1 │ Cholinergic Receptors

Receptor type	Signal transduction mechanism	Target cell	Effect on target cell
Nicotinic	Opens channels for sodium and potassium ions	Postganglionic cell body, chromaffin cells, skeletal muscle cells	Excitatory
Muscarinic	G protein–coupled; opens or closes specific ion channel	Effector organs of parasympathetic nervous system	Excitatory or inhibitory

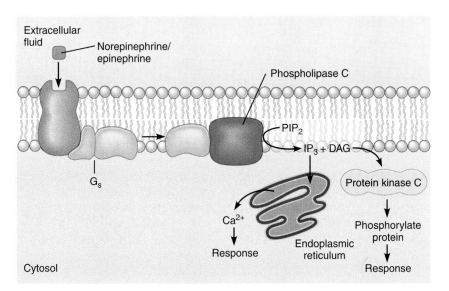

(a) Phosphatidylinositol biphosphate (PIP_2) and α_1 receptors

Figure 12.8 Signal transduction mechanism at effector organs of the sympathetic nervous system. (a) Binding of norepinephrine or epinephrine to an α_1 adrenergic receptor triggers activation of the phosphatidylinositol second messenger system. (b) Binding of norepinephrine or epinephrine to an α_2 adrenergic receptor triggers inhibition of the cAMP second messenger system, whereas binding to a β adrenergic receptor triggers excitation of the cAMP second messenger system.

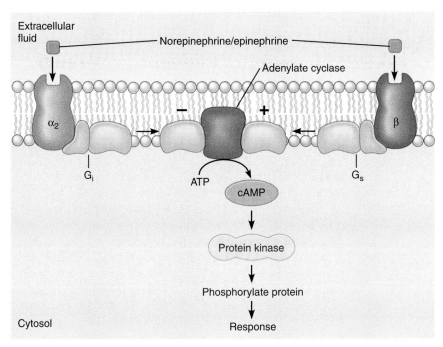

(b) cAMP and α_2 and β receptors

Table 12.2 | Adrenergic Receptors

Receptor type	Effector organ with receptor type	Relative affinities*	Signal transduction mechanism	Effect on effector organ†
α_1	Most vascular smooth muscle, pupils	NE > Epi	Activates IP_3	Excitatory
α_2	CNS, platelets, adrenergic nerve terminals (autoreceptors), some vascular smooth muscle, adipose tissue	NE > Epi	Inhibits cAMP	Excitatory
β_1	CNS, cardiac muscle, kidney	NE = Epi	Activates cAMP	Excitatory
β_2	Some blood vessels, respiratory tract, uterus	Epi ≫ NE	Activates cAMP	Inhibitory
β_3	Adipose tissue	NE = Epi	Activates cAMP	Excitatory

*NE = norepinephrine; Epi = epinephrine; > = greater than; ≫ = much greater than
†Effects are generalizations and not absolute.

Hypertension can be treated with drugs such as propranolol, which are called beta-blockers because they block the effects of norepinephrine or epinephrine at beta adrenergic receptors including those in the heart. Atropine, a muscarinic cholinergic antagonist, comes from the plant *Atropa belladonna,* so named because extracts from the plant were once used to dilate the pupils of women's eyes for cosmetic purposes (*belladonna* means "beautiful woman"). Atropine is sometimes used to dilate the pupils of the eye before an eye exam and is used to treat intestinal spasms and nausea.

Quick Test 12.2

1. What are the two neurotransmitters of the efferent branch of the peripheral nervous system, and which is the more common of the two?

2. What peripheral neurons release the less common of the two neurotransmitters in the efferent branch of the PNS?

3. What branch of the autonomic nervous system would be affected by administration of a nicotinic cholinergic antagonist, and what action would such a drug have?

Autonomic Neuroeffector Junctions

The synapse between an efferent neuron and its effector organ is called a *neuroeffector junction.* Synapses between autonomic postganglionic neurons and their effector organs differ from ordinary neuron-to-neuron synapses in that the postganglionic neurons do not have discrete axon terminals. Instead, neurotransmitters are released from numerous swellings located at intervals along the axons of these neurons, called **varicosities** (**Figure 12.9**a). Within these varicosities, neurotransmitters are

synthesized and then stored in vesicles (Figure 12.9b). The membrane of the axon contains the usual voltage-gated sodium and potassium channels that support propagation of action potentials. In addition, the membrane in the region of each varicosity contains voltage-gated calcium channels that open when an action potential reaches them.

The mechanism of neurotransmitter release from these varicosities is similar to the mechanism of transmitter release from an ordinary axon terminal (**Figure 12.10**). An action potential arriving at a varicosity opens voltage-gated calcium channels, which allows calcium to enter the cytoplasm and stimulates the release of neurotransmitter by exocytosis. However, unlike ordinary synapses, no discrete anatomical arrangement exists between the varicosity and the effector organ. First, because a single postganglionic axon has several varicosities, an action potential propagated along the axon triggers the release of neurotransmitter from all the varicosities. Second, because the distance between the varicosity and effector organ is greater than the width of a synaptic cleft, the neurotransmitter released from a varicosity diffuses over a greater area of the effector organ, binding to receptors on cells throughout the effector organ.

The effects of the neurotransmitter are terminated as they are in a typical neuron-to-neuron synapse: either diffusion of the neurotransmitter away from the receptors, or active reuptake of the neurotransmitter, or degradation of the neurotransmitter by enzymes. One such enzyme is *acetylcholinesterase,* which is located on the membrane of either the postganglionic neuron or the effector organ at cholinergic neuroeffector junctions. Following degradation of acetylcholine to acetate and choline by acetylcholinesterase, choline is actively transported back into the postganglionic varicosity and used to synthesize more

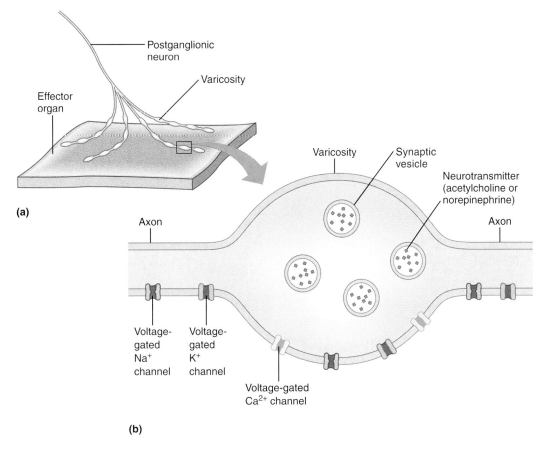

Figure 12.9 Neuroeffector junctions of the autonomic nervous system. (a) Near the effector organ the axon of a postganglionic neuron splits into branches bearing swellings called varicosities. (b) An enlarged view of a varicosity. Whereas voltage-gated sodium and potassium channels are present throughout the axon to support propagation of action potentials, voltage-gated calcium channels are concentrated at the varicosity. Neurotransmitter is stored in synaptic vesicles in the varicosity.

Where with respect to the varicosity depicted in Figure 12.9b are action potentials propagated: at the left, at the right, or on both sides?

acetylcholine. Monoamine oxidase, another degradative enzyme, is located at adrenergic neuroeffector junctions within mitochondria in postganglionic neurons, where it degrades catecholamines that have been actively transported back into the cells.

Regulation of Autonomic Function

Balance of activity levels between the two branches of the autonomic nervous system is necessary for the maintenance of homeostasis. Under resting conditions, the body expends less energy than at other times, with the primary energy expenditure going into the digestion and absorption of nutrients from recently ingested food. With low energy demands, the heart does not need to work as hard, because organs' demand for blood is low. Appropriately, the parasympathetic nervous system exerts dominant control over the body's organs under

these conditions. However, when the body is active, energy demands of the tissues increase. These demands must be met, even at the expense of suspending operation of the gastrointestinal organs. Under these conditions, parasympathetic activity decreases and sympathetic activity increases. The result is increased activity of the heart, which helps increase blood flow to the skeletal muscles. But how does the brain regulate the balance between parasympathetic and sympathetic activity to meet the body's changing demands?

Most changes in the activity of the autonomic nervous system are accomplished through the operation of **visceral reflexes**, automatic changes in the functions of organs that occur in response to changing conditions within the body. Consider, for example, the autonomic response that operates when a person stands up rapidly

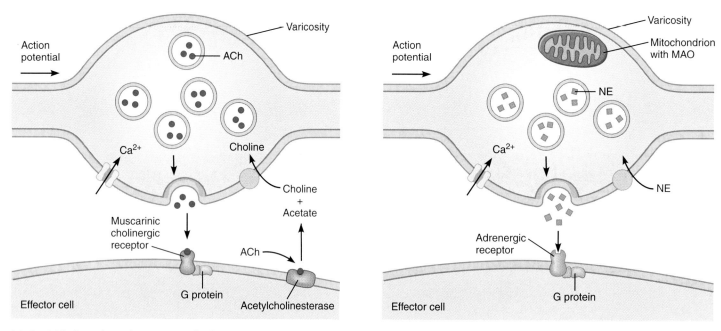

(a) Acetylcholine release from postganglionic neuron

(b) Norepinephrine release from postganglionic neuron

Figure 12.10 Neurotransmitter release from varicosities. (a) Release of acetylcholine from a varicosity. When an action potential depolarizes a varicosity, voltage-gated calcium channels open, allowing calcium ions to enter the cell and triggering the release of acetylcholine by exocytosis. Acetylcholine binds to muscarinic cholinergic receptors on the effector organ membrane, thereby activating a G protein. Once acetylcholine is degraded by acetylcholinesterase located on the effector organ, choline is actively transported back into the postganglionic varicosity and used to synthesize more acetylcholine. (b) Release of norepinephrine from a varicosity. When an action potential depolarizes the varicosity, voltage-gated calcium channels open, allowing calcium ions to enter the cell and trigger the release of norepinephrine by exocytosis. Norepinephrine binds to adrenergic receptors on the effector organ membrane, thereby activating a G protein. Some norepinephrine is actively transported back into the postganglionic neuron's varicosity, where it is degraded by monoamine oxidase (MAO) in mitochondria.

(**Figure 12.11**). A drop in blood pressure that occurs because blood pools in the lower limbs due to the force of gravity is detected by receptors in some of the major arteries (aorta and carotid arteries), and that information is then relayed to cardiovascular regulatory areas in the medulla oblongata through an afferent pathway. In response, these areas adjust their output to the sympathetic and parasympathetic nervous systems. Sympathetic output to the heart and blood vessels is enhanced, which brings blood pressure back to normal. At the same time, parasympathetic output to the heart is decreased, which allows the blood pressure to increase unopposed. (Recall that parasympathetic activity tends to slow the heart. This action tends to lower blood pressure.) Blood pressure typically returns to normal in just a few seconds, so the person is unaware of any changes that might have occurred. This reflex arc acts to prevent reduced blood flow to the brain and loss of consciousness when a person arises from lying down.

The major areas of the brain that regulate autonomic function include the hypothalamus, pons, and medulla oblongata (**Figure 12.12**). The hypothalamus

initiates the fight-or-flight response to elicit widespread activation of the sympathetic nervous system when a person is in danger or is otherwise excited. The hypothalamus also contains the regulatory centers for body temperature, food intake, and water balance, all of which are regulated in some way by autonomic efferent neurons. The medulla oblongata and pons contain cardiovascular and respiratory regulatory centers that control the heart, blood vessels, and smooth muscle in the respiratory airways and regulate the automatic breathing patterns that do not require conscious thought. These areas of the brain receive input from other brain regions, including the hypothalamus, cerebral cortex, and limbic system. They also receive afferent information needed for reflex control of visceral function, such as that just described concerning blood pressure. Other autonomic reflexes involving the brainstem include the pupillary light reflex (Chapter 10), the accommodation reflex (Chapter 11), the vomiting reflex (Chapter 21), and the swallowing reflex (Chapter 21). Some visceral reflexes are coordinated by neural circuits in the spinal cord and are thus called spinal reflexes. Examples

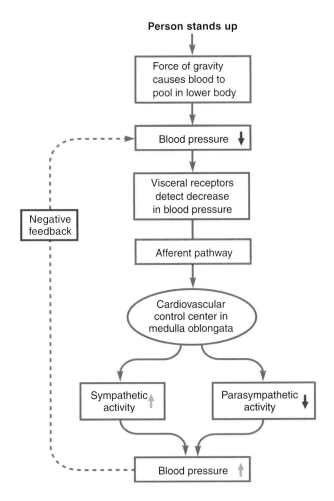

Person stands up

Force of gravity causes blood to pool in lower body

Blood pressure ↓

Visceral receptors detect decrease in blood pressure

Negative feedback

Afferent pathway

Cardiovascular control center in medulla oblongata

Sympathetic activity ↑

Parasympathetic activity ↓

Blood pressure ↑

Figure 12.11 Autonomic reflex response that controls blood pressure when a person stands up.

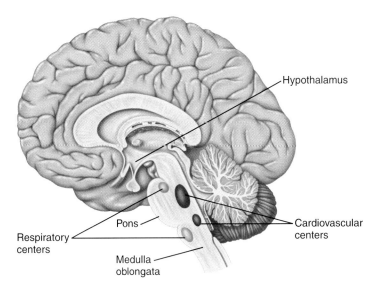

Figure 12.12 Areas of the brain that regulate autonomic function. This midsagittal section of the brain reveals the pons and medulla oblongata of the brainstem and the hypothalamus, all of which regulate autonomic function.

include the reflexes that control urination (Chapter 19), defecation (Chapter 21), erection (Chapter 22), and ejaculation (Chapter 22).

Recall from Chapter 10 that one area of the brain that influences activity in autonomic control centers is the limbic system, which is involved in the development of emotions. Emotions have a strong effect on activity of the autonomic nervous system and thus affect the functions of effector organs controlled by it. We are all aware of some of the more common autonomic responses to emotions, including a racing heartbeat, upset stomach ("butterflies"), blushing, fainting, and sweating.

The autonomic nervous system is also responsible for many of the symptoms associated with motion sickness, which is generally caused by a mismatch of sensory inputs, including those from the vestibular apparatus (which detects motion of the body), the visual system, and proprioceptors throughout the body (which detect the body's position in three-dimensional space). Common symptoms include nausea and sweating. The neural mechanisms mediating motion sickness are not fully understood, but communication between the vestibular system and autonomic control centers in the brainstem may be involved, with

both branches of the autonomic nervous system being excited. A common treatment for motion sickness is the drug scopolamine, a muscarinic cholinergic antagonist that can be administered as a tablet or via a patch.

Table 12.3 summarizes the effects of the two branches of the autonomic nervous system. We turn our discussion to the somatic nervous system in the next section.

Quick Test 12.3

1. What is a neuroeffector junction? In what respects are autonomic neuroeffector junctions different from ordinary neuron-to-neuron synapses? In what respects are they the same?

2. What structures in postganglionic neurons store and release neurotransmitters? Where are these structures located?

3. What are visceral reflexes, and how are they involved in the control of autonomic functions?

4. What areas of the CNS exert primary control over the autonomic nervous system?

The Somatic Nervous System

Unlike the autonomic nervous system, which controls the functions of many different types of effector organs, the somatic nervous system controls only one type of effector organ—skeletal muscle. Most skeletal muscles are connected to bones and thus function in the support and movement of the body. In addition, the somatic nervous system has only a single type of efferent neuron: motor

Table 12.3 ▮ Effects of Innervation by the Autonomic Nervous System

Organ system	PARASYMPATHETIC NERVOUS SYSTEM*	SYMPATHETIC NERVOUS SYSTEM	
	Effect	*Effect*	*Adrenergic receptor class*
Heart			
SA node	Decreases heart rate	Increases heart rate	β_1
AV node	Decreases conduction velocity	Increases conduction velocity	β_1
Force of contraction	Decreases (small effect)	Increases	β_1
Blood vessels			
Arterioles to most of body	None	Vasoconstriction	α_1
Arterioles to skeletal muscle	None	Vasoconstriction	α_1
		Vasodilation (epinephrine)	β_2
Arterioles to brain	None	None	
Veins	None	Vasoconstriction	α_1
		Vasodilation (epinephrine)	β_2
Lungs			
Bronchial muscle	Contraction	Relaxation	β_2
Bronchial glands	Stimulates secretion	Inhibits secretion	α
Digestive tract			
Motility	Increased	Decreased	$\alpha_1, \alpha_2, \beta_2$
Secretions	Stimulated	Inhibited	α_2
Sphincters	Relaxation	Contraction	α_1
Pancreas			
Exocrine glands	Stimulates secretion	Inhibits secretion	α
Endocrine glands	Stimulates secretion	Inhibits secretion	α_2
Salivary glands	Stimulates watery secretion	Stimulates mucus secretion	α_1
Kidneys			
Renin release	None	Stimulated	β_1

neurons, the neurons that innervate skeletal muscle. Most skeletal muscle is under voluntary control; that is, one can consciously decide to contract a muscle. Therefore, the somatic nervous system is also referred to as the *voluntary nervous system.*

Anatomy of the Somatic Nervous System

In the somatic nervous system, a single motor neuron travels from the central nervous system to a skeletal muscle cell (**Figure 12.13**); recall that two neurons are present in the autonomic nervous system pathway to an effector organ. Motor neurons originate in the ventral horn of the spinal cord (or the analogous brainstem nuclei) and receive input from multiple sources, including afferents (for spinal reflexes), the brainstem, and the cerebral cortex. A single motor neuron innervates many muscle cells (called **muscle fibers**), but each muscle fiber is innervated by only one motor neuron. A motor neuron plus all the mus-

cle fibers it innervates is a **motor unit** (**Figure 12.14**). When a motor neuron is activated, it stimulates all the muscle fibers in its unit to contract. The functional importance of motor units is discussed in Chapter 13.

The Neuromuscular Junction

Each branch of a motor neuron synapses with a skeletal muscle fiber at a single highly specialized central region of the fiber called the **neuromuscular junction** (**Figure 12.15**a). The axon terminals of the motor neuron, called **terminal boutons**, store and release acetylcholine, the only neurotransmitter in the somatic nervous system. Opposite these terminal boutons is a specialized region of the muscle fiber's plasma membrane called the **motor end plate**, which has invaginations containing large numbers of acetylcholine receptors. These receptors are of the nicotinic cholinergic variety, although pharmacological studies indicate that these nicotinic cholinergic receptors differ

Table 12.3 | *(continued)*

Organ system	PARASYMPATHETIC NERVOUS SYSTEM*	SYMPATHETIC NERVOUS SYSTEM	
	Effect	*Effect*	*Adrenergic receptor class*
Urinary bladder			
Bladder wall	Contraction	Relaxation (small effect)	β_2
Sphincter	Relaxation	Contraction	α_1
Male reproductive tract			
Blood vessels (erection)	Vasodilation	None	
Vas deferens and seminal vesicles (ejaculation)	None	Ejaculation	α_1
Female reproductive tract			
Uterus, nonpregnant	Unknown	Relaxation	β_2
Uterus, pregnant	Unknown	Contraction	α_1
Skin			
Sweat glands	Stimulates secretion	Stimulates secretion	α_1, muscarinic[†]
Piloerector muscles	None	Contraction (hairs stand up)	α_1
Eye			
Iris muscles (pupil size)	Contraction of circular muscle (pupillary constriction)	Contraction of radial muscle (pupillary dilation)	α_1
Ciliary muscles (accommodation)	Contraction for near vision	Relaxation for far vision (small effect)	β_2
Metabolism			
Liver	None	Stimulates glycogenolysis and gluconeogenesis	α_1, β_2
Adipose tissue	None	Stimulates lipolysis	β_3

*Receptor types for the parasympathetic nervous system are not given because *all* effector organs have muscarinic cholinergic receptors.
†Sympathetic postganglionic neurons to the sweat glands release acetylcholine as the neurotransmitter.

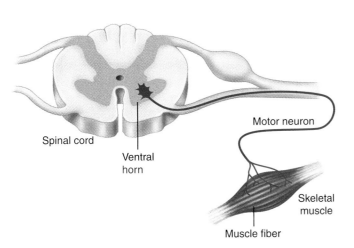

Figure 12.13 Anatomy of somatic nervous system pathways. The somatic nervous system consists of motor neurons, which originate in the ventral horns of the spinal cord and terminate on skeletal muscle fibers throughout the body.

Spinal cord
Ventral horn
Motor neuron
Skeletal muscle
Muscle fiber

Figure 12.14 Motor units. A motor unit consists of a motor neuron and all the muscle fibers it innervates. Whereas a single neuron innervates many muscle fibers, a given muscle fiber is innervated by one motor neuron only. Note that the muscle fibers within a given motor unit are scattered throughout the muscle.

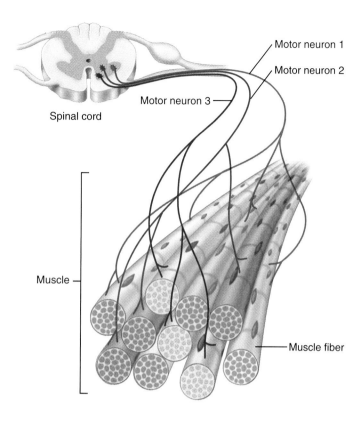

Motor neuron 1

Motor neuron 2

Motor neuron 3

Spinal cord

Muscle

Muscle fiber

somewhat from those found on postganglionic neurons in autonomic ganglia. Acetylcholinesterase, which is found between the invaginations of the motor end plate, terminates the excitatory signal, which allows the muscle fiber to relax.

Exercise Link

One of the most obvious ways that exercise training improves muscle function is an increase in the speed of contraction. But the speed of relaxation is just as important. Rapid clearance of acetylcholine from the neuromuscular junction is essential in preparing the muscle fiber to receive the next neuronal signal. One of the benefits Jane and Bill gained in their premarathon training was an increase in acetylcholinesterase at the motor end plates of the muscle fibers used in running.

The signal transmission mechanism at the neuromuscular junction is similar to that at excitatory neuron-to-neuron synapses (Figure 12.15b). When a motor neuron is activated by converging synaptic input, action potentials are propagated to the terminal boutons at the neuromuscular junctions of all muscle fibers in the motor unit. The resulting depolarization causes voltage-gated calcium channels in the boutons to open, allowing

calcium to enter the cytosol and triggering the release of acetylcholine by exocytosis. Acetylcholine diffuses across the synaptic cleft and binds to nicotinic cholinergic receptors at the motor end plate, causing cation channels to open. This allows sodium to flow into the muscle fiber and produces a depolarization called an **end-plate potential (EPP)** that is similar in many ways to the excitatory postsynaptic potentials (EPSPs) generated in neurons. The one major difference is that end-plate potentials are normally of sufficient magnitude to depolarize the muscle fiber to threshold, which generates an action potential and triggers contraction of the muscle fiber by mechanisms described in Chapter 13.

Whereas hormonal regulation of skeletal muscle *metabolism* occurs, only innervation by a motor neuron controls skeletal muscle *contraction*. Therefore, all neural communication to skeletal muscle is excitatory; that is, it stimulates the muscle to contract. For skeletal muscle to relax, the neural stimulation must decrease.

Clearly, normal signal transmission at the neuromuscular junction is necessary for normal control of skeletal muscle contraction. If transmission at the neuromuscular junction is altered, as occurs in the disease *myasthenia gravis*, then normal function of skeletal muscle is lost (see **Clinical Connections: Myasthenia Gravis**, p. 370).

The neuromuscular junction is also targeted by many of the toxins present in animal venoms. The venom of the

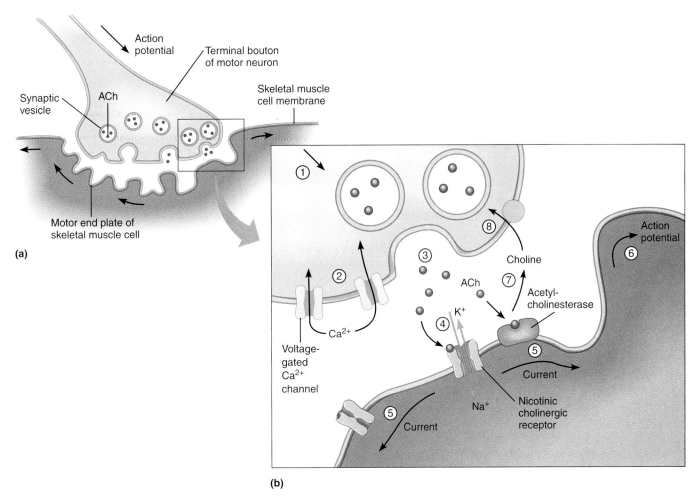

(a)

(b)

Figure 12.15 Functional anatomy of the neuromuscular junction. (a) Both the axon terminal of the motor neuron and the portion of the plasma membrane of the skeletal muscle called the motor end plate are specialized at the neuromuscular junction. (b) Communication at the neuromuscular junction. When an action potential arrives at the axon terminal of a motor neuron ①, voltage-gated calcium channels open and calcium enters the cytosol ②. The entry of calcium triggers the release by exocytosis of acetylcholine ③, which diffuses to and binds to nicotinic cholinergic receptors at the motor end plate, opening cation channels. Sodium enters the cell ④, producing an end-plate potential that generates currents throughout the plasma membrane of the skeletal muscle cell ⑤, depolarizing the membrane to threshold to generate an action potential ⑥. The action potential spreads along the skeletal muscle cell membrane, ultimately stimulating contraction. Acetylcholinesterase degrades acetylcholine to produce acetate and choline ⑦. Choline is actively transported into the terminal bouton ⑧, where it can be used to synthesize more acetylcholine.

black widow spider, for example, contains the toxin *latroxin,* which by stimulating the release of acetylcholine at the neuromuscular junction induces muscle spasms and rigidity. Because the respiratory muscles (which are skeletal muscles) are affected, this venom can cause respiratory failure and death by inducing spastic contractions of these muscles. The venom of the rattlesnake, by contrast, contains the toxin *crotoxin,* which has the opposite effect of latroxin. Crotoxin inhibits the release of acetylcholine, which induces flaccid paralysis of skeletal muscles. Another toxin with paralyzing effects on skeletal muscle is *curare,* a poison used by native South Americans on the tips of the darts used in blow guns (see **Discovery: Curare, p. 371**).

Quick Test 12.4

1. Describe two differences between the somatic nervous system and the two branches of the autonomic nervous system with regard to the anatomical arrangement of their efferent neurons. What are the efferent neurons of the somatic nervous system called?

2. What neurotransmitter is released by motor neurons at the neuromuscular junction? To what type of receptor does this neurotransmitter bind?

3. How many motor neurons innervate a single skeletal muscle fiber?

MYASTHENIA GRAVIS

All of us have experienced muscle fatigue, but for persons afflicted with myasthenia gravis, muscle fatigue is more than just a nuisance—it can be debilitating and sometimes fatal. *Myasthenia gravis* is a disease affecting transmission at neuromuscular junctions. Its victims are mostly women, and it usually strikes between the ages of 20 and 50. The defining characteristic of the disease is fatigue of unusually rapid onset and severity following the use of certain muscle groups. Because the muscles most frequently affected are those of the head, difficulties in speaking *(dysarthria)* and in swallowing *(dysphagia)* are common symptoms; drooping of the eyelids *(ptosis)*, shown in the photo, is also a common sign. The disease also frequently targets limb muscles, with resultant weakness in the arms and legs. In certain people, muscles used in breathing are affected, sometimes necessitating the use of a mechanical ventilator. Before the development of modern treatments, the mortality rate for those afflicted with myasthenia gravis was more than 30%, mostly as a result of respiratory problems.

Compared with other neuromuscular diseases, the time course of myasthenia gravis is unusual. Following onset of the disease, the severity of symptoms waxes and wanes from day to day or month to month, or even within the course of a single day. During certain periods the disease might even appear to be in complete remission.

Although the disease has been known to medical practitioners for at least the past few hundred years, the underlying mechanism remained a mystery until the mid-20th century, when techniques permitting the recording of electrical signals in nerves and muscles were developed. These techniques revealed that symptoms of myasthenia gravis were due to failure of motor neurons to excite muscle cells to contract. Even though motor neurons were capable of transmitting action potentials that were followed by end-plate potentials (EPPs) in muscle cells, these EPPs were often smaller than normal, particularly if neurons were repetitively stimulated. When this happened, muscle cell membrane potentials would often fail to reach threshold, resulting in a "dropped" contraction.

These observations suggested a number of possible mechanisms for the development of the disease. Perhaps a reduction in the amount of acetylcholine released by motor neurons was responsible, or maybe the problem was some deficiency in acetylcholine receptor function. We now know that myasthenia gravis is an *autoimmune* disease—a disease in which the immune system attacks proteins that are normal components of body tissues.

In myasthenia gravis, the immune system produces antibodies against acetyl-choline receptors at the neuromuscular junctions. These antibodies bind to nicotinic cholinergic receptors, triggering their removal from the plasma membrane and subsequent destruction by immune cells. The resulting decrease in the number of functional receptors on the cell surface impairs the ability of these muscle cells to respond to acetylcholine.

Current therapies attempt to blunt the immune system's action by reducing antibody levels in the blood. One approach is *thymectomy,* removal of the thymus gland, which plays an important role in immune function. This treatment reverses symptoms in about 50% of patients. *Plasmapheresis* is another approach. In this procedure, blood is removed from a patient (not all of it at once!). The *plasma* (liquid component of blood, which contains the antibodies) is separated from the cells, and the cells are then returned to the patient. The availability of these and other treatments has reduced the mortality rate for myasthenia gravis to near zero and has enabled most persons stricken with this disease to live normal lives.

This patient shows a drooping eyelid characteristic of Myasthenia gravis.

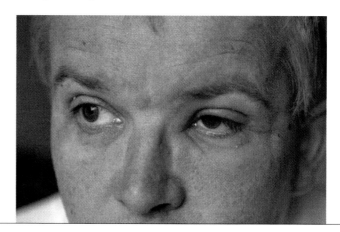

DISCOVERY

CURARE

Curare is an extract of a plant (*Chondrodendron tomentosum*) found in South America. Indians of that region crush and cook the roots and stems of the plant to produce a poison for the tips of their arrows and darts (see photo). (The word *curare* comes from the Indian word for poison.) Curare was valued by Indians because they used the poison in hunting game. When an animal was struck by a curare-laced arrow or dart, it would become paralyzed and eventually die from respiratory failure.

The effective component of curare is a compound called *tubocurarine,* which blocks communication at the neuromuscular junction. Tubocurarine binds to nicotinic cholinergic receptors, thereby preventing acetylcholine from binding. When this occurs, skeletal muscles are unable to contract even when action potentials are transmitted by the motor neurons that innervate them.

In the late 19th and early 20th centuries, curare was studied for its possible pharmaceutical benefits. Curare was first used as a skeletal muscle relaxant to supplement general anesthesia in the 1930s. Today, curare has numerous clinical uses, including dilating hollow organs such as the rectum and relaxing the throat, which enables easier examination. Curare is also used for relief of *spastic paralysis,* a type of paralysis that can result from excessive skeletal muscle activity.

Table 12.4 ∎ Properties of the Autonomic and Somatic Nervous Systems

Property	Autonomic: parasympathetic	Autonomic: sympathetic	Somatic
Origin	Brainstem or lateral horns of sacral spinal cord	Lateral horns of thoracic and lumbar spinal cord	Ventral horns of spinal cord
Neurons in pathway	Two (preganglionic and postganglionic)	Two (preganglionic and postganglionic)	One (motor neuron)
Effector organs	Cardiac muscle, smooth muscle, glands	Cardiac muscle, smooth muscle, glands, adipose tissue	Skeletal muscle
Neurotransmitters at neuroeffector junction	Acetylcholine	Norepinephrine	Acetylcholine
Receptor type at effector organ	Muscarinic cholinergic	Adrenergic (all classes)	Nicotinic cholinergic
Effects on effector organ	Either excitation or inhibition	Either excitation or inhibition	Excitation
Control	Primarily involuntary	Primarily involuntary	Primarily voluntary

SYSTEMS INTEGRATION Nervous System

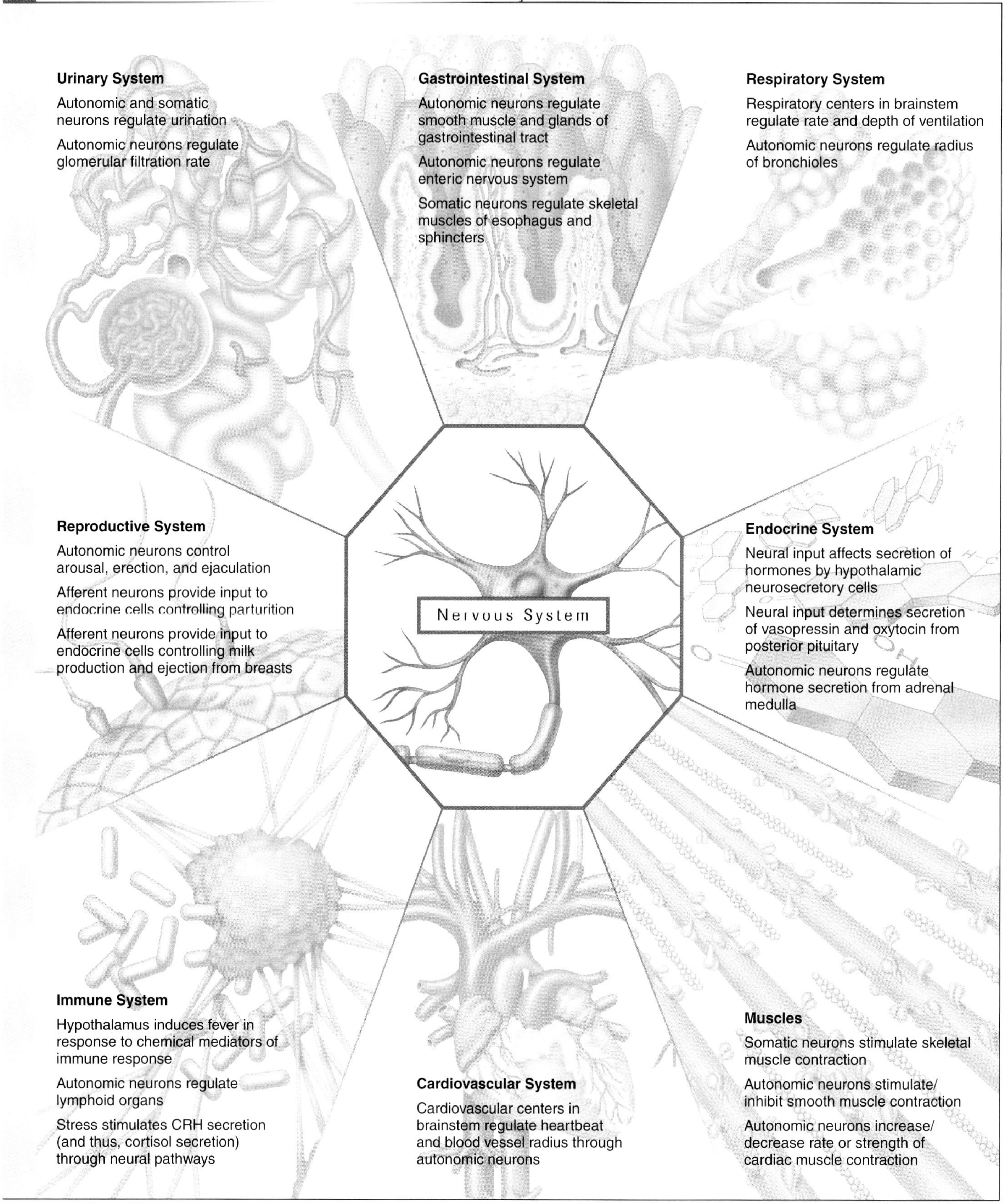

Urinary System

Autonomic and somatic neurons regulate urination

Autonomic neurons regulate glomerular filtration rate

Gastrointestinal System

Autonomic neurons regulate smooth muscle and glands of gastrointestinal tract

Autonomic neurons regulate enteric nervous system

Somatic neurons regulate skeletal muscles of esophagus and sphincters

Respiratory System

Respiratory centers in brainstem regulate rate and depth of ventilation

Autonomic neurons regulate radius of bronchioles

Reproductive System

Autonomic neurons control arousal, erection, and ejaculation

Afferent neurons provide input to endocrine cells controlling parturition

Afferent neurons provide input to endocrine cells controlling milk production and ejection from breasts

Nervous System

Endocrine System

Neural input affects secretion of hormones by hypothalamic neurosecretory cells

Neural input determines secretion of vasopressin and oxytocin from posterior pituitary

Autonomic neurons regulate hormone secretion from adrenal medulla

Immune System

Hypothalamus induces fever in response to chemical mediators of immune response

Autonomic neurons regulate lymphoid organs

Stress stimulates CRH secretion (and thus, cortisol secretion) through neural pathways

Cardiovascular System

Cardiovascular centers in brainstem regulate heartbeat and blood vessel radius through autonomic neurons

Muscles

Somatic neurons stimulate skeletal muscle contraction

Autonomic neurons stimulate/ inhibit smooth muscle contraction

Autonomic neurons increase/ decrease rate or strength of cardiac muscle contraction

CHAPTER SUMMARY

The Autonomic Nervous System, p. 354

There are two main branches of the efferent nervous system: the autonomic nervous system and the somatic nervous system. **Table 12.4** (p. 371) compares the properties of the two branches of the autonomic nervous system with those of the somatic nervous system. The autonomic nervous system includes the parasympathetic and sympathetic nervous systems, which innervate cardiac muscle, smooth muscle, glands, and adipose tissue. Effector organs are generally innervated by both the parasympathetic and sympathetic divisions, an arrangement termed dual innervation. The parasympathetic nervous system is most active during rest, whereas the sympathetic nervous system is most active during periods of activity or excitation and is responsible for the fight-or-flight response.

Pathways in the autonomic nervous system consist of two neurons that communicate between the CNS and the effector organ: preganglionic neurons and postganglionic neurons. Postganglionic neurons innervate the effector organ. The sympathetic nervous system also has an endocrine component because one set of preganglionic neurons innervates the adrenal medulla, stimulating the release of the hormone epinephrine. All preganglionic neurons contain the neurotransmitter acetylcholine. The parasympathetic postganglionic neurons also contain the neurotransmitter acetylcholine, but most sympathetic postganglionic neurons contain the neurotransmitter norepinephrine. The receptors for acetylcholine on postganglionic neurons are nicotinic cholinergic receptors, whereas the receptors for acetylcholine on effector organs in the parasympathetic nervous system are muscarinic cholinergic receptors. The receptors for norepinephrine and epinephrine on the effector organs in the sympathetic nervous system are adrenergic receptors.

The synapse between an efferent neuron and its effector organ is called a neuroeffector junction. At the neuroeffector junctions between autonomic postganglionic neurons and their effector organs, neurotransmitter is released diffusely from varicosities and then binds to receptors on the effector organ. The mechanism of release of neurotransmitter is similar to that for neuron-to-neuron synapses.

The autonomic nervous system is under involuntary control. Areas of the brain that influence autonomic activity include the brainstem, hypothalamus, and limbic system.

> **IP** Nervous II, Synaptic Transmission, pp. 3–11
> **IP** Nervous II, Ion Channels: pp. 4–5, 7
> **IP** Nervous I, The Membrane Potential, pp. 3–6, 11

The Somatic Nervous System, p. 365

The somatic division of the efferent nervous system consists of pathways composed of single motor neurons. Motor neurons originate in the ventral horn of the spinal cord and innervate skeletal muscle cells. A single motor neuron and the muscle cells it innervates is called a motor unit. The synapse between a motor neuron and a skeletal muscle fiber is called a neuromuscular junction. The motor neuron contains the neurotransmitter acetylcholine. The receptors in skeletal muscle are nicotinic cholinergic. Binding of acetylcholine to nicotinic cholinergic receptors at the motor end plate produces an end-plate potential, which ultimately causes the skeletal muscle fiber to contract.

> **IP** Nervous II, Synaptic Transmission, p. 11
> **IP** Muscular, The Neuromuscular Junction, pp. 1–10

EXERCISES

Multiple-Choice Questions

1. Effector organs of the autonomic nervous system include all of the following *except*
   a) the heart muscle.
   b) smooth muscle in the pupils of the eye.
   c) respiratory muscles.
   d) sweat glands.
   e) salivary glands

2. According to the concept of dual innervation by the autonomic nervous system, if sympathetic activity inhibits pancreatic secretions, then the parasympathetic nervous system should
   a) inhibit pancreatic secretions as well.
   b) stimulate pancreatic secretions.
   c) have no effect on pancreatic secretions.

3. The adrenal medulla
   a) contains sympathetic postganglionic neurons.
   b) is part of the brainstem.
   c) releases epinephrine into the blood.
   d) is part of the parasympathetic nervous system.
   e) is controlled by the somatic nervous system.

4. Which of the following cranial nerves does *not* contain parasympathetic preganglionic neurons?
   a) oculomotor (cranial nerve III)
   b) facial (cranial nerve VII)
   c) glossopharyngeal (cranial nerve IX)
   d) vagus (cranial nerve X)
   e) hypoglossal (cranial nerve XII)

5. Which of the following receptor types does *not* activate G proteins?
   a) nicotinic cholinergic
   b) muscarinic cholinergic
   c) α_1 adrenergic
   d) β_1 adrenergic
   e) β_3 adrenergic

6. The origin of spinal preganglionic neurons is the
   a) ventral horn of the spinal cord.
   b) dorsal horn of the spinal cord.
   c) lateral horn of the spinal cord.

7. The origin of motor neurons is the
   a) ventral horn of the spinal cord.
   b) dorsal horn of the spinal cord.
   c) lateral horn of the spinal cord.

8. Which of the following second messengers stimulates the release of calcium from intracellular stores?
   a) cAMP
   b) inositol triphosphate
   c) diacylglycerol

9. Which of the following is the location of the cardiovascular regulatory centers?
   a) thalamus
   b) hypothalamus
   c) limbic system
   d) pons
   e) medulla oblongata

10. How many motor neurons innervate a single skeletal muscle cell?
    a) zero
    b) one
    c) several
    d) hundreds
    e) millions

11. The motor end plate is
    a) the specialized synaptic terminal of the motor neuron.
    b) the specialized synaptic terminal of autonomic postganglionic neurons.
    c) the specialized region of skeletal muscle innervated by a motor neuron.
    d) the specialized region of an effector organ innervated by an autonomic postganglionic neuron.

12. Neurotransmitter is released from which portion of a postganglionic neuron?
    a) terminal bouton
    b) axon terminal
    c) varicosity
    d) cell body
    e) dendrites

Objective Questions

1. Which branch of the autonomic nervous system has longer preganglionic neurons?

2. Which branch of the autonomic nervous system is most active when the body is at rest?

3. The communication between preganglionic neurons and postganglionic neurons in the autonomic nervous system is one-to-one. (true/false)

4. What part of the adrenal gland secretes epinephrine?

5. Name the four cranial nerves that contain parasympathetic preganglionic neurons.

6. Autonomic nerves contain only efferent neurons. (true/false)

7. Which neurons in the peripheral nervous system are cholinergic?

8. Which neurons in the peripheral nervous system are adrenergic?

9. What enzyme catalyzes the formation of diacylglycerol and inositol triphosphate?

10. A decrease in cAMP is associated with what class of adrenergic receptor?

11. Activation of α adrenergic receptors usually produces (excitation/inhibition).

12. β_2 adrenergic receptors have a greater affinity for (epinephrine/norepinephrine).

13. The motor end plate has (nicotinic/muscarinic) cholinergic receptors.

14. Skeletal muscle can be excited to contract only; that is, it cannot be inhibited to relax. (true/false)

15. The enyzme that degrades acetylcholine in the synaptic cleft is called _____.

Essay Questions

1. Describe the different anatomical arrangements found in the two branches of the autonomic nervous

system and in the somatic nervous system.

2. Explain the concept of dual innervation.

3. Explain why sympathetic activation produces a more diffuse effect compared to activation of the parasympathetic nervous system.

4. Compare the signal transduction mechanisms for the different types of adrenergic receptors.

5. What areas of the brain regulate autonomic function?

Critical Thinking

1. If a patient has such a slow heart rate that not enough blood circulates to the brain, what type of receptor agonist could be administered to increase heart rate? What type of receptor antagonist could be administered? (See Table 12.3.)

2. Of the sympathetic preganglionic neurons and postganglionic neurons, one has myelinated axons whereas the other is unmyelinated. Based on the anatomy of the sympathetic nervous system in Figure 12.4, which neuron is myelinated and which is unmyelinated?

3. Explain how binding of norepinephrine to α_2 or β_1 receptors can produce an excitatory effect on the effector organ even though α_2 receptors inhibit production of cAMP whereas β_1 receptors stimulate production of cAMP.

4. Monoamine oxidase inhibitors are sometimes used to treat depression. What are some of the side effects you might expect with usage of this class of drug?

Find the answers to these exercises, and additional study tools,
at the Physiology Place (www.physiologyplace.com).

Muscle Physiology

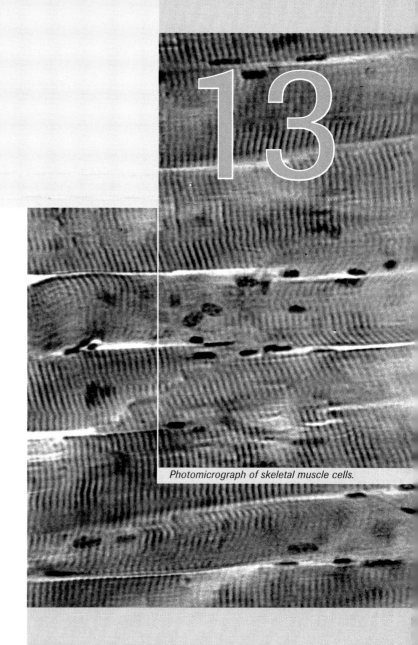

Photomicrograph of skeletal muscle cells.

W e have all heard the story of how Benjamin Franklin discovered electricity by flying a kite in a thunderstorm. Now you will hear about how Luigi Galvani discovered "animal electricity" by hanging dissected frogs on wire in a thunderstorm.

Luigi Galvani taught anatomy at the University of Bologna in Italy and served as professor of obstetrics in the Institute of Arts and Sciences from 1765 to 1797. In his teachings, he used frog dissections. In the process of preparing the dead frogs, he would hang the frog on a copper hook and then suspend the frog on an iron wire. He did this during a thunderstorm and saw the freshly dead frog contract its muscles. He moved the hanging frog carcass indoors (away from the electrical activity of the storm), and noticed that the muscles still contracted. This suggested to Galvani that frogs, and other animals, had their own form of electricity and that this "animal electricity" caused muscle contraction.

In further experiments, Galvani's assistant discovered that nerve tissue activated muscle tissue. While the assistant was dissecting a frog, he used a scalpel that was sitting near a machine that generated electricity. When he touched the scalpel to the sciatic nerve, the muscle contracted. Galvani concluded that electrostatic charge on the scalpel excited the nerve, and that nerves provide the electricity that causes muscle to contract. Although this is not precisely true, as you will learn in this chapter, Galvani's work led to the field of *electrophysiology,* the study of electrical activity in animals (or plants).

In Chapter 12, we learned about the efferent branch of the nervous system and its innervation of effector organs. The effector organ of the somatic nervous system is skeletal muscle, whereas smooth and cardiac muscle are effector organs of the autonomic nervous system. In this chapter, you will learn about these muscle types and how they, like neurons, are excitable tissues and generate action potentials similar to those described in Chapter 8. However, action potentials in muscle tissue have a different effect than those in neurons: They induce contractions and generate a force.

To perform their jobs properly, the muscles that move the body must be able to respond faithfully and quickly to commands from the nervous system. Indeed, a typical skeletal muscle cell can activate its contractile machinery within milliseconds of receiving a neural signal, and can turn it off nearly as quickly. It is this quickness of response that enables us to perform complicated motions.

We begin our study of muscle physiology with an examination of muscle anatomy, moving from the gross anatomical level to the molecular level, with special emphasis on the structures that generate and regulate contractile force. Then we examine contractile and regulatory mechanisms to see how they work. Although we concentrate on *skeletal muscle,* most of the basic principles apply to each of the three muscle types found in the body. At the end of the chapter we examine the special properties of the other two muscle types—*cardiac muscle* and *smooth muscle.*

Skeletal Muscle Structure

With few exceptions, **skeletal muscles**, such as the *biceps* of the arm, are connected to at least two bones. In contrast, the other muscles of the body—those found in internal organs, blood vessels, and certain other structures—are not connected to bones. The exceptions to this rule include certain skeletal muscles that are connected to the skin (as is the case with some facial muscles), to cartilage (for example, muscles of the larynx), or to other muscles (the external anal sphincter, for instance). Muscles are connected to bones by **tendons**, cords of elastic connective tissue that transmit force from the muscle to the bone.

Structure at the Cellular Level

The part of the muscle that generates force is called the *body,* the "meaty" part of the muscle (**Figure 13.1**). The body contains many bundles (called *fascicles*) of individual muscle cells, as well as connective tissue, blood vessels, and nerves. Each fascicle contains hundreds or thousands of muscle cells, which are called **muscle fibers** because of their elongated shape. Each muscle fiber runs the full length of the muscle (often running at a diagonal) and is encased in a sheath of connective tissue. Unlike most cells, which have a single nucleus, muscle fibers have many because each muscle fiber is formed during embryonic life from the fusion of several cells. These nuclei lie immediately below the muscle fiber's plasma membrane, which is called the **sarcolemma**.

A muscle fiber's semifluid cytoplasm, called sarcoplasm, is packed with mitochondria and hundreds of banded, rodlike elements called **myofibrils**, which contain the fiber's contractile machinery (**Figure 13.2**). Each myofibril is a bundle of overlapping thick and thin filaments made of the proteins *myosin* and *actin,* respectively. A saclike membranous network called the **sarcoplasmic reticulum** surrounds each of the myofibrils and is closely associated with other structures called **transverse tubules (T tubules)**, which are connected to the sarcolemma and penetrate into the cell's interior. The sarcoplasmic reticulum and T tubules play an important role in the activation of muscle contractions because they help transmit signals from the sarcolemma to the myofibrils, enabling a muscle cell to respond to neural input. The functions of the sarcoplasmic reticulum are to store calcium ions (Ca^{2+}) and to release them into the cytosol when the muscle cell is stimulated to contract. As we will see, these calcium ions are released in response to electrical signals that travel from the sarcolemma to the

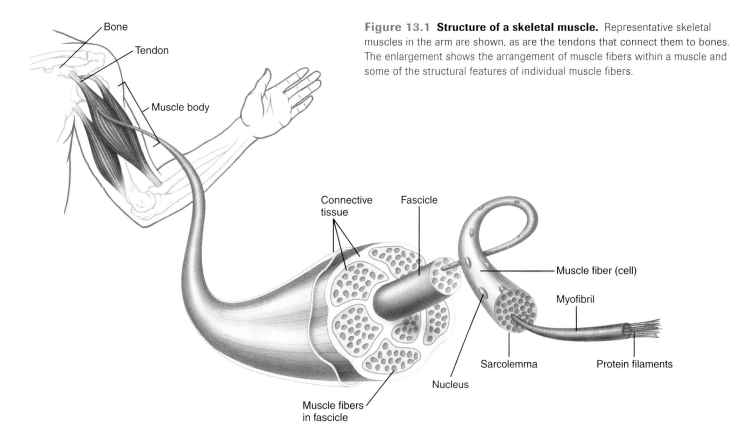

Figure 13.1 Structure of a skeletal muscle. Representative skeletal muscles in the arm are shown, as are the tendons that connect them to bones. The enlargement shows the arrangement of muscle fibers within a muscle and some of the structural features of individual muscle fibers.

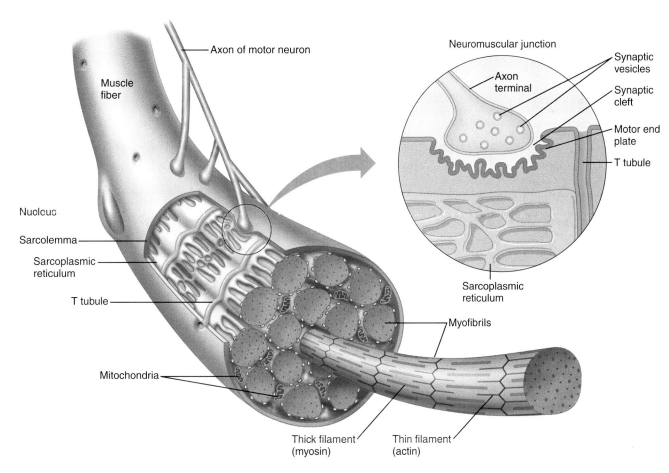

Figure 13.2 Structure of a skeletal muscle fiber. Major internal components of a muscle fiber are shown. A single myofibril in the muscle fiber has been extended and slightly enlarged to reveal the arrangement of thick and thin filaments within it. The enlarged view shows a magnified image of a neuromuscular junction.

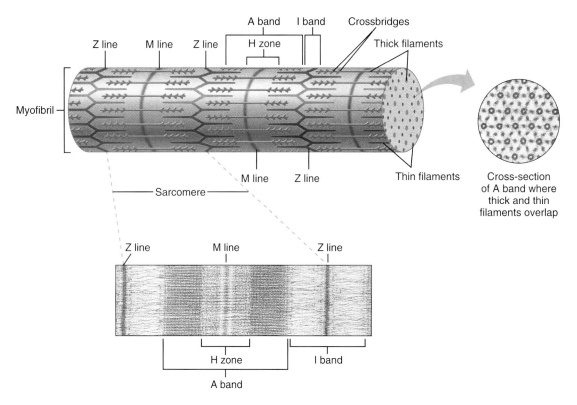

Figure 13.3 Sarcomere structure. A drawing of a myofibril, showing the regular arrangement of protein filaments within sarcomeres. The lower photomicrograph shows the banding pattern typical of striated muscle; the photomicrograph at right shows a cross section through the A band of a myofibril, in which the three-dimensional arrangement of thick and thin filaments can be seen clearly.

T tubules, and they serve as chemical messengers that carry these signals to the cell's interior, where the myofibrils are located.

Structure at the Molecular Level

When viewed under a microscope, skeletal muscle cells have a striped appearance, and for this reason this muscle (and also cardiac muscle) is often referred to as *striated muscle*. A close-up view shows that these striations are due to the orderly arrangement of protein fibers in the myofibrils called **thick filaments** and **thin filaments**, which run parallel to the muscle cell's long axis. Myofibrils are composed of a fundamental unit called a **sarcomere** that repeats over and over (**Figure 13.3**). Each sarcomere is bordered on either end by *Z lines,* which run perpendicular to the long axis and anchor the thin filaments at one end. The thick filaments in a sarcomere are connected by *M lines,* which also run perpendicular to the long axis.

Before the structure of the sarcomere was elucidated and its protein filaments identified, early investigators used the terms *A band, I band,* and *H zone* to designate certain regions in the muscle fiber's banding pattern (see Figure 13.3). The *A band,* the region that appears darkest under the microscope, spans the length of the thick filaments. The *H zone* is a region in the center of the A band that appears noticeably lighter. In all parts of the A band except the H zone, thin filaments are present along with the thick filaments and overlap them. (A cross section of the fiber reveals that six thin filaments surround each thick filament.) It is the absence of thin filaments in the H zone that gives it its lighter appearance. The *I band,* the brightest region of the muscle fiber, occupies the space between the A bands of adjacent sarcomeres. It contains only thin filaments and the Z line that connects them, and the absence of thick filaments accounts for the brightness of the I band.

The thin and thick filaments of the sarcomere are made up of two proteins called **actin** and **myosin**, respectively, which are referred to as *contractile proteins* because they constitute the machinery that generates contractile force. Just as myofibrils are made up of orderly, repeating structures, thick and thin filaments are also made up of structures arranged in an orderly, repeating fashion.

The basic components of each thin filament are actin monomers called G actin (G because they are globular proteins), each of which has a myosin-binding site. As we will see, the ability of actin and myosin to bind together under certain conditions is critical to a muscle's ability to generate force. G actins are linked together end

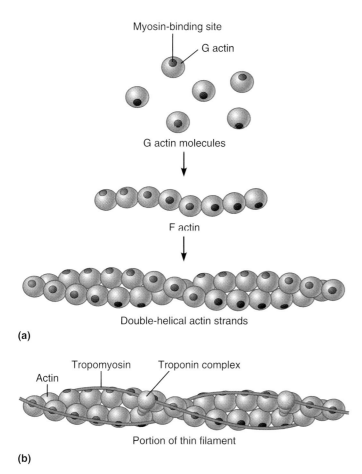

(a)

(b)

Figure 13.4 Structure of a thin filament. (a) The backbone of a thin filament consists of two strands of polymerized actin molecules wound together to form a double helix. Myosin-binding sites on individual actin molecules (G actin) are represented by dark dots. (b) A portion of a thin filament showing troponin and tropomyosin in their normal resting positions on the actin strands. Notice that actin's myosin-binding sites are covered by tropomyosin when a muscle cell is at rest.

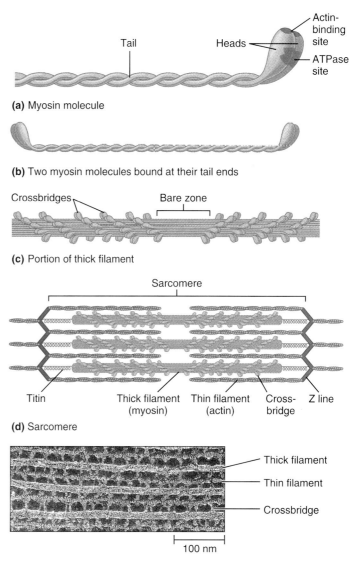

(a) Myosin molecule

(b) Two myosin molecules bound at their tail ends

(c) Portion of thick filament

(d) Sarcomere

(e) Electron micrograph of portion of sarcomere

Figure 13.5 Structure of a thick filament. (a) A myosin molecule, a dimer composed of two subunits wound together. Note the actin-binding and ATPase sites in the head region. (b) Two myosin molecules joined tail to tail. (c) A portion of a thick filament showing myosin heads (crossbridges) protruding at either end but not in the middle region (the bare zone). (d) A detailed view of a sarcomere showing the relative positions of thick and thin filaments and the protein titin, which anchors the thick filaments in place. (e) A photomicrograph of a sarcomere showing thick and thin filaments and crossbridges.

to end, like pearls in a necklace, to form strands called F actin (F because they are fibrous proteins). Two F actins are arranged in a double helix to form the actin strands found in thin filaments (**Figure 13.4**a).

Also present in thin filaments are two special proteins called *regulatory proteins* that enable muscle fibers to start or stop contracting: tropomyosin and troponin (Figure 13.4b). **Tropomyosin** is a long fibrous molecule that extends over numerous actin molecules in such a way that it blocks the myosin-binding sites in muscles at rest. **Troponin** is a complex of three proteins—one that attaches to the actin strand, another that binds to tropomyosin, and a third containing a site to which calcium ions can bind reversibly. As we will see, the binding of calcium to this site triggers muscle contraction by causing troponin to move tropomyosin aside, thereby exposing the myosin-binding sites on the actin molecules.

Each thick filament is made of hundreds of myosin molecules, each of which looks a bit like two golf clubs

wrapped around each other (**Figure 13.5**a). Each myosin molecule is a dimer consisting of two intertwined subunits, each having a long tail and a fat, protruding head. These heads are called **crossbridges** because under certain conditions (discussed shortly) they bridge the gap between the thick and thin filaments. Within a thick filament, the myosin molecules bind to each other at their tail ends so that their heads extend in opposite directions away from the center (Figure 13.5b). The tails of adjacent

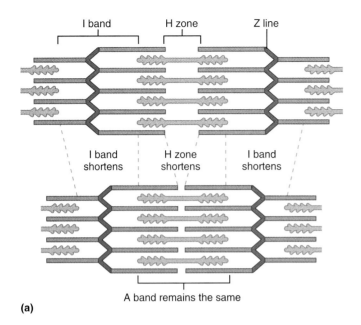

I band   H zone   Z line

I band shortens | H zone shortens | I band shortens

A band remains the same

(a)

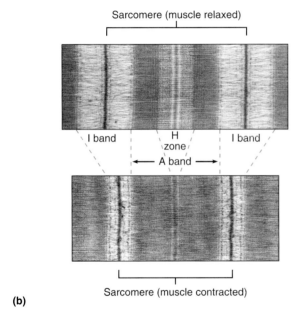

Sarcomere (muscle relaxed)

I band | H zone | I band

← A band →

Sarcomere (muscle contracted)

(b)

Figure 13.6 How changes in striation pattern are explained by the sliding-filament model of muscle contraction. (a) A schematic drawing and (b) photomicrographs showing the relative positions of the thick and thin filaments in sarcomeres in relaxed muscle (top) and contracted muscle (bottom).

myosin molecules are also arranged in a staggered fashion so that their heads protrude from the thick filament in an orderly helical pattern (Figure 13.5c). Because the middle of the thick filament is devoid of crossbridges, this region is appropriately called the *bare zone.*

The head is the "business end" of the myosin molecule because it is the part that actively generates a muscle's mechanical force. Each head possesses two sites that are critical to its force-generating ability: an *actin-binding*

site, which is capable of binding to the actin monomers in the thin filaments, and an *ATPase site,* which has enzymatic activity and hydrolyzes ATP (see Figure 13.5a).

Like thin filaments, thick filaments have additional proteins associated with them, most notably *titin,* an extraordinarily elastic protein that can be stretched to more than three times its unstressed length (Figure 13.5d). Strands of titin extend along each thick filament from the M line to each Z line, anchoring the thick filaments in their proper positions relative to the thin filaments. When an external stretching force is applied to a muscle, titin strands (see Figure 13.5d) elongate as the sarcomeres lengthen, and these strands begin to exert an opposing force, just as a spring resists stretching. When the external force is removed, this opposing force pulls the Z lines and thick filaments closer together and causes the sarcomeres to shorten, allowing the titin strands to spring back to their original length. As this occurs, individual muscle fibers shorten, as does the whole muscle.

Quick Test 13.1

1. Define the following terms: *muscle fiber, myofibril, Z line, sarcomere, crossbridge.*

2. Name and describe the locations and general functions of the two contractile proteins and the two regulatory proteins present in sarcomeres.

3. What two functions are performed by the heads of myosin molecules?

The Mechanism of Force Generation in Muscle

Throughout the body, function follows form from the cellular level to organs. This is nowhere more apparent than in skeletal muscle. As you will see in the next section, the regular arrangement of thick and thin filaments is the key to muscle contraction.

The Sliding-Filament Model

When physiologists first discovered the presence of actin and myosin in myofibrils, they thought that muscle contractions were caused by shortening of the proteins themselves. As advances in microscopy occurred, researchers discovered that during muscle-cell contraction the A band does not change in length, but the I bands and the H zone shorten (**Figure 13.6**). Given that the A band spans the length of the thick filaments in a sarcomere, this means the thick filaments do not change length when the muscle cell contracts. Researchers realized that the shortening of the I bands (which contain only thin filaments) occurred not because thin filaments contract, but because they slide past the thick filaments, moving deeper into the H zone and decreasing its width. As this occurs, adjacent A bands move closer together, which decreases the width of the I bands. The end result is that the Z lines at either end of a sarcomere move closer together, and thus the sarcomere shortens. As sarcomeres shorten, myofibrils also shorten, as do muscle fibers and ultimately whole muscles. In other words, muscles contract because the thick and thin filaments of the myofibrils slide past each other. Appropriately, this is called the **sliding-filament model** of muscle contraction.

The Crossbridge Cycle: How Muscles Generate Force

During muscle contraction, the mechanism that drives the sliding of thick and thin filaments past one another is called the **crossbridge cycle** (**Figure 13.7**). At the heart of this mechanism is an oscillating, back-and-forth motion of myosin crossbridges that is powered by ATP hydrolysis. Coupled with this is cyclic binding and unbinding of crossbridges to the thin filaments, which occurs in such a way that the motion of the crossbridges pulls the thin filaments toward the center of the sarcomere.

The back-and-forth movement of crossbridges is due to changes in the conformation (shape) of the myosin molecules. These conformational changes not only cause the heads to change position, but also alter both their ability to bind to actin monomers in the thin filaments and the *energy content* of the myosin molecules. One conformation of myosin is referred to as the *high-energy form,* which is indicated in step 5 of Figure 13.7. Myosin heads go into this conformation after they hydrolyze ATP. It is called the *high-energy form* because the myosin molecule stores energy that is released in the hydrolytic splitting of ATP. Myosin heads go into the other conformation, called the *low-energy form,* after the stored energy is released to drive the movement of the thin filaments.

Each crossbridge cycle involves the following five steps (see Figure 13.7):

① *Binding of myosin to actin.* We start with myosin in its energized form; that is, ADP and P_i (inorganic phosphate) are bound to the ATPase site of the myosin head. In this state, myosin has a high affinity for actin, and the myosin head binds to an actin monomer in the adjacent thin filament. This step can occur only in the presence of calcium, for reasons described below.

② *Power stroke.* The binding of myosin to actin triggers the release of the P_i and ADP from the ATPase site. During this process, the myosin head pivots toward the middle of the sarcomere, pulling the thin filament along with it, and the myosin molecule goes into its low-energy state.

③ *Rigor.* In its low-energy form, myosin and actin are tightly bound together, a condition called *rigor* (after *rigor mortis,* the stiffening of the body that occurs after death because the crossbridge cycle gets stuck at this step due to a lack of ATP; rigor mortis continues until enzymes leaked by disintegrating cellular components begin to break down the myofibrils).

④ *Unbinding of myosin and actin.* A new ATP enters the ATPase site on the myosin head, triggering a conformational change in the head, which decreases the affinity of myosin for actin, so the myosin detaches from the actin.

⑤ *Cocking of the myosin head.* Soon after it binds to myosin's ATPase site, ATP is split by hydrolysis into ADP and P_i, which releases energy. Some of the energy is captured by the myosin molecule as it goes into its high-energy conformation. Although ATP has been hydrolyzed at this point, the end-products of the reaction (ADP and P_i) remain bound to the ATPase site. If calcium is present, the cycle will start over at step 1.

Although a given crossbridge generates force only part of the time while it is active (during the power stroke), a muscle cell generates force continually during a contraction, because many crossbridges go through the cycle simultaneously but out of phase ("out of step") with each other. Thus at any given time, some crossbridges are starting the cycle, others are finishing it, and still others are at various stages in between. To see the significance of this, consider what happens when you walk: Your legs move back and forth (as crossbridges do during a muscle contraction), but your body moves forward in a smooth, continuous fashion. As one leg moves backward, it pushes you forward; this is the "power stroke." When you lift that leg up and move it forward, it does not propel you but instead simply returns to its original position; let us call this the "return stroke." When you walk, you move forward smoothly because one leg goes through the power stroke while the other is going through the return stroke.

Because the crossbridges at opposite ends of a thick filament are oriented in opposite directions from each other (see Figures 13.3 and 13.5), the power strokes of crossbridges at opposite ends of the thick filament move in opposite directions, pulling the thin filaments on either side of the A band in toward the center and causing the sarcomere to shorten. When the crossbridge cycle stops and the contraction ends, the thin filaments passively slide back to their original position. During a contraction, each myosin head completes only about five crossbridge cycles per second, but because each thick filament has several hundred heads, thousands of power strokes can occur each second. For this reason, sarcomeres—and entire muscle fibers—can shorten very rapidly, in many cases taking less than a tenth of a second to contract fully.

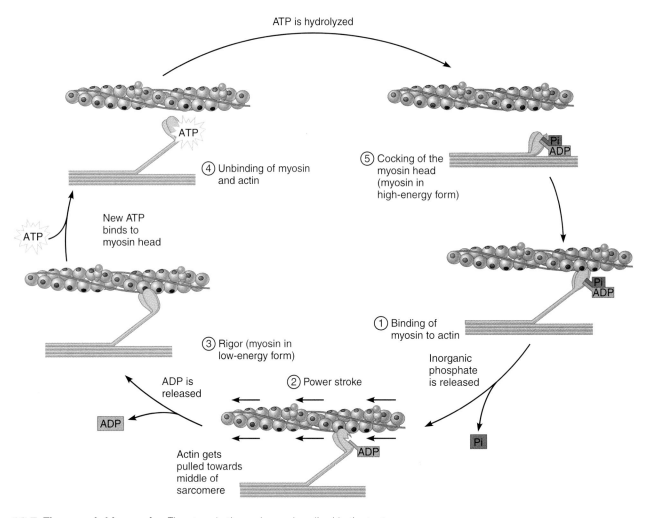

Figure 13.7 The crossbridge cycle. The steps in the cycle are described in the text.

In theory the crossbridge cycle could go on indefinitely, so long as there is a continual supply of ATP and actin sites are available for binding with myosin. This does not happen, so what initiates the cycle, and what keeps the cycle from occurring when a muscle is relaxed? In the next section we see that the answer to both questions lies in the actions of the regulatory proteins troponin and tropomyosin.

Quick Test 13.2

1. Which of the following shortens during a muscle contraction: thick filaments, thin filaments, A bands, I bands, H zones, sarcomeres? (Choose all that apply.)

2. To what does the sliding-filament model refer?

3. When ATP is hydrolyzed, myosin crossbridges change conformation. Do they go into the high-energy form or the low-energy form?

4. What triggers the power stroke of the crossbridge cycle?

Excitation-Contraction Coupling: How Muscle Contractions Are Turned On and Off

We saw in Chapter 12 that the central nervous system ultimately controls skeletal muscle contractions, with *motor neurons* delivering to the muscles commands telling them when and when not to contract. We saw that input from motor neurons always has an excitatory effect on muscle cells and serves to trigger contraction of those cells. Like neurons, muscle cells are *excitable,* meaning that they are capable of generating action potentials if their plasma membranes are depolarized to a sufficient degree. When a muscle cell receives input from a motor neuron, the cell depolarizes and fires an action potential that then stimulates contraction. The sequence of events that links the action potential to the contraction is referred to as **excitation-contraction coupling (Figure 13.8).**

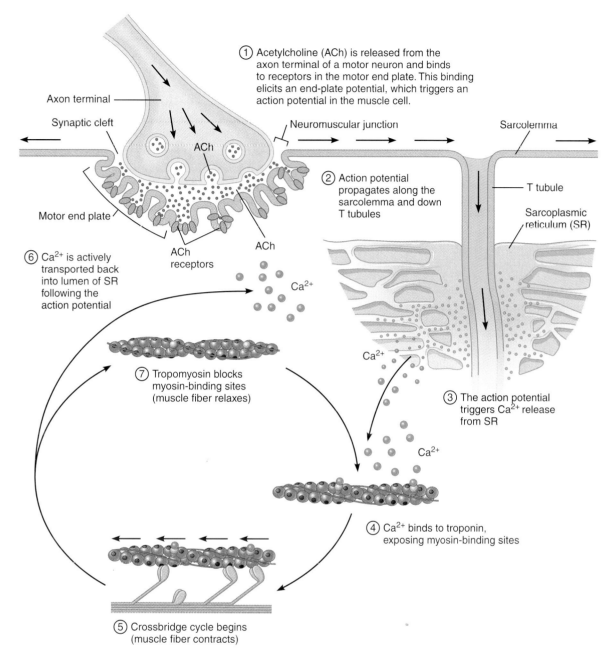

① Acetylcholine (ACh) is released from the axon terminal of a motor neuron and binds to receptors in the motor end plate. This binding elicits an end-plate potential, which triggers an action potential in the muscle cell.

Axon terminal

Synaptic cleft

Neuromuscular junction

Sarcolemma

ACh

Motor end plate

② Action potential propagates along the sarcolemma and down T tubules

T tubule

Sarcoplasmic reticulum (SR)

ACh receptors

ACh

⑥ Ca^{2+} is actively transported back into lumen of SR following the action potential

Ca^{2+}

Ca^{2+}

⑦ Tropomyosin blocks myosin-binding sites (muscle fiber relaxes)

③ The action potential triggers Ca^{2+} release from SR

Ca^{2+}

④ Ca^{2+} binds to troponin, exposing myosin-binding sites

⑤ Crossbridge cycle begins (muscle fiber contracts)

Figure 13.8 Events in excitation-contraction coupling. Contraction of a skeletal muscle fiber is initiated and maintained by the arrival of action potentials at the axon terminal of a motor neuron. Upon the cessation of action potentials and the transport of calcium back into the SR, the contraction stops and the muscle fiber relaxes.

What enzyme breaks down acetylcholine after it is released?

The Role of the Neuromuscular Junction in Excitation-Contraction Coupling

We have seen that the connection between a motor neuron and a muscle cell, referred to specifically as a *neuromuscular junction,* is fundamentally no different from an "ordinary" synapse between two neurons in the nervous system. The motor neuron (the presynaptic cell) transmits an action potential and secretes the neurotransmitter *acetylcholine* upon its arrival at the axon terminal (step 1, Figure 13.8). After release, acetylcholine diffuses to the plasma membrane of the muscle cell (the postsynaptic cell), where it binds to specific receptors, triggering a change in ion permeability that results in a depolarization.

Acetylcholinesterase

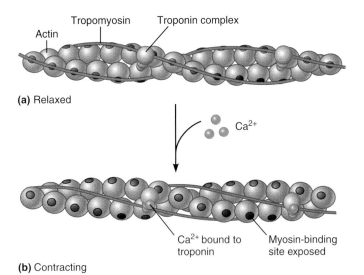

(a) Relaxed

Ca²⁺

Ca²⁺ bound to troponin   Myosin-binding site exposed

(b) Contracting

Figure 13.9 Actions of troponin and tropomyosin in excitation-contraction coupling. (a) In relaxed muscle, tropomyosin covers up actin's myosin-binding sites, which prevents the crossbridge cycle from occurring. (b) Following their release from the sarcoplasmic reticulum, calcium ions bind to troponin, causing a conformational change in the troponin complex that shifts tropomyosin's position on the actin filament and exposes the myosin-binding sites.

Despite its similarity to an ordinary synapse, recall that a neuromuscular junction has many characteristics that make it special (see Chapter 12). Although a motor neuron typically branches and innervates more than one muscle cell, each muscle fiber receives input from only one motor neuron. At the neuromuscular junction, the motor neuron's terminal boutons fan out over a wide area of the sarcolemma. Opposite these boutons is a specialized region of the sarcolemma called the *motor end plate,* which is highly folded and contains a high density of acetylcholine receptors (see Figure 13.8). An action potential in the motor neuron triggers the release of acetylcholine from each of its many terminal boutons, which causes many acetylcholine receptors to become activated. As a consequence, the resulting depolarization (called an *end-plate potential*) is much larger than an ordinary postsynaptic potential—so large, in fact, that it is always above threshold and triggers an action potential in the muscle cell. Thus, if an action potential occurs in a motor neuron, it is always followed by an action potential in the muscle cells it innervates (step 2, Figure 13.8).

Once an action potential is initiated in a muscle cell, it propagates through the entire sarcolemma and also travels through the T tubules, which are essentially extensions of the sarcolemma. As the action potential travels through the T tubules, it triggers the release of calcium from the nearby sarcoplasmic reticulum (step 3, Figure 13.8). This calcium then serves as the signal that initiates the crossbridge cycle and, hence, contraction of the muscle cell.

The Roles of Calcium, Troponin, and Tropomyosin in Excitation-Contraction Coupling

When a muscle cell is relaxed, the concentration of calcium in the cytosol is very low, and little binding occurs between calcium and troponin. Troponin is in its normal (resting) conformation and because of this, tropomyosin is positioned on the thin filaments in such a way that it blocks actin's myosin-binding sites, so the crossbridge cycle cannot occur (**Figure 13.9**a). The cytosolic calcium level is normally low because the membrane of the sarcoplasmic reticulum (SR) is equipped with pumps that actively transport calcium ions from the cytosol into the SR. Due to the action of these pumps, the SR is able to accumulate calcium against a concentration gradient and therefore function as a calcium storehouse.

In addition to calcium pumps, the membrane of the SR contains voltage-gated calcium channels that are normally closed, which prevents calcium inside the SR from leaking out. When an action potential travels through the T tubules, however, it causes these channels to open briefly, allowing calcium to flow out into the cytosol. The end result is a rise in the cytosolic calcium concentration.

Although the calcium channels that allow Ca²⁺ out of the SR are voltage-gated, it is an unusual kind of voltage-gating because the electrical signal that triggers gating occurs in the membrane of the T tubule—not the membrane of the SR itself. An action potential in the T tubule can trigger the release of Ca²⁺ from the SR because adjacent T tubules and SR membranes are physically linked by proteins called *foot structures* (or *ryanodine receptors*) that bridge the gap between them and also function as calcium channels (**Figure 13.10**). Where the foot structures come into contact with the T-tubule membranes are other proteins (called *dihydropyridine receptors,* or DHP receptors) that function as voltage sensors. When an action potential travels through the T tubules, these voltage sensors react (presumably by undergoing a conformational change) and in so doing transmit a signal directly to the foot structures with which they are in contact. This signal triggers the opening of calcium channel pores within the foot structures, which allows Ca²⁺ to flow out of the SR. As the ions enter the cytosol, some of them bind to specific sites on other SR calcium channels and cause them to open. In this manner the initial release of calcium triggers the release of even more calcium from the SR.

As the cytosolic calcium concentration rises, some of it binds to one of the three proteins making up each troponin complex (step 4 in Figure 13.8, and Figure 13.9b), which then undergoes a conformational change that causes tropomyosin to shift out of its normal resting position, thereby exposing the myosin-binding sites on the actin monomers. With the myosin heads of the thick filament now able to bind to actin, the crossbridge cycle can begin (step 5, Figure 13.8), and the sarcomere contracts. The design of the SR and T-tubule network surrounding the myofibrils permits nearly simultaneous delivery of calcium to all sarcomeres of a muscle fiber, so the sarcomeres contract in unison, as does the entire muscle fiber.

A muscle cell stops contracting when it no longer receives input from its motor neuron, and action potentials no longer occur in the sarcolemma. When an action potential triggers the release of calcium from the SR, this release does not continue indefinitely because as the cytosolic calcium concentration rises, calcium ions begin to bind to certain sites on the SR calcium channels, causing them to close. (These sites are distinct from those that trigger channel opening and have a lower affinity for calcium, so they do not come into play until cytosolic calcium has risen to a sufficiently high level.) The closure of these channels turns off the release of calcium and enables the active transport of calcium back into the SR (an ongoing process) to clear calcium from the cytosol (step 6 in Figure 13.8), which causes the calcium concentration to fall. Because the binding of calcium to troponin is reversible, this concentration change causes calcium to dissociate from troponin, which allows both troponin and tropomyosin to revert to their original positions (step 7 in Figure 13.8). The number of exposed sites on the actin filament therefore decreases, leading to a decline in the number of active crossbridges. Eventually, as calcium concentration returns to normal, the muscle contraction ends.

Muscle Cell Metabolism: How Muscle Cells Provide ATP to Drive the Crossbridge Cycle

Muscles are able to begin contracting within a fraction of a second after receiving neural input and are capable of sustaining contractile activity for a time. For example, think of a competitive sprinter, whose muscles begin to work near maximum capacity just after the starting gun goes off and continue to do so to the end of the race. For muscles to perform in this manner, it is necessary that muscle cells always have a readily available supply of ATP, even when the demand for ATP increases suddenly and rapidly. Energy sources for contracting muscle are shown in **Figure 13.11** and described below.

The Role of the Creatine/ Creatine Phosphate System

When a muscle fiber is resting, its demand for ATP is small, but when it is signaled to contract, the demand for ATP soars. At rest, a muscle cell contains a small store of ATP, but it cannot rely on this ATP for long once it begins contracting. To keep from depleting its ATP supply, it must gear up ATP production to keep pace with the increased rate of utilization.

The ATP that powers muscle contraction is produced in muscle cells, as in other cells, by substrate-level phosphorylation and oxidative phosphorylation. When a cell's rate of ATP utilization increases, the concentration

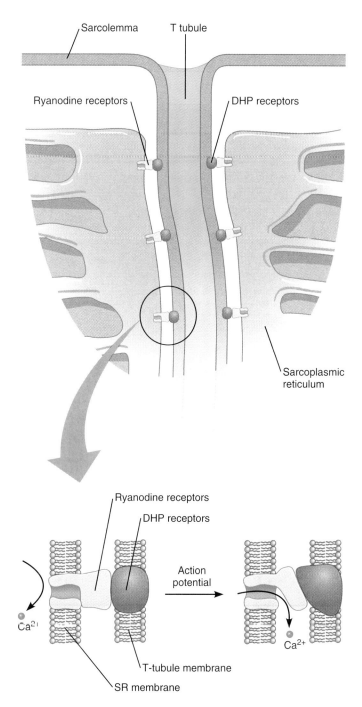

Figure 13.10 Gating of sarcoplasmic reticulum calcium channels. Voltage-sensitive DHP receptors, located in the T tubules, are coupled to ryanodine receptors in the SR membrane. When an action potential travels down T tubules and is detected by DHP receptors, calcium channels associated with ryanodine receptors open.

of ATP inside the cell falls and the concentration of ADP rises. These and other changes then stimulate the enzymes that control ATP-producing reactions, such that ATP is generated at a higher rate. Although this occurs once a muscle cell begins to contract, these reactions need a few seconds to come up to speed. To ensure a

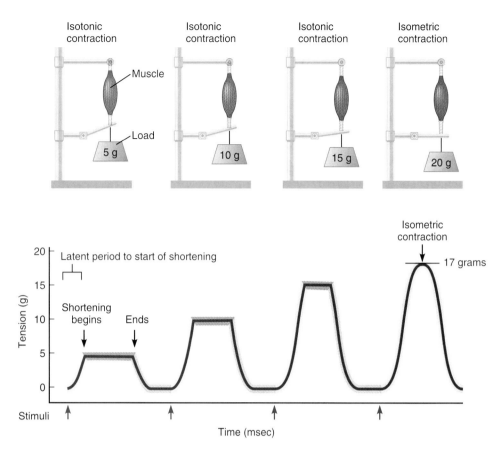

Figure 13.14 Effect of load on peak tension in an isotonic twitch. The responses to stimulation of a muscle subjected to four different loads are plotted. The muscle contracts isotonically when the loads are 5, 10, or 15 grams, and in each of these cases muscle tension plateaus at a level equal to the load. When the load is 20 grams, however, the muscle is unable to shorten because the load exceeds the maximum tension it can generate in a single twitch (17 grams), and thus the muscle contracts isometrically. Note that as load increases, the duration of muscle shortening decreases, and that the latent period before the onset of muscle shortening increases.

If a 16-gram load were placed on the muscle, would the contraction be isotonic or isometric?

shorten when it contracts (Figure 13.13b). Note that the curve for the isotonic twitch shows a distinct plateau, indicating that the force is constant for a period of time (hence the name *isotonic,* which means "same tension"). It is during this plateau phase that the muscle shortens and the load moves. Before the plateau phase the force increases but the muscle does not shorten because it is not yet generating enough force to lift the load. Only when the force becomes equal to the load does the muscle begin to shorten. This force remains constant so long as the load is moving, but eventually the muscle starts to relax and the load starts to fall. When the load comes to rest, the plateau phase ends and the force begins to decline.

Unlike an isometric twitch, an isotonic twitch is not an all-or-nothing event—its size and shape depend on the size of the load that is placed on the muscle. When the load is increased, for example, greater tension is needed to

overcome it, and for this reason the force tracings show plateaus at higher tension levels (**Figure 13.14**). At the same time, the time lag (latent period) between the stimulus and the beginning of muscle shortening (the start of the plateau) also increases, because it takes the muscle longer to develop the force required to move the load. When the load exceeds the amount of force the muscle can generate, the muscle cannot move it and therefore contracts isometrically. Under these conditions, the force tracing has the rounded peak characteristic of an isometric twitch (Figure 13.14, far right).

When a muscle contracts isometrically, its sarcomeres shorten even though the whole muscle does not. This is possible because the sarcomeres (collectively referred to as the muscle's **contractile component, CC**) do not extend the entire length of each muscle fiber and therefore do not transmit force directly to the ends of the cells. Instead, the force is transmitted through certain cellular components that connect the myofibrils to the ends of the cells and then through connective tissue that anchors

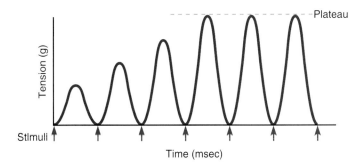

Figure 13.15 Treppe. In response to sufficiently frequent repetitive stimuli (arrows), peak tension in a muscle contracting isometrically rises with each of the first few twitches but eventually reaches a plateau.

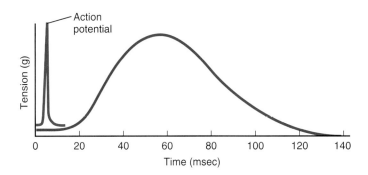

Figure 13.16 Duration of an isometric twitch. The twitch of a muscle fiber lasts significantly longer than the action potential that stimulated it.

What term is given to the series of events occurring in the time interval between an action potential and the rise in muscle tension?

the ends of the cells in place and extends through the tendons. Those parts of a muscle (or muscle cell) that do not actively generate force but only serve to passively transmit force to the ends of the muscle are collectively referred to as the muscle's **series elastic component (SEC)**. When a muscle contracts isometrically, the CC shortens and stretches the SEC, causing it to pull on the ends of the muscle. In so doing, the SEC lengthens as the CC shortens, giving an overall length change of zero.

Note that although muscles do contract isometrically in the body, they rarely (if ever) contract in a strictly isotonic fashion—that is, with a constant force as they shorten. When you walk, run, lift objects, or otherwise move such that contracting muscles shorten, your nervous system continually adjusts its input to the muscles to ensure that the muscular force is appropriate for the intended activity. This is quite different from the laboratory situation depicted in Figures 13.13 and 13.14, in which a single invariant stimulus is delivered to a muscle. Furthermore, when a muscle contracts in the body, the load placed on the muscle is rarely constant (as it is in Figures 13.13 and 13.14), even when a constant weight is being lifted. This is the case because the positions of the joints change as you move, and this alters the loads placed on muscles. You can easily demonstrate this by attempting to hold a 10-pound weight steady in your hand under two conditions: with your elbow close to your body and bent at a 90-degree angle, and with your arm extended straight out in front of you. Without a doubt you will find it more difficult to hold the weight steady in the second instance because the extended position of the arm puts greatly increased stress (that is, *load*) on certain muscles of the arm and shoulder. (To learn more about the physics of muscle activity, see **Toolbox: Skeletal Muscles at Work**.)

Factors Affecting the Force Generated by Individual Muscle Fibers

The force generated in a muscle depends on two factors: (1) the force generated in individual muscle fibers and (2) the number of muscle fibers contracting. The force generated in individual muscle fibers depends on the number

of active crossbridges that bind to actin. More active crossbridges generate more force, just as the force exerted on the rope in tug-of-war increases when more participants are added. In this section we examine factors affecting the number of active crossbridges, and thus the force generated by the contraction of individual muscle fibers: frequency of stimulation, fiber diameter, and changes in fiber length.

Frequency of Stimulation

Isometric muscle twitches are in fact reproducible, all-or-nothing events only if a muscle is stimulated at a frequency low enough to ensure that twitches are well separated in time. At higher frequencies, the rate of calcium release from the SR into the cytosol exceeds the rate of calcium active transport from the cytosol back into the SR, resulting in an increase in the peak tension. The more calcium in the cytosol, the more calcium bound to troponin, which moves the tropomyosin to expose more myosin-binding sites on actin. The more myosin-binding sites exposed, the more crossbridges can participate in crossbridge cycling and the greater the tension development. As the stimulation frequency increases, muscles move from twitch contractions to *treppe*, to *summation*, and finally to *tetanus*.

Treppe occurs at a frequency of muscle stimulation where independent twitches follow one another closely such that the peak tension rises in a stepwise fashion with each twitch, until eventually it reaches a constant level (**Figure 13.15**). The cause of treppe is unknown, but it may result from an increase in cytosolic calcium between twitches or from a "warming" of the muscle fiber that occurs with work.

Summation and tetanus occur at greater frequencies of stimulation due to twitches overlapping in time. Compared to an action potential, which takes at most a few milliseconds to complete, a muscle twitch is fairly slow, taking anywhere from tens to hundreds of milliseconds to complete (**Figure 13.16**). Because of this, a muscle fiber

SKELETAL MUSCLES AT WORK

Almost all skeletal muscles are connected to at least two bones by tendons. When a muscle contracts, one of these bones typically moves while the other remains relatively stationary. A muscle's point of attachment to the stationary bone is called the *origin,* whereas its point of attachment to the movable bone is the *insertion* [figure (a)]. Depending on a muscle's function and location, its tendons may be short or long. The biceps muscle, for instance, is connected to the bones of the arm by short tendons, but the calf muscle called the *gastrocnemius* is connected to the heel bone by a long tendon (the *Achilles tendon*), an arrangement that frees the ankle of both bulk and weight. Similarly, long tendons permit the muscles that control the fingers to reside in the forearm, which is useful because a muscle-bound hand would be detrimental to finger dexterity.

A muscle can actively exert force only by contracting, which in the case of skeletal muscle means *pulling,* not pushing, on a bone. Nevertheless, it is clear that you can use muscular force to move the elbow joint, for example, in opposite directions: You can *flex* the forearm (that is, reduce the angle of the elbow joint) by contracting the biceps muscle, and you can *extend* the forearm (increase the angle of the elbow joint) by contracting the triceps muscle, which is located opposite the biceps in the arm [figure (b)]. The biceps and triceps are examples of muscles that are *antagonistic*—each exerts force in a direction that opposes the action of the other. (Most other skeletal muscles are arranged in antagonistic groups as well.) To flex the forearm, the biceps generates force actively while the triceps relaxes and

stretches passively—that is, in response to forces exerted on it by the action of the biceps. To extend the forearm, the triceps contracts actively, and the biceps stretches passively. At times, simultaneous contraction of antagonistic muscle groups is useful. Thus when the biceps and triceps are stimulated to contract at the same time, as when you "brace yourself" to receive a package whose weight you are unsure of, the elbow joint stiffens to resist motion, such that the forearm remains stationary.

Bones function as levers. When you use your muscles to lift a weight or move a part of your body, the force your muscles must generate is actually much greater than you might think. For instance, to hold a 15-kg weight (about 33 pounds) in the palm of your hand, the biceps in your arm generates an impressive 105 kg (about 232 pounds) of force! The reason relates to the fact that the insertion of the biceps is located close to the elbow joint (about 5 cm away), whereas the hand is located farther from the joint (about 35 cm away) [figure (c)]. To lift the weight, the biceps pulls upward on the bones of the lower arm, causing the hand to exert an upward force against the weight. (When the weight is being held steady or is being lifted at a constant speed, the magnitude of this force equals the downward force of gravity acting on the weight.) Because the insertion of the biceps is close to the elbow joint, which acts as a pivot, the muscle must exert a force significantly greater than the force exerted by the hand [figure (d)].

To see why the biceps must exert more force, it is instructive to consider what happens when the weight is being lifted, as shown

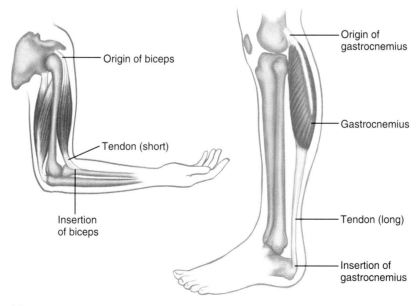

Origin of biceps

Tendon (short)

Insertion of biceps

Origin of gastrocnemius

Gastrocnemius

Tendon (long)

Insertion of gastrocnemius

(a)

(continued)

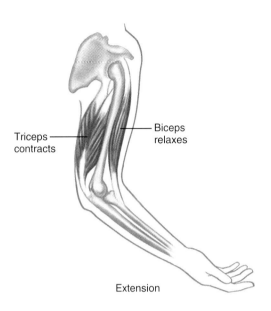

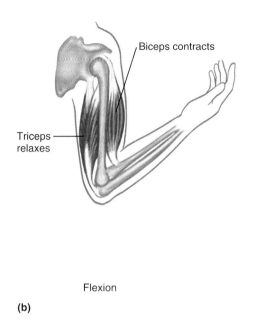

Flexion   Extension

(b)

in figure **(e)**. Whenever a force is applied to an object, causing it to move a certain distance, a quantity of *work* is performed:

$$work = force \times distance$$

In this case, the biceps performs a certain amount of work to move the arm, and the hand performs an equal quantity of work to move the weight. As the forearm pivots at the elbow, the hand and the biceps' insertion both move through an arc and travel a certain distance, but the hand travels farther because its arc has a greater radius, as shown in the figure. Because the hand and biceps perform the same amount of work, the product of force exerted and distance traveled is the same for both. Given that

the insertion of the biceps travels a shorter distance than the hand, it follows from the previous equation that the biceps must exert a greater force than the hand.

Using these principles, we can determine how much force the biceps must generate in order to lift a given weight. Since the length of an arc is proportional to its radius, the distance the hand or the insertion of the biceps travels is proportional to its distance from the elbow joint. Therefore, if the hand exerts a force F_1 and is located a distance R_1 from the elbow, it performs a quantity of work proportional to the product $F_1 \times R_1$ [figure **(f)**]. Likewise, if the biceps exerts a force F_2 and its insertion is located a distance R_2 from the elbow, the work it performs is

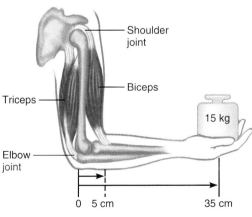

(c) Distance from elbow

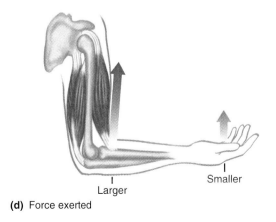

(d) Force exerted

SKELETAL MUSCLES AT WORK *(continued)*

proportional to the product $F_2 \times R_2$. Because the hand and biceps perform equal amounts of work, the force of the hand and biceps are related by the following expression:

$$F_1 \times R_1 = F_2 \times R_2$$

or

$$F_2 = F_1 \times (R_1/R_2)$$

Using values given previously, we find that when the hand is lifting a 15-kg weight, the biceps exerts a force given by

$$F_2 = 15 \text{ kg} \times (35 \text{ cm/5 cm}) = 105 \text{ kg}$$

This is also true when the hand is holding the weight steady.

In the example above, the bones of the forearm act as a lever that pivots at the elbow. The biceps pulls upward on this lever, while the weight in the hand pushes down on it. To hold the weight steady (or lift it at constant speed), the biceps must exert a force greater than the downward force of the weight because its *lever arm*—the distance between the lever's pivot point and the point at which the force is applied—is shorter than that of the weight. As a reflection of this, the shorter lever arm is said to put the biceps at a *mechanical disadvantage*. Consequences of the lever arm effect may be familiar to you: If you have ever had to change an automobile tire, for instance, you know that it is easier to loosen the lug nuts with a longer wrench than with a shorter wrench. The shorter wrench puts you at a mechanical disadvantage, such that you have to exert more force to loosen the nuts.

Even though the lever arm effect puts muscles at a mechanical disadvantage with respect to the force they must generate, the situation is actually advantageous in another way: As a muscle contracts it shortens with a certain velocity, but the limb (or other body part) to which it is attached moves significantly faster. If, for example, the biceps shortens at a rate of 2 cm/sec, the hand moves at a rate seven times as great—14 cm/sec. Thus the lever arm effect makes it possible for the biceps and other muscles to move body parts more quickly than the muscles themselves can shorten. This is important because it not only makes activities such as running and throwing possible but also enables us to move out of danger quickly in times of emergency.

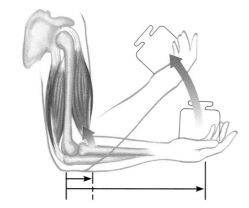

(e) Radius of arc (distance from elbow)

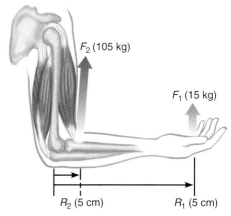

(f) Radius of arc (distance from elbow)

can have several action potentials in the time it takes to complete one twitch. When a muscle is stimulated repetitively such that additional action potentials arrive before twitches can be completed, the twitches superimpose on one another, yielding a force greater than that of a single twitch; this process is called **summation** (**Figure 13.17**). Summation happens whenever twitches occur at a frequency such that calcium cannot be removed from the cytosol as rapidly as it is released from the SR. Calcium removal is necessary for relaxation; thus the muscle fiber cannot relax completely between twitches.

At higher frequencies of stimulation, summation reaches a peak called **tetanus**. (The term *tetanus* also refers to a disease in which toxins produced in a bacterial infection cause motor neurons to stimulate muscle contraction inappropriately; see **Clinical Connections: Tetanus**, p. 397.) In *unfused* (or incomplete) *tetanus,* the force has small oscillations with brief periods of relaxation between peaks (Figure 13.17). The peaks are reached when calcium levels are great enough to saturate troponin, exposing all myosin-binding sites on actin. At even greater frequencies, calcium levels are great enough to continually saturate

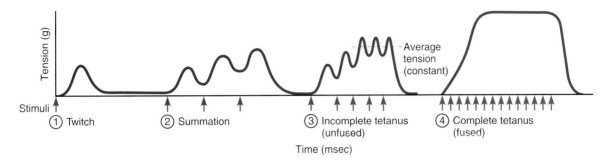

Figure 13.17 Effects of high stimulus frequency: summation and tetanus. In response to repetitive stimuli (arrows) delivered close together in time, muscle twitches superimpose in summation. A train of more frequent stimuli causes tension to rise with each twitch until the muscle reaches incomplete tetanus, characterized by a plateau composed of individually distinguishable twitches. Still greater stimulus frequency produces complete tetanus, in which force increases swiftly and smoothly to a plateau in which individual twitches are no longer distinguishable.

troponin such that all myosin-binding sites on actin are continually exposed, resulting in a plateau called *fused* (or complete) *tetanus* (Figure 13.17). If the stimulus intensity is increased still further, tetanic tension (that is, the tension generated during tetanus) increases, but only up to a point; further increases in frequency beyond this point yield no further increases in force. Under these conditions the muscle is generating all the force it can, which is referred to as *maximum tetanic tension*.

Fiber Diameter

We accept as a fact of everyday life that some muscles have an inherent ability to generate more force than others. Why else would we equate a weightlifter's bulging muscles with superior strength? The inherent ability of a muscle to generate force is referred to as the muscle's *force-generating capacity,* which is usually assessed by measuring maximum tetanic tension or peak tension in an isometric twitch.

The force-generating capacity of a muscle fiber depends on both the number of crossbridges in each sarcomere and the geometrical arrangement of the sarcomeres. Other things being equal, a muscle whose sarcomeres contain more crossbridges can generate more force, just as the force exerted on the rope in tug-of-war increases when more participants are added. In addition, a muscle that has more sarcomeres—and hence more thick and thin filaments—arranged in parallel can generate more force than a muscle with fewer sarcomeres arranged in parallel. Because the number of thick and thin filaments per unit of cross-sectional area does not vary significantly from one muscle to another, a fiber's *diameter* is the crucial variable that determines its force-generating capacity. The greater the fiber diameter, the greater its cross-sectional area, and the more force it can generate. This is why a weightlifter's bulging muscles are stronger than the average person's slimmer muscles.

Note that although the number of parallel sarcomeres strongly affects a muscle's force-generating capacity, the force-generating capacity does *not* depend on the number of sarcomeres that are joined in series (end to end). This means that two muscles, one longer than the other but otherwise identical, have the same force-generating capacity. Consider a chain with a weight dangling from one end: Assuming that the weight of the chain itself is negligible, each of its links exerts on its neighbors a force that is equal to the force exerted on the chain by the weight. This will be true regardless of the number of links in the chain.

Changes in Fiber Length

Although a muscle fiber's length does not affect its force-generating capacity insofar as it reflects the number of sarcomeres in series, *changes* in a fiber's length do influence its ability to generate force. For each fiber, maximum force-generating capacity occurs over a certain range of lengths. When a fiber either shortens beyond or is stretched beyond this optimum range, its force-generating capacity decreases, because such changes in length alter the length of individual sarcomeres and reduce their ability to generate force.

Figure 13.18 shows how the force of contraction, measured as percentage of maximum tetanic tension or peak tension in an isometric twitch, varies with a muscle's length. The graph is an example of a *length-tension curve,* and its shape is a consequence of the sliding-filament model and the nature of the crossbridge cycle. We have seen that the force of a muscle contraction is related to the number of active crossbridges—the greater this number, the greater the force—and that crossbridge activity requires that myosin crossbridges be able to bind to actin. If we start with a muscle fiber at optimum length and then stretch it, the tension generated by the fiber decreases linearly as its length increases (segment b–c in Figure 13.18). The reason is that the degree of overlap between thick and thin filaments decreases as sarcomeres lengthen, and crossbridges that are not overlapped by thin filaments cannot bind to actin and therefore cannot generate force.

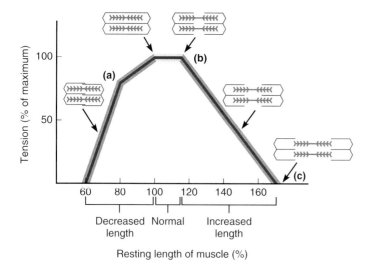

Figure 13.18 A length-tension curve. The curve plots the tension developed by a muscle as a function of the muscle's resting length. The center bracket indicates the normal range of muscle lengths *in situ*. Drawings of sarcomeres indicate changes in the relative positions of thick and thin filaments as muscle length changes.

If the muscle is significantly shorter than its optimum length, tension declines in two stages as the length decreases. At first tension decreases relatively slowly, but once the length decreases beyond a *certain* point (to the left of point a in Figure 13.18), tension decreases more rapidly. The decrease in tension that occurs with decreasing length is not the result of any change in the force generated by crossbridges, because the crossbridges are completely overlapped by thin filaments, and all are active. However, as sarcomeres shorten beyond their optimum length, the thin filaments at opposite ends of the sarcomere begin to overlap each other, which interferes with their movement. Then, as the sarcomeres shorten beyond point a in Figure 13.18, the Z lines eventually come into contact with thick filaments, so most of the force generated by crossbridges is exerted on the sarcomere itself instead of being transmitted to the ends of the muscle fiber.

The constant force-generating capacity of a muscle fiber at the peak of the length-tension curve in Figure 13.18 occurs for two reasons: All crossbridges are overlapped by thin filaments and therefore are capable of generating force, and sarcomeres are not short enough to allow thin filaments to come into contact, so interference between them does not occur.

The fact that muscles shorten when they contract does not mean that muscles in the body compromise their force-generating capacity when they contract. Note the bracket in Figure 13.18 indicating the normal operating range of muscles *in situ* (in the body), which shows that muscles *in situ* are always close to their optimum lengths, even when they have been maximally shortened or stretched. The reason is that the movement of a muscle is constrained by the bones to which it is attached; muscles can lengthen or

shorten only so far because the possible range of joint angles is limited by the architecture of the skeleton. As a result, muscles usually operate within the range of lengths in which they can generate maximum force.

■ ■ ■

Based on what we have seen concerning summation of contractions, we know that the action potential frequency in a muscle's motor units has a direct bearing on the force the muscle generates. The most force a single muscle fiber can develop by contracting isometrically is the maximum tetanic tension, which for most muscle fibers is only about five times greater than peak tension in a single twitch. Given that muscular tension can vary over several orders of magnitude (for example, the forces required to hold a paper clip compared to that required to hold a chair), it is clear that variation in action potential frequency can account for only a small fraction of the range of forces a muscle can generate. We see next that the force generated depends also on the number of muscle fibers contracting.

Regulation of the Force Generated by Whole Muscles

When a muscle contracts, only rarely do *all* of its fibers actively generate force. Some motor units are active, but the fibers in other motor units simply "go along for the ride," passively shortening in response to forces generated by actively contracting fibers. When larger forces are needed, the nervous system can activate some of these "extra" fibers, thereby increasing the total number of active fibers. Indeed, the nervous system exerts most of its control over muscular force by varying the number of active motor units; variation in the frequency of stimulation of individual fibers plays a secondary role. An increase in the number of active motor units is called **recruitment**.

Recruitment

We have seen that within a muscle, fibers belonging to a given motor unit are intermixed with fibers from other motor units. But not all motor units are created equal; they often differ in size, with some having relatively more fibers and others having relatively few. **Figure 13.19**a shows two motor units (X and Y) in a muscle, the fibers of which have identical force-generating properties and have been stimulated to give maximum tetanic tension; motor unit X contains five fibers, whereas motor unit Y contains seven fibers. When motor unit X is stimulated to contract, it generates five times the force of a single fiber because its fibers are working together in parallel; likewise, stimulation of motor unit Y results in a force seven times that of a single fiber, and stimulation of both motor units produces a force 12 times greater than that of a single fiber (Figure 13.19b). Because a muscle may contain hundreds of motor units, muscular tension can be

TETANUS

Just a few decades ago, U.S. schoolchildren faced the very real possibility of contracting any one of several severe, often-fatal diseases that have since largely disappeared from public consciousness thanks to widespread immunization programs. One of these diseases—*tetanus,* characterized by severe muscle spasms and convulsions—now afflicts fewer than 100 Americans each year.

Tetanus results from infection with *Clostridium tetani,* an anaerobic bacterium whose spores are found in soil and animal feces. The disease most commonly results from contamination of deep puncture wounds or burns, although it can occur after relatively minor wounds. Deep wounds are most typically involved because the anaerobic bacteria grow well in the oxygen-poor conditions found in such wounds. Symptoms of the disease appear after an incubation period of between 2 days and 50 days, although 5–10 days is most common.

Symptoms of the disease result from the action of tetanus neurotoxin secreted by the bacteria at the wound site. The toxin exerts its effects after reaching the CNS, where it binds to synaptic terminals and blocks the transmission of signals that normally inhibit the activity of motor neurons. As a consequence, motor neurons become hyperexcitable, leading to inappropriate stimulation of skeletal muscles. Depending on the extent of the toxin's spread, effects may be localized to muscles in the vicinity of the wound or

may involve muscles all over the body. Local effects include soreness and increased muscle tone in the affected area, whereas systemic effects include spastic movements and intermittent convulsions.

The most frequent early systemic symptom of tetanus is stiffness of the jaw. Commonly, stiffness spreads to other areas including the neck, arms, or legs and may be accompanied by sore throat, headache, fever, restlessness, irritability, or difficulty in swallowing. As the disease progresses, muscle spasms may cause a patient to experience difficulty opening the jaw *(trismus),* which accounts for the disease's common name—*lockjaw.* Other facial muscles may become affected, causing the patient to take on a bizarre facial expression characterized by a fixed smile and raised eyebrows *(risus sardonicus).* Death, if it comes, normally results from asphyxiation due to spasm of laryngeal or thoracic muscles.

The low incidence of tetanus in the United States is largely a result of mass immunization programs involving preschool children. Immunity is conferred by injection of inactivated tetanus toxin, toxoid, which stimulates the body to produce antibodies against the toxin. The tetanus vaccine is usually combined with vaccines against diphtheria and pertussis (whooping cough) to form a *DPT shot.* To maintain immunity, booster shots must be given every ten years. For unimmunized individuals who sustain a puncture wound, some protection can be conferred

by injection of antitoxins (antibodies produced against the toxin in vaccinated humans or animals). The mortality rate for unimmunized persons who contract the disease is estimated to be about 60%.

Barring effective prophylaxis, measures that can manage the symptoms include administration of drugs that interfere with neuromuscular transmission and thus reduce or eliminate muscle spasms. In all cases, adequate air flow to the lungs must be ensured, usually by the insertion of a breathing tube into a surgically created opening in the trachea. If neuromuscular block is strong enough to interfere with normal breathing, the patient must be placed on an artificial ventilator.

Stiffness of the jaw is an early systemic symptom of tetanus.

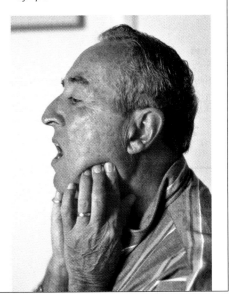

varied over a wide range merely by varying the number of active motor units.

Muscles differ in regard to the numbers of motor units they contain, from a handful in the muscles that control movements of the eyes to hundreds in larger muscles such as the biceps. Within a given muscle, the various motor units differ in both the numbers of fibers they possess and in the diameter and strength of those fibers. The fibers within any given motor unit tend to fall within a narrow range of sizes, with some consisting mostly of small fibers and others consisting mostly of

large fibers. Furthermore, motor units that have larger fibers also tend to have *more* fibers.

The Size Principle

When a muscle is called upon to generate small forces, generally only the smaller motor units come into play; when larger forces are needed, larger motor units are recruited. This correspondence between the size of motor units and the order of recruitment is known as the **size principle**. In addition, when contractions are sustained over a long time,

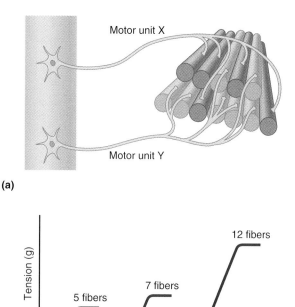

(a)

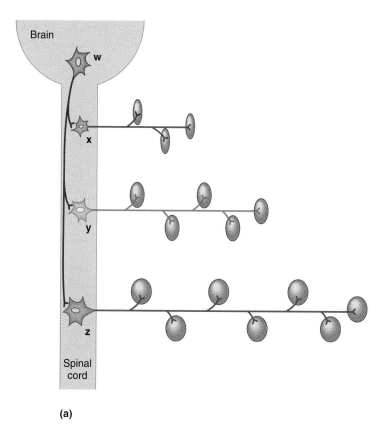

(b)

Figure 13.19 Increases in force generation with recruitment of motor units. (a) Motor units X and Y, which possess five fibers and seven fibers, respectively. (b) Tension developed by a single fiber, by motor unit X, by motor unit Y, and by motor units X and Y together.

(a)

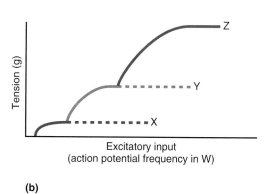

(b)

Figure 13.20 The size principle. (a) The anatomical relationship of three motor units (X, Y, and Z) of increasing size to an excitatory neuron W within the CNS. (b) As the frequency of action potentials in neuron W increases, the order of motor units activated proceeds from smallest (X) to largest (Z).

motor units are activated asynchronously—as one becomes active, another ceases its activity. In this manner the total force of the muscle is maintained at a constant level without overworking any of the individual motor units.

The fact that motor units differ in size has practical implications for precise control of muscular force. As a general rule, fine control is easier when muscular forces are small, because only the smaller motor units are recruited. Smaller variations in force are possible under these conditions because the recruitment of additional motor units causes only small increases in the total number of active fibers. In contrast, when large forces are involved, only larger increments in force are possible because larger motor units are recruited.

The basis for the size principle is not only that motor units vary in size, but that the *motor neurons* that control them also vary in size. Larger motor units are controlled by motor neurons with larger-than-average cell bodies and axon diameters, whereas smaller motor units are controlled by neurons with smaller-than-average cell bodies and axon diameters. This distribution of neuron sizes has important consequences for muscular control. For complicated reasons, larger cells are harder to depolarize to threshold; more excitatory synaptic input (a higher action potential frequency in the presynaptic cell) is required to induce a larger neuron to fire. Thus, when gradually increasing

synaptic input is delivered to a set of motor neurons, the small neurons will fire first and the large ones last.

This idea is illustrated schematically in **Figure 13.20**, in which small, medium, and large motor units (X, Y, and Z, respectively) are controlled by excitatory input from a neuron (W) located in the CNS. Note that as the action potential frequency in neuron W increases, the motor units become active in order of increasing size. Note as well that as each motor unit is recruited, the force increases in step-wise increments that reflect both the increased number of fibers in the larger units and the larger size of those fibers.

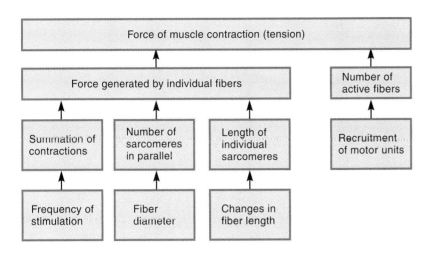

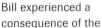

Figure 13.21 **Factors affecting the force developed by whole muscles.**

Now we can see that the force generated by whole muscles is affected by a combination of the factors acting on individual fibers and the number of fibers that are active (**Figure 13.21**).

But more than just the force of muscle contraction is important in movement; the speed with which muscles contract is important as well.

Velocity of Shortening

To determine a muscle's velocity of shortening, the muscle is stimulated to contract isotonically, and while it does so the distance it shortens is plotted over time, usually while different loads are placed on it. The results of these measurements, shown in **Figure 13.22**, reveal three effects: (1) The latent period of shortening (the time between the stimulus and the beginning of shortening) increases with increasing load (see Figure 13.14); (2) the duration of shortening (the time during which the muscle is shorter than its resting length) decreases with increas-

ing load; and (3) the velocity of shortening decreases with increasing load.

The velocity of shortening is defined as the *rate of change* of the distance shortened, which is the slope of each curve in Figure 13.22. (Because the slope changes continually throughout the period of shortening, it is customary to use the initial slopes of the curves as the measures of shortening velocity.) When velocity of shortening is plotted as a function of the load, the result is a load-velocity curve like that in **Figure 13.23**. Note that as the load increases, the velocity of shortening gradually decreases, eventually reaching zero when the load is equal to or greater than the maximum tension that can be generated by the muscle, and that velocity of shortening is greatest when no load is placed on the muscle. These observations are in accord with our everyday experience: We know that one cannot lift a box full of books as quickly as one can lift an empty box.

Measurements of shortening velocity have proven to be informative because they have revealed that muscles

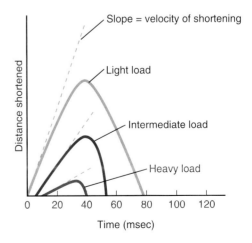

Figure 13.22 **The effect of load on muscle shortening.** The distance a muscle shortens while contracting isotonically is plotted over time for three different loads. Dashed lines represent the initial slopes of the curves, a measure of the muscle's velocity of shortening.

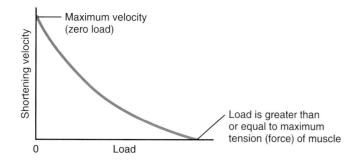

Figure 13.23 A load-velocity curve. The curve plots the velocities of shortening (as determined in Figure 13.22) over a range of loads.

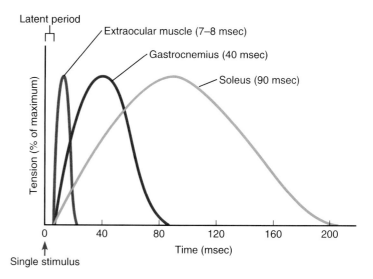

Figure 13.24 Differences in speed of contraction for three selected muscles. The curves represent isometric twitches; times in parentheses are those required for each muscle to develop maximum tension following a single stimulus (arrow). To facilitate comparison, the curves have been adjusted to make their peaks the same height, even though these muscles develop different maximum tensions.

differ in ways other than force-generating capacity. Researchers have found, for example, that certain types of muscle fibers can shorten faster than others. The bases for such differences are discussed in the next section.

> ### Quick Test 13.4
>
> 1. What is a twitch? How does an isometric twitch differ from an isotonic twitch?
> 2. Define the following terms: *summation* (of contractions), *contractile component, series elastic component, tetanus, recruitment, length-tension curve.*
> 3. What is the size principle?
> 4. What is the meaning of the term *velocity of shortening*?

Types of Skeletal Muscle Fibers

Although all skeletal muscle fibers are fundamentally alike with respect to the mechanisms of excitation-contraction coupling and force generation, there are significant differences among them in terms of how quickly they can contract and how they produce most of their ATP. In this section we examine these differences and their effects on muscle performance in everyday life.

Differences in Speed of Contraction: Fast-Twitch Fibers and Slow-Twitch Fibers

When different muscles are stimulated to contract isometrically, some take longer to reach peak tension than others (**Figure 13.24**). The reason is these muscles contain different populations of fibers. Some muscles (such as the soleus muscle of the leg) contain mostly *slow-twitch fibers,* which contract relatively slowly. In other muscles (such as the extraocular muscles, which control eye movements) the predominant fibers are *fast-twitch fibers,* which contract relatively quickly. In still other muscles (such as the gastrocnemius of the leg) the proportion of slow-twitch and fast-twitch fibers is intermediate. Differences between the

two fiber types are seen not only in isometric contractions but in isotonic contractions as well. Fast-twitch fibers attain peak isometric tension sooner than slow-twitch fibers and also have higher maximum shortening velocities when they contract isotonically (compared to slow-twitch fibers of similar length).

The difference between fast-twitch and slow-twitch fibers is based not on their size or shape, but instead on the type of myosin present in their thick filaments. So-called *fast myosin* has the inherent ability to hydrolyze ATP at a faster rate than *slow myosin,* and this ATPase rate has been found to correlate strongly with a fiber's speed of contraction. The higher ATPase rate of fast myosin implies that this form of myosin can complete more crossbridge cycles per second, which means that sarcomeres shorten faster, all else being equal.

Differences in the Primary Mode of ATP Production: Glycolytic Fibers and Oxidative Fibers

Even though all muscle fibers have the ability to produce ATP by both oxidative phosphorylation and substrate-level phosphorylation, they differ in their capacities for doing so and are grouped into two general categories on this basis. *Glycolytic fibers* have high cytosolic concentrations of glycolytic enzymes and therefore can generate ATP rapidly via glycolysis (substrate-level phosphorylation); these fibers have a relatively low capacity for generating ATP via oxidative phosphorylation because they contain relatively few mitochondria, where oxidative

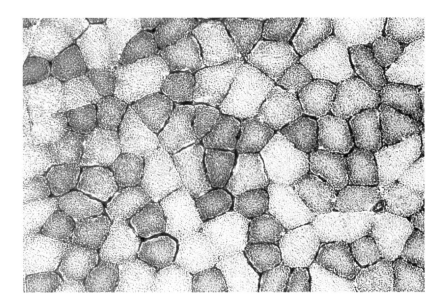

Figure 13.25 Oxidative and glycolytic muscle fibers. A micrograph of a cross section through a skeletal muscle showing numerous muscle fibers. The muscle has been prepared with a stain that causes mitochondria to appear purple and capillaries to appear black. Smaller fibers stain more deeply, indicating a higher density of mitochondria, which is typical of oxidative fibers. Note that capillary density is highest around these fibers.

phosphorylation occurs. By contrast, *oxidative fibers* are rich in mitochondria and have a high capacity for producing ATP via oxidative phosphorylation. However, these fibers contain relatively low concentrations of glycolytic enzymes and therefore have a low glycolytic capacity. Both fiber types are found in all muscles of the body, but their proportions vary among muscles.

The distinction between glycolytic and oxidative fibers goes beyond differences in glycolytic enzyme content or numbers of mitochondria. Oxidative fibers are generally of smaller diameter and well-supplied with capillaries, whereas glycolytic fibers are of larger diameter and are surrounded by fewer capillaries (**Figure 13.25**). This makes sense given that oxidative fibers have a higher capacity for utilizing oxygen and therefore rely more heavily on rapid oxygen delivery for proper function. A rich capillary supply ensures rapid oxygen delivery to the interstitial fluid surrounding the fibers, and the fibers' small diameter minimizes the distance oxygen must diffuse to reach the mitochondria.

Another difference between the fiber types is that oxidative fibers contain an oxygen-binding protein known as **myoglobin**, whereas glycolytic fibers lack it. Myoglobin, like hemoglobin (the oxygen-carrying protein in red blood cells), is a reddish molecule that binds oxygen reversibly. Its function is to serve as an oxygen buffer—to store a supply of oxygen that can be released whenever the oxygen concentration inside cells declines (as can happen when a muscle contracts strongly and compresses nearby blood vessels, thereby interrupting the blood supply). Because this oxygen store is limited, however, it can supply adequate amounts of oxygen for only a short time before it must be replenished, which occurs when blood flow is restored and the oxygen concentration rises. Because myoglobin imparts a reddish-brown color to oxidative fibers, these fibers are often referred to as *red muscle*. In contrast, glycolytic fibers, which lack

myoglobin and this reddish color, are referred to as *white muscle*. Familiar examples of red and white muscle are the "dark" and "white" meat found in chicken.

Glycolytic fibers produce ATP less efficiently than oxidative fibers because fewer ATP molecules are synthesized per unit of fuel consumed. However, glycolytic fibers are better able to produce ATP when oxygen availability is low because glycolysis does not require oxygen. When glycolytic fibers are active and producing ATP at a high rate, lactic acid is produced as a by-product, because these cells have a low oxidative capacity in addition to their high glycolytic capacity. One consequence of these differing capacities is that pyruvate, the end-product of glycolysis, is generated faster than it can be consumed and therefore accumulates in these cells. As it accumulates, it is converted to lactic acid, which has been implicated as a cause of muscle fatigue (discussed shortly). For this reason, glycolytic fibers fatigue more rapidly than oxidative fibers. In contrast, oxidative fibers produce little lactic acid so long as they are supplied with adequate amounts of oxygen, and as a consequence they are more resistant to fatigue. Lactic acid does not normally accumulate in these cells because their high oxidative capacity enables them to convert pyruvate to acetyl CoA as fast as it is produced.

Slow Oxidative, Fast Oxidative, and Fast Glycolytic Fibers

We have seen that skeletal muscle fibers can be classified as fast-twitch fibers or slow-twitch fibers on the basis of their contractile speeds, and as glycolytic fibers or oxidative fibers on the basis of their metabolic capacities. Not surprisingly, various combinations of contractile speed and oxidative or glycolytic capacity are possible. Indeed, three major classes of skeletal muscle fibers have been identified: *slow oxidative fibers, fast*

Table 13.1 | Properties of Skeletal Muscle Fiber Types

	Slow oxidative (red)	Fast oxidative (red)	Fast glycolytic (white)
Oxidative capacity	High	High	Low
Glycolytic capacity	Low	Intermediate	High
Speed of contraction	Slow	Intermediate	Fast
Myosin ATPase activity	Low	Intermediate	High
Mitochondrial density	High	High	Low
Capillary density	High	High	Low
Myoglobin content	High	High	Low
Resistance to fatigue	High	Intermediate	Low
Fiber diameter	Small	Intermediate	Large
Force-generating capacity	Low	Intermediate	High

glycolytic fibers, and the relatively rare *fast oxidative fibers.* Muscles generally contain all three fiber types, but in different proportions.

As their name implies, **slow oxidative fibers** contain slow myosin and have a high oxidative capacity, producing most of their ATP by oxidative phosphorylation. **Fast glycolytic fibers** contain fast myosin and have a high glycolytic capacity, producing most of their ATP through glycolysis. **Fast oxidative fibers** have a high oxidative capacity and contain fast myosin. (Actually, the myosin ATPase activity in these fibers is intermediate between the slowest and fastest myosin.)

Although there is no direct connection between a muscle fiber's type and its ability to generate force, there is an indirect connection because the three fiber types also differ in diameter. Slow oxidative fibers are the smallest in diameter and are therefore capable of generating only small forces. Fast glycolytic fibers have the largest diameter and generate the highest forces, whereas fast oxidative fibers are intermediate in terms of diameter and force-generating capacity. Properties of the three types of fibers are summarized in **Table 13.1**.

Response of the Three Fiber Types to Exercise

Even though the different types of fibers are intermixed in skeletal muscles, all three types are not typically utilized every time a muscle contracts. For one thing, fiber types are segregated into different motor units, such that a given motor unit usually contains fibers of only one type. Furthermore, because a correlation exists between the size of a motor unit and the type of fiber it contains, the different fiber types are recruited in a specific order (in accordance with the size principle) as muscle tension increases.

Slow oxidative fibers are recruited first because these fibers are located in the smaller motor units. Fast oxidative fibers are recruited next because these fibers are found in intermediate motor units. Fast glycolytic fibers are the last

to be recruited because they are found in the larger motor units. These fibers are not usually recruited unless a muscle is generating a large amount of force, as occurs in high-intensity exercises such as weight lifting or sprinting.

Resistance to Fatigue

It is not uncommon for competitive cyclists to ride all day, covering well over 100 miles in the process. By comparison, even Olympic-class weight lifters cannot lift at or near their maximum capacity for more than a few seconds. This difference is due to the fact that muscles differ in their ability to resist **fatigue**, a decline in a muscle's ability to maintain a constant force of contraction in the face of long-term, repetitive stimulation. Although fatigue eventually sets in after any kind of muscular activity, it generally occurs more quickly when a muscle is stimulated at higher frequencies and when larger forces are generated (**Figure 13.26**).

Although the precise causes of muscle fatigue are not fully understood, different types of exercise are known to induce fatigue for different reasons. In high-intensity exercise, glycolytic muscle fibers are recruited and, as previously mentioned, have a tendency to generate lactic acid because of their low oxidative capacity. As a consequence, a rapid buildup of lactic acid in muscles is common in high-intensity exercise. When contractions are strong and sustained, an additional factor can come into play: Strong contractions can compress vessels that supply blood to the muscles, which interrupts or reduces the muscles' blood supply; the resulting decrease in oxygen delivery causes muscle cells to produce larger amounts of lactic acid, which by lowering intracellular pH can alter enzyme activities and interfere in a variety of metabolic processes. In low-intensity exercise, by contrast, lactic acid accumulation is not generally a problem because few glycolytic fibers are recruited; the active fibers are mostly oxidative fibers, which have little tendency to produce lactic acid.

The cause of fatigue in low-intensity exercise, which takes a longer time to develop, is thought to be linked to the depletion of energy reserves, glycogen in particular. Full recovery from this type of fatigue requires about 24 hours, whereas recovery from the more rapid-onset type of fatigue requires only minutes or a few hours.

Very high intensity exercise can induce *neuromuscular fatigue*, which occurs when motor units are stimulated to contract at high frequencies, as occurs when large forces are generated. When cells are stimulated to contract strongly over long periods, repeated firing of motor neurons can deplete synaptic terminals of acetylcholine, which ultimately causes failure of neuromuscular transmission.

In addition to these mechanisms of fatigue, which affect either a muscle cell's ability to generate force or the nervous system's ability to trigger muscle contractions, fatigue also has a psychological component that resists any purely physiological explanation. It is widely accepted among athletes that performance is influenced by mental state as well as by physical condition. Often, the athletes who run the fastest or the longest are the ones with the strongest "will to win." Competitors with less desire are more likely to succumb to discouragement when fatigue sets in and muscles begin to ache; as a consequence, they may put forth less effort.

Long-Term Responses of Muscles to Exercise

Athletes go into training not simply to hone the skills needed in their particular sport but also to change the body's physical condition so that it is fitter to perform those skills. Such changes involve, among other things, changes in muscles' cellular architecture that result from regular exercise over time. As a result of these changes, the muscles' capacity to generate force and resist fatigue is altered.

As every athlete knows, different types of exercise affect muscles in different ways, and thus it is important to choose an exercise regimen that can achieve the desired results. For example, an aspiring marathon runner, who requires great endurance for her sport, should train using long-duration, low-intensity exercises ("aerobic" exercise) such as jogging. In contrast, an aspiring boxer, who desires bigger and stronger muscles, would do well to include short-duration, high-intensity exercise in his regimen.

Aerobic exercise increases the oxidative capacity of muscle fibers, thereby increasing their resistance to fatigue. As a result of such training, some fast glycolytic fibers are effectively converted to fast oxidative fibers. (However, because exercise does not alter the type of myosin present in muscle fibers, slow-twitch fibers remain slow and fast-twitch fibers remain fast.) Changes include increases in the size and number of mitochondria within the fibers, and an increase in the number of capillaries surrounding the fibers. In addition, the average diameter of the fibers decreases, which facilitates the movement of oxygen into the cells but also decreases the cells' force-generating capacity.

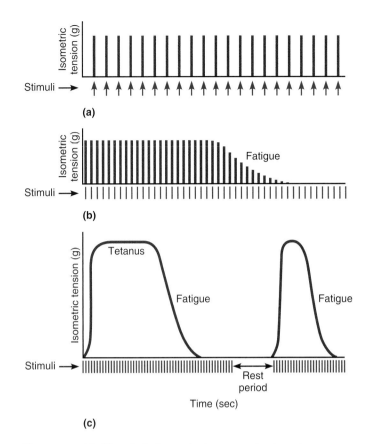

(a)

(b)

(c)

Figure 13.26 Muscle fatigue. In these records of tension developed by a muscle in response to repetitive stimuli (arrows) the time scale is compressed, so individual twitches appear as spikes. **(a)** Stimulation at a relatively low frequency produces little, if any, fatigue. Note that peak tension is constant over time. **(b)** Stimulation at a higher frequency produces fatigue, apparent in the decline in peak tension near the end of the record. **(c)** A muscle stimulated at a frequency high enough to yield maximum tetanic tension fatigues even more rapidly. After a recovery period, tension returns to previous levels upon resumption of stimulation, but fatigue sets in more rapidly than before.

In contrast, high-intensity exercise decreases the oxidative capacity of muscle fibers and increases their glycolytic capacity, thereby converting a portion of the fast oxidative fibers into fast glycolytic fibers. Changes include decreases in the size and number of mitochondria, increases in the concentration of glycolytic enzymes, and increases in average fiber diameter. However, the declining oxidative capacity of these fibers reduces their resistance to fatigue. As fibers grow, new myofibrils are synthesized, which enables the fibers to generate more force; reflecting the increase in average fiber diameter, entire muscles also become bulkier and more massive. Note that muscle growth is not due to the addition of new fibers, because muscle fibers are *postmitotic*—that is, they cannot divide to form new cells. Even though new fibers can be generated from immature precursors called *satellite cells*, this normally happens only when fibers die and must be replaced.

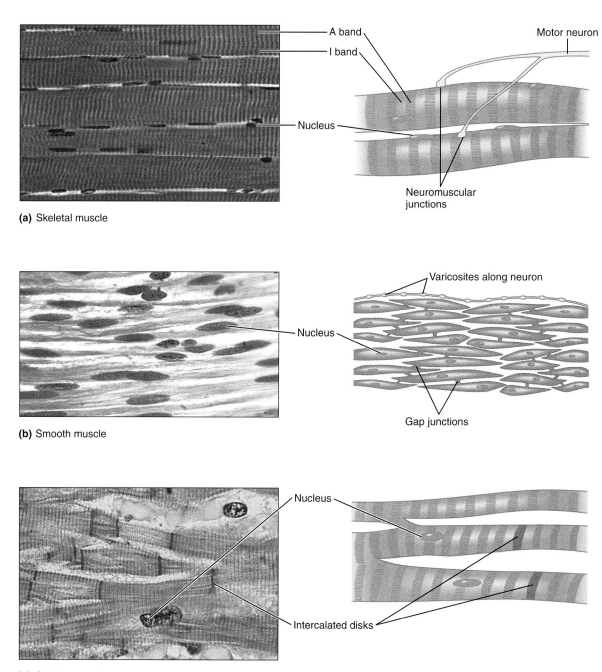

(a) Skeletal muscle

(b) Smooth muscle

(c) Cardiac muscle

Figure 13.27 Photomicrographs (left) and schematic diagrams (right) of the three types of muscle in the body.

The neurons that control skeletal muscles and those that control smooth muscle belong to different branches of the efferent nervous system. What are these branches called?

The somatic and autonomic nervous systems, respectively

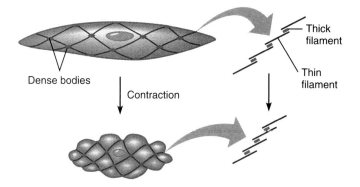

Figure 13.28 Oblique arrangement of thick filaments and thin filaments in a smooth muscle cell. The relative positions of the filaments are shown for a relaxed cell (top) and a contracting cell (bottom).

Smooth and Cardiac Muscle

When people hear the word *muscle,* most automatically think of *skeletal muscle,* which has been our sole focus so far in this chapter. However, two other types of muscle in the body—*smooth muscle* and *cardiac muscle*—perform their functions without attracting much notice. We now shift our focus to these muscles, and to the special properties that make them different from skeletal muscle. For easy comparison, the three types of muscle are shown in **Figure 13.27**. The specific features of smooth muscle and cardiac muscle that are apparent in this diagram will be discussed in the next two sections.

Smooth Muscle

Smooth muscle, which gets its name from the fact that it lacks the striations characteristic of skeletal and cardiac muscle and therefore appears uniformly bright under the light microscope (see Figure 13.27), is the type of muscle found in internal organs, blood vessels, and other structures that are not under voluntary control. Functions performed by these muscles are many and varied and depend on the organ in which they are located. In the gastrointestinal tract, for example, smooth muscle contractions mix the ingested food with digestive secretions and propel it from one location to another. In blood vessels, smooth muscle regulates blood flow to organs and tissues by causing the vessels to constrict or dilate.

Like skeletal muscle, smooth muscle has thick and thin filaments and generates force through the crossbridge cycle. However, the filaments are not arranged in sarcomeres, which accounts for the lack of striations. Even though thick and thin filaments are arranged in parallel with each other, as in skeletal muscle, they tend to run obliquely in various directions, which means that contraction occurs along several axes (**Figure 13.28**). *Dense bodies,* points of attachment between these filaments and connective tissue inside the cells, serve to transmit contractile force to the cell's exterior.

The Mechanism of Excitation-Contraction Coupling

Smooth muscle contractions are regulated by intracellular calcium, but the sarcoplasmic reticulum is not as extensive as in skeletal muscle. Moreover, much of the calcium that triggers contractions comes from outside the cells, because when the cell is depolarized, voltage-gated calcium channels in the plasma membrane open and allow calcium to flow in. (Depolarization also triggers the release of calcium from the sarcoplasmic reticulum.)

Cytosolic calcium activates the crossbridge cycle in smooth muscle, but it does so differently than in skeletal muscle because contractions are not regulated by the troponin-tropomyosin system. Instead, contractions in smooth muscle are triggered when calcium binds reversibly to **calmodulin**, a cytosolic protein that regulates many processes in almost all cells of the body (see p. 150). This binding triggers a conformational change that enables the calcium-calmodulin complex to bind to and activate an enzyme called **myosin kinase** (or *myosin light-chain kinase*). The activated kinase then catalyzes the phosphorylation of myosin crossbridges, which activates them and initiates crossbridge activity. Crossbridge cycling then proceeds essentially as shown in Figure 13.7 on page 382. Note that in smooth muscle the calcium signal that triggers crossbridge activity targets *myosin* filaments, whereas in skeletal muscle it targets *actin* filaments because this is where troponin and tropomyosin are located.

Termination of the crossbridge cycle in smooth muscle requires more than just removal of calcium from the cytosol, because the phosphate groups attached to myosin are covalently bound and therefore do not dissociate readily. For this reason, termination of the crossbridge cycle in smooth muscle requires the action of phosphatase enzymes, which by removing phosphate groups inactivate myosin. Because these phosphatases (which are continually active) compete with myosin kinase, activation of myosin occurs only when enough calcium is present to activate myosin kinase to a degree sufficient to overcome the action of the phosphatases.

Because the mechanism of excitation-contraction coupling is entirely different from that in skeletal muscle, it takes a lot longer to initiate and also to terminate smooth muscle contractions. This is no real handicap because

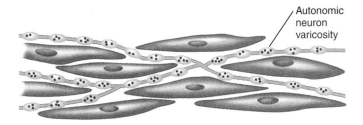

(a) Multi-unit smooth muscle

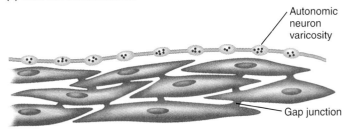

(b) Single-unit smooth muscle

Figure 13.29 Multi-unit and single-unit smooth muscle.

smooth muscle isn't exactly "built for speed" anyway; myosin ATPase activity in smooth muscle is anywhere from 10 to 100 times lower than in skeletal muscle, which means that smooth muscle contraction is an inherently slow process.

Neural Regulation of Contraction

Unlike skeletal muscle, which is regulated by motor neurons, smooth muscle is regulated by *autonomic neurons.* A given smooth muscle cell may be regulated by sympathetic neurons, parasympathetic neurons, or frequently both. Whereas motor neurons always exert excitatory effects on skeletal muscle cells, the effect of autonomic input to smooth muscle cells may be excitatory or inhibitory, depending on whether the neurons in question are sympathetic or parasympathetic; these two types of neuron almost always affect a given smooth muscle cell in opposite ways. For instance, intestinal smooth muscle contracts in response to parasympathetic input and relaxes in response to sympathetic input. Furthermore, whether a muscle cell contracts or relaxes depends on where it is located. Thus, for example, sympathetic input induces relaxation of smooth muscle in the intestine but contraction of smooth muscle in most blood vessels. These opposite responses are due to differences in the type of neurotransmitter receptors found on the muscle cells, not to differences in the type of neurotransmitter released, because sympathetic or parasympathetic neurons release the same neurotransmitters (norepinephrine and acetylcholine, respectively) at virtually all their target tissues.

Another difference between skeletal and smooth muscle concerns the specificity of neural connections. In skeletal muscle, neural input is delivered to each cell individually because motor neurons are connected to specific cells via neuromuscular junctions. In contrast, smooth muscle cells do not receive neural input by "personal telegram" but instead by "bulk mailing," because an autonomic neuron does not make synaptic connections to specific cells. Instead, neurotransmitter is released from varicosities (swellings) located at intervals along the axon and diffuses over a relatively long distance to large groups of cells (see Figure 13.27b). Consequently, neighboring muscle cells tend to contract or relax together. This synchronized activity is also promoted by the presence in most smooth muscle tissue of *gap junctions* that allow ions (and other small molecules) to move from one cell to another, so an electrical signal initiated in one cell spreads to neighboring cells.

Skeletal and smooth muscles also differ in regard to the electrical signal generated in response to neural input. Whereas skeletal muscle cells always respond to neural input with an action potential that triggers a reproducible twitch, in smooth muscle cells slow but twitchlike contractions may be elicited in response to action potentials, but this is not necessarily so. The membrane potential of most smooth muscle cells varies in a graded fashion in response to neural input, depolarizing if the input is excitatory and hyperpolarizing if it is inhibitory, which causes the contractile force to increase or decrease in a graded fashion. Thus contractions in smooth muscle need not be elicited by action potentials. In fact, some smooth muscle cells do not have action potentials at all. When action potentials do occur, they are not usually followed by real twitches (that is, individual contractions, one in response to each action potential), just greater tension.

Some types of smooth muscle cells are able to actively exert tension even in the absence of external stimulation, because resting calcium levels are high enough to maintain a constant low level of crossbridge activity. This resting tension is referred to as *tone.* (Skeletal muscles also exhibit some degree of basal tone, but in their case it is due to a constant low level of neural stimulation.) Many smooth muscle cells have the ability to contract in response to hormones and other chemical agents, independent of neural input. Furthermore, some smooth muscle cells exert active tension in response to mechanical stretch.

Single-Unit and Multi-Unit Smooth Muscle Smooth muscle tissue varies in both the degree to which muscle cells are connected by gap junctions and the pattern of innervation. In some places, most smooth muscle cells are not connected by gap junctions but instead are largely separate and richly supplied with neurons; smooth muscle of this type is referred to as **multi-unit smooth muscle** (**Figure 13.29**a). Multi-unit smooth muscle occurs in the large respiratory airways and large arteries, where the number of active smooth muscle cells may be large or small depending on the circumstances. In other locations, smooth muscle cells are extensively linked by gap junctions, such that electrical signals originating in a few cells are transmitted to the rest of the cells; such smooth muscle is innervated by relatively few neurons and is referred to as **single-unit smooth muscle** (Figure 13.29b). Example of organs containing single-unit smooth muscle are the gastrointestinal tract and the uterus, in which large groups of cells contract synchronously.

Pacemaker Activity   In some cases, smooth muscle cells exhibit spontaneous depolarizations that occur on a regular basis and that may or may not be accompanied by action potentials. These depolarizations are called **pacemaker potentials**, and the cells that generate them are called **pacemakers**. In single-unit smooth muscle, pacemaker activity causes cells to contract and relax in unison. Although the frequency or amplitude of electrical signals in pacemakers can be influenced by neural activity, pacemaker signals occur even in the absence of any neural influence.

Cardiac Muscle

Cardiac muscle is similar to skeletal muscle in that it is striated (see Figure 13.27c), has the same sarcomere structure, and has contractions that are regulated by the troponin-tropomyosin system. Cardiac muscle cells are similar to smooth muscle cells in that they are extensively connected by gap junctions, such that an action potential, once initiated, travels throughout the entire cell network.

Cardiac action potentials are broad and last for hundreds of milliseconds, making them quite different from the spiky action potentials that occur in skeletal muscle and most neurons and last for 1–2 milliseconds. The relatively long duration of cardiac action potentials is significant: Because they last nearly as long as it takes for cardiac muscle cells to contract and relax (**Figure 13.30**), summation of cardiac muscle contractions cannot occur, even when the action potential frequency is high and the heart is beating rapidly. This is good, because summation would be detrimental to the heart's pumping action in that the heart would not be able to relax completely and fill with blood between contractions.

Certain heart muscle cells that are concentrated in two regions known as the *sinoatrial* and *atrioventricular nodes* exhibit pacemaker activity. The heartbeat is triggered by action potentials originating in pacemaker cells and does not depend on neural input; this is why a heart removed from the body continues to beat, even though all neural connections to the heart have been severed. (Because the signals that trigger the heartbeat originate within the heart muscle itself, the heart's contractile activity is said to be *myogenic,* whereas the contractile activity of skeletal muscle is said to be *neurogenic.*) However, the autonomic nervous system does regulate the heart muscle by modulating the frequency and force of heart muscle contractions. Properties of cardiac muscle and their importance to the heart's function are discussed further in Chapter 14. **Table 13.2** compares the properties of skeletal, smooth, and cardiac muscle.

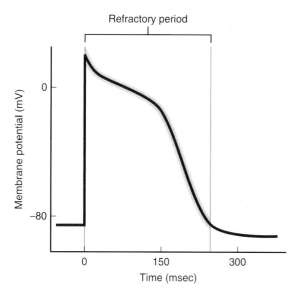

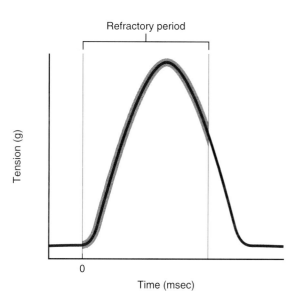

Figure 13.30 Relative durations of contractions and action potentials in cardiac muscle. The common time scale for the two graphs allows comparison of an action potential (above) with the tension developed in response to it (below) in a cardiac muscle cell.

Quick Test 13.6

1. How does smooth muscle differ from skeletal muscle with respect to the arrangement of thick and thin filaments? With respect to the mechanism of excitation-contraction coupling?

2. What is the difference between multi-unit smooth muscle and single-unit smooth muscle?

3. What does a pacemaker cell do, and in what types of muscle are they found?

4. What prevents summation of cardiac muscle contractions, and what is the significance of this phenomenon?

Table 13.2 | Comparison of Skeletal, Smooth, and Cardiac Muscle

Property	Skeletal	Smooth	Cardiac
Striations (sarcomeres)	Yes	No	Yes
Actin and myosin	Yes	Yes	Yes
Level of control	Voluntary	Involuntary	Involuntary
Neural input	Somatic	Autonomic	Autonomic
Neuroeffector junction	Neuromuscular junction–specific	Varicosities–diffuse	Varicosities–diffuse
Hormonal control	None	Several depending on location	Epinephrine
Source of calcium	SR	SR and ECF	SR and ECF
Regulatory protein that binds calcium	Troponin	Calmodulin	Troponin
Gap junctions	No	Yes (single unit)	Yes
Pacemaker activity	No	Yes (single unit)	Yes
Myosin ATPase activity	Fastest	Slowest	Intermediate
Recruitment	Yes	Yes (multi-unit)	No

▮ CHAPTER SUMMARY

Skeletal Muscle Structure, p. 376

Most skeletal muscles are connected to bones by tendons and contain numerous elongated cells (muscle fibers) that generate contractile force using energy from ATP hydrolysis. Within muscle fibers are myofibrils that contain the contractile machinery. The sarcoplasmic reticulum surrounds the myofibrils, stores calcium ions, and is closely associated with transverse (T) tubules, which penetrate into the cell interior from the sarcolemma. Skeletal and cardiac muscle are striated, reflecting the orderly arrangement of thick and thin filaments in the myofibrils, which are made up of fundamental force-generating units (sarcomeres) joined end to end. Thick and thin filaments contain the contractile proteins myosin and actin, respectively. The heads (crossbridges) of myosin molecules are responsible for generating the motion that drives contraction and possess two important sites: an actin-binding site and an ATPase site. Two regulatory proteins (troponin and tropomyosin) present on the thin filaments serve to initiate and terminate contractions.

[IP] Muscular, Anatomy Review: Skeletal Muscle Tissue, pp. 4–11

[IP] Muscular, Sliding Filament Theory, pp. 3–15

The Mechanism of Force Generation in Muscle, p. 380

When a muscle contracts, thick and thin filaments slide past one another. This sliding is driven by the crossbridge cycle, in which the motion of crossbridges is coupled to their cyclic binding and unbinding to actin molecules in adjacent thin filaments. In skeletal muscle, each fiber receives input from one motor neuron, which branches and innervates more than one fiber. An action potential in a motor neuron triggers the release of acetylcholine, which binds to receptors in the muscle fiber's motor end plate. The result is an electrical signal (end-plate potential) that triggers an action potential in the sarcolemma. This is followed by propagation of the action potential through the T tubules, release of Ca^{2+} from the sarcoplasmic reticulum, binding of Ca^{2+} to troponin, movement of tropomyosin away from actin's myosin-binding sites, and initiation of the crossbridge cycle.

The Mechanics of Skeletal Muscle Contraction, p. 388

A motor neuron plus the muscle fibers it innervates constitutes a motor unit. When a motor neuron fires an action potential, all fibers in the motor unit contract together. The mechanical response of a motor unit to a single action potential is a twitch, which is reproducible in size. Twitches can be isometric, in which case the muscle generates force but does not shorten, or isotonic, in which case the muscle shortens. The force generated by an entire muscle is determined by both the force generated by individual fibers (which depends on the frequency of stimulation, fiber diameter, and changes in fiber length) and the number of fibers that are active. Stimulation at high frequencies causes summation of twitches, such that the force eventually reaches a plateau (tetanus).

SYSTEMS INTEGRATION Muscles

Urinary System

Smooth muscle in afferent and efferent arterioles regulates glomerular filtration

Smooth muscle in wall of urinary bladder contracts during micturition to drive urine flow

Smooth and skeletal muscle form sphincters that regulate micturition

Gastrointestinal System

Smooth muscle in the wall of the gastrointestinal tract mixes and propels chyme

Muscular sphincters regulate the flow of chyme at various places in the gastrointestinal tract

Skeletal muscles are used in chewing and swallowing

Respiratory System

Action of diaphragm and other skeletal muscles works to expand lungs during inspiration

Smooth muscles in small airways and pulmonary blood vessels regulate ventilation and perfusion of alveoli

Muscles

Reproductive System

Smooth muscle contractions propel sperm or eggs through the reproductive tract

Smooth muscle contractions in the uterus propel the fetus through the birth canal during parturition

Endocrine System

Contraction of cardiac muscle drives blood flow, which delivers hormones to target tissues

Immune System

Smooth muscle in the walls of larger lymphatic ducts propels lymph flow

Skeletal muscle contractions aid in the flow of lymph

Cardiovascular System

Cardiac muscle generates arterial pressure, which drives blood flow throughout the body

Vascular smooth muscle regulates resistance of blood vessels, which is important in the control of blood pressure

Skeletal muscle contractions aid in the flow of blood through the venous system back to the heart

Nervous System

Smooth muscle in cerebral vasculature adjusts the distribution of blood flow to different regions, according to changes in brain activity

Muscles in the eyes and ears help adapt them to changing conditions

The central nervous system regulates muscular force by varying both the action potential frequency in motor neurons and the number of active motor units (recruitment). As muscular force increases, motor units are recruited in order of increasing size, a phenomenon referred to as the size principle.

IP Muscular, Sliding Filament Theory, pp. 17–19

IP Muscular, Contraction of Whole Muscle, pp. 1–16

IP Muscular, The Neuromuscular Junction, pp. 1–16

Types of Skeletal Muscle Fibers, p. 400

Skeletal muscles contain different types of fibers in various proportions.

Fast-twitch fibers and slow-twitch fibers differ in their speed of contraction, which is related to the type of myosin they contain. Glycolytic fibers synthesize most of their ATP via glycolysis and generate lactic acid, which makes them fatigue rapidly. Oxidative fibers synthesize ATP mostly via oxidative phosphorylation and are more resistant to fatigue.

IP Muscular, Muscle Metabolism, pp. 3–23

Smooth and Cardiac Muscle, p. 405

Smooth muscle is found in internal organs and other structures that are not under voluntary control and is regulated by autonomic neurons.

Contractions are triggered by the binding of Ca^{2+} to calmodulin, which activates myosin kinase, resulting in the phosphorylation of myosin cross-bridges.

In cardiac muscle, contractions are triggered by action potentials initiated in pacemaker cells. Action potentials travel from cell to cell through gap junctions, so the entire network of cells contracts as a unit.

IP Muscular, Anatomy Review: Skeletal Muscle Tissue, p. 3

IP Cardiovascular, Anatomy Review: The Heart, p. 6–7

EXERCISES

Multiple-Choice Questions

1. When a muscle cell is relaxed and intracellular ATP levels are normal, a crossbridge will remain in which of the following states?
   a) bound to actin and in the low-energy form
   b) bound to actin and in the high-energy form
   c) in the high-energy form, with ADP and P_i bound to it
   d) in the high-energy form, with ATP bound to it
   e) in the low-energy form with nothing bound to it

2. During a muscle contraction, which of the following does *not* change length?
   a) the distance between Z lines
   b) the width of I bands
   c) the width of A bands
   d) none of the above

3. Which of the following would tend to *reduce* the concentration of lactic acid that accumulates in a muscle cell as a result of contractile activity?
   a) increasing the concentration of glycolytic enzymes
   b) decreasing the oxygen supply to the cell
   c) increasing the diameter of the cell
   d) increasing the number of mitochondria in the cell
   e) all of the above

4. Which of the following statements is a valid generalization regarding the properties of smooth muscle?
   a) Neurotransmitters can either excite or inhibit smooth muscle contraction, but any given neurotransmitter is always excitatory or inhibitory, regardless of where the muscle is located.
   b) A given smooth muscle cell can respond to more than one type of neurotransmitter.
   c) Smooth muscle cells are generally unresponsive to neurotransmitters of all types.
   d) Smooth muscle cells can respond to neural input from the somatic or autonomic nervous systems.
   e) none of the above

5. Which of the following is *not* a determinant of whole muscle tension?
   a) the number of muscle fibers contracting
   b) the tension produced by each contracting fiber
   c) the proportion of each motor unit that is contracting at any given time
   d) the extent of fatigue
   e) the frequency of action potentials in the motor neurons

6. In an isotonic contraction,
   a) muscle length shortens.
   b) muscle tension exceeds the force of the load.
   c) the load is moved.

d) a and c
   e) all of the above

7. Which of the following is true for the excitation-contraction coupling of *all* muscle types (skeletal, cardiac, and smooth)?
   a) An action potential causes calcium levels in the cytosol to increase.
   b) Calcium binds to troponin.
   c) Thick and thin filaments slide past each other.
   d) a and c
   e) all of the above

8. During contraction of a skeletal muscle fiber,
   a) the thick filaments contract.
   b) the thin filaments contract.
   c) the A band becomes shorter.
   d) the I band becomes shorter.
   e) all of the above

9. Which of the following statements concerning the characteristics of different types of muscle fibers is *false*?
   a) The higher the myosin ATPase activity, the faster the speed of contraction.
   b) Muscles that have high glycolytic capacity and large glycogen stores are more resistant to fatigue.
   c) Oxidative types of muscle fibers contain myoglobin.
   d) Oxidative fibers have a richer blood supply.
   e) Larger diameter fibers can produce greater tension.

10. Which of the following muscle types contain gap junctions?
    a) skeletal muscle
    b) smooth muscle
    c) cardiac muscle
    d) a and b
    e) b and c

Objective Questions

1. In skeletal muscle, when calcium is released from the sarcoplasmic reticulum it binds to (troponin/tropomyosin) to initiate the crossbridge cycle.

2. When a muscle fiber contracts, the I bands shorten. (true/false)

3. Glycolytic fibers generate more force than oxidative fibers because they are larger in diameter. (true/false)

4. (Glycolytic/Oxidative) fibers contain high concentrations of the oxygen-binding protein myoglobin.

5. The plasma membrane of a muscle cell is also known as the _____.

6. During muscle contraction, ATP hydrolysis is catalyzed by (myosin head groups/actin monomers).

7. During an (isometric/isotonic) muscle contraction, a muscle develops contractile force but does not change in length.

8. The velocity of contraction of a muscle fiber is directly related to its (diameter/myosin ATPase activity).

9. A reduction in the number of active crossbridges is responsible for a decrease in force-generating capacity of a muscle fiber that is significantly (longer/shorter) than its optimum length.

10. (Oxidative/Glycolytic) muscle fibers are more resistant to fatigue.

11. According to the size principle, the force-generating capacity of a muscle fiber increases in direct proportion to its length. (true/false)

Essay Questions

1. Compare and contrast mechanisms of excitation-contraction coupling in striated muscle and smooth muscle. Be sure to include a description of the mechanisms responsible for termination of muscle contractions.

2. Compare and contrast skeletal muscle and smooth muscle with respect to regulation of contractile activity by the nervous system.

3. Describe the relationship between contractile force and action potential frequency in skeletal muscle, and explain how the situation differs in cardiac muscle.

4. Explain the size principle and how it relates to the amount of tension developed by a skeletal muscle.

5. Discuss how oxidative muscle fibers differ from glycolytic fibers, and explain how each of these factors relates to the ability of a fiber to resist fatigue.

6. Describe the role of creatine kinase in muscle cell metabolism.

Critical Thinking

1. A physiologist measures peak twitch tension for a series of muscle contractions performed with different loads. If peak tension is plotted as a function of load, the results should look something like

2. Following the contraction phase of a muscle twitch, cytoplasmic calcium concentration declines, and calcium ions dissociate from troponin. When the calcium concentration returns to its resting level, ATP hydrolysis ceases because
   a) ATP concentration goes to zero.
   b) low calcium levels cause the energy of ATP hydrolysis to become positive, favoring ATP synthesis.
   c) ATP binding sites on myosin molecules lose their ability to catalyze ATP hydrolysis.
   d) the number of available (unoccupied) myosin ATPase sites declines to virtually zero.

3. Two muscle fibers (X and Y) generate the same isometric peak twitch tension, but fiber X contains myosin that hydrolyzes ATP at a high rate, whereas fiber Y has myosin that hydrolyzes ATP more slowly. Properties of these fibers are consistent with which of the following sets of curves?

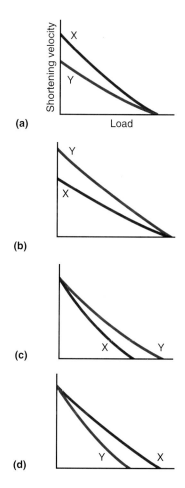

4. Suppose you have a beaker containing a suspension of purified myosin molecules and ATP in a saline solution. Based on your understanding of events occurring during the crossbridge cycle,
   a) ATP hydrolysis should occur.
   b) ATP hydrolysis will occur if actin is added to the solution.
   c) ATP hydrolysis will occur if calcium is added to the solution.
   d) ATP hydrolysis will occur if calcium and troponin are added to the solution.

5. During aerobic activities, glucose and fatty acids are delivered to skeletal muscle cells by the bloodstream. Based on what you learned in Chapter 3, explain the sources of glucose and fatty acids for contracting muscle and describe the metabolic pathways by which they provide energy.

6. Glycolysis under anaerobic conditions provides approximately 5% of the energy that could be produced under aerobic conditions. Thus, one might assume that anaerobic exercises (such as weight training) would be a better way to lose weight than aerobic exercise. Explain why this is not true.

Find the answers to these exercises, and additional study tools, at the Physiology Place (www.physiologyplace.com).

The Cardiovascular System: Cardiac Function

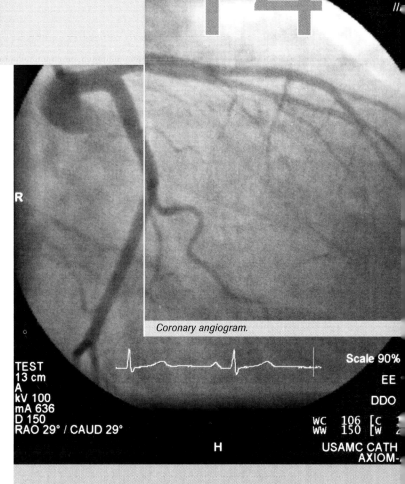

14

USAMC CATH
AXIOM-
VA20F 0205

Coronary angiogram.

TEST
13 cm
A
kV 100
mA 636
D 150
RAO 29° / CAUD 29°

Scale 90%

EE

DDO

WC 106 [C
WW 150 [W

H

USAMC CATH
AXIOM-

The photo on this page is of an angiogram displaying an electrocardiogram (ECG) trace of a healthy heart. You have likely seen an ECG in the movies or on television hospital shows, with the ECG trace fluctuating violently when a patient is in trouble, or "flatlining" when the patient's heart stops beating! The trace is a recording of the electrical activity of a heart. In this chapter, you will learn what the waves in the trace represent and how the electrical activity of the heart drives the rhythmic contraction that efficiently propels blood throughout the body. You will understand why the circulation of blood is so critical, and why any interruption of that blood flow is potentially life-threatening. If blood flow were to suddenly stop, you would experience no more than a few seconds of consciousness before blacking out—and then you would have but a few minutes to live. Clearly, the heart must perform its functions continuously and nearly flawlessly for every minute of every day that you live.

This chapter focuses on the inner workings of the heart—a wondrous muscle that beats approximately 3 billion times in an average lifetime.

In Chapter 13, we learned about muscle and how it contracts, ending with a discussion of cardiac muscle. We now expand on cardiac muscle and the organ in which it is found, the heart. The function of the heart is to pump blood into the blood vessels. The blood vessels contain another type of muscle we learned about in Chapter 13, smooth muscle. Both cardiac and smooth muscle are under control by the autonomic nervous system, described in Chapter 12. In this chapter and the next, you will learn about how the autonomic nervous system regulates cardiovascular function by controlling cardiac and smooth muscle.

An Overview of the Cardiovascular System

The ability of cells to exchange materials with their immediate environment is an absolute requirement of life. When exchanging materials with interstitial fluid, each cell relies on diffusion to bring needed materials, such as oxygen and nutrients, to it and to carry unwanted materials, such as carbon dioxide and other wastes, away. Because these materials ultimately come from or go into the *external* environment, which is at a considerable distance from most body cells, diffusion alone cannot provide all cells with what they need as quickly as they need it; as a mechanism of transport, it is too slow. Providing much more efficient transport is the purpose of the cardiovascular system, the subject of this and the next two chapters.

The **cardiovascular system** consists of three components: (1) the **heart**—a muscular pump that drives the flow of blood through blood vessels; (2) **blood vessels**—conduits through which the blood flows; and (3) **blood**—a fluid that circulates around the body, carrying materials to and from the cells.

At first glance, the function of the cardiovascular system is simple: The heart pumps blood through blood vessels to various organs, and the blood carries oxygen and nutrients to tissues and removes carbon dioxide and other wastes. However, the heart is more than a pump because it performs sensory and endocrine functions that help regulate cardiovascular variables such as blood volume and pressure. The blood vessels are not just conduits for blood but are also important sensory and effector organs that regulate blood pressure and the distribution of blood to various parts of the body. The blood not only carries nutrients and wastes but also transports hormones from one part of the body to another and thus serves as a communications link acting in conjunction with the nervous system. As we will see, it is not possible to fully understand how one part of the system works—the heart, for instance—without understanding how the rest of the system works. Furthermore, regulation of the cardiovascular system involves interactions with several other organ systems, including the nervous system, the endocrine system, and the kidneys.

Before we examine the workings of the heart, the central focus of this chapter, we discuss the major components of the cardiovascular system and their functions.

The Heart

The heart is a muscular organ whose function is to generate the force that propels blood through the blood vessels. The heart contains four chambers: two upper chambers, called **atria** (singular: *atrium*) that receive the blood that comes back to the heart from the vasculature, and two lower chambers, called **ventricles**, that receive blood from the atria and generate the force that pushes the blood away from the heart and through the blood vessels (**Figure 14.1**).

The heart can be functionally separated into left and right halves: the atrium and ventricle on the left side of the heart constitute the *left heart;* the atrium and ventricle on the right side constitute the *right heart.* The atria and ventricles on either side of the heart are separated by a

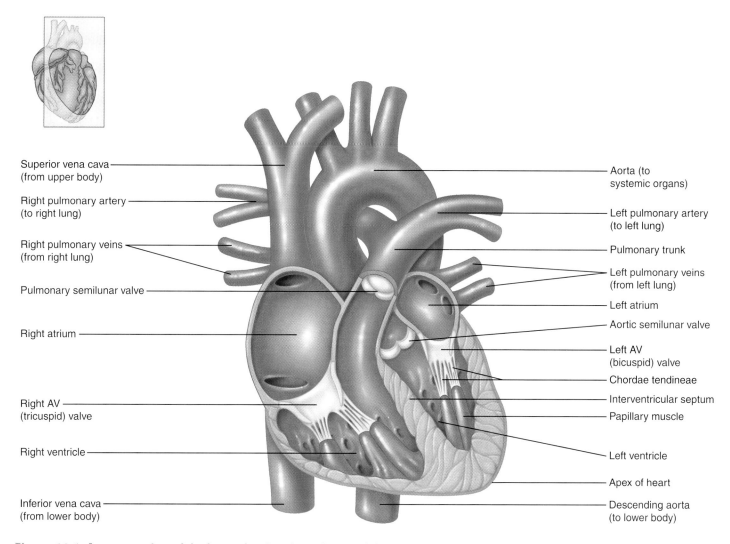

Superior vena cava
(from upper body)

Right pulmonary artery
(to right lung)

Right pulmonary veins
(from right lung)

Pulmonary semilunar valve

Right atrium

Right AV
(tricuspid) valve

Right ventricle

Inferior vena cava
(from lower body)

Aorta (to
systemic organs)

Left pulmonary artery
(to left lung)

Pulmonary trunk

Left pulmonary veins
(from left lung)

Left atrium

Aortic semilunar valve

Left AV
(bicuspid) valve

Chordae tendineae

Interventricular septum

Papillary muscle

Left ventricle

Apex of heart

Descending aorta
(to lower body)

Figure 14.1 **A cutaway view of the heart showing the atria, ventricles, atrioventricular valves, and connections to major blood vessels.**

wall called the *septum* that prevents blood in the left heart from mixing with blood in the right heart. The portion separating the left and right atria is referred to as the *interatrial septum;* the portion separating the left and right ventricles is the *interventricular septum.* Just as the heart has left and right sides, it also has a "top" and a "bottom." The wider upper pole (end) of the heart is known as the *base;* the narrower lower pole is the *apex.* In this chapter, you will learn about the anatomy of the heart and how it works efficiently as two pumps to supply all organs with the nutrients and oxygen they need.

Blood Vessels

When blood moves through the body, it travels in a circular pattern through a system of blood vessels that carry it from the heart to the various organs and then back to the heart again. This system of blood vessels is generally referred to as the **vasculature**. As blood flows away from the heart, blood vessels branch repeatedly,

becoming more numerous and smaller in diameter, just as the limbs of a tree become smaller and more numerous as you move from the trunk to the outer branches. As blood flows back to the heart, the vessels conveying it converge, becoming less numerous and larger in diameter, just as the limbs of a tree do as one moves from the outer branches to the trunk. When blood leaves the heart, it is transported to the body's organs and tissues in relatively large vessels called **arteries**, which branch repeatedly within the organs and tissues. The smallest arteries branch into still smaller vessels called **arterioles**, which carry blood to the smallest vessels, called **capillaries**. From the capillaries, blood moves to larger vessels called **venules**, which lead to still larger vessels called **veins**, which carry blood back to the heart. Because blood moves from the heart into the vasculature and then from the vasculature back to the heart, the cardiovascular system is a *closed* system. The structure and function of blood vessels are discussed further in Chapter 15.

Blood

Although blood is a fluid, nearly half its volume is composed of cells. The most numerous cells are **erythrocytes**, also known as *red blood cells.* These cells contain *hemoglobin,* a protein that carries oxygen and carbon dioxide. The presence of hemoglobin in these cells gives them their characteristic red color. The remainder of the cells are **leukocytes**, or *white blood cells,* which come in a variety of types and help the body defend itself against invading microorganisms. Also present are **platelets**, which are not cells but instead cell fragments that play an important role in blood clotting. The liquid portion of the blood, called **plasma**, is made up of water containing dissolved proteins, electrolytes, and other solutes. The composition of blood is discussed in greater detail in Chapter 16.

Quick Test 14.1

1. What is the liquid portion of the blood called? What two types of cells are found in the blood?

2. What are the five types of blood vessels found in the vasculature?

3. What are the "receiving chambers" of the heart called? The "pumping chambers"?

The Path of Blood Flow Through the Heart and Vasculature

Blood follows an essentially circular path as it travels through the body. Although the sheer number of blood vessels makes the structure of the cardiovascular system complex, the layout of the system is simple in concept. In this section we look at the path of blood flow through the cardiovascular system.

Series Flow Through the Cardiovascular System

The general pattern of blood flow through the cardiovascular system is shown in **Figure 14.2**. In the diagram you can see that the circulatory system consists of two divisions: the **pulmonary circuit**, which consists of all blood vessels within the lungs and also those connecting the lungs with the heart, and the **systemic circuit**, which encompasses the rest of the blood vessels in the body. Note that these two divisions are supplied with blood by different sides of the heart. The *right* heart supplies blood to the pulmonary circuit, whereas the *left* heart supplies blood to the systemic circuit. Notice that blood on one side of the heart never mixes with blood on the other side. Thus the heart is actually two separate pumps housed within a single organ.

The pulmonary and systemic circuits both possess dense networks of capillaries called *capillary beds,* where

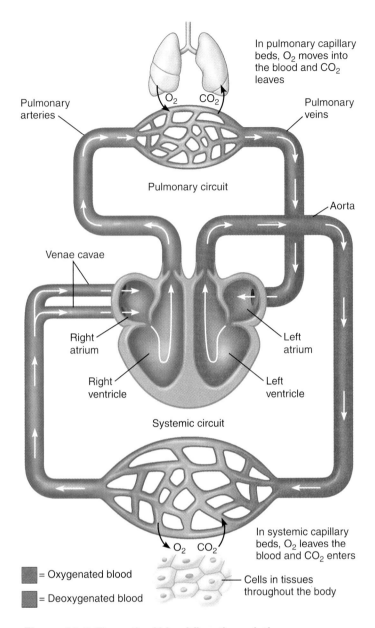

Figure 14.2 The path of blood flow through the cardiovascular system. The pulmonary and systemic circuits and major blood vessels connecting with the heart are shown. Arrows indicate direction of blood flow.

exchange of nutrients and gases (oxygen and carbon dioxide) takes place. In pulmonary capillaries, oxygen (O_2) moves into the blood from air in the lungs while carbon dioxide (CO_2) leaves the blood. When it leaves pulmonary capillaries, the blood is relatively rich in oxygen and is called *oxygenated* blood. Capillary beds of the systemic circuit are located in all organs and tissues except the lungs. In these organs and tissues, cells consume oxygen and generate carbon dioxide, so as blood travels through systemic capillaries, oxygen leaves the blood and carbon dioxide enters. Blood leaving these capillaries is called *deoxygenated* blood because it is relatively oxygen poor.

When oxygenated blood becomes deoxygenated or vice versa, its color actually changes. Oxygenated blood is bright

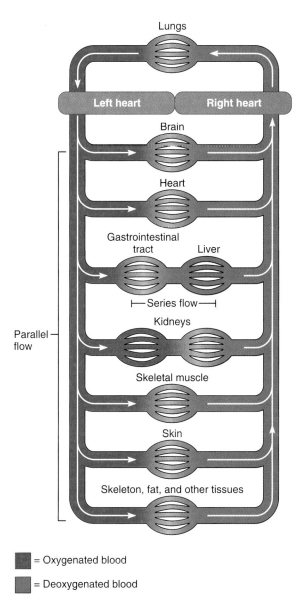

Figure 14.3 Blood flow patterns in the cardiovascular system. Parallel flow through organs in the system circuit.

= Oxygenated blood

= Deoxygenated blood

red, whereas deoxygenated blood is a darker red. Despite the fact that both forms of blood are red, the terms *red blood* and *blue blood* are frequently used to denote oxygenated and deoxygenated blood, respectively. Deoxygenated blood is called *blue* because it imparts a bluish color to veins visible beneath the skin. Oxygenated and deoxygenated blood are indicated by red and blue colors in Figure 14.2.

As blood flows through the cardiovascular system, it travels through the pulmonary and systemic circuits in an alternating fashion, returning to the heart each time (see Figure 14.2). Let us follow the path of blood flow step-by-step, starting in the left ventricle:

1. The left ventricle pumps oxygenated blood into the **aorta**, a major artery whose branches carry blood to capillary beds of all organs and tissues in the systemic circuit.

2. Blood becomes deoxygenated in systemic tissues and then travels back to the heart in the **venae cavae** (singular: *vena cava*), two large veins that carry blood into the right atrium. The *superior* vena cava carries blood from parts of the body above the diaphragm, whereas the *inferior* vena cava carries blood from parts below the diaphragm.

3. From the right atrium, blood passes through the tricuspid valve into the right ventricle.

4. The right ventricle pumps blood into the pulmonary trunk, which almost immediately branches into the **pulmonary arteries**, which carry deoxygenated blood to the lungs. Note that the pulmonary arteries are the only arteries in the body carrying deoxygenated blood. They are called *arteries* because they carry blood *away* from the heart.

5. Blood becomes oxygenated in the lungs and then travels to the left atrium in the **pulmonary veins**. These are the only veins in the body carrying oxygenated blood and are called *veins* because they carry blood *toward* the heart.

6. From the left atrium, blood passes through the bicuspid valve into the left ventricle, which is where we started. The whole cycle then repeats.

The above path describes how blood flows through the pulmonary and systemic circuits *in series* with each other; that is, for the cardiovascular system as a whole, blood must pass through the two circuits in sequence before it can return to the starting point. However, it is important to realize that blood flows through both circuits simultaneously; that is, the right heart is pumping blood to the lungs at the same time that the left heart is pumping blood to the systemic organs. If we look at blood flow within either the systemic or pulmonary circuit, we see a different pattern—*parallel flow.*

> **Quick Test 14.2**
>
> 1. Which organs are supplied by blood flowing through the *pulmonary* circuit? The *systemic* circuit?
>
> 2. Arrange the order of the following terms so that they correctly describe the path of blood flow through the body: *left ventricle, pulmonary arteries, bicuspid valve, aortic semilunar valve, pulmonary semilunar valve, right ventricle, aorta, pulmonary veins, vena cava, tricuspid valve, right atrium, left atrium.*

Parallel Flow Within the Systemic or Pulmonary Circuit

Figure 14.3 shows why blood flow in the systemic circuit is called *parallel flow* (and why the systemic organs are said to be *in parallel* with one another). In the

CORONARY ARTERIES

Although the heart lumen is full of blood, this blood is too distant from most heart cells to provide them with the oxygen and nutrients they need. Therefore, blood vessels must deliver blood to the heart, as they do to other organs. The blood vessels that do this compose the *coronary circulation.*

Two coronary arteries, called the left and right coronary arteries, branch off the base of the aorta as shown in figure (a).

The *left coronary artery* branches into the *left anterior descending artery,* which supplies blood to the interventricular septum and the anterior of both ventricles, and the *circumflex artery,* which supplies blood to the left atrium and posterior of the left ventricle. The smaller *right coronary artery* supplies blood to the right atrium and posterior walls of the right ventricle.

The heart requires high levels of oxygen, even at rest, because cardiac muscle

has a limited capacity for anaerobic glycolysis. In fact, blood flow per tissue mass through the coronary circulation is higher for the heart than most other organs. Unlike other tissues, the heart has decreased blood flow during systole and increased blood flow during diastole, because when cardiac muscle contracts, the coronary arteries are compressed, increasing resistance and thereby decreasing blood flow.

The coronary circulation is subject to both intrinsic and extrinsic regulations. Local metabolites such as carbon dioxide and oxygen affect the arterioles of the heart, as they do those of the systemic circulation (discussed further in Chapter 15). However, coronary arterioles are also affected by *adenosine,* a vasodilator produced by the breakdown of ATP in cardiac cells. During increased activity, blood flow to the heart can increase fourfold. Coronary arterioles are also affected by the sympathetic nervous system. Smooth muscle of the coronary arterioles has both α and β adrenergic receptors. Thus, although norepinephrine would tend to cause vasoconstriction of arterioles during increased sympathetic activity by binding to α receptors, epinephrine tends to cause vasodilation by binding to β receptors, and vasodilation normally predominates.

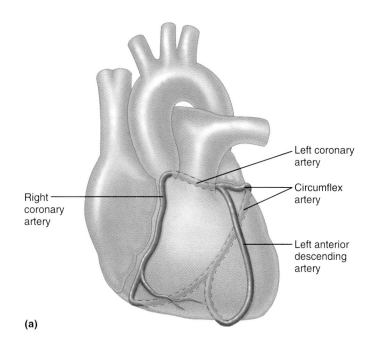

Right coronary artery

Left coronary artery

Circumflex artery

Left anterior descending artery

(a)

systemic circuit, blood does not flow from one organ directly to the next. Instead, blood travels through the aorta and the arteries that branch off it to reach only one organ at a time before flowing through veins that converge to either the superior or inferior vena cava. Moreover, the pattern of blood flow *within* organs, including the lungs in the pulmonary circuit, is also parallel because arteries branch to arterioles, which branch to capillaries, and so on.

Note in Figure 14.3 that the *heart* is in parallel with the other organs in the systemic circuit. Even though the heart pumps a large volume of blood, the blood within the heart's chambers does not supply the heart muscle with significant quantities of oxygen or nutrients. Instead, the heart muscle obtains most of its nourishment

from blood via the *coronary arteries,* which branch off the aorta near its base and run through the heart muscle (see **Discovery: Coronary Arteries**).

The parallel arrangement of organs in the systemic circuit confers two distinct advantages. First, because each organ is fed by a separate artery, each receives fully oxygenated blood—that is, blood that has not been depleted of oxygen as a result of having already flowed through another organ. Thus, for instance, when a muscle contracts and extracts oxygen from the blood, the muscle's increased metabolic rate does not deprive other organs of fully oxygenated blood because blood leaving the muscle returns to the heart and is reoxygenated in the lungs before it returns to the systemic circuit. Second, because blood reaches the organs

(continued)

A decrease in cardiac blood flow to levels insufficient to provide adequate oxygen and to remove metabolites is called *myocardial ischemia* and can result in chest pain called *angina pectoris*. In some cases the ischemia is temporary, or *acute*, caused by vascular spasms of the coronary arteries or increased activity of the heart. In more severe cases, the ischemia is prolonged, or *chronic*, often caused by *atherosclerosis*, narrowing of the arteries due to the buildup of plaques. Chronic myocardial ischemia can lead to *myocardial infarction (MI)*, also known as a heart attack. An MI is generally caused by a blood clot lodging in a narrowed coronary artery. An MI causes irreversible damage to cardiac muscle—contractile cells die and are replaced by scar tissue. Interesting, the cells do not die during the ischemic episode; they die during *reperfusion*, when blood flow is resumed. The precise mechanisms of reperfusion injury are not understood, but involve the presence of free radicals.

Many things in addition to myocardial ischemia can produce chest pain. For example, acid reflux from the stomach to the esophagus produces similar symptoms. How do doctors determine if a patient suffers from myocardial ischemia? A variety of tests help determine if an ischemic episode has occurred. Indicators of cardiac ischemia include elevated levels of

cardiac muscle enzymes and troponin in the blood, an elevated S-T segment in an ECG, and pain that has been *referred* to the left arm and shoulder. If ischemia is indicated, *cardiac catheterization* is often performed to determine if there is a blockage of an artery, and if so, to try removing the block using *balloon angioplasty*. During cardiac catheterization, a dye is injected into the coronary arteries and its flow through the arteries is observed using X rays. Figure (b) shows an angiogram of a heart with blockage of the right coronary artery. To remove the blockage, a small deflated balloon is inserted into the artery and then inflated to

expand the artery and push the plaque back against the walls of the artery.

A person at risk of having an MI is often put on a low-cholesterol, low-fat diet (to reduce the risk of atherosclerosis) and takes an aspirin a day (to decrease the likelihood of forming a blood clot). Those who have already suffered an MI are often given nitroglycerin (a vasodilator) to have available in case of another episode. Other drugs used to help prevent a second MI include digitalis (decreases heart rate so it requires less energy), calcium channel blockers (decrease likelihood of coronary vascular spasms), and beta blockers (decrease heart rate and cardiac contractility).

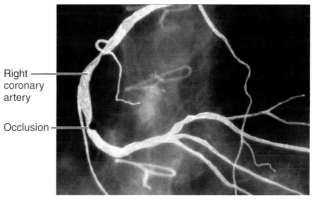

(b) Angiogram of a heart showing blockage of the right coronary artery.

via parallel paths, blood flow to the organs can be independently regulated, enabling blood flow to be adjusted to match the constantly changing metabolic needs of organs. At any given instant, blood flow can be increased to more active organs and decreased to less active organs.

Although parallel flow is the norm for the systemic circuit, in some exceptions to the rule, blood flows in series between two capillary beds (Figure 14.3). We already saw one of the exceptions in Chapter 6, with the path of blood flow between the hypothalamus and anterior pituitary. Capillary beds in the hypothalamus are in series with those in the anterior pituitary, connected together by *portal* veins. A *portal circulation* is one in which blood flows from one capillary bed to another before returning

to the heart. Other portal circulations exist between the intestines and liver (Chapter 21) and within the kidneys (Chapter 19).

Anatomy of the Heart

The heart is located within a membranous sac called the *pericardium* and is centrally located in the *thoracic cavity* (chest cavity) just above the *diaphragm,* a muscular partition that separates the thoracic cavity from the *abdominal cavity* (**Figure 14.4**). It is about the size of a fist, weighing approximately 300–350 grams in males and 250–300 grams in females.

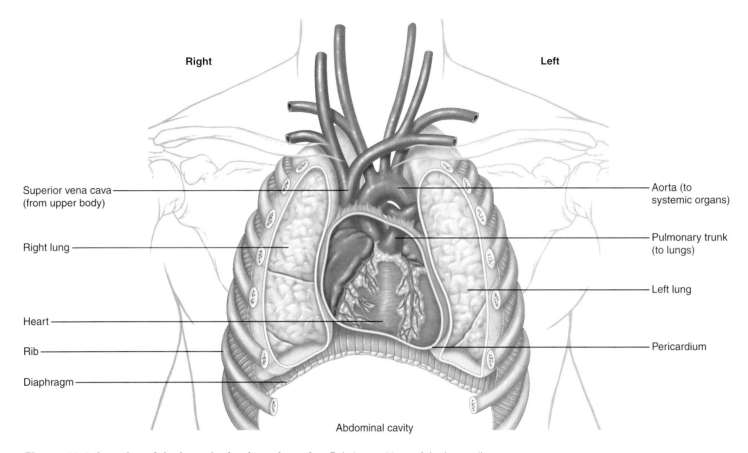

Figure 14.4 Location of the heart in the thoracic cavity. Relative positions of the heart, rib cage, and diaphragm. Also shown are the lungs and the major blood vessels connecting to the heart.

The heart generates the force that propels blood flow through the vasculature. To carry out this function, the walls of the heart consist primarily of *cardiac muscle*. To ensure that blood flows in the correct direction through the heart, four valves prevent backflow. In the next section, we look at the anatomy of the heart wall and the valves within the heart.

Myocardium and the Heart Wall

The heart wall consists of three layers: an outer layer of connective tissue called the **epicardium**, a middle layer of cardiac muscle called the **myocardium**, and an inner layer of epithelial cells called the **endothelium**. (The endothelial layer extends throughout the entire cardiovascular system.)

The heart's pumping action is conferred by the rhythmic contraction and relaxation of the myocardium. When muscle in the wall of an atrium or ventricle contracts, the wall moves inward and squeezes the blood in the chamber. This squeezing increases the pressure within the chamber and forces the blood out. When the muscle relaxes, the chamber expands and fills with blood.

Note in Figure 14.1 that ventricular muscle is substantially thicker than atrial muscle. This reflects the fact that the ventricles pump blood over relatively long distances through the vasculature (and not just into the next chamber, as the atria do), so they must work harder to pump a given volume of blood. Note as well that the ventricular muscle is much thicker on the left side than it is on the right (**Figure 14.5**). The thicker muscle enables the left ventricle to develop greater pressure than the right ventricle. It is important for the left ventricle to develop greater pressure because it pumps blood to all the organs in the body except the lungs, whereas the right ventricle pumps blood only to the lungs. The pressure required to pump blood at a given rate through the whole body is greater than that required to pump blood at the same rate through the lungs.

What we call the *heartbeat* is actually a wave of contraction that sweeps through heart muscle fibers (cells) in an orderly, coordinated fashion. The atria contract first, driving blood into the ventricles; then the ventricles contract, driving blood to the organs. Although the entire heart muscle functions as a unit, atrial muscle (the *atrial myocardium*) and ventricular muscle (the *ventricular*

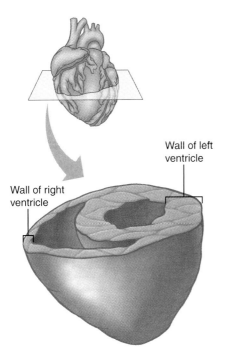

Figure 14.5 Right and left ventricle muscle thickness. The greater thickness of the muscle of the left ventricle allows it to generate the force necessary to pump blood throughout the body.

myocardium) are physically anchored to and separated by a layer of fibrous connective tissue called the *fibrous skeleton* of the heart.

Valves and Unidirectional Blood Flow

Because the heartbeat is a cycle, pressure within the chambers changes with the cycle. The pressure differences are what drive blood flow from atria to ventricles and from ventricles to the arteries. However, it is critical that blood flow down its pressure gradient only when that gradient favors movement in the normal direction, from atria to ventricle and from ventricle to artery.

The heart has four valves that keep blood flowing in the proper direction within the heart itself and between the heart and the arteries connected directly to it (the *aorta* and the *pulmonary trunk*) (see Figure 14.1). The atrium and ventricle on each side are separated by **atrioventricular valves (AV valves)**, which permit blood to flow from the atrium to the ventricle but not in the opposite direction. AV valves open or close in response to cyclic changes in pressure that occur with every heartbeat (**Figure 14.6**). When atrial pressure is higher than ventricular pressure, the valves open; when ventricular pressure becomes higher than atrial pressure, the valves close. The AV valve on the left consists of two flaps or *cusps* of connective tissue and is thus called the **bicuspid valve**

or **mitral valve**. The right AV valve has three cusps and is called the **tricuspid valve**.

When a ventricle contracts, the increased ventricular pressure exerts an upward force against the AV valve. Because of this force, there is a potential danger that one or more valve cusps could be pushed into the atria, a condition called *prolapse*. If this were to happen, the edges of the cusps would no longer meet properly when the valve closes, and the valve would not be able to seal completely. Prolapse of the AV valves is normally prevented because the valve cusps are held in place by strands of connective tissue (known as the *chordae tendineae*) that extend from the edges of the cusps to *papillary muscles*, which protrude from the ventricular wall. During ventricular contraction, the papillary muscles also contract, which exerts tension on the chordae tendineae. The chordae tendineae pull downward on the valve cusps, thereby enabling the AV valves to seal properly while resisting the upward force of ventricular pressure. The cusps of these valves (and also those of the *semilunar valves*, discussed next) are anchored at their bases to rings of connective tissue formed by the fibrous skeleton.

In addition to the AV valves, other valves, called **semilunar valves**, are located between the ventricles and arteries. The **aortic semilunar valve** (or *aortic valve*) is located between the left ventricle and the aorta, and the **pulmonary semilunar valve** *(pulmonary valve)* is located between the right ventricle and the pulmonary trunk. The function of these valves is similar to that of the AV valves—to permit blood to flow forward while preventing it from flowing backward (**Figure 14.7**). The aortic and pulmonary valves open when ventricular pressure is greater than arterial pressure (when the ventricles contract). This allows blood to leave the ventricles and enter the arteries. When the ventricles relax and ventricular pressure becomes lower than arterial pressure, the valves close, thus preventing blood from flowing back into the ventricles from the arteries.

Quick Test 14.3

1. What are the three layers of the heart wall?

2. Which chamber of the heart has the thickest walls? Explain why.

3. Where are the atrioventricular valves located? What is their function? What is the location and function of the semilunar valves?

Electrical Activity of the Heart

For the heart to adequately pump blood through the circulatory system, the cardiac muscle must contract in a highly synchronized manner; first the contraction of both

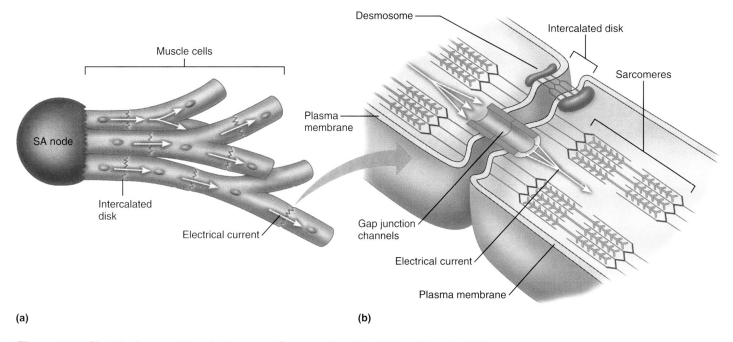

Figure 14.8 Electrical connections between cardiac muscle cells. (a) An action potential generated spontaneously in cells of the SA node spreads to adjacent muscle cells by means of electrical current passing through gap junctions in intercalated disks. (b) A schematic view of the junction between two adjacent muscle cells showing a gap junction and a desmosome.

Besides cardiac muscle, what other type of muscle possesses gap junctions?

neighbors by gap junctions, which permit electrical current to pass in the form of ions from one cell to another across their plasma membranes. In the heart, gap junctions are concentrated in structures called **intercalated disks**, which form the junctions between adjacent muscle fibers (**Figure 14.8**). Intercalated disks also contain large numbers of *desmosomes,* areas in which protein fibers link adjacent cells together, forming a physical bond between them that resists mechanical stress. This is important because it enables the myocardium to resist stretching, which occurs every time the heart fills with blood, and also because it enables the myocardium to withstand the tension that is generated every time the muscle cells contract.

Initiation and Conduction of an Impulse During a Heartbeat

The sequence of electrical events that normally triggers the heartbeat occurs as follows (**Figure 14.9**):

①　An action potential is initiated in the SA node. From the SA node, impulses travel to the AV node by way of *internodal pathways*—systems of conduction fibers that run through the walls of the atria. As these signals move through the internodal pathways, they also spread through the bulk of the atrial muscle by way of *interatrial pathways.*

②　The impulse is conducted to cells of the AV node, which transmit action potentials less rapidly than

other cells of the conduction system. As a result, the impulse is momentarily delayed by about 0.1 second (called the *AV nodal delay*) before moving onward.

③　From the AV node, the impulse travels through the atrioventricular bundle, also known as the **bundle of His** (pronounced "hiss"), a compact bundle of muscle fibers located in the interventricular septum. The AV node and bundle of His are the only electrical connection between the atria and the ventricles, which are otherwise separated by the fibrous skeleton.

④　The signal travels only a short distance through the atrioventricular bundle before it splits into left and right *bundle branches,* which conduct impulses to the left and right ventricles, respectively.

⑤　From the bundle branches, impulses travel through an extensive network of branches referred to as **Purkinje fibers**, which spread through the ventricular myocardium from the apex upward toward the valves. From these fibers, impulses travel through the rest of the myocardial cells.

Control of the Heartbeat by Pacemakers

Although the SA node and the AV node are both capable of generating spontaneous action potentials, the heartbeat is almost always triggered by impulses originating from the SA node. The AV node rarely initiates contractions for

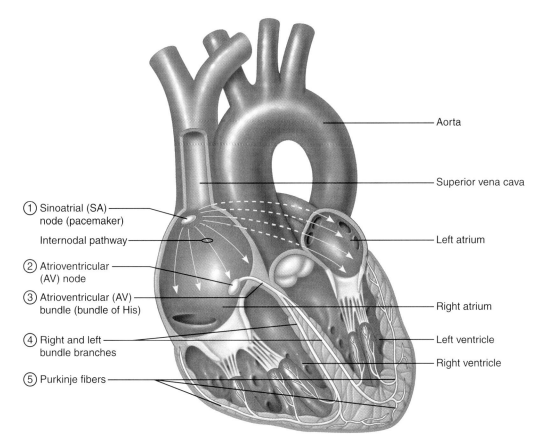

Figure 14.9 The conduction system of the heart. The pathway of impulse conduction through the heart is indicated in this longitudinal section.

two reasons. The first is that action potentials originating in the SA node travel through the AV node on their way to the ventricles. When this happens, cells in the AV node go into a *refractory period*, during which they cannot generate their own action potentials. The second reason is that the SA node has a higher "beat frequency" than the AV node. This means that if we were to measure the action potential frequency in isolated cells of either node under normal resting conditions, we would see that SA node cells fire more frequently than AV node cells—about 70 impulses/minute for the SA node, as opposed to 50 impulses/minute for the AV node. Thus the AV node rarely has a chance to fire an action potential because the SA node always "beats it to the punch."

However, if the SA node fails to fire an action potential or if it slows down dramatically—that is, if it "misses a beat," so to speak—the AV node *will* initiate action potentials, which travel through the conducting system and trigger ventricular contraction in the normal manner. The AV node can also take over control of the heartbeat if conduction between the nodes is blocked or slowed down for some reason. In these circumstances, the AV node functions as an emergency backup system that keeps the ventricles beating. If for some reason the AV node is unable to drive ventricular contraction, the heart has yet another backup system: Certain cells in the Purkinje

fibers (sometimes referred to as *idioventricular pacemakers*) can take over. However, the firing frequency of these cells is only 30–40 impulses per minute.

Spread of Excitation Through the Heart Muscle

As impulses propagate through the heart muscle, they travel in an orderly pattern as a kind of wavefront—a "wave of excitation." As this wave of excitation spreads, contraction of the muscle follows. The pattern of excitation is shown in **Figure 14.10**.

The wave of excitation starts at the SA node and then spreads outward through the atria. The wave then "funnels" through the atrioventricular bundle by way of the AV node, which acts as a kind of bottleneck due to the relative slowness of impulse conduction in this region. This delay is essential for efficient cardiac function; it allows the wave of excitation to spread completely through the atria before it reaches the ventricles, thus ensuring that atrial contraction is complete before ventricular contraction starts. Given that the function of atrial contraction is to drive blood into the ventricles, if no such delay occurred, ventricular contraction would work against the pumping action of the atria.

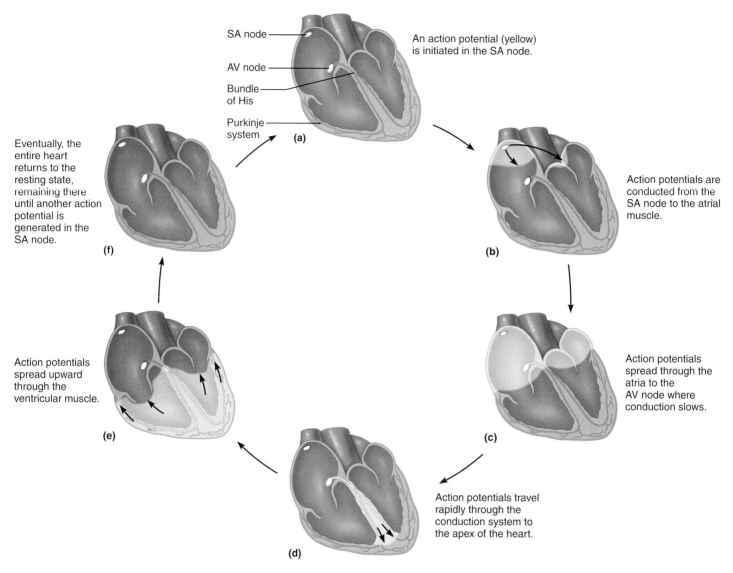

SA node

AV node

Bundle of His

Purkinje system

(a)

An action potential (yellow) is initiated in the SA node.

(b)

Action potentials are conducted from the SA node to the atrial muscle.

(c)

Action potentials spread through the atria to the AV node where conduction slows.

(d)

Action potentials travel rapidly through the conduction system to the apex of the heart.

(e)

Action potentials spread upward through the ventricular muscle.

(f)

Eventually, the entire heart returns to the resting state, remaining there until another action potential is generated in the SA node.

Figure 14.10 The spread of action potentials through the heart. The sequence of electrical excitation during a single heartbeat, starting with (a) depolarization of the SA node and ending with (f) the return of the heart to the resting state.

Once impulses reach the bundle branches and the Purkinje fibers, they are carried relatively quickly to the lower portion of the ventricles. From there, the wave of excitation fans out through the entire ventricular muscle. Thus ventricular contraction begins at the apex and spreads upward. This makes sense when you consider that blood exits the ventricles from the top (see Figure 14.1, p. 415). Ventricular contraction is thus reminiscent of how one should squeeze a tube of toothpaste—from the bottom up.

The Ionic Basis of Electrical Activity in the Heart

We now know that the heartbeat is triggered by action potentials that originate in pacemaker cells and propagate through the heart muscle in an orderly, predictable fash-

ion. Here we examine the cellular mechanisms responsible for generating these electrical signals, beginning with events occurring in the membrane of pacemaker cells.

Electrical Activity in Pacemaker Cells

A cardiac contractile cell fires an action potential only when it is depolarized to threshold by a stimulus. Normally this stimulus is a circulating electrical current that originates in neighboring cells that are firing action potentials. We have seen that this current enters the cell through gap junctions that connect it with its neighbors. After entering, the current exits the cell by passing through the plasma membrane, and in doing so it triggers depolarization.

Recall that pacemaker cells are different because they can fire action potentials in the absence of any external stimulus, and do so in a regular, periodic fashion. A pacemaker cell is able to fire action potentials spontaneously because it does not have a steady resting potential. After an action potential, a pacemaker cell immediately begins to depolarize slowly and continues to do so until its membrane potential reaches threshold, which triggers another action potential (**Figure 14.11**). Following this, the membrane potential returns to about −60 to −70 mV and then begins another round of slow depolarization until another action potential is triggered. The slow depolarizations or "ramps" that lead up to each action potential are referred to as **pacemaker potentials**.

In pacemaker cells and other cardiac muscle cells, electrical signals are caused by changes in plasma membrane ion permeability brought about by the opening and closing of specific types of ion channels, just as in any other type of cell. To understand how these permeability changes affect the membrane potential, recall the following rule, which we first encountered in Chapter 8: *As a membrane's permeability to a particular ion increases relative to that of other ions, the membrane potential moves toward the equilibrium potential of that ion.* In cardiac muscle cells, the most important permeability changes involve sodium, potassium, and calcium ions (Na^+, K^+, and Ca^{2+}, respectively). Ion concentrations in cardiac muscle cells are similar to those in other cells—the intracellular fluid is rich in potassium but poor in sodium and calcium compared to extracellular fluid. Thus the equilibrium potential of potassium is negative, whereas the equilibrium potentials of sodium and calcium are both positive. (Approximate values are $E_K = -94$ mV, $E_{Na} = +60$ mV, and $E_{Ca} = +130$ mV.) Therefore, increased sodium or calcium permeability (P_{Na} or P_{Ca}) tends to make the membrane potential become more positive, whereas increased potassium permeability (P_K) tends to make it become more negative.

In pacemaker cells, electrical signals are triggered by changes in P_K, P_{Na}, and P_{Ca}, as shown in Figure 14.11. The slow depolarization that occurs in the early stages of the pacemaker potential is due to closing of potassium channels and opening of so-called *funny channels*. Potassium channels open during repolarization of the action potential, and then close when the membrane returns to its polarized state. Funny channels, so named because investigators noticed that they had some unusual characteristics, open after the cell repolarizes and allow sodium and potassium ions to cross the plasma membrane. With potassium channels closed and funny channels open during the early stages of the pacemaker potential, potassium movement out of the cell decreases, whereas sodium movement into the cell increases, causing the initial depolarization.

The funny channels are open for only a brief time, closing when the membrane potential approaches −55 mV, approximately 5 mV short of the threshold needed to generate an action potential. However, the initial depolarization triggers the opening of voltage-gated calcium channels called *T-type channels*. This raises P_{Ca}, which depolarizes the cell even further. Although the T-type channels stay open for only a short time before

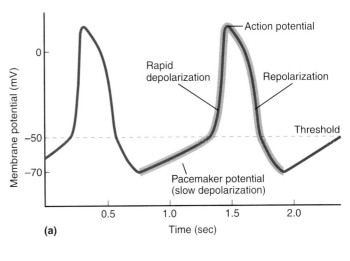

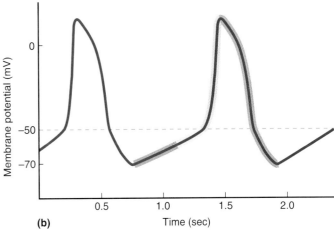

Figure 14.11 Electrical activity in a pacemaker cell.
(a) A recording of the membrane potential showing action potentials and pacemaker potentials. (b) Changes in membrane permeability responsible for potential changes in pacemaker cells. During the initial spontaneous depolarization (orange), P_K is decreasing and P_{Na} is increasing. During the latter spontaneous depolarization (yellow), P_{Ca} is increasing and P_{Na} is decreasing. During the rapid depolarization (green), P_{Ca} is increasing more. During the repolarization phase (pink), P_{Ca} is decreasing and P_K is increasing.

inactivating, the resulting depolarization triggers the opening of a second population of voltage-gated calcium channels *(L-type channels)*, which stay open longer and inactivate only slowly. The result is a large increase in P_{Ca} that produces the rapid depolarization characteristic of the upswing of the action potential. (As these calcium channels open, they also allow some sodium to flow into the cell, which increases P_{Na} and adds to the depolarizing effect.) This depolarization triggers the opening of potassium channels and, consequently, a rise in P_K that occurs shortly after the increase in P_{Ca} and acts to pull the membrane potential back down. The resulting fall in potential removes the stimulus for calcium channel opening, allowing these channels to begin closing. This reduces P_{Ca} and decreases the flow of calcium into the cell, which works along with the increase in P_K to

Table 14.1 | Ionic Bases of the Autorhythmic Cell Action Potential

Autorhythmic cell potential change	Ion channel gating	Ion movement
Pacemaker potential Initial period of spontaneous depolarization to subthreshold	Funny channels open	Sodium moves in, potassium moves out
Latter period of spontaneous depolarization to threshold	T-type calcium channels open	Calcium moves in
Rapid depolarization phase of action potential	L-type calcium channels open	Calcium moves in
Repolarization phase of action potential	Potassium channels open	Potassium moves out

repolarize the membrane and terminate the action potential. **Table 14.1** summarizes the function of ion channels in autorhythmic cells.

Because the heart initiates its own action potentials, it does not require neural input to trigger its contractions. However, as discussed later in this chapter, autonomic neurons do exert control over the rate and force of these contractions.

Electrical Activity in Cardiac Contractile Cells

Action potentials in cardiac contractile cells from different regions of the heart vary in regard to shape (time course and level of depolarization) and speed of propagation. These differences occur because contractile cells vary in the type and number of ion channels they possess, as well as in their physical dimensions. Despite these differences, two important events characterize most cardiac action potentials: (1) During a typical cardiac action potential, P_K *decreases* due to the action of a certain type of voltage-gated potassium channel that *closes* in response to depolarization. (Recall that in pacemaker cells and most other excitable tissues, P_K increases during an action potential because these tissues contain voltage-gated potassium channels that open in response to depolarization.) (2) During a cardiac action potential, depolarization causes the opening of voltage-gated *calcium* channels, which not only affects the membrane potential but also is instrumental in triggering muscle cell contractions.

The majority of ventricular muscle cells, which make up the bulk of the myocardium, are unlike pacemaker cells in that they have stable resting potentials. They also have longer-lasting action potentials with a distinctive shape that can be divided into several phases (designated 0–4), as shown in **Figure 14.12**. Permeabil-

ity changes occurring during each of these phases are described next:

⓪ Phase 0: Phase 0 of the cardiac action potential is similar to the upswing of a neuronal action potential and is caused by similar events: Depolarization of the membrane triggers the opening of voltage-gated sodium channels, which raises P_{Na} and increases the flow of sodium ions into the cell. Consequently, the membrane potential becomes more positive, which triggers the opening of more sodium channels, additional increases in P_{Na}, more depolarization, and so on. The result is a rapid rise in membrane potential that peaks at between +30 and +40 mV.

① Phase 1: The sodium channels that were opened in phase 0 start to inactivate, which reduces P_{Na}. This

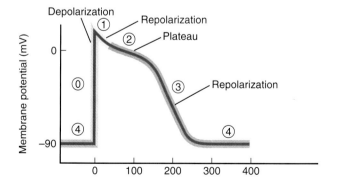

Figure 14.12 The cardiac action potential. An action potential recorded from a ventricular muscle cell. P_{Na} increases during phase 0 (green) and decreases during phase 1 (yellow). P_{Ca} increases and P_K decreases during phase 2 (orange), and then P_{Ca} decreases and P_K increases during phase 3 (purple). During phase 4 (blue), all ion channels are in their resting state (P_K high, P_{Ca} and P_{Na} low).

Table 14.2 | Ionic Bases of the Contractile Cell Action Potential

Phase of contractile cell action potential	Ion channel gating	Ion movement
0 Rapid depolarization	Sodium channels open	Sodium moves in
1 Small repolarization	Sodium channels inactivate	Sodium movement in decreases
2 Plateau	Potassium inward rectifier channels close Calcium L-type channels open	Potassium movement out decreases Calcium moves in
3 Repolarization	Potassium delayed rectifier channels open Calcium L-type channels close	Potassium moves out Calcium movement in decreases
4 Resting potential	Potassium channels (both type) open Sodium and calcium channels still closed	Potassium moves out Little sodium or calcium moves in

decreases the flow of sodium into the cell and causes the membrane potential to fall toward more negative values due to continued movement of potassium ions out of the cell. The membrane potential drops only a small amount, however, because the depolarization of the membrane that began in phase 0 has set into motion two additional events that are also occurring at this time: (1) the closing of voltage-gated potassium channels (known as *inward rectifier channels*), which reduces P_K and decreases the flow of potassium out of the cell; and (2) the opening of voltage-gated calcium channels (L-type channels), which raises P_{Ca} and increases the flow of calcium into the cell. Both of these changes act to depolarize the membrane, thereby counteracting the effect of sodium channel inactivation.

② Phase 2: During this phase, which is also referred to as the *plateau phase*, most of the potassium channels that were closed in phase 1 stay closed, so that P_K remains lower than its resting value. At the same time, most of the calcium channels that opened in phase 1 remain open, and P_{Ca} remains elevated. The lowered P_K and elevated P_{Ca} both act to keep the membrane in its depolarized state.

③ Phase 3: During this phase P_K increases, partly because of the action of a second population of potassium channels similar to those in neurons (called *delayed rectifier channels*), which open in response to depolarization. These channels begin to open during phases 1 and 2 but do not exert a significant influence on the membrane potential until phase 3 because they open slowly. As P_K rises, the flow of potassium out of the cell increases, which pulls the membrane potential down toward more negative values. Furthermore, this fall in potential removes the stimulus that kept the inward rectifier channels

closed in phase 2, and as a consequence these channels begin to open, which raises P_K even further. The fall in potential also removes the stimulus that kept the calcium channels open during phase 2 and allows them to begin closing, which lowers P_{Ca} and reduces the flow of calcium into the cell. This works hand in hand with the increase in P_K to repolarize the membrane, thereby terminating the action potential.

④ Phase 4: During this phase, which corresponds to the *resting potential*, P_K, P_{Na}, and P_{Ca} are at their resting values. Because P_K is much greater than P_{Na} or P_{Ca} under these conditions, the membrane potential is around −90 mV, which is close to the equilibrium potential of potassium.

Table 14.2 summarizes the role of ion channels in the contractile cell action potential.

Excitation-Contraction Coupling in Cardiac Contractile Cells

The mechanism by which a cardiac action potential stimulates contraction is similar to that for skeletal muscle in many respects, but also has similarities to smooth muscle (**Figure 14.13**). The stimulus that triggers an action potential in the cardiac muscle cell is current coming through gap junctions ①. An action potential spreads through the plasma membrane and down T tubules ②, causing voltage-sensitive calcium channels on the sarcoplasmic reticulum to open and release calcium into the cytosol ③. The action potential also triggers opening of voltage-gated calcium channels on the plasma membrane, allowing calcium to enter the cell. (This increased permeability to calcium is also responsible for the plateau phase of the action potential.) The calcium that enters during the plateau phase acts on the voltage-sensitive channels that trigger calcium release from the

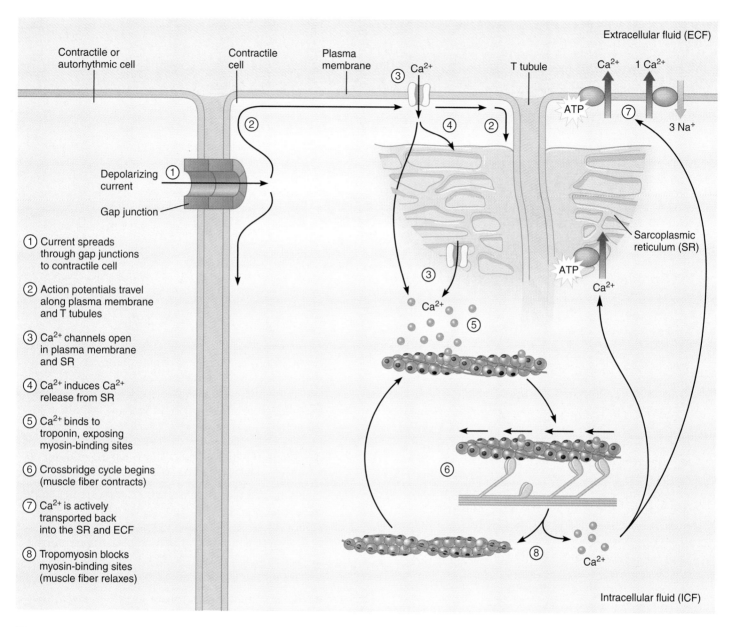

Figure 14.13 Excitation–contraction coupling in cardiac muscle. Cardiac muscle is excited by the spread of depolarizing current through gap junctions. The depolarization triggers the opening of calcium channels in the plasma membrane and sarcoplasmic reticulum. Calcium binds to troponin, enabling the crossbridge cycle to occur. To terminate contraction, calcium is pumped out of the cytosol into the sarcoplasmic reticulum and interstitial fluid.

sarcoplasmic reticulum and stimulates them to stay open longer ④. As a result, the sarcoplasmic reticulum releases more calcium with each action potential. This phenomenon is known as *calcium-induced calcium release*.

Cytosolic calcium triggers contraction of cardiac muscle in the same way it triggers contraction of skeletal muscle: Calcium binds to troponin, shifting tropomyosin off of the myosin-binding sites on actin ⑤, and crossbridge cycling occurs ⑥. Most of the calcium that binds to troponin (95%) comes from the sarcoplasmic reticulum, with only 5% coming from the extracellular fluid. Relaxation of cardiac muscle re-

quires removal of calcium from the cytosol ⑦, which occurs by three mechanisms: (1) As in skeletal muscle, a Ca^{2+}-ATPase located in the membrane of the sarcoplasmic reticulum actively transports calcium from the cytosol into the lumen of the sarcoplasmic reticulum. (2) Cardiac muscle also has a Ca^{2+}-ATPase located in the plasma membrane that actively transports calcium from the cytosol into the interstitial fluid. (3) Cardiac muscle also has a Na^+-Ca^{2+} exchanger in the plasma membrane that actively transports calcium out of the cell by countertransport with sodium. Without calcium bound to troponin, tropomyosin shifts back over the myosin-binding sites on actin and the muscle fiber relaxes ⑧.

Recording the Electrical Activity of the Heart with an Electrocardiogram

The **electrocardiogram** (**ECG** or EKG; the *K* is for the German form of the word, *elektrokardiograph*) is a noninvasive means for monitoring the electrical activity of the heart. To understand the interpretation of ECGs, it is important to remember that the recorded electrical events cause the contraction of cardiac muscle. Physicians use ECG recordings to determine whether problems exist in the electrical activity of the heart. ECG recordings do not give information about mechanical problems of the heart unless those problems result from an electrical problem. An ECG, for example, would not be useful for identifying a mechanical malfunction of a valve.

The ECG is a record of the overall spread of electrical current through the heart as a function of time during the cardiac cycle. The ECG is usually recorded by means of electrodes placed on the skin. The concept of recording ECGs is similar to that for recording EEGs described in Chapter 10; namely, electrical activity generated in nervous or muscle tissue spreads through the body because body fluids function as conductors. The more synchronized the activity, the larger the amplitude of signals that are recorded at a distance from the source. Because the electrical activity of the heart is highly synchronized, relatively large amplitude electrical potentials that correspond to distinct electrical phases of the heart can be detected at the surface of the skin.

A Dutch physiologist, Willem Einthoven, developed the technique of ECG recordings. The procedure for standard ECG recording is based on an imaginary equilateral triangle surrounding the heart. The triangle is expanded until its corners fall on the right arm, left arm, and left leg, a pattern known as Einthoven's triangle (**Figure 14.14**). Electrodes placed on the skin at the corners of the triangle are connected in pairs to a voltage-measuring device such as an oscilloscope or chart recorder. Certain pairs of electrodes are referred to as *leads* and are designated by Roman numerals. One electrode in each lead is designated as the positive electrode, the other as the negative.

Each specific lead detects the difference in the surface electrical potential between the positive and negative electrodes. Lead I detects the potential at the left arm minus that at the right arm; lead II detects the potential at the left leg minus that at the right arm; lead III detects the potential at the left leg minus the left arm. The direction of the recorded waveforms (up or down) depends on whether the difference in potential between the two electrodes is positive (which gives an upward deflection) or negative (which gives a downward deflection).

The waveforms recorded with a standard lead II ECG and the action potential of a ventricular contractile cell are shown in **Figure 14.15**. ECGs are recorded on

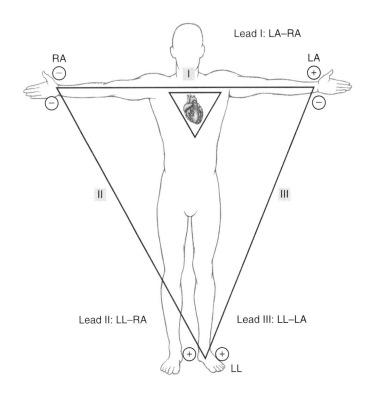

Figure 14.14 Einthoven's triangle. Three electrodes are placed on limbs to form an equilateral triangle around the heart.

chart paper at a rate of 25 mm/sec with an amplitude of 1 mV/cm. The ECG normally shows three characteristic waveforms: (1) The **P wave** is an upward deflection occurring as a result of atrial depolarization. (2) The **QRS complex** is a series of sharp upward and downward deflections as a result of ventricular depolarization; it is correlated with phase 0 of the ventricular contractile cell action potential. (3) The **T wave** is an upward deflection caused by ventricular repolarization; it is correlated with phase 3 of the ventricular contractile cell action potential. Atrial repolarization is generally not detected in an ECG recording because it occurs at the same time as the QRS complex. Between the waves, a normal ECG trace consists of a horizontal line, called the *isoelectric line,* because no changes in electrical activity are occurring.

It is important to note that although the phases of the ECG are due to action potentials traveling through the heart muscle, the ECG is *not* simply a recording of an action potential. During the heartbeat, cells fire action potentials at different times, and the ECG reflects *patterns* of action potential firing in the entire *population* of cells that make up the heart muscle.

In addition to waves, certain intervals and segments can provide important information about the function of the heart. The *P-Q* or *P-R interval* occurs between the onset of the P wave and the onset of the QRS complex and is an estimate of the time of conduction through the AV node. The *Q-T interval* is the time from the onset of the QRS complex to the end of the T wave and is an estimate of the time the ventricles are contracting, called

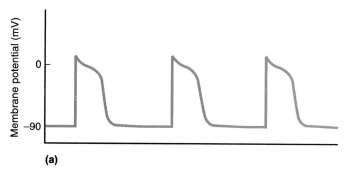

(a)

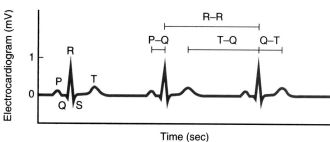

Component	Amplitude (mV)	Duration (sec)
P wave	0.2	0.10
QRS complex	1.0	0.08–0.12
T wave	0.2–0.3	0.16–0.27
P–Q interval	N/A	0.12–0.21
Q–T interval	N/A	0.30–0.43
T–Q segment	N/A	0.55–0.70
R–R interval	N/A	0.85–1.00

(b)

Figure 14.15 Electrical activity of the heart. (a) Recording of the membrane potential in a ventricle contractile cell. (b) Recording of a lead II ECG. The table gives normal values for ECG waves, intervals, and segments.

ventricular systole. The *T-Q segment* is the time from the end of the T wave to the beginning of the QRS complex and is an estimate of the time the ventricles are relaxing, called ventricular diastole. The *R-R interval* is the time between the peaks of two successive QRS complexes; it represents the time between heartbeats. Heart rate can be determined by dividing 60 seconds by the R-R interval. If the R-R interval is 1 second, for example, then the heart rate is 60 beats per minute.

Figure 14.16 compares normal ECGs to those recorded during examples of abnormal electrical activity of the heart, called *cardiac arrhythmias*. Abnormal SA nodal firing can cause either a sinus tachycardia, which is

Normal

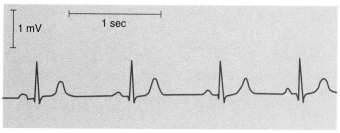

Sinus tachycardia (with inverted T wave)

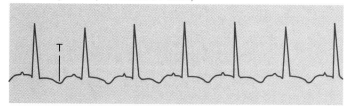

Sinus bradycardia

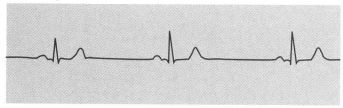

Heart block, 3rd degree

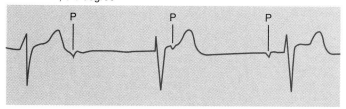

Pre-mature atrial contraction (PAC)

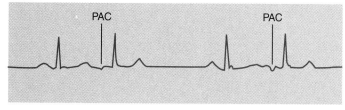

Ventricular fibrillation

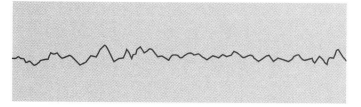

Figure 14.16 Lead II ECGs showing various arrhythmias.
25 mm on the horizontal = 1 sec; 1 cm on the vertical = 1 mV. Some data courtesy of C. V. Massey, University of South Alabama.

an abnormally fast resting heart rate (greater than 100 beats/min), or a sinus bradycardia, which is an abnormally slow resting heart rate (less than 50 beats/min). Altered conduction through the AV node can cause various degrees of heart block: During first-degree heart block, conduction through the AV node is slowed, causing a longer delay in AV nodal conduction (an increased P-Q interval). During second-degree heart block, conduction through the AV node does not always occur. If conduction does not occur, the ventricles do not depolarize (resulting in the absence of a QRS complex and T wave) and thus do not contract. The 1:1 relationship between atrial contractions and ventricular contractions, therefore, is lost. During third-degree heart block, conduction through the AV node does not occur at all, causing complete dissociation of atrial and ventricular contractions. The atria contract at the rate of SA nodal discharge, but the ventricles contract at the rate of Purkinje fiber discharge, which is only about 30–40 times per minute. This slow rate of ventricular contraction is insufficient to supply the body generally with the oxygen and nutrients required, and thus third-degree block can be deadly.

Sometimes the heart is depolarized by an electrical stimulus arising outside the normal conduction pathway. Because cardiac muscle cells are connected by gap junctions, an abnormal depolarization will spread throughout the heart, causing an extra contraction called an *extrasystole*. If the depolarization occurs in an atrium, then a premature atrial contraction (PAC) occurs, and conduction follows through the AV node, causing contraction of the ventricle. If the depolarization occurs in a ventricle, then a premature ventricular contraction (PVC) occurs, and no atrial contraction precedes it. PACs and PVCs are generally of little clinical significance unless they occur at high frequencies.

More serious arrhythmias are fibrillations, which occur when the heart muscle no longer has synchronized depolarization. In atrial fibrillation, atrial muscle fibers depolarize independently, so atrial contraction is inefficient in pumping blood to the ventricle. Atrial fibrillation results in weakness and light-headedness due to decreased blood flow, but as long as the ventricles still contract at a sufficient rate, atrial fibrillation is generally not deadly because contraction of the atria contributes little to ventricular filling (most ventricular filling is passive). Ventricular fibrillation, by contrast, can cause death within minutes. When ventricular muscle cells depolarize independently, the ventricles can no longer efficiently pump the blood out to the tissues, including the brain. Clinicians must quickly *defibrillate* the ventricular muscle to keep the person alive. Defibrillation is often done by passing a large current through the chest wall to the heart such that the externally applied current depolarizes all the muscle cells at the same time, returning synchronous electrical activity to the heart.

Quick Test 14.4

1. Define the following terms: *autorhythmicity, pacemaker cell, conduction fiber, pacemaker potential.*

2. Under normal conditions, which controls the heartbeat—the SA node or the AV node? Explain why.

3. Arrange the order of the following terms so that they describe the normal path of electrical impulses through the heart: *Purkinje fibers, atrioventricular bundle, AV node, bundle branches, SA node, ventricular muscle, atrial muscle.*

4. The entry of calcium into a ventricular muscle cell helps to maintain depolarization of the membrane during the plateau phase of the action potential, but this calcium also performs what other important function?

5. Match the terms *P wave, QRS complex,* and *T wave* with the following events: ventricular depolarization, ventricular repolarization, atrial depolarization.

The Cardiac Cycle

The **cardiac cycle** includes all the events associated with the flow of blood through the heart during a single complete heartbeat. In our discussion we concentrate on the following aspects of the cardiac cycle: (1) the various phases in the pumping action of the heart, often called the *pump cycle;* (2) periods of valve opening and closure; (3) changes in atrial, ventricular, and aortic pressure, which reflect contraction and relaxation of the heart muscle; (4) changes in ventricular volume, which reflect the amount of blood entering and leaving the ventricle during each heartbeat; and (5) the two major heart sounds.

The relationships among the various aspects of the cardiac cycle are depicted in **Figure 14.17**. (The pressure graphs pertain to the left heart only; the graphs for pressures in the right heart are similar, except that the peak pressures are lower.)

The Pump Cycle

Because the cardiac cycle involves the events of one heartbeat, a complete cycle involves both ventricular contraction and ventricular relaxation. As a result, the cycle can be divided into two major stages: **systole**, the period of ventricular contraction, and **diastole**, the period of ventricular relaxation. (Even though the atria also undergo periods of contraction and relaxation—termed atrial systole and diastole, respectively—we use the terms *systole* and *diastole* to refer to ventricular events.)

We begin our examination of the cardiac cycle in the middle of diastole, a time at which the atria and ventricles are completely relaxed:

1. *Ventricular filling:* During mid-to-late diastole (phase 1 in Figure 14.17), blood returning to the heart via the

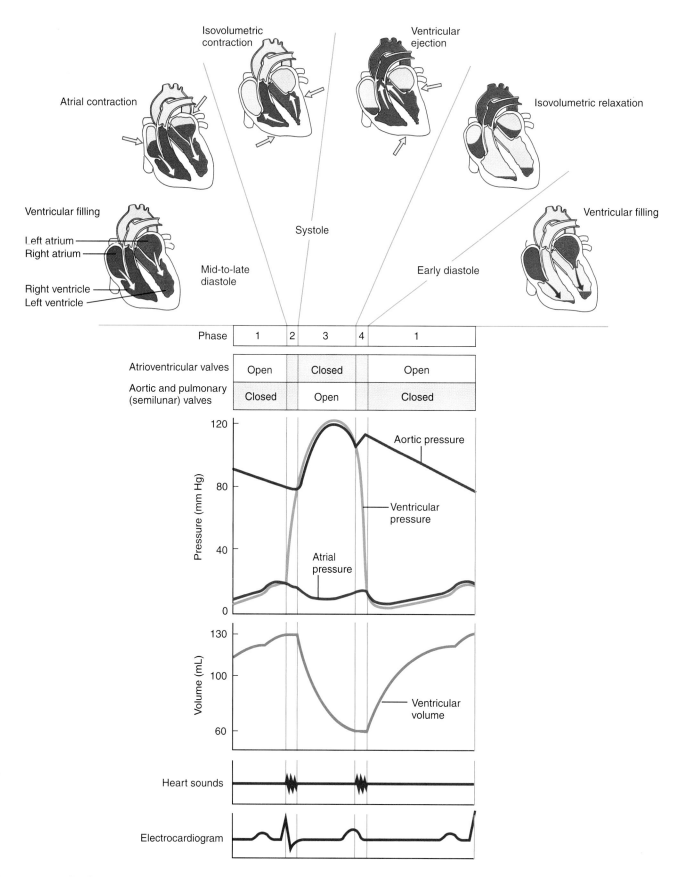

Figure 14.17 Cardiac cycle. All values are given in reference to the left heart. Diastole corresponds to phases 4 and 1, whereas systole corresponds to phases 2 and 3. Note the correlations between changes in pressure gradients across valves and the open/closed state of the valve and the changes in ventricular volume. Heart sounds are correlated with the closing of the valves. ECG waves correlate to the mechanical events of the heart: the P wave precedes atrial contraction (evident as an increase in atrial pressure), the QRS complex precedes ventricular contraction (evident as the rapid increase in ventricular pressure), and the T wave precedes ventricular relaxation (evident as the rapid decrease in ventricular pressure).

systemic and pulmonary veins enters the relaxed atria and passes through the AV valves and into the ventricles under its own pressure; that is, the pressure in the veins is sufficiently high to drive the flow of blood into the heart (called **venous return**). As the ventricles fill, the pulmonary and aortic (semilunar) valves are closed because ventricular pressure is lower than that in the aorta and pulmonary arteries.

Late in diastole (*at the end of phase 1*), the atria contract, driving more blood into the ventricles. Shortly thereafter, the atria relax and systole begins. This entire phase of blood entering the ventricle is called **ventricular filling**.

2. *Isovolumetric contraction:* At the beginning of systole (phase 2), the ventricles contract, which raises the pressure within them. When ventricular pressure exceeds atrial pressure (which occurs very early in systole), the AV valves close; the semilunar valves remain closed because ventricular pressure is not yet high enough to force them open. At this point, no blood is flowing into or out of the ventricles because all the valves are closed. Thus even though the ventricles are contracting, the volume of blood within them remains constant, so phase 2 is termed **isovolumetric contraction** (iso = "same"). Phase 2 ends when the ventricular pressure is great enough to force open the semilunar valves so that blood can leave the ventricles.

3. *Ventricular ejection:* In the remainder of systole (phase 3), blood is ejected into the aorta and pulmonary arteries through the open semilunar valves, and ventricular volume falls. During the exit of blood from the ventricles, referred to as **ventricular ejection**, ventricular pressure rises to a peak and then begins to decline. When it falls below aortic pressure, the semilunar valves close, ending ejection (and systole) and marking the beginning of diastole.

4. *Isovolumetric relaxation:* At the onset of early diastole (phase 4), the ventricular myocardium is relaxing. Some blood is present in the ventricles, and it is still under pressure because it takes time for the tension in the ventricular muscle to wane. Ventricular pressure is simultaneously too low to keep the semilunar valves open and too high to allow the AV valves to open. Because all valves are closed and the volume of blood remains constant within the relaxing ventricles, phase 4 is referred to as **isovolumetric relaxation**.

Once ventricular pressure becomes low enough to permit the AV valves to open again, blood enters the ventricles from the atria. This marks the beginning of phase 1, and the pump cycle begins once again.

The durations of systole and diastole are not equal. For a heart beating at the normal resting rate of 72 beats per minute (one beat every 0.8 second), most of the cardiac cycle (about 65%, or 0.5 second) is spent in diastole; systole lasts only about 0.3 second. This longer

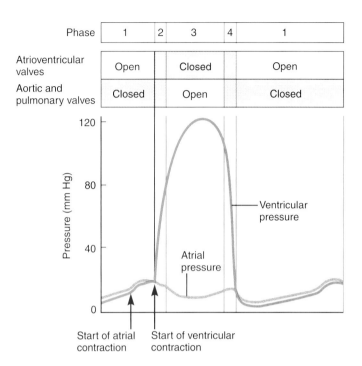

Figure 14.18 Ventricular pressure during the cardiac cycle. Note that the AV valves close when ventricular pressure exceeds atrial pressure, which occurs immediately after the ventricles start to contract.

diastole gives the heart adequate time to fill with blood, which is essential for efficient pumping, and it also gives the heart muscle more time to relax, which helps prevent fatigue.

Now that we have seen this overview, we examine the cardiac cycle in detail.

Atrial and Ventricular Pressure

By convention, cardiovascular pressures (the pressure of blood in the chambers of the heart or in the vasculature) are given in millimeters of mercury (mm Hg). Recall that atmospheric pressure is also measured in millimeters of mercury, and that normal atmospheric pressure at sea level is 760 mm Hg. In cardiovascular physiology, all pressures are given relative to atmospheric pressure, which is taken to be *zero*. Thus, when physiologists say that blood is at a pressure of 100 mm Hg, they mean that it is 100 mm Hg *above* atmospheric pressure.

The rise in atrial pressure that occurs in late diastole indicates the beginning of atrial contraction (see Figure 14.17). This rise in pressure is small and short-lived, however, and is followed by a series of similar small increases at various times throughout the cardiac cycle. Because these changes in pressure are of little significance to overall cardiac function, we will not discuss them further.

In mid-diastole (phase 1), ventricular pressure stays very low until the end of the phase, when an abrupt but small rise occurs (**Figure 14.18**). This rise is due to atrial

CHAPTER SUMMARY

An Overview of the Cardiovascular System, p. 414

The cardiovascular system includes the heart, blood, and blood vessels. The heart, which acts as a pump that drives the flow of blood, is a muscular organ possessing four chambers: the left and right atria, which receive blood as it returns to the heart from the vasculature, and the left and right ventricles, which pump blood away from the heart through the vasculature. The blood vessels (arteries, arterioles, capillaries, venules, and veins) function as conduits for blood flow. Blood, which consists of a liquid (plasma) in which the other components (erythrocytes, leukocytes, and platelets) are suspended, acts as a medium that carries oxygen and nutrients to the body's cells while carrying away carbon dioxide and other waste products.

IP Cardiovascular, Anatomy Review: Blood Vessels p. 6

IP Cardiovascular, Anatomy Review: The Heart p. 4

IP Cardiovascular, Cardiac Action Potential pp. 4–5

IP Cardiovascular, Cardiac Cycle p. 3

The Path of Blood Flow Through the Heart and Vasculature, p. 416

The vasculature is divided into a pulmonary circuit, which supplies blood to the lungs, and a systemic circuit, which supplies blood to all the other organs and tissues of the body. In the pulmonary circuit, blood becomes oxygenated and gives up carbon dioxide; in the systemic circuit it becomes deoxygenated and picks up carbon dioxide. Blood is ejected from the right ventricle through the pulmonary semilunar valve into the pulmonary trunk, which divides into left and right pulmonary arteries, carrying blood to the lungs. The pulmonary veins carry blood away from the lungs and deliver it to the left atrium. From there, blood moves into the left ventricle. The left ventricle pumps blood

into the aorta, which delivers it to the systemic organs and tissues. Blood returns to the heart by way of the venae cavae, which carry it to the right atrium. From there, the blood enters the right ventricle. In contrast to the series flow of blood through the right and left sides of the heart, blood flow through the systemic circuit is parallel flow, with different arteries supplying fresh blood to different organs. Furthermore, the branching of blood vessels ensures that each capillary bed receives fresh blood.

IP Cardiovascular, Anatomy Review: The Heart, pp. 6–7

Anatomy of the Heart, p. 419

The heart is located in the thoracic cavity and is surrounded by a pericardial sac. Most of the heart consists of the myocardium. Valves in the heart ensure unidirectional flow of blood: Atrioventricular valves allow blood to flow from atrium to ventricle and semilunar valves allow blood to flow from ventricle to artery (left ventricle to aorta and right ventricle to pulmonary trunk).

Electrical Activity of the Heart, p. 421

The heart muscle fibers that make up the heart's conduction system are specialized to initiate action potentials and conduct them rapidly through the myocardium. Contractions of the heart are triggered on a regular basis by action potentials initiated by pacemaker cells concentrated in certain regions of the myocardium. Normally the heartbeat is driven by pacemakers in the sinoatrial (SA) node, located in the upper right atrium. Following each action potential, pacemaker cells exhibit slow, spontaneous depolarizations (pacemaker potentials) that eventually depolarize the membrane to threshold and trigger the next action potential. In most cardiac contractile cells, action potentials are characterized by a broad plateau phase that largely results from an in-

crease in the cell membrane's calcium permeability; the flow of calcium into the cells is important in triggering heart muscle contractions.

The heart's electrical activity can be recorded using electrodes placed on the skin surface, yielding an electrocardiogram (ECG), which consists of three phases: a P wave, corresponding to atrial depolarization; a QRS complex, corresponding to ventricular depolarization; and a T wave, corresponding to ventricular repolarization.

IP Cardiovascular, Intrinsic Conduction System, pp. 3–4

IP Cardiovascular, Cardiac Action Potential, p. 3

The Cardiac Cycle, p. 433

The cardiac cycle is divided into two distinct periods: diastole (ventricular relaxation), during which ventricular filling occurs; and systole (ventricular contraction), during which the exit of blood from the ventricles (ejection) occurs. Aortic pressure varies throughout the cardiac cycle; it rises to a maximum (systolic pressure, SP) during systole and falls to a minimum (diastolic pressure, DP) during diastole. The average pressure throughout the cycle, which represents the driving force for blood flow through the systemic circuit, is the mean arterial pressure (MAP). Ventricular volume falls to a minimum at the end of systole (end-systolic volume, ESV) and rises to a maximum at the end of diastole (end-diastolic volume, EDV). The difference between these volumes is the stroke volume (SV), the volume pumped by each ventricle in a single heartbeat.

IP Cardiovascular, Cardiac Cycle, pp. 4–19

IP Cardiovascular, Factors That Affect Blood Pressure, p. 14

IP Cardiovascular, Cardiac Output, p. 6

IP Cardiovascular, Intrinsic Conduction System, pp. 5–6

Cardiac Output and Its Control, p. 438

The volume of blood pumped by each ventricle per minute is the cardiac output (CO), which depends on the heart rate (HR) and stroke volume: $CO = HR \times SV$. The heart is regulated by sympathetic and parasympathetic neurons and hormones (extrinsic control), and by factors operating entirely within the heart (intrinsic control). Heart rate, determined by the firing frequency of the SA node, is entirely under extrinsic control. Stroke volume is under extrinsic and intrinsic control and is affected by three major factors: ventricular contractility, which is regulated by sympathetic neurons and epinephrine; end-diastolic volume, which depends on preload; and after-load, which depends on arterial pressure. The influence of end-diastolic volume on stroke volume is the basis of Starling's law of the heart, an example of intrinsic control of cardiac function.

IP Cardiovascular, Cardiac Output, pp. 4–9

IP Cardiovascular, Blood Pressure Regulation, pp. 4–9

▮ EXERCISES

Multiple-Choice Questions

1. Minimum aortic pressure during the cardiac cycle is attained
   a) immediately after closure of the aortic semilunar valve.
   b) immediately before opening of the aortic semilunar valve.
   c) immediately before opening of the atrioventricular valves.
   d) in mid-diastole.
   e) at the end of systole.

2. The first heart sound occurs when the atrioventricular valves close, and thus it marks
   a) the end of the ejection period.
   b) the start of the ejection period.
   c) the start of systole.
   d) the start of isovolumetric contraction.
   e) c and d are both true

3. If you know end-diastolic volume, the only other thing you need to know to determine stroke volume is
   a) afterload.
   b) ventricular contractility.
   c) end-systolic volume.
   d) heart rate.
   e) cardiac output.

4. As a result of the Starling effect, stroke volume should increase following an increase in
   a) mean arterial pressure.
   b) heart rate.
   c) sympathetic activity.
   d) afterload.
   e) preload.

5. Sympathetic and parasympathetic input to the SA node influences
   a) ventricular filling time.
   b) ventricular contractility.
   c) afterload.
   d) atrial contractility.
   e) all of the above

6. Which of the following contains deoxygenated blood?
   a) the right ventricle
   b) the left ventricle
   c) pulmonary veins
   d) the aorta
   e) a and c are both true

7. Which of the following is *not* normally apparent in the ECG?
   a) atrial depolarization
   b) atrial repolarization
   c) ventricular depolarization
   d) ventricular repolarization
   e) none of the above

8. The second heart sound occurs when the semilunar valves close, and thus it marks
   a) the end of the ejection period.
   b) the start of the ejection period.
   c) the start of systole.
   d) the start of isovolumetric contraction.
   e) c and d are both true

9. The QRS complex of the ECG is due to
   a) atrial depolarization.
   b) atrial repolarization.
   c) ventricular depolarization.
   d) ventricular repolarization.
   e) opening of the AV valves.

10. As a wave of action potentials travels from the atria to the ventricles, it is momentarily delayed by about 0.1 second as a result of slow conduction through
    a) the SA node.
    b) the AV node.
    c) the atrioventricular bundle.
    d) the left and right bundle branches.
    e) Purkinje fibers.

11. Which of the following is most likely to cause a *decrease* in the stroke volume of the left ventricle?
    a) an increase in mean arterial pressure
    b) an increase in end-diastolic pressure
    c) an increase in end-diastolic volume
    d) an increase in the activity of sympathetic nerves to the heart
    e) an increase in central venous pressure

12. Left ventricular pressure and aortic pressure are virtually identical during
    a) isovolumetric contraction.
    b) isovolumetric relaxation.
    c) diastole.
    d) systole.
    e) the ejection period.

Objective Questions

1. Heart rate is normally determined by the action potential frequency in the (SA/AV) node.

2. According to the Starling effect, stroke volume should increase if end-diastolic volume (increases/decreases).

3. Heart rate is determined entirely by the inherent action potential frequency in cells of the SA node, with no external influences. (true/false)

4. Blood flow through the systemic circuit is driven by contractions of the (right/left) ventricle.

5. The valve located at the junction between the left ventricle and the aorta is an example of a(n) (atrioventricular/semilunar) valve.

6. (Isovolumetric contraction/Ejection) comes immediately after diastole.

7. Maximum aortic pressure during the cardiac cycle is called (diastolic/systolic) pressure.

8. Under normal conditions, pressures in the left and right ventricles are equal during systole. (true/false)

9. Stroke volume and _____ completely determine cardiac output.

10. If end-diastolic volume does not change but end-systolic volume decreases, stroke volume (increases/decreases).

11. If end-diastolic volume does not change but end-systolic volume decreases, ejection fraction (increases/decreases).

12. If sympathetic and parasympathetic inputs are constant and end-diastolic volume increases, contractility of the ventricular myocardium increases. (true/false)

13. The period of relaxation of the heart muscle is known as _____.

14. The (P/T) wave of the ECG corresponds to ventricular repolarization.

15. Action potentials generated by pacemaker cells are called *pacemaker potentials*. (true/false)

Essay Questions

1. Discuss autonomic regulation of cardiac function. Include in your discussion a description of the effects of autonomic activity on the rate and force of ventricular contraction. Feel free to use Starling curves to clarify your discussion.

2. Describe the process of action potential propagation through the heart. Include a description of the role of pacemaker cells and gap junctions in cardiac electrical activity.

3. Discuss the interplay of the various influences on stroke volume. Be sure to include a discussion of the Starling effect and the influence of autonomic neurons. Use graphs or charts in your explanation if you feel that it is appropriate.

4. Clarify the distinction between ventricular *contractility* and *force of contraction*. Use graphs or charts in your explanation if applicable.

Critical Thinking

1. Jane trained for the marathon for 6 months. Jane's resting heart rate is 50 beats/min and her blood pressure is 105/75. Jane's roommate, Sue, watches a lot of TV and has a resting heart rate of 85 beats/min and blood pressure of 135/85. Both women have the same resting cardiac output of 5.0 liters/min. Determine the resting stroke volume and mean arterial pressure. Based on your findings, whose heart is beating more efficiently?

2. Cardiac arrhythmias are often treated with verapamil, an L-type calcium channel blocker. Explain the effects such calcium blockers would have on the heart. Include effects on conduction and contraction. What types of arrhythmias would these drugs be used to treat?

3. Damage to a valve can often be detected by auscultation with a stethoscope because valve damage causes murmurs. Mitral valve prolapse results from a defective valve and allows blood to flow backward. Where is the mitral valve? The murmur caused by mitral valve prolapse would be associated with what heart sound? Describe the changes in cardiac function that can result in severe cases of mitral valve prolapse.

4. In the ECG trace below, label all the P, QRS, and T waves. What abnormality is readily apparent? Based on your knowledge of the conduction pathway of the heart, determine where the abnormality is in the pathway.

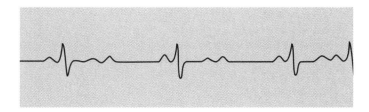

Find the answers to these exercises, and additional study tools, at the Physiology Place (www.physiologyplace.com).

The Cardiovascular System: Blood Vessels, Blood Flow, and Blood Pressure

15

Blood pressure is the driving force behind blood flow.

Have you ever stood up too fast and felt dizzy? Most people have experienced this sensation, called *orthostatic hypotension,* at one time or another. When a person stands up, the force of gravity pulls blood toward the legs and away from the upper parts of the body. This decreases the amount of blood returning to the heart, which—as we learned in the last chapter— decreases stroke volume and thus cardiac output. With less blood flowing into the vasculature, blood pressure decreases, called *hypotension.* Because pressure is the driving force for blood flow, the amount of blood reaching the brain may be decreased, causing the symptoms of dizziness.

As you will learn in this chapter, arterial blood pressure and blood flow to the organs are regulated to deliver adequate supplies of oxygen and nutrients to organs and tissues under most circumstances. So though you may sometimes feel dizzy when you stand up, usually the cardiovascular system rapidly adapts and maintains adequate blood flow to the brain so that your body can change position without detriment.

In Chapter 14, we learned how the heart pumps blood into the vasculature and how cardiac output is regulated by autonomic control of heart rate and stroke volume. We now turn to the vasculature, and how blood flow through the vasculature is regulated. We start with a discussion of the physical laws that govern the flow of blood through the vasculature. We then discuss the types of blood vessels that make up the vasculature and their roles in cardiovascular function. In the last part of this chapter, we describe blood pressure and how it is regulated by autonomic control of the vasculature and the heart.

Physical Laws Governing Blood Flow and Blood Pressure

In a certain sense the vasculature is an elaborate system of pipes that runs through the body, so the fundamental physical laws that describe the flow of any liquid through a system of pipes also pertain to blood flow in the cardiovascular system. The rule that is pertinent to

our discussion here states that the flow rate of a liquid (the volume flowing per unit of time) through a pipe is directly proportional to the difference between the pressures at the two ends of the pipe (the *pressure gradient*) and inversely proportional to the resistance of the pipe:

$$\text{flow} = \text{pressure gradient/resistance} = \Delta P/R$$

The quantity ΔP, the size of the pressure gradient, represents the driving force that *pushes* the flow of liquid through a pipe; the quantity R, the resistance, is a measure of the various factors that *hinder* the flow of liquid through a pipe.

This rule is so crucial to our understanding of blood flow that it is the starting point for all our discussions pertaining to flow, pressure, and resistance in the cardiovascular system. It is so universally applicable that it applies to liquids flowing in a single pipe or blood vessel, or in a system of pipes or blood vessels, no matter how complicated. As we will see in Chapter 17, it even pertains to the flow of air into and out of the lungs.

In this section we examine the general principles governing how pressure gradients and resistance affect blood flow in individual vessels and networks of vessels. First we consider pressure gradients.

Pressure Gradients in the Cardiovascular System

When you inflate a balloon, it expands because the pressure the air exerts on the inside of the balloon is greater than the pressure the air exerts on the outside. If you remove your fingers from the nozzle of the balloon, air rushes out for the same reason—air pressure is greater inside the balloon than outside. Whenever there is a difference in pressure between two locations, the pressure gradient drives the flow from a region of higher pressure to one of lower pressure, or *down the pressure gradient*.

Air flowing out of a balloon, like the flow of blood through the cardiovascular system, is an example of *bulk flow*. Regardless of whether the flowing medium is a gas or a liquid, the driving force for bulk flow is always a pressure gradient, and the direction of flow is always down the gradient from a region of greater pressure to a region of lower pressure. This rule applies to blood flow and to all other examples of bulk flow that occur in the body, such as the flow of air into and out of the lungs.

The Role of Pressure Gradients in Driving Blood Flow

As we saw in Chapter 14, the primary function of the heart is to generate the pressure that drives the flow of blood through the vasculature. Strictly speaking, however, it is not absolute pressure that drives blood flow, but rather a pressure gradient. By pumping blood into the arteries, the heart raises mean arterial pressure, which creates a difference in pressure between the arteries and veins that drives the flow of blood.

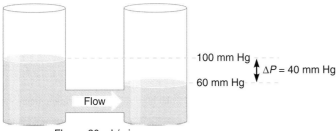

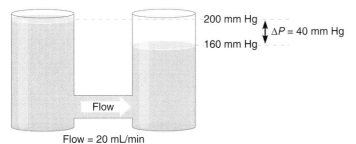

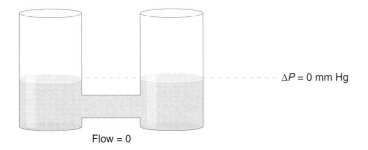

Figure 15.1 A model that relates blood flow to the pressure gradient. A single blood vessel is represented by a tube connecting two reservoirs, in which the depth of liquid determines the pressure. **(a)** The difference in the pressures in the two reservoirs produces a pressure gradient (Δ*P*) of 40 mm Hg, creating a flow of 20 mL/min, as indicated by the arrow. **(b)** When the levels of the liquid are the same in both reservoirs, the pressure gradient is zero, and hence flow is zero. **(c)** When the levels in both reservoirs are raised such that Δ*P* remains at 40 mm Hg, flow remains at 20 mL/min, indicating that the pressure gradient, not absolute pressure, determines flow.

Figure 15.1 shows a useful model for explaining the relationship between pressure and flow in blood vessels. In the diagram, a tube or "blood vessel" connects two large reservoirs containing liquid. The vertical distance from the vessel to the surface of the liquid, the so-called *hydrostatic column*, determines the pressure at either end of the vessel. (The pressure also depends on the *density* of the liquid—its mass per unit volume—which we assume is constant.) A higher hydrostatic column corresponds to a greater pressure. (You can feel the effect of a hydrostatic

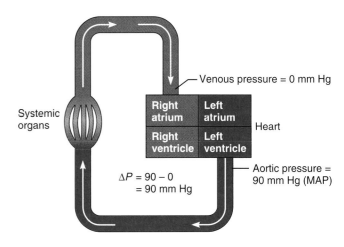

Figure 15.2 A pressure gradient is the driving force for blood flow. Aortic pressure averages about 90 mm Hg, whereas the pressure in the vena cava is close to zero where it joins the heart. This creates a pressure gradient of 90 mm Hg, which represents the overall driving force that pushes the flow of blood through the systemic circuit. Note that this pressure gradient is virtually identical to the mean arterial pressure (MAP).

column by diving underwater. The deeper you go, the more pressure the water exerts on your body.)

When the liquid level is different on the two sides (Figure 15.1a), a pressure gradient exists, and the liquid flows at a rate of 20 mL/min through the tube from the high-pressure side (100 mm Hg) to the low-pressure side (60 mm Hg), or down the pressure gradient (Δ*P* = 40 mm Hg). As a result, the level on one side drops while that on the other side rises. Eventually, the levels become equal (Figure 15.1b), and flow stops because a pressure gradient no longer exists.

Figure 15.1c illustrates that the rate of flow through the tube depends only on the *difference* between the pressures at either end, not the absolute pressure. When the liquid level is raised on both sides, the pressure on both sides increases. However, if the difference between the levels remains constant (here, Δ*P* still equals 40 mm Hg), the flow does not change (flow still equals 20 mL/min).

Pressure Gradients Across the Systemic and Pulmonary Circuits

In the systemic circuit, *mean arterial pressure* (*MAP,* the average pressure in the aorta throughout the cardiac cycle) is about 90 mm Hg. At the other end of the circuit, in the large veins in the thoracic cavity that lead to the right atrium, the pressure—known as the **central venous pressure (CVP)**—is approximately 2–8 mm Hg. The difference between the mean arterial pressure and central venous pressure is the pressure gradient that drives blood flow through the systemic circuit. Because central venous pressure is so small, we can say that *the pressure gradient (ΔP) driving blood flow through the systemic circuit is virtually identical to the mean arterial pressure* (**Figure 15.2**).

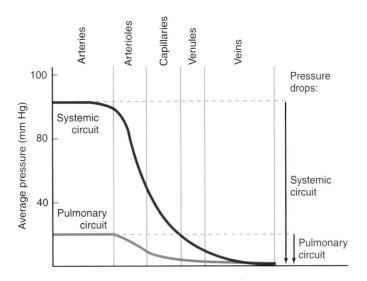

Figure 15.3 Pressures and pressure drops in the pulmonary and systemic circuits. The pressure drop represents the pressure gradient for blood flow through the circuit.

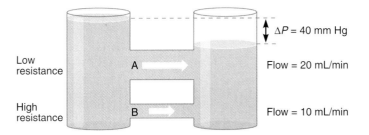

Figure 15.4 The effect of resistance on flow. In this model, two blood vessels are depicted as two tubes (A and B) connecting the reservoirs. Even though the pressure gradient is the same (40 mm Hg) for both vessels, blood flows through vessel B at a lower rate than through vessel A. Because it has a smaller diameter, vessel B has a higher resistance than vessel A.

> *If the pressure difference decreases to 20 mm Hg, what will be the flow rate in tube A? In tube B?*

Blood flow through the pulmonary circuit is also driven by a pressure gradient—the difference between the pressure in the pulmonary arteries and the pressure in the pulmonary veins. However, this pressure gradient is smaller than the one that drives flow through the systemic circuit because pulmonary arterial pressure is lower than aortic pressure (**Figure 15.3**). During the cardiac cycle, pulmonary arterial pressure averages about 15 mm Hg, as opposed to about 90 mm Hg in the aorta. Pulmonary venous pressure, like central venous pressure, is close to zero.

Resistance in the Cardiovascular System

In Chapter 14 we saw that the blood flow through the pulmonary circuit is identical to that through the systemic circuit (about 5 liters per minute at rest). But if the pressure gradient in the pulmonary circuit is lower than that in the systemic circuit, how is it possible to have the same blood flow in the two circuits? The answer can be deduced from the flow rule, flow = $\Delta P/R$: The pulmonary circuit offers less *resistance* (because of its physical characteristics), so a smaller pressure gradient can achieve the same flow.

Here we examine in turn the factors that determine the resistance of individual blood vessels and of networks of vessels such as the systemic and pulmonary circuits.

Resistance of Individual Blood Vessels

The resistance of any tube (including a blood vessel) is a measure of the degree to which the tube hinders or resists the flow of liquid through it. From the flow rule, it's apparent that for a given pressure gradient, a vessel with higher resistance yields a lower flow (**Figure 15.4**). Put another way, for a given pressure gradient, blood flow is greater when resistance is lower because it is easier for blood to flow (**Discovery: Permeability and Resistance**).

If you have ever drunk liquid through a drinking straw, you have experienced the effects of resistance. You likely noticed that it is easier to drink through a wide straw than through a narrow one, and that it is easier to drink through a short straw than through a long one. You also know that it's more difficult to drink a milkshake through a straw than it is to drink a soda. Resistance, then, depends on the physical dimensions of the tube and the properties of the fluid flowing through it, namely the tube's *radius* and *length*, and the fluid's *viscosity* ("thickness" or "syrupiness"). The effect of these factors on blood flow are described below and mathematically in the **Toolbox: Poiseuille's Law**, p. 456.

1. *Vessel radius.* Changes in resistance to blood flow in the cardiovascular system almost always result from changes in the radii of blood vessels: As radius decreases, resistance increases. A decrease in blood vessel radius is called **vasoconstriction**; an increase in vessel radius is called **vasodilation**.

2. *Vessel length.* Even though longer vessels have greater resistance than shorter ones (all else being equal), changes in vascular resistance are rarely due to changes in vessel length; vessels do not change length except as a person grows.

3. *Blood viscosity.* Vascular resistance increases as viscosity increases, but blood viscosity does not change appreciably under normal conditions. The major determinant of blood viscosity is the concentration of cells and proteins in the blood; as either increases, blood viscosity increases.

Resistance of Blood Vessel Networks: Total Peripheral Resistance

Although thus far we have limited our consideration to the resistance of individual blood vessels, a *network* of

In examining membrane transport in Chapter 4, we encountered Fick's law, which relates the flux to the concentration gradient for a substance crossing a membrane by simple diffusion. In simple diffusion, a concentration gradient of a given magnitude produces a larger or smaller flux, depending on a variable called *permeability*, which in turn depends on properties of the diffusing substance and of the membrane through which the substance is moving. Permeability therefore relates a flow of something (the flux) to a driving force (the concentration gradient).

In this chapter we explore another relationship between a flow and a driving force. In this case the flow is the bulk flow of a fluid, such as blood, and the driving force is a pressure gradient. The parameter relating them is the *resistance*. Resistance, like permeability, depends on properties of the substance that is flowing and of the structure through which the substance is moving. The following table compares bulk flow and simple diffusion.

	Bulk flow	**Simple diffusion**
Driving force	Pressure gradient (ΔP)	Concentration gradient (ΔC)
Flow	Volume of fluid moving through a tube or vessel per unit time (flow)	Number of molecules crossing a membrane per unit time (flux)
Variable relating flow to driving force	Resistance of blood vessel (R)	Permeability of membrane (P)
Mathematical relationship	Flow = $\Delta P/R$	Flux = $PA(\Delta C)$ (A = membrane area)
Direction of flow	From higher to lower pressure (down the pressure gradient)	From higher to lower concentration (down the concentration gradient)
Characteristics of moving substance that affect resistance or permeability	Fluid's viscosity	Diffusing molecules' size shape, and lipid solubility
Characteristics of vessel or membrane affecting resistance or permeability	Blood vessel's radius and length	Membrane's thickness and lipid composition

blood vessels (such as the systemic or pulmonary circuits, the vasculature within an organ, or even a single capillary bed) also has a resistance. For blood vessel networks, the rules governing flow, pressure, and resistance are fundamentally the same as for individual vessels: For any network, the total flow increases in proportion to the pressure gradient along the network and decreases as the resistance of the network increases.

As we might expect, the resistance of a vascular network depends on the resistances of all the individual blood vessels it contains. Any factor that causes the resistance of individual vessels in a network to increase or decrease also tends to cause a respective increase or decrease in the resistance of the network as a whole. From this, it follows that *vasoconstriction anywhere within a network of blood vessels tends to increase the resistance of the network, whereas vasodilation anywhere within a network tends to decrease the resistance of the network.* In the systemic circuit, the combined resistances of all the blood vessels within the circuit is known as **total peripheral resistance (TPR)**.

Relating Pressure Gradients and Resistance in the Systemic Circulation

We can express the relationship among pressure, resistance, and flow in the systemic circuit by making some substitutions in the flow rule, $F = \Delta P/R$. First, because all the blood that flows from the heart goes through the systemic circuit, the flow is equal to the volume of blood flowing through the circuit each minute, or cardiac output (CO). Next, we have already seen that this flow of blood is driven by the pressure gradient represented by the difference between mean arterial pressure (MAP) and central venous pressure, and that this pressure gradient is virtually identical to MAP. And finally, we know that the

POISEUILLE'S LAW

A fluid flowing through a tube or blood vessel encounters resistance, some of which is due to frictional forces acting between the fluid and the walls of the tube or vessel. Friction within the fluid itself also contributes to the resistance, which is why some fluids, such as molasses, flow more slowly than others, such as water. The speed at which a fluid moves varies from one location to another within the moving liquid. Therefore, layers of fluid moving at different speeds rub against each other, which creates friction and dissipates energy. For a fluid moving smoothly through a cylindrical tube, the resistance (R) is given by the following equation, which is called *Poiseuille's law:*

$$R = 8L\eta/\pi r^4$$

where L is the length of the tube, η is the viscosity of the fluid, and r is the tube's internal radius. Note that the resistance is strongly affected by the internal diameter of the tube because it depends on the *fourth power* of the radius. Therefore, if one tube is half as wide as another of the same length, its resistance is 16 times as great!

resistance in the systemic circuit is total peripheral resistance (TPR). Accordingly, substituting these variables into the flow rule yields

CO = MAP/TPR

Now that we understand the physics governing blood flow, we describe the different components of the vasculature and their role in blood flow and exchange of material between the blood and interstitium.

Quick Test 15.1

1. What is a pressure gradient? Where are pressure gradients present in the vasculature?
2. Define *mean arterial pressure*. Define *central venous pressure*. Which is greater? What is the significance of the difference between these two pressures?
3. Define *resistance*. What are some factors that affect resistance in the vasculature?

Overview of the Vasculature

Blood vessels are classified according to whether they carry blood away from or to the heart, and according to size. Arteries and smaller arterioles carry blood from the heart and to capillaries, which are drained by venules and then larger veins, which return the blood to the heart (**Figure 15.5**). The arterioles, capillaries, and venules can only be seen with the aid of a microscope, and therefore are called the **microcirculation**. All blood vessels possess a hollow interior called the *lumen*, through which blood flows; the lumen of all blood vessels is lined by a layer of epithelium called the endothelium. Surrounding the lumen is a wall that varies in thickness and composition from one vessel to another.

The smallest of all blood vessels, capillaries, consist of a layer of endothelial cells and a basement membrane; the walls of all other blood vessels contain various amounts of *smooth muscle* and fibrous and/or elastic connective tissue (**Figure 15.6**). Within the fibrous connective tissue are extracellular fibers made of a protein called *collagen*, which lends tensile strength to vessel walls, enabling them to withstand the pressure of blood within them without rupturing. Elastic connective tissue contains fibers of a highly stretchable extracellular protein called *elastin*, which enables blood vessels to expand or contract as the pressure of blood within them changes.

We next examine the structure and function of the various types of blood vessels, beginning with arteries.

Arteries

Arteries conduct blood away from the heart and toward the body's tissues. The largest artery, the aorta, has an internal diameter of about 12.5 mm and a wall that is 2 mm thick. The smaller arteries that branch off the aorta have internal diameters ranging from 2 mm to 6 mm and a wall thickness of about 1 mm, and these branch into yet smaller diameter arteries. The larger arteries provide little resistance to blood flow and therefore serve as a rapid conduit for blood to travel. The walls of large arteries contain large amounts of elastic and fibrous tissue, enabling arteries to withstand relatively high blood pressures, which are higher in these vessels than anywhere else in the vasculature.

As the arteries branch into smaller arteries, the amount of elastic tissue in the walls decreases while the amount of smooth muscle increases. Arteries less than 0.1 mm (100 microns) in diameter lose most of their elastic properties and are sometimes called *muscular arteries*. The smooth muscle enables regulation of the radius of small arteries in a manner similar to that described for arterioles shortly.

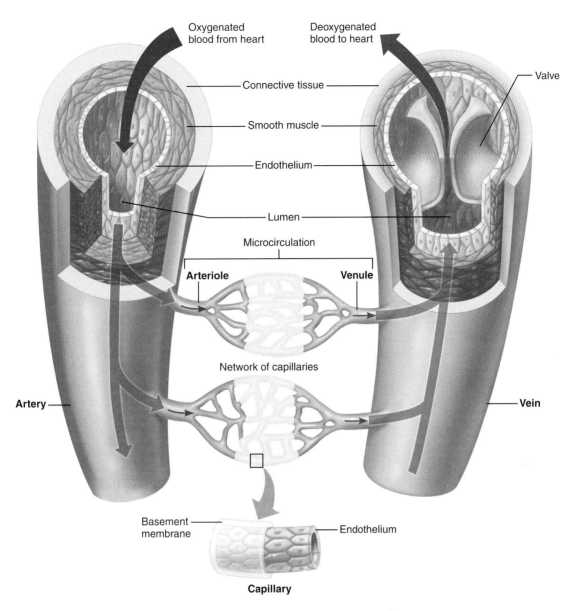

Figure 15.5 The relationships of blood vessels according to size and the direction of blood flow. The arrow at the upper left represents oxygenated blood (red) arriving from the heart, whereas the arrow at the upper right represents deoxygenated blood (blue) returning to the heart. Note the differences in lumen diameter and wall thickness between arteries and veins, and the presence of one-way valves in the vein.

Arteries: A Pressure Reservoir

The thickness of arterial walls, coupled with the relative abundance of elastic tissue, gives arteries both a certain stiffness and the ability to expand and contract as the blood pressure rises and falls with each heartbeat. This combination of stiffness and flexibility enables arteries to perform one of their major functions—acting as *pressure reservoirs* (storage sites for pressure) to ensure a continual, smooth flow of blood through the vasculature even when the heart is not pumping blood (diastole).

Figure 15.7 illustrates the concept of a pressure reservoir. As the arterial walls are expanding due to increased volume during systole, the elastin fibers act

much like a spring being stretched. This elastic force is stored such that during diastole, when no more blood is entering the arteries, the walls passively recoil inward, propelling blood forward. Therefore, blood moves through the vasculature continuously, propelled during systole by the ejection of blood from the heart and during diastole by the elastic recoil of the arterial walls. Although blood flow is continuous, palpitation of arteries, such as the radial artery in the wrist, reveals a pulse. The pulse is caused by a pressure wave that travels along the arteries in response to blood being pushed into the arteries during systole, causing the arterial walls to expand.

To serve as a pressure reservoir, arteries must have low **compliance**, which is a measure of the relationship

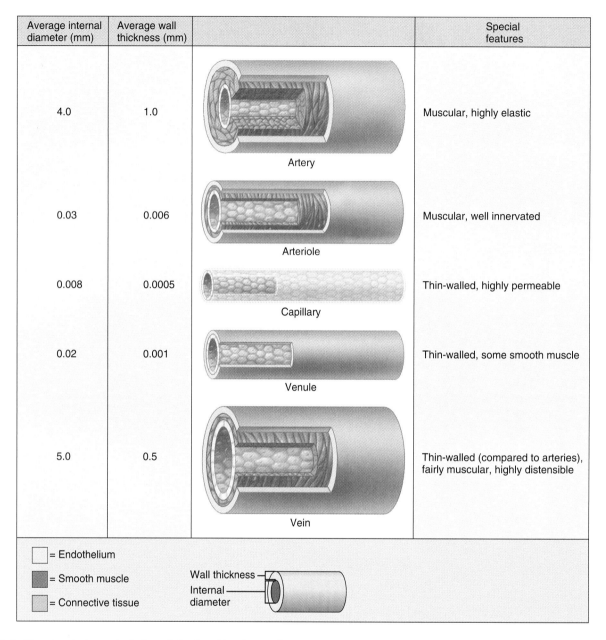

Average internal diameter (mm)	Average wall thickness (mm)		Special features
4.0	1.0	Artery	Muscular, highly elastic
0.03	0.006	Arteriole	Muscular, well innervated
0.008	0.0005	Capillary	Thin-walled, highly permeable
0.02	0.001	Venule	Thin-walled, some smooth muscle
5.0	0.5	Vein	Thin-walled (compared to arteries), fairly muscular, highly distensible

☐ = Endothelium

■ = Smooth muscle

▨ = Connective tissue

Wall thickness
Internal diameter

Figure 15.6 Structural characteristics of the five blood vessel types.

between pressure and volume changes (**Toolbox: Compliance**). In vessels with low compliance, such as arteries, a small increase in blood volume causes a large increase in blood pressure (or a large increase in pressure causes only a small degree of expansion of the blood vessel walls). Therefore, when the heart ejects blood into the arteries during systole and causes them to expand, the resulting rise in pressure is greater than it would be if arteries' compliances were higher. The low compliance of arteries is a function of the elasticity of the vessel walls.

Arterial Blood Pressure

As blood is ejected from the ventricle into the aorta, pressure in the aorta increases to almost that of the ventricle. Pressure in the aorta does not stay elevated, however, because during diastole, blood quits flowing into the aorta yet continues to flow out, which causes a slow decline in arterial blood pressure to a minimum just prior to the next systole. The pressure in the aorta is called the **arterial blood pressure**. Recall from Chapter 14 that ventricular pressure decreases to 0 mm Hg, whereas arterial pressure remains elevated. The pressure in the arteries during diastole is due to the elastic recoil property described above. As the arterial walls recoil inward, they exert a force on the blood, increasing the pressure. Because arterial blood pressure varies with the cardiac cycle, the maximum pressure that occurs during systole is called the systolic pressure, and the minimum pressure that occurs during diastole

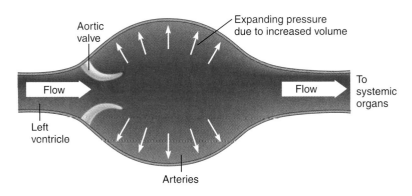

(a) Systole

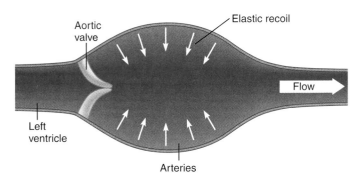

(b) Diastole

Figure 15.7 The role of arterioles as a pressure reservoir. Flow arrows indicate the movement of blood into and out of the arteries. Pressure arrows indicate the expansion and recoil of the arterial walls. **(a)** During systole, the pressure on the blood in the left ventricle is greater than that in the arteries, causing blood to flow through the aortic valve into the arteries. Blood entering the arteries causes their volume to increase, stretching the walls. **(b)** During diastole, the arterial walls recoil inward, pushing blood through the vasculature. Blood cannot flow back into the ventricle because the aortic valve is closed.

▓ TOOLBOX

COMPLIANCE

A blood vessel (indeed, any hollow structure) tends to expand as the pressure inside it rises, and to contract as the pressure falls. Strictly speaking, it is not just the pressure *inside* that determines whether a vessel expands or contracts, but rather the *difference* between the pressure inside and the pressure outside. This pressure difference is called the *distending pressure* (or *transmural pressure*). When the pressure inside a vessel is greater than the pressure outside, the distending pressure is positive, and a net outward force acts on the wall of the vessel, tending to make it expand. As the distending pressure increases, the volume of the vessel increases. If the pressure inside a vessel is less than the pressure outside, the distending pressure is negative, and a net inward force acts on the wall and tends to compress the vessel.

The *compliance* of a vessel is strictly defined as the change in volume per unit change in distending pressure. Mathematically, this is expressed as

$$\text{compliance} = \Delta V/\Delta(P_{\text{inside}} - P_{\text{outside}})$$

where ΔV is the change in volume, and P_{inside} and P_{outside} are pressures inside and outside, respectively. The term within the parentheses is the distending pressure.

The following diagram, which is a pressure-volume curve, graphically depicts the concept of compliance:

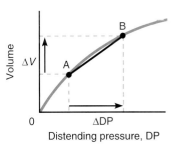

An increase in the distending pressure in a vessel, as when one goes from point A to point B in the diagram, produces a certain increase in the vessel's volume. The increase in volume is given by the vertical distance between the two points. The compliance is therefore the change in vertical distance (volume) divided by the change in horizontal distance (distending pressure). In other words, the compliance of the vessel over the range between point A and B is the *slope* of the line joining the two points.

is called the diastolic pressure. The average arterial pressure during the cardiac cycle is the mean arterial pressure (MAP).

When you go to a physician to have your blood pressure taken, what is actually measured is an *estimate* of the arterial pressure. Because there is no convenient way to measure aortic pressure directly, pressure is usually measured in the *brachial artery*, which runs through the upper arm. Pressure measured in this manner is close to aortic pressure because the brachial artery is not far from the heart and is also at about the same height as the aorta. (Blood pressure tends to be lower in upper regions of the body and higher in lower regions due to the force of gravity acting on blood.) For convention, blood pressure is usually recorded as systolic pressure over diastolic pressure (in mm Hg). For a healthy resting adult, the systolic pressure is less than 120 mm Hg and the diastolic pressure is less than 80 mm Hg. Using 120 and 80 mm Hg as examples of systolic and diastolic pressures, respectively, the blood pressure is recorded as 120/80, and the mean arterial pressure is 93 mm Hg (**Discovery: Measuring Arterial Blood Pressure**, p. 462).

> **Quick Test 15.2**
>
> 1. Name the three types of tissue commonly found in blood vessel walls. Compare the wall thickness of the different blood vessels.
>
> 2. Explain how blood can continue to flow through the vasculature during diastole.
>
> 3. Define *compliance*. How does the compliance of arteries contribute to their function as a pressure reservoir?

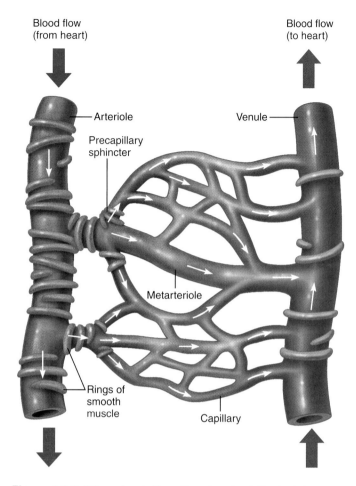

Figure 15.8 Microcirculation. The microcirculation includes arterioles, metarterioles, capillaries, and venules. Direction of blood flow is indicated by arrows within vessels. Smooth muscle is located in the walls of the arterioles and metarterioles and as precapillary sphincters. Contraction or relaxation of the smooth muscle increases or decreases, respectively, resistance to blood flow.

Arterioles

Arterioles are part of the microcirculation, illustrated in **Figure 15.8**. Arterioles lead either into capillary beds or into *metarterioles*, which then lead into capillary beds. The walls of arterioles contain little elastic material but have an abundance of circular smooth muscle that forms rings around the arterioles. Because this smooth muscle can contract or relax, changing the diameter of arterioles, arterioles are best known as the site where resistance to blood flow can be regulated.

Arterioles and Resistance to Blood Flow

We learned that blood flow depends on the pressure gradient and the resistance to blood flow (flow = $\Delta P/R$). The arterioles are the blood vessels that provide the greatest resistance to blood flow; in fact, over 60% of total peripheral resistance is attributable to arterioles. (Although individual capillaries have a smaller radius than arterioles, there are so many capillaries that their total cross-sectional area is

greater than that of arterioles, so together the capillaries have less resistance.) This is illustrated in **Figure 15.9**, which shows that as blood flows from arteries to veins, pressure decreases gradually. A difference in pressure across any portion of the vasculature is also called the *pressure drop* across that part. The largest pressure drop occurs along the arterioles: In the systemic circuit, blood enters them at an average pressure of about 90 mm Hg and leaves them at a pressure of about 40 mm Hg. The reason for this large pressure drop is related to the high resistance of arterioles. In contrast, the pressure drops in the larger vessels (the arteries and veins) are quite small, such that the pressure in these vessels is nearly uniform.

Arterioles not only provide the greatest resistance to blood flow, but their resistance can be regulated. In fact, the major function of arterioles is to serve as points of control for regulating resistance to blood flow, which serves two functions: (1) controlling blood flow to individual capillary beds, and (2) regulating mean arterial pressure.

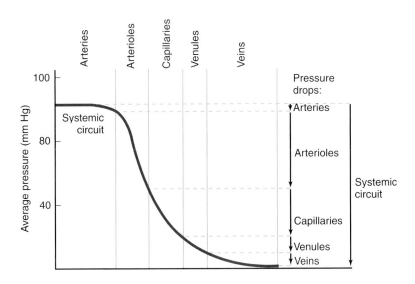

Figure 15.9 Pressures in the vasculature. Distances along the horizontal axis represent the relative distances blood travels in the various portions of the vasculature. Representative blood pressures and pressure drops for the various vessel types in the systemic circulation. Note that the greatest decrease in pressure occurs as blood travels through the arterioles.

Resistance is regulated by the contraction and relaxation of the circular smooth muscle. Arteriolar smooth muscle is single-unit smooth muscle, which has cells linked together by gap junctions such that they contract in unison, that is, as a unit. It also has pacemaker cells that spontaneously depolarize. Because of their spontaneous depolarizations, arteriolar smooth muscle is partially contracted in the absence of any external factors, called **arteriolar tone**. External factors can either increase or decrease the contractile state. When arteriolar smooth muscle contraction increases, the radius of arterioles decreases (vasoconstriction) and resistance increases, causing a decrease in blood flow. When the arteriolar smooth muscle relaxes, the radius of arterioles increases (vasodilation) and resistance decreases, causing an increase in blood flow (**Figure 15.10**).

Both intrinsic and extrinsic control mechanisms alter the contractile state of arteriolar smooth muscle. Intrinsic controls include a variety of local metabolites that regulate blood flow to match the needs of the cells in the region. Extrinsic controls include both the autonomic nervous system and hormones. Unlike intrinsic control, extrinsic controls are for the purpose of regulating mean arterial pressure.

Intrinsic Control of Blood Flow Distribution to Organs

Blood does not flow equally to all organs but instead gets distributed among the organs based on need. The distribution of blood flow is analogous to the problem of regulating water usage in a neighborhood: Whereas it is the job of the utility company to provide adequate water pressure for all houses in the system, the regulation of the flow of water to any individual house can be accomplished only by the people who live there, because they are the only ones who know of their moment-to-moment need for water. Similarly, extrinsic control mechanisms provide adequate arterial pressure for all organs in the

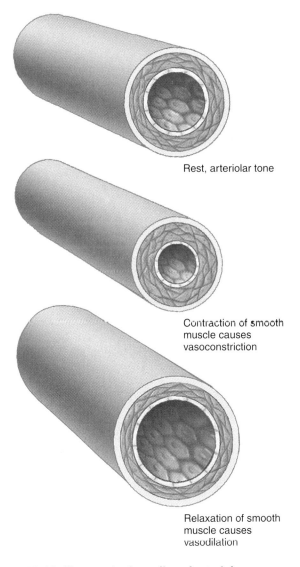

Rest, arteriolar tone

Contraction of smooth muscle causes vasoconstriction

Relaxation of smooth muscle causes vasodilation

Figure 15.10 Changes in the radius of arterioles.

■ **DISCOVERY**

MEASURING ARTERIAL BLOOD PRESSURE

The most common method for measuring blood pressure is based on the phenomenon of turbulence and the noise that accompanies it. Under most circumstances, blood moves through the cardiovascular system via *laminar* (streamlined) *flow,* similar to water that slowly leaves a faucet in a thin, silent stream. In laminar flow within a blood vessel, all the blood is flowing smoothly along the length of the vessel [see figure (a)], and blood flowing in the center of the vessel moves faster than blood nearer the wall because friction between the moving blood and the wall slows the blood down.

However, when a liquid is forced to move through a tube quickly enough, its flow becomes *turbulent*, just as water leaving a faucet more quickly begins to flow roughly and noisily. In turbulent flow [see figure (b)], movement of blood in a vessel is disrupted into eddies or whirls in which the blood flows in different directions and speeds in different places within the vessel. These variations increase the resistance and create vibrations that can be heard.

When technicians take your blood pressure, they use a device called a *sphygmomanometer*, which consists of an inflatable cuff and a pressure-measuring device that displays air pressure inside the cuff, and they use a stethoscope placed over the brachial artery to listen for sounds produced by turbulent blood flow.

To measure blood pressure, the technician places the cuff around the upper arm and inflates it to increase the cuff pressure. This pressure is transmitted through the tissue of the arm to the brachial artery, which runs to the lower arm. The technician increases cuff pressure until it is above systolic arterial pressure, which causes the artery to collapse, stopping blood flow. With no blood flowing through the artery, no sounds can be heard through the stethoscope.

Next, the technician opens a valve to slowly let air out of the cuff, allowing cuff pressure to fall. When cuff pressure drops to where it is just slightly below systolic arterial pressure, the artery opens briefly with each heartbeat, because the pressure inside the artery is higher than that outside it, forcing the vessel open. When this happens, blood flows through the artery in a turbulent fashion because it is forced through a narrow opening. This turbulence creates sounds, called *Korotkoff sounds*, that can be heard through the stethoscope. When the Korotkoff sounds first appear, the technician notes the cuff pressure and records it as systolic arterial pressure.

The technician continues to let air out of the cuff, which further lowers cuff pressure. As a consequence, the artery remains open for longer periods during each heartbeat, but flow remains turbulent. Eventually, cuff pressure falls just be-

low diastolic arterial pressure, from which point the artery stays open throughout the entire cardiac cycle because pressure inside the artery is always higher than that outside it. Under these conditions, blood flow returns to its laminar pattern and the Korotkoff sounds disappear. The technician notes the cuff pressure where sounds first disappear and records it as the diastolic arterial pressure.

The events involved in blood pressure measurement are illustrated in diagram (c); the straight line represents cuff pressure, and the wavy line represents arterial pressure.

The blood pressure is recorded as systolic pressure (SP) over diastolic pressure (DP), SP/DP. Average normal values for a healthy individual are 110/70. From the blood pressure measurement, pulse pressure and mean arterial pressure can be determined.

Pulse pressure (PP) is the difference between systolic pressure and diastolic pressure:

$$PP = SP - DP$$

Using average numbers for a healthy adult, the pulse pressure is

$$PP = 110 \text{ mm Hg} - 70 \text{ mm Hg}$$
$$= 40 \text{ mm Hg}$$

In older people, a pulse pressure that is abnormally high may indicate *hardening*

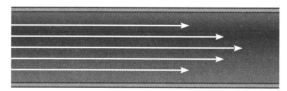

(a) Laminar flow

(b) Turbulent flow

systemic circuit, but only an individual organ or tissue can really "know" how much blood it needs at any given time, and so it regulates its own blood flow through local control. The factors that determine blood flow to organs are illustrated in **Figure 15.11**.

In Figure 15.11a, we consider three organs arranged in parallel such that blood flow for each organ is driven by the same pressure gradient: the difference between mean arterial pressure and central venous pressure, which effectively equals MAP (central venous pressure is almost 0). Suppose

DISCOVERY

(continued)

of the arteries, a condition in which the arteries become thickened and more rigid, which decreases their ability to stretch.

Mean arterial pressure (MAP), the average pressure occurring in the arteries during one cardiac cycle, is estimated by the following expression:

$$MAP = \frac{SP + (2 \times DP)}{3}$$

For a normal healthy adult, this becomes

$$MAP = \frac{110 + (2 \times 70)}{3} = 83.3 \text{ mm Hg}$$

Notice that the mean arterial pressure is *not* obtained simply by taking the average of the systolic and diastolic pressures, which is 90 mm Hg. Instead, MAP is a weighted mean, in which diastolic pressure is given twice the weight of systolic pressure. The reason has to do with the

changes in aortic pressure as seen in the shape of the pressure wave: During a single cardiac cycle, aortic pressure is near its maximum for a relatively short period and is closer to the minimum about twice as long. Any change away from this pattern affects the mean arterial pressure—even if systolic and diastolic pressure do not change. For this reason, the equation for mean arterial pressure should be regarded as only an estimate.

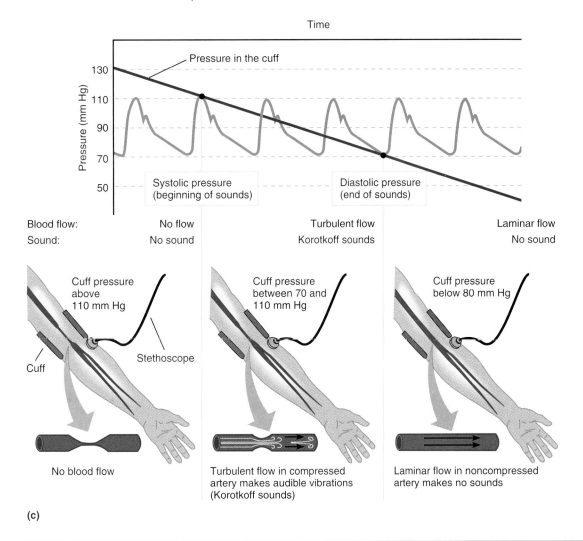

(c)

---

the flows through organs A, B, and C are 1.5 liters/min, 1.0 liter/min, and 0.5 liter/min, respectively (Figure 15.11b). Given that all these flows are driven by the same pressure gradient, the differences in flow must result from differences in resistance, according to the following rule:

organ blood flow = MAP/organ resistance

On the basis of this relationship, the flows depicted in Figure 15.11b indicate that resistance is lowest in organ A, intermediate in organ B, and highest in organ C.

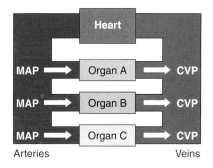

(a)

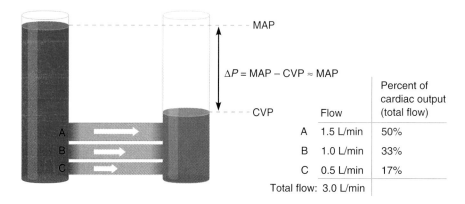

(b)

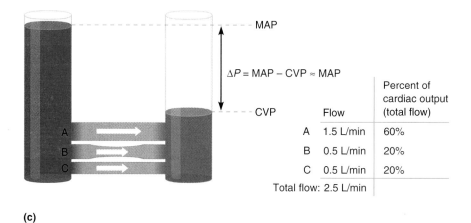

(c)

Figure 15.11 The effects of pressure gradients and resistance on blood flow to organs. (a) Blood flow through three parallel organs. The pressure gradient driving flow is the difference between MAP and CVP. (b) Differences among blood flows to organs are due to differences in resistance in the organs. (c) Given constant MAP and CVP, an increase in resistance in one organ (organ C) reduces flow to that organ alone. Note that this results in a reduction in total flow and in a change in the distribution of blood flow.

In part (c), how do the resistances of tubes B and C compare?

They are the same.

The effect of organ resistance on the *distribution* of blood flow is illustrated by comparing parts (b) and (c) of Figure 15.11. In both situations MAP is constant; the resistances, and thus the blood flows, of organs A and C are unchanged. But the resistance of organ B is higher in Figure 15.11c, which results in a decrease in blood flow compared to that in Figure 15.11b. Note that because the flow to organ B decreases while the flows to organs A and C remain unchanged, total flow declines, and organ B's *share* of cardiac output decreases while the *shares* of organs A and C increase.

These findings have an important implication: Changes in the distribution of blood flow to organs—that is, changes in the percentage of cardiac output supplied to each organ—are due to changes in the vascular resistance of individual organs.

The vascular resistance of an organ (or tissue) is altered by the contraction or relaxation of smooth muscle in arterioles (and small arteries). Arterioles can vasoconstrict or vasodilate in response to intrinsic factors, thereby changing the resistance to blood flow. Intrinsic control mechanisms regulate not only the distribution of blood flow among the organs, but also the distribution of blood flow *within* organs. These mechanisms are responsible for the sharing of flow among capillary beds.

Intrinsic control mechanisms are especially important in regulating blood flow to the heart, brain, and skeletal muscles for two reasons: (1) Neural mechanisms play only a minor role in regulating blood flow to the heart and brain, and (2) metabolic activity in the heart and skeletal muscles can vary greatly, changing their demands for oxygen and nutrients. Even though the metabolic activity of the brain as a whole is relatively constant, activity varies from region to region within the brain. These activity patterns change frequently, depending on what a person is doing at any given moment. Intrinsic mechanisms work on a continual basis to increase blood flow to regions that are becoming more active while reducing flow to those whose activity is decreasing.

Given that intrinsic control of organ blood flow is accomplished through relaxation or contraction of smooth muscle in arterioles, which control the flow through individual capillary beds within an organ or tissue, what senses whether blood flow is adequate or not? The answer is *vascular smooth muscle* itself. In the following sections we examine the responses of vascular smooth muscle in response to four factors: changes in metabolic activity, changes in blood flow, stretch of arteriolar smooth muscle, and local chemical messengers.

Regulation in Response to Changes in Metabolic Activity: Active Hyperemia

Vascular smooth muscle cells in arterioles are sensitive to conditions in extracellular fluid and respond to changes in the concentrations of a wide variety of chemical substances, including oxygen, carbon dioxide, potassium ions, hydrogen ions, and others. These changes in concentration occur as a result of metabolic activity. Arteriolar smooth muscle either contracts of relaxes depending on whether concentrations of particular substances rise or fall. The general rule of thumb is this: *Changes associated with increased metabolic activity generally cause vasodilation, whereas changes associated with decreased metabolic activity induce vasoconstriction.* (An important exception to this rule occurs in the pulmonary vasculature, as we see in Chapter 18.)

We can demonstrate how this rule works by considering the oxygen and carbon dioxide concentrations in tissue as an example (**Figure 15.12**). When blood flow is matched to the tissue's metabolic needs, the concentrations of oxygen and carbon dioxide in the tissue are in a steady state: The rate at which oxygen enters the tissue from the blood equals the rate at which it is consumed by the cells, and the rate of carbon dioxide entering the blood equals the rate at which it is produced by the cells. Now suppose that the metabolic rate increases, such that the rates of oxygen consumption and carbon dioxide production rise, causing a decrease in tissue oxygen (called *hypoxia*) and an increase in tissue carbon dioxide. Initially, blood flow is insufficient to keep up with metabolic demand, a condition called **ischemia**. The decrease in oxygen and the increase in carbon dioxide both act on arteriolar smooth muscle, causing it to relax. When the muscle relaxes, vascular resistance in the tissue drops, and blood flow in that region increases. This increase in blood flow following an increase in metabolic activity is termed **active hyperemia** (hyperemia is a general term for a higher-than-normal rate of blood flow). As a result of this increase in blood flow, the oxygen delivery to the tissue and carbon dioxide removal from the tissue increase, and eventually a new steady state is achieved. Note that the changes in blood flow are a result of a direct effect of the reduced oxygen and elevated carbon dioxide on the arterioles themselves, and that no nerves or hormones are involved. Thus, active hyperemia is an example of *intrinsic* control.

The same mechanism decreases blood flow if the local metabolic rate falls. As metabolic activity drops, oxygen levels tend to increase and carbon dioxide levels tend to decrease, which causes vasoconstriction of the arterioles, thereby increasing resistance to blood flow.

Regulation in Response to Changes in Blood Flow: Reactive Hyperemia

In the previous examples, blood flow increased in response to changes in metabolic activity; tissue oxygen and metabolite concentrations can also change as a result of changes in *blood flow*. For example, if blood flow is blocked or reduced below adequate levels for any reason (such as occlusion of a blood vessel), the oxygen concentration falls and the carbon dioxide level rises because rates of oxygen consumption and carbon dioxide production exceed rates of delivery and removal, respectively. Both of these changes induce vasodilation and a reduction in vascular resistance, which

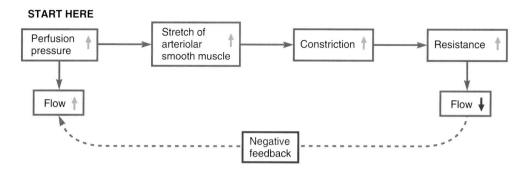

START HERE

Figure 15.14 The myogenic response to changes in perfusion pressure. Changes occurring in response to a rise in perfusion pressure.

in perfusion pressure also increases stretch of arteriolar walls, which also induces vasoconstriction and a reduction in blood flow.

Regulation by Locally Secreted Chemical Messengers

Contractile activity of vascular smooth muscle is also affected by a variety of chemical substances, most of which are secreted by blood vessel endothelial cells or by cells in surrounding tissues (**Table 15.1**). One such substance is *nitric oxide*, which is released on a continual basis by endothelial cells in arterioles and acts on smooth muscle to promote vasodilation. Substances produced by inflamed tissues, such as *bradykinin* and *histamine*, stimulate nitric oxide synthesis. The resulting increase in blood flow accounts for the redness of inflamed areas. Another potent vasodilator is *prostacyclin*, an eicosanoid that functions in preventing blood clots (Chapter 16). *Adenosine* is

Table 15.1 ▮ Local Vasoactive Substances and Their Actions on Vascular Smooth Muscle

Substance	Source	Effect on vascular smooth muscle
Oxygen	Delivered to tissues by blood; consumed in aerobic metabolism	Vasoconstriction
Carbon dioxide	Generated in aerobic metabolism	Vasodilation
Potassium ions	Released from cells (particularly in muscle) as a result of repeated depolarization occurring during activity	Vasodilation (vasoconstriction at high concentrations)
Acids (hydrogen ions)	Generated during anaerobic metabolism (lactic acid) and by reaction of carbon dioxide with water (carbonic acid)	Vasodilation
Adenosine	Released by cells in certain tissues in response to hypoxia	Vasodilation
Nitric oxide	Released by endothelial cells on a continuous basis and in response to various chemical signals	Vasodilation
Bradykinin	Generated from a precursor protein (kininogen) by action of an enzyme (kallikrein) secreted by cells in certain tissues in response to various chemical signals	Vasodilation
Endothelin-1	Released by endothelial cells in response to various chemical signals and mechanical stimuli	Vasoconstriction
Prostacyclin	Released by endothelial cells in response to various chemical signals and mechanical stimuli	Vasodilation

an important vasodilator in the *coronary arteries*. Among the substances that promote vasoconstriction is *endothelin-1*, which is produced by endothelial cells.

We just learned that several intrinsic mechanisms exist to regulate blood flow to individual organs. We now look at an example of how these factors affect organ blood flow during exercise.

Independent Regulation of Blood Flow During Exercise

Independent regulation of organ blood flow is illustrated in **Figure 15.15**, which shows cardiac output and blood flow to the various organs at rest and during heavy exercise. At rest, cardiac output (CO) is 5 liters per minute; during exercise, CO rises dramatically to 25 liters per minute, a fivefold increase.

We can see that blood flow to organs is independently regulated by comparing the proportions of total blood flow each organ receives at rest and during exercise. (A comparison of these proportions or relative *shares* of cardiac output enables us to identify what is termed the *distribution* of blood flow to organs.) If blood flow to organs were not regulated independently, the proportion of CO each organ receives would remain constant, and blood flow to every organ would rise as cardiac output rises during exercise.

Whereas skeletal muscle and skin combined receive 18–26% of CO at rest, during exercise the proportion of CO these organs receive increases to 80–85%. At the same time, blood flow to the liver and gastrointestinal tract declines from 20–25% of CO at rest to 3–5% of CO during exercise. Clearly, when the body makes a transition from rest to exercise, blood flow is diverted away from the liver and gastrointestinal tract (where metabolic demand is low) and toward the muscles and skin, which have high metabolic demands during exercise. The increased blood supply to muscles provides the oxygen and nutrients needed to generate contractile force; increased flow to the skin both helps to dissipate excess heat at the body surface and supplies sweat glands with the energy and water they need in order to cool the body via the evaporation of sweat. Blood flow is distributed according to need due to the local controls of arteriolar radius described above.

Extrinsic Control of Arteriole Radius and Mean Arterial Pressure

We now return to our flow equation to determine how control of arteriole radius affects mean arterial pressure:

$$\text{flow} = \Delta P/R \quad \text{or} \quad \text{CO} = \text{MAP}/\text{TPR}$$

Solving for mean arterial pressure:

$$\text{MAP} = \text{CO} \times \text{TPR}$$

As you can see in this equation, a change in either CO or TPR affects MAP. In this section, we describe the extrinsic

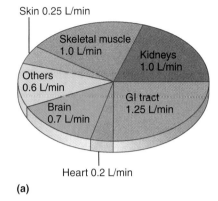

Resting blood flow
Cardiac Output (CO) = 5 L/min

Skin 0.25 L/min
Skeletal muscle 1.0 L/min
Kidneys 1.0 L/min
Others 0.6 L/min
GI tract 1.25 L/min
Brain 0.7 L/min
Heart 0.2 L/min

(a)

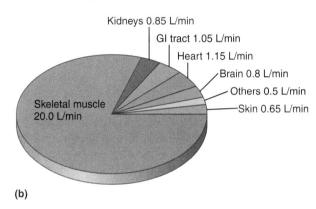

Exercise blood flow
Cardiac Output (CO) = 25.0 L/min

Kidneys 0.85 L/min
GI tract 1.05 L/min
Heart 1.15 L/min
Brain 0.8 L/min
Others 0.5 L/min
Skin 0.65 L/min
Skeletal muscle 20.0 L/min

(b)

Figure 15.15 Blood flow patterns in the cardiovascular system during rest and exercise. (a) Rest. (b) During heavy exercise. Cardiac output increases 5-fold during exercise. Blood flow to skeletal muscle increases 20-fold, flow to the heart increases 6-fold, and flow to the skin increases 2.5-fold.

control of arteriole radius as a factor that changes MAP. A later section in this chapter will more thoroughly cover the regulation of MAP.

Sympathetic Control of Arteriolar Radius

The sympathetic nervous system innervates the smooth muscle of most arterioles. During periods of increased sympathetic nerve activity, norepinephrine binds to α adrenergic receptors on arteriolar smooth muscle and activates the phosphatidylinositol biphosphate second messenger system. The end result is vasoconstriction, which increases TPR and thus increases MAP.

In addition to having α receptors, certain arteriolar smooth muscle (particularly that of skeletal and cardiac muscle) has β_2 adrenergic receptors. Norepinephrine released from sympathetic postganglionic fibers does not normally bind to these receptors. However, epinephrine

Table 15.2 | Extrinsic Controls of Arteriole Radius

Extrinsic factor	Change in radius	Effect on MAP
Sympathetic nerves	Vasoconstriction	Increase
Epinephrine	Depends on receptor type α Adrenergic: Vasoconstriction β₂ Adrenergic: Vasodilation	Increase (dominant effect is at α receptors)
Vasopressin	Vasoconstriction	Increase
Angiotensin II	Vasoconstriction	Increase

secreted from the adrenal medulla in response to sympathetic activity can bind to both α and β₂ receptors. Binding to α receptors, whether by norepinephrine or epinephrine, causes vasoconstriction as described above. However, binding of epinephrine to β₂ receptors activates the cAMP second messenger system, resulting in vasodilation and a decrease in resistance to blood flow.

Because epinephrine binds to both α and β₂ receptors, its effects on the vasculature are not readily apparent. When epinephrine is present at lower concentrations, it binds primarily to β₂ receptors and promotes vasodilation, because it has greater affinity for β₂ receptors than for α receptors. At higher concentrations, epinephrine binds to both types of receptors, and for this reason the hormone's effect on vascular resistance depends on which receptor type predominates. Because α receptors outnumber β₂ receptors in most locations, high concentrations of epinephrine usually promote vasoconstriction (the same effect as sympathetic neural input). Because β₂ receptors exert the predominant effect on blood vessels in cardiac and skeletal muscle, vasodilation occurs in these tissues in response to epinephrine. At the same time that this vasodilation promotes increased blood flow to these tissues, vasoconstriction decreases blood flow elsewhere. The significance of this increased blood flow is clear given that lots of epinephrine is released during the fight-or-flight response, which adapts the body for vigorous physical exercise. In such exercise, the workload of the heart and skeletal muscles increases, and greater blood flow is required to meet the increased metabolic demand (see Figure 15.15).

The parasympathetic nervous system does not innervate arteriolar smooth muscle except that of the external genitalia, where it causes vasodilation (see Chapter 22).

Hormonal Control of Arteriolar Resistance

In addition to epinephrine, two other hormones regulate arteriolar resistance. These hormones cause vasoconstriction and increase MAP.

Vasopressin (ADH) Vasopressin is a hormone secreted by the posterior pituitary gland (see Chapter 6). Because it acts on the kidneys to limit urine output, vasopressin is also known as *antidiuretic hormone.* (Increased urine flow is called *diuresis.*) Along with this effect, and more to the point here, it also promotes vasoconstriction in most tissues—hence the name *vasopressin.* (Effects that tend to raise blood pressure are referred to as *pressor* effects.)

Angiotensin II Angiotensin II is a protein derived from a precursor called *angiotensinogen,* which is always present in the plasma. The generation of angiotensin II from angiotensinogen is a two-step process: Angiotensinogen is first converted to *angiotensin I* by *renin,* an enzyme secreted by the kidneys (see Chapter 20). Angiotensin I is then converted to angiotensin II by *angiotensin converting enzyme,* which is present on the inner surface of blood vessels in many parts of the body, particularly in the lungs. One of the many effects of angiotensin II is to promote vasoconstriction, thereby increasing TPR and MAP.

Table 15.2 summarizes extrinsic factors that regulate arteriole radius.

Quick Test 15.3

1. Define *vasodilation* and *vasoconstriction.* What effect do they have on resistance to blood flow?

2. Define the following terms: *ischemia, hyperemia, flow autoregulation, perfusion pressure.*

3. What is the difference between active hyperemia and reactive hyperemia?

4. Explain the different functions of intrinsic and extrinsic controls of arteriole radius. Name some factors that affect radius.

Capillaries, the Lymphatic System, and Venules

Capillaries are the primary site where exchange of nutrients and waste products occurs between blood and tissue. In this section, we describe the functional anatomy of capillaries, how blood flow into capillary beds is under local controls, and mechanisms of exchange between blood and tissue.

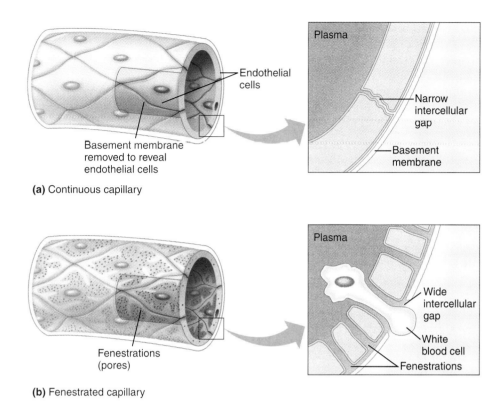

(a) Continuous capillary

(b) Fenestrated capillary

Figure 15.16 Two types of capillaries. (a) A continuous capillary, featuring narrow, water-filled gaps between endothelial cells. (b) A fenestrated capillary, which possesses pores (fenestrations) that penetrate through endothelial cells, in addition to intercellular gaps between endothelial cells. In some fenestrated capillaries, gaps between endothelial cells are large enough to permit blood cells to move through them.

Capillary Anatomy

Capillaries, which are the smallest and most numerous blood vessels in the body, range from 5 to 10 μm in diameter (1 μm is one *micrometer,* one-millionth of a meter). Numbering around 10 billion, capillaries are also the vessels with the thinnest walls (about 0.5 μm). The thinness of capillary walls—they consist of a single layer of endothelial cells and a basement membrane—facilitates the capillaries' primary function: to permit the exchange of materials between cells in tissues and the blood. The thinness of capillary walls permits small molecules (for example, oxygen, carbon dioxide, monosaccharides, sugars, amino acids, and water) to enter and leave capillaries readily, promoting efficient material exchange. The exchange of materials across capillaries is also facilitated by the large number of them: The body's total capillary surface area available for the exchange of materials likely exceeds 6000 square meters, and most body cells are within 1 mm of a capillary.

In most regions of the body, small solutes readily enter and leave the bloodstream by simple diffusion across capillary walls. (A notable exception is in the brain, where the blood-brain barrier limits the diffusion of solutes across capillary walls; see p. 267.) However, the permeability of capillaries varies from region to region because

capillaries differ with regard to the physical properties of their walls. On the basis of these anatomical differences, capillaries are grouped into two major classes: continuous capillaries and fenestrated capillaries (**Figure 15.16**).

Continuous Capillaries

In *continuous capillaries* (Figure 15.16a), which are more common, the endothelial cells are joined together such that the spaces between them are relatively narrow. These capillaries are highly permeable to substances having small molecular sizes and/or high lipid solubility (such as oxygen, carbon dioxide, and steroid hormones) and are somewhat less permeable to small water-soluble substances (such as sodium, potassium, glucose, or amino acids). The permeability of continuous capillaries to proteins and other macromolecules is very low because these substances can neither readily cross membranes of endothelial cells nor easily penetrate the gaps between cells.

Fenestrated Capillaries

In *fenestrated capillaries* (Figure 15.16b), the endothelial cells possess relatively large pores *(fenestrations)* that are wide enough to allow proteins and other large molecules to pass through. In some fenestrated capillaries, the gaps between endothelial cells are also wide enough that large proteins and in some cases entire cells can pass through easily. For this reason, fenestrated capillaries are highly permeable not only to small molecules (whether lipid-soluble or water-soluble) but to macromolecules as well. Fenestrated capillaries are found most in organs whose functions depend on the rapid movement of materials across capillary walls, including the kidneys, liver, intestines, and bone marrow. In the liver, the presence of fenestrated capillaries allows newly synthesized proteins such as albumin or clotting factors to enter the plasma. In bone marrow, fenestrated capillaries allow newly formed blood cells to enter the circulation.

Local Control of Blood Flow Through Capillary Beds

The simplest way to regulate the exchange of material across capillary walls is to regulate the amount of blood flowing through a particular vascular bed. This is done through local control of smooth muscle located in various areas of the microcirculation. We already learned about the regulation of blood flow through arterioles. Next, we describe metarterioles and precapillary sphincters that regulate blood flow through certain capillaries.

Most tissues contain metarterioles, which are structurally intermediate between arterioles and capillaries; instead of the continuous layer of smooth muscle that surrounds arterioles, metarterioles possess isolated rings of smooth muscle that act as gatekeepers at strategic points (see Figure 15.8). Unlike arterioles, which direct blood into the interbranching vessels in a capillary bed, metarterioles serve as bypass channels or *shunts* by directly connecting arterioles to venules. The presence of these shunts allows blood to continue flowing from arterioles to venules, bypassing capillaries.

Like that of arterioles, the smooth muscle of metarterioles can contract or relax to increase or decrease, respectively, resistance to blood flow. Because metarterioles are the bypass, when their resistance to blood flow is high, blood flow through the capillary beds increases, whereas when their resistance is low, blood flow through the capillary beds decreases. The smooth muscle of metarterioles is under local control by metabolites, similar to that of the arterioles. Whether or not metarterioles receive sympathetic innervation, as arterioles do, is still unclear.

Blood flow through capillaries is also regulated by smooth muscle that surrounds capillaries on the arteriole end, called **precapillary sphincters** (see Figure 15.8). Contraction of precapillary sphincters constricts the capillaries, increasing their resistance to blood flow. Precapillary sphincters are only affected by local controls, that is, metabolites produced based on the metabolic activity of the tissue around the sphincters. An increase in metabolites, such as carbon dioxide, causes relaxation of the sphincters and increases blood flow through the capillaries, whereas a decrease in carbon dioxide causes contraction of the sphincters and decreases blood flow. Blood that enters the capillaries can participate in exchange with interstitial fluid through mechanisms described next.

Movement of Material Across Capillary Walls

Movement of material across capillary walls serves two purposes: (1) exchange of material between blood and cells and (2) normal distribution of the extracellular fluid.

Exchange Across Capillary Walls

Mechanisms of transport across capillary walls differ for different substances, depending on their molecular sizes and degree of lipid solubility (**Figure 15.17**). Most small solutes, whether lipid- or water-soluble, move across capillary walls by simple diffusion. Capillaries have high permeability to lipid-soluble substances because these substances readily diffuse through cell membranes and therefore can easily cross endothelial cells. Continuous capillaries show a somewhat lower permeability to small water-soluble solutes because these substances are mostly restricted to moving through the water-filled gaps between the endothelial cells. The permeability to proteins and other macromolecules is negligible because these substances are too large to pass through any but the largest water-filled pores and are thus prevented from passing through endothelial cells or around them. This is not true of *all* proteins, however, because certain proteins (referred to as *exchangeable proteins*) are selectively transported across endothelial cells by a slow, energy-requiring process known as *transcytosis*. In this process, capillary endothelial cells engulf proteins in the plasma by endocytosis. The proteins are then ferried across the cells by vesicular transport and released by exocytosis into the interstitial fluid on the other side.

The direction of movement across capillary walls depends on the electrochemical gradient for a specific substance (except the exchangeable proteins). Because nutrients and oxygen tend to be plentiful in the blood and needed by cells, these substances diffuse from blood to tissue. Waste products and carbon dioxide, however, are produced by the cells and therefore are more abundant in the tissues. These substances diffuse from tissue to blood.

Bulk Flow Across Capillary Walls

Because capillary walls are freely permeable to water and small solutes, fluid can move from blood to interstitial fluid (called **filtration**) or from interstitial fluid to blood (called **absorption**) based on pressure gradients. The purpose of this *bulk flow* is not to provide nutrients and remove waste products but to maintain balance between the extracellular compartments: interstitial fluid and plasma. A shift in fluid from plasma to interstitial fluid causes a swelling of tissues, called **edema**.

The forces that drive the movement of fluid into and out of capillaries are called **Starling forces**, and include

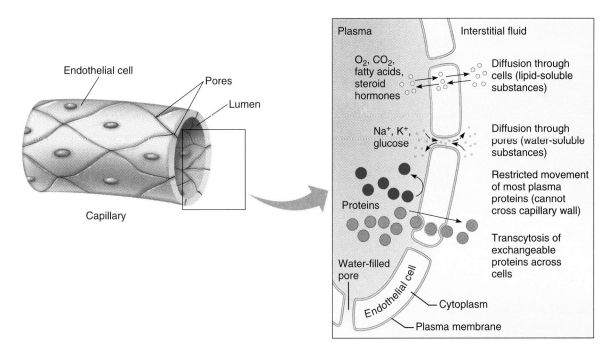

Figure 15.17 Exchange of materials across the wall of a continuous capillary. Lipid-soluble substances are able to diffuse through endothelial cells, whereas passage of small water-soluble substances is restricted to water-filled pores. Although most proteins in the plasma are unable to cross the wall, certain proteins cross the wall via transcytosis.

Is the movement of glucose across capillary walls an active or a passive process?

the following: (1) *capillary hydrostatic pressure* (P_{CAP}) due to the hydrostatic pressure of fluid inside the capillary, (2) *interstitial fluid hydrostatic pressure* (P_{IF}) due to the hydrostatic pressure of fluid outside the capillary, (3) *capillary osmotic pressure* (π_{CAP}) due to the presence of nonpermeating solutes inside the capillary, and (4) *interstitial fluid osmotic pressure* (π_{IF}) due to the presence of nonpermeating solutes outside the capillary. The four Starling forces are illustrated in **Figure 15.18**a and described below.

Hydrostatic Pressures Whenever a hydrostatic pressure exists across a semipermeable barrier such as a capillary wall, the pressure tends to move water across the wall. The capillary hydrostatic pressure favors filtration and is equal to the blood pressure in the capillaries, which varies along the length of the capillary, because the blood pressure decreases as blood flows from the arteriole end (P_{CAP} = 38 mm Hg) of the capillary to the venule end (P_{CAP} = 16 mm Hg) (Figure 15.18b). The interstitial fluid hydrostatic pressure favors absorption and is due to the presence of fluid outside the capillary walls driving the fluid into the capillary. This pressure does not vary along the length of capillaries and is usually low (P_{IF} = 1 mm Hg).

Osmotic Pressures Recall that when a semipermeable barrier separates two solutions with different concentrations of nonpermeating solute, water tends to flow from the side where the solute concentration is lower to the side where the solute concentration is higher. The osmotic pressure of a solution is a measure of its solute concentration, and it increases as the solute concentration increases. Put another way, the presence of a nonpermeating solute exerts an osmotic pressure that tends to draw water to the side where it is present in greater concentration. Only nonpermeating solutes exert an osmotic pressure because permeating solutes, such as glucose, are generally at equal concentrations across the capillary wall. The primary nonpermeating solutes in plasma are proteins. Osmotic pressure that is exerted by proteins is referred to as **oncotic pressure**. Thus the capillary osmotic pressure is also called the capillary oncotic pressure.

The capillary oncotic pressure is the osmotic force exerted by proteins in the plasma. Plasma proteins draw water into the capillary, and therefore exert a force for absorption. Under normal conditions, the concentration of proteins in the plasma is 6–8 grams per 100 mL, which exerts an osmotic force of 25 mm Hg (Figure 15.18b). Although the capillary oncotic pressure is affected by any movement of water into or out of capillaries that might

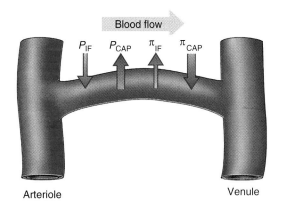

(a)

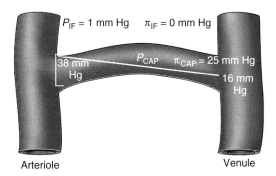

Arteriole end

Filtration pressure:	Absorption pressure:
P_{CAP} = 38 mm Hg	π_{CAP} = 25 mm Hg
π_{IF} = 0 mm Hg	P_{IF} = 1 mm Hg
38 mm Hg	26 mm Hg

NFP = Filtration pressure
– Absorption pressure

= 38 mm Hg – 26 mm Hg = 12 mm Hg

Venule end

Filtration pressure:	Absorption pressure:
P_{CAP} = 16 mm Hg	π_{CAP} = 25 mm Hg
π_{IF} = 0 mm Hg	P_{IF} = 1 mm Hg
16 mm Hg	26 mm Hg

NFP = Filtration pressure
– Absorption pressure

= 16 mm Hg – 26 mm Hg = –10 mm Hg

(b)

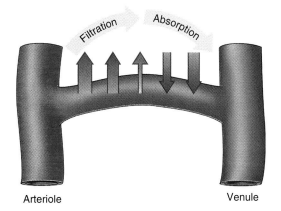

(c)

occur as blood flows through them, for the sake of simplicity we will ignore this and assume that the oncotic pressure of plasma does not change as blood flows from one end of the capillary to the other.

The interstitial fluid oncotic pressure is the osmotic force exerted by proteins in the interstitial fluid, which tends to draw fluid out of the capillary and into the interstitial fluid. However, relatively few proteins are in the interstitial fluid, and thus the oncotic pressure of interstitial fluid is approximately 0 mm Hg.

Net Filtration Pressure The direction of fluid flow across the wall of a capillary is determined by the **net filtration pressure** at any given location, that is, the difference in the filtration pressures and the absorption pressures. Of the four Starling forces, capillary hydrostatic pressure and interstitial fluid osmotic pressure favor filtration, whereas interstitial fluid hydrostatic pressure and capillary osmotic pressure favor absorption. The net filtration pressure (NFP) can be defined as:

$$NFP = \text{filtration pressure} - \text{absorption pressure}$$
$$= (P_{CAP} + \pi_{IF}) - (\pi_{CAP} + P_{IF})$$

When the sign of the net filtration pressure is positive, filtration occurs; when it is negative, absorption occurs.

Recall that while three of the Starling forces are relatively constant at any given location, the capillary hydrostatic pressure varies along the length of the capillary. Thus, the net filtration pressure also varies along the length of the capillary, being higher at the arteriole end and lower at the venule end. Figure 15.18b shows the four Starling forces and the net filtration pressure at two points along the capillary: the arteriole end and the venule end. In this example, the capillary hydrostatic pressure is 38 mm Hg at the arteriole and 16 mm Hg at the venule end. Interstitial fluid hydrostatic pressure is 1 mm Hg, capillary colloid osmotic pressure is 25 mm Hg,

Figure 15.18 Starling forces across capillary walls. A schematic representation of an arteriole leading into a capillary bed (represented by a single tube) that drains into a venule is shown. **(a)** Of the four Starling forces, capillary hydrostatic pressure (P_{CAP}) and interstitial fluid oncotic pressure (π_{IF}) favor filtration, whereas interstitial fluid hydrostatic pressure (P_{IF}) and capillary oncotic pressure (π_{CAP}) favor absorption. **(b)** Average values for the Starling forces. Note that capillary hydrostatic pressure decreases as blood flows from the arteriole end of the capillary (P_{CAP} = 38 mm Hg) to the venule end (P_{CAP} = 16 mm Hg). The diagonal line represents the decrease in P_{CAP}. **(c)** Net filtration pressure across the capillary wall is indicated by arrows. Note that filtration occurs near the arteriole end whereas absorption occurs near the venule end.

Would an increase in the plasma protein concentration tend to favor increased filtration or absorption?

Table 15.3 | Forces Affecting the Movement of Fluid Across Capillary Walls

Force	Definition	Direction of force	Approximate value
Capillary hydrostatic pressure, P_{CAP}	Hydrostatic pressure exerted by the presence of fluid inside the capillary	Filtration	16–38 mm Hg
Interstitial fluid hydrostatic pressure, P_{IF}	Hydrostatic pressure exerted by the presence of fluid outside the capillary	Absorption	1 mm Hg
Capillary oncotic pressure, π_{CAP}	Osmotic force due to presence of proteins in plasma	Absorption	25 mm Hg
Interstitial fluid oncotic pressure, π_{IF}	Osmotic force due to presence of proteins in interstitial fluid	Filtration	0 mm Hg
Net filtration pressure, NFP	**Difference between forces for filtration and absorption**	**If positive: filtration** **If negative: absorption**	**2 mm Hg**

and interstitial fluid osmotic pressure is 0 mm Hg. Based on these data, we can determine the net filtration pressure at the arteriole and venule ends of the capillary:

$$\text{Arteriole end: NFP} = (P_{CAP} + \pi_{IF}) - (\pi_{CAP} + P_{IF})$$
$$= (38 + 0) - (25 + 1)$$
$$= 12 \text{ mm Hg}$$

$$\text{Venule end: NFP} = (P_{CAP} + \pi_{IF}) - (\pi_{CAP} + P_{IF})$$
$$= (16 + 0) - (25 + 1)$$
$$= -10 \text{ mm Hg}$$

At the arteriole end the NFP is 12 mm Hg, which favors filtration; at the venule end, the NFP is −10 mm Hg, which favors absorption. Thus, as blood enters a capillary bed, fluid moves out, but near the end of the capillary bed, *most* of the fluid returns to the blood (Figure 15.18c). It is important to note that not all the fluid that leaves the blood by bulk flow is returned before the blood leaves the capillary bed.

Table 15.3 summarizes the forces for fluid movement across capillary walls.

Factors Affecting Filtration and Absorption Across Capillaries

The rate at which fluid is filtered or absorbed across capillary walls is influenced by any factor that alters the relative sizes of the Starling forces and, hence, the net filtration pressure. Increased filtration is favored by an increase in the capillary hydrostatic pressure or interstitial fluid oncotic pressure or by a decrease in interstitial fluid hydrostatic pressure or capillary oncotic pressure. Changes in the opposite direction, of course, favor an increase in the rate of fluid absorption.

Under normal conditions, the total blood volume of fluid that filters out of an individual's capillaries amounts to about 20 liters per day—more than six times total plasma volume! About 17 liters is absorbed into capillaries in the same period of time, giving a net volume of 3 liters per day that is filtered, which is roughly equal to an individual's entire plasma volume. If this volume of fluid shifts from the plasma to the interstitial space every day, why don't tissues swell from edema? Why doesn't blood volume decrease? The answer is that the 3 liters or so of filtered fluid is picked up from the interstitium and returned to the cardiovascular system by the *lymphatic system*, which is described in the next section.

The balance between filtration and absorption can be altered as a result of certain pathological conditions, or even as a result of everyday occurrences. When a person stands up, for example, capillary hydrostatic pressure increases in the lower parts of the body because the column of blood raises the hydrostatic pressure in the lower arterioles and veins. Any increase in pressure, whether it be at the arteriolar end of a capillary or at the venular end, tends to raise capillary blood pressure, which increases capillary hydrostatic pressure and causes an increase in the rate of filtration.

An increase in capillary filtration, with accompanying tissue swelling, also occurs in response to certain types of injuries. When the skin is cut or abraded, for instance, the affected part becomes swollen within a few minutes. In part this results from damage to capillaries, which allows protein-rich fluid to leak out. Consequently, the concentration of proteins in the interstitial fluid rises, which raises its oncotic pressure. In addition, certain cells in the injured area release a chemical called *histamine*, which increases the permeability of capillary walls to proteins, thereby increasing the leakage of proteins into the interstitium.

Other conditions that promote increased capillary filtration and edema include liver, kidney, and heart

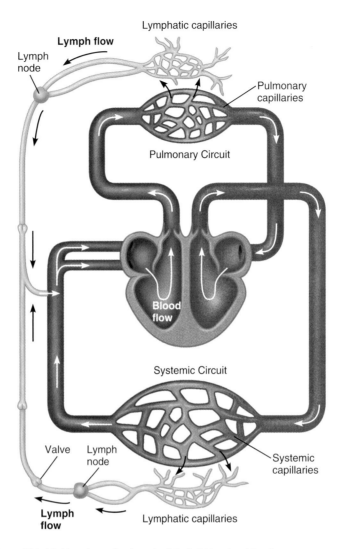

Figure 15.19 The lymphatic system. This highly schematic view depicts fluid leaving blood capillaries and entering lymphatic capillaries in the lungs and systemic tissues. This fluid, called lymph, flows through lymphatic ducts and lymph nodes to veins in the systemic circuit.

disease. Because most plasma proteins are manufactured by the liver, damage to the liver can result in a decrease in plasma protein concentration, which lowers plasma oncotic pressure. Certain forms of damage to the kidneys interfere with their ability to eliminate excess water (and solutes) in the urine, which results in the accumulation of excess fluid in the body. As a consequence, blood volume increases and blood pressure rises throughout the cardiovascular system. This increase in pressure raises capillary hydrostatic pressure, which increases filtration. Damage to the kidneys can also cause them to eliminate significant quantities of plasma proteins in the urine. (Normally, only negligible quantities of protein are voided in the urine.) This loss of protein triggers increased capillary filtration because it reduces plasma oncotic pressure. Damage to the heart can result in pulmonary edema, the accumulation of fluid in the lungs (**Clinical Connections: Heart Failure**, p. 478).

Lymphatic System

Even though approximately 3 liters of fluid leak out of the capillaries each day, the tissues do not normally swell because this fluid enters the **lymphatic system**, a network of vessels (often referred to as *ducts*) that courses throughout the body (**Figure 15.19**). Once the fluid gets into the lymphatic system, it is carried through the ducts and is eventually returned to the cardiovascular system. The lymphatic system is sort of a silent partner to the cardiovascular system, because although fluid continually moves through it, this fluid (called **lymph**) often goes unnoticed because it is clear, with only a slight yellow tinge to it. If you have ever scraped your knee slightly and noticed clear fluid oozing out of it, this is mostly lymph.

Fluid enters the lymphatic system by way of small, blind-ended ducts called **lymphatic capillaries**, the walls of which have large pores that allow water, small solutes, and even proteins and larger particles to pass through. As a consequence, any fluid that routinely leaks out of "ordinary" blood capillaries can easily move from the interstitium into the lymphatic system. From the lymphatic capillaries, fluid moves through a series of successively larger ducts called **lymphatic veins** until it eventually reaches one of the two ducts that drain into the bloodstream near the jugular veins, the *right lymphatic duct* and the *thoracic duct*. In this manner, filtered fluid is returned to the cardiovascular system. The lymphatic capillaries and veins have valves that allow fluid to enter the capillaries and the lymph to move through the veins in only one direction, toward the right lymphatic and thoracic ducts. The flow of lymph through the lymphatic veins to the ducts is similar to the flow of blood through "ordinary" veins, described shortly.

At certain points in the lymphatic system, lymph passes through structures known as **lymph nodes**. In lymph nodes, any particles that may be present in the lymph, including bacteria or other foreign matter, are filtered out and removed by phagocytic cells called **macrophages**. Lymphocytes and other cells of the immune system also congregate in the lymph nodes, making them important sites for the cellular interactions that are an integral part of the immune response (see Chapter 23).

Venules

Capillaries come together to form vessels called venules. Venules are slightly smaller than arterioles, averaging about 20 μm in diameter, but their walls are much thinner than those of arterioles and contain little or no smooth muscle (see Figure 15.6). In fact, the smaller of the venules resemble capillaries more than arterioles in that they consist of a single layer of endothelium that is porous, allowing for exchange between blood and interstitium. Thus, exchange between blood and interstitium occurs in capillaries and the small venules.

Quick Test 15.4

1. How do fenestrated capillaries differ from continuous capillaries?

2. To what does the term *oncotic pressure* refer?

3. Of the following changes, which would tend to cause an *increase* in the rate at which fluid is filtered from capillaries? Choose all that apply: a decrease in plasma oncotic pressure; a decrease in interstitial fluid oncotic pressure; an increase in venous pressure; an increase in plasma protein concentration.

4. What is lymph? What are its origins?

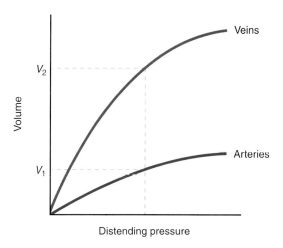

Figure 15.20 Curves showing how the volume of blood contained in arteries and veins varies with the pressure inside them. Comparison of the two curves shows that at a given pressure, veins hold more blood (V_2) than arteries (V_1).

Veins

Venules come together to form larger veins. Veins have roughly the same diameter as arteries but have walls about one-half as thick. A typical vein has an internal diameter of 5 mm but a wall thickness of only 0.5 mm. The largest veins, the venae cavae, are even larger in diameter than the aorta (30 mm, as opposed to 12.5 mm) but have a wall thickness of only 1.5 mm compared to 2 mm for the aorta. The relative thinness of the walls of veins reflects the fact that blood pressure in the veins is significantly lower than in arteries. The walls of veins are similar to those of arteries in that they contain smooth muscle and elastic and fibrous connective tissue (see Figure 15.6).

Unlike any other blood vessels in the body, veins are equipped with one-way valves that permit blood to flow toward the heart but prevent it from flowing back toward organs and tissues (see Figure 15.5). These valves are present in veins located outside the thoracic cavity *(peripheral veins)* but are absent from veins located within the thoracic cavity *(central veins)*. The significance of this difference will become apparent when we discuss the action of the *respiratory pump*.

Veins: A Volume Reservoir

Unlike arteries, which function as pressure reservoirs, veins function as *volume reservoirs*, a property that is also related to vessel compliance. Because veins are thin-walled and easily stretched, they have high compliance—a relatively small increase in the pressure within veins causes a relatively large degree of expansion (increase in volume). Put another way, veins can accommodate a large increase in blood volume in response to a small increase in blood pressure and therefore are good at storing volume. As a result of their high compliance, veins can hold a larger volume of blood than arteries can at a given pressure (**Figure 15.20**). In fact, the veins in the human

HEART FAILURE

The term *heart failure* refers to any change in the heart's condition that reduces its ability to maintain an adequate cardiac output. Most commonly this is a result of a chronic decrease in ventricular contractility. (Recall that the contractility of a ventricle is a measure of its ability to generate force at a given end-diastolic volume.) However, heart failure can occur even if ventricular contractility is normal or above normal. Obstruction of an atrioventricular valve, for example, can cause heart failure because it slows ventricular filling, which reduces end-diastolic volume. Alternatively, chronic *hypertension* or *stenosis* (constriction) of the aorta may induce overgrowth and thickening *(hypertrophy)* of the ventricular myocardium due to the increased workload placed on the heart. The increase in ventricular thickness makes the left ventricle harder to stretch (that is, it reduces ventricular compliance), which tends to reduce the end-diastolic volume. Either way, a decrease in end-diastolic volume causes stroke volume and cardiac output to fall by virtue of the Starling effect. This constitutes heart failure because it interferes with the ability

of the heart to maintain an adequate cardiac output.

In heart failure, one or both sides of the heart may be affected. Signs of heart failure can thus depend on which side is failing. If one side of the heart fails, it can cause changes that place stress on the other side of the heart, causing it to fail as well. For this and other reasons, the initial cause of heart failure can be difficult to unravel. One thing is certain, however: Once the heart begins to fail, it sets into motion a complex chain of events that only worsens the condition.

What might cause a heart to fail in the first place? Sometimes heart failure is caused by a heart attack or *myocardial infarction,* in which the death of muscle cells leads to weakening of the heart. Abnormalities in electrical conduction can also lead to weakening the heart. In third-degree *heart block,* for example, impulse conduction to the ventricles is prevented, which causes a dramatic reduction in heart rate. The decrease in heart rate causes the end-diastolic volume to become chronically elevated because ventricular filling time increases. When this happens, the myocardium

becomes overstretched and weakened. Bacterial and viral infections can also induce heart failure by weakening the myocardium.

Once the heart muscle weakens, it generates less contractile force, which reduces stroke volume and cardiac output. If this happens in the left heart, mean arterial pressure falls. Signs include weakening of the pulse, fatigue, and *cyanosis*—a bluish tinge to the skin indicating inadequacy of the oxygen supply. Another sign of left heart failure is *pulmonary edema* (accumultion of fluid in the lungs) due to a rise in pressure on the venous side of the pulmonary circuit. (The rise in venous pressure causes edema because it causes pressure in pulmonary capillaries to increase, which promotes filtration of fluid across capillary walls.) The resulting accumulation of fluid in the lungs can impair gas exchange, causing shortness of breath or even respiratory distress (gasping for air) in severe cases.

To understand why ventricular failure causes a rise in venous pressure, we must consider how venous pressure is affected by the pumping action of the heart. To

body contain a substantially greater volume of blood than do the arteries (**Figure 15.21**), even though the pressure within veins is much lower than that within arteries.

The volume reservoir function of veins is not merely a curiosity; it has important practical consequences. When one donates blood or loses a fraction of total blood volume for any reason, most of the lost blood volume comes from the veins. As the volume of blood in the veins decreases, however, the accompanying drop in venous pressure is relatively small, owing to the fact that veins have high compliances. This is critical because the driving force for venous return is the difference in venous pressure (specifically, central venous pressure) and right atrial pressure. Therefore, the veins can lose a substantial volume of blood before the drop in venous pressure becomes large enough to cause a significant decrease in venous return and, hence, cardiac output. (Recall that venous pressure is an important determinant of end-diastolic volume, which influences the stroke volume and cardiac output by virtue of the Starling effect; see p. 443.)

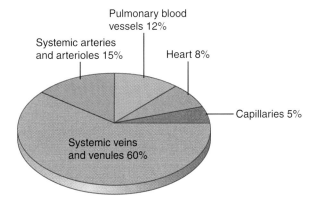

Figure 15.21 Distribution of blood volume in the various portions of the cardiovascular system. Percentages indicate proportions of total blood volume.

(continued)

begin, imagine that the heart has stopped pumping blood. Under these conditions, only a small pressure (a few mm Hg) would exist in the cardiovascular system due to blood filling up blood vessels and stretching their walls. However, the pressure would be the same in the arteries, veins, and everywhere else, so no blood flow would occur. Now imagine that the heart were to suddenly resume its pumping action. In doing so it would receive blood from veins and push it into the arteries, which would effectively transfer a certain fraction of the total blood volume from the veins to the arteries. This transfer of blood would raise mean arterial pressure and lower venous pressure, as you can see in diagram (a) where the resumption of pumping occurs at the arrow.

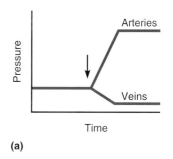

(a)

Arterial pressure rises more than venous pressure falls because the arteries have a lower compliance than the veins. In heart failure, the heart's ability to pump blood is reduced, which causes a portion of the total blood volume to shift back into the veins from the arteries. This reduces arterial pressure and also raises venous pressure, as occurs at the arrow in diagram (b):

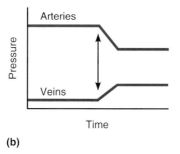

(b)

In heart failure, pressure can rise in either the systemic veins or the pulmonary veins, depending on which side of the heart is affected. Failure of the left ventricle *(left-sided failure)* causes pressure to rise in the pulmonary veins because they carry blood to the left side of the heart. In contrast, failure of the right ventricle (*right-sided* failure) causes pressure to

rise in the vena cava and other systemic veins. In left-sided failure, however, this is only the beginning of the problem. The fall in mean arterial pressure that accompanies left-sided failure usually triggers a multitude of reflex responses, including a reduction in salt and water excretion by the kidneys. This leads to accumulation of excess fluid in the body *(fluid retention)*, which raises blood volume. Because the veins act as volume reservoirs, most of this excess volume ends up on the venous side of the circulation, which increases venous pressure even further.

Edema, a common sign of heart failure, is a consequence of increased venous pressure. As venous pressure rises, the increase in pressure is transmitted to vessels upstream, including capillaries. When the left side of the heart fails, edema occurs in the lungs because the pressure increases in pulmonary capillaries. When the right side of the heart fails, it tends to cause edema in systemic tissues because pressure rises in systemic capillaries. Edema is most noticeable in parts of the body that hang down, such as the wrists and ankles. When edema occurs, the condition is referred to as *congestive heart failure*.

The existence of a volume reservoir in the veins is important for another reason, too: In exercise and other circumstances in which it is desirable to increase the cardiac output or maintain it at an adequate level, a number of mechanisms come into play to force much of the blood volume out of the veins and toward the heart, thereby promoting increased ventricular filling and cardiac output. These mechanisms act by increasing central venous pressure.

Factors That Influence Central Venous Pressure and Venous Return

Central venous pressure has an important, though indirect, influence on mean arterial pressure, and therefore it affects the flow of blood to all systemic organs. Because the difference between CVP and right atrial pressure is the driving force for venous return, the greater the CVP, the greater the end-diastolic volume and thus stroke volume, cardiac output, and ultimately mean arterial pressure.

In this section we examine four factors that affect CVP and thus indirectly affect blood flow to organs: the skeletal muscle pump, the respiratory pump, blood volume, and venomotor tone.

Exercise Link

While Bill and Jane were running the marathon, several factors were contributing to diminished venous return and therefore reduced central venous pressure, including increased blood flow to skeletal muscle and skin, upright posture, and fluid loss through sweat. Fortunately, the compensatory processes that are described next became functional during exercise and played important roles in maintaining central venous pressure.

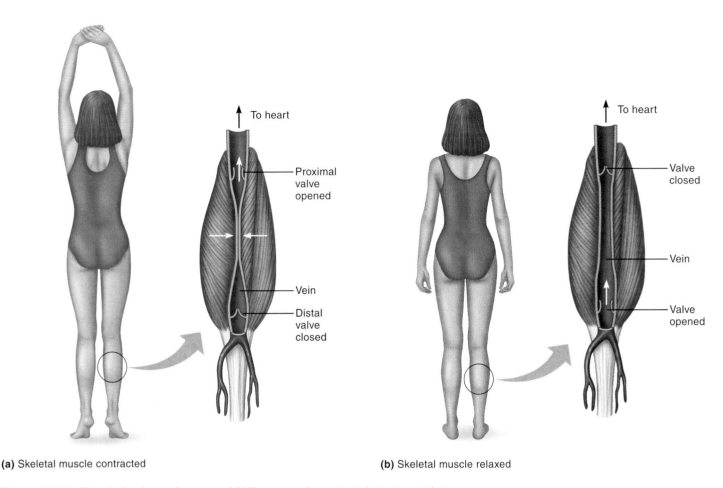

(a) Skeletal muscle contracted

(b) Skeletal muscle relaxed

Figure 15.22 The skeletal muscle pump. (a) When a muscle contracts, it presses against veins, driving blood toward the heart (left). (b) When the muscle relaxes, backward flow is prevented by closure of one-way valves in the veins.

The Skeletal Muscle Pump

We have seen that peripheral veins contain one-way valves that allow blood to flow forward toward the heart but prevent it from flowing backward. When skeletal muscles contract, they press against veins traveling between them, which raises the venous pressure of blood. This increased pressure forces the more distal valves to close, preventing blood from flowing backward, and forces the more proximal valves to open, allowing blood to flow toward the heart (**Figure 15.22**a). When the muscles relax and the pressure drops, the reduced pressure allows the distal valves to open, so that blood can flow forward into the previously compressed vein, and also causes the proximal valve to close, thereby preventing blood from flowing away from the heart (Figure 15.22b). By alternately contracting and relaxing, muscles act as "pumps" that help drive blood toward the central veins, which raises central venous pressure. For this reason, any exercise that involves rhythmic muscle contractions, such as walking or running, promotes an increase in venous return, increased stroke volume, and increased cardiac output.

The Respiratory Pump

Just as exercise helps to move blood back to the heart through the action of the skeletal muscle pump, the vigorous respiratory movements that accompany exercise also help to move blood back to the heart. We call the effect of respiratory movements on venous return the *respiratory pump*.

The respiratory pump works in the following manner: When you inhale, your diaphragm pulls downward and your rib cage expands, which lowers pressure in the thoracic cavity and raises pressure in the abdominal cavity. This action creates a pressure gradient that promotes the movement of blood from abdominal veins to the central veins located in the thoracic cavity, thereby increasing central venous pressure. When you exhale, thoracic pressure rises and abdominal pressure falls. This creates a pressure gradient that would tend to favor the backward movement of blood from the central veins to the abdominal veins, but such backward flow is prevented by the closure of valves in the abdominal veins. Instead, the rise in thoracic pressure drives the forward movement of blood from the central veins to the heart, thereby promoting increased venous return and cardiac output.

Blood Volume

The body's total blood volume has an important influence on mean arterial pressure through its effect on central venous pressure. The relationship between blood volume and central venous pressure is a simple one: An increase in blood volume produces an increase in venous pressure, and a decrease in blood volume produces a decrease in venous pressure. If blood volume falls as a result of bleeding or dehydration or for any other reason, central venous pressure falls, as do venous return and end-diastolic volume. The resulting fall in cardiac output causes a decrease in mean arterial pressure. Conversely, a rise in blood volume has the opposite effect, tending to increase mean arterial pressure.

Certain forms of **hypertension** (elevated mean arterial pressure), for example, are due to failure of the kidneys to excrete adequate amounts of salt and water, which results in the retention of excess fluid in the body. This excess fluid causes blood volume to increase, which raises the mean arterial pressure.

Because arterial pressure is strongly affected by blood volume, control of blood volume is an important part of blood pressure regulation. A fall in blood volume activates reflex mechanisms that act to reduce the kidneys' output of water in the urine, which helps the body to conserve water, thereby maintaining blood volume and venous pressure. At the same time, activation of thirst centers in the brain induces a person to drink fluids. Absorption of these fluids by the gastrointestinal tract serves to restore lost blood volume, thereby maintaining adequate venous pressure. These reflex mechanisms that regulate blood volume generally involve the endocrine system and act relatively slowly and thus regulate blood pressure over the long term (hours or days). (Long-term regulation of blood pressure by the kidneys is discussed in greater detail in Chapter 20.)

Under certain circumstances the high compliance of veins actually works to the detriment of the heart's pumping action by leading to **venous pooling**—accumulation of blood in veins. When a person stands up, for instance, the force of gravity increases the pressure on the blood in the lower veins of the body, causing those veins to expand and enabling the volume of blood within them to increase. This pooling of venous blood is detrimental to the heart's pumping action because it reduces venous return; instead of returning to the heart, much of the blood entering the veins remains there. Thus venous pooling reduces central venous pressure by reducing the volume of blood in the central veins, and it therefore lowers arterial pressure in the same manner as does a reduction in blood volume.

A drop in mean arterial pressure upon standing (referred to as *orthostatic hypotension*) may cause a person to feel dizzy due to a decrease in blood flow to the brain, but reflex mechanisms normally quickly compensate for it under normal conditions. The presence of aggravating factors, such as dehydration or a failing heart, may cause a person to faint upon standing. In such a case, fainting is actually advantageous because once a person has fallen over, blood that had pooled in the veins in the legs moves toward the central veins, just as water flows out of a glass when it is tipped over. This increases central venous pressure, which promotes an increase in venous return and an increase in cardiac output. The rise in cardiac output raises mean arterial pressure, which helps restore blood flow to the brain.

Venomotor Tone

The smooth muscle in the walls of veins contracts or relaxes in response to input from autonomic nerves and certain chemical agents. In particular, venous smooth muscle contains α adrenergic receptors and activity of the sympathetic nervous system triggers increased contractile activity, with a resulting rise in tension referred to as **venomotor tone**.

An increase in venomotor tone has two effects: (1) Constriction of veins raises the venous pressure, which forces blood to the central veins and then to the heart, briefly increasing stroke volume, and (2) increased wall tension reduces venous compliance, which raises central venous pressure and produces a sustained increase in stroke volume. Therefore, an increase in venomotor tone promotes a rise in cardiac output and mean arterial pressure. Changes in venomotor tone are an important component of the reflexes that regulate arterial pressure. When arterial pressure falls as a result of blood loss, for example, activity in sympathetic neurons increases, which acts to raise arterial pressure.

Figure 15.23 summarizes the ways these factors influence central venous pressure. This diagram illustrates that increases in muscle pump activity, respiratory pump activity, blood volume, or venomotor tone all act to raise central venous pressure and therefore tend to raise mean arterial pressure.

Quick Test 15.5

1. Does an increase in central venous pressure tend to increase or decrease cardiac output? Explain.

2. Of the following, which would tend to *increase* central venous pressure? Choose all that apply: standing up; an increase in blood volume; contraction and relaxation of skeletal muscle; decreased venomotor tone.

3. Compare the compliance of arteries and veins, and explain the functional significance.

Mean Arterial Pressure and Its Regulation

We have seen that two factors influence blood flow to an organ: mean arterial pressure and the organ's resistance. Given that mean arterial pressure influences blood flow to all organs in the systemic circuit, it becomes clear why matching blood flow to each organ's needs requires that

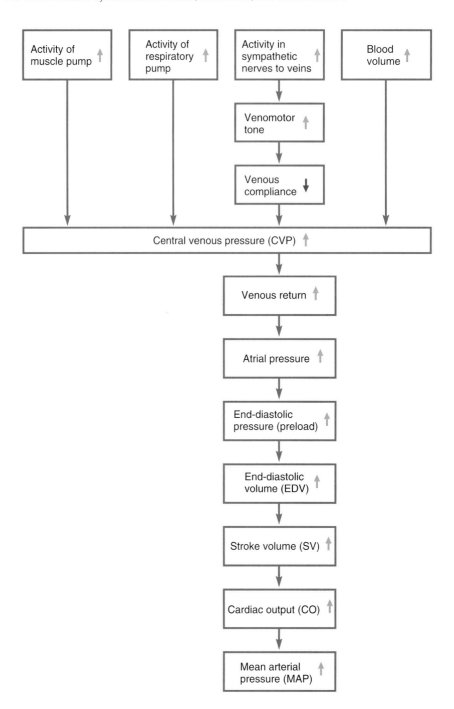

Figure 15.23 Factors affecting central venous pressure and thus mean arterial pressure.

the cardiovascular system maintain adequate MAP: Any decline in MAP tends to compromise blood flow to all the systemic organs. We now further explore the determinants of MAP and its regulation.

Determinants of Mean Arterial Pressure: Heart Rate, Stroke Volume, and Total Peripheral Resistance

We previously saw that one variation of the flow rule is MAP = CO × TPR. Recalling from Chapter 14 that

cardiac output is determined by heart rate (HR) and stroke volume (SV), and substituting these terms into the expression, we see that

$$MAP = HR \times SV \times TPR$$

Thus, mean arterial pressure is completely determined by three factors: (1) heart rate, (2) stroke volume, and (3) total peripheral resistance.

The previous expression indicates that mean arterial pressure should rise following an increase in heart rate or stroke volume (which tend to increase cardiac output) or

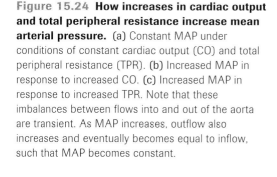

Figure 15.24 How increases in cardiac output and total peripheral resistance increase mean arterial pressure. (a) Constant MAP under conditions of constant cardiac output (CO) and total peripheral resistance (TPR). (b) Increased MAP in response to increased CO. (c) Increased MAP in response to increased TPR. Note that these imbalances between flows into and out of the aorta are transient. As MAP increases, outflow also increases and eventually becomes equal to inflow, such that MAP becomes constant.

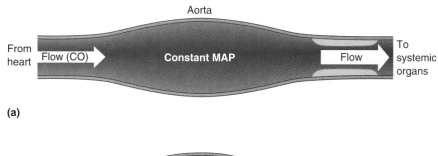

Aorta

From heart — Flow (CO) — Constant MAP — Flow — To systemic organs

(a)

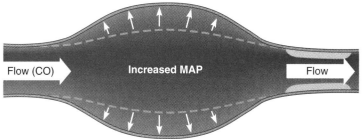

Flow (CO) — Increased MAP — Flow

An increase in cardiac output . . .   . . . leads to an increase in the volume of blood contained in the aorta and an increase in mean arterial pressure . . .   . . . when total peripheral resistance remains the same.

(b)

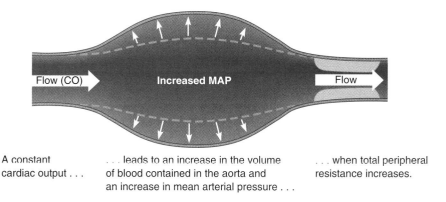

Flow (CO) — Increased MAP — Flow

A constant cardiac output . . .   . . . leads to an increase in the volume of blood contained in the aorta and an increase in mean arterial pressure . . .   . . . when total peripheral resistance increases.

(c)

total peripheral resistance. We can get an intuitive feeling for why increases in cardiac output or total peripheral resistance should cause MAP to rise by looking at **Figure 15.24**, which shows blood flow (wide arrows) into and out of the aorta in different circumstances. Total peripheral resistance is shown schematically as a constriction at the distal end of the aorta, where the blood flows out.

When mean arterial pressure is steady (Figure 15.24a), blood flows into the aorta at the same rate as it flows out, such that the volume of blood contained within it does not change. Because the flow into the aorta is cardiac output, the flow out of the aorta under these conditions equals cardiac output. (The flows shown in Figure 15.24 represent *average* blood flows over several cardiac cycles. Within a single cardiac cycle, the flows into and out of the aorta differ from one another as aortic pressure cycles up and down; see p. 434.)

Figure 15.24b shows how mean arterial pressure would change given a sudden increase in cardiac output (due to an increase in heart rate and/or stroke volume), with total peripheral resistance remaining constant. Under these conditions, blood flows into the aorta faster than it flows out, so that blood volume within the aorta increases and the vessel expands. This expansion stretches the wall of the aorta, causing it to exert a large inward force on the blood, so that the pressure of the blood rises. Thus an increase in heart rate or stroke volume causes mean arterial pressure to rise. (If cardiac output remains elevated, the rising aortic pressure causes the flow of blood out of the aorta to increase. Eventually, mean arterial pressure reaches a steady level, with rates of blood flow into and out of the aorta being equal.)

Figure 15.24c shows how mean arterial pressure would change when a sudden increase in total peripheral resistance occurs while cardiac output remains constant. The increase in total peripheral resistance reduces the flow of blood out of the aorta, so that blood flows in faster than it flows out, and mean arterial pressure rises.

We now look at how the body regulates mean arterial pressure by regulating heart rate, stroke volume, and cardiac output.

Regulation of Mean Arterial Pressure

Control of mean arterial pressure is accomplished by extrinsic regulatory mechanisms—that is, mechanisms involving the control of organs and tissues by the nervous and endocrine systems. In this section we concentrate on understanding the *short-term* regulation of arterial pressure, which takes place over a time span of seconds to minutes. *Long-term* regulation of arterial pressure, which is discussed briefly in this chapter but explained more fully in Chapter 20, involves control of the blood volume by the kidneys and takes place over minutes to days.

To see how short-term and long-term regulation work together, consider this example: When a person loses blood, the resulting decrease in blood volume causes a fall in central venous pressure, which causes a decrease in venous return, end-diastolic volume, and cardiac output, and ultimately a decrease in mean arterial pressure. As we will see, this drop in arterial pressure usually triggers a number of neural and hormonal responses that act within seconds to raise the arterial pressure back toward its normal level. At the same time, the drop in arterial pressure also triggers a decrease in the rate of urine output by the kidneys, which helps the body to maintain the blood volume by conserving water. In contrast, when a person drinks excess fluids, blood volume and mean arterial pressure rise above normal. This rise in arterial pressure triggers neural and hormonal responses that act quickly to lower the pressure, returning it to near normal. At the same time, the rise in pressure triggers an increase in the rate of urine output by the kidneys, which rids the body of excess water, thereby returning blood volume to normal.

Neural Control of Mean Arterial Pressure

When the body is at rest, extrinsic regulatory mechanisms work to keep mean arterial pressure at a constant level. If for some reason arterial pressure drops, regulatory responses "kick in" to bring it back up to the normal level; if mean arterial pressure rises, regulatory responses work to bring it back down. Thus mean arterial pressure is a *regulated variable* that is regulated by *negative feedback control*. As we saw in Chapter 1, *sensors* monitor the regulated variable in any mechanism involving negative feedback control; the sensors for monitoring mean arterial pressure are called *arterial baroreceptors*, which we examine next.

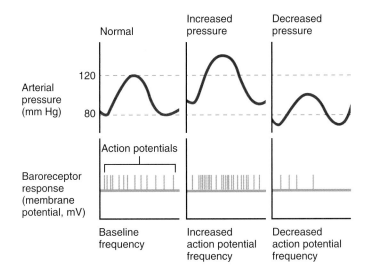

Figure 15.25 Response of arterial baroreceptors to changes in arterial pressure. Top trace in each panel shows an arterial pressure wave; vertical lines in the lower traces represent action potentials recorded from a carotid sinus baroreceptor.

Here, a change in the intensity of a stimulus (arterial pressure) triggers a change in the frequency of action potentials in afferent neurons. What is the general term for this phenomenon?

Arterial Baroreceptors: Sensors of Mean Arterial Pressure

A **baroreceptor** is a general term for a type of sensory receptor neuron in blood vessels and the heart that responds to changes in pressure within the cardiovascular system. (Recall that a *baro*meter is an instrument that monitors atmospheric pressure.) *Arterial* baroreceptors respond specifically to the stretching that occurs during pressure changes in arteries. The sensory endings of arterial baroreceptors are embedded within arterial walls. When arterial pressure (or more precisely *distending pressure*) rises, the arteries expand, stretching the walls of the arteries and the sensory endings of the baroreceptors within them, and inducing depolarization. Depolarization triggers action potentials, which are then conducted to the central nervous system by the baroreceptors' axons. Increased pressure induces greater stretch of the sensory endings, which produces greater depolarization and an increase in action potential frequency (**Figure 15.25**).

Arterial baroreceptors are found in two locations: the *aortic arch*, the curved portion of the aorta located close to where it emerges from the heart, and the *carotid sinuses* of the *carotid arteries*, which are located in the neck (**Figure 15.26**). Thus arterial baroreceptors are also called *sino-aortic baroreceptors*. These baroreceptors are strategically placed because pressure in the aorta affects blood flow to every organ in the systemic circuit, and because pressure in the carotid arteries affects blood flow to the brain,

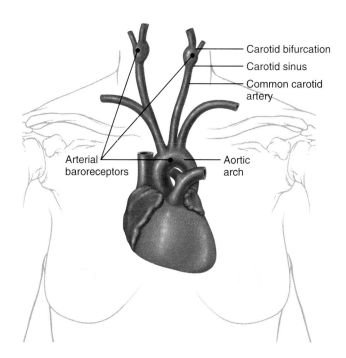

Figure 15.26 Arterial baroreceptors. Arterial baroreceptors are located in the aortic arch and carotid sinuses.

which is exceedingly sensitive to any reduction in its blood supply.

Arterial baroreceptors are important in the regulation of mean arterial pressure because they relay pressure information to the central nervous system, which exerts control over cardiovascular function via autonomic neurons projecting to the heart and blood vessels. When mean arterial pressure changes, activity in autonomic neurons changes, which causes the heart and blood vessels to alter their function in response. Baroreceptor input to the nervous system also triggers changes in the secretion of several hormones that target the heart and blood vessels.

Cardiovascular Control Center of the Medulla Oblongata

Neural control of mean arterial pressure is orchestrated primarily by the medulla oblongata. It possesses a diverse set of neural networks encompassing several nuclei, called the *cardiovascular control center*, that regulate different aspects of cardiovascular function (**Figure 15.27**). The cardiovascular control center is able to assess various indicators of cardiovascular performance (such as arterial pressure) and decide whether this performance is sufficient to meet the body's current needs. If it is not, the cardiovascular control center instructs the cardiovascular system to make appropriate adjustments by sending output to effectors via autonomic nerves.

Information from a variety of sensory receptors projects to the cardiovascular control center. Foremost among these are the arterial baroreceptors, which inform the center about current pressures in the aortic arch and carotid sinus. Other receptors include *low-pressure baroreceptors* in the right atrium and large systemic veins (also

called *volume receptors*), which monitor venous pressure, and *chemoreceptors* in the brain and carotid arteries that monitor concentrations of oxygen, carbon dioxide, and hydrogen ions in arterial blood. Functions of these receptors are discussed shortly. Proprioceptors in skeletal muscle and joints, which sense body movement and position, and other receptors of various types in internal organs throughout the body also provide input; these receptors are important in the cardiovascular response to exercise.

The cardiovascular control center receives input not only from sensory receptors but also from higher brain areas, including the cerebral cortex and hypothalamus. The hypothalamus is important in orchestrating the fight-or-flight responses of the cardiovascular system, and it also regulates the resistance of blood vessels in the skin in response to changes in body temperature. These changes in resistance control blood flow through the skin, which helps to regulate the rate of heat loss from the body. The precise nature of cortical influences on cardiovascular function is not fully understood, but these influences are thought to be wide ranging. The cortex is involved in the cardiovascular changes that occur in response to pain and emotional states (such as the rise in blood pressure that frequently accompanies anxiety) and in exercise. The cortex also exerts continual control over cardiovascular function by modulating the cardiovascular control center's responses to its sensory inputs.

Autonomic Inputs to Cardiovascular Effectors   The cardiovascular control center integrates the information received from various sources described above and determines what adjustments in the cardiovascular system are needed. To make these adjustments, the center communicates to the autonomic nervous system, affecting the levels of activity in sympathetic and parasympathetic nerves to the heart and blood vessels.

Major autonomic innervations of the cardiovascular system include the following (Figure 15.27): (1) sympathetic and parasympathetic nerves to the sinoatrial node, which control heart rate; (2) sympathetic nerves to the ventricular myocardium, which control ventricular contractility; (3) sympathetic nerves to arterioles and other resistance vessels, which control vascular resistance, and (4) sympathetic nerves to veins, which control venomotor tone. Clearly, the primary neural influence on cardiovascular function is *sympathetic;* parasympathetic input comes into play only at the SA node.

The Baroreceptor Reflex

Most of us have probably felt the effects of baroreceptors in action. We have already seen that blood pressure can fall when a person stands, causing a sensation of dizziness. Within seconds, however, the dizziness goes away because the pressure has risen to nearly normal. This occurs because the fall in arterial pressure is detected by arterial baroreceptors, which trigger an increase in sympathetic activity and a decrease in parasympathetic

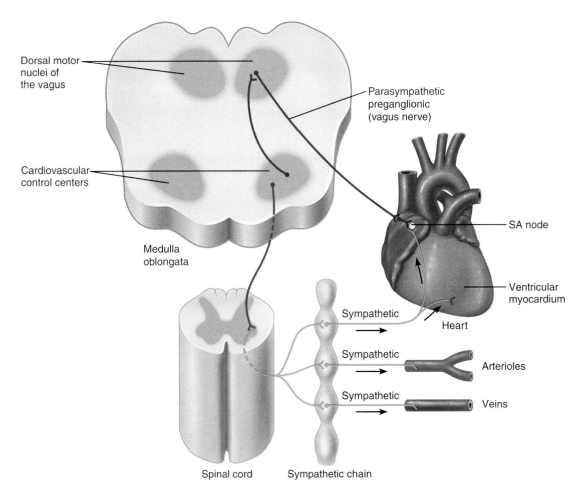

Figure 15.27 Major neural pathways in the control of cardiovascular function.
Cardiovascular function is regulated primarily by the cardiovascular control center of the medulla oblongata, which relays signals to the heart and blood vessels by way of sympathetic and parasympathetic neurons.

Does the sympathetic chain belong to the central nervous system or to the peripheral nervous system?

activity, which brings about increases in heart rate, myocardial contractility, and vascular resistance. This sequence of events is called the **baroreceptor reflex**.

Figure 15.28 illustrates the baroreceptor reflex in response to a decrease in mean arterial pressure. The drop in pressure is detected by the arterial baroreceptors and this information is transmitted to the cardiovascular control center, which triggers a decrease in parasympathetic activity and an increase in sympathetic activity to compensate. The decrease in parasympathetic activity and increase in sympathetic activity to the SA node increases the frequency of action potentials in the SA node and causes heart rate to increase. An increase in sympathetic activity to the ventricular myocardium increases cardiac contractility, which causes an increase in stroke volume. The increase in sympathetic activity to the veins also increases stroke volume by increasing venous return. An increase in sympathetic activity to arterioles causes vasoconstriction, which increases total peripheral resistance.

Increases in heart rate, stroke volume, and total peripheral resistance all tend to increase mean arterial pressure.

Figure 15.29 illustrates in graphical form the baroreceptor response to a drop in blood volume due to hemorrhage. When blood volume falls, the end-diastolic volume decreases, and the stroke volume and cardiac output also fall. At the same time that cardiac output drops, mean arterial pressure falls. The drop in arterial pressure then activates the baroreceptor reflex, which drives the pressure back up to near its initial value by causing an increase in heart rate, total peripheral resistance, and stroke volume. Notice in Figure 15.29 that heart rate and total peripheral resistance increase to greater than resting values, but the stroke volume, which was decreased in the first place, returns to less than normal. The pressure does not return all the way to its original level because if it did, baroreceptors

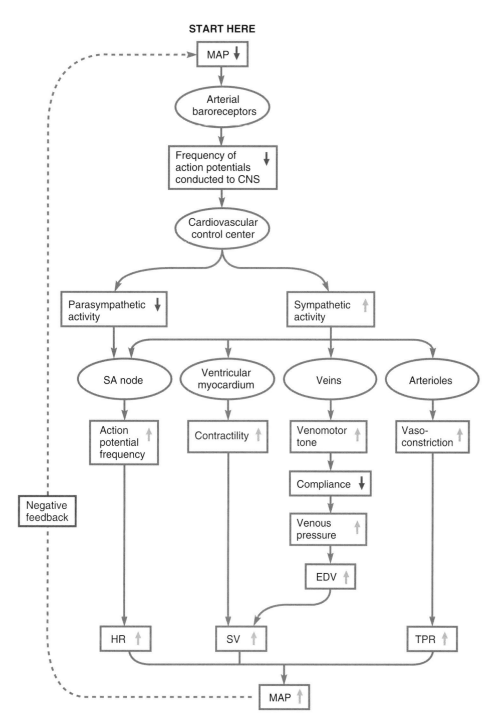

Figure 15.28 The events in the baroreceptor reflex in response to a drop in mean arterial pressure.

would relay signals to the cardiovascular control center, telling it that the pressure was normal; in that case the compensatory response to blood loss would shut down, and the pressure would drop again. Thus a small error signal is necessary for keeping the response activated and compensating for the blood loss.

Because a drop in arterial pressure triggers a subsequent compensatory rise, the baroreceptor reflex acts by negative feedback to keep pressure constant. The same principle holds when arterial pressure rises. In this case

the change in pressure triggers responses in the opposite direction, producing a decline in pressure.

The action of the baroreceptor reflex raises an interesting question: If baroreceptor reflexes work to keep mean arterial pressure constant at normal levels, why do certain people suffer hypertension? The reason is that hypertension is a chronic condition that develops slowly over long periods of time. The accompanying gradual rise in arterial pressure causes baroreceptors to lose their sensitivity such that they reset at a new, higher pressure, which effectively becomes

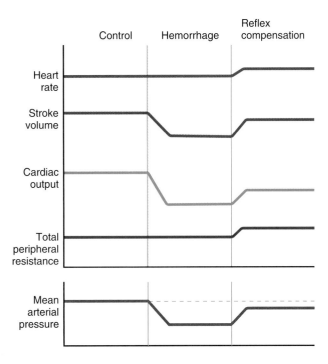

Figure 15.29 Baroreceptor-mediated responses to hemorrhage. Changes in cardiovascular variables are indicated relative to normal values before hemorrhage occurs (the "control" condition in the graph). The delay between hemorrhage and onset of reflex compensation is exaggerated for clarity. Note that reflex compensation raises mean arterial pressure only to near-normal levels.

In this example, which of the following decreases following hemorrhage: preload, afterload, or both?

"normal." Under these conditions the baroreceptors still function to regulate arterial pressure, but they act to maintain it at a level higher than it would normally be. Once this reset has occurred, baroreceptors cannot correct the problem (**Clinical Connections: Hypertension**, p. 490).

When all is considered, the *immediate* danger posed by low arterial pressure *(hypotension)* is far greater than that posed by hypertension, because a low mean arterial pressure acts to reduce blood flow to all systemic organs, which can compromise their function and even permanently damage them. This is not to say that hypertension is not dangerous, but hypertension usually takes years to kill, whereas hypotension can kill in minutes. Thus the most important function of the baroreceptor reflex is to counteract potentially dangerous reductions in organ blood flow. However, we are now faced with a puzzle: If a fall in mean arterial pressure triggers an increase in total peripheral resistance, which tends to *reduce* organ blood flow, then how is blood flow maintained?

When a baroreceptor reflex triggers an increase in sympathetic activity, the resistance in most, but not *all*, organs increases. In particular, resistance in the brain and

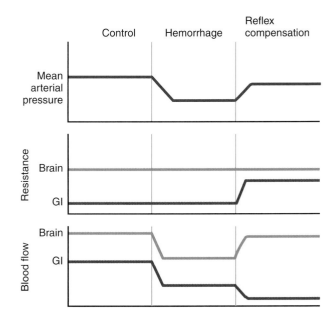

Figure 15.30 Effects of a hemorrhage-induced baroreceptor reflex on mean arterial pressure, on vascular resistance, and on blood flow in the brain and gastrointestinal tract. Increased vascular resistance in the GI tract results in lower-than-normal blood flow to these organs. By contrast, virtually no change occurs in vascular resistance in the brain; as a result, blood flow to this vital organ returns to near normal as the reflex restores MAP to near normal.

heart are affected very little by sympathetic influences and are therefore not altered significantly by baroreceptor reflex. When a baroreceptor reflex triggers a compensatory rise in mean arterial pressure, it acts to maintain blood flow to the heart and brain—but it does so *at the expense of other organs*, whose blood flow actually decreases as a result of vasoconstriction. The end result is that blood flow is shifted *away from* certain organs (such as the skin and gastrointestinal system), and *toward* other organs (such as the brain and heart) that are more vital in the short-term survival of the body (**Figure 15.30**).

Exercise Link

As Bill and Jane ran the marathon, their cardiovascular systems made similar adjustments. Blood flow increased (1) to their legs, providing oxygen and energy to the working muscles; (2) to their skin, transferring heat out of their bodies; (3) and to their heart muscles, supporting the work of increased cardiac output. As compensation, blood flow was directed away from the liver, kidneys, and gastrointestinal tract, the functions of which were not directly or immediately necessary to support exercise.

Note that when a drop in arterial pressure is due to a loss of blood volume, the baroreceptor reflex is only a "quick fix." As long as the drop in volume is not so great that the body cannot compensate for it, the baroreceptor reflex will keep arterial pressure high enough to ensure survival until the lost volume can be replaced by drinking fluids (or by blood transfusion, if necessary). Long-term regulation of arterial pressure is accomplished by controlling blood volume, which depends on adjusting the balance between fluid intake and excretion. (Temporary compensation for a decrease in blood volume can also occur through shifts in extracellular fluid: Fluid moves from interstitial fluid to plasma due to the decrease in capillary hydrostatic pressure, which results in a decrease in filtration across capillary walls.)

When individuals lose a lot of blood, the baroreceptor "quick fix" can sometimes get them into a "real fix" because the baroreceptor reflex reduces blood flow to most organs. Eventually these organs require restoration of normal flow; otherwise they will be damaged by either the reduced availability of oxygen and nutrients or the accumulation of toxic metabolic by-products. Because local controls also influence arteriolar smooth muscle, when tissues are deprived of an adequate blood supply for a prolonged period, local chemical changes cause vascular smooth muscle to relax. Unless lost blood volume is restored quickly (within one or two hours in the case of severe hemorrhage), these local influences will override the influence of sympathetic vasoconstrictor nerves on vascular smooth muscle, and vascular resistance will begin to fall. As a result, arterial pressure will fall, despite the actions of compensatory mechanisms to raise it. When this happens, blood flows to the heart and brain fall. This condition, known as *circulatory shock*, may become irreversible, in which case it progresses inexorably to death.

Hormonal Control of Mean Arterial Pressure

Arterial baroreceptors exert control over cardiovascular function not only via the baroreceptor reflex but also by regulating the secretion of the hormones epinephrine, vasopressin, and angiotensin II, which work hand in hand with the autonomic nervous system to regulate mean arterial pressure.

Recall that epinephrine is released in response to sympathetic nerve activity to the adrenal medulla. Low arterial pressure is a stimulus for epinephrine secretion, although evidence suggests that secretion is enhanced only when the drop in pressure is relatively severe. Epinephrine affects both cardiac output (described in Chapter 14) and total peripheral resistance. The effects of epinephrine on cardiac function mirror the actions of sympathetic input: At the SA node, epinephrine increases the action potential frequency of pacemaker cells, which increases heart rate. In the myocardium, epinephrine increases cardiac contractility, which increases stroke volume. In both cases, these effects are brought about as a result of epinephrine binding to β_1 receptors in cardiac tissue. These are the same receptors that bind norepinephrine, which explains the similarity between epinephrine's actions and those of the sympathetic nervous system. At the vasculature, epineph-

rine has mixed effects consisting of vasoconstriction in most vascular beds, but it can cause vasodilation in skeletal and cardiac muscle. Under most circumstances, epinephrine causes an increase in total peripheral resistance and increases blood pressure. Thus, epinephrine tends to increase mean arterial pressure by increasing heart rate, stroke volume, and total peripheral resistance.

Vasopressin and angiotensin II both cause vasoconstriction, thereby increasing total peripheral resistance and mean arterial pressure. Vasopressin secretion is regulated by a variety of factors, including the level of activity in arterial baroreceptors. When arterial pressure falls, vasopressin release is enhanced, which promotes an increase in mean arterial pressure. Vasopressin also reduces urine output by the kidneys to maintain plasma volume. Recall that angiotensin II is produced in response to renin secretion from the kidneys. When arterial pressure falls, the release of renin is stimulated both directly by reduced arterial pressure and via activity of sympathetic nerves to the kidneys. As a result of the increase in renin secretion, the plasma concentration of angiotensin I rises, and this is followed by an increase in the concentration of angiotensin II. Angiotensin II increases mean arterial pressure in a number of different ways including promoting vasoconstriction, reducing urine output by the kidneys, and stimulating thirst.

Extrinsic factors involved in the control of mean arterial pressure are summarized in **Table 15.4**.

Control of Blood Pressure by Low-Pressure Baroreceptors (Volume Receptors)

In addition to the arterial baroreceptors, which monitor systemic arterial pressure, other baroreceptors monitor pressures elsewhere in the cardiovascular system, specifically on the low-pressure side of the circulation. Particularly important are baroreceptors in the walls of large systemic veins and in the walls of the right atrium. These receptors function in the same way as arterial baroreceptors in that they have receptor endings that respond to stretch, but because of their locations they monitor pressure on the *venous* side of the systemic circulation and therefore act directly to detect changes in *blood volume*. Because the venous side of the circulation acts as a volume reservoir, low-pressure baroreceptors are frequently referred to as *volume receptors*. Additional baroreceptors located in the pulmonary vasculature act indirectly to monitor systemic venous pressure.

Because arterial pressure is influenced by venous pressure, the functions of low-pressure and arterial baroreceptors are intertwined; levels of activity in these receptors frequently vary in the same direction, because when venous pressure increases or decreases, arterial pressure also tends to increase or decrease, respectively. In addition, low-pressure baroreceptors exert many of the same actions as arterial baroreceptors. When venous pressure falls, for example, low-pressure baroreceptors trigger an increase in both sympathetic nerve activity and vasopressin secretion.

HYPERTENSION

Hypertension (high blood pressure) affects at least 60 million people in the United States (1 in 6 adults). Because of the dangers of hypertension and its prevalence, in 2003 the National Heart, Lung, and Blood Institute released new clinical guidelines for its prevention, detection, and treatment. These guidelines state that a normal blood pressure (systolic/diastolic) is <120/ <80 mm Hg; prehypertension is 120–139/ 80–89 mm Hg; stage 1 hypertension is 140–159/90–99 mm Hg; and stage 2 hypertension is ≥160/≥100 mm Hg.

There are two main forms of hypertension, *primary* and *secondary hypertension*. There is also an acute form of hypertension, called *labile hypertension*, that is a transient increase in blood pressure associated with stress. Primary and secondary hypertension are considerably more dangerous and described here.

Primary hypertension (or *essential hypertension*) accounts for 90–95% of all hypertensive cases. The precise cause of primary hypertension in an individual cannot be established. However, the disease is known to be associated with certain risk factors, including obesity, high cholesterol levels, smoking, and genetic disposition.

In secondary hypertension the elevated blood pressure is secondary to another disease. In certain forms of secondary hypertension, the cause is on the *arterial side* of the circulation; mean arterial pressure is elevated because cardiac output or total peripheral resistance is inappropriately high. In other cases the problem is on the *venous side;* mean arterial pressure is elevated because blood volume is inappropriately high. Secondary hypertension includes *renal hypertension*, associated with kidney disease, and *endocrine hypertension*, associated with inappropriate secretion of a hormone.

In renal hypertension, the primary cause is a disorder of kidney function. Kidney disease may cause failure to excrete normal amounts of salt and water, leading to fluid retention and expansion of blood volume. Sometimes the problem is occlusion of blood flow to a kidney or within a kidney, which can trigger inappropriately high rates of renin release. Abnormally high levels of renin results in overproduction of angiotension II in the plasma, which stimulates vasoconstriction and thereby increases peripheral resistance and also stimulates the kidneys to retain salt and water.

In endocrine hypertension, a hormone is inappropriately secreted by an endocrine gland. For example, a tumor of the adrenal medulla (called a *pheochromocytoma*) can oversecrete epinephrine, which stimulates increased cardiac output and total peripheral resistance, both of which elevate blood pressure. Other causes of secondary hypertension include sleep apnea, cirrhosis of the liver, smoking, and stress.

Hypertension is closely associated with another disease, *atherosclerosis* or hardening of the arteries. In atherosclerosis a fatty plaque builds up in the walls of arteries, decreasing the elasticity of arterial walls and narrowing the lumen, as shown in the photograph. The decreased lumen increases resistance, thereby contributing to hypertension. However, hypertension also produces damage to the walls of the arteries, predisposing them for atherosclerosis. Thus there is a vicious cycle in the relationship between hypertension and atherosclerosis, with each facilitating the development of the other. Atherosclerosis is a dangerous condition because blood flow through the narrowed arteries is hindered and can lead to stroke or heart attack.

In addition to atherosclerosis, hypertension has several adverse effects on the

---

Quick Test 15.6

1. What is a baroreceptor? Where are arterial baroreceptors located? Where are low-pressure baroreceptors located? What is the baroreceptor reflex?

2. What area of the brain contains the cardiovascular control center?

3. Indicate whether each of the following autonomic nervous activities increases or decreases when arterial pressure falls: sympathetic nervous activity, parasympathetic nervous activity, heart rate, myocardial contractility, vascular resistance (in most tissues), venomotor tone.

4. Name two organs whose vascular resistance is generally unaffected by baroreceptor reflexes. Why is this beneficial?

5. Describe the effects of the following hormones on mean arterial pressure: epinephrine, vasopressin, and angiotensin II.

Other Cardiovascular Regulatory Processes

In addition to baroreceptor control of mean arterial blood pressure, other processes control cardiovascular function. Some of these, such as the cardiovascular response to exercise, act to alter the pattern of blood flow so that the body can adapt to a particular set of circumstances. Others, such as *respiratory sinus arrhythmia*, have no known functional significance. What follows is a sampling of these other regulatory processes.

Respiratory Sinus Arrhythmia

Respiratory sinus arrhythmia is a rhythmic variation in heart rate in which inspiration is accompanied by increases in sympathetic activity and heart rate, whereas expiration is accompanied by increases in parasympathetic activity and a decrease in heart rate. Although not

(continued)

cardiovascular system. Because elevated arterial blood pressure increases the workload on the heart, it can increase the likelihood of a *myocardial infarction*, or heart attack. It can also lead to heart failure, because it increases afterload and can chronically elevate end-diastolic volume. Hypertension-induced damage to blood vessels can also lead to kidney failure or loss of vision.

Because it is painless, those affected by hypertension can remain unaware of it for years—while it causes irreversible damage to the cardiovascular system and other organs. Fortunately, hypertension is easily detectable during routine blood pressure checks. (Repeated measurements over time are necessary for a firm diagnosis, however.) Unfortunately, the condition can be permanently reversed only on rare occasions because the cause is unknown in the vast majority of cases (primary hypertension). When the cause is known (secondary hypertension), the disease underlying the hypertension can be treated.

Regardless of the cause, hypertension cannot be "cured" or compensated for by baroreceptor reflexes or any of the normal mechanisms that regulate blood pressure. The reason is that baroreceptors are part

of the problem: Under conditions of chronically elevated pressure, the baroreceptors reset, and regulatory mechanisms work to maintain the new, high pressure. Thus hypertension is mostly controlled through chronic medication or behavioral means.

Treatments for hypertension include diuretics, which promote increased excretion of salt and water by the kidneys, and specific antihypertensive drugs, such as beta blockers and calcium channel blockers. *Beta blockers* reduce cardiac output by interfering with the ability of epinephrine and norepinephrine to bind to beta

receptors, thereby reducing the stimulatory influence of these agents on the heart. *Calcium channel blockers* reduce the flow of calcium into vascular smooth muscle cells across the plasma membrane, which reduces vasomotor tone and lowers peripheral resistance. Other drugs, called *angiotensin converting enzyme (ACE) inhibitors*, reduce plasma levels of angiotensin II by blocking the enzyme that catalyzes its formation from angiotensin I. Other treatment options include exercise, which has been shown to lower resting blood pressure, and weight reduction.

Plaque in the wall of an artery.

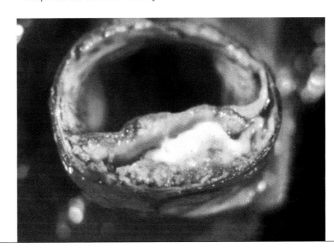

---

completely understood, the changes in sympathetic and parasympathetic activity influence the sinoatrial node. It is seen in only some people and appears to be more pronounced in children.

Respiratory sinus arrhythmia has at least one known practical consequence: Competitive archers and target shooters use it to aid their accuracy. If one takes a deep breath and then breathes out while aiming, respiratory sinus arrhythmia slows the heart rate, enabling the shooter to fire between heartbeats and thus avoid even a slight movement of the body that could throw aim off.

Chemoreceptor Reflexes

As we see in Chapter 18, control of respiration is governed by *chemoreceptors* located in the carotid sinuses and the brain. These chemoreceptors are neurons specialized to monitor concentrations of carbon dioxide (and oxygen

when it falls to extremely low levels) in the blood. When carbon dioxide rises, chemoreceptors stimulate breathing, so that more carbon dioxide is eliminated. These receptors also influence cardiovascular function, but the directions and sizes of these responses are difficult to predict, because some cardiovascular effects are primary (that is, due to stimulation of chemoreceptors themselves), whereas others are secondary (that is, due to changes in respiration triggered by stimulation of chemoreceptors).

One of the primary cardiovascular effects of chemoreceptor stimulation when arterial carbon dioxide levels rise is to trigger a decrease in heart rate and an increase in peripheral resistance. These responses seem counterproductive, because when the concentration of carbon dioxide in the blood rises, blood flow to tissues should be *increased* in order to maintain the same rate of carbon dioxide removal. However, when viewed from another

Table 15.4 | Factors Involved in Extrinsic Control of Mean Arterial Pressure

Target organ or tissue	Neural or hormonal factor	Factor's effect on target	Influence on mean arterial pressure
Heart			
Sinoatrial node	Sympathetic nerves	↑HR	↑MAP
	Parasympathetic nerves	↓HR	↓MAP
	Epinephrine	↑HR	↑MAP
Ventricular myocardium	Sympathetic nerves	↑Contractility (↑SV)	↑MAP
	Epinephrine	↑Contractility (↑SV)	↑MAP
Arteriolar smooth muscle (most tissues)	Sympathetic nerves	Vasoconstriction (↑TPR)	↑MAP
	Epinephrine	Vasoconstriction or vasodilation, depending on concentration and location	Variable
	Vasopressin	Vasoconstriction (↑TPR)	↑MAP
	Angiotensin II	Vasoconstriction (↑TPR)	↑MAP
Venous smooth muscle	Sympathetic nerves	↑Venomotor tone	↑MAP
	Epinephrine	↑Venomotor tone	↑MAP

perspective, these responses are entirely appropriate. Due to the decrease in heart rate, oxygen consumption by the heart muscle itself is diminished, which tends to conserve oxygen. Although the fall in heart rate also tends to reduce mean arterial pressure, the increase in peripheral resistance offsets this effect. As a result, blood flow to the brain is maintained at normal or near-normal levels.

The chemoreceptor-mediated increase in peripheral resistance may also be adaptive in another way: When arterial carbon dioxide levels rise, the concentration of carbon dioxide rises in tissues all over the body. Given that vascular smooth muscle is sensitive to local carbon dioxide concentrations, such a rise in arterial carbon dioxide should trigger vasodilation in many tissues, leading to a potentially dangerous drop in peripheral resistance and mean arterial pressure. By stimulating an increase in peripheral resistance, chemoreceptor reflexes work to protect against this possibility.

Thermoregulatory Responses

The ability to control heat loss through the skin is an essential component of the body's ability to regulate its own temperature. As we saw in Chapter 1, the body's response to changes in temperature is mediated by a *thermoregulatory center* in the hypothalamus that receives input from *thermoreceptors*, heat-sensitive neurons found at various locations throughout the body. The thermoregulatory center receives from these thermoreceptors information concerning skin and core temperatures.

Under normal conditions, there is a significant level of activity in sympathetic nerves that project to blood vessels in the skin. When the body's heat content increases, the resulting rise in temperature decreases the level of sympathetic activity in nerves supplying the skin, which induces relaxation of vascular smooth muscle. (Over time, other factors that promote vasodilation are thought to come into play.) As a result, blood vessels dilate, the skin's vascular resistance decreases, and blood flow to skin increases. In addition, under conditions of heat stress, sweat glands produce bradykinin, whose vasodilatory effects tend to decrease vascular resistance in the skin. As a result of the increase in blood flow, the rate of heat loss through the skin increases, which helps to counteract the rise in body temperature. A decrease in the heat load has the opposite effect: Increased activity in the skin's sympathetic nerves stimulates contraction of smooth muscle in the blood vessels, which increases the resistance of blood vessels in the skin and decreases the blood flow through the skin. As a consequence, blood is diverted away from the skin and toward the deeper structures of the body, so that less heat is lost through the skin.

Because the influence of body temperature on skin blood vessels takes precedence over other reflex controls in most circumstances, changes in skin resistance can sometimes be dangerously maladaptive. For instance, when a person overexerts on a very hot day, blood volume decreases due to excessive sweating, thereby decreasing blood pressure. The baroreceptor reflex triggers constriction of blood vessels throughout the body, including those in the skin. If that person then becomes overheated, the thermoregulatory center takes control, inducing dilation of skin blood vessels to increase heat loss. The resulting decrease in vascular resistance in skin vessels tends to cause a decrease in total peripheral resistance. As a consequence, arterial pressure may fall, counteracting the baroreceptor reflex, which should be acting to raise total peripheral resistance in order to maintain arterial pressure.

Responses to Exercise

Exercise is accompanied by profound alterations in cardiovascular function. **Figure 15.31** shows that jogging increases cardiac output from the resting value of 5 liters per minute to over 11 liters per minute; in highly trained athletes, cardiac output can reach 35 liters per minute. Heart rate also rises from the resting average of 72 beats per minute to about 135 beats per minute. Exercise also causes dramatic changes in blood flow: Flow to skeletal muscles (and to cardiac muscle and skin as well) increases while flow to the liver and the gastrointestinal tract decreases. Clearly, these changes are beneficial in several ways: (1) Delivery of oxygen and nutrients to cardiac muscle and active skeletal muscles is increased, which is appropriate, given the increase in metabolic activity in these organs; (2) oxygen and nutrients are conserved by curtailing delivery to tissues for which the need for nutrients is not as acute; and (3) increased blood flow to the skin aids the body in getting rid of excess heat generated during exercise.

In exercise, the drop in vascular resistance in skin and muscle is not quite compensated for by the increase in the resistance of other organs, so total peripheral resistance drops. Blood pressure *rises* slightly, however, due to the increase in cardiac output. Note in Figure 15.31 that stroke volume increases markedly despite the fact that only a modest increase in end-diastolic volume occurs (in light exercise). This indicates that the increase in stroke volume is due not to the Starling effect but to an increase in ventricular contractility resulting from an increase in sympathetic nervous activity and increases in the levels of circulating epinephrine.

What triggers these responses to exercise? In the central nervous system, cortical and limbic regions of the brain exert a direct influence on the output of sympathetic and parasympathetic neurons, resulting in increased sympathetic activity and decreased parasympathetic activity to the heart that account for the increase in heart rate and ventricular contractility. In addition, there is an increase in sympathetic activity in the digestive system and other organs (resulting in vasoconstriction) and

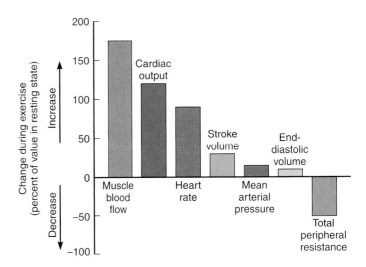

Figure 15.31 Cardiovascular responses to light exercise.
Note that the decrease in TPR allows MAP to remain nearly constant despite the huge increase in CO.

a *decrease* in sympathetic activity to the skin (which promotes vasodilation). This change in sympathetic input to the skin is a thermoregulatory response triggered by the rise in body temperature that accompanies exercise. Vasodilation in the skin is also promoted by bradykinin released from sweat glands, whose activity increases in exercise. The changes in sympathetic activity are facilitated by the cardiovascular control center, which receives input from various types of receptors in muscles, including chemoreceptors responsive to local chemical factors and mechanoreceptors and proprioceptors sensitive to muscular physical activity. The increase in blood flow to skeletal and cardiac muscles is largely a result of the action of local chemical factors on vascular smooth muscle.

Other mechanisms also promote increased venous return in exercise, thereby facilitating the rise in cardiac output. One such mechanism is the skeletal muscle pump; another is the respiratory pump, whose activity increases in exercise because breathing becomes deeper and faster. A third mechanism is an increase in venomotor tone, which results from an increase in sympathetic activity to the veins.

Quick Test 15.7

1. What is respiratory sinus arrhythmia?

2. What is a chemoreceptor? What cardiovascular changes are triggered when chemoreceptors detect a drop in arterial oxygen levels?

3. When body temperature rises, does vascular resistance in the skin increase or decrease? How does this change help the body regulate its temperature?

4. Name two organs in which blood flow increases during exercise.

CHAPTER SUMMARY

Physical Laws Governing Blood Flow and Blood Pressure, p. 452

The flow of blood through any vessel or network of vessels depends on the pressure gradient (ΔP) and the resistance (R) of the vessel or network: flow $= \Delta P/R$. The overall pressure gradient driving flow through the systemic circuit is the difference between mean arterial pressure (MAP) and central venous pressure (CVP), which is virtually identical to mean arterial pressure. The main influence on vascular resistance is vessel radius. The combined resistances of all blood vessels in the systemic circuit is the total peripheral resistance (TPR). When describing flow across the systemic circuit, the flow rule can be written as CO = MAP/TPR.

> **IP** Cardiovascular, Factors that Affect Blood Pressure, pp. 3–8

Overview of the Vasculature, p. 456

All blood vessels possess a lumen and are lined by a layer of endothelial cells. Their walls contain varying amounts of smooth muscle and connective tissue, which are critical to their specific functions.

> **IP** Cardiovascular, Anatomy Review: Blood Vessel Structure and Function, pp. 3–19

Arteries, p. 456

Arteries, which have thick walls that enable them to withstand the high pressure of blood within them, have relatively low compliance. Arterial walls are elastic, allowing them to expand during systole and then recoil inward during diastole. Because of this elastic recoil, arteries function as a pressure reservoir that maintains blood flow throughout the cardiac cycle.

Arterioles, p. 460

Arterioles contain relatively large amounts of smooth muscle, which enables them to expand or contract, thereby regulating blood flow through capillary beds. Arterioles are important in regulating the distribution of

cardiac output to the organs and in controlling mean arterial pressure.

Regulation of the distribution of blood flow among the various systemic organs is achieved through intrinsic control of organ vascular resistance. Blood flow through any given systemic organ is determined by mean arterial pressure and that organ's vascular resistance: organ blood flow = MAP/organ resistance. The resistance of an organ or tissue can change in response to variations in the metabolic activity of that organ or tissue because arteriolar smooth muscle is sensitive to local concentrations of chemicals produced or consumed in metabolism, including oxygen and carbon dioxide. Chemical changes associated with increased metabolic activity lead to vasodilation, decreased resistance, and increased blood flow (active hyperemia). The resistance of an organ can also change in response to local variations in blood flow. If blood flow becomes insufficient to meet metabolic demands (ischemia), local mechanisms induce vasodilation and a resultant increase in blood flow (reactive hyperemia). In those tissues in which vascular smooth muscle is responsive to stretch, an increase in perfusion pressure causes arterioles to stretch, which stimulates vasoconstriction and a reduction in blood flow. A response of this type is termed a myogenic response.

Extrinsic controls of arteriole radius (and therefore total peripheral resistance) regulates mean arterial pressure: MAP = CO × TPR. Extrinsic factors include the autonomic nervous system and hormones (epinephrine, vasopressin, and angiotensin II).

Capillaries, the Lymphatic System, and Venules, p. 470

Capillaries have the thinnest walls of all the blood vessels and are highly permeable to water and small solutes. Their primary function is to permit exchange of materials between the blood and the tissues. The movement of fluid across capillary walls is driven

by the net filtration pressure, which depends on the Starling forces. Most of the fluid that is filtered from capillaries is returned to the cardiovascular system by absorption. Excess filtrate is returned to the cardiovascular system by the lymphatic system. Venules are thin walled and participate in material exchange.

Veins, p. 477

Veins are large thin-walled vessels. Most veins possess valves that permit blood to flow toward the heart but not away from it. Veins have a high compliance and function as volume reservoirs. The pressure in the veins of the thoracic cavity is called central venous pressure. Central venous pressure influences arterial pressure because it affects venous return, end-diastolic volume, stroke volume, and cardiac output. As CVP rises or falls, cardiac output and MAP also tend to rise or fall, respectively. Factors affecting CVP include: activity of the skeletal muscle pump, activity of the respiratory pump, blood volume, and venomotor tone (which is regulated by sympathetic input to the veins).

Mean Arterial Pressure and Its Regulation, p. 481

In order to supply the organs and tissues with adequate blood flow, the driving force for flow (MAP) must be maintained. MAP is controlled by both short-term and long-term extrinsic regulatory mechanisms. Whereas short-term regulation is achieved through neural and hormonal control of cardiovascular function, long-term regulation is achieved through control of blood volume, which involves the kidney. Short-term regulation includes the baroreceptor reflex, a negative feedback system whereby baroreceptors detect changes in MAP, relay this information to the cardiovascular control center, which then acts on the autonomic nervous system to exert appropriate control over cardiovascular function. Autonomic control of MAP is accomplished through (1) sympathetic and parasympathetic input to the SA node, which controls heart rate; (2) sympathetic input to the myocardium, which controls ventricular contractility and stroke volume; and (3) sympathetic input to arteriolar smooth muscle in most tissues, which regulates total peripheral resistance.

IP Cardiovascular, Factors that Affect Blood Pressure, pp. 3–8

Other Cardiovascular Regulatory Processes, p. 490

The cardiovascular system is subject to a variety of regulatory processes that function in specific situations. For example, cardiovascular function is influenced by activity in arterial chemoreceptors, which monitor concentrations of oxygen and carbon dioxide in arterial blood. Regulation of blood flow to the skin, which is controlled by sympathetic nerves to skin blood vessels, is important in

body temperature regulation. Cardiovascular responses to exercise are largely achieved through changes in the activity of autonomic nerves to the heart and blood vessels, changes that are orchestrated by cortical and limbic brain regions. Blood flow to the heart and skeletal muscle is also regulated by local factors operating within these tissues.

EXERCISES

Multiple-Choice Questions

1. Total peripheral resistance is
   a) the combined resistance of all organs in the body.
   b) the resistance of capillaries located in distal body parts.
   c) the combined resistance of all organs in the systemic circuit.
   d) the combined resistance of all the blood vessels within an organ or tissue.
   e) the resistance to blood flow through the heart.

2. Central venous pressure increases
   a) when blood volume decreases.
   b) as a result of venous pooling.
   c) as a result of an increase in venomotor tone.
   d) when a person stands up.
   e) all of the above

3. If total peripheral resistance is constant, an increase in mean arterial pressure could be the result of
   a) vasoconstriction in the systemic circuit.
   b) an increase in sympathetic input to the ventricular myocardium.
   c) a decrease in blood volume.
   d) a and b
   e) none of the above

4. Which of the following tends to promote edema in systemic tissues?
   a) a decrease in the concentration of plasma proteins
   b) an increase in pressure in the vena cava
   c) an increase in arterial pressure
   d) leakage of proteins from capillaries into the interstitial fluid
   e) all of the above

5. Which of the following tends to cause a decrease in ventricular end-diastolic volume?
   a) an increase in central venous pressure
   b) an increase in skeletal muscle pump activity
   c) a decrease in filling time
   d) an increase in blood volume
   e) an increase in venomotor tone

6. Knowing stroke volume, heart rate, and mean arterial pressure provides sufficient information to determine
   a) total peripheral resistance.
   b) cardiac output.
   c) combined blood flows to all systemic organs.
   d) a and b
   e) all of the above

7. Which of the following blood vessels possess valves that prevent blood from flowing backward?
   a) arteries
   b) arterioles
   c) capillaries
   d) venules
   e) veins

8. Exchange of oxygen and carbon dioxide between blood and respiring tissues occurs across the walls of
   a) arteries.
   b) arterioles.
   c) capillaries.
   d) veins.
   e) all of the above

9. Where is the greatest proportion of total blood volume at rest?
   a) heart
   b) arteries
   c) arterioles

   d) capillaries
   e) veins

10. Which of the following tends to cause a decrease in mean arterial pressure?
    a) a drop in total peripheral resistance
    b) an increase in stroke volume of the left ventricle
    c) an increase in heart rate
    d) an increase in venous return
    e) an increase in sympathetic activity

11. Lymphatic capillaries differ from blood capillaries in that
    a) lymphatic capillaries have a lower permeability to water.
    b) lymphatic capillaries have a lower permeability to small solutes.
    c) lymphatic capillaries are blind ended.
    d) lymphatic capillaries are not connected to any other vessels.
    e) all of the above

12. Which of the following tends to cause increased venous pooling?
    a) a decrease in venomotor tone
    b) a decrease in the osmotic pressure of plasma proteins
    c) a decrease in mean arterial pressure
    d) exercise
    e) dehydration

13. Moment-to-moment changes in total peripheral resistance are normally due to changes in
    a) the lengths of blood vessels in the systemic circuit.
    b) the radius of certain blood vessels in the systemic circuit.
    c) the viscosity of blood.
    d) the mean arterial pressure.
    e) the cardiac output.

In this chapter, we cover the different components of blood and how they participate in the overall function of blood. We start off discussing the liquid portion of blood, called **plasma**, and its role in the transport of proteins, hormones, electrolytes, organic nutrients, and waste products. Next, the cellular components of blood are described, including **erythrocytes** (red blood cells) that transport oxygen and carbon dioxide, **leukocytes** (white blood cells) that defend the body against pathogens, and **platelets** (cell fragments) that are critical in the formation of blood clots to prevent the loss of blood.

Overview of the Composition of Blood: The Hematocrit

The total volume of blood in a normal healthy adult is about 5.5 liters and consists mostly of plasma (approximately 3 liters) and erythrocytes (approximately 2.5 liters), but also includes leukocytes and platelets (**Figure 16.1**). The fractional contribution of erythrocytes to the blood is called the **hematocrit** (abbreviation, *hct*), which is determined by centrifuging a sample of blood in a tube and is usually presented as a percentage (**Figure 16.2**). When blood is centrifuged, the elements of blood are separated based on density. Because erythrocytes are denser than other elements of the blood, they are pulled to the bottom of the tube. Plasma, the least dense component, remains at the top. Between these two layers rests a thin layer of leukocytes and platelets (less than 1% of total blood volume) called the *buffy coat*.

To determine the hematocrit, the heights of the erythrocyte column and the whole blood column are measured and then the hematocrit is calculated by determining the percentage of whole blood that consists of erythrocytes, as follows:

$$\text{hematocrit} = \frac{\text{height of erythrocyte column}}{\text{height of whole blood column}} \times 100$$

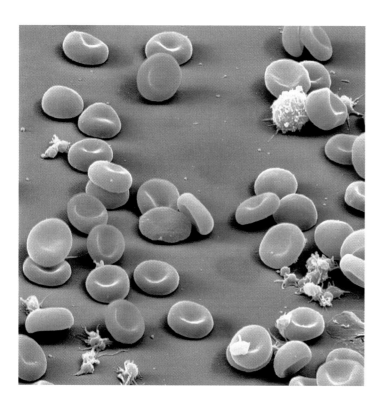

Figure 16.1 Cellular components of the blood. (Electron micrograph about 3000×)

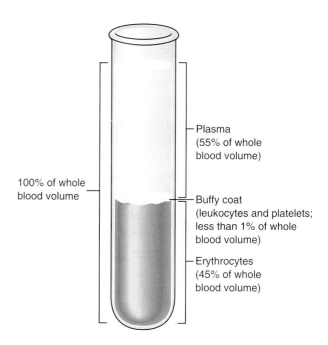

Figure 16.2 Determination of hematocrit. When a blood sample is centrifuged, erythrocytes are pulled to the bottom of the tube because they are denser than other blood elements.

The hematocrit is a useful clinical measure because it indicates whether a person has a normal complement of erythrocytes. For men, the normal range of the hematocrit is 42–52, meaning that erythrocytes occupy 42–52% of blood volume. For women, the normal range of the hematocrit is 37–47. A low hematocrit indicates a lower than normal concentration of erythrocytes in the blood. A high hematocrit indicates a higher than normal concentration of erythrocytes in the blood, called *polycythemia*. Polycythemia is a normal adaptive response in low-oxygen environments, such as occurs at high altitudes. For example, men living at an altitude of approximately 15,000 feet in the Andes often have hematocrits as high as 60. Further, the hematocrit increases within two weeks for people who travel to the Andes.

Now that we know the major components in blood, we will describe each component in detail. We start with the plasma, then discuss erythrocytes, leukocytes, and lastly platelets.

Plasma

Plasma is an aqueous solution in which a great variety of solutes are dissolved. These solutes include proteins, small nutrients (such as glucose, lipids, and amino acids), metabolic waste products (such as urea and lactic acid), gases (oxygen, carbon dioxide, nitrogen, and others), and electrolytes (such as sodium, potassium, and chloride). Although the proteins are the most abundant solutes in the plasma by weight, the smaller solutes are generally present in higher concentrations.

With respect to small solutes—that is, solutes other than proteins—the composition of plasma is very similar to that of interstitial fluid. This similarity occurs because capillary walls (which separate plasma from interstitial fluid) are highly permeable to small solutes, which allows these solutes to move freely between plasma and interstitial fluid. With respect to proteins, however, plasma and interstitial fluid differ greatly in composition; the concentration of proteins in the plasma is significantly greater than that in interstitial fluid. This difference in concentration is maintained by the low permeability of capillary walls to proteins, which limits proteins' ability to move out of the plasma.

Plasma proteins are categorized into three main groups: *albumins, globulins,* and *fibrinogen.* The albumins, which are synthesized by the liver, are the most abundant plasma proteins and make a large contribution to the osmotic pressure of plasma, which affects the movement of fluid across capillaries. The globulins encompass a large number of different proteins that transport lipids, steroid hormones, and other substances in the blood; play a critical role in the blood's ability to form clots; and are important in defending the body against foreign substances. Fibrinogen is synthesized by the liver and is a key substance in the formation of blood clots. **Serum** is plasma

Table 16.1 ▌ Components of Plasma

Component	Description and importance
Water	Makes up 90% of plasma volume; provides dissolving and suspending medium for solutes and formed elements
Solutes	
Proteins	Accounts for 8% of plasma (by weight); most are synthesized by liver
Albumin	60% of plasma proteins; largely responsible for plasma oncotic pressure
Globulins	36% of plasma proteins; include clotting proteins, antibodies secreted by certain leukocytes during the immune response, and proteins that bind to lipids, fat-soluble hormones, and metal ions to transport these substances in the blood
Fibrinogen	Important in the formation of blood clots
Others	Enzymes, hormones, and antibacterial proteins
Nitrogenous waste products	By-products of metabolism, such as urea, uric acid, and creatinine
Organic nutrients	Materials absorbed from the intestines and used by cells throughout the body; includes glucose and other simple sugars, amino acids, fatty acids, glycerol, triglycerides, cholesterol, and vitamins
Electrolytes	
Cations	Sodium, potassium, calcium, magnesium (important in neuromuscular signaling), and trace metals (important in normal enzyme activity)
Anions	Chloride (important in neuromuscular signaling), bicarbonate, and phosphate (important in maintenance of normal plasma pH)
Respiratory gases	Oxygen and carbon dioxide; most oxygen and some carbon dioxide is bound to hemoglobin in erythrocytes; a significant fraction of carbon dioxide is found in the plasma in the form of bicarbonate

from which fibrinogen and other clotting proteins have been removed. Major components of the plasma are listed in **Table 16.1**.

Erythrocytes

Erythrocytes (red blood cells) are the most abundant cells in the blood, numbering about 5 million per cubic millimeter (mm^3) of blood. Among cells of the body, erythrocytes are unique in that they lack nuclei, mitochondria, and other organelles, such as ribosomes, that are necessary

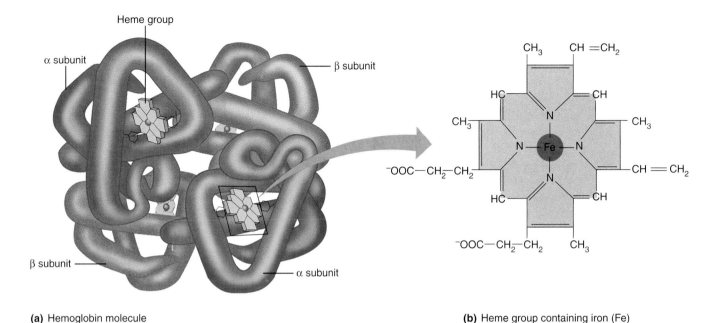

(a) Hemoglobin molecule

(b) Heme group containing iron (Fe)

Figure 16.3 Hemoglobin. (a) The hemoglobin molecule consists of four protein subunits, designated as either α or β, each covalently bound to a heme group containing iron. (b) Chemical structure of a heme group.

for manufacturing proteins. Erythrocytes are shaped like disks and are about 7 μm in diameter and a little more than 2 μm thick (see Figure 16.1). They are often described as *biconcave disks* because they are indented on both sides.

The shape of erythrocytes is due to the presence of a cytosolic protein called *spectrin*. Spectrin is a fibrous protein that forms a network linked to the plasma membrane. This *spectrin net* is flexible, giving erythrocytes the ability to bend and flex as necessary to move through capillaries that are sometimes smaller in diameter than the erythrocytes. In addition to its flexibility, the biconcave shape gives erythrocytes a large surface area, which makes them suitable for exchange. This is critical because, as we see next, the erythrocytes transport oxygen and carbon dioxide in blood for exchange with body cells and lung tissue.

Oxygen and Carbon Dioxide Transport

The major function of erythrocytes is to transport oxygen and carbon dioxide in the blood. Such transport is essential for the delivery of oxygen from the lungs to respiring cells, and for the delivery of carbon dioxide from respiring cells to the lungs, where it is eliminated from the body. Erythrocytes have a high capacity for carrying these gases because they contain in their cytoplasm two proteins: *hemoglobin* and *carbonic anhydrase*. **Hemoglobin** binds, and thus transports, oxygen and carbon dioxide, whereas **carbonic anhydrase** is essential for the transport of carbon dioxide only. The transport of these gases

is covered in detail in Chapter 18, but here we give a general description of these proteins and their importance to erythrocytes.

Hemoglobin's Reversible Binding of Oxygen and Carbon Dioxide

Hemoglobin is the most abundant protein in erythrocytes, with over 250 million hemoglobin molecules per erythrocyte. Hemoglobin is composed of four polypeptide chains of two types (two alpha and two beta), each of which has an iron-containing ring structure known as a *heme group* (**Figure 16.3**). This iron is the site to which a molecule of oxygen binds. Because there are four heme groups in a hemoglobin molecule, each hemoglobin can bind four oxygens. Carbon dioxide binds reversibly to amino acids within the polypeptide chains. The iron in hemoglobin is oxidized and gives hemoglobin a red color, which is responsible for the red color of erythrocytes and blood.

Carbonic Anhydrase and the Carbon Dioxide–Bicarbonate Reaction

Although less abundant, carbonic anhydrase is another protein found in erythrocytes that is critical in the transport of gases. Carbonic anhydrase is an enzyme that catalyzes the reversible conversion of carbon dioxide and water to carbonic acid:

$$CO_2 + H_2O \xrightleftharpoons{\text{carbonic anhydrase}} H_2CO_3$$

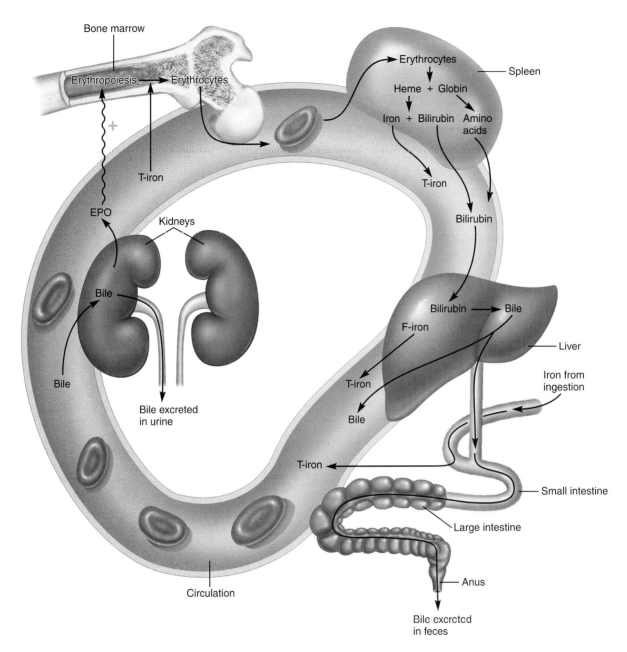

Figure 16.4 Life cycle of erythrocytes. The production and breakdown of erythrocytes involves several organs. EPO = erythropoietin, T-iron = transferrin, and F-iron = ferritin.

The pathway continues with the reversible dissociation of carbonic acid to yield a hydrogen ion and a bicarbonate ion:

$$H_2CO_3 \rightleftharpoons H^+ + HCO_3^-$$

The entire pathway can be written as follows:

$$CO_2 + H_2O \overset{\text{carbonic anhydrase}}{\rightleftharpoons} H_2CO_3 \rightleftharpoons H^+ + HCO_3^-$$

Thus, carbon dioxide can be converted to free hydrogen ions, which affects the pH of blood. In Chapter 18, you will learn how this reaction is critical to the transport of carbon dioxide in blood and the key role carbon dioxide has in acid-base balance of the blood.

Life Cycle of Erythrocytes

Once erythrocytes are released into the bloodstream, they remain there for only about 120 days. They have no nucleus (no DNA) nor any organelles, and thus they cannot undergo cell division. Therefore, new erythrocytes are produced on a regular basis, at a rate of approximately 2–3 million per second or 200 billion per day! The bone marrow has the enormous task of producing these erythrocytes by a process called **erythropoiesis**, while the spleen removes the old erythrocytes from the blood. **Figure 16.4** shows the organs involved in the synthesis and breakdown of erythrocytes.

Erythrocyte Production

All blood cells develop from precursor cells called *hematopoietic* (blood-forming) *stem cells,* located in the bone marrow (**Figure 16.5**). Whereas erythrocytes and most leukocytes come to full maturity in the bone marrow, T lymphocytes must migrate to the *thymus gland* (located in the thoracic cavity above the heart) before they develop to maturity. The development of a particular type of blood cell depends on cytokines called *hematopoietic growth factors (HGFs)*. The HGF that stimulates erythrocyte production is **erythropoietin**. The HGFs involved in leukocyte production include **colony-stimulating factors** and **interleukins**.

Erythropoietin is released from certain cells in the kidney in response to low oxygen levels in blood. Erythropoietin travels in the bloodstream to the bone marrow, where it triggers differentiation of pluripotent cells to erythrocytes. Several levels of differentiation are needed to produce erythrocytes. During differentiation, erythrocytes produce hemoglobin and lose their nuclei and organelles. The last cell stage prior to developing into the mature erythrocyte is the *reticulocyte*, a red blood cell with some ribosomes still present in the cytoplasm, giving the cell a weblike, or *reticular*, appearance. Normally, only erythrocytes are released into the bloodstream. However, under conditions of rapid erythrocyte synthesis, such as would occur following a severe hemorrhage, some reticulocytes enter the bloodstream as well.

Dietary requirements for erythrocyte production are iron, folic acid, and vitamin B_{12}. The iron is needed to synthesize hemoglobin. Although the liver stores some iron and some of the iron released from old erythrocytes is recycled, maintaining adequate hemoglobin levels in the blood (13–18 grams per 100 mL in men, and 12–16 grams per 100 mL in women) requires iron in the diet. Folic acid and vitamin B_{12} are necessary for the synthesis of DNA. Although deficiencies of folic acid or vitamin B_{12} affect cell division of all cells, their effects on erythropoiesis are most notable because of the rapid production of erythrocytes.

A shortage of the dietary elements for erythropoiesis can produce either a reduction in the amount of hemoglobin per cell or a reduction in the number of erythrocytes in the blood, either of which reduces the oxygen carrying capacity of the blood, a condition called **anemia**. If the condition is due to a lack of iron in the diet, it is called *iron-deficiency anemia*, and the erythrocytes are characteristically smaller than normal due to a decrease in the amount of hemoglobin per erythrocyte. If the condition is due to a lack of vitamin B_{12}, it is called *pernicious anemia* and is characterized by larger than normal erythrocyte size but a decrease in the number of erythrocytes. Anemia can also result from hemorrhage (bleeding) or an abnormally high rate of erythrocyte destruction (hemolysis). Symptoms of anemia include chronic fatigue and shortness of breath upon exertion. For more about anemia, see **Clinical Connections: Anemia** (p. 506).

Filtering and Destruction of Erythrocytes by the Spleen

The spleen is a lymphoid organ that stores blood cells and removes old erythrocytes from the circulation (see Figure 16.4). Erythrocytes have a life span of approximately 120 days. Some old erythrocytes hemolyze in the bloodstream, but most are engulfed by macrophages in the spleen and, to a lesser extent, in the liver. When macrophages destroy erythrocytes, the hemoglobin is also catabolized. After iron is removed, the resulting heme is converted to **bilirubin**, a yellow compound. Bilirubin is then released into the bloodstream, where it gives plasma its yellowish tinge. The bilirubin travels to the liver where it is catabolized further. Most products of bilirubin catabolism are secreted in the bile to the small intestine and ultimately excreted in the feces, imparting a brownish tinge to the feces. Other bilirubin metabolites are released into the bloodstream where they travel to the kidneys and are eliminated in the urine, imparting a yellowish color to the urine. During certain conditions (liver diseases, blockage of the bile duct, or excessive hemolysis of erythrocytes in the blood), plasma bilirubin levels are elevated, which is called *jaundice* and causes the skin and the whites of the eyes to take on a yellowish color.

The iron that was released by hemoglobin catabolism is recycled to form new hemoglobin. Iron is transported in blood bound to a protein called *transferrin*. Transferrin picks up iron from the gastrointestinal tract or from the spleen and transports the iron to the red bone marrow for erythrocyte synthesis, or to the liver where some iron can be stored bound to the protein *ferritin*. Some iron is also stored bound to ferritin in the spleen and in cells lining the small intestine.

Quick Test 16.1

1. What is the hematocrit, and how is it determined?

2. How do the concentrations of small solutes in the plasma compare to their concentrations in interstitial fluid? How do the two fluids compare in regard to their protein concentrations?

3. What is hemoglobin? Why is it important to the function of red blood cells?

4. What reaction does the enzyme carbonic anhydrase catalyze?

Leukocytes

Compared to erythrocytes, leukocytes (white blood cells) are far less numerous in the blood, numbering only about 4000–10,000 per cubic millimeter (see **Table 16.2** for the distribution of erythrocytes and leukocytes in blood). Leukocytes are nucleated and possess all the normal cellular machinery, and they are thus the only fully functional cells in the blood. Unlike erythrocytes, leukocytes

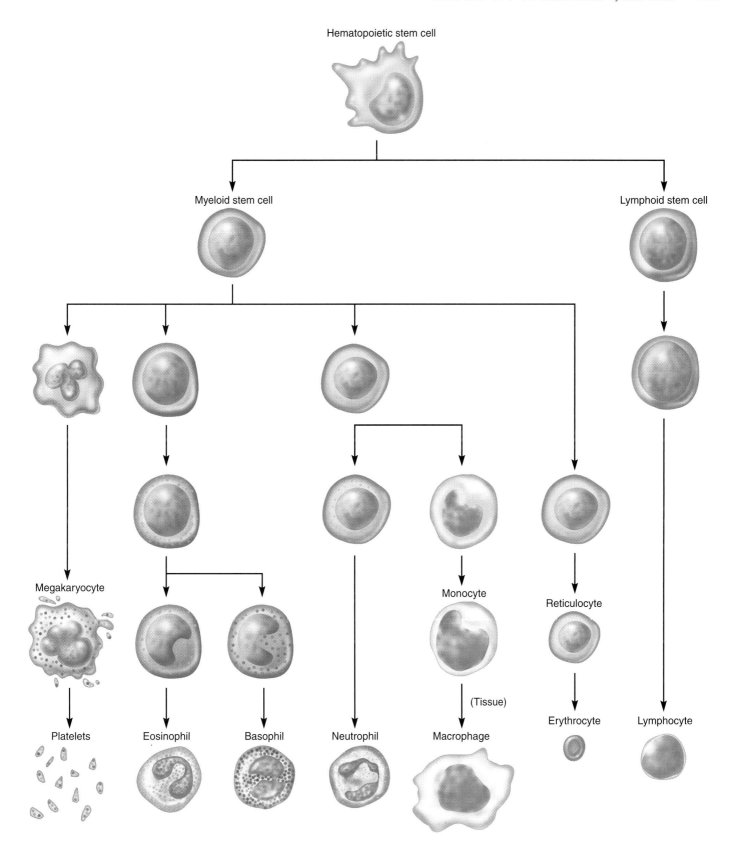

Figure 16.5 Production of blood cells and platelets. All blood cells are derived from hematopoietic stem cells. During the first level of differentiation, two cell lines are generated: myeloid stem cells, which are precursors for most blood cells, and lymphoid cells, which are precursors to lymphocytes.

Table 16.2 | Cellular Composition of Blood

Component	Amount per microliter (mm³)	Diameter (μm)	Anatomical features	Primary function
Erythrocytes	5,000,000	7–8	No nucleus; no organelles; biconcave disk	Transport O_2 and CO_2
Leukocytes	4000–10,000			Defend body against pathogens
Neutrophils	3000–7000	10–14	Multilobed nucleus; red and blue staining granules	Phagocytosis of foreign material
Eosinophils	100–400	10–14	Bilobed nucleus; red staining granules	Kill parasites
Basophils	20–50	10–12	Multilobed nucleus; blue staining granules	Secrete chemical mediators in inflammation and allergic reactions
Monocytes	100–700	14–24	Large oval nucleus; no granules	Phagocytosis; mature into macrophages in tissue
Lymphocytes	1500–3000	5–17	Large nucleus; little cytoplasm; no granules	B cells—secrete antibodies T cells—secrete cytokines that support immune response of other cells; secrete factors that kill infected or tumor cells
Platelets	250,000	2–4	Cytoplasmic fragments; granules	Hemostasis

are normally found not only in the bloodstream but also in other tissues of the body. The presence of leukocytes outside blood vessels results from their mobility, which allows them to squeeze through pores in capillaries and migrate through tissues. This ability to migrate is important to their function, which is to defend the body against invading microorganisms and other foreign materials.

There are five major types of leukocytes, each with a particular role in immunity (see **Table 16.3**). Three of these cell types—neutrophils, eosinophils, and basophils—are called *granulocytes* because they have prominent protein-containing vesicles in their cytoplasm known as *cytoplasmic granules*. The others—monocytes and lymphocytes—do not have prominent granules and thus are sometimes called *agranulocytes*. Each leukocyte type is traditionally identified by its reaction with Wright's stain, which reveals its distinctive nuclear shape and cytoplasmic color.

Neutrophils

Neutrophils constitute about 50–80% of all leukocytes and are capable of one of the most important defense activities in the body: phagocytosis. As a phagocyte ("eating cell"), a neutrophil engulfs and digests microorganisms, abnormal cells, and foreign particles present in blood and tissues. Only a few other cell types (eosinophils, monocytes, and macrophages) are capable of phagocytosis. Newly produced neutrophils circulate in the blood for 7–10 hours and then migrate into the tissues, where they

live for only a few days. During an infection the number of circulating neutrophils increases dramatically, providing not only assistance in defence but also a clinical indication of an ongoing infection.

Eosinophils

About 1–4% of all leukocytes are **eosinophils**. Like neutrophils, eosinophils are phagocytic, but their main contribution in defense is in attacking parasitic invaders that are too large to be engulfed. Eosinophils mount an attack against these parasites by attaching to their bodies and discharging toxic molecules from their cytoplasmic granules. Unfortunately, eosinophil defense—and the overall immune response to such parasites—is weak. Eosinophil responses can sometimes be harmful because the toxic molecules they release can also damage normal tissues and may trigger allergic reactions.

Basophils

Basophils are nonphagocytic cells that are thought to defend against larger parasites; they may operate much like eosinophils by releasing toxic molecules that damage invaders. However, basophils also release histamine, heparin, and other chemicals that contribute significantly to allergic reactions such as hay fever. Although basophils constitute less than 1% of all leukocytes, they have a significant effect in people with allergies.

Table 16.3 | Leukocytes and Their Roles in Immunity*

Leukocyte morphology	Functions in immunity
Neutrophil	Engulf microorganisms, abnormal cells, and foreign particles by phagocytosis
Eosinophil	Secrete enzymes that kill parasites; contribute to tissue damage in allergic reactions
Basophil	Secrete chemical mediators of inflammation and allergic reactions
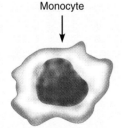 Monocyte ↓  Macrophage	Secrete cytokines; engulf microorganisms by phagocytosis
Lymphocyte	Plasma cells (mature form of B cells) secrete antibodies Helper T cells secrete cytokines that activate multiple cell types: cytotoxic T cells secrete factors that lead to the death of infected cells and tumor cells Null cells called natural killer cells secrete factors that lead to the death of infected cells and tumor cells

*Morphological features shown are characteristic of preparation with Wright's stain.

Monocytes

Monocytes, which make up about 2–8% of the leukocytes, are important in phagocytic defense. New monocytes circulate in the blood for only a few hours before they migrate into tissues, where they become five to ten times larger and develop into very active phagocytic cells known as **macrophages** ("big eaters").

Some macrophages, called "wandering macrophages," migrate throughout body tissues, whereas others, known as "fixed macrophages," remain at particular sites. Macrophages are especially abundant in connective tissue, in the wall of the gastrointestinal tract, in the alveoli of the lung, and in the walls of certain blood vessels in the liver (where they are known as Kupffer cells) and in the spleen, two sites in which they phagocytose abnormal, dead, and dying erythrocytes. Macrophages are also prevalent in regions of the spleen and lymph nodes where they are most likely to contact infectious agents circulating in the blood and lymphatic fluid.

Lymphocytes

Lymphocytes constitute about 20–40% of all leukocytes in the blood, and about 99% of all cells found in interstitial fluid. Lymphocytes are of three major types: *B lymphocytes (B cells), T lymphocytes (T cells),* and *null cells,* so called because they lack cell membrane components that are characteristic of B cells and T cells. Most null cells are large, granular lymphocytes known as *natural killer (NK) cells.* The various lymphocytes function in the immune system, as described in Chapter 23.

> **Quick Test 16.2**
>
> 1. List the five major types of leukocytes, and describe one function of each.
> 2. Define *phagocytosis.* What types of cells are capable of phagocytosis?
> 3. Name the three major types of lymphocytes.

Platelets and Hemostasis

Platelets are colorless cell fragments that arise when portions of large bone-marrow cells called *megakaryocytes* break off. They are smaller than erythrocytes and contain mitochondria, smooth endoplasmic reticulum, and cytoplasmic granules, but no nucleus. Platelets number from 100,000 to 500,000 per cubic millimeter and are important in triggering the sequence of events that leads to the formation of blood clots.

Blood vessels get damaged frequently, leading to internal and/or external bleeding. Although bleeding is usually minor, a lack of **hemostasis**—mechanisms to stop the bleeding—would mean even a superficial cut

ANEMIA

Anemia, defined as a decrease in the oxygen carrying capacity of blood, is generally associated with a low hematocrit, which can result from either a decrease in the number of erythrocytes or a decrease in the size of erythrocytes. However, a person can also have anemia if the hematocrit is normal but each red cell contains less than the normal concentration of hemoglobin. Because most of the oxygen transported in blood is bound to hemoglobin in erythrocytes, a decrease in erythrocyte abundance or size is associated with low levels of hemoglobin. There are six general categories of anemia:

1. *Nutritional anemias* are caused by a dietary deficiency, most commonly an iron deficiency. Because iron is a component of hemoglobin, an iron deficiency decreases hemoglobin synthesis. The number of erythrocytes is normal, but less hemoglobin is present in each erythrocyte, causing the erythrocytes to be smaller. Less hemoglobin means a decrease in the oxygen carrying capacity of blood.

Another essential nutrient for normal oxygen transport is folic acid, which is required for the synthesis of thymine, one of the bases in DNA. A lack of folic acid affects all cells of the body that undergo rapid cell division, because DNA replication is required for cell division. Given that each second the body produces 2–3 million erythrocytes by the rapid cell division of certain precursor cells, insufficient folic acid in the diet results in fewer erythrocytes, each of which is larger than normal. Large erythrocytes, however, tend to be more fragile and have a shorter life span.

2. *Pernicious anemia* is caused by a deficiency of intrinsic factor, which is required for the absorption of vitamin B_{12} in the intestinal tract. Because vitamin B_{12}, like folic acid, is required for the synthesis of thymine, this anemia has the same characteristics as anemia caused by a folic acid deficiency.

3. *Aplastic anemia* is caused by a defect in the bone marrow, the primary site of erythrocyte and leukocyte production; the result is a deficiency of both classes of blood cells. Fewer erythrocytes causes anemia, and fewer leukocytes impairs the body's defense against pathogens such as bacteria and viruses.

4. *Renal anemia* is associated with decreased production of the hormone erythropoietin due to a pathological state in the kidneys. Because erythropoietin stimulates the synthesis of erythrocytes in the bone marrow, a decrease in erythropoietin production results in fewer erythrocytes in the blood.

5. *Hemorrhagic anemia* is caused by the rapid loss of blood. Within a few days following a severe hemorrhage, fluid intake and the shifting of body fluids between compartments increases the blood volume back to normal. However, erythrocyte production takes a few weeks, so even as blood volume returns to normal, the hematocrit remains low until new erythrocytes can be produced.

6. *Hemolytic anemia* is caused by the rupture, or *hemolysis,* of an excessive number of erythrocytes. An increase in hemolysis can be caused by infections such as malaria, or by defects in the erythrocytes, such as occurs in *sickle cell anemia.*

Sickle cell anemia is a hereditary disease caused by a defect in the gene that codes for hemoglobin. Individuals who inherit the gene from both parents have *sickle cell disease,* whereas those who inherit the gene from only one parent have *sickle cell trait.* About 1 in 12 African-Americans have sickle cell trait. A single base in the gene coding for the beta chain is wrong, causing it to code for the wrong amino acid. (Specifically, the sixth amino acid in the beta chain is normally glutamate, but is replaced by valine.) This single amino acid substitution causes the hemoglobin molecules in an erythrocyte

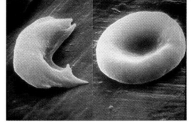

A sickled (left) and normal (right) red blood cell.

to polymerize (link to one another) when oxygen levels are low, making the erythrocyte sickle-shaped and fragile (see the photo). Such erythrocytes also can adhere to blood vessels, producing blockages of blood flow.

Symptoms of sickle cell disease are indicative of the anemia caused by excessive hemolysis of the fragile erythrocytes and blockage of blood vessels. The blockage causes the tissue in the area to become *hypoxic* (low P_{O_2}), often resulting in pain. Other symptoms include swelling of the extremities, decreased immune function, gallstones, and kidney failure. Individuals with sickle cell disease are also more prone to strokes. (Sickle cell disease symptoms may appear in individuals with sickle cell trait if they are subjected to conditions of low oxygen.) Therapy for sickle cell disease is limited to treating the symptoms. Pain is treated with *analgesics* (pain blocking drugs), and the diet is often supplemented with folic acid.

Diseases can often be associated with a particular ethnic group or a geographical location. The predominance of sickle cell disease in African-Americans may have developed as a protection against malaria, which is caused by a parasite spread by mosquitoes. The parasites invade erythrocytes, where they proliferate. Sickle cell disease actually protects against malarial infections because the malaria parasite enhances the "sickling" of erythrocytes in individuals with sickle cell disease. Once "sickled," the erythrocytes are more fragile and are recognized as abnormal cells by the spleen. The spleen destroys these erythrocytes and the malaria parasite within them.

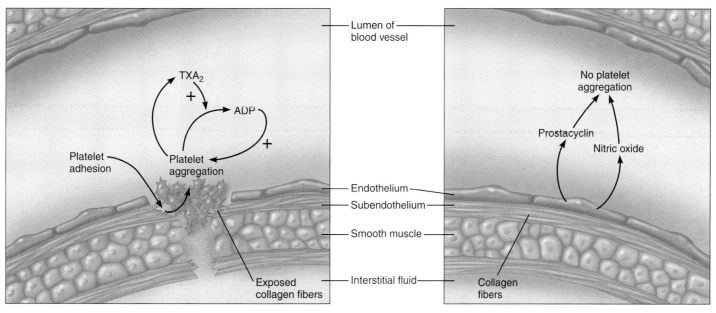

(a) Damaged blood vessel endothelium

(b) Normal blood vessel endothelium

Figure 16.6 Formation of a platelet plug. (a) Platelets adhere to collagen fibers at the site of vessel damage. Adhered platelets secrete ADP to stimulate platelet aggregation at the site of adhesion, causing more platelets to secrete ADP. Aggregated platelets also release thromboxane A$_2$ (TXA$_2$) into the lumen of the blood vessel, which stimulates further platelet aggregation. (b) Healthy endothelial cells secrete nitric oxide (NO) and release prostacyclin, both of which inhibit platelet aggregation.

could cause a person to bleed to death. The process of hemostasis occurs in three steps: vascular spasm, formation of a platelet plug, and formation of a blood clot, or **thrombus**. Each of these steps is described next.

Vascular Spasm

When a blood vessel is damaged, intrinsic mechanisms trigger a constriction called a *vascular spasm,* which increases resistance to blood flow. Damage also tends to activate the sympathetic nervous system, which causes further vasoconstriction. With less blood flowing to the area of damage, blood loss is minimized. However, decreasing blood loss is not sufficient; blood loss must be stopped altogether.

Platelet Plug

Platelets, also called *thrombocytes,* are non-nucleated fragments of megakaryocytes. Platelets possess granules containing a variety of substances that can be secreted into the plasma, including ADP, serotonin, epinephrine, and a variety of chemicals that participate in the formation of a blood clot. Platelets also are "sticky" under certain circumstances, allowing them to adhere to surfaces, especially those of damaged blood vessels.

Both the formation of platelet plugs and the subsequent blood clot require the presence of platelets and a fairly large set of specific plasma proteins. In platelet plug formation, the key protein is **von Willebrand factor (vWf)**, which is secreted by megakaryocytes, platelets, and endothelial cells lining blood vessels. Although vWf is present in the plasma at all times, it accumulates at the site of vessel damage.

The first step in platelet plug formation is *platelet adhesion*, which occurs when blood vessel damage exposes tissue underlying the endothelium (the layer of endothelial cells that line a vessel), called *subendothelial tissue.* When blood contacts subendothelial tissue, vWf binds to collagen fibers in the subendothelium, triggering the binding of platelets to vWf, which anchors platelets in place. Contact with vWf also activates platelets, changing their metabolism and surface properties (making them "sticky") and stimulating the secretion of certain products.

Two of the secretory products of activated platelets— serotonin and epinephrine—cause vasoconstriction, which increases resistance to blood flow and minimizes blood loss. A third secretory product of activated platelets, ADP, causes *platelet aggregation*, the second step in platelet plug formation. ADP stimulates morphological changes in the platelets that cause them to adhere to one another such that they form a mass, or aggregate. Aggregated platelets secrete more ADP, which stimulates further platelet aggregation, thereby providing a positive feedback loop that increases the rate of platelet plug formation (**Figure 16.6**a). ADP also stimulates the production of **thromboxane A$_2$ (TXA$_2$)**, which further supports platelet aggregation.

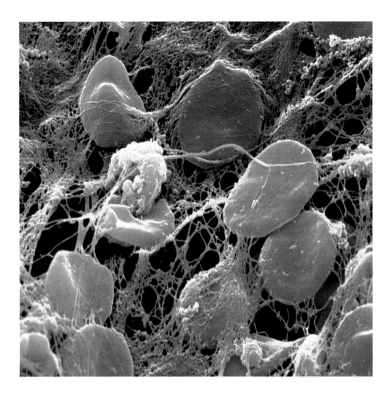

Figure 16.7 A fibrin clot.

TXA$_2$ is formed from a phospholipid, **arachidonic acid**, located in the membrane of platelets. TXA$_2$ has many roles in hemostasis, including stimulation of platelet aggregation, stimulation of ADP secretion (providing more positive feedback for platelet plug formation), and vasoconstriction (reducing blood flow to the area).

Platelet plug formation is limited to the area of blood vessel damage to prevent unnecessary blood clots that, by occluding blood flow, might deprive tissue of its needed nutrients and allow waste products to accumulate. Platelet plugs do not form on normal endothelium because healthy endothelial cells continuously release **prostacyclin** (also called prostaglandin I$_2$) and nitric oxide, both of which inhibit platelet aggregation (Figure 16.6b). Whereas aggregated platelets convert arachidonic acid to TXA$_2$ to facilitate platelet plug formation, healthy endothelial cells convert arachidonic acid to prostacyclin to inhibit platelet plug formation.

Platelets contain high concentrations of the contractile proteins actin and myosin. As platelets aggregate and form a plug, they contract to increase the "tightness" of the plug.

Formation of a Blood Clot

Integral to the formation of a blood clot is a plasma protein called **fibrin**, which is necessary for the blood to coagulate, or be converted into a gel that traps erythrocytes and plugs the damage to the blood vessel and thus pre-

vents blood loss. For this reason blood clots are also called *fibrin clots* (**Figure 16.7**). Fibrin clot formation is secondary to platelet plug formation because it requires that a number of steps occur on phospholipids located on the surface of activated platelets, and because secretory products of aggregated platelets are necessary for clot formation.

The formation of a fibrin clot requires a sequence of reactions called the **coagulation cascade**. During this cascade, plasma proteins called **coagulation factors**, which are always present in the plasma in their inactive form, undergo a series of proteolytic activations resulting from hydrolysis of certain peptide bonds. Most coagulation factors are designated by Roman numerals; the number of the factor provides no insight to function or location in the coagulation cascade, as these coagulation factors were numbered based on the order of discovery. A lowercase "a" following the Roman numeral indicates activation of the factor. Most of the activated coagulation factors function as proteolytic enzymes for the next step of the cascade, although some serve as cofactors.

The ultimate step of clot formation is conversion of a filamentous plasma protein, *fibrinogen,* into its active form, fibrin. Conversion of fibrinogen to fibrin is a proteolytic reaction catalyzed by **thrombin**, the active form of another coagulation factor, prothrombin. Once formed, fibrin molecules adhere to each other, forming a loose meshwork of strands. The meshwork is then stabilized by formation of covalent linkages between strands, a reaction catalyzed by another coagulation factor, *factor XIII* (also called fibrin-stabilizing factor). Like fibrin and thrombin, factor XIII is present in the plasma in an inactive form and must be activated to XIIIa before participating in clot formation.

Central to formation of the stable fibrin meshwork is activation of thrombin (**Figure 16.8**). Factor Xa converts prothrombin (factor II) to thrombin. Thrombin has several roles in clot formation, including converting fibrinogen to fibrin, activating factor XIII, and providing positive feedback for its own activation. Although not shown in the figure, thrombin also contributes to additional platelet aggregation and stimulates platelets to secrete several products.

Two pathways lead to the activation of thrombin: an intrinsic pathway involving coagulation factors and other necessary chemicals that are already present in the plasma, and an extrinsic pathway involving some coagulation factors present in damaged tissue adjacent to the site of vessel damage. Both clotting pathways are generally activated simultaneously because rarely is vessel damage not accompanied by damage to other tissues. The stimuli that initiate the pathways, however, differ.

The intrinsic pathway starts when circulating factor XII (also called Hageman factor) is activated by contact with substances in the subendothelium, including collagen and phospholipids (**Figure 16.9**). Activation of factor XII starts a cascade of reactions that leads to activation of factor X, to activation of thrombin, and ultimately to fibrin formation. The extrinsic pathway starts when tissue damage allows **tissue factor** (factor III) to contact the plasma and react with inactive factor VII to form a complex, which activates factor VII. The complex of factor VIIa and tissue factor then activates factor X, leading to the activation of thrombin. Even though the intrinsic and extrinsic pathways start from separate places, they eventually merge at activation of factor X to form a common pathway.

Only some of the several other chemicals in blood that participate in clot formation are shown in Figure 16.9. At several steps, Ca^{2+} (factor IV) and *platelet factor 3 (PF3)* are needed. PF3, located on the surface of activated platelets, is a phospholipid necessary for activation of thrombin by the intrinsic pathway.

Factors Limiting Clot Formation

Like platelet aggregation, fibrin clot formation is limited to the immediate vicinity of the damaged area because certain proteins found in plasma and on the surface of endothelial cells act as **anticoagulants**, chemicals that inhibit blood clotting or coagulation (**Discovery: Leeches and Bloodletting**, p. 511). During the initial phase of clotting, **tissue factor pathway inhibitor** is secreted by healthy endothelial cells and inhibits the extrinsic pathway. Additionally, healthy endothelial cells se-

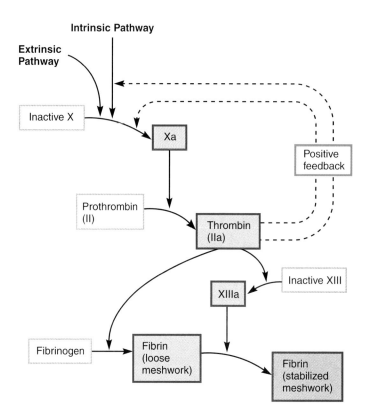

Figure 16.8 Role of thrombin in forming fibrin clot. Thrombin is activated by the action of factor Xa on prothrombin and provides positive feedback for its own synthesis. Once active, thrombin converts fibrinogen to fibrin and activates factor XIII.

crete a molecule called *thrombomodulin*, which binds to thrombin, forming a complex. When it is a part of the thrombomodulin-thrombin complex, thrombin cannot convert fibrinogen to fibrin; instead it activates **protein C**, another plasma protein that continuously circulates in the plasma. Protein C is an anticoagulant that inhibits both the intrinsic and extrinsic pathways. Thus thrombin not only promotes clotting at injury sites, it also indirectly inhibits clotting in healthy tissue.

Once formed, fibrin clots are eventually dissolved by **plasmin**, a protein derived from the plasma protein *plasminogen*. Plasmin dissolves clots by enzymatically breaking down fibrin. Plasminogen is converted to plasmin by *plasminogen activators* secreted by a variety of cell types. One example of a plasminogen activator is *tissue plasminogen activator (TPA)*, which is secreted by endothelial cells during clot formation. Fibrin activates TPA, which subsequently converts plasminogen to plasmin.

The Role of Coagulation Factors in Clot Formation Disorders

Most coagulation factors are synthesized in the liver, which releases them into the plasma in their inactive form. *Serum* is plasma from which these coagulation factors have been removed.

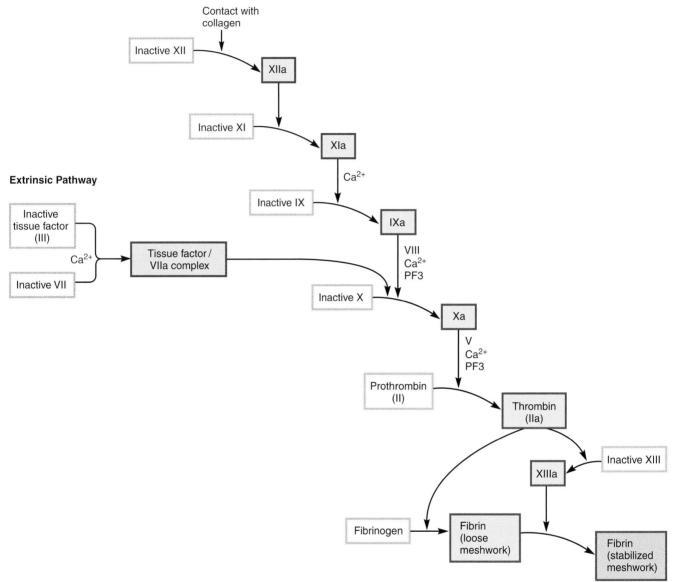

Figure 16.9 The intrinsic and extrinsic coagulation pathways. Formation of a fibrin clot is a sequence of reactions involving several coagulation factors, platelet factors (such as PF3), and calcium. All components of the intrinsic pathway are located in the blood. The extrinsic pathway requires a factor not in the blood: tissue factor (factor III).

Several coagulation factors are essential for the formation of a fibrin clot. The lack of any essential factor impairs clot formation and thus results in excessive bleeding. Hemophilia, a genetic disorder caused by the deficiency of the gene for a specific coagulation factor, is most commonly due to a deficiency of factor VIII in the blood. Another genetic bleeding disorder, von Willebrand's disease, is characterized by reduced levels of vWf, which interferes with platelet plug formation. However, because vWf also serves as a plasma carrier for factor VIII, the absence of vWf causes factor VIII to be less stable and to break down more quickly, leading to lower factor VIII levels in the bloodstream. Excessive bleeding can also occur if vitamin K is lacking in the diet, because the liver requires vitamin K for the synthesis of many of the proteins necessary in blood clotting.

Aspirin as an Anticoagulant

Aspirin, one of the most commonly used analgesics, has received a lot of attention lately for another of its actions: It prevents blood clotting. Aspirin is one of many anticlotting drugs given to people who are susceptible to stroke or coronary artery disease. Formation of blood clots in

LEECHES AND BLOODLETTING

Leeches are invertebrates known for their ability to suck blood. From ancient times to the early 20th century, leeches (or lances) were used in *bloodletting*, a procedure involving the removal of blood from a patient in order to treat disease. In 1094, some barbers adopted the practice of bloodletting and became known as "barber surgeons." Barber surgeons were trained not only to cut hair but also to perform surgery, extract teeth, and let blood. During the bloodletting procedure, the patient would hold onto a pole to make the veins more prominent. After the procedure, the bandages would be wrapped around the pole and placed outside the store for advertisement. This image of a striped pole is now familiarly known as a "barber's pole."

In the 1800s, bloodletting was used for surgery before the advent of anesthesia. If enough blood was removed, the patient would faint, due to a low mean arterial pressure. In the 1900s, the practice of bloodletting came to be considered primitive, and virtually came to a stop.

Today, however, the practice of using leeches is growing once again. Leeches are currently being used to increase blood flow to specific regions of the body in treatments of ischemia, pain, and inflammation. The medical value of a leech lies in its saliva, which contains an anesthetic, a vasodilator, and an anticoagulant. The anesthetic prevents the patient from feeling the presence of the leech, allowing the leech to stay attached longer. The vasodilator increases blood flow to the leech attachment site. The anticoagulant prevents the formation of a blood clot.

Medicinal leeches drawing blood and injecting vasodilators and anticoagulants.

cerebral arteries (in the brain) or coronary arteries (in the heart) can have serious deleterious effects, including death. Given at low dosages, aspirin acts as an anticoagulant by inhibiting the formation of TXA_2, decreasing platelet aggregation and platelet plug formation. At high dosages, however, aspirin decreases formation of prostacyclin, which actually increases the likelihood of clot formation. Clinical studies have shown that low doses of aspirin both reduce the incidence of subsequent heart attacks and decrease the severity of damage when given after a heart attack.

Quick Test 16.3

1. What are platelets? What are some of their secretory products? Explain the role of platelets in forming a blood clot.

2. Name two chemicals that prevent formation of a platelet plug in healthy blood vessels.

3. Describe the role of thrombin in hemostasis.

4. Define *whole blood*, *plasma*, and *serum*.

Summary of Factors Influencing Organ Blood Flow

Figure 16.10 is a chart summarizing the many concepts we have encountered in this and the two preceding chapters. This chart can help you recall and understand the many factors that bear on the cardiovascular system's primary function, which is to maintain adequate blood flow to the body's organs. The chart is also an emphatic reminder that all these factors are interconnected, a truth it illustrates in a way that no words ever could. The chart can help you understand relationships between certain variables. For example, the fact that arrows from two boxes labeled *heart rate* and *stroke volume* lead to the box labeled *cardiac output* should remind you that cardiac output is determined by these two variables ($CO = HR \times SV$), a concept discussed in Chapter 14. If you have difficulty understanding any portion of the chart, review the appropriate sections in the three cardiovascular chapters.

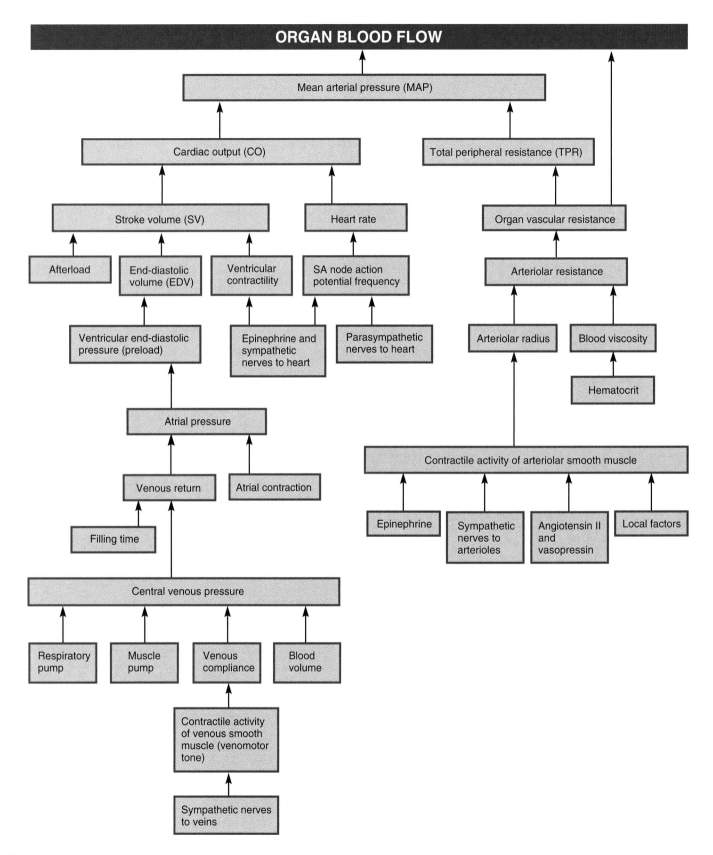

Figure 16.10 The relationships of the factors that affect the delivery of blood to systemic organs by the cardiovascular system.

Urinary System

Blood delivers O_2 and nutrients to respiring tissues

Blood removes CO_2 and other wastes from respiring tissues

Blood acts as source of raw material for urine formation

Blood pressure drives kidney filtration

Blood delivers hormones important for kidney function and carries others from kidney to targets

Gastrointestinal System

Blood delivers O_2 and nutrients to respiring tissues

Blood removes CO_2 and other wastes from respiring tissues

Blood carries absorbed nutrients to rest of body

Blood delivers gastrointestinal hormones to targets

Respiratory System

Blood picks up O_2 from air in lungs and releases CO_2

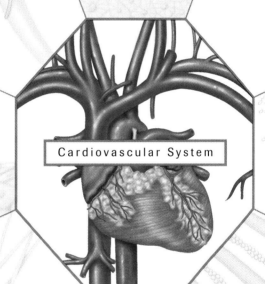

Cardiovascular System

Reproductive System

Blood delivers O_2 and nutrients to respiring tissues

Blood removes CO_2 and other wastes from respiring tissues

Blood delivers gonadotropic hormones to gonads and carries sex hormones to rest of body

Endocrine System

Blood delivers O_2 and nutrients to respiring tissues

Blood removes CO_2 and other wastes from respiring tissues

Blood carries hormones from glands to target tissues

Immune System

Blood delivers O_2 and nutrients to respiring tissues

Blood removes CO_2 and other wastes from respiring tissues

Capillary filtration produces lymphatic fluid

Blood carries chemical messengers of immune system and transports lymphocytes

Nervous System

Blood delivers O_2 and nutrients to respiring tissues

Blood removes CO_2 and other wastes from respiring tissues

Muscles

Blood delivers O_2 and nutrients to respiring tissues

Blood removes CO_2 and other wastes from respiring tissues

CHAPTER SUMMARY

Overview of the Composition of Blood: The Hematocrit, p. 498

The hematocrit is determined by centrifuging blood to separate it into components: plasma, erythrocytes, leukocytes, and platelets. The hematocrit is the percentage of blood volume composed of erythrocytes and is approximately 42–45. Plasma is the most abundant component, making up approximately 55% of blood volume. Leukocytes and platelets together constitute less than 1% of total blood volume.

Plasma, p. 499

Plasma is the liquid component of blood, consisting of water and dissolved solutes such as proteins and electrolytes. Plasma proteins include albumin, globulins, and fibrinogen. Major electrolytes in plasma include sodium, calcium, and chloride. Potassium and bicarbonate are also present in plasma, but to a lesser extent. The plasma is important in the transport of hormones, nutrients, waste products, clotting proteins, and antibacterial proteins.

Erythrocytes, p. 499

Erythrocytes, or red blood cells, are small and contain neither a nucleus nor any organelles. Erythrocytes function in the transport of oxygen and carbon dioxide between the lungs and tissues. Erythrocytes contain hemoglobin, a protein that binds and thus transports both oxygen and carbon dioxide in blood, and carbonic anhydrase, an enzyme that converts carbon dioxide to bicarbonate and hydrogen ions. Erythrocytes are produced in the bone marrow in response to erythropoietin secreted from the kidneys. Production of erythrocytes requires iron, a necessary element of hemoglobin. Erythrocytes survive in the blood approximately 120 days and are then removed by the spleen and broken down. As the hemoglobin is catabolized, most of the iron is recycled for synthesis of new hemoglobin. A decrease in the oxygen carrying capacity of blood is called anemia and can be caused by a decrease in the amount of hemoglobin per erythrocyte or a decrease in the number of circulating erythrocytes.

Leukocytes, p. 502

Leukocytes, or white blood cells, are larger than erythrocytes, but much sparser in blood. The leukocytes function in our immune system, protecting the body against foreign matter. There are five types of leukocytes in the blood: neutrophils, eosinophils, basophils, monocytes, and lymphocytes. Although leukocytes travel in blood, they migrate to the tissues where they carry out their defense functions. In the tissues, monocytes develop into macrophages. Neutrophils, monocytes, and macrophages (and to a lesser extent eosinophils) are phagocytes, cells that engulf foreign matter or other debris to remove it from the body.

Platelets and Hemostasis, p. 505

Mechanisms for stoppage of bleeding (hemostasis) include vascular spasm (blood vessel constriction), platelet plug formation, and blood clot formation, which occur in response to blood vessel damage. In platelet plug formation, platelets aggregate around the site of damage, forming a physical barrier to blood leakage. During this process, platelets also become activated, which sets the stage for clot formation. In clot formation, fibrinogen (a soluble plasma protein) is transformed by proteolytic cleavage into fibrin, which forms a fibrous mesh around the platelet plug. This transformation is set into motion following a cascade of reactions (the coagulation cascade) involving activated platelets and several clotting factors in the plasma. The coagulation cascade can be initiated by an intrinsic pathway, which involves only components present in plasma, or an extrinsic pathway, which involves factors present in tissues outside blood vessels. Spread of the blood clot beyond the site of damage is prevented by substances secreted by undamaged tissues and by other mechanisms.

EXERCISES

Multiple-Choice Questions

1. Which of the following is *not* found in plasma?
   a) albumin
   b) sodium
   c) glucose
   d) hemoglobin
   e) potassium

2. Which component of blood makes up most of the blood volume?
   a) plasma
   b) erythrocytes
   c) leukocytes
   d) platelets

3. What organ synthesizes new erythrocytes?
   a) liver
   b) spleen
   c) kidney
   d) bone marrow
   e) heart

4. Which of the following classes of leukocytes functions in phagocytosis?
   a) neutrophils only
   b) basophils only
   c) lymphocytes only
   d) neutrophils and basophils
   e) basophils and lymphocytes

5. Which class of leukocytes develops into macrophages in tissue?
   a) neutrophils
   b) eosinophils
   c) basophils
   d) monocytes
   e) lymphocytes

6. Which class of leukocytes have granules that stain preferentially with a red dye?
   a) neutrophils
   b) eosinophils
   c) basophils
   d) monocytes
   e) lymphocytes

7. What cytokine stimulates erythrocyte production?
   a) ADH
   b) erythropoietin
   c) von Willebrand factor
   d) epinephrine
   e) renin

8. Which class of blood cell is the most abundant?
   a) erythrocytes
   b) neutrophils
   c) basophils
   d) monocytes
   e) lymphocytes

9. Contact of blood with collagen triggers
   a) platelet aggregation.
   b) activation of the intrinsic clotting cascade.
   c) activation of the extrinsic clotting cascade.
   d) both a and b are true
   e) all of the above are true

10. What is the function of plasmin?
    a) stimulate vasoconstriction of damaged blood vessels
    b) inhibit platelet aggregation in healthy blood vessels
    c) dissolve blood clots
    d) activate thrombin
    e) activate the extrinsic clotting pathway

Objective Questions

1. Plasma with clotting factors removed is called _____.

2. Iron can be stored in the liver bound to _____.

3. During the clotting cascade, _____ converts fibrinogen to fibrin.

4. (B lymphocytes/T lymphocytes) develop into antibody-secreting cells.

5. Catabolism of hemoglobin by the spleen produces (bilirubin/bile).

6. _____ is an enzyme that converts carbon dioxide and water into carbonic acid.

7. At low doses, aspirin inhibits production of _____, thereby decreasing platelet aggregation.

Essay Questions

1. Describe the components of erythrocytes and explain how they are necessary for erythrocytes to carry out their function.

2. Describe the life cycle of erythrocytes. Be sure to include a description of the roles of the various organs needed for erythrocyte synthesis and breakdown.

3. Describe the body's mechanisms for preventing blood loss following damage to a blood vessel.

Critical Thinking

1. The percentage of blood that is composed of erythrocytes affects blood viscosity (see Chapter 15). Explain how an increase in the hematocrit at higher altitudes can affect blood flow and blood pressure.

2. Aspirin is often prescribed at low doses to people at risk of suffering a heart attack or stroke. Explain the rationale for this prescription.

17 The Respiratory System: Pulmonary Ventilation

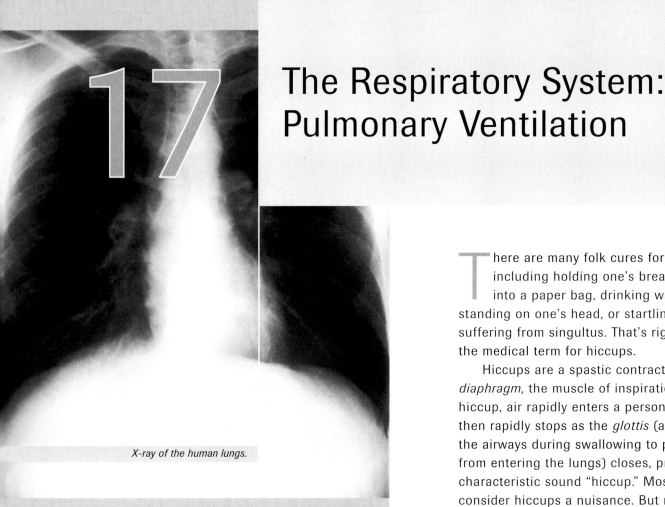

X-ray of the human lungs.

There are many folk cures for *singultus* including holding one's breath, rebreathing into a paper bag, drinking water while standing on one's head, or startling a person suffering from singultus. That's right, singultus is the medical term for hiccups.

Hiccups are a spastic contraction of the *diaphragm,* the muscle of inspiration. During a hiccup, air rapidly enters a person's lungs and then rapidly stops as the *glottis* (a flap that closes the airways during swallowing to prevent food from entering the lungs) closes, producing the characteristic sound "hiccup." Most of us consider hiccups a nuisance. But many people suffer from intractable hiccups, which are bouts of hiccups that last over 48 hours and could be indicative of an underlying disease. In pregnant women, hiccups can be more frequent due to the pressure the fetus puts on the diaphragm.

Some of the folk cures are based on physiology of the respiratory system. For example, holding one's breath or rebreathing into a bag increases the amount of carbon dioxide in the blood, which may inhibit the neural pathway for hiccups (which are similar to pathways that control breathing). Startling a person alters his breathing pattern, which may also affect neural pathways for hiccups. As for standing on one's head and drinking water, that could make one choke, which would definitely alter respiration and possibly hiccups.

OBJECTIVES

- Compare internal respiration to external respiration, and describe the processes occurring in each.

- Describe the major structures of the respiratory system, and list the functions of each.

- Describe the anatomy of the respiratory membrane, and explain how the structure facilitates the diffusion of gases.

- Describe the anatomy of alveoli. Explain the roles of type I cells, type II cells, and alveolar macrophages in respiratory function.

- Explain the function of pulmonary surfactant.

- Describe the mechanics of breathing. Name the muscles of respiration. List the different pulmonary pressures and explain their roles in ventilation.

- Describe the roles of lung compliance and airway resistance in ventilation.

- List the different lung volumes and capacities. Explain the clinical applications of lung volumes, forced vital capacity, and forced expiratory volume.

Before You Begin

Make sure you have mastered the following topics:

1. *Diffusion, p. 110*
2. *Pulmonary circulation, p. 416*
3. *Capillary anatomy, p. 471*
4. *Flow, pressure gradients, and resistance, p. 452*
5. *Compliance, p. 457*

In Chapter 3, we learned that our cells produce most of the ATP they need through oxidative phosphorylation, which requires oxygen. During energy metabolism, carbon dioxide is also produced. In Chapters 14, 15, and 16, we learned how blood flow to the tissues provides our cells with the nutrients they need, including oxygen, and removes waste products, including carbon dioxide. We now examine the respiratory system, which functions in delivering oxygen to blood and eliminating carbon dioxide from blood.

We all know that the depth of breathing changes with changing circumstances. When you exercise, for example, you breathe more deeply than while resting, and when something startles you, you might hold your breath for a moment before breathing a deep "sigh of relief." In either case, a deeper breath brings more air into the lungs, where oxygen in inhaled air moves into the blood, and where carbon dioxide moves out of the blood and leaves the body in exhaled air. In this chapter we explore *pulmonary ventilation*—the movement of air into and out of the lungs.

After a brief overview of respiratory function we examine the functional anatomy of the respiratory system. Then we explore how the muscles of respiration produce several different pulmonary pressures, and the roles of those pressures in pulmonary ventilation. Next we examine the factors that affect the rates of air flow during pulmonary ventilation. Finally, we conclude by considering the bases for and clinical significances of various lung volumes and capacities.

Overview of Respiratory Function

The respiratory system is so named because its function is **respiration**, the process of gas exchange. This exchange of gases occurs at two levels, termed internal respiration and external respiration (**Figure 17.1**). Whereas **internal respiration** (or cellular respiration) refers to the use of oxygen within mitochondria to generate ATP by oxidative phosphorylation, and the production of carbon dioxide as a waste product, **external respiration** refers to the exchange of oxygen and carbon dioxide between the atmosphere and body tissues, which involves both the respiratory and circulatory systems.

External respiration encompasses four processes:

1. **Pulmonary ventilation**, the movement of air into the lungs (inspiration) and out of the lungs (expiration) by bulk flow

2. Exchange of oxygen and carbon dioxide between lung air spaces and blood by diffusion

3. Transportation of oxygen and carbon dioxide between the lungs and body tissues by the blood

4. Exchange of oxygen and carbon dioxide between the blood and tissues by diffusion

This chapter focuses on the first process, pulmonary ventilation. The next chapter will cover the remaining three aspects of external respiration.

In addition to its main function—respiration—the respiratory system also performs several other functions, including (1) contributing to the regulation of acid-base balance in the blood, (2) enabling vocalization, (3) participating in defense against pathogens and foreign particles in the airways, (4) providing a route for water and heat losses (via the expiration of air that was moistened and warmed during inspiration), (5) enhancing venous return (through the respiratory pump), and (6) activating certain plasma proteins as they pass through the pulmonary circulation.

Anatomy of the Respiratory System

The major organs of the respiratory system are the *lungs*, which are located in the thoracic cavity. Each lung is divided into lobes; the right lung has three lobes

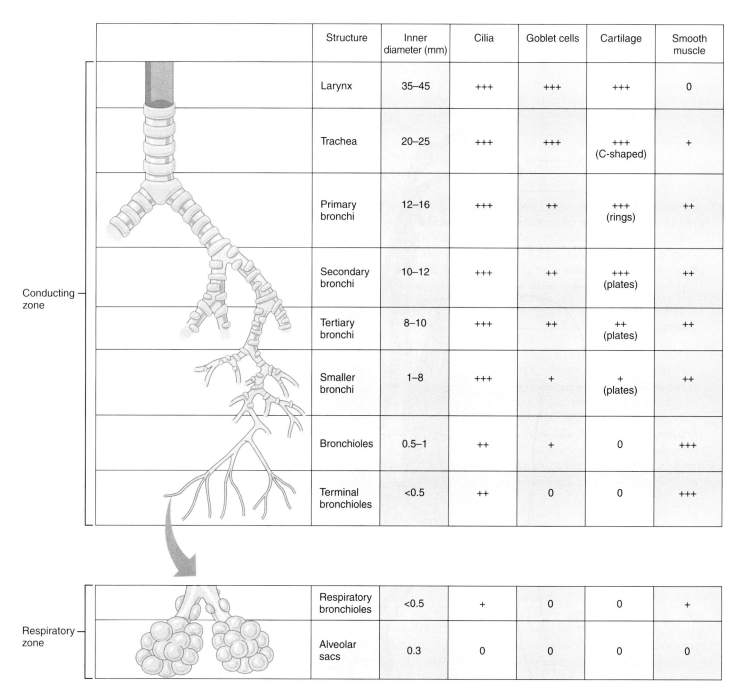

	Structure	Inner diameter (mm)	Cilia	Goblet cells	Cartilage	Smooth muscle
Conducting zone	Larynx	35–45	+++	+++	+++	0
	Trachea	20–25	+++	+++	+++ (C-shaped)	+
	Primary bronchi	12–16	+++	++	+++ (rings)	++
	Secondary bronchi	10–12	+++	++	+++ (plates)	++
	Tertiary bronchi	8–10	+++	++	++ (plates)	++
	Smaller bronchi	1–8	+++	+	+ (plates)	++
	Bronchioles	0.5–1	++	+	0	+++
	Terminal bronchioles	<0.5	++	0	0	+++
Respiratory zone	Respiratory bronchioles	<0.5	+	0	0	+
	Alveolar sacs	0.3	0	0	0	0

Figure 17.3 Anatomical features of the conducting and respiratory zones of the respiratory tract. 0 indicates not present, + indicates sparse, + + indicates present, and + + + indicates abundant.

After it enters the thoracic cavity, the trachea divides into left and right **bronchi** (singular: *bronchus*) that conduct air to each lung. Like the trachea, the bronchi contain cartilage; however, the cartilage forms rings around the entire circumference of the bronchus. Within each lung, the bronchi divide into smaller tubes called **secondary bronchi**; three secondary bronchi conduct air to the lobes of the right lung, and two secondary bronchi conduct air to the lobes of the left lung. The cartilage in secondary bronchi is less abundant than that in the primary bronchi and occurs as plates.

Each secondary bronchus divides into smaller *tertiary bronchi,* which in turn divide into successively smaller bronchi such that approximately 20–23 orders of branching occurs. The extensive branching ultimately results in approximately 8 million tubules, the smallest less than 0.5 mm in diameter.

Once the tubules become less than 1 mm in diameter, they are called **bronchioles** ("little bronchi"). Unlike the larger bronchi, bronchioles have no cartilage and are thus capable of collapsing. To help prevent collapse, the walls of bronchioles contain elastic fibers. The bronchioles divide further into **terminal bronchioles**, the final and smallest component of the conducting zone.

The primary function of the conducting zone is to provide a passageway through which air can enter and exit the respiratory zone, where gas exchange occurs. The conducting zone holds approximately 150 mL of air and is considered "dead space," because the air does not participate in gas exchange with blood. The functional significance of the dead space is described later. As air travels through the conducting zone, its temperature is adjusted to body temperature and humidified to keep the respiratory tract moist.

The conducting zone is lined by an epithelium that changes in composition as the tubules become smaller in diameter. The epithelium lining the larynx and trachea (and to a lesser extent, the bronchi) contains numerous *goblet cells;* also abundant in the epithelium throughout the conducting zone are *ciliated cells* (**Figure 17.4**). **Goblet cells** secrete a viscous fluid called *mucus,* which coats the airways and traps foreign particles in inhaled air; the **cilia** (hairlike projections) of the *ciliated cells* beat in a whiplike fashion to propel the mucus containing the trapped particles up toward the glottis and then into the pharynx, where the mucus is then swallowed. This process, called the *mucus escalator,* prevents mucus from accumulating in the airways and clears trapped foreign matter. Accumulation of mucus in the airways increases the likelihood of infections such as bronchitis and pneumonia because it promotes retention and growth of bacteria. Because cilia are easily paralyzed by tobacco smoke, smoking disables the mucus escalator, so that mucus and trapped debris accumulate in the airways and can be cleared only by coughing; this is one reason for the familiar "smoker's cough." At levels below the bronchioles, phagocytic cells called macrophages engulf foreign matter in the interstitial space and on the surface of the epithelium.

In addition to changes in the epithelium, other tissue changes occur as the airways become smaller. As previously noted, cartilage is plentiful in the walls of the trachea and bronchi, but it becomes less abundant as the diameter of the bronchi decreases, until in the bronchioles it is completely absent. Smooth muscle is sparse in the trachea and bronchi but increases in abundance as the airways become smaller. The lack of cartilage and the presence of circular smooth muscle within the bronchioles enable these airways to change their diameter; such changes alter the resistance to air flow, just as the actions of circular smooth muscle in arterioles enables them to alter resistance to blood flow.

The Respiratory Zone

We saw in Chapter 4 that the rate at which substances diffuse across a membrane increases as membrane surface area increases and as membrane thickness decreases. The arrangement of structures in the respiratory zone, the site of gas exchange, maximizes surface area and minimizes thickness, such that the diffusion of oxygen and carbon dioxide between air and blood is facilitated.

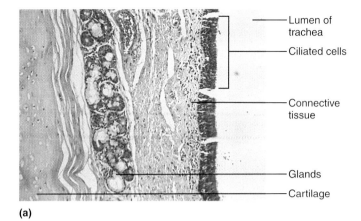

(a)

Lumen of trachea
Ciliated cells
Connective tissue
Glands
Cartilage

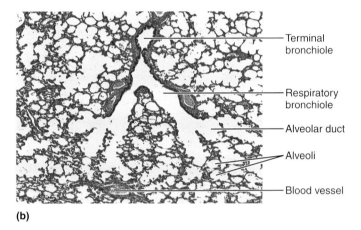

(b)

Terminal bronchiole
Respiratory bronchiole
Alveolar duct
Alveoli
Blood vessel

Figure 17.4 Respiratory tract epithelia. (a) A photomicrograph of tracheal epithelium, located in the conducting zone. (b) A photomicrograph of respiratory bronchioles and alveoli, located in the respiratory zone.

Past the site where the terminal bronchioles of the conducting zone branch, the respiratory zone begins (**Figure 17.5**). The first respiratory zone structures, **respiratory bronchioles**, terminate in *alveolar ducts,* which lead to **alveoli**, the primary structures where gas exchange occurs(Figure 17.5a). Most alveoli occur in clusters called **alveolar sacs**, which resemble clusters of grapes; some alveoli open off of respiratory bronchioles.

Adjacent alveoli are not completely independent structures. Because they are connected by *alveolar pores* (Figure 17.5), air flows between alveoli, allowing equilibration of pressure within the lungs.

Alveolar structure facilitates diffusion of gases between blood and air. The wall of an alveolus consists primarily of a single layer of epithelial cells called **type I cells** overlying a basement membrane (Figure 17.5c). Recall from Chapter 15 that a capillary wall consists of a single layer of endothelial cells and an underlying basement membrane. In many places in the lungs, the alveolar epithelial cells and the endothelial cells of the nearby capillaries are so close together that their basement membranes are fused (Figure 17.5d). Together, the capillary and the alveolar wall form a barrier, called the **respiratory membrane**,

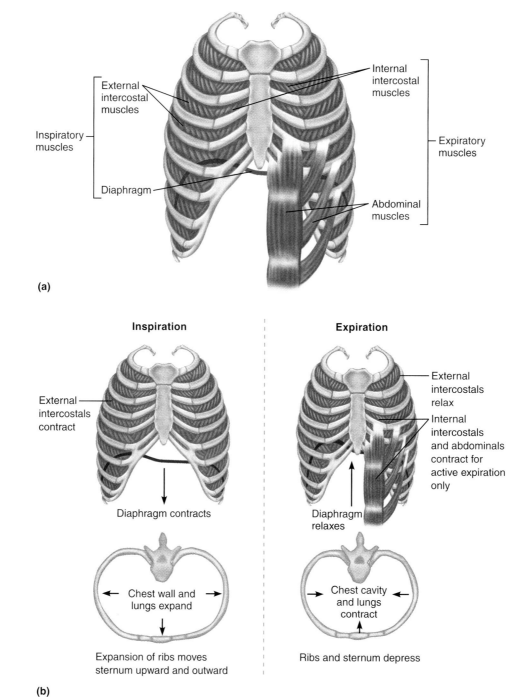

(a)

Inspiration

External intercostals contract

Diaphragm contracts

Chest wall and lungs expand

Expansion of ribs moves sternum upward and outward

Expiration

External intercostals relax

Internal intercostals and abdominals contract for active expiration only

Diaphragm relaxes

Chest cavity and lungs contract

Ribs and sternum depress

(b)

Figure 17.11 Muscles of ventilation. (a) Locations of inspiratory and expiratory muscles. Notice the opposite orientation of the external and internal intercostal muscles. (b) Actions of muscles of ventilation. When inspiratory muscles contract (at left), the chest wall expands, causing the lungs to expand. Quiet expiration (at right) occurs passively by relaxation of the muscles of inspiration, which allows the lungs and chest wall to recoil to their original positions. Active expiration requires contraction of the muscles of expiration, while the muscles of inspiration relax.

ventilation (**Figure 17.11**a). The diaphragm and the external intercostal muscles are the primary inspiratory muscles, whereas the internal intercostals and abdominal muscles are the primary expiratory muscles, although expiration is primarily a passive process not requiring any muscle contraction.

Inspiration

The process of inspiration depicted in Figure 17.11b is initiated by neural stimulation of the inspiratory muscles (**Figure 17.12**). These skeletal muscles are stimulated to contract by the release of acetylcholine at the neuromuscular junction, as described in Chapter 12. Contraction of the diaphragm causes it to flatten and move downward; meanwhile, contraction of the obliquely oriented external intercostals causes the ribs to pivot upward and outward, expanding the chest wall. These combined actions increase the volume of the thoracic cavity. Other muscles of the neck (scalenes and sternocleidomastoids) and chest region (pectoralis minor) play subsidiary roles in inspiration, especially during forceful inspiration.

As the chest wall expands, it pulls outward on the intrapleural fluid, causing the intrapleural pressure to decrease. This decrease in intrapleural pressure causes an increase in the transpulmonary pressure, or the difference between the intrapleural pressure and the intra-alveolar pressure ($P_{alv} - P_{ip}$; **Figure 17.13**). An increase in transpulmonary pressure due to a decrease in P_{ip} creates a larger distending pressure across the lungs, so the lungs (alveoli) expand with the chest wall. When the lungs expand, pressure in the alveoli decreases to less than atmospheric pressure, so air flows into the alveoli by bulk flow, and continues to flow in until the pressure in the alveoli increases to atmospheric pressure. Stronger contractions of the inspiratory muscles produce a greater expansion of the thoracic cavity, making intrapleural pressure even more negative and creating a greater transpulmonary pressure, resulting in greater lung expansion and a deeper inspiration; that is, a larger volume of air moves into the lungs.

Expiration

During quiet breathing, expiration is a *passive* process in that it does not require muscle contraction. At the end of an inspiration, the chest wall and lungs are expanded by muscle contraction. By simply relaxing these muscles, which occurs when motor neurons to the inspiratory muscles stop firing, the elastic chest wall and lungs recoil to their resting position. As the chest wall and lungs recoil, the volume of the lungs decreases, causing alveolar pressure to increase to values greater than atmospheric pressure. Air flows out (expiration occurs) due to the pressure gradient until the volume in the lungs equals the FRC.

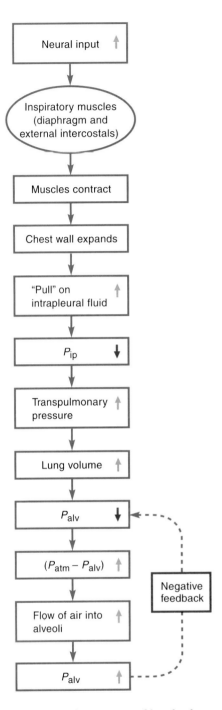

Figure 17.12 Events in the process of inspiration.

A more forceful expiration can be produced by contraction of the expiratory muscles in a process called *active expiration*. Contraction of the expiratory muscles produces a greater and more rapid decrease in the volume of the thoracic cavity, which increases intra-alveolar pressure and causes a greater pressure gradient for air flow out of the alveoli.

Intra-alveolar pressure is the pressure of air within the alveoli. Intrapleural pressure is the pressure of the intrapleural fluid. Transpulmonary pressure is the difference between the intrapleural pressure and the intra-alveolar pressure. Because the lungs and the chest wall are elastic, between breaths the chest wall tends to recoil outward, and the lungs to recoil inward. These forces tend to separate the chest wall from the lungs, creating a negative intrapleural pressure.

Inspiration and expiration are driven by differences in atmospheric and intra-alveolar pressures. These pressure gradients are created when the volume of the lungs is changed. Inspiration is caused by contraction of the diaphragm and the external intercostal muscles; when these muscles contract, the volume of the thoracic cavity increases. As the thoracic cavity expands, the intrapleural pressure decreases, creating a force that expands the lungs as the chest wall expands. Intra-alveolar pressure decreases below atmospheric pressure, and inspiration occurs. During quiet breathing, expiration occurs when the chest wall and lungs return passively to their original positions. Active expiration involves contraction of the internal intercostals and abdominal muscles.

IP Respiratory, Pulmonary Ventilation, pp. 3–13

Factors Affecting Pulmonary Ventilation, p. 530

The rate of air flow into or out of the lungs is determined by the magnitude of the pressure gradient driving the flow and airway resistance. Lungs have a high compliance; that is, they are easily stretched to increase lung volume for inspiration. Airway resistance depends primarily on the radius of the tubules of the respiratory tract. Airway resistance generally is low but can be affected by breathing mechanics, the autonomic nervous system, chemical factors, and pathological states.

IP Respiratory, Pulmonary Ventilation, pp. 14–18

Clinical Significance of Respiratory Volumes and Air Flows, p. 534

Lung volumes and capacities can be measured using spirometry. Lung volumes include tidal volume, inspiratory reserve volume, expiratory reserve volume, and residual volume. The lung capacities include inspiratory capacity, vital capacity, functional residual capacity, and total lung capacity. Other lung measurements take into account the rate of air flow. The forced vital capacity is the amount of air a person can expire following a maximum inspiration, expiring as forcefully and rapidly as possible. The forced expiratory volume is a measure of the percentage of the forced vital capacity that can be exhaled within a certain time frame.

Minute ventilation is the total amount of air that flows into or out of the respiratory system in a minute. Minute alveolar ventilation is a measure of the volume of fresh air reaching the alveoli each minute, which is minute ventilation corrected for dead space volume. To increase minute alveolar ventilation, it is more efficient to increase tidal volume than respiration rate.

▌▌ *EXERCISES*

Multiple-Choice Questions

1. Which of the following is a component of internal respiration?
   a) ventilation
   b) transport of oxygen in the blood
   c) diffusion of carbon dioxide from tissues to blood
   d) diffusion of oxygen from blood to tissues
   e) oxidative phosphorylation

2. Which of the following is *not* a function of the conducting zone of the respiratory system?
   a) humidifying the air
   b) adjusting the air to body temperature
   c) exchanging gases between respiratory system and blood
   d) secreting mucus
   e) protecting the lungs from inhaled particles

3. The smallest airways in the conducting zone are
   a) terminal bronchioles.
   b) respiratory bronchioles.
   c) alveolar ducts.
   d) alveolar sacs.
   e) bronchi.

4. Surfactant is secreted by
   a) goblet cells.
   b) alveolar macrophages.
   c) type I cells.
   d) type II cells.
   e) ciliated cells.

5. The product of tidal volume and breathing frequency gives
   a) respiration rate.
   b) total lung capacity.
   c) alveolar ventilation.
   d) minute ventilation.
   e) dead space volume.

6. When all muscles of respiration are relaxed and alveolar pressure is zero, lung volume is equal to
   a) residual volume.
   b) vital capacity.
   c) functional residual capacity.
   d) tidal volume.
   e) total lung capacity.

7. Which of the following statements describes the lungs at the functional residual capacity?
   a) Atmospheric, intra-alveolar, and intrapleural pressures are all equal.
   b) The lungs tend to collapse due to their elastic properties.
   c) The chest wall tends to collapse due to its elastic properties.
   d) a and c are both true
   e) all of the above are true

8. Which of the following factors decreases airway resistance?
   a) activation of the parasympathetic nervous system
   b) epinephrine
   c) histamine

9. Pulmonary surfactant
   a) prevents collapse of alveoli.
   b) prevents small alveoli from joining with larger alveoli.
   c) increases lung compliance.
   d) a and c are both true
   e) all of the above are true

10. Which of the following muscles contracts during quiet expiration?
    a) diaphragm
    b) internal intercostals
    c) external intercostals
    d) none of the above is true
    e) all of the above are true

Objective Questions

1. Contraction of the diaphragm increases the rate of air flow during forced expiration. (true/false)

2. During inspiration, transpulmonary pressure (increases/decreases).

3. During inspiration, intrapleural pressure becomes (more/less) negative.

4. If airway resistance increases, a (higher/lower) intra-alveolar pressure is required to produce a given rate of air flow during expiration.

5. Pulmonary surfactant (increases/decreases) the surface tension of water.

6. Pulmonary surfactant (increases/decreases) lung compliance.

7. Dead space volume is the volume of air in the (conducting zone/respiratory zone).

8. (Obstructive/Restrictive) lung diseases are characterized by increased airway resistance.

9. _____ cells secrete mucus.

10. The (internal/external) intercostals are muscles of inspiration.

Essay Questions

1. Describe the various ways in which the structure of the respiratory system is adapted to facilitate gas exchange.

2. Describe the balance between distending pressures and recoil forces acting on the lungs and chest wall at FRC. Include in your discussion an explanation of why intrapleural pressure is subatmospheric.

3. Explain the difference between minute ventilation and minute alveolar ventilation. Include a discussion of the effect of tidal volume on alveolar ventilation.

4. Describe the role of surfactant in decreasing the surface tension in alveoli. What are some of the consequences of decreasing surface tension?

5. Describe the mechanics of ventilation. What muscles contract for inspiration? What muscles contract for expiration? How does the elasticity of the lungs and chest wall affect ventilation?

Critical Thinking

1. Which of the curves in the following graph illustrates a condition of greater compliance, A or B?

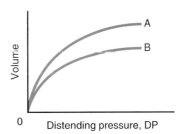

2. Determine your resting respiration rate by counting how many times you breathe in one minute. Assume a tidal volume of 500 mL and estimate your dead space volume as 1 mL per pound of body weight (that is, if you weighed 150 pounds, your dead space volume would be 150 mL). Calculate your minute ventilation and minute alveolar ventilation.

   During exercise, oxygen demand increases. Your respiratory system compensates by increasing minute alveolar ventilation. Calculate your minute alveolar ventilation assuming that your respiration rate increases 50% from what it was at rest. Then calculate your minute alveolar ventilation with your resting respiration rate but with an increase in tidal volume of 50% (from 500 mL to 750 mL). Does an increase in respiration rate or tidal volume more efficiently increase minute alveolar ventilation? Explain.

3. According to Boyle's law, pressure is inversely proportional to volume. In Figure 17.13, breath volume is shown to increase until the end of inspiration. Which pressure, intra-alveolar or intrapleural, follows Boyle's law? Explain.

4. Black widow spiders release a venom called latrotoxin that enhances acetylcholine release. The venom is deadly because it interferes with breathing. Explain how.

Find the answers to these exercises, and additional study tools, at the Physiology Place (www.physiologyplace.com).

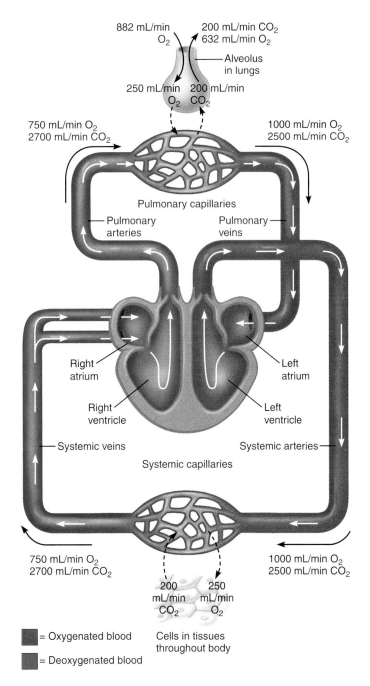

Figure 18.1 Movements of oxygen and carbon dioxide in pulmonary and systemic tissues during rest. In capillary beds in the lungs, oxygen diffuses from alveoli to blood and carbon dioxide diffuses from blood to alveoli. In systemic capillary beds, oxygen diffuses from blood to the cells and carbon dioxide diffuses from cells to the blood.

Under resting conditions, how much blood is pumped to the lungs each minute?

5 liters

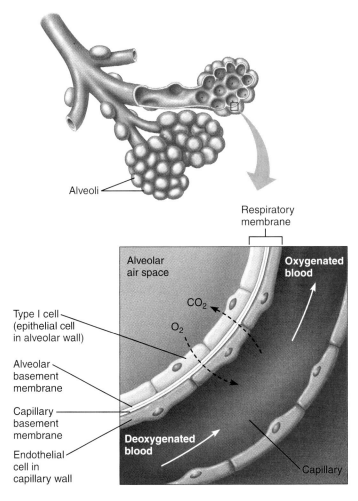

Figure 18.2 The respiratory membrane. The exchange of oxygen and carbon dioxide between alveoli and blood occurs by simple diffusion across the respiratory membrane, which is composed of a single layer of type I epithelial cells lining the alveolus, a single layer of endothelial cells lining the capillary, and the alveolar and capillary basement membranes.

The movement of oxygen and carbon dioxide between alveolar air and blood occurs by diffusion down concentration gradients. Oxygen is at a higher concentration in the alveoli and diffuses into the blood, whereas carbon dioxide is at a higher concentration in the blood and diffuses into the alveoli. In Chapter 4 we learned that oxygen and carbon dioxide are able to cross cell membranes by simple diffusion; we also learned that the rate of transport in simple diffusion is proportional to the magnitude of the concentration gradient, the surface area of the membrane through which it is moving, and the permeability of the membrane to that particular substance. **Figure 18.2** shows the structure of the respiratory membrane, which is composed of three layers: type I epithelial cells in the alveolar wall and endothelial cells in the capillary wall "sandwiched" around their fused basement membranes. The respiratory membrane provides a large surface area of very thin membrane, favoring fast rates of diffusion for oxygen and carbon dioxide between alveolar air and blood.

According to the ideal gas law, for any pure gas, $PV = nRT$, where P = pressure, V = volume, n = number of moles, R = universal gas constant, and T = temperature. If V and T are constant, then the pressure exerted by a gas is directly proportional to the number of moles of the gas. If 1 mole of helium produces a pressure of 100 mm Hg, for example, then 2 moles of helium at the same volume and temperature would exert a pressure of 200 mm Hg.

What about a mixture of two gases? If we start with 1 mole of helium at 100 mm Hg, then add 1 mole of oxygen, what will happen to the pressure? Note that the ideal gas law does not depend on the kind of gas; all gases behave identically so far as pressure is concerned. Therefore, the pressure of the mixture would be 200 mm Hg, just as if we had added another mole of helium.

In 1801, the English chemist John Dalton formulated a law regarding mixtures of gases. Dalton's law states that the pressure exerted by a mixture of gases is equal to the sum of the pressures exerted by the individual gases occupying the same volume alone. The pressure exerted by an individual gas in a mixture is called the *partial pressure* of that gas. Therefore, the total pressure of a mixture of gases is a sum of the partial pressures exerted by each gas in the mixture:

$$P_{total} = P_1 + P_2 + P_3 + \cdots + P_n$$

where n is the total number of gases in the mixture.

Quick Test 18.1

1. Under normal conditions, if respiring tissues consume oxygen at a rate of 250 mL/min, at what rate does oxygen diffuse from the alveoli to blood?

2. If the cells of the body consume 300 mL of oxygen per minute and produce 270 mL of carbon dioxide per minute, then what is the respiratory quotient?

3. Which blood vessel contains blood with the greater concentration of oxygen, a pulmonary artery or a pulmonary vein?

Diffusion of Gases

We have seen that the overall movements of oxygen and carbon dioxide between lungs and systemic tissues occur by diffusion down concentration gradients. In this section we look more closely at how these gradients are established and find that they are affected by the *partial pressures* and *solubilities* of oxygen and carbon dioxide.

Partial Pressure of Gases

We saw in Chapter 17 that the pressure of a gas depends on its temperature and the number of gas molecules contained in a given volume. (The exact relationship was described in Toolbox: Boyle's Law and the Ideal Gas Law, p. 531.) A gas (air, for instance) is often a mixture of more than one type of molecule. The total pressure of such a gas is the sum of the pressures of the individual gases that make up the mixture:

$$P_{total} = P_1 + P_2 + P_3 + \cdots + P_n$$

where n is the number of gases. In any gas mixture, the **partial pressure** of an individual gas is the proportion of the pressure of the entire gas that is due to the presence of the individual gas (see **Toolbox: Partial Pressures and Dalton's Law**). For example, if helium and nitrogen are mixed together in equal proportions and the pressure of the mixture is 500 mm Hg, then half the pressure (250 mm Hg) is exerted by helium, and half by nitrogen. In this example, 500 mm Hg is the *total* pressure of the mixture, and 250 mm Hg is the *partial pressure* of helium or nitrogen.

The partial pressure of a gas mixture is determined by two factors: (1) the *fractional concentration* of that gas, which is the quantity of that gas (moles) relative to the total quantity of gas in the mixture, and (2) the total pressure exerted by the gas mixture. To find the partial pressure of a gas, one multiplies these two factors together.

Air is composed almost entirely of two gases: nitrogen and oxygen. (Other gases such as carbon dioxide, argon, neon, helium, and methane are found in only minute amounts in air.) The amount of water vapor in air depends on the humidity. The total pressure of air can be described as the sum of the partial pressures of the gases found in air, plus the pressure of water vapor, as follows:

$$P_{air} = P_{N_2} + P_{O_2} + P_{H_2O}$$

On a molar basis, nitrogen is the most abundant gas in air, making up 79% of the molecules in air. Oxygen is the next most abundant at 21%. Depending on the humidity, water may become a critical component of air, decreasing the contribution of the other two gases to the total pressure of air. At sea level, air pressure is 760 mm Hg. Assuming zero humidity, the partial pressures of the two primary gases in air are

$$P_{N_2} = 0.79 \times 760 \text{ mm Hg} = 600 \text{ mm Hg}$$

$$P_{O_2} = 0.21 \times 760 \text{ mm Hg} = 160 \text{ mm Hg}$$

PULMONARY EDEMA

Pulmonary edema, the accumulation of excess fluid in the lungs, is a fairly common but dangerous disorder. The condition is marked by signs of respiratory distress (including *tachypnea,* or rapid, shallow breathing) and by low oxygen levels in the tissues *(hypoxia),* which may be evident as a bluish coloration *(cyanosis)* of the skin and mucous membranes.

Pulmonary edema occurs in two stages: *interstitial edema,* in which excess fluid accumulates in the interstitial spaces in lung tissue, and *alveolar edema,* in which fluid accumulates in the alveoli. In severe cases, fluid may even move into the airways, in which case the affected person may cough up a frothy foam. Pulmonary edema interferes with breathing in two ways: (1) by increasing the distance gases must diffuse to move between alveolar air and capillary blood, which impedes gas exchange, and (2) by interfering with the action of pulmonary surfactant, which causes a decrease in lung compliance and thus an increase in the work of breathing.

Pulmonary edema is similar to systemic edema (described in Chapter 15). The most common cause of pulmonary edema is increased hydrostatic pressure in the pulmonary capillaries. Left heart failure (see Chapter 15, p. 478) can cause a backup of blood in the pulmonary circulation, which increases the pressure in the pulmonary veins. As pressure builds up in the pulmonary capillaries, fluid is pushed out of the capillaries and into the interstitial space of the lungs.

Treatment of pulmonary edema is critical because it is a life-threatening situation. The symptoms of pulmonary edema are treated by administering oxygen and diuretics (medications that increase fluid output by the kidneys). However, the cause of the pulmonary edema must be determined and treated appropriately once the symptoms are stabilized.

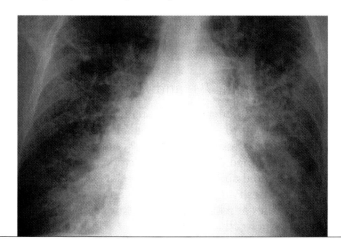

X-ray of a person with pulmonary edema.

can alter normal functioning of the body, including the nervous system. (A review of terms used in respiratory physiology is provided in **Table 18.2.**)

Quick Test 18.4

1. What three factors determine alveolar P_{O_2} and P_{CO_2}?

2. When alveolar ventilation increases, what happens to alveolar P_{O_2} and P_{CO_2}?

3. What happens to alveolar P_{CO_2} when a person hyperventilates? What happens to alveolar P_{CO_2} when a person hypoventilates?

4. If alveolar P_{O_2} decreases and P_{CO_2} increases, what happens to arterial P_{O_2} and P_{CO_2}?

Transport of Gases in the Blood

The previous section explained how gas is exchanged between air and blood, and between tissue and blood. This section explains how the gas is transported within the blood. We saw that gases are not very soluble in blood. For example, when arterial P_{O_2} is at the normal value of 100 mm Hg, blood contains only about 3 mL of dissolved oxygen per liter of blood. To deliver dissolved oxygen to tissues at the normal resting rate of approximately 250 mL/min, cardiac output would have to be about 83 liters/min if the total oxygen concentration in the blood were this low. Because the normal resting cardiac output is approximately 5 liters/min, oxygen must be transported in blood by a more efficient means than merely being dissolved in plasma. The same is true for carbon dioxide. First we examine how blood transports oxygen, and then carbon dioxide.

Oxygen Transport in the Blood

Transport of oxygen in the blood has a special need: The transport mechanism must be readily reversible such that oxygen enters the blood in the lungs and leaves the blood in other tissues of the body. **Hemoglobin,** a protein found in erythrocytes, has a unique structure that allows oxygen to do just that.

Table 18.2 ▏ Some Terms Used in Respiratory Physiology

Term	Definition
Hyperpnea	An increase in ventilation to meet an increase in the metabolic demands of the body
Dyspnea	Labored or difficult breathing
Apnea	Temporary cessation of breathing
Tachypnea	Rapid, shallow breathing
Hyperventilation	A condition in which ventilation exceeds the metabolic demands of the body
Hypoventilation	A condition in which ventilation is insufficient to meet the metabolic demands of the body
Hypoxia	A deficiency of oxygen in the tissues
Hypoxemia	A deficiency of oxygen in the blood
Hypercapnia	An excess of carbon dioxide in the blood
Hypocapnia	A deficiency of carbon dioxide in the blood

Oxygen Transport by Hemoglobin

Every liter of arterial blood contains about 200 mL of oxygen. Approximately 3 mL of this oxygen (1.5%) is dissolved in the plasma or in the cytosol of erythrocytes, and only this dissolved oxygen contributes to the P_{O_2} in blood. The remaining 197 mL of oxygen (98.5%) is transported bound to hemoglobin. Although the bound oxygen does not contribute to the P_{O_2}, it is in equilibrium with the dissolved oxygen, and thus the amount of oxygen bound to hemoglobin is a function of the P_{O_2}.

As we saw in Chapter 16, hemoglobin consists of four subunits, each of which contains a *globin* (globular polypeptide chain) and a *heme group* that contains iron. The heme groups are each capable of binding one oxygen molecule, so each hemoglobin molecule can carry four oxygen molecules. The complex of hemoglobin and bound oxygen is called *oxyhemoglobin;* a hemoglobin molecule without any oxygen is called *deoxyhemoglobin.*

In the lungs, as oxygen molecules move from alveolar air to capillary blood, they bind to hemoglobin (**Figure 18.6**a); when the blood reaches respiring tissues, oxygen molecules dissociate from the hemoglobin and diffuse to the cells (Figure 18.6b). For hemoglobin to function in oxygen transport, it is critical that it binds the oxygen *reversibly;* that is, tightly enough so that it can pick up large quantities of oxygen in the lungs, but not so tightly that it cannot release the oxygen into the respiring tissues later on.

The binding or release of oxygen depends on the P_{O_2} in the fluid surrounding hemoglobin. High P_{O_2} facilitates the binding of oxygen with hemoglobin, whereas low P_{O_2} facilitates release of oxygen from hemoglobin. The reaction of oxygen with hemoglobin can be written as

$$Hb + O_2 \rightleftharpoons Hb{\cdot}O_2$$

where Hb is deoxyhemoglobin, O_2 is the dissolved oxygen in blood, and $Hb{\cdot}O_2$ is oxyhemoglobin. The law of mass action states that an increase in the concentration of the reactants drives the reaction to the right, resulting in the generation of more product. Therefore, as oxygen levels in the pulmonary capillaries increase, more oxyhemoglobin is formed. Conversely, as oxygen levels in the systemic capillaries decrease, the reaction is driven to the left to release oxygen from the hemoglobin.

Note, however, that because hemoglobin can bind up to four oxygen molecules, the previous equation could be written as follows:

$$Hb \underset{O_2 \downarrow}{\rightleftharpoons} Hb{\cdot}O_2 \underset{O_2 \downarrow}{\rightleftharpoons} Hb{\cdot}(O_2)_2 \underset{O_2 \downarrow}{\rightleftharpoons} Hb{\cdot}(O_2)_3 \underset{O_2 \downarrow}{\rightleftharpoons} Hb{\cdot}(O_2)_4$$

The law of mass action is still in effect in that the more oxygen available, the more oxyhemoglobin is formed. When all oxygen-binding sites on a hemoglobin molecule are occupied, the hemoglobin molecule is said to be 100% saturated.

When hemoglobin is 100% saturated, 1 gram (g) of hemoglobin carries 1.34 mL of oxygen. The normal concentration of hemoglobin in the blood is 12–17 g/dL, or an average of 150 g/liter. Therefore, the oxygen-carrying capacity of the hemoglobin in blood is about 200 mL of oxygen per liter of blood (1.34 mL/g × 150 g/liter). At the normal arterial P_{O_2} of 100 mm Hg, hemoglobin is at approximately 98% of its oxygen-carrying capacity (is 98% saturated) (**Figure 18.7**a). When cardiac output is 5 liters per minute, the blood supplies almost 1000 mL of oxygen to respiring tissues each minute (5 liters of blood per minute times 200 mL of O_2 in a liter of blood equals 1000 mL of O_2 per minute). Because respiring tissues need only about 250 mL of O_2 per minute, only 25% of the oxygen diffuses into respiring cells, which means that 75% of the binding sites on hemoglobin are still occupied when blood leaves the tissue, which occurs at a P_{O_2} of 40 mm Hg (Figure 18.7b). Therefore, in mixed venous blood under resting conditions, hemoglobin is still 75% saturated.

Anemia is a decrease in the oxygen-carrying capacity of blood. There are many causes of anemia, including a deficiency or defect in hemoglobin (see Clinical Connections: Anemia, p. 506). With less functioning hemoglobin in the blood, the oxygen-carrying capacity is decreased, and tissues may not be supplied with the oxygen they need, even when the P_{O_2} of blood is normal. Therefore, people suffering from anemia tire more easily.

The Hemoglobin-Oxygen Dissociation Curve

The relationship between P_{O_2} and hemoglobin saturation just described can be summarized in the hemoglobin-oxygen dissociation curve, a plot of the percent saturation

7. Suppose that alveolar P_{O_2} = 100 mm Hg and P_{CO_2} = 60 mm Hg. Which of the following is true?
   a) pH will be less than normal.
   b) Percent saturation of hemoglobin by oxygen will be below normal.
   c) Bicarbonate concentration will be above normal.
   d) a and c are both true
   e) all of the above

8. Suppose a person's arterial P_{O_2} and P_{CO_2} are normal (P_{O_2} = 100 mm Hg; P_{CO_2} = 40 mm Hg). Which of the following would most likely stimulate an increase in ventilation?
   a) a decrease in P_{O_2} to 90 mm Hg
   b) a decrease in P_{CO_2} to 35 mm Hg
   c) an increase in P_{O_2} to 110 mm Hg
   d) an increase in P_{CO_2} to 45 mm Hg

9. A rise in arterial P_{CO_2} triggers an increase in ventilation by stimulating both central and peripheral chemoreceptors. The response of central chemoreceptors is due to
   a) diffusion of carbon dioxide into brain extracellular fluid, which stimulates chemoreceptors directly.
   b) diffusion of hydrogen ions into brain extracellular fluid, which stimulates chemoreceptors directly.
   c) diffusion of carbon dioxide into brain extracellular fluid, which reacts with water to form hydrogen ions, which stimulates chemoreceptors directly.
   d) diffusion of carbon dioxide into brain extracellular fluid, which reacts with water to form bicarbonate ions, which stimulates chemoreceptors directly.
   e) direct stimulation by hydrogen ions in arterial blood.

10. When a person exercises, ventilation increases to meet the demands of more active tissues. This is an example of
    a) hyperventilation.
    b) hypoventilation.
    c) hypoxia.
    d) apnea.
    e) hyperpnea.

11. The normal ratio of bicarbonate concentration to carbon dioxide concentration in arterial blood is
    a) 1:5.
    b) 5:1.
    c) 10:1.
    d) 20:1.
    e) 1:20.

12. Which of the following can hemoglobin bind and transport in blood?
    a) oxygen
    b) carbon dioxide
    c) hydrogen ions
    d) a and c are both true
    e) all of the above

13. Which of the following areas of the brain contain inspiratory neurons?
    a) the dorsal respiratory group only
    b) the ventral respiratory group only
    c) both the dorsal and ventral respiratory groups
    d) neither the dorsal nor ventral respiratory groups

Objective Questions

1. Under normal conditions, the rate at which oxygen is brought into the alveoli in inspired air is (the same as/greater than/less than) the rate at which it is consumed in respiring tissues.

2. Under resting conditions, tissues normally extract (exactly half/more than half/less than half) of the oxygen that is delivered to them in arterial blood.

3. The amount of carbon dioxide in systemic arterial blood is less than 50% of that in mixed venous blood. (true/false)

4. When the P_{CO_2} of the blood increases, the concentration of bicarbonate (increases/decreases), and the concentration of hydrogen ions (increases/decreases).

5. The enzyme that catalyzes the conversion of carbon dioxide to carbonic acid is _____.

6. As the pH of the blood increases, the affinity of hemoglobin for oxygen (increases/decreases).

7. When a person hypoventilates, the P_{CO_2} of arterial blood (increases/decreases).

8. A decrease in alveolar ventilation would be expected to cause a(n) (increase/decrease) in arterial P_{O_2}, and a(n) (increase/decrease) in arterial P_{CO_2}.

9. Hemoglobin with carbon dioxide bound to it is called _____.

10. In gas exchange in both the lungs and respiring tissues, oxygen and carbon dioxide always move down their partial pressure gradients. (true/false)

11. (Central/peripheral) chemoreceptors respond directly to hydrogen ions produced during metabolism.

12. Coughing is triggered by stimulation of pulmonary _____ receptors.

13. In respiratory acidosis, arterial P_{CO_2} is (higher/lower) than normal.

14. An increase in the P_{CO_2} of alveolar air would be expected to trigger local (bronchoconstriction/bronchodilation) in airways.

15. An increase in the P_{O_2} of alveolar air would be expected to trigger local (vasoconstriction/vasodilation).

Essay Questions

1. Explain how changes in blood P_{CO_2} affect loading and unloading of oxygen in the lungs and in respiring tissues.

2. Sketch a hemoglobin-oxygen dissociation curve, and explain how it is affected by pH and P_{CO_2}. Include in your explanation changes in affinity and shifts in the hemoglobin-oxygen dissociation curve.

3. Describe what happens to oxygen and carbon dioxide as blood travels through the circulatory system. Start in the left ventricle and finish in the left atrium.

4. Suppose that alveolar P_{O_2} and P_{CO_2} are normal. If a sudden increase occurs in tissue metabolic activity and CO_2 production, but no change in minute alveolar ventilation occurs, then what would you expect to happen to arterial P_{CO_2}, mixed venous P_{CO_2}, and alveolar P_{CO_2}?

5. Describe how chemoreceptors work to keep arterial P_{CO_2} constant. Include an explanation of how arterial P_{CO_2} affects both central and peripheral chemoreceptors.

Critical Thinking

1. Carbon monoxide is a poison because it binds to hemoglobin at the heme group, preventing oxygen from binding. Draw a hemoglobin-oxygen dissociation curve for the normal condition and then with carbon monoxide present.

2. Chronic exposure to high altitude results in several adaptations to enhance oxygen delivery to tissues, one of which is an increase in the hematocrit or *polycythemia*. Based on what you learned about blood flow in

Chapter 15, describe the deleterious effects of polycythemia.

3. Barbiturate overdoses cause respiratory depression. Describe what happens to the partial pressures of oxygen and carbon dioxide in arterial blood following an overdose. What happens to the pH? Will a respiratory acidosis or alkalosis result? What would you do to treat somebody following an overdose?

4. Premature babies often do not secrete surfactant at adequate levels. Describe the consequences of low surfactant and appropriate treatments.

Find the answers to these exercises, and additional study tools, at the Physiology Place (www.physiologyplace.com).

19

The Urinary System: Renal Function

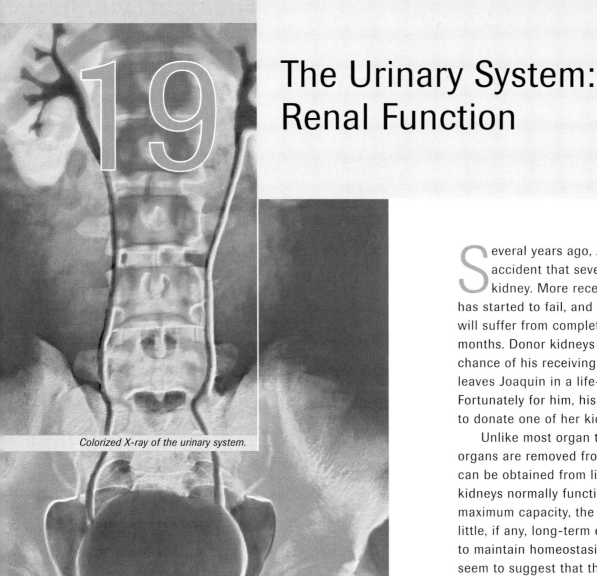

Colorized X-ray of the urinary system.

S everal years ago, Joaquin was in an accident that severely damaged his left kidney. More recently, his right kidney has started to fail, and his doctors predict that he will suffer from complete renal failure within six months. Donor kidneys are scarce, and the chance of his receiving one is unlikely. This leaves Joaquin in a life-threatening situation. Fortunately for him, his sister Elena has agreed to donate one of her kidneys to him.

   Unlike most organ transplants, in which donor organs are removed from the deceased, kidneys can be obtained from living donors. Because the kidneys normally function far below their maximum capacity, the loss of one kidney has little, if any, long-term effect on the body's ability to maintain homeostasis. Although this might seem to suggest that the kidneys do not work very hard, the contrary is true: The kidneys filter the entire plasma volume (approximately 3 liters) every 22 minutes, and they are essential to maintaining the normal extracellular fluid environment that bathes the cells of the body. Without medical intervention, the loss of both kidneys would result in death within a couple of weeks.

OBJECTIVES

- Identify and describe the functions of the following structures in the urinary system: nephron, glomerulus, renal tubule, collecting duct, ureter, bladder, and urethra.

- Describe how the urinary excretion of solutes and water influences the volume and composition of plasma, and identify other processes that affect plasma volume and composition.

- Explain how the basic renal exchange processes of filtration, secretion, and reabsorption affect the rate at which materials are excreted in the urine.

- Define the following terms: *filtered load, glomerular filtration rate, clearance, transport maximum,* and *renal threshold.*

- Describe the events that occur during micturition.

Before You Begin

Make sure you have mastered the following topics:

1. *Movement of molecules across epithelia, p. 126*
2. *Movement of molecules across capillary walls, p. 472*
3. *Starling forces, p. 472*

In Chapters 14, 15, and 16, we learned how the cardiovascular system delivers nutrients and removes wastes from our cells. In Chapters 17 and 18, we learned how one waste product, carbon dioxide, was eliminated from the body. We now turn to the kidneys and their role in filtering blood to eliminate wastes and other solutes that are present in blood in excess. In this chapter we describe basic renal processes and the formation of urine. In Chapter 20 we will describe the regulation of these processes to maintain homeostasis.

Functions of the Urinary System

Of all the organs of the body, the kidneys are perhaps the most misunderstood and underappreciated. Because the kidneys filter the blood and produce **urine**, a fluid that is eliminated from the body, many people think that the sole function of the kidneys is to clear the blood of waste products. Such a view, however, overlooks the many other vital functions performed by these amazingly versatile organs. Although urine does contain metabolic by-products and other substances properly described as "wastes," it also contains water and solutes (such as sodium and potassium) that must be maintained at certain levels in the plasma and other body fluids. The significance of this is that the rate at which these materials are eliminated from the body, or *excreted,* by the kidneys has a significant impact on the volume and composition of these fluids and therefore is highly regulated according

to the body's needs. The kidneys perform the following primary functions:

1. *Regulation of plasma ionic composition.* By increasing or decreasing the excretion of specific ions in the urine, the kidneys regulate the concentration of these ions in the plasma. Ions whose concentrations are regulated by the kidneys include sodium (Na^+), potassium (K^+), calcium (Ca^{2+}), magnesium (Mg^{2+}), chloride (Cl^-), bicarbonate (HCO_3^-), and phosphates (HPO_4^{2-} and $H_2PO_4^-$).

2. *Regulation of plasma volume.* By controlling the rate at which water is excreted in the urine, the kidneys regulate plasma volume, which has a direct effect on total blood volume, and thus on blood pressure.

3. *Regulation of plasma osmolarity.* Because the kidneys vary the rate at which they excrete water relative to solutes, they have the ability to regulate the osmolarity (solute concentration) of the plasma.

4. *Regulation of plasma hydrogen ion concentration (pH).* By regulating the concentration of bicarbonate and hydrogen ions in the plasma, the kidneys partner with the lungs to regulate the pH of the blood.

5. *Removal of metabolic waste products and foreign substances from the plasma.* Because the kidneys excrete wastes and other undesirable substances in the urine, they clear the plasma of waste products and eliminate them from the body. These materials include metabolic by-products such as urea and uric acid that are generated during protein and nucleic acid catabolism, respectively, as well as foreign substances such as food additives, drugs, or pesticides that enter the body from the external environment.

Because a free exchange of water and small solutes occurs between the plasma and interstitial fluid throughout most of the body, as the kidneys regulate the volume and composition of the plasma they also regulate the volume and composition of interstitial fluid. In addition, changes in the interstitial fluid affect the intracellular fluid. Thus the kidneys ultimately control the volume and composition of all the body's fluids. As we will see, the ability of the kidneys to form urine, and thus to perform their primary functions, hinges on their ability to filter and process large quantities of solutes and water.

The kidneys perform several secondary functions as well. The kidneys are endocrine organs because they secrete the hormone erythropoietin (which stimulates erythrocyte production by the bone marrow) and the enzyme renin (which is necessary for the production of angiotensin II, a hormone important in regulating salt and water balance for long-term control of blood pressure). The kidneys are also necessary for the activation of vitamin D_3 (ultimately to $1,25(OH)_2$ vitamin D_3, or calcitriol), an important factor in regulating blood calcium and phosphate levels. Furthermore, the kidneys can

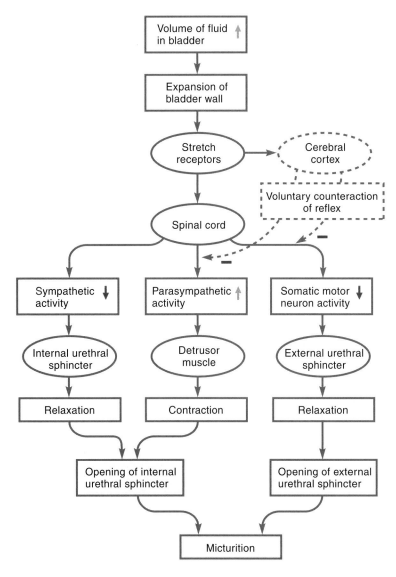

Figure 19.22 Elements of the micturition reflex. Voluntary control over the reflex is indicated by the dashed arrows.

sphincters coupled with the contraction of the detrusor muscle allows the bladder to empty.

In older children and adults, the micturition reflex can be overridden by voluntary control. Signals from the stretch receptors that detect filling of the bladder are transmitted in ascending pathways to the cerebral cortex, giving rise to the conscious sensation of fullness of the bladder and serving as a signal to activate descending pathways that *inhibit* the parasympathetic neurons controlling the detrusor muscle, and to

activate descending pathways that *excite* the motor neurons controlling the external urethral sphincter, allowing postponement of the micturition reflex if needed. Such postponement cannot be continued indefinitely, however, because continued filling of the bladder leads to greater excitation of stretch receptors. Eventually the level of activity in these neurons becomes high enough to trigger the micturition reflex despite conscious efforts to the contrary, triggering uncontrollable urination.

In addition to voluntary control to prevent micturition, one can initiate micturition through voluntary relaxation of the external urethral sphincter and the lowering of the pelvic floor. This lowering both causes the bladder to drop, which pulls open the internal urethral sphincter, and stretches the bladder wall, which activates the stretch receptors and induces the micturition reflex. Contraction of the diaphragm and abdominal muscles can increase the volume of urine voided by increasing the pressure in the abdominal cavity, which increases the pressure on the bladder for micturition.

> **Quick Test 19.4**
>
> 1. Name three structural differences between the epithelial cells of the proximal tubule and the distal tubule. The structure of which tubule is more favorable for exchange?
>
> 2. Define the following terms: *filtered load, clearance,* and *micturition.*
>
> 3. The clearance of what two substances can be used to estimate the GFR? The clearance of what substance can be used to estimate renal plasma flow?
>
> 4. If the clearance of molecule X is greater than the GFR, was X reabsorbed or secreted in the renal tubules?
>
> 5. Describe the innervation of the bladder, internal urethral sphincter, and external urethral sphincter. Voluntary inhibition of micturition is mediated through innervation to which of these structures?

Figure 19.23 is a chart summarizing many of the concepts covered in this chapter. Note that regulation of GFR, reabsorption, and secretion determine the plasma volume and solute concentration. Whereas the mechanisms for regulating GFR were described in this chapter, those for regulating reabsorption and secretion of water and solutes are described in Chapter 20.

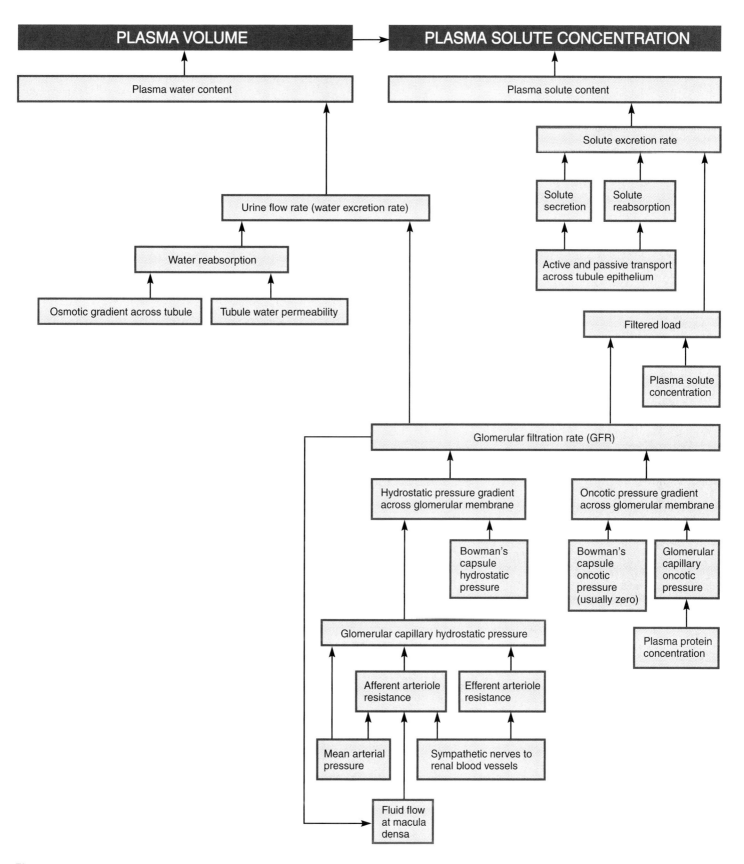

Figure 19.23 Role of renal processes in determining plasma volume and composition.

CHAPTER SUMMARY

Functions of the Urinary System, p. 579

The primary function of the kidneys is to filter the blood to regulate the ionic composition, osmolarity, volume, and pH of plasma, and to remove metabolic waste products and foreign substances from the plasma. Urine is formed in this process.

IP Urinary System, Anatomy Review, pp. 1–3

Anatomy of the Urinary System, p. 580

The urinary system consists of the kidneys, ureters, bladder, and urethra. The functional units of the kidneys are nephrons, consisting of Bowman's capsule, proximal tubule, descending loop of Henle, ascending loop of Henle, and distal tubule. The distal tubule drains into a collecting duct.

Filtration occurs at the renal corpuscle, which includes Bowman's capsule and the glomerulus. The glomerular filtrate resembles plasma in composition except that it lacks proteins. As the filtrate moves through the nephron, its volume and composition change as a result of the reabsorption and secretion of water and solutes. Reabsorbed materials move from the tubular fluid in the lumen of the tubule to the peritubular fluid that surrounds the tubule, and then into the plasma of peritubular capillaries that surround the tubule. Secretion moves material in the opposite direction, from the plasma into the filtrate.

The kidneys receive a large proportion of the cardiac output via the renal artery. An afferent arteriole leads into each glomerulus, and an efferent arteriole leaves the glomerulus. The efferent arteriole branches into either peritubular capillaries or vasa recta, which drain into veins that eventually lead to the renal vein. The juxtaglomerular apparatus, which consists of the macula densa cells in the distal tubule

and granular cells in the walls of the afferent and efferent arterioles, is important in the regulation of glomerular filtration and in salt and water reabsorption.

IP Urinary System, Glomerular Filtration, p. 3

IP Urinary System, Anatomy Review, pp. 5, 9–20

Basic Renal Exchange Processes, p. 585

Glomerular filtration is driven by the four Starling forces that contribute to the glomerular filtration pressure: (1) the glomerular capillary hydrostatic pressure, (2) the hydrostatic pressure inside Bowman's capsule, (3) the oncotic pressure of plasma in glomerular capillaries, and (4) the oncotic pressure of fluid in Bowman's capsule. The glomerular filtration pressure and the presence of fenestrations in the glomerular capillaries and slit pores in the epithelium of Bowman's capsule favor the bulk flow of protein-free fluid between blood and the lumen of Bowman's capsule. The normal glomerular filtration rate is approximately 125 mL/min.

The filtration fraction is the percentage of renal plasma flow that is filtered; on average it is approximately 20%. The filtered load is the quantity of a certain solute that is filtered at the glomerulus. For a solute that is freely filtered, the filtered load equals the product of the GFR and the plasma concentration of the solute.

Under normal conditions, the glomerular filtration rate is regulated to stay nearly constant by three intrinsic control mechanisms: (1) myogenic regulation of smooth muscle in the afferent arteriole, (2) tubuloglomerular feedback, and (3) mesangial cell contraction. Extrinsic control of GFR includes sympathetic nervous control of smooth muscle in the afferent and efferent arterioles.

When solutes are transported across the tubular epithelium by car-

rier proteins during reabsorption or secretion, the transport is subject to a transport maximum, which occurs when the concentration of solute is great enough to saturate the carrier proteins. The plasma concentration at which the solute appears in the urine is called the renal threshold.

IP Urinary System, Glomerular Filtration, pp. 1–14

Regional Specialization of the Renal Tubules, p. 595

The proximal tubule is specialized to reabsorb large quantities of solutes and water, and returns most filtered material to the bloodstream. In contrast, the distal tubule and collecting duct are specialized for the regulation of transport, which is important in controlling the volume and composition of the plasma. The transport of water and many solutes in the distal tubule and collecting duct is regulated by hormones.

IP Urinary System, Early Filtrate Processing, pp. 1–13

IP Urinary System, Late Filtrate Processing, pp. 1–14

Excretion, p. 597

The rate at which a substance is excreted in the urine is determined by three factors: the rate at which it is filtered at the glomerulus, the rate at which it is reabsorbed, and the rate at which it is secreted. If the amount of solute excreted per minute is less than the filtered load, then the solute was reabsorbed in the renal tubules. If the amount of solute excreted per minute is greater than the filtered load, then the solute was secreted in the renal tubules.

Clearance is a measure of the volume of plasma from which a substance is completely removed or "cleared" by the kidneys per unit time. The clearance of inulin or creatinine can be used to estimate the GFR. The clearance of PAH can be used to estimate renal plasma flow and thus renal blood flow.

The fluid that remains in the renal tubules after filtration, reabsorption, and secretion is excreted as urine. The fluid drains from the collecting ducts into the renal pelvis and then into the ureter. Wavelike contractions of smooth muscle in the wall of the ureter propel the urine toward the bladder. The bladder stores the urine until it is excreted during micturition. Micturition is under both reflex and voluntary control. The micturition reflex is triggered by stretching of the bladder wall.

EXERCISES

Multiple-Choice Questions

1. Which structure of the urinary system stores urine until it is excreted?
   a) kidneys
   b) bladder
   c) ureter
   d) urethra
   e) gallbladder

2. Which of the following is *not* one of the mechanisms by which a solute can be exchanged between the plasma and the renal tubules?
   a) glomerular filtration
   b) secretion
   c) excretion
   d) reabsorption

3. What type of specialized junction connects epithelial cells lining the renal tubules?
   a) gap junctions
   b) tight junctions
   c) desmosomes
   d) intercalated disks
   e) slit pores

4. Which of the following does *not* favor a large glomerular filtration rate?
   a) slit pores
   b) fenestrations
   c) high glomerular hydrostatic pressure
   d) high resistance in the afferent arteriole
   e) high resistance in the efferent arteriole

5. Most reabsorption of water and solutes occurs in the
   a) proximal tubule.
   b) descending limb of the loop of Henle.
   c) ascending limb of the loop of Henle.
   d) distal tubule.
   e) collecting duct.

6. In which of the following are microvilli most abundant?
   a) Bowman's capsule
   b) glomerular capillaries
   c) distal tubule
   d) proximal tubule
   e) collecting duct

7. Which of the following would occur if mean arterial pressure increased from 95 mm Hg to 125 mm Hg?
   a) Glomerular filtration rate would increase due to the increased glomerular capillary hydrostatic pressure.
   b) Glomerular filtration rate would decrease due to increased Bowman's capsule hydrostatic pressure.
   c) Glomerular filtration rate would not change due to autoregulation.
   d) Glomerular filtration rate would not change due to activation of the sympathetic nervous system.

8. The normal fasting plasma glucose concentration is 100 mg/dL, and the renal threshold is 300 mg/dL. If the plasma concentration doubles to 200 mg/dL, then
   a) the rate at which glucose is reabsorbed will double.
   b) the capacity of the renal tubule for transporting glucose will be exceeded.
   c) urinary water excretion will increase.
   d) glucose clearance will increase.
   e) the filtered load of glucose is halved.

9. A substance S is freely filterable and is excreted at a rate (in moles/min) that is lower than the filtered load. On the basis of this information alone, which of the following is the *most precise* conclusion that can justifiably be drawn regarding the kidneys' processing of S?
   a) S is neither reabsorbed nor secreted.
   b) S is definitely reabsorbed and may be secreted.
   c) S is definitely secreted and may be reabsorbed.
   d) S is definitely both reabsorbed and secreted.

10. Which of the following observations would enable you to definitely conclude that a substance X is being secreted?
    a) The clearance of X is greater than the GFR.
    b) The concentration of X in the urine is greater than its concentration in the plasma.
    c) The concentration of X in the plasma is decreasing over time.
    d) a or c
    e) any of the above

11. Micturition occurs in response to
    a) relaxation of the detrusor muscle.
    b) contraction of the internal and external urethral sphincters.
    c) activation of parasympathetic neurons to the bladder.
    d) activation of somatic motor neurons to the bladder.

Objective Questions

1. The (ureter/urethra) carries urine from the bladder to the outside of the body.

2. Urinary excretion is the elimination of urine from the bladder. (true/false)

3. The (afferent/efferent) arteriole carries blood toward the glomerulus.

4. The combination of a glomerulus and the surrounding Bowman's capsule is called a(n) _____ _____.

5. The hydrostatic pressure in glomerular capillaries is (higher/lower) than that in most capillaries of the body.

6. The glomerular filtration rate tends to (increase/decrease) as the concentration of proteins in the plasma increases.

7. Autonomic neurons regulate contraction of the (internal/external) urethral sphincter.

8. The glomerular filtration pressure is synonymous with the hydrostatic pressure inside glomerular capillaries. (true/false)

9. The filtered load of a solute is determined by its plasma concentration and the (glomerular filtration rate/urine flow rate).

10. If the clearance of a substance is greater than the glomerular filtration rate, then that substance must have undergone (reabsorption/secretion) in the renal tubules.

11. The clearance of (PAH/creatinine) is approximately equal to the renal plasma flow rate.

12. Substances that are reabsorbed move into the (peritubular capillaries/tubule lumen).

13. An increase in the flow rate through the macula densa causes a(n) (increase/decrease) in the glomerular filtration rate.

14. Glucose reabsorption occurs primarily in the (proximal tubule/distal tubule).

Essay Questions

1. Describe the pathway of filtrate flow from the renal corpuscle to elimination from the body.

2. Explain why the glomerular filtration rate does not change with a moderate decrease in mean arterial pressure.

3. Compare the mechanisms of active and passive reabsorption of solute.

4. Explain the concept of transport maximum. Why does glucose appear in the urine of diabetic patients?

5. Explain why the proximal tubules are considered mass absorbers.

6. In your own words, describe what the term *clearance* means.

Critical Thinking

1. Suppose that sodium is being excreted in the urine at a rate of 0.5 mmole/min, and that sodium is present in the plasma at a concentration of 150 mM. If the GFR is 125 mL/min, what is the filtered load of sodium? What is the sodium clearance? Is sodium being reabsorbed or secreted by the renal tubules?

2. In many types of kidney diseases, proteins can leak across the glomerular wall, increasing the colloid oncotic pressure of Bowman's capsule and decreasing the colloid oncotic pressure of the plasma. Calculate the glomerular filtration pressure given the following information from a patient with acute renal failure: glomerular hydrostatic pressure = 60 mm Hg, Bowman's capsule hydrostatic pressure = 20 mm Hg, glomerular oncotic pressure = 23 mm Hg, and Bowman's capsule oncotic pressure = 3 mm Hg. Is filtration across the kidneys greater or less than normal? Does urine flow rate increase or decrease?

3. Compare the responses of smooth muscle in the afferent arterioles and the detrusor muscles to stretch.

Find the answers to these exercises, and additional study tools, at the Physiology Place (www.physiologyplace.com).

The Urinary System: Fluid and Electrolyte Balance

20

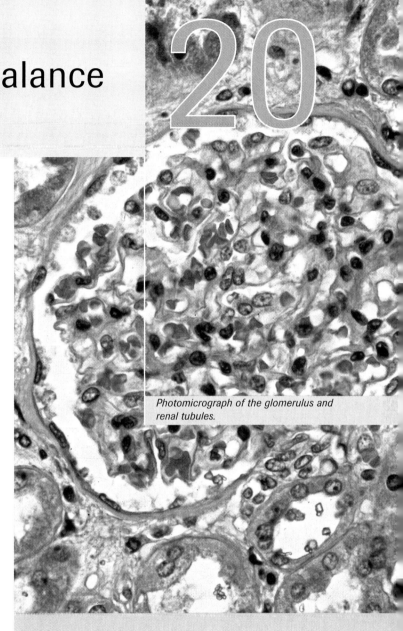

Photomicrograph of the glomerulus and renal tubules.

W e all have different eating and drinking habits. For example, some people need a large glass of water in the morning to quench their thirst; others need a couple cups of coffee in the morning to get them going. The immediate outcome is the same, excretion of large volumes of water in the urine, but the consequences differ.

When you drink a large volume of water, it isn't long before your body eliminates the excess water consumed to maintain normal plasma volume and osmolarity. Likewise, when you consume salty food, such as french fries, your body retains fluid and you feel thirsty. These are part of negative feedback systems to maintain normal fluid and electrolyte composition of the plasma. However, caffeine is a drug that disrupts your body's homeostasis by causing your body to eliminate water inappropriately. Ultimately, drinking caffeine makes you thirsty to replace the water lost shortly after consumption.

In this chapter, you will learn about the mechanisms of water and solute movement into and out of the kidney tubules and the hormonal regulation of these movements. You will also learn why caffeine causes an increase in water excretion, even if plasma volume and osmolarity are normal.

In Chapter 19, we learned about the anatomy of the kidneys and the processes of filtration, reabsorption, secretion, and excretion. In this chapter, we will learn about the hormonal regulation of these processes for a specific solute or water in order to maintain normal fluid and electrolyte composition, which is critical to maintaining normal blood pressure and normal functioning of cells.

The Concept of Balance

To maintain homeostasis, the human body must be kept in balance. To be in balance in this context means that what comes into the body and what is produced by the body must equal the sum of what is used by the body and what is eliminated from the body, as shown in the following equation:

$$\text{input} + \text{production} = \text{utilization} + \text{output}$$

The kidneys play a key role in regulating fluid and electrolyte balance and acid-base balance.

Factors Affecting the Plasma Composition

The kidneys exert control over the volume and composition of plasma by regulating its solute and water content. The volume and composition depend on each other and must be maintained within narrow limits. The volume of plasma is determined almost entirely by its water content because solutes make only a negligible direct contribution to the volume. However, the amount of solute in plasma indirectly affects plasma volume because changes in plasma osmolarity (solute concentration) can cause water to shift between the plasma and other body fluid compartments. Plasma volume has an important influence on the body's homeostasis because it affects mean arterial pressure. As we will see later in this chapter, mean arterial pressure is regulated over the long run by changes in plasma volume.

The solute and water content of the plasma is affected both by the movement of materials into and out of the body and by the movement of materials between different compartments within the body. **Figure 20.1** shows pathways for the movement of water and solutes small enough to permeate capillary walls and cell membranes and thus to move freely into and out of the plasma. The plasma and interstitial fluid compartments are combined in this diagram to emphasize their similarity in composition, which is due to the free exchange of water and small solutes between them.

As Figure 20.1 shows, the plasma can gain or lose materials by exchange with cells or with extracellular connective tissue, such as the bone matrix. When the bone is resorbed, for example, calcium and phosphates are released into the plasma, which raises the concentrations of these substances; conversely, the deposition of calcium and phosphates into bone lowers their plasma concentrations. Figure 20.1 shows three other routes by which the plasma and the external environment can exchange materials: The plasma can (1) either gain or lose materials by exchange with the lumen of the gastrointestinal tract, (2) either gain or lose materials by exchange with the lumens of the renal tubules, and (3) lose materials through sweating, hemorrhage, or respiration. (The plasma normally gains oxygen through respiration, however.)

Solutes and water are absorbed into the plasma from the gastrointestinal tract and also move from the plasma to the lumen of the tract in the form of saliva, bile, pancreatic juice, and other gastrointestinal secretions. Under normal conditions, the absorptive capacity of the gastrointestinal tract is sufficient to recover virtually 100% of all secreted materials as well as those that enter the tract via ingestion (eating or drinking). The rate at which solutes and water are lost by elimination from the gastrointestinal tract is very small compared to the rate at which these substances are lost through the excretion of urine by the kidneys. Consequently, *the transport of materials across the wall of the gastrointestinal tract normally amounts to a net gain of solutes and water by the body.*

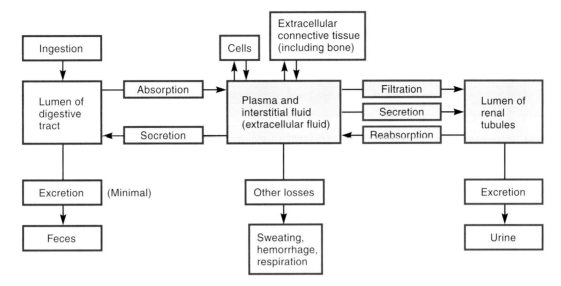

Figure 20.1 Material exchanges affecting plasma content. Materials (water and solutes) enter the body through ingestion into the digestive tract. Once in the body, materials are exchanged among the digestive tract, the plasma (and the interstitial fluid), cells of the body, connective tissue, and the renal tubules. The highlighted areas represent exchanges between plasma and the renal tubules.

Small solutes and water move from the plasma to the lumen of the renal tubules by glomerular filtration and secretion. Solutes and water are returned to the plasma from the renal tubules by reabsorption. Since not all this material is reabsorbed, *the transport of materials across the walls of the renal tubules amounts to a net loss of water and solutes by the body.* These lost materials are excreted in the urine.

Solute and Water Balance

When solutes and water enter and exit the plasma at the same rate, the plasma's volume and composition do not change, and it is said to be in balance. Changes in volume and/or composition occur when materials enter the plasma faster than they exit or vice versa (**Figure 20.2**). When a substance enters the body faster than it exits, the substance is said to be in a state of *positive balance;* under these conditions the quantity of that substance in the plasma tends to increase, unless the substance enters cells or is metabolized within the body. If a substance leaves the body faster than it enters, the substance is in a state of *negative balance;* under these conditions the quantity of that substance in the plasma tends to decrease.

For certain substances whose plasma concentrations are controlled by specific regulatory mechanisms, a state of positive balance or negative balance can exist with little or no change in the plasma concentration. When you eat a meal containing glucose, for example, your body goes into a state of positive glucose balance. The absorption of glucose from the lumen of the gastrointestinal tract causes a rise in the plasma glucose level, but this rise

is transient because it triggers insulin secretion and other hormonal changes. Insulin causes an increase in cellular glucose uptake which, combined with other hormonally triggered responses, quickly lowers the blood glucose concentration, eventually restoring it to normal. Once the glucose enters cells, it is catabolized for energy or converted to glycogen or fats for storage.

In Chapter 19 we saw that the kidneys filter 180 liters of plasma per day. Approximately 70% of the filtered water and sodium is reabsorbed in the proximal tubule in the absence of any regulation. However, the body can

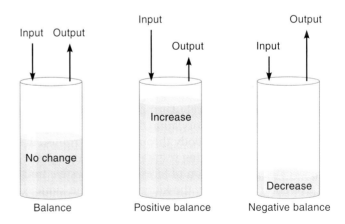

Figure 20.2 The concept of balance. A substance is in balance in the body when input equals output. A substance is in positive balance when input exceeds output, causing a net increase in the amount of that substance. A substance is in negative balance when output exceeds input, causing a net decrease in the amount of that substance.

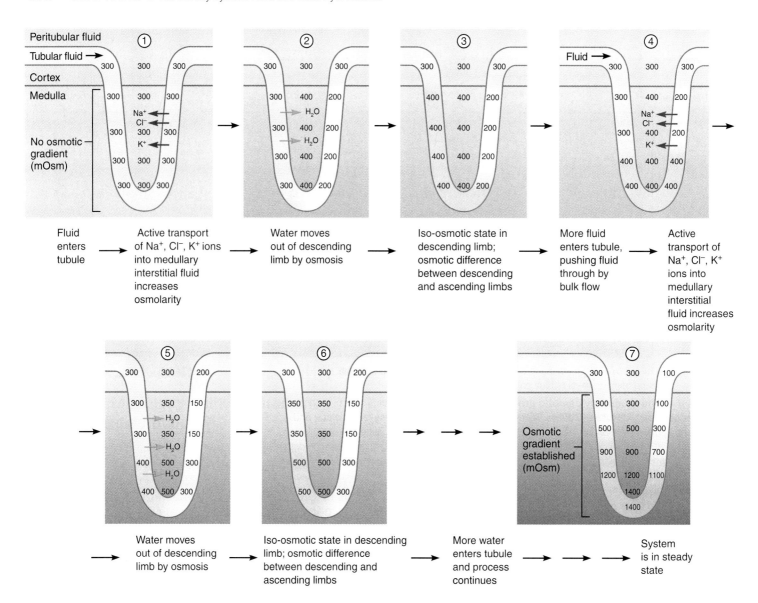

Figure 20.7 How the countercurrent multiplier establishes the medullary osmotic gradient. ① Initially all fluids are iso-osmotic at 300 mOsm. Active transport of solutes (Na^+, Cl^-, K^+) from the ascending limb of the loop of Henle into the medullary interstitial fluid increases the osmolarity of the interstitial fluid and decreases the osmolarity of the tubular fluid in the ascending limb. ② The increased osmolarity of the medullary interstitial fluid draws water from the lumen of the descending limb of the loop of Henle into the interstitial fluid, ③ increasing the osmolarity of the tubular fluid in the descending limb. ④ More tubular fluid then enters the loop of Henle, pushing the fluid farther into the renal tubules. The process of solute transport from the ascending limb ⑤ followed by water movement from the descending limb, ⑥ increasing the osmolarity of the tubular fluid in the descending limb, followed by more tubular fluid entering the loop of Henle, repeats until ⑦ the medullary osmotic gradient is established.

regions of the medulla with higher osmolarity, water leaves the capillaries by osmosis, and solutes enter the plasma by diffusion, which would tend to reduce the osmolarity of the interstitial fluid if left unchecked. This process continues to the tip of the vasa recta due to the increasing osmolarity of the medullary interstitial fluid. However, as the blood flows back toward the cortex, the direction of the osmotic gradient across the capillary walls

reverses, so water moves into the plasma and solutes move into the interstitial fluid, which tends to raise the osmolarity of the interstitial fluid. As a result, the osmolarity of the interstitial fluid stays relatively constant and the osmolarity of plasma leaving the renal medulla in the vasa recta capillaries is almost equal to that of the plasma entering the renal medulla.

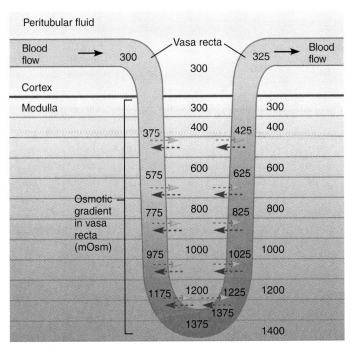

---▸ Water movement
◂--- Solute flow

Figure 20.8 **How the vasa recta prevents the dissipation of the medullary osmotic gradient.** As the vasa recta capillaries accompany the loops of Henle through the medulla, plasma water losses and solute gains on the way into the medulla are counteracted by plasma water gains and solute losses on the way out of the medulla.

Water Reabsorption in the Distal Tubule and Collecting Duct

Recall that 70% of the water filtered from plasma at the renal corpuscle is reabsorbed in the proximal tubule. Approximately 20% of the filtered water is reabsorbed in the distal tubule, and most of the remaining 10% is reabsorbed in the collecting ducts. In the initial portion of the distal tubule, the lumenal fluid (100–200 mOsm) is hypo-osmotic to the peritubular fluid (300 mOsm). As fluid moves down the collecting duct, the osmolarity of the lumenal fluid is always less than the increasing osmolarity of the medullary interstitial fluid, so water moves from the lumen of the collecting duct to the medullary interstitial fluid, and from there into the plasma (that is, water is reabsorbed) when the wall of the collecting duct is permeable to water.

The epithelial cells lining the late distal tubule and collecting duct are connected by tight junctions such that water cannot pass between cells from peritubular fluid to tubular fluid or vice versa. In addition, the lipid bilayers of these cells' plasma membranes are not permeable to water. The ability of water to cross the plasma membrane (and therefore the epithelial layer) depends on the presence of water channels or pores, called **aquaporins**, in

the plasma membrane of principal cells. Aquaporin-3 is present in the basolateral membrane of principal cells at all times, whereas aquaporin-2 is present in the apical membrane only in the presence of the hormone ADH (discussed in the next section).

The effects of the medullary osmotic gradient and water permeability of the late distal tubule and collecting duct on water reabsorption are shown in **Figure 20.9**. Basically, the more permeable these tubules, the greater the water reabsorption.

In Figure 20.9a, the walls of the late distal tubule and collecting duct are impermeable to water. As tubular fluid of 100 mOsm enters the late distal tubule, an osmotic gradient for water movement exists, but because of the impermeable membrane, water cannot move. The osmotic gradient gets larger as the tubular fluid at 100 mOsm travels down the collecting duct toward the renal pelvis, but water still cannot cross the impermeable membrane. The final result is excretion of a large volume of urine with low osmolarity.

Figure 20.9b shows how the kidneys can conserve water when the late distal tubule and collecting duct are made highly permeable to water. In the early portion of the collecting duct, the tubular fluid is initially hypo-osmotic to the cortical interstitial fluid, and water is reabsorbed. As the collecting duct leaves the cortex, the tubular fluid is iso-osmotic with the interstitial fluid at 300 mOsm. As the fluid moves down the collecting duct, water continues to be reabsorbed from the collecting duct into the medullary interstitial space such that the fluid in the collecting duct always remains very nearly iso-osmotic with the medullary interstitial fluid; eventually the fluid reaches an osmolarity of 1400 mOsm at the end of the collecting duct. Tubular fluid osmolarity can never exceed the medullary interstitial fluid osmolarity because water will stop moving across the wall once the osmolarity inside the tubule becomes equal to that outside the tubule. Therefore, the maximum osmolarity of urine is 1400 mOsm. Because those solutes that are not 100% reabsorbed must be excreted in the urine and because there is an upper limit on the osmolarity of urine, a minimum volume of water must be excreted to eliminate the solutes. This volume is the **obligatory water loss**, which is approximately 440 mL of water per day under normal conditions (see **Discovery: Why We Can't Drink Seawater**, p. 619).

The length of the loop of Henle determines the maximum concentration of urine. Longer loops of Henle can form a larger medullary osmolarity gradient by the countercurrent multiplier and thereby allow greater water reabsorption. Camels, for example, have longer loops of Henle than humans and can generate urine with a concentration of 2800 mOsm. Australian hopping mice, which have the longest loops of Henle of any known species, can concentrate urine to 9800 mOsm. Because of their strong ability to conserve water, Australian hopping mice are subjected to very small obligatory water loss and can survive with little water to drink.

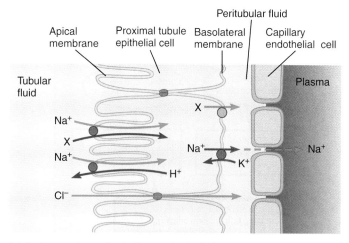

(a) Sodium reabsorption in the proximal tubule

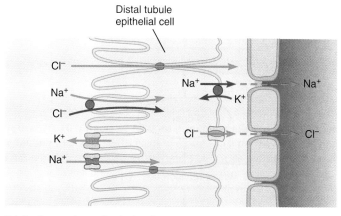

(b) Sodium reabsorption in the distal tubule

Figure 20.13 Mechanisms of sodium reabsorption in the proximal and distal tubules. (a) Sodium reabsorption in the proximal tubule. Sodium is actively transported across the basolateral membrane by the Na^+/K^+ pump. Sodium moves across the apical membrane either by cotransport with an organic molecule (X) such as glucose or an amino acid, or by countertransport with another ion such as hydrogen. Note that chloride follows sodium reabsorption. (b) Sodium reabsorption in the distal tubule. Again, sodium is actively transported across the basolateral membrane by the Na^+/K^+ pump. Sodium moves across the apical membrane either by cotransport with chloride ions or through sodium channels. Potassium secretion from peritubular fluid to the tubule lumen sometimes accompanies sodium reabsorption.

Quick Test 20.2

1. Name two stimuli for ADH release from the posterior pituitary. Where does ADH act in the kidneys, and what does it do?

2. Describe the effects of ADH on principal cells.

Sodium Balance

Because sodium is the primary solute in extracellular fluid, it must be regulated if normal osmolarity is to be maintained. In addition, because the electrochemical gradient for sodium across plasma membranes is critical to the function of excitable cells, it is important that plasma sodium levels be regulated. An increase in plasma sodium levels above normal, called *hypernatremia*, is often accompanied by water retention and an increase in blood pressure. A decrease in plasma sodium levels below normal, called *hyponatremia*, is associated with low plasma volume and hypotension.

Sodium is freely filtered at the glomerulus and undergoes tubular reabsorption, but it is not secreted. Regulation of plasma sodium occurs at the level of reabsorption. We first look at renal handling of sodium. Next, we discuss the two hormones that function in the regulation of sodium reabsorption: aldosterone and atrial natriuretic peptide, or ANP.

Mechanisms of Sodium Reabsorption in the Renal Tubule

In all tubular segments where sodium is reabsorbed, sodium ions are actively transported. This active reabsorption is driven by Na^+/K^+ pumps located in the basolateral membrane of renal tubule epithelial cells. (Recall that these pumps utilize energy from ATP hydrolysis to transport sodium and potassium ions against their electrochemical gradients.) Because the active transport of sodium out of the epithelial cell keeps its concentration low in the intracellular fluid, sodium passively enters the cell from the tubular lumen across the apical membrane. Even though this latter step is passive, the overall movement of sodium ions across the cell (that is, from tubular fluid to peritubular fluid) is active because it depends on the active transport of sodium across the basolateral membrane. The active transport of sodium affects the movement of water across the renal tubules by osmosis.

In the proximal tubule, the entry of sodium into the tubule epithelial cells is carried out by transport proteins in the apical membrane that couple sodium movement to the flow of other solutes. **Figure 20.13**a shows two such pathways for sodium entry into cells: (1) cotransport with solutes such as glucose and amino acids (designated X in the figure), and (2) countertransport with hydrogen ions. In the first of these processes, energy released by the passive entry of sodium is harnessed to drive the flow of glucose or amino acids against their electrochemical gradients as they enter the epithelial cell. These solutes then exit the cell passively across the basolateral membrane. Given that the transport of glucose and amino acids from the tubular fluid to peritubular fluid requires an active step, the reabsorption of these solutes is active and requires energy. The ultimate source of this energy is ATP hydrolysis, because the reabsorption

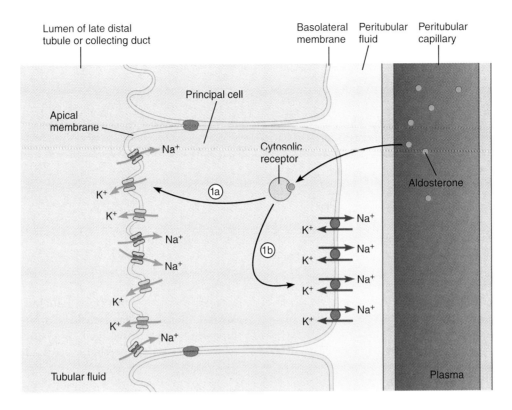

Figure 20.14 Effects of aldosterone on principal cells of the distal tubules and collecting ducts. After binding to its receptor, aldosterone ⓐ stimulates both the opening of sodium channels and potassium channels and the synthesis of new channels on the apical membrane, and ⓑ stimulates the synthesis and insertion of more Na^+/K^+ pumps on the basolateral membrane.

What class of hormone is aldosterone, and in what location in principal cells are receptors for the hormone located?

of these solutes is coupled to the flow of sodium, which ultimately depends on the ATP-driven Na^+/K^+ pumps. Other transport proteins couple the passive entry of sodium to the active secretion of hydrogen ions into the tubule lumen, a process that is important in acid-base regulation.

Figure 20.13b shows the mechanism of active sodium reabsorption in the distal tubule. The process involves passive movement of sodium across the apical membrane into the tubule epithelial cell and active transport of sodium across the basolateral membrane out of the epithelial cell and into the peritubular fluid. In this regard, sodium reabsorption in the distal tubule resembles that in the proximal tubule. However, the two processes differ with regard to the mechanism of sodium transport across the apical membrane. In the distal tubule, sodium enters the epithelial cell by two means: by cotransport with chloride ions, and by facilitated diffusion through sodium channels.

Sodium reabsorption in the distal tubule is also often coupled to potassium secretion, which, along with the cotransport of sodium with chloride, minimizes changes in the electrical potential that exists across the walls of

the tubules. If the electrical potential is to be maintained, the reabsorption of a cation such as sodium must be balanced by reabsorption of anions (and to a lesser extent by secretion of other cations). Because chloride and bicarbonate are the most abundant anions in the tubular fluid, they account for the bulk of the anions that are reabsorbed with sodium. Potassium and hydrogen ions are the most common cations secreted as sodium ions are reabsorbed.

The Effects of Aldosterone

Aldosterone is a steroid hormone released from the adrenal cortex that regulates both the reabsorption of sodium and the secretion of potassium. Here we focus on the role of aldosterone in sodium reabsorption.

Aldosterone binds to cytosolic receptors in principal cells of the late distal tubules and collecting ducts, where it has several effects (**Figure 20.14**). Aldosterone increases the number of open sodium channels and potassium

A steroid hormone; in the cytosol

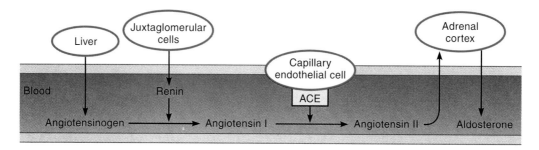

Figure 20.15 The renin-angiotensin-aldosterone system. The liver and the juxtaglomerular cells secrete and release angiotensinogen and renin, respectively, into the blood, where renin cleaves amino acids from angiotensinogen to form angiotensin I. Angiotensin converting enzyme (ACE) located on certain endothelial cells in capillaries cleaves amino acids from angiotensin I to form angiotensin II, which travels in the bloodstream to the adrenal cortex, where it stimulates the release of aldosterone into the blood.

channels in the apical membrane, both by causing existing channels to open and by stimulating the synthesis of new channels. Aldosterone also stimulates the synthesis of Na^+/K^+ pumps, which increases the concentration of Na^+/K^+ pumps in the basolateral membrane. Through these actions, aldosterone increases sodium reabsorption and potassium secretion simultaneously; it cannot affect one without affecting the other.

Of the factors that control aldosterone release, the one that is most important in the control of sodium reabsorption is the renin-angiotensin-aldosterone system (RAAS).

The Renin-Angiotensin-Aldosterone System

Recall from Chapter 19 that where the distal tubule travels close to the afferent and efferent arterioles, these structures form the juxtaglomerular apparatus. Within the walls of the afferent arteriole are granular cells that secrete **renin**; within the walls of the distal tubule are the macula densa cells, which can detect changes in the sodium and chloride concentrations of and the flow of the tubular fluid. When sodium concentration in the tubular fluid decreases, renin secretion increases. Although often called a hormone, renin is actually a proteolytic enzyme.

Once renin is released from the granular cells into the bloodstream, it starts a series of reactions that lead to the release of aldosterone (**Figure 20.15**). Renin acts on another protein that is always present in the plasma, *angiotensinogen,* which like most plasma proteins is secreted by the liver. Renin cleaves off some amino acids from angiotensinogen, converting it to angiotensin I. As angiotensin I molecules circulate in the bloodstream, they encounter another proteolytic enzyme called **angiotensin converting enzyme (ACE)**, which is bound to the inner surfaces of capillaries throughout the body and is particularly abundant in the capillaries of the lungs. ACE cleaves off some amino acids from angiotensin I, converting it to angiotensin II. In addition to acting as a vasoconstrictor that is important in the regulation of

mean arterial pressure (see Chapter 15), angiotensin II has another key role: the stimulation of aldosterone release from the adrenal cortex. (Angiotensin II also acts in the hypothalamus, where it stimulates ADH release and thirst.)

Figure 20.16 summarizes the four mechanisms whereby angiotensin II increases mean arterial pressure: (1) Angiotensin II stimulates vasoconstriction of systemic arterioles, which by increasing the total peripheral resistance increases mean arterial pressure. (2) Angiotensin II stimulates the adrenal cortex to secrete aldosterone, which by increasing sodium reabsorption causes water reabsorption to increase. (3) Angiotensin II stimulates the posterior pituitary to secrete ADH, which by increasing water reabsorption minimizes fluid loss and maintains plasma volume, thereby maintaining mean arterial pressure. (4) Angiotensin II activates hypothalamic neurons to stimulate thirst and fluid intake, which by increasing plasma volume increases mean arterial blood pressure.

Renin release is controlled by a variety of stimuli related to blood pressure (**Figure 20.17**). Because the RAAS tends to increase blood pressure, a decrease in blood pressure is a primary stimulus for renin release. Specifically, a decrease in afferent arteriolar pressure triggers renin release, because granular cells are directly sensitive to the degree of stretch of the afferent arteriole. An increase in renal sympathetic nerve activity also stimulates renin release by direct input to granular cells. The sympathetic nervous system is activated during the baroreceptor reflex response to a decrease in blood pressure (see Chapter 15). A large decrease in mean arterial pressure also decreases the glomerular filtration rate, which when coupled with the continual reabsorption of sodium and chloride in the proximal tubule and ascending loop of Henle, decreases sodium and chloride levels in the distal tubule. A decrease in sodium and chloride in the distal tubules is detected by macula densa cells of the tubule, which secrete a chemical signal that stimulates renin release from the juxtaglomerular cells.

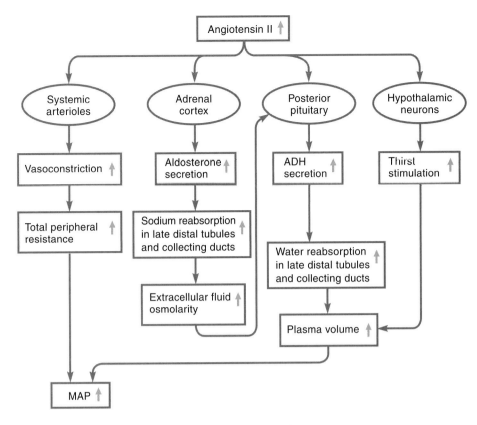

Figure 20.16 **Mechanisms by which angiotensin II increases mean arterial pressure.**

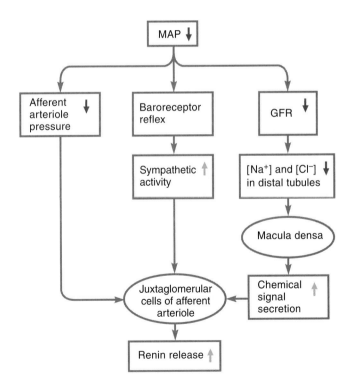

Figure 20.17 **Mechanisms by which decreases in mean arterial pressure stimulate renin release.**

What is renin's RAAS function in the plasma?

It cleaves amino acids off angiotensinogen to form angiotensin I.

Atrial Natriuretic Peptide

Atrial natriuretic peptide (ANP) is secreted by cells in the atria of the heart in response to distension of the atrial wall, which occurs when plasma volume has increased. ANP increases sodium excretion by increasing the glomerular filtration rate and by decreasing sodium reabsorption (**Figure 20.18**). ANP causes dilation of the afferent arteriole and constriction of the efferent arteriole, which by increasing glomerular capillary pressure increases the glomerular filtration rate and increases the filtered sodium load. ANP decreases sodium reabsorption directly by decreasing the number of open sodium channels in the apical membrane of the principal cells. In addition, ANP decreases secretion of both renin and aldosterone.

Quick Test 20.3

1. Briefly explain how an increase in the secretion of renin stimulates the secretion of aldosterone. How does an increase in aldosterone secretion affect sodium reabsorption?

2. Where are the cells that secrete renin located? Name three stimuli for renin secretion.

3. How does atrial natriuretic peptide affect sodium reabsorption?

4. Describe the effects of aldosterone on principal cells.

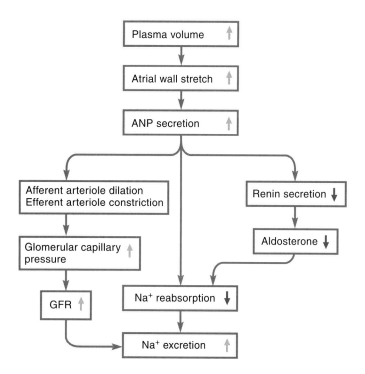

Figure 20.18 Mechanisms by which secretion of atrial natriuretic peptide increases sodium excretion in response to increased plasma volume.

Potassium Balance

The gradient that results from high potassium concentrations in the intracellular fluid and low potassium concentrations in the extracellular fluid is critical to the function of excitable cells. An increase in plasma potassium levels is called *hyperkalemia*. Some common symptoms of hyperkalemia include cardiac arrhythmias, muscle weakness and cramps, dizziness, nausea, and diarrhea. A decrease in plasma potassium levels is called *hypokalemia*. Common symptoms of hypokalemia include cardiac arrhythmias, muscle weakness and tenderness, hypotension, confusion, alkalosis, and shortness of breath.

Renal Handling of Potassium Ions

In the kidneys, potassium is freely filtered at the glomerulus and undergoes both reabsorption and secretion in the tubules. Normally, the amount of potassium reabsorbed is greater than the amount secreted; that is, the net effect is reabsorption. In fact, most of the potassium filtered is reabsorbed.

Unlike water and sodium, whose plasma levels are regulated by varying the amounts that are reabsorbed from the renal tubules, the plasma concentration of potassium is regulated by varying the amounts that are secreted into the renal tubules. As with sodium, renal handling of potassium ions varies within the renal tubules (**Figure 20.19**). Potassium ions are reabsorbed in the proximal tubule and secreted in the late distal tubule and collecting duct; the secretion is regulated.

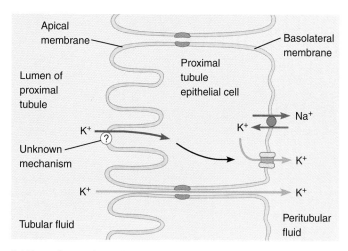

(a) Potassium reabsorption in the proximal tubule

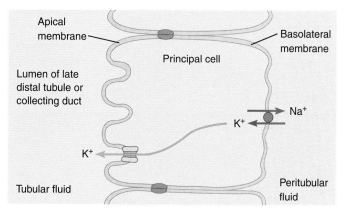

(b) Potassium secretion in the principal cells of the late distal tubule and collecting duct

Figure 20.19 Potassium transport in renal tubules. (a) In the proximal tubule, potassium is reabsorbed because of the presence of potassium channels in the basolateral membrane. (b) In principal cells of the distal tubule and collecting duct, potassium is secreted because of the presence of potassium channels in the apical membrane.

In the proximal tubule (Figure 20.19a), potassium is reabsorbed by the following mechanisms: Potassium ions move from the peritubular fluid into the tubule epithelial cell via the Na^+/K^+ pump located on the basolateral membrane; potassium ions also move from the tubular fluid into the epithelial cell by some as yet unknown mechanism. Once inside the epithelial cell, potassium ions move through potassium channels in the basolateral membrane into the peritubular fluid. Therefore, most potassium entering the tubule epithelial cell (whether it originated in the peritubular fluid or in the tubular fluid) moves into the peritubular fluid and then into the plasma. In addition, potassium ions can move between cells from tubule lumen to the peritubular fluid and then into the plasma.

In principal cells of the late distal tubule and collecting duct (Figure 20.19b), potassium is secreted by the

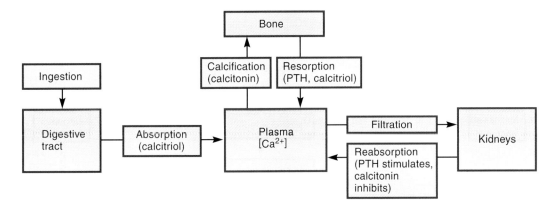

Figure 20.20 Routes of calcium exchange. Calcium can enter the plasma by absorption from the digestive tract or resorption of bone. Calcium leaves the plasma by calcification of bone or excretion in the urine. The amount of calcium excreted in the urine is regulated by varying the rate of calcium reabsorption.

following mechanism: As in the proximal tubule, potassium ions move from the peritubular fluid into the epithelial cell via the Na^+/K^+ pump in the basolateral membrane. Unlike epithelial cells in the proximal tubule, however, principal cells have their potassium channels in the apical membrane, allowing potassium ions to move out of the epithelial cell into the tubular fluid of the distal tubule and collecting duct.

Regulation of Potassium Secretion by Aldosterone

Potassium secretion is regulated by aldosterone. Recall that this hormone increases both the number of Na^+/K^+ pumps on the basolateral membrane in principal cells lining the late distal tubules and collecting ducts, and the number of potassium channels in the apical membrane. The increase in Na^+/K^+ pumps causes greater potassium movement into the epithelial cells, which is followed by greater movement of potassium ions through apical potassium channels and into the lumen of the tubules, resulting in greater excretion of potassium in the urine.

As discussed previously, aldosterone secretion is regulated by the renin-angiotensin-aldosterone system, whereby angiotensin II stimulates aldosterone release from the adrenal cortex. However, high plasma potassium levels also directly stimulate aldosterone secretion by acting on secretory cells in the adrenal cortex. The aldosterone released then increases potassium secretion, which brings plasma potassium levels toward normal.

Calcium Balance

Calcium is critical to the function of most cells: It triggers exocytosis of chemical messengers, stimulates secretion of various substances, stimulates muscle contraction, and

increases the contractility of the heart and blood vessels. Calcium is also an important component of the bone and teeth. An increase in plasma calcium, called *hypercalcemia,* has widespread effects on the body, including muscle weakness and atrophy, lethargy, behavioral changes, hypertension, constipation, and nausea. A decrease in plasma calcium, called *hypocalcemia,* causes numbness and tingling sensations, muscle cramps and spasms, exaggerated reflexes, and hypotension.

Plasma calcium concentration is regulated through the interaction of a number of organs, including the kidneys, digestive tract, bone, and skin (**Figure 20.20**). Calcium can be added to the plasma from bone and absorbed via the digestive tract, and it can be removed from the plasma by bone and the kidneys. Even though most of the calcium in the body (99%) is located in the bones, this calcium is not permanently fixed in the bone. The bone actually provides a reservoir of calcium such that when plasma calcium levels are low, the plasma can obtain calcium via a process called *resorption,* during which bone is broken down to liberate calcium ions. Conversely, when plasma calcium levels are high, calcium can be deposited into bone (see **Clinical Connections: Osteoporosis,** p. 631). The body also obtains calcium from ingested food. Although the absorption of most substances by the gastrointestinal tract is not regulated, calcium absorption is regulated according to the needs of the body.

Renal Handling of Calcium Ions

Calcium is transported in blood both bound to carrier proteins and free in the plasma. Calcium that is free in the plasma is freely filtered at the glomerulus. Normally, 99% of the filtered calcium is reabsorbed as the tubular fluid moves through the renal tubules. Approximately 70% of the filtered calcium is reabsorbed in the proximal

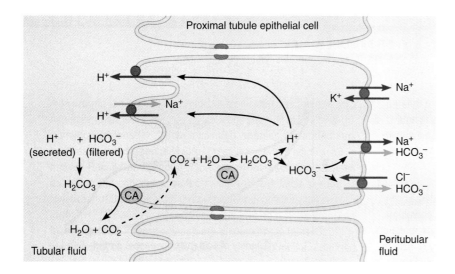

Figure 20.26 Bicarbonate reabsorption and hydrogen ion secretion in the proximal tubule. Filtered bicarbonate ions combine with secreted hydrogen ions to form carbonic acid, which is converted to water and carbon dioxide by carbonic anhydrase on the apical membrane. The carbon dioxide diffuses into the epithelial cell, where intracellular carbonic anhydrase catalyzes the conversion of carbon dioxide and water to carbonic acid, which then dissociates into bicarbonate and hydrogen ions. The hydrogen ions are secreted by countertransport with sodium ions, whereas the bicarbonate ions are reabsorbed by cotransport with sodium ions and by countertransport with chloride ions.

Renal Compensation

The third line of defense against changes in blood pH is the renal system, which takes hours or even days to compensate for changes in pH. The kidneys regulate the pH of arterial blood by regulating the renal excretion of hydrogen ions and bicarbonate, and by producing new bicarbonate, according to the following rules: *If the hydrogen ion concentration in the blood increases, the kidneys increase hydrogen ion secretion and bicarbonate reabsorption and synthesize new bicarbonate; if hydrogen ion concentration in the blood decreases, the kidneys decrease hydrogen ion secretion and bicarbonate reabsorption.* The secretion of hydrogen ions is coupled to the reabsorption or synthesis of bicarbonate ions, as described shortly.

Several substances critical to renal compensation for acid-base disturbances are filtered at the glomerulus, including CO_2, H^+, HCO_3^-, and $H_2PO_4^-$. The fates of these substances vary in the different segments of the renal tubules.

Renal Handling of Hydrogen and Bicarbonate Ions in the Proximal Tubule

In the proximal tubule, bicarbonate reabsorption is coupled to hydrogen ion secretion (**Figure 20.26**). In epithelial cells of the proximal tubule, several carrier proteins required for movement of hydrogen or bicarbonate ions are located on either the basolateral or apical membrane. Which transporters are active depends on the pH of the extracellular fluid.

The basolateral membrane contains three transporters: (1) Na^+/K^+ pumps that transport sodium ions out of the cell and into the peritubular fluid while transporting potassium ions into the cell, (2) Na^+/HCO_3^- cotransporters that transport both sodium and bicarbonate ions out of the cell and into the peritubular fluid, and (3) HCO_3^-/Cl^- countertransporters that transport chloride ions into the cell and bicarbonate ions into the peritubular fluid.

The apical membrane contains two transporters: (1) Na^+/H^+ countertransporters that transport sodium ions into the cell and hydrogen ions out of the cell and into

the tubular fluid, and (2) H^+ pumps that use ATP to transport hydrogen ions into the tubular fluid.

The enzyme carbonic anhydrase (CA), which is located in the cytosol and on microvilli of the apical membrane of the epithelial cell, catalyzes the following reversible reaction:

$$CO_2 + H_2O \rightleftharpoons H_2CO_3$$

The membrane-bound carbonic anhydrase converts carbonic acid (which comes from filtered bicarbonate ions, as described shortly) to carbon dioxide in the lumen of the proximal tubule. The carbon dioxide then diffuses into the epithelial cell, where it is converted back to carbonic acid by carbonic anhydrase. The carbonic acid then dissociates by the following reversible reaction:

$$H_2CO_3 \rightleftharpoons H^+ + HCO_3^-$$

The hydrogen ion formed inside the epithelial cell by this reaction is secreted into the lumen of the tubules by either countertransport with sodium ions or active transport by the H^+ pumps. The intracellular concentration of sodium is kept low by the Na^+/K^+ pumps on the basolateral membrane. In the lumen of the tubule, hydrogen ions combine with filtered bicarbonate to form carbonic acid. The carbonic anhydrase located on microvilli catalyzes the conversion of carbonic acid to carbon dioxide and water. The carbon dioxide can then diffuse into the epithelial cell, as previously described.

The bicarbonate ion formed inside the epithelial cell by the carbonic anhydrase–catalyzed reaction moves from the epithelial cell into the peritubular fluid by either cotransport with sodium or countertransport with chloride. The Na^+/HCO_3^- cotransporter functions in reabsorption of both sodium and bicarbonate. The net effect for the bicarbonate ion is reabsorption, because bicarbonate is moved from the lumen of the tubules into the peritubular fluid, as follows: A bicarbonate ion in the lumen is converted to a carbon dioxide molecule that moves from the lumen into the epithelial cell, where it is converted back into a bicarbonate ion that moves into the peritubular fluid, where it can diffuse into the blood.

Overall, these actions in the proximal tubule produce three primary effects: (1) Under normal conditions, approximately 80–90% of the filtered bicarbonate is reabsorbed, (2) hydrogen ions arc sccrctcd, and (3) sodium is reabsorbed.

Renal Handling of Hydrogen and Bicarbonate Ions in the Late Distal Tubule and Collecting Duct

In certain epithelial cells of the late distal tubule and collecting duct, called intercalated cells, secretion of hydrogen ions is coupled to the synthesis of new bicarbonate ions (**Figure 20.27**). The intercalated cells lining the distal tubules and collecting ducts have different membrane proteins than the epithelial cells lining the proximal tubules.

The basolateral membrane contains (1) HCO_3^-/Cl^- countertransporters that move bicarbonate out of the cell and into the peritubular fluid while moving chloride ions into the cell, and (2) chloride channels that allow the chloride to diffuse back into the peritubular fluid. The apical membrane contains (1) H^+ pumps that utilize ATP to transport hydrogen ions out of the cell and into the tubular fluid, and (2) K^+/H^+ countertransporters that transport potassium ions into the cell and hydrogen ions into the tubular fluid.

The enzyme carbonic anhydrase is located in the cytosol of the intercalated cell. Carbon dioxide levels in the epithelial cell increase from either cellular metabolism or diffusion from the peritubular fluid into the cell. Inside the cell, carbonic anhydrase converts carbon dioxide and water to carbonic acid, which then dissociates into H^+ and HCO_3^-, as follows:

$$CO_2 + H_2O \rightleftharpoons H_2CO_3 \rightleftharpoons H^+ + HCO_3^-$$

The reaction is driven to the right by the removal of hydrogen ions and bicarbonate ions from the epithelial cell. Hydrogen ions are removed through transport by the H^+ pump or in exchange for K^+ by the countertransporter, resulting in secretion of hydrogen ions. Bicarbonate ions are removed by countertransport with chloride ions across the basolateral membrane into the peritubular fluid, and then into the plasma. However, this bicarbonate is not being reabsorbed; because it was never in the lumen of the renal tubules but instead was produced by the epithelial cell, this bicarbonate is considered *new* bicarbonate.

The secreted hydrogen ions decrease the pH of the tubular fluid. However, the pH of the tubular fluid, and therefore urine, is limited to a minimum of 4.5, at which point hydrogen ion secretion stops. To minimize decreases in urine pH, secreted hydrogen ions are buffered. Recall that hydrogen ions secreted in the proximal tubule were buffered by filtered bicarbonate ions. In the distal tubule and collecting ducts, however, very little bicarbonate

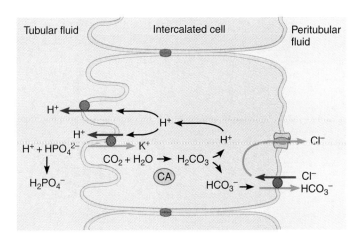

Figure 20.27 Bicarbonate synthesis and hydrogen ion secretion by intercalated cells of the distal tubule and collecting duct. Carbonic anhydrase in the cytosol converts carbon dioxide (which is either metabolically produced or has diffused into the cell from the peritubular fluid) into carbonic acid, which dissociates into hydrogen ions and bicarbonate ions. The hydrogen ions are secreted by a H^+ pump or a K^+/H^+ countertransporter; the bicarbonate ions are transported into the peritubular fluid by a HCO_3^-/Cl^- countertransporter.

remains in the lumen of the tubules, as most of it was reabsorbed. Recall as well that phosphate ions are freely filtered by the glomerulus. In the lumen of the distal tubules and collecting ducts, hydrogen ions are buffered by phosphates according to the following equation:

$$HPO_4^{2-} + H^+ \rightleftharpoons H_2PO_4^-$$

Overall, these actions in the distal tubules and collecting ducts produce two primary effects: Newly formed bicarbonate ions are added to the plasma, and hydrogen ions are secreted into the tubular fluid.

Role of Glutamine in Renal Compensation During Severe Acidosis

The mechanisms just described generally compensate for increases in hydrogen ion concentration produced by normal daily activities. However, these mechanisms are insufficient to compensate for large increases in the plasma hydrogen ion concentration. Under conditions of severe acidosis, a third renal mechanism contributes to compensation (**Figure 20.28**).

In the proximal convoluted tubule, glutamine is transported from both the tubular fluid and the peritubular fluid into the epithelial cells. Catabolism of glutamine in the epithelial cells generates bicarbonate ions and ammonia (NH_3), as follows:

$$glutamine \rightarrow HCO_3^- + NH_3$$

The bicarbonate moves into the peritubular fluid by either cotransport with sodium or countertransport with a chloride ion. This bicarbonate is not being reabsorbed, however; because it was never in the tubular fluid, *new*

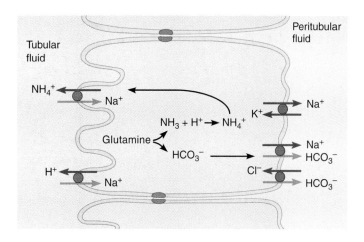

Figure 20.28 Bicarbonate production and hydrogen secretion by glutamine metabolism in the proximal tubule. Glutamine is catabolized to bicarbonate ions and ammonia. The bicarbonate ions are transported into the peritubular fluid by a Na^+/HCO_3^- cotransporter or a HCO_3^-/Cl^- countertransporter. The ammonia binds a hydrogen ion to form ammonium, which is secreted by countertransport with sodium ions.

bicarbonate is being added to the blood. The ammonia is converted to ammonium (NH_4^+) by the following reaction:

$$NH_3 + H^+ \rightarrow NH_4^+$$

This ammonium is transported into the tubular fluid by countertransport with sodium ions and is eventually excreted.

The overall effect of these actions is that a new bicarbonate ion is added to the blood, and a hydrogen ion is secreted in the form of ammonium.

Compensation for Acid-Base Disturbances

Recall from Chapter 18 that the Henderson-Hasselbalch equation describes the relationship of plasma pH to the ratio of bicarbonate and carbon dioxide levels in blood:

$$pH = 6.1 + \log[HCO_3^-]/[CO_2]$$

Given that blood pH must be maintained at 7.4, this equation can be solved for the ratio of bicarbonate to carbon dioxide, as follows:

$$7.4 = 6.1 + \log[HCO_3^-]/[CO_2]$$

$$1.3 = \log[HCO_3^-]/[CO_2]$$

$$[HCO_3^-]/[CO_2] = 20$$

Therefore, for plasma pH to be normal, the ratio of bicarbonate to carbon dioxide must be 20:1. The respiratory and renal systems work together to control this ratio; the respiratory system controls carbon dioxide levels, and the kidneys regulate bicarbonate levels.

In acidosis, the ratio of bicarbonate to carbon dioxide decreases to less than 20:1, either because of a decrease in bicarbonate or an increase in carbon dioxide. In alkalosis, the ratio of bicarbonate to carbon dioxide is greater than 20:1, either because of an increase in bicarbonate or a decrease in carbon dioxide.

We turn now to a description of the four types of acid-base disturbances, and how the body compensates for them.

Respiratory Acidosis

Respiratory acidosis is caused by hypoventilation, in which ventilation is less than that needed by the body. Carbon dioxide increases in the plasma, decreasing the pH:

$$CO_2 + H_2O \underset{\longleftarrow}{\longrightarrow} \overset{\text{add}}{\uparrow} H_2CO_3 \underset{\longleftarrow}{\longrightarrow} \uparrow H^+ + \uparrow HCO_3^-$$

Hypoventilation can be caused by lung diseases, depression of the respiratory center in the brainstem, or diseases that affect respiratory muscles. As a result of hypoventilation, arterial P_{CO_2} increases, decreasing the ratio of bicarbonate to carbon dioxide. To bring the ratio (and therefore pH) back to normal, the kidneys compensate by increasing the secretion of hydrogen ions and the reabsorption of bicarbonate ions. The lungs cannot compensate because that is where the problem developed initially (unless the hypoventilation was voluntary and not pathological, in which case the lungs can compensate by returning ventilation to normal).

Respiratory Alkalosis

Respiratory alkalosis is caused by hyperventilation, in which ventilation is greater than that needed by the body. Causes of hyperventilaton include fever and anxiety. As a result of the hyperventilation, arterial P_{CO_2} decreases, increasing the ratio of bicarbonate to carbon dioxide:

$$CO_2 + H_2O \underset{\longrightarrow}{\overset{\text{remove}}{\longleftarrow}} \downarrow H_2CO_3 \underset{\longrightarrow}{\longleftarrow} \downarrow H^+ + \downarrow HCO_3^-$$

To bring the ratio back to normal, the kidneys compensate by decreasing the reabsorption of bicarbonate ions and by secreting fewer hydrogen ions. The lungs cannot compensate because that is where the problem developed initially (unless the hyperventilation was voluntary).

Metabolic Acidosis

Metabolic acidosis is caused by an increase in acids in the plasma from sources other than carbon dioxide:

$$CO_2 + H_2O \underset{\longrightarrow}{\longleftarrow} H_2CO_3 \underset{\longrightarrow}{\longleftarrow} \overset{\text{add}}{\uparrow} H^+ + \downarrow HCO_3^-$$

Causes of metabolic acidosis include diarrhea, which results in loss of bicarbonate through elimination of the intestinal contents; diabetes mellitus, which by increasing fat metabolism causes a buildup of keto acids; strenuous exercise, which increases lactic acid production; and renal

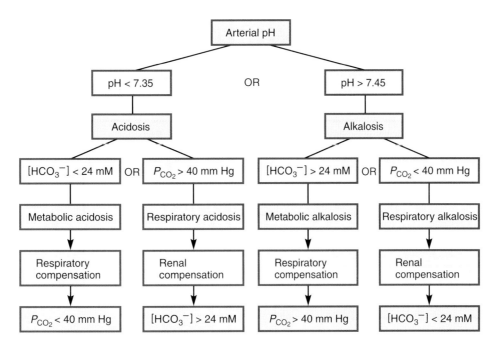

Figure 20.29 Summary of acid-base disturbances and compensation.

failure. Compensation for metabolic acidosis includes an increase in ventilation and, if the kidneys are not part of the initial problem, an increase in production of new bicarbonate by the kidneys and an increase in excretion of H^+.

In respiratory compensation, an increase in hydrogen ion levels in the plasma activates peripheral chemoreceptors, which reflexively increase ventilation. The increase in ventilation decreases arterial P_{CO_2}, which drives the conversion of bicarbonate and hydrogen ions to more carbon dioxide, decreasing the free hydrogen ion levels in blood. However, the lungs cannot completely compensate because the increased ventilation decreases arterial P_{CO_2}, which decreases the stimulatory influence of carbon dioxide on central chemoreceptors.

In renal compensation, the production of new bicarbonate is essential to replace bicarbonate that was lost via two processes: (1) when bicarbonate was used to buffer excess acid, and (2) when plasma bicarbonate levels were decreased during respiratory compensation as described above.

Metabolic Alkalosis

Metabolic alkalosis is caused by a decrease in acids in the plasma from sources other than carbon dioxide:

$$CO_2 + H_2O \xrightleftharpoons{} H_2CO_3 \xrightleftharpoons{\overset{\overset{\text{remove}}{\Uparrow}}{}} \downarrow H^+ + \uparrow HCO_3^-$$

Causes of metabolic alkalosis include vomiting, which results in a loss of acidic gastric contents, and ingestion of alkaline drugs such as sodium bicarbonate (baking soda) or antacids. Compensation for metabolic alkalosis involves both the lungs and kidneys.

In respiratory compensation, decreases in hydrogen ion levels in the plasma remove a stimulatory effect on peripheral chemoreceptors, which reflexively decreases ventilation. The decrease in ventilation increases arterial P_{CO_2}, which combines with water to produce bicarbonate and hydrogen ions, increasing free hydrogen ion levels in blood. As in metabolic acidosis, the lungs cannot completely compensate because the decreased ventilation increases arterial P_{CO_2}, which activates the central chemoreceptors and increases ventilation.

In renal compensation, the kidneys excrete more bicarbonate ions and fewer hydrogen ions. Ridding the blood of bicarbonate shifts the equilibrium to the right, causing more carbon dioxide to react with water to form carbonic acid, which then dissociates and increases the plasma concentration of hydrogen ions.

Evaluation of Acid-Base Disturbances

Diagnosis of acid-base disturbances involves measuring plasma levels of hydrogen ion concentration (pH), carbon dioxide (P_{CO_2}), and bicarbonate. The different acid-base disturbances can be diagnosed as follows (**Figure 20.29**):

■ A decrease in pH coupled with a decrease in plasma bicarbonate levels indicates that a metabolic acidosis is occurring. Because the lungs compensate for a metabolic acidosis by increasing ventilation, P_{CO_2} levels decrease.

■ A decrease in pH coupled with an increase in P_{CO_2} indicates that a respiratory acidosis is occurring. Because the kidneys compensate for a respiratory acidosis by increasing bicarbonate reabsorption, plasma bicarbonate levels increase.

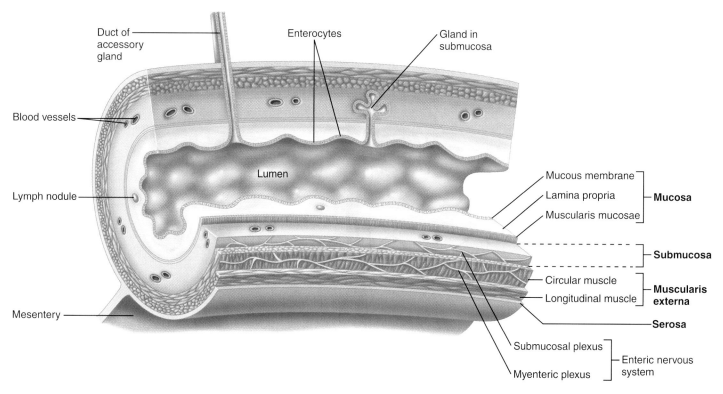

Figure 21.3 Generalized structure of the gastrointestinal wall, depicting the four major layers and the tissues within them.

barrier separating the lumen from the body's internal environment. Some enterocytes are classified as *absorptive cells* because they are specialized for *absorption,* the transport of nutrients and other materials from the lumen to the bloodstream. Other enterocytes are classified as *exocrine cells* because they secrete materials such as fluids and enzymes into the lumen (which is outside the body, as explained in Chapter 1). Among these cells are *goblet cells,* which secrete **mucus** (a sticky, viscous fluid containing glycoproteins called *mucins*) throughout the length of the GI tract; mucus forms a coating that protects the lining against abrasion and substances in the lumen that may attack tissue. Still other enterocytes are *endocrine cells,* which secrete hormones into the bloodstream. As we will see, these hormones play an important role in the regulation of digestive function.

The **lamina propria** is a layer of connective tissue underlying the mucous membrane. Contained within this layer are small blood vessels, nerves, and lymphatic vessels that communicate with larger nerves and vessels in still deeper tissue layers. The lamina propria also contains lymphoid tissue, including *lymph nodules* and *Peyer's patches,* that is important in defending the body against bacteria, which are plentiful in the lumen of the intestines.

The **muscularis mucosae** is a thin layer of smooth muscle that serves primarily to contract the mucosa into folds, which stirs the lumenal contents and promotes contact with the mucosal surface. Within this layer are

longitudinal muscle fibers, which run parallel to the tract's long axis, and *circular* muscle fibers, which run around the tract's circumference.

The Submucosa The submucosa is a thick layer of connective tissue that provides the GI tract with much of its distensibility and elasticity, enabling it to tolerate a large degree of stretch without sustaining damage. This layer also contains many of the tract's larger blood and lymphatic vessels. At its outer border is a network of nerve cells known as the **submucosal plexus** *(Meissner's plexus),* which communicates with another nerve cell network in the muscularis externa called the **myenteric plexus** *(Auerbach's plexus)* (see Figure 21.3); together these nerve plexuses make up what is called the **enteric nervous system** (or *intrinsic nervous system*).

The enteric nervous system is an elaborate network of sensory neurons, motor neurons, and interneurons located within the wall of the GI tract. It is capable of regulating many GI functions independent of external influences. Input to the enteric nervous system comes from receptors located in the GI tract and autonomic neurons. Output from the enteric nervous system goes to effector cells (smooth muscle, exocrine glands, and endocrine glands) located within the GI tract and plays an important role in the control of digestive function.

The Muscularis Externa The muscularis externa is largely responsible for the motility of the GI tract and contains two separate layers of smooth muscle: an inner layer of circular muscle and an outer layer of longitudinal muscle. The circular muscle layer consists of single-unit smooth muscle capable of generating spontaneous depolarizations, called *slow-wave potentials*. The circular muscle makes up the bulk of the muscularis externa. Contractions of the circular muscle layer decrease the diameter of the lumen of the GI tract. The longitudinal muscle layer consists of multi-unit smooth muscle that depends on neural input for contraction. Contraction of the longitudinal muscle layer shortens the GI tract. Coordinated contractions of the circular and longitudinal muscles propels the lumenal contents through the GI tract and mixes the contents with secretions that help digest food particles. Motility also promotes contact of digestive end-products with the mucosal epithelium, thereby increasing the absorption efficiency.

The Serosa   The serosa, the outermost layer of the gastrointestinal wall, consists of an inner layer of fibrous connective tissue, which provides structural support, and an outer layer of epithelial tissue called the **mesothelium**, which secretes a watery lubricating fluid that makes it easier for organs to slide past one another. The mesothelium (along with a layer of underlying connective tissue) is continuous with the **mesenteries**, a system of clear, thin membranes that interconnects most of the abdominal organs and houses nerves and blood vessels running to them. The mesenteries help to anchor the organs in place and are continuous with the **peritoneum**, a membrane lining the inside of the abdominal cavity.

Quick Test 21.1

1. Name the four major layers that make up the wall of the GI tract.

2. Name the three major cell types in the mucous membrane, and briefly describe the function of each.

3. Name the two major divisions of the enteric nervous system, and give their locations.

4. Define the following terms: *digestion, absorption, secretion, motility, GI tract, accessory glands, longitudinal muscle, circular muscle, mesenteries, peritoneum,* and *enterocyte.*

Functional Anatomy of Gastrointestinal Tract Organs

Now we turn our attention to the individual organs of the GI tract, progressing in order from the upper end of the tract to the lower end. We begin with the mouth, pharynx, and esophagus.

The Mouth, Pharynx, and Esophagus   The *mouth* or *oral cavity*, the beginning of the GI tract, is where food enters and where the processes of mechanical breakdown and

digestion begin (**Figure 21.4**). In the mouth, food is chewed (a process called *mastication*) to decrease the size of food particles and to mix them with a secretion called *saliva.* Saliva lubricates the food and contains an enzyme called *salivary amylase,* which begins the digestion of carbohydrates by breaking down starch and glycogen.

From the mouth, the food-saliva mixture is propelled by the tongue into the pharynx (commonly known as the *throat*), a common passageway for food and air. From the pharynx, the passageways for food and air diverge. Whereas air enters the larynx and trachea via the glottis and proceeds toward the lungs, food enters the esophagus, which runs parallel and dorsal to the trachea.

The **esophagus** is a muscular tube whose primary function is to conduct food from the pharynx to the stomach. Unlike the trachea, it is thin-walled and pliant, so that it can easily stretch to accommodate food as it is swallowed; when food is not present, however, it is normally collapsed. The esophagus is unusual among the organs of the GI tract in that its wall contains both skeletal muscle (in the upper third of its length) and smooth muscle.

The movement of food from the pharynx to the esophagus is regulated by the **upper esophageal sphincter** (see Figure 21.4), a ring of skeletal muscle surrounding the esophagus at its upper end. (A **sphincter** is generally defined as a ring of muscle that surrounds an orifice and regulates the passage of material through it by altering its diameter.) At the esophagus's lower end is the **lower esophageal sphincter**, a ring of smooth muscle that regulates the flow of food from the esophagus to the stomach. Both of these sphincters are normally closed, and open only when food is being swallowed. The lower esophageal sphincter prevents the contents of the stomach, which are acidic, from entering the esophagus. However, backflow of stomach contents into the esophagus (*gastric reflux*) can on occasion occur and produce *heartburn,* a burning sensation in the chest caused by irritation of the esophageal lining.

The Stomach   An important function of the **stomach**, a J-shaped sac located beneath the diaphragm, is to store food after it is swallowed and to release it into the small intestine. The lining of the stomach contains glands (called **gastric glands**) that secrete a watery fluid called **gastric juice** into the lumen. Contractile activity of smooth muscle in the stomach's wall pulverizes food into smaller particles and mixes it with gastric juice, forming a mixture called **chyme**.

The stomach has three major anatomical regions (see Figure 21.4): a domed upper portion called the *fundus,* which extends above the lower esophageal sphincter; a middle region called the *body,* which accounts for the bulk of the stomach's volume; and a lower region called the *antrum,* which is narrower and smaller in volume. Contractions of the antrum propel chyme from the stomach into the small intestine, a process called *gastric*

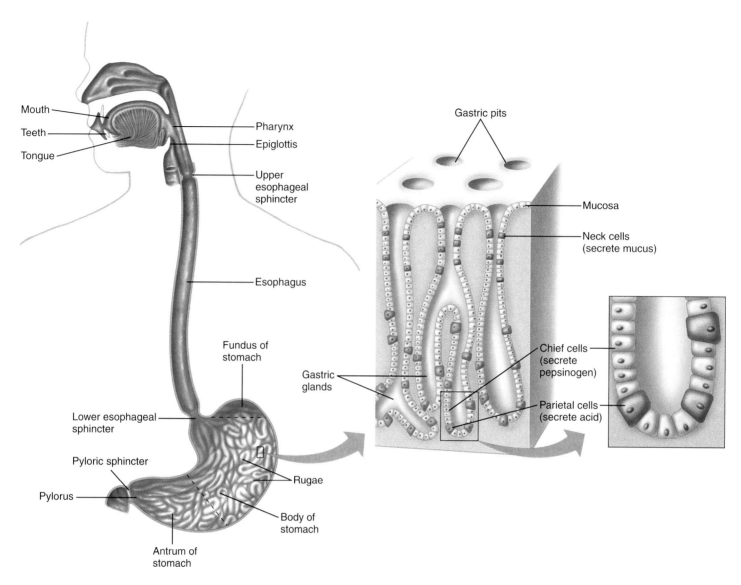

Figure 21.4 Anatomy of the mouth, pharynx, esophagus, and stomach. Enlarged views show gastric pits and gastric glands.

emptying. As chyme exits the stomach, it passes through a narrow passage called the **pylorus** on its way to the small intestine. The flow of chyme through the pylorus is regulated by a surrounding ring of smooth muscle called the **pyloric sphincter**, which opens and closes with each cycle of stomach contraction, such that chyme exits the stomach in spurts.

The stomach wall differs anatomically and functionally in the different regions. In the fundus, the wall is thin and easily expands to accommodate increases in volume following a meal. In the body of the stomach, the gastric mucosa is thrown into longitudinal folds called rugae, which flatten as the stomach expands. These two anatomical specializations allow the stomach to expand approximately 20 times from its empty volume (50 mL) to its full volume (1000 mL). The antrum has the thickest muscle layer and produces strong contractions responsible for gastric mixing and emptying. Both the fundus and

body contain **gastric pits**, which contain cells that secrete the products found in gastric juice.

Gastric pits contain a variety of secretory cells, including both exocrine and endocrine cells. In the neck, or upper region, of the pits, **neck cells** secrete mucus. Deeper in the pits are gastric glands, which contain the following cells: (1) **chief cells**, which secrete **pepsinogen**, the precursor for a proteolytic enzyme called **pepsin**; (2) **parietal cells**, which secrete both hydrogen ions (as HCl) to acidify the stomach contents and **intrinsic factor**, which is necessary for the absorption of vitamin B_{12}; and (3) **G cells**, which secrete the hormone **gastrin**.

The lumen of the stomach is the only locale in the GI tract where the contents are acidic. In fact, the pH of stomach contents can go as low as 2, which is equivalent to a 10 mM solution of hydrochloric acid! This acidity is necessary for converting pepsinogen into its active form, pepsin; it is also useful because it helps denature proteins

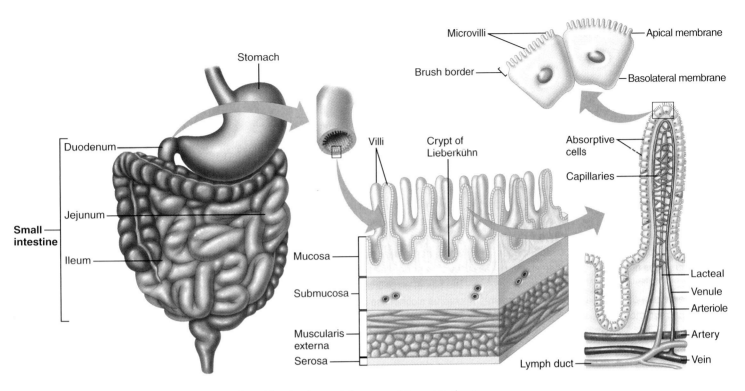

Figure 21.5 Anatomy of the small intestine. The duodenum, jejunum, and ileum are shown in relation to other GI tract organs; the enlarged views show the structure of villi and microvilli in the duodenum.

in the food and kills many foodborne bacteria and thus protects against certain illnesses.

The stomach's lining is protected against the potential harmful effects of acid and pepsin by a surface covering of mucus and bicarbonate, called the **gastric mucosal barrier**. Neck cells in the gastric pits secrete mucus, whereas surface epithelial cells secrete both mucus and bicarbonate. If the gastric mucosal barrier is penetrated by acids, such as salicylic acid (aspirin), an ulcer can result (**Clinical Connections: Ulcers**, p. 654).

The Small Intestine From the stomach, chyme travels to the **small intestine**, a coiled tube about 2.5–3 meters long that is the primary site for the digestion of all nutrients in food. The small intestine is also where most of the ingested nutrients, water, vitamins, and minerals (inorganic ions such as sodium, potassium, and calcium) are absorbed.

On the basis of subtle anatomical distinctions, the small intestine is divided into three major regions (**Figure 21.5**): an initial portion called the **duodenum**, which begins at the pylorus and extends for approximately 30 cm; a middle portion called the **jejunum**, which extends for about another 1 meter; and a terminal portion called the **ileum**, which extends approximately 1.5 meters and joins the colon.

In the duodenum, chyme is mixed with a watery secretion from the pancreas called **pancreatic juice**, which contains a wide variety of digestive enzymes and is rich in

bicarbonate. Bicarbonate neutralizes the acid in the chyme when it exits the stomach, which is necessary because the enzymes in pancreatic juice function at the normal pH of the small intestine (which is slightly basic) but not at an acidic pH. In addition to pancreatic juice, the duodenum also receives **bile**, a fluid secreted by the liver that contains bicarbonate and *bile salts*, which aid in the digestion of fats, as described later.

As nutrients in the chyme are broken down by enzymes, digestive end products are released into solution and absorbed by cells in the mucosal epithelium. These simultaneous processes of digestion and absorption begin in the duodenum and continue to completion in the remainder of the small intestine. Unless an unusually large quantity of food has been ingested, absorption is typically completed within the first 20% of the small intestine's length, or before the chyme has reached the ileum. Thus the small intestine has a large excess capacity for absorbing nutrients, indicating that its absorptive mechanisms are highly efficient.

The small intestine's absorptive efficiency is attributable in part to the fact that the mucosal surface is highly folded (see Figure 21.5). These folds, called **villi** (singular: *villus*), increase the surface area by roughly a factor of 10, compared to what the surface area would be if the small intestine were simply a smooth cylinder on the inside. Each villus houses structures crucial to the absorption of nutrients, including a network of *capillaries* and a blind-ended lymphatic vessel called a **lacteal**. After

CHAPTER SUMMARY

Overview of Gastrointestinal System Function, p. 647

The primary purpose of the gastrointestinal system is to extract needed materials from ingested food and deliver them to the bloodstream for distribution to cells throughout the body. Because most nutrient molecules are too large to be transported in the bloodstream, they must be broken down to smaller molecules by enzymes in the lumen of the digestive tract (digestion). The resultant digestion products are transported into the bloodstream (absorption). To aid in these processes, fluids and enzymes are transported into the lumen of the tract (secretion), and muscular activity in the wall (motility) propels the lumenal contents from one digestive organ to the next.

Functional Anatomy of the Gastrointestinal System, p. 647

The gastrointestinal system comprises the gastrointestinal (GI) tract, or digestive tract, and accessory glands. Throughout most of its length, four layers make up the wall of the tract: (1) the mucosa, which lines the lumen; (2) the submucosa, an underlying layer of connective tissue containing numerous nerves and blood vessels; (3) the muscularis externa, which contains circular and longitudinal smooth muscle; and (4) the serosa, composed of connective tissue and an outermost mesothelium. The mucosa contains the mucous membrane, an epithelial layer containing secretory, absorptive, and endocrine cells. Within the wall is the enteric nervous system, consisting of two divisions: the submucosal plexus and the myenteric plexus. Organs of the GI tract include the mouth, esophagus (which conducts food to the stomach), stomach (which holds the food and mixes it with secretions to form chyme), small intestine (the primary site of digestion and absorption),

colon (which absorbs water and electrolytes and stores feces), rectum, and anus (a passage leading to the outside). The flow of material between organs is regulated by sphincters. Accessory glands include the salivary glands (which secrete saliva), the pancreas (which secretes pancreatic juice containing enzymes and bicarbonate), and the liver (which secretes bile and processes absorbed nutrients).

Digestion and Absorption of Nutrients and Water, p. 659

Digestion of starch and glycogen begins in the mouth with the action of salivary amylase and is continued in the small intestine by pancreatic amylase. Brush border enzymes in the small intestine reduce carbohydrates to monosaccharides, which are transported across the mucosal epithelium and diffuse into the bloodstream. Protein digestion begins with the action of pepsin in the stomach and is continued in the small intestine by pancreatic enzymes (including trypsin, chymotrypsin, and carboxypeptidase, secreted as inactive zymogens that are activated upon entry into the GI tract) and membrane-bound enzymes (including aminopeptidase). Most proteins are reduced completely to amino acids, which are transported into the bloodstream. Dietary fats (mostly triglycerides) are reduced to fatty acids and monoglycerides by pancreatic lipases. This process is aided by bile salts, which emulsify fat droplets. Products of fat digestion enter epithelial cells by simple diffusion and then are reassembled into triglycerides, which are transported (along with other lipids) into the lymphatic system in the form of chylomicrons, a type of lipoprotein. Water absorption is secondary to solute absorption and is driven by an osmotic gradient. Vitamins and minerals are absorbed chemically unaltered.

General Principles of Gastrointestinal Regulation, p. 667

Gastrointestinal regulatory mechanisms maximize the efficiency of digestion and absorption but generally do not act to maintain homeostasis. Gastrointestinal function is regulated by short reflex and long reflex pathways involving the enteric nervous system, autonomic nervous system and hormones (including gastrin, secretin, cholecystokinin, and glucose-dependent insulinotropic peptide). The enteric nervous system receives input both from the autonomic nervous system and from mechanoreceptors, chemoreceptors, and osmoreceptors that monitor conditions in the digestive tract.

Gastrointestinal Secretion and Its Regulation, p. 671

Saliva secretion is controlled by autonomic input to the salivary glands and is coordinated by the medullary salivary center. Gastric secretion of acid and pepsinogen is influenced by cephalic-phase, gastric-phase, and intestinal-phase stimuli and controlled by neural and hormonal reflexes. Pancreatic secretion is also controlled by neural and hormonal signals (primarily secretin and cholecystokinin). Bile secretion by the liver is stimulated by secretin and by cholecystokinin, which also stimulates contraction of the gallbladder.

Gastrointestinal Motility and Its Regulation, p. 675

Gastrointestinal smooth muscle contractions are triggered by slow waves generated by pacemaker cells. Nerves and hormones generally influence the strength of contractions, but not the frequency. The stomach and intestines exhibit motility patterns that change depending on conditions in the lumen.

▮▮ *E X E R C I S E S*

Multiple-Choice Questions

1. What do sodium, fatty acids, and vitamin A have in common?
   a) They are not enzymatically modified prior to absorption into the bloodstream.
   b) They cross the apical membranes of enterocytes by simple diffusion.
   c) They are transported into blood capillaries in the villi.
   d) They are all hydrophilic.
   e) all of the above

2. Blockage of the flow of bile into the duodenum interferes with the digestion of which of the following?
   a) carbohydrates only
   b) lipids only
   c) proteins only
   d) carbohydrates and lipids only
   e) carbohydrates and proteins only

3. Which of the following is an accurate statement regarding the various phases of gastrointestinal control?
   a) The autonomic nervous system is involved in cephalic-phase regulation only.
   b) The autonomic nervous system is involved in gastric-phase regulation only.
   c) The autonomic nervous system is involved in intestinal-phase regulation only.
   d) The autonomic nervous system is involved in gastric-phase and intestinal-phase regulation only.
   e) The autonomic nervous system is involved in all three phases of regulation.

4. Failure of the salivary glands to secrete amylase would make it impossible to digest which of the following?
   a) proteins
   b) fats
   c) disaccharides
   d) starch
   e) none of the above

5. Which of the following tends to inhibit acid secretion by the stomach?
   a) an increase in the osmolarity of duodenal contents
   b) entry of stomach acid into the duodenum
   c) the arrival of food in the stomach
   d) a and b are both true
   e) all of the above

6. Which of the following best illustrates the phenomenon of potentiation?
   a) Bile secretion is normally stimulated by both secretin and CCK.
   b) Secretin and CCK act primarily on different parts of the hepatic secretory apparatus.
   c) Secretin and CCK both stimulate bile duct cells to secrete fluid.
   d) The secretion of fluid by bile duct cells is greater when both CCK and secretin are present, compared to when either hormone is present alone (at the same concentration).

7. The enzyme enterokinase is directly or indirectly responsible for the proper functioning of
   a) bile salts.
   b) lipases.
   c) trypsin.
   d) chymotrypsin.
   e) both trypsin and chymotrypsin.

8. Which of the following is an example of a zymogen?
   a) enterokinase
   b) chymotrypsinogen
   c) salivary amylase
   d) cholecystokinin
   e) all of the above

9. Increases in gastric motility are generally accompanied by increases in ileal motility because of the
   a) gastroileal reflex.
   b) ileogastric reflex.
   c) gastrocolic reflex.
   d) colonocolonic reflex.
   e) intestino-intestinal reflex.

10. Which of the following is a digestion product of amylase?
    a) maltose
    b) glucose
    c) sucrose
    d) lactose
    e) fructose

Objective Questions

1. Through the action of pancreatic amylase alone, starch could be broken down to a form that could be absorbed completely. (true/false)

2. The glands that secrete acid in the stomach are examples of accessory glands. (true/false)

3. Receptive relaxation is an example of (cephalic-phase/gastric-phase) control of stomach function.

4. The (submucosal/myenteric) nerve plexus is located within the muscularis externa.

5. The _____ is the outermost layer of the gastrointestinal wall, consisting of the mesothelium and an underlying layer of connective tissue.

6. When food is present in the small intestine, contraction of the gallbladder is stimulated by (secretin/cholecystokinin).

7. Bile salts are necessary for the proper functioning of pancreatic (proteases/lipases).

8. The lamina propria is located within the (mucosa/submucosa).

9. Gastrin (stimulates/inhibits) the secretion of acid by the stomach.

10. In the stomach, pepsinogen is secreted by (chief cells/parietal cells).

11. The small intestine is periodically swept of its contents by (segmentation/migrating motility complexes).

12. The term *enterohepatic circulation* refers to the conduction of blood from the intestine to the liver via the hepatic portal vein. (true/false)

13. Nerves and hormones exert their effects on gastrointestinal motility primarily by altering the (frequency/amplitude) of slow waves.

14. Chewing is under voluntary control but is also controlled by reflex neural pathways. (true/false)

15. Disaccharides are broken down to monosaccharides by (pancreatic enzymes/brush border enzymes).

Essay Questions

1. Describe the process by which ingested triglycerides are digested, absorbed, and transported into the bloodstream. Indicate which of the steps are adaptations to triglycerides' hydrophobic nature, and explain why. Compare these steps to the corresponding processes pertaining to the digestion and absorption of hydrophilic substances such as proteins or carbohydrates.

2. When chyme moves from the stomach to the small intestine, a number of stomach functions are altered. Identify these functions, describe the changes

that occur, and explain the regulatory mechanisms that bring about these changes. Explain why these changes make sense in light of the gastrointestinal system's ultimate function.

3. Describe the structural adaptations of the small intestine that increase its capacity for absorbing nutrients.

4. Describe the processes of segmentation, peristalsis, haustration, and mass movement, and explain their roles in digestive function. What is the apparent function of the muscularis mucosae?

5. Describe the various mechanisms that regulate the secretion of acid and pepsinogen by the stomach. Compare these mechanisms to those that regulate gastric motility.

Critical Thinking

1. Pantoprazole (Protonix) is a fairly new drug that blocks the active transport of hydrogen ions in the stomach. Explain the consequences of taking this drug and what disorder(s) it might be useful to treat. What would be some of the negative side effects?

2. In 1996, the FDA approved a new fat substitute called olestra. Olestra consists of a sucrose backbone with six fatty acids attached. Olestra is neither digested nor absorbed in the GI tract. Explain why olestra can cause diarrhea and deficiencies of fat-soluble vitamins.

3. A normal flora of bacteria resides in the GI tract in a symbiosis with the human host. These bacteria can digest cellulose and other products. In severe cases of diarrhea, many of these bacteria are lost. Physicians often prescribe a diet that includes cottage cheese or yogurt. What benefit does this diet give the patient?

4. The body composition (fat mass versus fat-free mass) of Americans includes a greater proportion of fat than ever before, and the trend is toward continued increases in fat. The medical community is concerned with this trend because obesity predisposes people to many diseases (hypertension and diabetes, for example). New diets come out on a regular basis. Describe the physiological basis behind some of these new diets.

Find the answers to these exercises, and additional study tools, at the Physiology Place (www.physiologyplace.com).

The Reproductive System

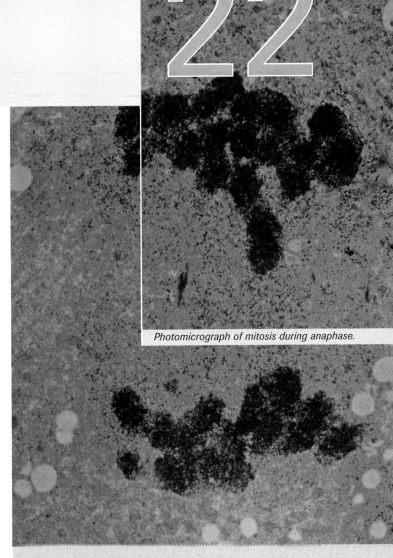

Photomicrograph of mitosis during anaphase.

Lucy and Lisa are sisters separated by only one year in age. However, Lucy is tall, with dark brown hair and blue eyes, and wears glasses because she is near-sighted. Meanwhile, Lisa is petite, has blonde hair and brown eyes, and has perfect vision. Given that Lucy and Lisa have the same mother and father, how can the two sisters exhibit such different characteristics? Answering this question requires an understanding of the reproductive process and how genetic makeup is determined.

In this chapter, we will learn about the process of reproduction, from copulation to fertilization to fetal growth to parturition and lactation. We will also learn what differentiates males from females, and how these differences contribute to the reproductive process.

Before You Begin

Make sure you have mastered the following topics:

1. *Hypothalamic tropic hormones, p. 165*

2. *Negative feedback, p. 11*

3. *Mitosis, p. 56*

Throughout this book, we have been emphasizing the concept of homeostasis: the maintenance of a constant internal environment by the interworkings of the organ systems. Here, we diverge from that concept. The reproductive system's primary function is not survival of the individual but survival of the species. We begin our study of the reproductive system with a general overview of reproductive physiology, emphasizing similarities between the male and female systems and how they work together to create offspring. In subsequent sections we concentrate on sex-specific aspects of reproductive function.

An Overview of Reproductive Physiology

Reproduction in humans is *sexual,* meaning that offspring are produced as a result of the mating of parents of different sexes. Each offspring inherits a unique combination of genes from both parents and therefore develops its own unique set of characteristics. This ability to create new combinations of genes is a hallmark of sexual reproduction, and it is one reason for the great diversity of living things in nature.

The Role of Gametes in Sexual Reproduction

In sexual reproduction, each parent produces cells called **gametes**, each of which contains a copy of half of his or her genetic material. Male gametes are known as **spermatozoa** (or simply *sperm*); female gametes are **ova** (singular: *ovum*) or *eggs*. Gametes have half the number of chromosomes that most cells of the body have.

Most cells of the body are described as **diploid** because they contain 2 sets of 23 chromosomes (designated $2n$). However, gametes are described as **haploid** because they contain only a single set of 23 chromosomes (designated as n). As a result of mating, gametes from each parent may fuse together to produce a new cell, a phenomenon known as **fertilization**. (Fertilization in humans is frequently referred to as *conception.*) This new cell, known as a **zygote**, has the potential to develop into a completely new individual and is diploid, having received half of its chromosomes from one parent and half from the other.

Gametes are generated from a pool of specialized, relatively undifferentiated precursor cells known as *germ cells* in a process called **gametogenesis**. In this process, the diploid germ cells undergo a series of cell divisions that ultimately reduces the number of chromosomes in each cell by half (that is, from $2n$ to n).

A single sperm or ovum contains 23 chromosomes, 22 autosomes, and one sex chromosome. Each chromosome is distinguishable from the others on the basis of shape and possesses its own characteristic set of genes. **Sex chromosomes** are the chromosomes that determine an individual's sex and are of two types—the Y or male chromosome and the X or female chromosome. **Autosomes** are the chromosomes other than the sex chromosomes. During fertilization, genetic material from a sperm and an egg combines to yield a total of 46 chromosomes (44 autosomes plus two sex chromosomes), the normal complement of chromosomes found in most cells of the body. Unlike the sex chromosomes, autosomes always come in matching pairs, the members of which are similar in size and shape and possess genes governing the same characteristics. As a reflection of this, the members of each pair of autosomes are said to be *homologous.* In the zygote and all other diploid cells that arise from it, one member of each pair of homologous chromosomes is inherited from the egg and is therefore of *maternal* origin (that is, comes from the mother), whereas the other member of the pair is inherited from the sperm and is therefore of *paternal* origin (that is, comes from the father).

The fact that autosomes come in pairs means that for every autosomal gene, there is a corresponding gene on the homologous chromosome that governs the same trait or characteristic. Thus autosomal *genes* come in pairs, one each from the father and the mother. The genes in each pair, though similar in many important respects, are not necessarily identical because genes generally come in a variety of different versions or *alleles.* The allele inherited from the mother may be the same as or different from the allele inherited from the father. A familiar example involves a gene that determines eye color. One version of the gene (one allele) codes for blue eyes, whereas another allele codes for brown eyes. Because each person has two genes for eye color, it is possible for a person to have two "blue" alleles, two "brown" alleles, or one "blue" allele and one "brown" allele.

The rule that chromosomes come in matching pairs does not apply to the sex chromosomes. The Y chromosome is significantly smaller than the X chromosome and has a different shape. Furthermore, most of the genes present on the X chromosome are not matched by corresponding genes on the Y chromosome, and vice versa. Because many genes that are present on the X but absent from the Y chromosome are necessary for life, every individual must inherit at least one X chromosome. Females inherit two X chromosomes, whereas males inherit one Y and one X.

Gene Sorting and Packaging in Gametogenesis: Meiosis

In both males and females, all the gametes an individual will ever produce are ultimately derived from a relatively small set of diploid germ cells. At some point in life these cells undergo mitosis to produce a colony or *clone* of daughter cells, each possessing an exact copy of all 46 of the individual's chromosomes. Subsequently these cells undergo **meiosis**, a series of two cell divisions following only a single replication of DNA, thereby generating daughter cells with half the normal chromosome number. Eventually these cells become mature sperm or ova.

In meiosis, an individual's chromosomes are sorted and packaged in such a way that each of the final daughter cells receives one sex chromosome plus one chromosome from each of the 22 pairs of autosomes. The sorting and packaging that occurs in meiosis proceeds in the following series of steps (**Figure 22.1**):

① The process begins with a diploid germ cell containing 46 chromosomes ($2n$), which in the figure are represented by four bars of two different colors that indicate whether they are paternal or maternal. Homologous chromosomes are indicated by their similar lengths. (For simplicity, the germ cell's nuclear membrane is not shown.)

② In a process similar to what occurs prior to mitosis, the nuclear membrane breaks down, and the original DNA is replicated, yielding exact copies of all 46 chromosomes ($2n \times 2$). The two copies of each chromosome (called *sister chromatids*) remain joined together at a structure called the centromere.

③ Homologous chromosomes begin to group together in pairs such that genes on paternal chromosomes line

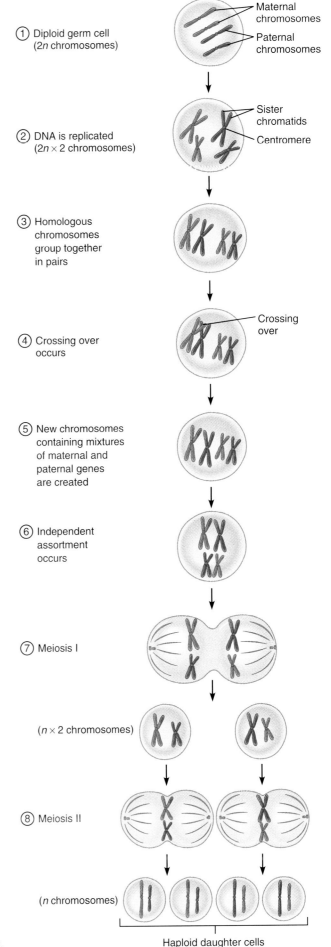

Figure 22.1 **Meiosis.** Details of the steps of meiosis are described in text.

up opposite the corresponding genes on maternal chromosomes.

④ Portions of homologous chromosomes begin to overlap and exchange segments with one another, a phenomenon known as *crossing over.*

⑤ Crossing over results in new chromosomes that contain mixtures of maternal and paternal genes. Because crossing over occurs randomly, the number of possible new chromosomes is extremely large.

⑥ Pairs of homologous chromosomes line up along a plane bisecting the cell in such a way that a random mixture of maternal and paternal chromosomes is present on either side of the cell. This random grouping of maternal and paternal chromosomes is known as *independent assortment.*

⑦ The first meiotic division *(meiosis I)* occurs. The cell divides in two, and each daughter cell receives one chromosome from each homologous pair. Due to independent assortment, each daughter cell receives a different combination of maternal and paternal chromosomes. Because the chromosomes are duplicates consisting of two sister chromatids, each cell receives half the normal chromosome number but two copies of every chromosome ($n \times 2$). Thus each cell receives two copies of each autosome (either the maternal or paternal version) and two copies of either an X or a Y chromosome.

⑧ The second meiotic division *(meiosis II)* occurs. Each of the cells generated in meiosis I divides in two, yielding a total of four cells. In this process, sister chromatids separate, and each of the new haploid daughter cells receives one sister chromatid from each chromosome. Consequently, each of the four cells receives single copies of 23 chromosomes ($n \times 1$), which are eventually enclosed within a new nuclear membrane (not shown).

Because of crossing over and independent assortment, a given parent can produce a large number of genetically different gametes, even though they are all derived from the same gene pool. Accordingly, each child of a given set of parents normally inherits its own unique set of genes.

Components of the Reproductive System

The *reproductive system* of an individual encompasses all organs involved in mating, gametogenesis, or other functions directly involved in the production of offspring. Reproductive system organs include the gonads and the accessory reproductive organs.

The **gonads** *(testes* in the male and *ovaries* in the female) are considered the primary reproductive organs because they perform two functions that ultimately govern all reproductive activity: They produce gametes, and they secrete sex hormones, a variety of steroids that promote gametogenesis, growth and maintenance of reproductive

organs, development of secondary sex characteristics, and various other effects throughout the body. The testes secrete a class of sex hormones known as **androgens** (notably *testosterone*); the ovaries secrete **estrogens** (such as *estradiol*) and **progesterone**.

Note that even though androgens and estrogens are commonly referred to as "male" and "female" sex hormones, respectively, they are actually present in both sexes, but in different relative amounts. Androgens are more abundant in males, whereas estrogens are more abundant in females. As noted in Chapter 6, sex hormones are produced not only by the gonads but also by the adrenal cortex. In women the adrenal cortex is the primary source of androgens, which are responsible for promoting the sex drive in either sex.

The *accessory reproductive organs* include structures specialized to perform additional functions required for proper reproductive activity, such as transporting gametes from one place to another and providing nourishment for gametes once they are produced. Accessory organs include (1) the organs of the **reproductive tract**, a system of interconnecting passageways through which gametes are transported after leaving the gonads, and (2) various glands that secrete fluids into the reproductive tract.

Events Following Fertilization

Sperm are transferred from the male to the female during the act of mating, known as **copulation**. Ova, in contrast, remain in the female's body, where they are fertilized by the arriving sperm. After fertilization, the ovum undergoes many cell divisions, eventually giving rise to billions of cells. As the cells increase in number, they also differentiate and begin to organize into distinct tissues according to instructions encoded in their genes. Over a period of approximately nine months, these cells and tissues gradually develop into a functional human organism. During this time the developing human is carried within its mother's body (a condition referred to as **gestation** or *pregnancy*), an arrangement that provides it with nourishment and protection from the outside environment. In the first two months after conception the developing human is called an **embryo**; from that point on it is referred to as a **fetus**. As the fetus develops, it eventually acquires the capacity to live outside its mother's body and separates from it in a process known as **parturition** (birth). Next we examine the events in embryonic and fetal development that determine the offspring's gender.

Sex Determination

Because a female's germ cells possess two X chromosomes (and thus no Y chromosome), every egg she ever produces possesses an X chromosome. In contrast, a male produces some sperm with an X chromosome and others with a Y chromosome. For this reason, the sex of the offspring is determined entirely by the genetic makeup of the sperm, not the egg: When the sperm bears an X chromosome, the fertilized egg inherits two Xs (one from the mother and one from the father) and develops into a female; when the sperm bears a Y chromosome, the egg inherits one Y and one X and therefore develops into a male. That males and females are born in approximately equal numbers is a consequence of the fact that on average about 50% of all sperm cells possess a Y chromosome. Although a person's sex is normally determined by the genes he or she inherits, the genes themselves do not confer the full complement of male or female traits; instead they simply determine whether a fetus will develop ovaries or testes. The role of genes in deciding a person's sex is referred to as **sex determination**.

In early life the embryo possesses primitive gonads that have the potential to become either ovaries or testes. Whether a fetus develops ovaries or testes is determined by the presence or absence in the embryo of a gene located on the Y chromosome—the **srY gene**—which codes for a protein called *testis-determining factor.* (srY stands for *s*ex-determining *r*egion of the *Y* chromosome.) When the embryo inherits a Y chromosome, the srY gene instructs the primitive gonads to become testes; when the embryo inherits two X chromosomes, the srY gene is absent, so the primitive gonads develop into ovaries. Thus the primitive gonads develop into ovaries "by default"—that is, unless they are instructed by the srY gene to do otherwise.

Sex Differentiation

In the first few weeks of development, the embryo is said to be *sexually indifferent* and possesses rudimentary female and male reproductive systems, called **Wolffian ducts** *(mesonephric ducts)* and **Müllerian ducts** *(paramesonephric ducts),* respectively (**Figure 22.2**). These structures have the potential to give rise to all reproductive organs except the gonads, the *external genitalia* (sex organs that are externally visible) and, in females, the vagina (the initial portion of the reproductive tract, which leads to the outside of the body).

The development of gonads in the embryo (sex determination) sets the stage for the development of other sexual characteristics. Whether these characteristics are male or female depends on the presence or absence of two hormones normally secreted by the testes—*testosterone* and another hormone called **Müllerian inhibiting substance (MIS)**. If these hormones are present (and target tissues are able to respond to them normally), the fetus develops as a male; if these hormones are absent, it develops as a female. The role of these hormones in the development of sexual characteristics is referred to as **sex differentiation**.

When testes are present and functioning normally, embryonic tissues are exposed to both testosterone and MIS. Testosterone acts on the Wolffian ducts to promote the development of male reproductive organs, whereas MIS promotes the regression and eventual disappearance of the Müllerian ducts, thereby preventing the develop-

ment of female organs. At the same time, testosterone acts on certain other embryonic tissues to promote the development of male external genitalia (such as the penis and scrotum, which are discussed shortly). In many target tissues, however, it is not testosterone that triggers these changes; instead, an enzyme in these target tissues converts testosterone to dihydrotestosterone, which then binds to receptors and triggers the target tissue response.

When the embryo develops ovaries instead of testes, the absence of testosterone causes the Wolffian ducts to regress, thereby preventing the development of male organs. In addition, the absence of MIS allows the Müllerian ducts to develop into female organs. The lack of testosterone also eliminates the influences that would otherwise stimulate the development of male external genitalia and allows female external genitalia to develop instead. The roles of sex determination and sex differentiation in the development of males and females are depicted in **Figure 22.3**.

Patterns of Reproductive Activity over the Human Life Span

Humans are not born with the ability to reproduce but instead acquire this ability during **puberty**, a period of sexual maturation that typically begins sometime between the ages of 10 and 14 and extends to the mid- to late teens. In the years before puberty the reproductive organs are immature and incapable of generating sperm or ova. In addition, a person's outward appearance is childlike, with little obvious distinction between the bodies of boys and girls. During puberty the reproductive organs mature, gametogenesis begins, and other physiological changes occur throughout the body. In the process, males and females develop various *secondary sex characteristics*—the external features that distinguish the sexes from each other, such as the growth of facial hair in men and the widening of the hips relative to the shoulders in women.

> ### *Exercise Link*
>
>
>
> Some other secondary sex characteristics may have affected Bill and Jane's marathon performance. Testosterone promoted increased muscle mass development in Bill, allowing him to run faster (at least for a short time). Estrogen influenced some of the enzymes that liberate and oxidize carbohydrates and fats for energy. As a result, Jane used her energy stores in a way that may have provided her with greater stamina.

Although both men and women acquire full reproductive function during puberty, the subsequent patterns of reproductive activity are very different for men and women. Activity of the female reproductive system exhibits cyclic variations known as the *menstrual cycle,* which averages about one month (28 days) in length, during which usually

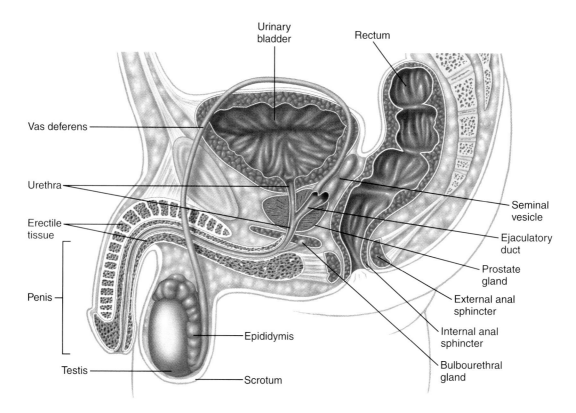

Figure 22.4 The male reproductive system, as seen in a sagittal section.

connective tissue. Each testis (**Figure 22.5**a) is divided internally into 250–300 compartments, each containing a set of thin, highly coiled hollow tubes known as **seminiferous tubules**, where sperm are produced. In the spaces between the tubules (Figure 22.5b) are clusters of cells known as **Leydig cells** *(interstitial cells),* which are responsible for the synthesis and secretion of testosterone and other androgens.

The wall of a seminiferous tubule (Figure 22.5c) contains an outer layer of smooth muscle cells and an inner layer of epithelial cells called **Sertoli cells**, whose primary function is to nurture sperm and control their development. Sperm cells in various stages of development are located in the spaces between the Sertoli cells, and mature sperm are located in the lumen of the seminiferous tubules, which is filled with fluid. The smooth muscle layer, which is separated from the Sertoli cells by a basement membrane, exhibits peristaltic contractions that help propel the sperm and fluid through the tubule.

Each Sertoli cell is joined to its neighbors by tight junctions, which limit the diffusion of materials between cells. The cells and the tight junctions that connect them form a barrier (called the *blood-testis barrier*) that isolates the lumenal fluid from the fluid bathing the cells on the other side (referred to as the *basal compartment*). The basal compartment communicates freely with the blood and is similar in composition to normal interstitial fluid, whereas the lumenal fluid has a different composition.

The significance of the blood-testis barrier is that by maintaining this difference in fluid composition, the barrier ensures that when sperm move from the basal compartment to the lumenal compartment at a certain stage of their development, the new fluid environment they encounter in the lumen fulfills their developmental requirements. The blood-testis barrier is also important because it prevents possible attack of sperm cells by the male's immune system.

Sertoli cells perform a number of other functions crucial to sperm development and transport. They are responsible for secreting the lumenal fluid, which serves as a medium for sperm development and transport, and for transporting nutrients to the developing sperm cells. The Sertoli cells also manufacture and secrete *androgen-binding protein,* which by binding androgens reversibly acts as an "androgen buffer" that helps maintain in the lumenal fluid a steady concentration of androgens, which must be present for sperm development to take place. Additionally, Sertoli cells play a critical role in orchestrating germ cells' responses to testosterone and *follicle stimulating hormone (FSH),* the hormones that control their development. These hormones are both necessary for sperm production, but they do not target the developing sperm themselves; instead they target the Sertoli cells, which then secrete various chemical messengers that actually stimulate the germ cells to grow and develop. Sertoli cells are also responsible for the secretion of a hormone called *inhibin*

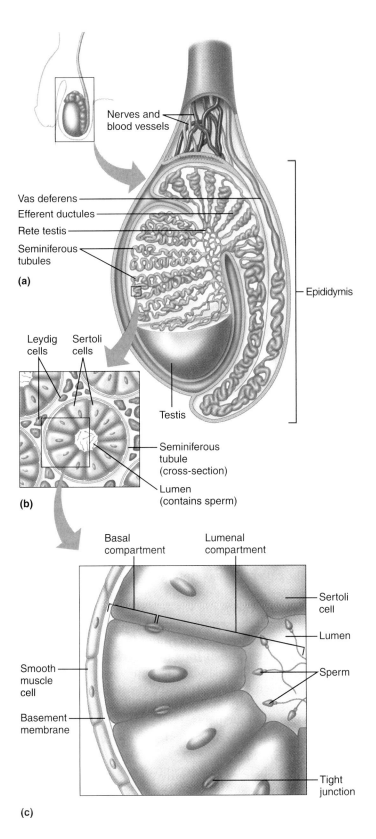

Nerves and blood vessels

Vas deferens
Efferent ductules
Rete testis
Seminiferous tubules

(a)

Epididymis

Leydig cells

Sertoli cells

Testis

Seminiferous tubule (cross-section)

Lumen (contains sperm)

(b)

Basal compartment

Lumenal compartment

Sertoli cell

Lumen

Sperm

Smooth muscle cell

Basement membrane

Tight junction

(c)

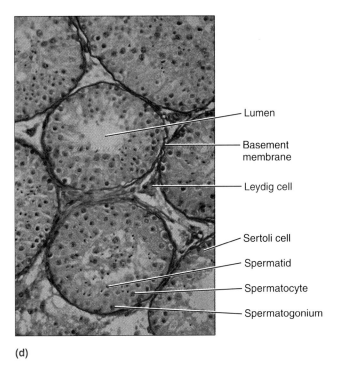

Lumen

Basement membrane

Leydig cell

Sertoli cell

Spermatid

Spermatocyte

Spermatogonium

(d)

Figure 22.5 The testis and associated structures. (a) A sagittal section of the testis and epididymis showing the seminiferous tubules and other ducts that carry sperm away from the testis. (b) An enlargement showing seminiferous tubules in cross section and Leydig cells located in the spaces between the tubules. (c) A portion of a seminiferous tubule showing structure of the wall. This diagram is highly schematized to emphasize essential features. (d) Photomicrograph of a cross section through seminiferous tubules.

Table 22.2 | Selected Actions of Androgens in Males

Stimulate spermatogenesis

Promote development of secondary sex characteristics during puberty and maintenance of these characteristics in adult life

Increase sex drive

Promote protein synthesis in skeletal muscle

Stimulate growth hormone secretion, which permits bone growth during adolescence

Promote development of male reproductive structures during embryonic life

Testosterone and other androgens normally limit their own secretion through negative feedback. Specifically, they act both on the hypothalamus, to suppress the secretion of GnRH, and on the anterior pituitary, making it less responsive to GnRH. As a result, testosterone tends to suppress the release of FSH and LH, which in turn tends to suppress subsequent testosterone secretion. Sertoli cells also secrete a protein hormone called **inhibin**, which suppresses the release of FSH (but not the release of LH) by the anterior pituitary.

Rates of sex hormone secretion in males are fairly constant and are not subject to the cyclic variations that occur in females during the menstrual cycle. Throughout a male's reproductive life the hypothalamus releases GnRH in bursts occurring at approximately two-hour intervals, which causes rates of FSH and LH secretion to rise during bursts and fall between bursts. Despite these fluctuations, however, *average* plasma levels of FSH and LH (and hence levels of testosterone) are fairly constant from day to day. Significant variations over the long term do occur, however; testosterone levels rise dramatically during puberty, reach a peak sometime in the third decade of life, and subsequently decline slowly.

In prepubescent males, testosterone levels are relatively low, reflecting low rates of GnRH and gonadotropin secretion. However, GnRH secretion rises dramatically at the beginning of puberty in response to a change in brain activity that alters neural input to the hypothalamus. (The precise nature of this change in brain activity is presently unknown.) The increase in GnRH triggers a similar rise in gonadotropin secretion, which stimulates growth and maturation of the testes as well as testosterone secretion.

The rise in plasma testosterone levels that occurs during puberty stimulates further maturation of the reproductive system and other widespread changes throughout the body. For example, testosterone stimulates the secretion of growth hormone by the anterior pituitary, which in turn stimulates the bone growth responsible for the dramatic increase in stature that occurs during this period. Testosterone's actions also trigger the

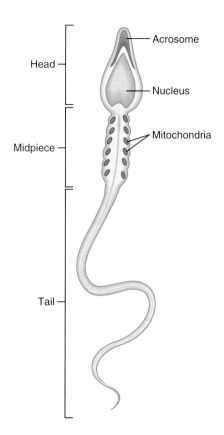

Figure 22.7 Anatomy of a sperm cell.

development of secondary sex characteristics during puberty, including growth of facial hair; growth of coarse body hair in the pelvic region and armpits; increased protein synthesis in muscle that leads to increased muscle size and strength, particularly in the upper body; growth of the larynx, which causes the voice to deepen; and secretion of a thick oil by *sebaceous glands* (oil glands) in the skin. Actions of androgens in males are summarized in **Table 22.2**.

Sperm and Their Development

Fully developed sperm cells possess three distinct segments (**Figure 22.7**): a bulblike *head,* a short cylindrical *midpiece,* and a long, slender *tail* (or *flagellum*). Within the head are the chromosomes and a large vesicle called an **acrosome**, which contains enzymes and other proteins that enable the sperm to fuse with the egg during fertilization. The tail contains complex machinery made up of proteins that hydrolyze ATP and convert the released chemical energy into motion. The resulting whiplike movements of the tail propel the sperm forward in a swimming motion. The midpiece anchors the tail to the head and contains mitochondria, which produce the ATP required for motility.

Spermatogenesis starts near the basement membrane in the seminiferous tubules with undifferentiated germ

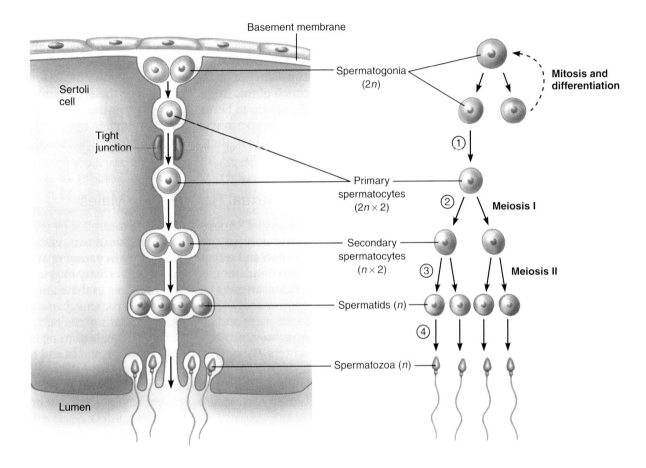

Figure 22.8 Spermatogenesis. As germ cells migrate from the basal compartment to the lumen of a seminiferous tubule (depicted at left), they undergo various stages of development (depicted at right). The dashed arrow indicates that some spermatogonia do not differentiate following mitosis and remain as spermatogonia.

Over a male's lifetime, does the number of spermatogonia normally increase, decrease, or stay the same?

cells called **spermatogonia** (**Figure 22.8**). As spermatogenesis proceeds, germ cells gradually migrate toward the lumen of the tubule in the spaces between Sertoli cells. At a certain point the developing germ cells move from the basal compartment to the lumenal compartment by passing through the tight junctions, which open up temporarily to let the cells pass through. As a result of this migration, cells at different stages of development are characteristically found at certain locations in the tubule; the older, more mature cells are nearer the lumen, whereas the newer, less differentiated cells are nearer the basement membrane.

Each male is born with a finite number of spermatogonia, but these cells undergo mitosis repeatedly, giving males the capacity to produce sperm for an indefinite period following puberty. One of the daughter cells produced by each mitotic division differentiates further and eventually becomes a mature sperm; the other daughter cell does not differentiate and remains a spermatogonium. Therefore, the total number of spermatogonia does

not change, ensuring that this pool of potential sperm is never depleted.

To become a mature sperm, a spermatogonium goes through the following stages of development (see Figure 22.8, at right): ① The spermatogonium's chromosomes are replicated, and the cell differentiates to become a *primary spermatocyte,* which has 46 duplicated chromosomes possessing two sister chromatids apiece. (In the figure, the chromosome number is designated as $2n \times 2$ to indicate that the chromosomes are duplicated.) ② The primary spermatocyte goes through meiosis I to yield two *secondary spermatocytes,* which possess 23 duplicated chromosomes ($n \times 2$). ③ Secondary spermatocytes undergo meiosis II to become *spermatids,* which have 23 single chromosomes (n) but have not developed the mature sperm's characteristic features. ④ Spermatids differentiate to become spermatozoa (n), which possess the

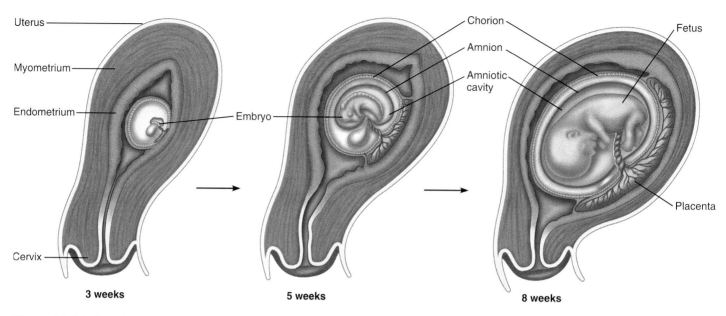

Figure 22.21 Growth and development of the embryo and the structures that support it. Note that at about eight weeks the developing human is termed a fetus.

surrounding cells develop into an epithelial tissue called the **amnion** (or *amniotic sac;* see Figure 22.21), which eventually fuses with the chorion to form a single membrane that surrounds the developing embryo. The fluid contained within the amnion (called *amniotic fluid*) is similar in composition to normal extracellular fluid; it bathes the embryo and forms a physical barrier between it and the uterine wall that cushions the developing embryo against possible physical trauma and protects it against sudden changes in temperature.

As development proceeds, the embryo undergoes a series of dramatic transformations in structure and function. Within the inner cell mass, cells begin to differentiate into distinct tissues that will ultimately give rise to all the organ systems found in adults. Indeed, the pace of change is so rapid that rudimentary organ systems are already discernible within the first few weeks after fertilization. Over time these primitive organ systems gradually acquire more and more of their adult characteristics; different systems develop at different rates. As this is occurring, the embryo's external appearance is also rapidly changing, so that by the end of the eighth week of life, it has a recognizably human form, complete with clearly discernible limbs as well as a head and face. From this point onward, the developing human is a *fetus.*

Formation of the Placenta

In its earlier stages of development, the embryo is able to obtain nutrients through the breakdown of endometrial tissue, and sufficient quantities of oxygen by diffusion from nearby blood vessels. However, as the embryo

grows, its need for oxygen increases such that it can no longer rely on diffusion from the mother's tissues as a mechanism of delivery; likewise, its demand for nutrients increases such that it can no longer obtain sufficient amounts from the surrounding tissue.

To obtain essential materials (and to rid itself of carbon dioxide and other wastes), the growing embryo or fetus relies on the placenta, a structure that permits the ready exchange of gases, nutrients, and other materials between the fetus's bloodstream and the maternal circulatory system. The placenta develops in the first few weeks after implantation in the region of contact between the embryo and the endometrium, and it is made of specialized, closely interwoven fetal and maternal tissues (**Figure 22.22**).

Placental development begins as the chorion sends out fingerlike projections called *chorionic villi.* The villi contain both capillaries of the fetal circulation and cells that secrete paracrines that alter the structure and function of the surrounding endometrial tissue, such that the villi become surrounded by sinuses containing maternal blood. As a result, maternal blood and fetal blood come into close proximity (which facilitates material exchange) without actually mixing. (Separation of maternal and fetal blood is important because it prevents attack of fetal tissues by the mother's immune system.) Maternal blood is delivered to the placental sinuses by the *uterine artery* and is returned to the mother's general circulation by the *uterine vein.* Fetal blood is carried to the placenta by the paired *umbilical arteries* and back to the fetus's general circulation by the *umbilical vein,* both of which are housed in

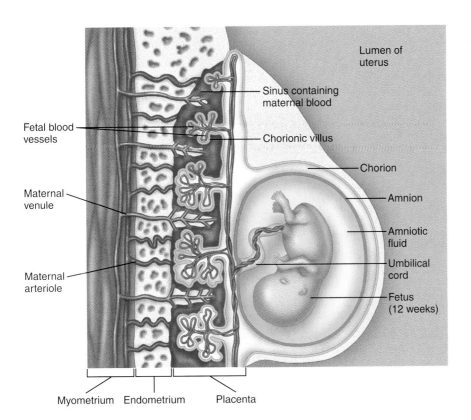

Figure 22.22 The placenta. The interweaving of fetal and maternal tissues enables the exchange of materials between the circulatory systems of mother and fetus.

Lumen of uterus

Sinus containing maternal blood

Fetal blood vessels

Chorionic villus

Chorion

Amnion

Maternal venule

Amniotic fluid

Umbilical cord

Maternal arteriole

Fetus (12 weeks)

Myometrium   Endometrium   Placenta

a ropelike structure called the **umbilical cord**, which extends between the placenta and the fetus. As we will see shortly, the placenta is not just an organ of exchange; it also secretes several hormones that are important in maintaining pregnancy and preparing the mother's body for the eventual birth of the child.

Hormonal Changes During Pregnancy

During the first three months of pregnancy, the corpus luteum is maintained by the actions of a hormone called **human chorionic gonadotropin (hCG)**, which is secreted by the chorionic portion of the placenta and exerts many of the same effects as LH. Under the influence of hCG, the corpus luteum continues to secrete large amounts of estrogens and progesterone, which exert several actions described later. Secretion of hCG begins at implantation, when trophoblast cells start to infiltrate the endometrial tissue, and then it rises for the next two months as the placenta grows (**Figure 22.23**). At the end of this period the secretion of hCG falls off dramatically, and as a result the corpus luteum begins to degenerate. Estrogen and progesterone levels remain high, however, and continue to rise until parturition because the placenta begins to secrete these hormones just as secretion by the corpus luteum is declining. Placental estrogens are actually manufactured from androgens (primarily

dehydroepiandrosterone, or DHEA) secreted by the adrenal cortex of the developing fetus. Placental progesterone is manufactured from cholesterol, which is delivered to the placenta by the maternal bloodstream.

During pregnancy, plasma levels of estrogen and progesterone, which rise after formation of the corpus

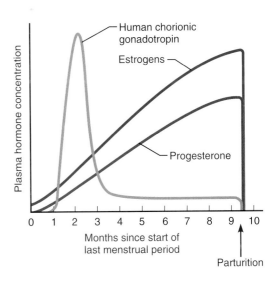

Human chorionic gonadotropin

Estrogens

Progesterone

Plasma hormone concentration

0 1 2 3 4 5 6 7 8 9 10

Months since start of last menstrual period

Parturition

Figure 22.23 Plasma concentrations of estrogen, progesterone, and human chorionic gonadotropin during pregnancy.

luteum, continue to rise until parturition, at which time estrogen and progesterone levels fall precipitously (Figure 22.23). These elevated hormone levels are important for maintaining the pregnancy and for preparing the mother's body for the eventual delivery of the fetus. In particular, estrogen promotes the following effects:

1. *Growth of duct tissue in the breasts.* In the milk-producing glands of the breasts (called **mammary glands**), estrogens stimulate the development of tissue in ducts that carry milk to the nipples. This prepares the breasts for **lactation** (milk production), which begins after parturition. Estrogens also promote the deposition of fatty tissue in the breasts, which causes them to enlarge.

2. *Prolactin secretion.* Estrogen stimulates the secretion by the anterior pituitary of prolactin, which promotes breast growth and prepares the mammary glands for lactation. Although prolactin and estrogens both work to build up the milk-producing apparatus during pregnancy, actual milk production does not occur yet because the high plasma levels of estrogens and progesterone during pregnancy suppress lactation.

3. *Growth and enhanced contractile responsiveness of uterine smooth muscle.* Estrogens stimulate the growth of uterine smooth muscle, thereby enhancing the ability of the uterus to contract and expel the fetus during parturition. Estrogens also work to promote increased contractile activity in uterine smooth muscle and to increase its responsiveness to *oxytocin,* which stimulates uterine contractions during birth. Despite the fact that estrogens enhance contractile activity by themselves, the uterine contractions that occur during pregnancy are infrequent and weak, because the contraction-promoting actions of estrogens and oxytocin are blocked by the elevated levels of progesterone during this period.

While estrogens are working to promote these effects, progesterone is also working to maintain the pregnancy and prepare the mother's body for parturition. In particular, progesterone promotes the following effects:

1. *Growth of glandular tissue in the breasts.* In the breasts, progesterone stimulates the growth of glandular tissue (as opposed to ductal tissue, whose growth is stimulated by estrogens). This glandular tissue is responsible for secreting the milk that is produced after parturition.

2. *Suppression of contractile activity in uterine smooth muscle.* As previously mentioned, progesterone suppresses uterine contractions, thereby counteracting the effects of estrogens and oxytocin on uterine smooth

muscle. Because contractions could prematurely expel an embryo or fetus from the uterus, this suppressive action is necessary for successful implantation and for the maintenance of pregnancy.

3. *Maintenance of secretory-phase uterine conditions.* Progesterone works during pregnancy to maintain the same secretory-phase uterine conditions that it promotes during the luteal phase of the ovarian cycle. This ensures hospitable conditions for embryonic and fetal growth.

Actions of estrogens and progesterone are summarized in **Figure 22.24**, which shows hormonal interactions occuring in early pregnancy, when the corpus luteum is still active.

In pregnancy, progesterone exerts negative feedback on the hypothalamus and anterior pituitary that keeps rates of LH and FSH secretion low. Consequently, no new dominant follicles appear and no LH surges occur, despite the fact that estrogen levels are elevated at this time. In addition, the placenta secretes a new hormone called **placental lactogen**, which has effects similar to some of the actions of growth hormone and prolactin. This hormone works hand in hand with estrogens and prolactin to promote not only breast growth but also the mobilization within the mother of energy stores needed to meet the increased demands arising from the growth of breasts and fetal tissue.

Quick Test 22.5

1. Define the following terms: *capacitation, acrosome reaction, polyspermy, morula, blastocyst, inner cell mass, trophoblast, implantation, placenta, chorion, chorionic villi, amnion, amniotic fluid,* and *umbilical cord.*

2. Where does fertilization occur? Implantation? What is the primary function of the placenta?

3. Estrogens and progesterone are secreted by what structure during the first three months of pregnancy? During the final six months?

4. For each of the following hormones, name the site from which it is secreted, and give a brief description of its function: human chorionic gonadotropin, prolactin, oxytocin, and placental lactogen.

Parturition and Lactation

Pregnancy normally lasts about nine months (40 weeks), during which time the fetus gains in size, weight, and maturity such that it acquires the ability to live outside its mother's body.

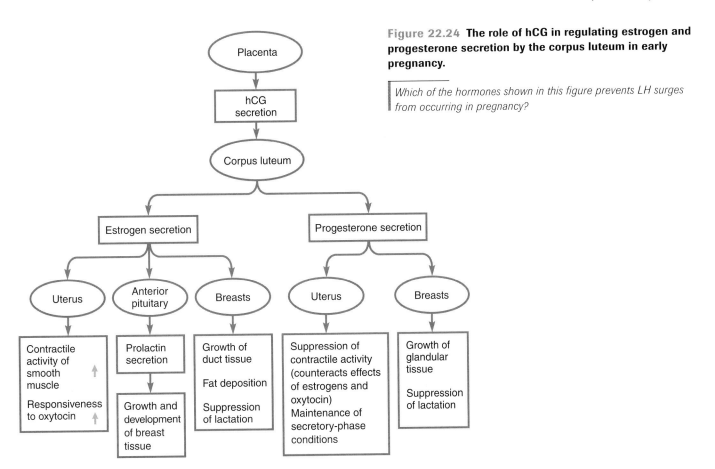

Figure 22.24 **The role of hCG in regulating estrogen and progesterone secretion by the corpus luteum in early pregnancy.**

Which of the hormones shown in this figure prevents LH surges from occurring in pregnancy?

Events of Parturition

In the last few weeks of pregnancy, weak and infrequent uterine contractions begin to appear. Over time these contractions gradually increase in strength and frequency. Because estrogens are known to promote contractile activity in uterine smooth muscle, it is likely that these contractions are triggered, at least in part, by the increased levels of estrogens relative to progesterone that exist in late pregnancy. Moreover, uterine smooth muscle is capable of initiating its own contractions in response to stretch, and as the fetus grows and the uterus expands, the resulting stretch of the smooth muscle may initiate contractile activity.

To facilitate delivery, the cervix undergoes a process called *ripening* in the weeks before parturition. During this process, the cervix becomes softer and more flexible as a result of the enzymatic breakdown of collagen fibers in its connective tissue. These changes make it easier for the cervical canal to enlarge for the fetus to pass through it.

Shortly before parturition the fetus, which is normally upside down, moves downward and comes into contact with the closed cervix (**Figure 22.25**a), so that the strong uterine contractions of parturition will push the fetus through the cervical canal headfirst. Moreover, the fetus's headfirst orientation enables the head, which is narrower than the rest of the body, to act as a wedge to force the cervical canal open. (Opening of the canal is called *cervical dilation*). In a small fraction of cases the legs or hips pass through the cervical canal first, a phenomenon called *breech birth;* because the hips are wider than the head, such a delivery is more difficult than normal and increases the time required. Breech birth also exposes the fetus to greater danger because it increases the risk that the umbilical cord may become compressed between the fetus and the wall of the birth canal, which chokes off the fetus's blood supply. For these reasons, *cesarean section* (surgical delivery through the abdominal wall) is often recommended when breech birth is imminent.

In the hours just before parturition, the amniotic sac typically ruptures, causing the leakage of amniotic fluid to the outside of the body. This event is often the first reliable signal to the mother that parturition is imminent. Soon afterward a series of strong uterine contractions

Progesterone

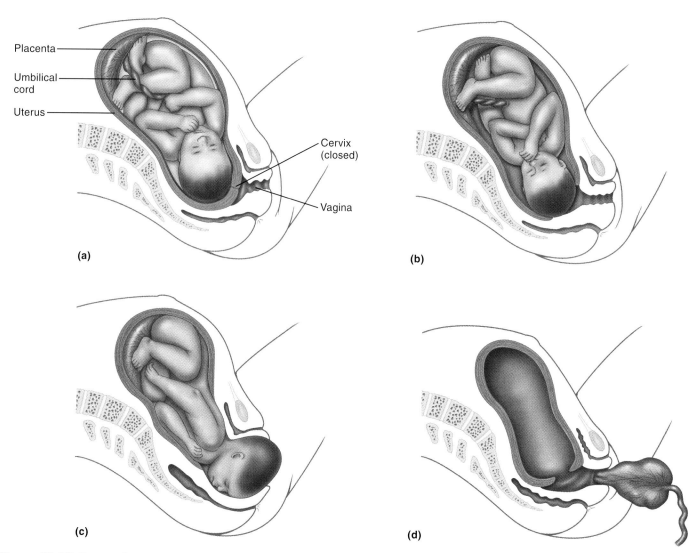

Figure 22.25 Events of parturition. (a) Position of the fetus in the uterus before parturition. (b) Dilation of the cervix. (c) Movement of the fetus through the birth canal (delivery). (d) Expulsion of the placenta following delivery.

(known as *labor*) begins. Each contraction starts at the top of the uterus and then travels downward through the smooth muscle as a wave. (Contractions are triggered by electrical signals that move from cell to cell through gap junctions.) The downward movement of these contractions pushes the contents of the uterus against the cervix, which eventually forces the cervical canal open. In labor, uterine contractions may be assisted by voluntary contractions of the mother's abdominal muscles, which increase pressure on the uterine contents.

Initially the contractions of labor are separated by intervals of 10–15 minutes, but contractions come closer together as labor progresses. The rising intrauterine pressure causes dilation of the cervical canal (Figure 22.25b), which eventually reaches a maximum diameter of about 10 centimeters. At that point the fetus moves into

the canal and within a few minutes emerges from the mother's body (Figure 22.25c). Blood vessels in the umbilical cord begin to constrict, and the placenta then separates from the endometrium. (The umbilical cord is usually tied off by birth attendants and then cut on the maternal side of the knot.) Finally, a wave of strong uterine contractions expels the placenta (now referred to as the *afterbirth*) outside the mother's body (Figure 22.25d).

Once the contractions of labor begin, the stretch sensitivity of uterine smooth muscle tends to perpetuate them. As the contractions push the fetus downward, the lower portion of the uterus is stretched, which stimulates more contractile activity. Uterine contractions are also promoted by oxytocin, whose release is stimulated in labor. The pressure exerted by the fetus against the uterine wall excites stretch receptors located in the lower portion of the uterus, which triggers the release of oxytocin from the posterior pituitary. Oxytocin enters the bloodstream and then stimulates uterine contractions both by direct action on muscle cells and by stimulating other cells in the myometrium to secrete prostaglandins, which act locally to promote contractile activity. The contraction-promoting effects of oxytocin are normally blocked during pregnancy by progesterone, which helps to keep uterine contractile activity low.

Although the precise nature of the signals that trigger parturition are incompletely understood, evidence suggests that they may originate from the fetus itself. The fetal portion of the placenta is known to secrete corticotropin releasing hormone (CRH), the same hormone that is secreted by the hypothalamus in adults and stimulates the release of ACTH from the anterior pituitary. In the fetus, placental CRH stimulates the release of ACTH, which stimulates the adrenal gland to secrete DHEA. This hormone is then converted by the placenta to estrogens, which enter the maternal bloodstream. The rising rates of placental CRH secretion are believed to be largely responsible for the rapidly rising estrogen levels that occur toward the end of pregnancy and help to initiate the events of labor. Because some placental CRH leaks into the maternal blood, it is also possible that rising CRH levels act as a direct trigger for parturition.

Lactation

The decrease in estrogens and progesterone that occurs in parturition is important not only because it helps to promote uterine contractions in labor, but also because it allows lactation to begin. In the first months after birth the infant has no teeth and cannot eat solid foods. For this reason, the milk produced by the mother's breasts can serve as a valuable source of early nutrition. For the first one or two days after birth, this milk is little more than a watery fluid (called *colostrum*) containing a vast array of proteins but few other nutrients. Subsequently, however, the milk becomes enriched with a variety of additional constituents, including fat, lactose (the same sugar present in cow's milk), growth factors and hormones (which promote tissue development in the infant), and antibodies, which confer on the infant some degree of immunity against bacteria and other pathogens. (In early life the infant is not able to make sufficient quantities of its own antibodies because its immune system is not yet fully developed.)

Milk is produced in the mammary glands by clusters of round, saclike glands called *alveoli* and is delivered by milk ducts to openings in the nipples (**Figure 22.26**a). To obtain this milk the infant simply sucks on the nipples

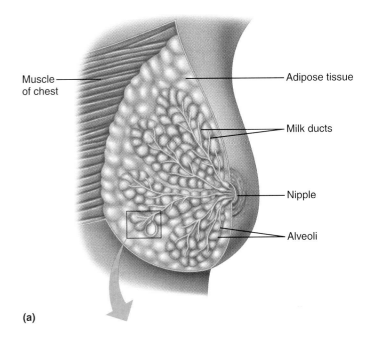

(a)

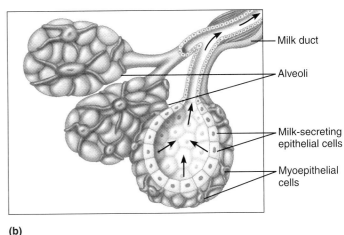

(b)

Figure 22.26 Mammary glands. (a) A cutaway view of a breast, showing the alveoli and milk ducts that constitute the mammary glands. (b) An enlargement showing milk-secreting epithelial cells and myoepithelial cells in alveoli.

(an act known as *suckling*), which triggers the flow of milk through the ducts (sometimes referred to as "letdown"). The alveoli are surrounded by *myoepithelial cells* (Figure 22.26b), which are flattened like ordinary epithelial cells but have the ability to contract like muscle cells. Suckling stimulates these cells to contract, which compresses the alveoli and forces the milk to flow through the ducts, a phenomenon called *milk ejection.*

The events by which suckling induces milk ejection are diagrammed in **Figure 22.27**. Suckling of the infant stimulates tactile receptors in the nipples, which project to the hypothalamus and excite neurosecretory cells that

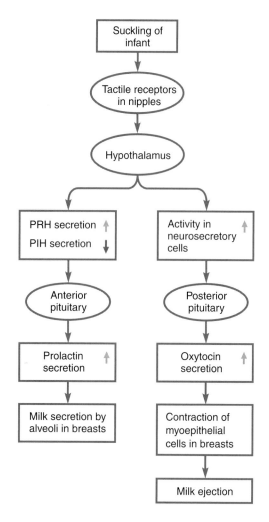

Figure 22.27 Hormonal regulation of lactation in response to suckling.

and inhibiting the release of prolactin inhibiting hormone (PIH). Both of these changes stimulate the anterior pituitary to secrete prolactin, which induces milk production by cells in the alveoli.

During puberty, rising estrogen and progesterone levels stimulate growth of the breasts and development of the milk-producing apparatus. Estrogens promote not only the development of tissue in the ducts but also the deposition of fatty tissue in the breasts, which results in visible breast enlargement; progesterone, by contrast, promotes development of the alveoli. In pregnancy, the breasts become enlarged and the milk-producing apparatus becomes fully developed as a result of the combined actions of increased plasma levels of estrogens, progesterone, prolactin, and placental lactogen. (Recall that prolactin release is stimulated by estrogen at this time.) No milk is actually produced, however, because the high estrogen and progesterone levels block the secretory activity of cells in the alveoli. After parturition the fall in estrogens and progesterone unblocks the stimulatory effect of prolactin on milk production and allows lactation to proceed.

> **Quick Test 22.6**
>
> 1. Define the following terms: *afterbirth, labor, colostrum, mammary glands,* and *alveoli.*
>
> 2. For each of the following descriptive phrases, indicate whether it pertains to oxytocin, prolactin, estrogens, or progesterone: (a) inhibits uterine contractions during pregnancy, (b) secretion is stimulated by estrogens, (c) promotes development of milk ducts and deposition of fat in breasts during puberty, (d) promotes development of alveoli during puberty, (e) is secreted in response to excitation of uterine stretch receptors, (f) is secreted in response to tactile stimulation of nipples, (g) stimulates milk production in response to suckling, (h) stimulates milk ejection in response to suckling.

extend to the posterior pituitary and secrete oxytocin. Oxytocin travels in the bloodstream and stimulates myoepithelial cells to contract. Suckling also stimulates the actual production of milk by stimulating the release of prolactin releasing hormone (PRH) by the hypothalamus

CHAPTER SUMMARY

An Overview of Reproductive Physiology, p. 686

Human reproduction involves the fundamental processes of gametogenesis, fertilization, pregnancy, and parturition. Reproductive ability is acquired during puberty, during which the reproductive organs mature,

gametogenesis begins, and secondary sex characteristics develop. Males are able to reproduce continuously through adulthood, but in females reproductive capacity is cyclic and is lost at menopause.

The male and female reproductive systems include the gonads (testes in

males and ovaries in females), which carry out gametogenesis and secrete sex hormones (androgens in males and estrogens and progesterone in females), and accessory reproductive organs, which include organs of the reproductive tract and various glands that secrete fluids into the tract.

SYSTEMS INTEGRATION   Reproductive System

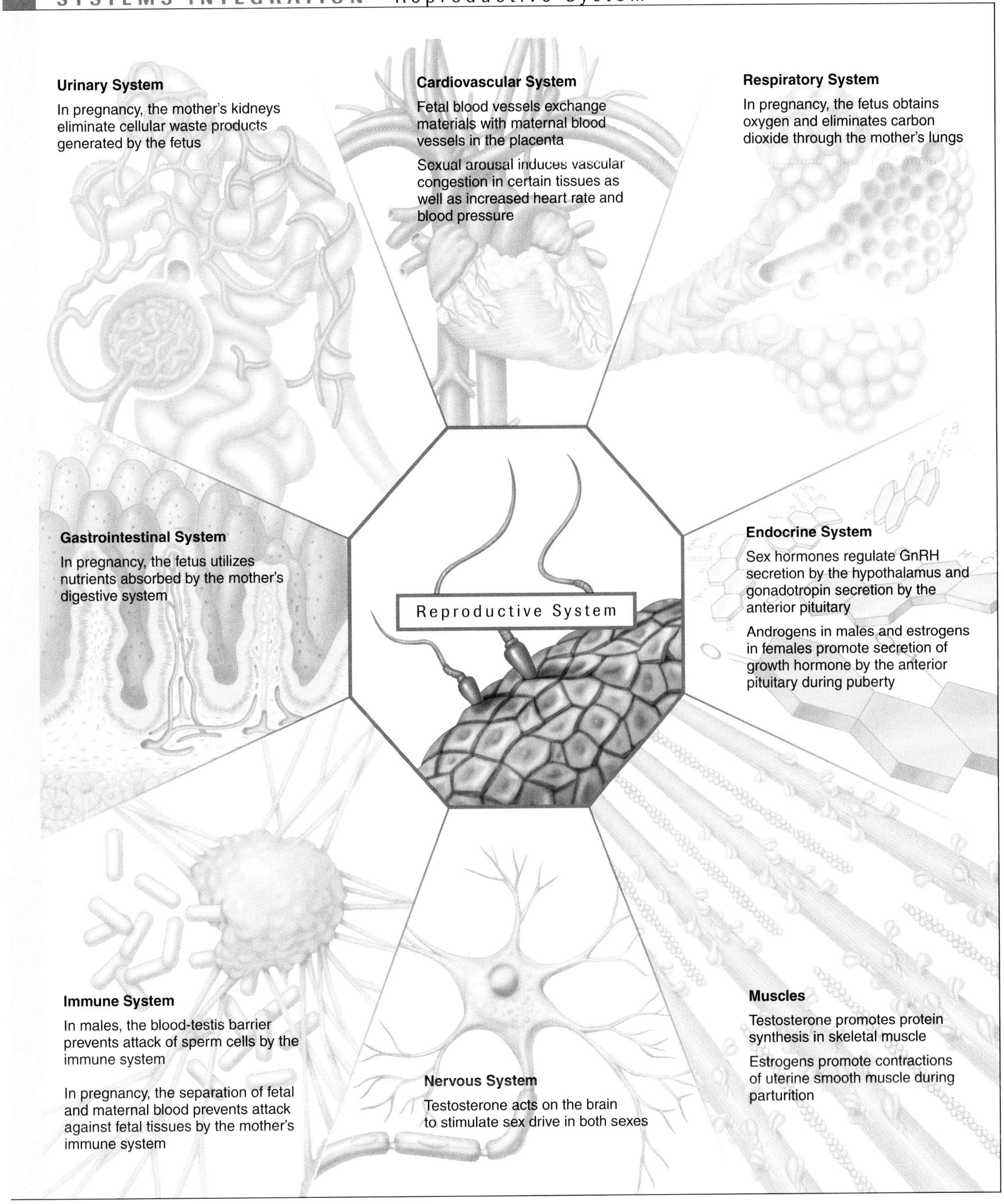

Urinary System

In pregnancy, the mother's kidneys eliminate cellular waste products generated by the fetus

Cardiovascular System

Fetal blood vessels exchange materials with maternal blood vessels in the placenta

Sexual arousal induces vascular congestion in certain tissues as well as increased heart rate and blood pressure

Respiratory System

In pregnancy, the fetus obtains oxygen and eliminates carbon dioxide through the mother's lungs

Gastrointestinal System

In pregnancy, the fetus utilizes nutrients absorbed by the mother's digestive system

Reproductive System

Endocrine System

Sex hormones regulate GnRH secretion by the hypothalamus and gonadotropin secretion by the anterior pituitary

Androgens in males and estrogens in females promote secretion of growth hormone by the anterior pituitary during puberty

Immune System

In males, the blood-testis barrier prevents attack of sperm cells by the immune system

In pregnancy, the separation of fetal and maternal blood prevents attack against fetal tissues by the mother's immune system

Nervous System

Testosterone acts on the brain to stimulate sex drive in both sexes

Muscles

Testosterone promotes protein synthesis in skeletal muscle

Estrogens promote contractions of uterine smooth muscle during parturition

The Male Reproductive System, p. 691

The male reproductive system includes the testes, external genitalia (the penis and scrotum), the reproductive tract (the epididymis, vas deferens, ejaculatory duct, and urethra), and accessory glands (seminal vesicles, bulbourethral glands, and the prostate gland). Sperm are formed in the testes in seminiferous tubules, which are lined by Sertoli cells. During sexual arousal a mixture of sperm and fluids (semen) is forcibly ejected from the penis through the urethra, an event called ejaculation.

Reproductive function in males is controlled by androgens and other hormones, including gonadotropins from the anterior pituitary (follicle stimulating hormone, FSH, and luteinizing hormone, LH) and gonadotropin releasing hormone (GnRH). Spermatogenesis and other Sertoli cell functions are stimulated by FSH. LH stimulates androgen secretion by Leydig cells. During reproductive life, androgen levels are fairly steady because they limit their own secretion through negative feedback control of GnRH and gonadotropin secretion.

The Female Reproductive System, p. 699

The female reproductive system includes the ovaries, the reproductive tract (the uterus, uterine tubes, and vagina), and the external genitalia (the mons pubis, labia majora, labia minora, vestibule, clitoris, and vestibular glands). The menstrual cycle, which lasts about 28 days, is marked by cyclic changes in pituitary and ovarian hormone secretion and begins with menstruation, the shedding of tissue and blood from the endometrium. Ova develop from a set of germ cells whose number is fixed at birth and do not become fully mature until fertilization has occurred.

Each ovary contains numerous follicles, each of which contains one oocyte. Follicles also contain granulosa cells, which nourish the oocyte, regulate its development, and secrete estrogens; and (in later stages of development) theca cells. The ovarian cycle is divided into a follicular phase, during which a dominant follicle is selected and develops into a Graafian follicle, and a luteal phase, during which the follicle is transformed into a corpus luteum. The follicular phase ends with release of the oocyte (ovulation), which then enters the nearby uterine tube. Coinciding with the ovarian cycle is a uterine cycle consisting of a menstrual phase, a proliferative phase, and a secretory phase.

In the follicular phase, FSH targets granulosa cells to stimulate growth and estrogen secretion; LH stimulates theca cells to secrete androgens, which granulosa cells then convert to estrogens. The estrogens promote oogenesis (along with FSH) and proliferative-phase uterine changes. In the late follicular phase, rising estrogen levels trigger an LH surge, which is responsible for ovulation and development of the corpus luteum. In the luteal phase the corpus luteum secretes estrogen and progesterone, which suppresses LH and FSH secretion and promotes secretory-phase uterine changes. In the absence of fertilization the corpus luteum degenerates, causing a drop in estrogen and progesterone levels that triggers menstruation.

Fertilization, Implantation, and Pregnancy, p. 711

After fertilization, which normally occurs in the uterine tube, the zygote is eventually transformed into a blastocyst, which implants in the endometrium. At the point of contact, embryonic and endometrial tissues develop into a placenta, which permits exchange of materials between the mother and the developing embryo. During pregnancy, estrogens and progesterone (secreted by the corpus luteum at first and then by the placenta) promote many effects, including growth and development of the mammary glands, secretion of prolactin by the anterior pituitary (which promotes breast growth and eventually lactation), and maintenance of secretory-phase uterine conditions.

Parturition and Lactation, p. 716

Parturition typically occurs 40 weeks after fertilization and is accompanied by a wave of strong uterine contractions (labor), dilation of the cervix, expulsion of the fetus from the uterus, and separation of the placenta (afterbirth) from the uterine wall. After delivery, nourishment is provided to the infant by milk secreted by the mammary glands. Suckling of the infant triggers secretion of prolactin, which promotes milk production, and of oxytocin, which promotes milk ejection.

∎ EXERCISES

Multiple-Choice Questions

1. In both males and females, gonadotropin secretion by the anterior pituitary is stimulated by
   a) inhibin.
   b) androgens.
   c) GnRH.
   d) FSH.
   e) GHRH.

2. In the embryo, testosterone promotes
   a) development of the primitive gonads into testes.
   b) regression of Müllerian ducts.
   c) development of Müllerian ducts into male reproductive organs.
   d) development of Wolffian ducts into male reproductive organs.
   e) expression of the srY gene.

3. The testes are housed in a structure called the
   a) prostate gland.
   b) scrotum.
   c) penis.

d) epididymis.

e) vas deferens.

4. In the first step of spermatogenesis, spermatogonia differentiate into cells called

a) spermatids.

b) primary spermatocytes.

c) secondary spermatocytes.

d) spermatozoa.

e) spermatophytes.

5. Cells in the ovaries secrete all of the following hormones *except*

a) estrogens.

b) progesterone.

c) androgens.

d) luteinizing hormone.

e) inhibin.

6. In oogenesis, meiosis I occurs

a) in early fetal life.

b) at birth.

c) just before ovulation.

d) after ovulation but before fertilization.

e) after fertilization.

7. During the early to mid-follicular phase of the ovarian cycle, granulosa cell functions are stimulated by

a) progesterone.

b) FSH.

c) LH.

d) GnRH.

e) estrogens.

8. In the late luteal phase, estrogen and progesterone levels fall due to

a) rupture of the dominant follicle.

b) degeneration of the corpus luteum.

c) an inhibitory effect of LH on secretory activity of the corpus luteum.

d) the inhibitory effect of inhibin on the secretory activity of granulosa cells.

e) all of the above.

9. In the uterine cycle, the proliferative phase is immediately preceded by the

a) menstrual phase.

b) secretory phase.

c) luteal phase.

d) follicular phase.

10. The placenta not only serves as an organ of exchange but also secretes all of the following hormones *except*

a) prolactin.

b) chorionic gonadotropin.

c) placental lactogen.

d) progesterone.

e) estrogens.

Objective Questions

1. In meiosis I, maternal and paternal chromosomes are segregated into separate daughter cells. (true/false)

2. The srY gene codes for (testosterone receptors/testis-determining factor), which determine(s) whether an embryo develops testes or ovaries.

3. In the absence of testosterone and MIS, the Müllerian ducts (persist/degenerate) in the embryo, and female structures eventually develop.

4. FSH and LH are classified as (sex hormones/gonadotropins).

5. GnRH, which is secreted by the hypothalamus, stimulates the secretion of both FSH and LH from the anterior pituitary. (true/false)

6. In the testes, androgens are secreted by (Sertoli cells/Leydig cells).

7. Spermatogenesis is stimulated by testosterone and (FSH/LH), which targets Sertoli cells.

8. The head of a sperm contains chromosomes and a(n) _____, a vesicle containing enzymes needed for fertilization.

9. Erection is accompanied by a(n) (increase/decrease) in the activity of sympathetic neurons projecting to blood vessels in the penis.

10. Once sperm are deposited in the female reproductive tract, they cannot fertilize the oocyte until they have undergone a process called _____.

11. Fertilization usually occurs in the (uterus/uterine tube).

12. The second half of the ovarian cycle is called the (luteal/follicular) phase.

13. In a follicle, the oocyte is surrounded by a layer of (granulosa/theca) cells that provide it nourishment and regulate its development.

14. (FSH/LH) stimulates theca cells to secrete androgens, which are converted to estrogens by granulosa cells.

15. The inner layer of the uterine wall is called the (endometrium/myometrium).

16. During the (proliferative/secretory) phase of the uterine cycle, the lining of the uterus thickens under the influence of rising estrogen levels.

17. Ovulation is triggered by (FSH/LH).

18. The corpus luteum secretes estrogens and (LH/progesterone).

19. In the late follicular phase, LH secretion is stimulated by (estrogens/progesterone).

20. Degeneration of the corpus luteum causes hormonal changes that trigger (ovulation/menstruation).

21. Before implantation, the morula develops into an _____, which consists of an inner cell mass contained within a hollow, fluid-filled outer cell layer.

22. Secretory-phase uterine conditions are promoted by (estrogen/progesterone), which inhibits gonadotropin secretion during the last half of the ovarian cycle.

23. During the last six months of pregnancy, estrogens and progesterone are secreted by the (ovaries/placenta).

24. During labor, strong uterine contractions are induced by (prolactin/oxytocin).

25. Suckling stimulates the release of _____, which promotes milk production by the breasts.

Essay Questions

1. Compare the hormonal regulation of reproductive function in males with that in females during the early to mid-follicular phase, drawing as many parallels as possible. Be sure to consider the actions of hormones as well as the mechanisms that regulate their secretion.

2. Compare the steps in spermatogenesis with those of oogenesis, drawing as many parallels as possible.

3. Describe all the events that must occur before a sperm deposited in the female reproductive tract can fertilize an oocyte released from an ovary.

4. Describe the events that occur during an LH surge, including the actions of LH on target cells. Be sure to include a description of the events that trigger the surge and prime the target tissue's responses to it.

5. Describe the processes that give rise to elevated rates of estrogen and progesterone secretion during pregnancy. Explain how these hormones promote the maintenance of pregnancy and prepare the body for the

eventual birth and nourishment of the infant.

Critical Thinking

1. After fertilization, some zygotes will have the following three sex chromosomes: XXY. Will this zygote develop? If so, will it develop into a male or female? Explain.

2. Endometriosis is a disease in which some of the endometrial tissue of the uterus grows outside the uterus and attaches to organs in the abdominal cavity. What are two ways by which endometrial tissue can get into the abdominal cavity? The endometrial tissue outside the uterus responds to hormones just like normal endometrial tissue does. What changes occur to this abnormal tissue during a female cycle and what symptoms develop from these changes?

3. Some males are sterile because their sperm cannot fertilize an oocyte. What are some possible defects in sperm that would prevent them from fertilizing an oocyte?

4. The genes for red photopigments and green photopigments are located on the X chromosome. Explain how this makes males more prone to red-green color blindness.

Find the answers to these exercises, and additional study tools, at the Physiology Place (www.physiologyplace.com).

The Immune System

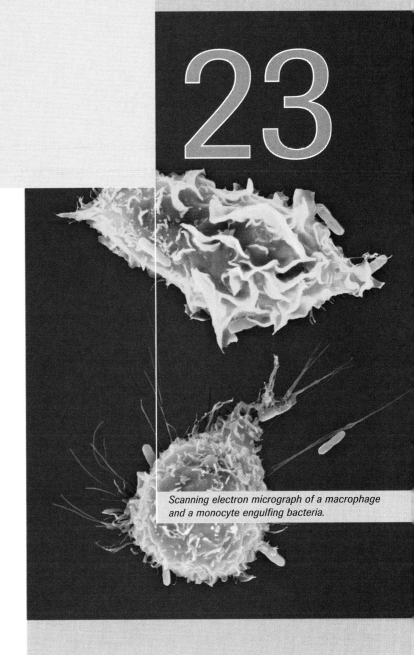

Scanning electron micrograph of a macrophage and a monocyte engulfing bacteria.

23

M ing is a bright first-year college student who is a champion off-road bicyclist. Usually in excellent health, Ming is surprised and dismayed when she develops a sore throat and a fever and feels so tired that she can't gather enough energy to so much as look at a textbook—or even at her bike! Her physician at the campus health center tells her she has a viral disease called *mononucleosis,* also known as "mono." She learns that there is no specific treatment for mono, but that in time she will feel better thanks to her *immune system*—the organs, tissues, circulating cells, and secreted molecules that resist and defeat infections. Like all of us, Ming understands the power of this system: When she had an infected cut, it healed; when she was ill with the flu, she got better; and when she was immunized against the bacterial infection diphtheria, she became protected from that disease for life.

▨ Identify the lymphoid organs and briefly describe their functions.

▨ Explain events that occur during inflammation and how each contributes to nonspecific defense. Describe how the skin and mucous membranes, interferons, natural killer cells, and the complement system contribute to nonspecific defense.

▨ Describe humoral immunity, or how B cells, through the production of antibodies, contribute to immune responses.

▨ Describe cell-mediated immunity, or how helper T cells and cytotoxic T cells contribute to immune responses.

▨ Explain how immunization can lead to protection from infectious disease.

▨ Discuss the major immunological issues regarding blood transfusion and organ transplantation.

▨ Explain how immune dysfunction can result in allergy, autoimmunity, or immunodeficiency.

Before You Begin

Make sure you have mastered the following topics:

1. *Phagocytosis, p. 125*

2. *Endocytosis and exocytosis, p. 125*

3. *Cytokines, p. 138*

4. *Composition of blood, p. 498*

5. *Lymphatic system, p. 476*

In the preceding chapters we have learned a great deal about how physiological systems function to maintain health. In this chapter we see how the body remains healthy even as disease-causing organisms and nonliving substances gain access to it via air, food, and water. **Immunity** refers to the immune system's capacity to protect individuals from disease by recognizing and eliminating potentially *pathogenic* (disease-generating) agents, including bacteria, bacterial toxins, viruses, parasites, and fungi. In addition to resisting foreign agents, the immune system disposes of unneeded components of the body, including aging cells and cellular debris present in diseased tissue; participates in healing wounds; and sometimes recognizes and eliminates mutant cells that may develop into cancer. As it seeks out and recognizes materials foreign to the body, the immune system also rejects tissues and cells that are not identical to "self" (that is, to the cells and tissues of the individual in question); such rejection poses the primary obstacle to organ transplantation. The presence in the body of such foreign and abnormal substances induces the immune system to develop an **immune response**, a complex series of physiological events that culminates in the destruction and elimination of these substances. Sometimes, however, immune responses are inappropriate and lead to disease. People with allergies, for example, experience exaggerated immune responses to otherwise benign foreign materials such as dust, pollen, or peanuts. In addition, some people suffer from *autoimmune diseases* in which a person's immune system attacks his or her own tissues and cells. Rheumatoid arthritis, diabetes mellitus, and multiple sclerosis are just a few examples of the many diseases that can result from autoimmune responses.

Anatomy of the Immune System

The immune system consists of two components: leukocytes (white blood cells), which are responsible for producing a wide range of immune responses, and **lymphoid tissues**, such as the bone marrow, thymus, spleen, lymph nodes, and tonsils, in which leukocytes develop, reside, and come into contact with foreign materials. We discuss each of these components in detail in the following sections.

Leukocytes

There are five major types of leukocytes, described briefly in Chapter 16. The granulocytes—neutrophils, eosinophils, and basophils—have cytoplasmic granules that contain secretory products; the release of these products is called *degranulation*. The agranulocytes—monocytes and lymphocytes—lack granules.

Phagocytes

Neutrophils, eosinophils, monocytes, and macrophages are all phagocytes—they can engulf foreign particles and microorganisms, thereby removing them from blood and tissues. Neutrophils are the most abundant of the leukocytes and are particularly important in fighting bacterial infections. An elevation in the number of neutrophils in blood is used clinically as a determinant of infection. In addition to phagocytosis, neutrophils release several cytokines involved in *inflammation*. Although eosinophils are also phagocytes, their primary role in the immune system is to defend against parasites by releasing toxic substances.

Monocytes are phagocytes in the blood, but they differentiate into macrophages in the tissue to phagocytose tissue foreign matter. Macrophages are five to ten times larger than monocytes and have greater activity (both phagocytic and secretory activity are enhanced). Macrophages are found in almost all tissues of the body, with some localized to certain tissue only (*fixed macrophages*) and others able to migrate to different tissues (*free* or *wandering macrophages*). Fixed macrophages include microglia (Chapter 10), alveolar macrophages (Chapter 17), *Kuppfer cells* in the liver, and *histiocytes* in connective tissue.

Lymphocytes

Lymphocytes are unique in that they provide the immune system with diversity, specificity, memory, and the ability to distinguish between self and nonself. Lymphocytes are of three major types: **B lymphocytes (B cells), T lymphocytes (T cells)**, and **null cells**, so called because they lack cell membrane components that are characteristic of B cells and T cells. Most null cells are large, granular lymphocytes known as **natural killer (NK) cells** (described shortly).

When B cells contact foreign or abnormal molecules known as **antigens**, they develop into **plasma cells**, which secrete antibodies. **Antibodies**, also known as **immunoglobulins**, are proteins present in the plasma and interstitial fluid that target specific antigens for destruction. If, for example, a B cell contacts the bacterium *Staphylococcus aureus*, it develops into a plasma cell that secretes antibodies that bind only to *S. aureus*. These antibodies do not damage the bacteria by themselves but instead mark them for destruction by various mechanisms that are described later in the chapter.

T cells, in contrast, do direct damage to foreign cells. T cells contact infected cells, mutant cells, and transplanted cells and then take several days to develop into active **cytotoxic T cells** that destroy the infected or abnormal cells. T cells kill by secreting molecules that form pores in the target cell's membrane; once it has a perforated membrane, the target cell succumbs to lysis, a process in which it fills with fluid and bursts. The roles of B lymphocytes and T lymphocytes in the body's immune response are discussed in greater detail later in the chapter.

Even though natural killer cells constitute only a small proportion of circulating lymphocytes, they are important in fighting viral infections. Viruses, unlike bacteria, must enter cells to reproduce, and by killing virus-infected cells NK cells limit the production of new viruses in the body. NK cells kill by a mechanism similar to that used by cytotoxic T cells but differ from them in how they recognize their target cells and by exhibiting immediate readiness, which enables them to respond well before B cells and T cells as an essential part of early immune responses. NK cells also recognize and kill mutant cells that may develop into cancer.

Lymphoid Tissues

All leukocytes (and erythrocytes as well) develop from precursor cells called *hematopoietic* (blood-forming) *stem cells,* located in the bone marrow. Whereas B lymphocytes and most other leukocyte types come to full maturity in the bone marrow, T lymphocytes must migrate to the thymus gland (located in the thoracic cavity above the heart) before they develop to maturity. Because the bone marrow and the thymus (and the fetal liver as well) are the sites of lymphocyte maturation, they are called **central lymphoid tissues** (**Figure 23.1**). After B cells and T cells reach maturity, they migrate from central lymphoid tissues to the various sites in the body where they are most likely to contact foreign substances. These sites, known as **peripheral lymphoid tissues**, include the spleen, lymph nodes, tonsils, adenoids, appendix, lymph nodules of the gastrointestinal tract, and regions in the lining of the gastrointestinal tract called *Peyer's patches,* which are essentially collections of B cells, T cells, and macrophages.

Each of the peripheral lymphoid tissues contains a dense network of cells that trap microorganisms and foreign particles, and each is located where it can ensnare invaders soon after they enter the body. While the spleen is collecting worn-out erythrocytes from the blood, it is also collecting bloodborne microorganisms and foreign particles. Once trapped in the spleen, these particles are eventually cleared by the actions of macrophages and lymphocytes. Similarly, microorganisms and particles carried in lymph are trapped by lymph nodes, which are found throughout the body where lymphatic vessels converge. Whereas the macrophage and lymphocyte networks of the spleen and lymph nodes filter blood and lymph, respectively, those of the tonsils and adenoids trap inhaled particles and microorganisms, and those of the appendix, lymph nodules, and Peyer's patches trap substances that enter the body in ingested food or water.

Quick Test 23.1

1. What types of leukocytes are phagocytes?

2. Name the two central lymphoid tissues. Why are they known as *central* lymphoid tissues?

3. List six peripheral lymphoid tissues. What general role do these tissues play in immunity?

Central Lymphoid Tissues **Peripheral Lymphoid Tissues**

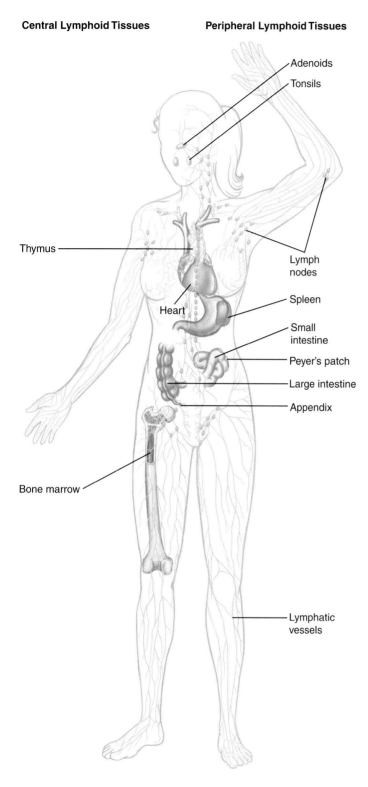

Figure 23.1 Lymphoid tissues. Lymphocytes reach maturity in the central lymphoid tissues and interact with foreign antigens in the peripheral lymphoid tissues. These lymphoid tissues are interconnected by blood vessels (not shown) and lymphatic vessels through which lymphocytes circulate.

A few days after infection with chicken pox a person experiences viremia, a high level of viruses in the blood. In which lymphoid tissue are these viruses most likely to contact lymphocytes?

Organization of the Body's Defenses

The body's defenses are categorized as either nonspecific or specific, depending on their selectivity for the invader and when the response occurs. The mechanisms that constitute **nonspecific defenses** defend against potentially harmful substances without regard to their precise identity. These mechanisms operate even before foreign material enters the body, in the form of the skin and mucous membranes—the body's first line of defense against infection. If those barriers are broken, internal nonspecific defenses immediately begin to operate. Nonspecific mechanisms also clear wounds and damaged tissue of debris and contribute to healing. **Specific immune responses**, in contrast, are highly selective (meaning they target specific substances) and come into play after nonspecific responses have already begun. These responses are specific because they are mediated by lymphocytes, which are uniquely designed to recognize particular substances and aid in their destruction. Unlike nonspecific responses, which always operate with about the same speed and effectiveness, specific responses strengthen with each exposure to a particular offending agent.

Nonspecific Defenses

Intact skin and the mucous membranes that line the digestive, respiratory, urinary, and reproductive tracts provide excellent initial barriers to most bacteria and viruses. Mucous membranes also produce viscous mucus, which bathes the surfaces of exposed epithelia and can trap foreign matter and potential pathogens (disease-causing agents). Microorganisms that enter the upper respiratory tract, for example, are often caught in mucus and are then coughed into the mouth and swallowed, an action that is enhanced by ciliated epithelial cells that line the trachea. In addition to acting as physical barriers, the skin and mucous membranes provide chemical defenses: Secretions from sebaceous glands (which secrete an oily product to the skin surface) and sweat glands give the skin a pH ranging from 3 to 5, which is acidic enough to prevent colonization by many pathogens. (Bacteria that are normally found on the skin are adapted to its acidic, relatively dry environment.)

As soon as a potential pathogen or foreign matter enters the body, internal nonspecific defenses are quickly initiated, even if the body has not been previously exposed to that pathogen or material. The body's internal nonspecific defenses include (1) **inflammation**, a complex series of events that culminates in the accumulation of proteins, fluid, and phagocytic cells in an area of tissue that has been injured or invaded by microorganisms; (2) **interferons**, a family of related proteins that are

Figure 23.2 Major events in local inflammation. Events are described in detail in the text.

> *Drugs known as antihistamines block the action of histamine. Explain how antihistamines diminish the symptoms of inflammation.*

secreted by leukocytes and virus-infected cells and can induce other cells to resist infection by the virus; (3) natural killer (NK) cells, which provide early defenses against virus-infected cells and cancer cells by recognizing and destroying them; and (4) the **complement system**, a set of plasma proteins that, once activated by a series of stepwise reactions, act to lyse foreign cells, especially bacteria. We consider each of these internal nonspecific defense mechanisms next.

Inflammation

Microbial invasion or damage to tissue triggers a complex series of events that rapidly lead to inflammation of the affected tissue. In fact, the tissue space, rather than the bloodstream, is the main site of defensive action against infection. Five major events occur in inflammation, typically in the following order: (1) nearby macrophages engulf debris and foreign matter, (2) nearby capillaries dilate and become more permeable to proteins and fluid, (3) foreign matter is contained, (4) additional leukocytes migrate into the region, and (5) recruited leukocytes continue to help clear the infection, mainly by phagocytosis. Note that these events are nonspecific; inflammation proceeds in much the same way regardless of which bacteria, virus, or injury triggered it. In the following subsections we examine the major events in inflammation by considering the body's response to an injury in which the skin is cut and abraded.

Phagocytosis of Pathogens and Debris by Nearby Macrophages Macrophages already present in affected tissues quickly detect bacteria introduced into the cut using receptor proteins on their surfaces that can bind to molecules on the surfaces of many different types of bacterial cells. The resulting attachment initiates phagocytosis (**Figure 23.2**a), and the macrophages begin to engulf the bacteria. Attaching to bacteria in this way also stimulates the macrophages to secrete cytokines, proteins that are secreted by cells in response to a stimulus (in this case, contact with bacteria) and affect the behavior of other nearby cells (see Chapter 5). As we will see, the cytokines secreted by such activated macrophages contribute to subsequent steps in inflammation (and to immune responses as well). The phagocytic activity of these macrophages is important in limiting the spread of bacteria early in infection, but these cells are usually too few in number to eliminate all the foreign cells and debris present at an injury site.

> *Inhibition of capillary dilation reduces redness and heat; inhibition of increased capillary permeability reduces edema.*

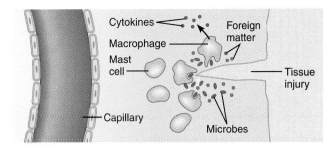

(a) Phagocytosis by nearby macrophages

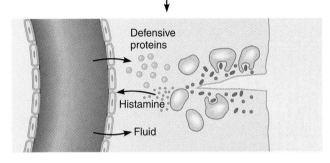

(b) Dilation and increased permeability of capillary

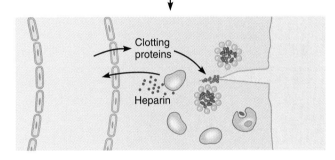

(c) Containment of bacteria and foreign matter

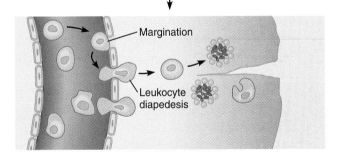

(d) Leukocyte proliferation and migration

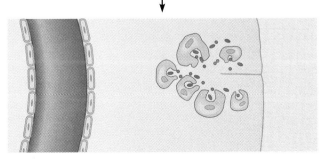

(e) Continued activity of recruited leukocytes

Dilation and Increased Permeability of Capillaries

Within minutes of bacterial invasion, nearby blood vessels dilate (Figure 23.2b), which increases local blood flow. The capillary walls also become more permeable, allowing proteins and fluid normally contained in the plasma to move into tissue spaces. The increased blood flow brings additional leukocytes and defensive proteins into the local circulation, and the increased capillary permeability allows these proteins to move into the tissues where they are needed. In addition, the leukocytes that gather in these dilated vessels also migrate from the blood into the tissue spaces by a process we examine shortly.

Both the *vasodilation* and the increased capillary permeability are induced by *histamine* released from damaged **mast cells**, which are cells that are dispersed throughout the body's connective tissues and are similar to the basophils in the blood. Because mast cells are especially numerous in submucosal tissue and the dermis (a tissue layer in the skin), they are prone to damage when the skin is injured and are thus poised to stimulate the early events of inflammation by releasing histamine.

The vascular changes mediated by histamine are ultimately responsible for the four characteristic symptoms of inflammation (from the Latin *inflammo,* "to set on fire"): redness, heat, edema (swelling), and pain. The increased blood flow both reddens the tissue (which is more or less apparent depending on skin tone) and makes it warmer. Both the histamine-induced increase in blood flow and capillary permeability contribute to the edema that follows. As the capillaries become engorged with blood, the resulting increase in hydrostatic pressure plus the increased interstitial oncotic pressure that accompanies the leakage of plasma proteins causes fluid to filter out of the capillaries and into the tissue spaces, resulting in edema. The edema exerts pressure against the surrounding tissue and skin, which contributes to pain, as does the production of pain-inducing chemicals such as *bradykinin* that stimulate nearby sensory neurons. Thus even though these vascular changes bring some discomfort, they also help to gather nonspecific defenses to the site of injury.

Containment of Foreign Matter

Early in the process mast cells and basophils also release the anticoagulant heparin, which temporarily suspends blood clotting and allows leukocytes unimpeded access to the area of tissue injury. In time, however, clotting factors that have leaked from the plasma into the tissue become active and form clots around clusters of bacteria, thereby inhibiting their spread within the body (Figure 23.2c). With the aid of other plasma proteins and also proteins released from damaged tissues, the process of clot formation continues and effectively walls off the region of damage and infection. Eventually the portion of the clot that is at the skin's surface dries and hardens, forming a scab.

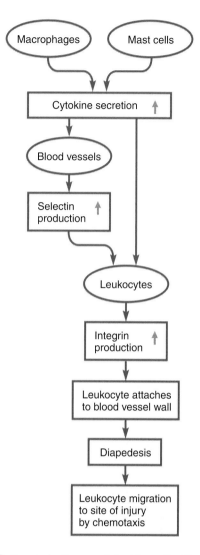

Figure 23.3 Events in the regulated transit of leukocytes from the bloodstream to tissues.

Leukocyte Migration and Proliferation

As previously mentioned, macrophages already present in the damaged tissue are the first to phagocytose bacteria and debris, but soon they are joined by other phagocytic cells. About an hour after the injury, neutrophils accumulate in great numbers within the affected tissue; about ten hours later, monocytes begin to move to the tissue, where they develop into large, active macrophages.

The signaling that tells leukocytes to move through blood vessel walls at the right location in the body is achieved by a process of regulated transit involving four events: margination, attachment, diapedesis, and chemotaxis (Figure 23.2d). *Margination,* the movement of leukocytes toward the blood vessel wall, begins when cytokines released from macrophages and mast cells at the site of injury induce nearby blood vessels to produce *adhesion molecules* called *selectins*. Selectins protrude from the vessel's interior wall and attach loosely to leukocytes as they pass by, stalling their forward progress. This gives leukocytes the chance to receive activating signals from other

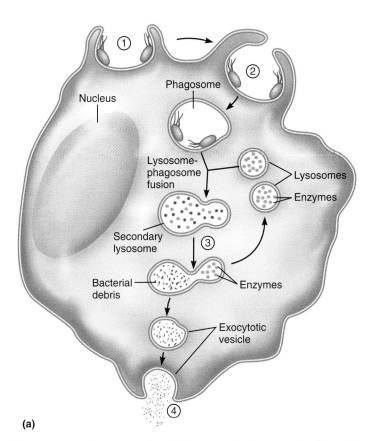

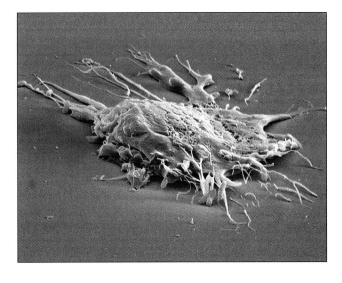

Figure 23.4 **Phagocytosis.** (a) The steps of phagocytosis: ① attachment, ② internalization, ③ degradation, and ④ exocytosis. (b) Electron photomicrograph of a macrophage phagocytosing bacteria.

cytokines, which, in turn, signal them to form other adhesion molecules called *integrins* that bind the cells tightly to the blood vessel wall. *Attachment* is soon followed by the cell's transit across the wall, known as *diapedesis* ("jump across"). In this step the leukocyte essentially "crawls" between endothelial cells of the blood vessel wall and through the basement membrane beneath. Once in the tissues, leukocytes move steadily and directly toward the point of injury, attracted by chemicals released from bacteria and injured tissues themselves, in a process called *chemotaxis*. Phagocytic leukocytes are thus delivered to the place where they are most needed: the original site of injury and bacterial invasion. Once at the site, phagocytes contact and engulf additional bacteria and debris. The process whereby the regulated transit of leukocytes occurs is depicted in **Figure 23.3**.

The hereditary disease called *leukocyte adhesion deficiency (LAD)* clearly demonstrates that the migration of phagocytic leukocytes from blood to infected tissues is an essential component of defense. In people with LAD, leukocytes possess faulty integrins, so the cells are unable to bind tightly to the blood vessel wall and therefore cannot cross into the tissues. Without the aid of additional phagocytes, tissue-borne infections are liable to fester and spread, and for these reasons people with this disease suffer from frequent, severe bacterial infections.

Note that the mass movement of neutrophils and monocytes from blood to tissues during inflammation does not diminish the number of circulating leukocytes. In fact, a common sign of bacterial infection is *leukocytosis,* a four- to fivefold increase in the number of circulating neutrophils. The reason for this increase is that the cytokines secreted by macrophages eventually reach the bone marrow, where they stimulate the proliferation and release of neutrophils and, later, monocytes into the circulation.

Continued Clearing of Infection by Recruited Leukocytes
The inflammatory processes described so far have brought defensive proteins (such as clotting factors) and phagocytic leukocytes (neutrophils and macrophages) to the site of injury. We have seen that phagocytes perform two important functions: They engulf bacteria and debris, and they secrete cytokines that further regulate inflammation and other defensive mechanisms. Let's take a closer look at these two important tasks.

Phagocytosis is achieved in four steps: attachment, internalization, degradation, and exocytosis (**Figure 23.4**). The first step, *attachment,* enables the phagocyte to distinguish between substances that should be engulfed and those that should not. The selectivity of a phagocyte for its target material is mainly determined by how well the two adhere. Phagocytes tend to bind to damaged and dead cells with irregular, rough surfaces, but not to healthy cells. Phagocytes also attach to many types of bacteria. Sometimes, attachment is enhanced by **opsonins**, proteins (including antibodies) that bind tightly to the foreign material and make it easier for the phagocyte to engulf it. Phagocytosis is greatly enhanced by this binding of opsonins, a process known as *opsonization* ("to make tasty"); a macrophage can engulf and destroy a substance about 4000 times faster when it is coated with antibodies.

As soon as the phagocyte has attached to its prey, *internalization* occurs. In less than 0.01 second, the phagocyte's plasma membrane extends outward around the site of attachment and surrounds the material, enclosing it within a large intracellular vesicle called a *phagosome*. The phagosome moves toward the cell interior and fuses with a *lysosome*, which contains a variety of digestive enzymes. *Degradation* of the phagocytosed material then occurs within the enlarged, enzymatically active organelle, now called a *secondary lysosome*. Subsequently the phagocyte uses some of the harmless bacterial debris (amino acids, for example) and eliminates others by *exocytosis*.

Attachment of a phagocyte to foreign cells and debris not only initiates phagocytosis but also induces the phagocyte to accomplish its other major task: the secretion of cytokines that help prolong and coordinate the body's responses to infection and injury. These cytokines are varied, and exert a wide range of effects on the body. Some of them are called *interleukins* because they act as chemical messengers that send signals between (*inter-*) leukocytes (*-leukins*). We focus in particular on three cytokines secreted by activated macrophages: *interleukin-1 (IL-1), interleukin-6 (IL-6),* and *tumor necrosis factor-alpha (TNF-α).* These cytokines act individually and collectively to induce a number of changes, including synthesis of more adhesion molecules by blood vessel endothelial cells and release of more neutrophils from bone marrow.

In addition to these actions, IL-1, IL-6, and TNF-α also act on the hypothalamus to raise body temperature; that is, they function as *endogenous pyrogens*. They achieve this function by stimulating the hypothalamus to release prostaglandins, which in turn adjust the body's "thermostat" setting, raising the temperature above normal. Although a high fever can be dangerous, a moderate one is generally thought to benefit host defense; elevated temperatures are thought to decrease the rates of bacterial and viral replication, and to accelerate phagocytosis and other defensive reactions.

The cytokines IL-1, IL-6, and TNF-α also stimulate liver cells to produce *acute phase proteins,* a group of proteins having a wide range of antibacterial and inflammatory effects and whose plasma concentrations increase soon after an infection begins. One of these is *C-reactive protein,* which binds to the surface of many types of

bacteria. A bacterium bound by C-reactive protein molecules is opsonized, and therefore is more susceptible to phagocytosis.

Finally, IL-1 helps induce the proliferation and differentiation of B lymphocytes and T lymphocytes. B cell differentiation leads to the production of antibodies that mark selected foreign substances for destruction, and T cell differentiation leads to *cell-mediated immunity,* in which certain types of T cells kill specific abnormal or infected body cells.

Interferons

A second nonspecific defense mechanism prevents the spread of viruses within the body by interfering with virus replication. This defense is provided by a group of related proteins appropriately named interferons. The secretion of two interferons, called *interferon-α* and *interferon-β* (beta), from virus-infected cells signals to the surrounding cells the presence of the virus and induces the cells' resistance.

The viral nucleic acid that accumulates in a virus-infected cell during viral replication stimulates the cell to synthesize and secrete interferons. Even though the infected cell will likely die, the interferons it secretes bind to nearby healthy cells, initiating in them a series of intracellular changes that cause the cells to become more resistant to the virus. This virus-resistant state is conferred by the presence of RNA-degrading enzymes and protein synthesis inhibitors in the cytoplasm. Because these enzymes and inhibitors are potentially dangerous to the cell itself, they come into play only after viral infection and last only a short time. Interferon-induced cells are thus poised to block the production of new viruses, but they can do so for a limited time only. Both the production of interferon and the virus-resistant state it induces are nonspecific; virtually any viral nucleic acid can induce interferon production, and interferon-induced resistance can defeat virtually any viral infection.

A third type of interferon, *interferon-γ* (gamma), is secreted not by virus-infected cells but by active T cells and NK cells, and it has a broader range of effects. In addition to inhibiting viral replication, interferon-γ enhances phagocytosis in macrophages, boosts antibody production in B cells, and helps to activate natural killer cells and cytotoxic T cells, both of which kill virus-infected cells and cancer cells. In addition, by inhibiting cell division, interferon-γ has a direct effect on cancer cells, suppressing the growth of tumors.

Natural Killer Cells

We have already encountered the third nonspecific defense mechanism: natural killer (NK) cells. NK cells function in early nonspecific defense by recognizing the general features of infected or abnormal cells and releasing **perforins**, proteins that form pores in the

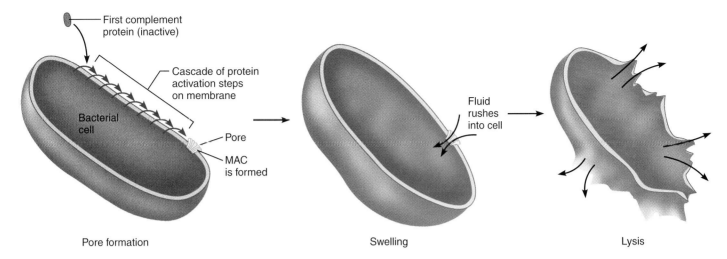

Pore formation   Swelling   Lysis

Figure 23.5 Actions of the complement system. The complement system can be triggered when the first in a series of complement proteins binds to the surface of a bacterial cell. This mode of complement activation is nonspecific because it occurs on any of a broad range of bacterial types. After a series of activation steps, some of the complement proteins form a membrane attack complex (MAC), which inserts into the cell membrane as a protein-lined channel or pore. Fluid rushes into the cell, causing it to swell and burst (undergo lysis).

infected or abnormal cell's membrane. The pores cause an increase in permeability of the membrane to ions and water. As fluid moves into the cell, the cell swells and eventually lyses.

In the case of virus-infected cells, NK cells can act on them without detecting the virus itself, and NK cells are ready to act immediately. NK cells thus provide an essential, general defense in the early stages of viral infection, and of tumor growth as well. These cells are also brought into play during immune responses because their activities can be enhanced by the cytokine *interleukin-2 (IL-2)* and by antibodies, which are produced by T cells and plasma cells, respectively, during specific responses.

The Complement System

The fourth and final nonspecific defense mechanism, the complement system, is so named because it completes or fulfills ("complements") the actions of specific antibodies. However, the system can also act in the absence of antibodies. The system consists of about 30 plasma proteins that act to destroy invading microorganisms, especially bacteria. The first in a series of complement protein reactions occurs in association with a bacterium and leads to a cascade of activation steps in which each component activates the next in the series (**Figure 23.5**). The cascade occurs at the bacterial surface and ends with the development of a *membrane attack complex (MAC)*, a collection of pore-forming proteins that pierce the bacterial membrane. The insertion of these channels into the membrane causes the cell to lose its integrity, so it fills with fluid, swells, and then bursts (Figure 23.5). Such complement-mediated lysis is the primary way by which antibody-coated bacteria are killed.

The complement cascade can be activated in two ways: (1) by binding directly to carbohydrates present on the surface of a broad range of bacterial cells (known as the *alternative pathway*), or (2) by binding to antibodies that are already attached to bacterial cells (known as the *classical pathway*). The alternative pathway is important in nonspecific, early responses to infection because it occurs in the absence of antibodies (that is, before they are secreted in response to an infection). The classical pathway, by contrast, becomes important only in the later stages of an infection because it requires antibodies. Note that whereas the alternative pathway is nonspecific, the classical pathway is specific because it involves antibodies. This is another example in which nonspecific and specific mechanisms converge; here, complement is essential to effective antibody action because antibodies can target a microorganism for destruction but cannot directly kill it.

Although the complement cascade is often viewed in terms of its final outcome—lysis of bacteria—a number of activated complement proteins also contribute to the development of inflammation. Because complement proteins are activated in the vicinity of the bacteria and accumulate there, they are appropriately positioned to assist in antibacterial defense. Indeed, some complement proteins act in chemotaxis, guiding phagocytes into the area; others bind to nearby mast cells and induce them to release histamine, which causes vasodilation and increased capillary permeability. Finally, one specific protein of the complement system, known as C3b, is generated in great quantity in the cascade and coats bacterial surfaces; because it acts as an opsonin, it enhances phagocytosis of these bacteria.

Specific Defenses: Immune Responses

In the previous section we saw that the skin and mucous membranes constitute the body's first line of defense; that inflammation, interferons, NK cells, and the complement system form a second line of nonspecific defense; and that these mechanisms respond rapidly to injury or exposure to foreign material or infective agents, even if such exposure is the initial one. But nonspecific defenses are not always completely successful in eliminating foreign materials; fortunately, the body has an exquisite and powerful mechanism in reserve: the immune response.

Suppose that the body's nonspecific defenses were not completely effective in responding to the cut and abrasion we previously discussed. While these defenses continue to operate, bacteria, bacterial fragments, and other foreign molecules gain access to the bloodstream and become trapped in the netlike architecture of the spleen. Likewise, bacteria present in interstitial fluid are carried into lymphatic vessels and eventually into the netlike lymph nodes, which swell and become tender. In the spleen, lymph nodes, and other lymphoid tissues, the bacteria come into contact with B lymphocytes and T lymphocytes, thereby inducing these cells to generate efficient and selective immune responses that work throughout the body to eliminate the invaders.

B lymphocytes and T lymphocytes each generate a particular kind of immune response. B lymphocytes develop into plasma cells that secrete antibodies, the actions of which bring about **humoral immunity**, so called because antibodies circulate in the blood and lymph, body fluids long ago called "humors." In contrast, certain T lymphocytes develop into active cytotoxic T cells, which bind to and kill abnormal body cells. This type of immune response constitutes **cell-mediated immunity**, so called because cytotoxic T cells must come into direct contact with their targets in order to act on them.

The circulating antibodies of the humoral response defend mainly against bacteria, toxins, and viruses present in body fluids. In contrast, the T cells of the cell-mediated response are active against bacteria and viruses that are hidden within infected body cells. Moreover, the cell-mediated response operates in the body's reaction to transplanted tissue and cancer cells, both of which are perceived as foreign.

Features of Immune Responses

About 2400 years ago, Thucydides of Athens described how those sick and dying of plague were attended by others who had recovered, "for no one was ever attacked a second time." We have all made similar observations: Those of us who experienced many childhood diseases do not worry about getting them again because exposure to each disease confers lifelong immunity to that disease. In this section we explore the nature of specific immune responses by addressing the following questions: Why does a measles infection generate a response to that disease but to no other disease (a phenomenon known as *specificity*)? How is it that the immune system is already poised to defend against a first exposure to chicken pox (which relates to a property known as *diversity*)? How do we acquire our lifelong immunity to chicken pox after the first exposure (which depends on the immune system's *memory*)? Finally, why does the immune system respond against foreign microbes but not against the body's own tissues (a phenomenon known as *self-tolerance*)? As we will see, these special features of the immune response—specificity, diversity, memory, and self-tolerance—derive from the nature of B lymphocytes and T lymphocytes.

Specificity B cells and T cells bind and respond to foreign or abnormal molecules known as *antigens.* Antigens (a term arising from *anti*body *gene*rators) are typically complex protein or polysaccharide components of viruses, bacteria, fungi, protozoa, parasitic worms, pollen, transplanted tissue, and tumor cells. Each antigen has a unique structure and contains different recognition sites called **epitopes** or *antigenic determinants,* each of which can be detected by specific lymphocytes, which then target that invader for destruction.

B cells and T cells are able to recognize specific antigens because they possess antigen-binding proteins called *antigen receptors* on their surfaces. All of the receptors on a particular B cell or T cell bind to certain antigens only and thus resemble other receptors that bind specific ligands, such as neurotransmitter receptors and hormone receptors. The antigen receptors on B cells and T cells are similar to receptors on antibodies.

A typical antibody is a Y-shaped molecule consisting of four protein chains: two identical *heavy chains* and two identical *light chains* that are joined by disulfide bridges (**Figure 23.6**a). Each antibody molecule possesses *variable regions (V)* and *constant regions (C).* The constant regions consist of portions of the heavy and light chains that make up the tail of the Y. There are five different types of heavy chains, which are the basis of five classes of antibodies, each with its own mechanism of eliminating antigen. The classes of antibodies and their mechanisms of action are described later in the text.

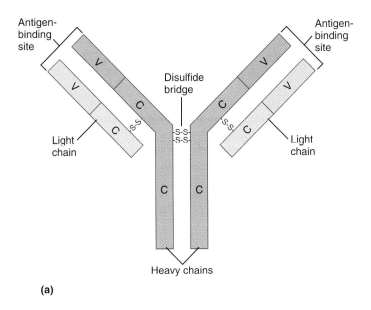

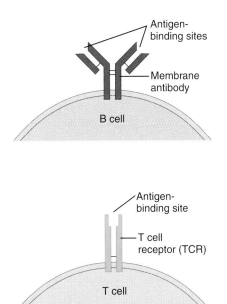

Figure 23.7 B cell and T cell antigen receptors. Antigen receptors on B cells are called membrane antibodies (or membrane immunoglobulins); antigen receptors on T cells are called T cell receptors (TCRs). Each lymphocyte has about 100,000 identical receptors specific for a particular antigen.

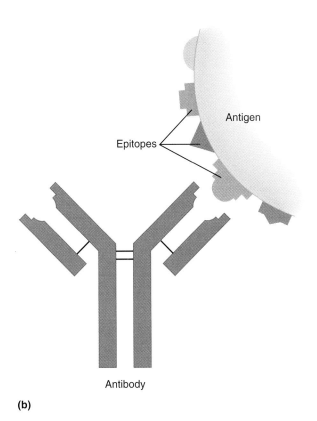

Figure 23.6 The basis of antigen–antibody specificity. (a) Each antibody molecule consists of two identical heavy chains and two identical light chains linked by disulfide bonds. The combined structures of the variable (V) regions form the antigen-binding sites of the molecule, which vary from antibody to antibody. (b) A schematic representation of how the complementary physical structures of an antigen's epitope and the antibody's antigen-binding site constitute antigen–antibody specificity.

The variable regions of an antibody, so called because the amino acid sequences in these regions vary extensively from antibody to antibody, consist of portions of heavy and light chains that make up the top of the Y. The two identical variable regions form two identical antigen-binding sites. Thus each antibody molecule is capable of binding two epitopes of the same kind.

The interaction between the antigen-binding site and its antigen is similar to that between an enzyme and its substrate (see Toolbox: Ligand-Protein Interactions, p. 73), which also shows specificity: The unique shape of the binding site allows for a close fit between the antibody and its antigen (Figure 23.6b), such that multiple noncovalent bonds can form between chemical groups on the respective molecules.

The antigen receptors on B cells are similar to antibody molecules except that the receptors are bound to the plasma membrane (**Figure 23.7**), whereas antibodies are secreted into the extracellular fluid. For this reason, B cell antigen receptors are often called *membrane antibodies* or *membrane immunoglobulins*. A B cell's membrane antibodies have the same specificity as the antibodies it later secretes as a plasma cell.

The antigen receptors on T cells, called *T cell receptors (TCRs)*, come from the same family of proteins that includes antibodies but are different in structure from antibodies (Figure 23.7). Moreover, unlike antibodies, T cell receptors act only as cell surface receptors for antigen. They are never secreted.

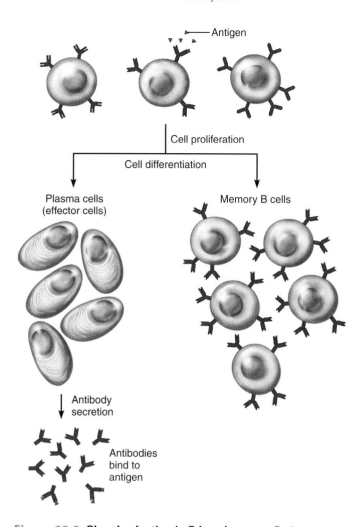

Figure 23.8 Clonal selection in B lymphocytes. Each lymphocyte produces a randomly generated set of identical antigen receptors. When a foreign antigen and a lymphocyte's receptors are sufficiently complementary in structure, binding induces the lymphocyte to proliferate and differentiate into a population (clone) of short-lived effector lymphocytes (in this case, plasma cells) and a clone of long-lived memory lymphocytes. Note that clonal selection also occurs in T lymphocytes.

Diversity   A single T lymphocyte or B lymphocyte has about 100,000 antigen receptors, all with the same specificity. Therefore, each B cell and T cell can detect just a few of the millions of possible antigens that might gain entry to the body. The particular antigen receptor molecules a given lymphocyte produces are determined by random genetic events that occur early in the development of the lymphocyte. These early events, which are unique to each lymphocyte and occur prior to any contact with foreign antigen, generate a phenomenal array of B lymphocytes and T lymphocytes in the body, each bearing antigen receptors of a particular specificity. With this diversity of lymphocytes, the immune system has the capacity to recognize and respond to millions of different antigens—even to antigens as yet unseen in the universe!

The specificity of B cell and T cell receptors explains the specificity of the B cell and T cell responses. When a particular microorganism invades the body, it interacts with and activates only those lymphocytes that have receptors specific for the antigens it possesses. When the virus that causes chicken pox invades the body, for example, only lymphocytes specific for chicken pox antigens are activated to proliferate (by successive cell divisions) and differentiate. In other words, the foreign antigen triggers an immune response against itself. This antigen-driven activation of lymphocytes is called **clonal selection**, and is absolutely necessary for immune responses (**Figure 23.8**). Lymphocyte differentiation gives rise to two populations (clones) of cells: **effector cells** (such as plasma cells), which are short-lived cells that combat the same antigen that stimulated their production, and **memory cells**, which are long-lived cells bearing membrane receptors specific for the same antigen. Each antigen, by binding to specific receptors, selectively activates a tiny fraction of cells from the body's diverse array of lymphocytes. This relatively small number of selected cells then gives rise to clones of thousands of cells, all specific for and dedicated to eliminating that particular microorganism.

Memory   The antigen-induced lymphocyte changes that occur when a person is first exposed to an antigen constitute a **primary immune response**. In the primary response, antigen-selected B cells and T cells proliferate and differentiate into effector cells (antibody-producing plasma cells and cytotoxic T cells, respectively) about 10–17 days after exposure to the antigen. Often, a person becomes ill during this time, because it takes only a few days for most viruses or bacteria to cause symptoms. Eventually, however, the symptoms of illness diminish and disappear as antibodies and cytotoxic T cells help clear the offending agent from the body.

Once someone has suffered through an infection, he or she has likely become immune to further infection by that same microorganism. The basis of this so-called *acquired immunity* is that upon subsequent exposures to that same antigen, the response (called a **secondary immune response**) is much faster (only 2–7 days), greater in magnitude, and more prolonged than a primary response (**Figure 23.9**). The secondary response results from the existence of immunological memory, which is due to the fact that each exposure to an antigen gives rise not only to effector cells but also to clones of long-lived memory T cells and memory B cells. Upon subsequent exposure to the same antigen, memory cells are poised to quickly proliferate and differentiate into still more memory and effector cells.

Self-Tolerance   Given that the diverse repertoire of lymphocyte specificities is randomly generated, how is it that an individual's B cells and T cells do not react to the body's proteins? The answer is that as B cells and T cells mature in the bone marrow and thymus, their antigen receptors are in effect tested for their potential to

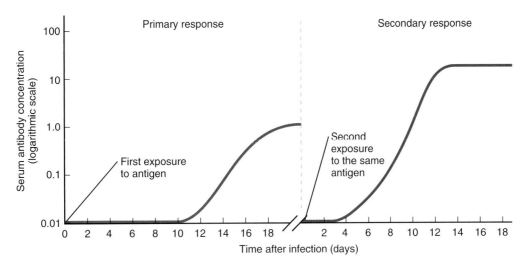

Figure 23.9 Primary and secondary responses to a given antigen. In secondary responses, serum antibody concentrations rise sooner, reach a higher level, and remain elevated longer than those stimulated by the initial exposure to the antigen.

If a person's initial exposure to antigen X coincided with a secondary exposure to antigen Y, what kind of response to antigen X would occur?

recognize and react against self. In general, those lympho-cytes whose receptors have the potential to react with self-molecules are either rendered nonfunctional or undergo *apoptosis*, the self-destructive events that accompany *pro-grammed cell death*. As a result of these processes, only those lymphocytes that are reactive against foreign *(nonself)* mol-ecules remain. This capacity to distinguish self from non-self, known as *self-tolerance*, continues to develop even as the lymphocytes migrate to lymphatic organs. Failure of self-tolerance can lead to various autoimmune diseases (discussed later).

Quick Test 23.3

1. What is humoral immunity? What lymphocyte type is responsible for this kind of immunity?

2. What is cell-mediated immunity? What lymphocyte type is responsible for this kind of immunity?

3. List the four key features of immune responses, and provide a brief explanation of each.

4. What is the basis for the stronger and more rapid immune response upon secondary exposure to an antigen?

Humoral Immunity

Now that we've discussed the characteristics of immune responses, we examine in greater detail the first type of immunity: humoral immunity. We begin by considering the role of B lymphocytes.

The Role of B Lymphocytes in Antibody Production

Millions of B lymphocytes with different specificities nor-mally circulate in the blood and lymph or reside in pe-ripheral lymphoid tissues. When one of the B cells binds with an antigen, the cell responds in the two ways de-picted in Figure 23.8. First, the B cell is stimulated to pro-liferate, which increases the number of B cells with the same specificity; second, the cells differentiate, such that some become long-lived memory B cells and others be-come short-lived antibody-synthesizing effector cells called *plasma cells*. A plasma cell secretes about 2000 anti-body molecules per second over its life span of 4–7 days. These antibodies then circulate throughout the blood and lymph for several weeks, binding to the same antigens that stimulated their production, thereby marking them for destruction by phagocytosis or complement-mediated lysis.

The antigens that evoke the production of both plasma cells and memory B cells are known as *T-dependent antigens* because they do so only with help from a special kind of T cell called a **helper T cell**. We examine the pre-cise functions of helper T cells shortly; we mention them here because they influence B cell activation. When helper T cells respond to specific antigens, they secrete many cytokines, including one called *interleukin-2 (IL-2)*. Together, IL-2 and T-dependent antigens (mainly pro-teins) induce the proliferation of B cells. By contrast, polysaccharide antigens, such as those found on many bacterial surfaces, can activate B cells without T cell help; these antigens are known as *T-independent antigens*. Be-cause they contain long arrays of repeated subunits, these polysaccharides bind with many antigen receptors on the

A primary response

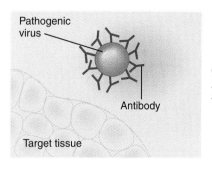

(a) Neutralization:
Antibodies block the activity of a pathogen.

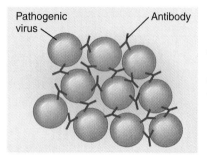

(b) Agglutination:
Multiple pathogens are aggregated by antibody molecules.

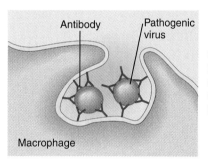

(c) Opsonization:
Pathogens bound by antibodies are more efficiently engulfed by phagocytes.

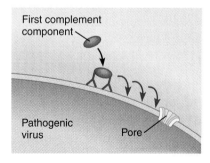

(d) Complement activation:
Antibodies bound to pathogens activate the complement cascade, resulting in lysis of the cell.

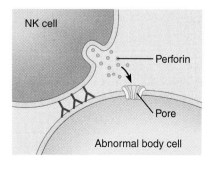

(e) Enhanced NK cell activity:
Abnormal body cells that are bound by antibodies are recognized by NK cells and are subsequently lysed.

Figure 23.10 Antibody-mediated mechanisms of antigen disposal.

B cell surface, providing a strong enough stimulus to induce the B cell to proliferate even in the absence of IL-2. However, without IL-2, all of the proliferating cells differentiate into antibody-secreting plasma cells; none develop into memory B cells. Therefore, even repeated exposures to a particular T-independent antigen produce primary responses only, because the antigen alone never generates immunological memory. Thus, B cell responses to T-independent antigens take 10–17 days to occur, and the quantity of antibodies produced is smaller than that produced in a secondary response. Specific events in B cell activation and the development of humoral immunity are discussed in the following sections.

Next we will see how antibodies bind to and mediate the disposal of microorganisms and foreign materials. Note that even though antibodies are generated in the immune response, they also enhance nonspecific defenses by binding to and thus marking specific microorganisms to focus nonspecific defenses on the invader at hand. A typical bacterium can be coated with as many as 4 million antibody molecules!

Antibody Function in Humoral Immunity

We have seen that an antibody has two functions: First it binds specifically to an antigen, and then it aids in the inactivation or disposal of that antigen. Whereas the antigen-binding sites are responsible for the recognition of antigen, the tail of the Y-shaped antibody molecule is responsible for the mechanisms by which it mediates antigen disposal. There are five classes of antibodies: IgG, IgM, IgA, IgE, and IgD. (*Ig* stands for immunoglobulin.)

The way in which an antibody aids in antigen disposal depends on its class. The structures and functions of each of the five immunoglobulin classes are summarized in **Table 23.1**. All classes of immunoglobulins can mediate the simplest forms of antigen attack—neutralization and agglutination—whereas particular classes specialize in opsonization of the antigen, activation of the complement system, and stimulation of natural killer cells. The functions of IgG are shown in **Figure 23.10**.

Neutralization

In **neutralization** the antibody blocks an antigen's activity just by binding to it (Figure 23.10a). For example, antibodies can neutralize a virus simply by attaching to the molecules that the virus must use to infect its host cell. Similarly, antibodies that coat bacterial toxins (such as the toxin produced by *Clostridium tetani,* which causes the disease *tetanus*) can render them inactive.

Agglutination

Antigens are frequently neutralized and clumped together simultaneously by thousands of antibody molecules. This process, called **agglutination**, is possible because each

Table 23.1 ▏ The Five Major Classes of Antibody

Class of antibody	Structure	Features	Roles in antigen disposal
IgM	J (joining) chain	The most common class of antibody produced in the primary response to antigen	Neutralizes antigen Agglutinates antigen Activates complement
IgD		Important as an antigen receptor on B cells	Neutralizes antigen Agglutinates antigen
IgG		The most common class of antibody in the blood, and the major class of antibody produced in secondary responses. Crosses the placenta, so is important in fetal and newborn immunity.	Neutralizes antigen Agglutinates antigen Activates complement Opsonizes antigen Enhances NK cell activity
IgE		Involved in allergies such as hay fever	Neutralizes antigen Agglutinates antigen Binds to mast cells and basophils, causing them to release histamine
IgA	J (joining) chain	Crosses epithelial cells, so is present on mucosal surfaces and in breast milk; is important in immunity in newborns	Neutralizes antigen Agglutinates antigen

antibody molecule has at least two antigen-binding sites (Figure 23.10b). (Those belonging to the IgM and IgA classes have more than two.) IgG, for example, can bind to equivalent epitopes on two separate pathogens, linking them together. With numerous IgG molecules likewise binding to two microbes apiece, a large complex forms.

Opsonization

Once bound to antibodies, an antigen is effectively opsonized and thus rendered more susceptible to phagocytosis (Figure 23.10c). As previously mentioned, antibodies act as opsonins because they can bind to both antigens and phagocytic cells. IgG antibodies are specialized for opsonization because their tails bind to specific surface receptors on phagocytic cells. This binding triggers the phagocytes to engulf both the antibodies and their targeted prey.

Complement Activation

Both IgM and IgG antibodies can activate the complement system, which brings about the lysis of bacteria to

which the antibodies are bound (Figure 23.10d). Although the complement system can also be activated when certain complement proteins bind directly to some kinds of bacteria (the nonspecific, alternative pathway), the specific, classical pathway is effective against almost any bacterial cell that has been marked with antibodies. Regardless of the initiating event, not only do activated complement proteins bring about the destruction of bacteria, but many of these proteins also help to advance and regulate inflammation, another essential nonspecific defense.

Enhanced Activation of Natural Killer Cells

Finally, IgG antibodies can enhance the nonspecific killing action of natural killer cells (Figure 23.10e). Recall that NK cells broadly recognize abnormal features of tumor cells and virus-infected cells and then produce membrane-perforating molecules that lead to the lysis of these cells. Often these abnormal body cells also possess abnormal surface molecules that can stimulate the production of antibodies. The antibodies mark the cells for death—in this case, death by NK cells. Like phagocytes, NK cells have surface receptors that bind with the constant regions of IgG. Antibodies thus provide a link between a specifically targeted, abnormal cell and the NK cell, serving once again to focus nonspecific responses upon a particular foreign substance.

> ### Quick Test 23.4
>
> 1. Explain two differences between plasma cells and memory B cells.
> 2. What is the difference between T-dependent antigens and T-independent antigens? Why are no memory B cells generated in response to a T-independent antigen?
> 3. Draw an IgG molecule, and list the four other classes of antibody.
> 4. State five ways in which antibodies help eliminate antigen.

Cell-Mediated Immunity

We have seen that whereas antibodies defend against invaders and other antigens floating free in the blood and lymph, T cells make contact with and respond to any body cells that are infected or otherwise abnormal. Because T cells require direct contact between them and their targets, their responses are described as being *cell-mediated*. We begin our discussion of cell-mediated responses by examining the roles of T cells in this type of immunity.

Roles of T Lymphocytes in Cell-Mediated Immunity

There are three major types of T lymphocytes: helper T cells, cytotoxic T cells, and suppressor T cells. Helper T cells, the primary regulators of immune responses, operate indirectly by secreting cytokines that enhance the activity of B cells, cytotoxic T cells, suppressor T cells, and helper T cells themselves. In addition, helper T cells produce cytokines that enhance the actions of macrophages and NK cells, which are essential to nonspecific defenses. Cytotoxic T cells, in contrast, are directly responsible for cell-mediated immunity in that they kill cells infected by viruses or intracellular bacteria, and cells that are otherwise abnormal (such as cancer cells and transplanted cells). Suppressor T cells are not well understood, but they are thought to produce cytokines that suppress the activity of B cells, helper T cells, and cytotoxic T cells.

All three types of T cells have antigen receptors (T cell receptors or TCRs) that detect foreign antigens on body cells, but they do so only when these antigens are associated with a special class of normal self-proteins known as **major histocompatibility complex (MHC) molecules**. This type of recognition is unlike the way in which B cells recognize antigens in that B cells and the antibodies they secrete are able to bind to epitopes in their native forms (for instance, as they exist on the surface of a bacterium). In order for T cells to be activated, their antigen receptors must contact an MHC molecule on the surface of some other body cell that is bound to a small fragment of antigen. Thus MHC molecules must first bind to a fragment of foreign antigen that is present within a body cell and then must transport it to the cell surface, where it can be detected by T cells. This process is called **antigen presentation**. Before we examine T cell responses in greater detail, let's take a closer look at MHC molecules and their roles in T cell maturation, antigen presentation, and T cell activation.

MHC Molecules: Markers of Self

Each body cell is identified as "self" by a set of MHC molecules (in humans, known as *human leukocyte antigens*, or HLA molecules), of which there are two classes. **Class I MHC** molecules are found on the surfaces of all nucleated cells—that is, on almost every cell of the body; **class II MHC** molecules are found on only a few specialized cell types, including macrophages, activated B cells and T cells, and the cells that make up the interior of the thymus.

Although the two classes of MHC molecules function the same way in all people, each person's MHC molecules are unique to himself or herself; it is virtually impossible for the tissues of any two people (except those of identical twins) to have the same set of MHC molecules, or *HLA tissue type*. That MHC molecules differ from person to person explains why skin grafts and organs are usually rejected when transplanted from one person to another; in fact, the existence of MHC molecules was discovered during investigations of graft rejection. These proteins play a major part in determining whether transplanted tissue is accepted as self (histocompatible) or rejected as foreign.

The discovery of MHC molecules and their role in graft rejection was somewhat puzzling; why would markers have evolved that prevent us from sharing tissues? The answer is that this diversity of MHC molecules is adaptive to the survival of the human species as a whole. Each person's MHC molecules are capable of presenting some antigen fragments but not others. The existence of a range of different MHC molecules in the population ensures that if the population is infected with a given pathogenic microbe, at least some individuals will have MHC molecules capable of presenting its specific antigens, and those individuals will be able to mount an immune response to the invader and survive the infection. In reality, most of us are capable of mounting an attack against all the microbes we encounter in everyday life, because antigens are complex enough, and each person's set of MHC molecules are varied enough, such that each of us develops an effective immune response. Although the cells of

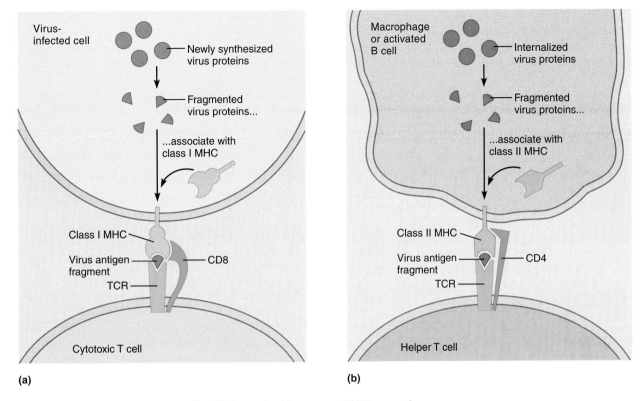

Figure 23.11 Presentation of antigens to T cells by major histocompatibility complex (MHC) molecules. (a) Class I MHC molecules, made by all nucleated cells of the body, capture fragments of viral (or bacterial) antigens synthesized within an infected cell and transport them to the cell surface. A cytotoxic T cell then binds to the infected cell through its T cell receptors (TCRs) and CD8 molecules. (b) Class II MHC molecules, made by macrophages and activated B cells, capture fragments of foreign antigens internalized by phagocytosis or receptor-mediated endocytosis, respectively, and transport them to the cell surface. A helper T cell then binds to the presenting cell through its TCR and CD4 molecules.

two different people may present different portions of a given antigen, both people will be able to mount a strong immune response to that antigen.

The Role of MHC Molecules in T Cell Maturation

If individuals are to mount an effective immune response, their T cells must have antigen receptors capable of making specific contacts with their own MHC molecules. This poses a particular challenge to the development of T cells, for the following reason: Because T cell receptors (like B cell receptors or membrane antibodies) are randomly generated to provide a great array of different lymphocytes, each with a unique specificity, developing T cells differ in their capacity to bind to an individual's own MHC molecules.

In a healthy immune system, only "desirable" T cells reach maturity and leave the thymus. As young T cells are developing in the thymus, they come into contact with thymic cells bearing high levels of the body's own class I and class II MHC molecules. Only T cells bearing receptors with appropriate affinity for self-MHC molecules continue to develop. Furthermore, T cells develop into two different types, depending on the class of MHC

molecules to which they bind: Those cells that bind to class I MHC molecules develop into cytotoxic T cells, whereas those that bind to class II MHC molecules develop into helper T cells.

The Role of MHC Molecules in Antigen Presentation and T Cell Activation

Each MHC molecule possesses a binding site that can bind to a variety of foreign antigen fragments. (In the absence of foreign antigen, MHC molecules bind to self-protein fragments, but the resulting combination does not normally induce an immune response.) As a newly synthesized MHC molecule makes its way to the surface of an infected or abnormal cell, it can capture an antigen fragment in its binding site and carry it out to the surface of the cell. How the MHC captures and presents the antigen and the recognition of this antigen by T cells varies for the two classes of MHC molecules.

Class I MHC molecules capture foreign or abnormal antigens synthesized within infected cells or tumor cells and present the antigen at the cell surface to cytotoxic T cells (**Figure 23.11**a). Exposure to the presented antigen stimulates the cytotoxic T cells to kill the unhealthy cells.

To assist in this process, cytotoxic T cells have a surface protein called CD8 that binds to class I MHC molecules, enhancing the association between a cytotoxic T cell and abnormal target cell. Thus, cytotoxic T cells are also called CD8 cells.

Class II MHC molecules capture foreign antigens that have been taken into cells through phagocytosis or receptor-mediated endocytosis and present the antigen at the cell surface to helper T cells (Figure 23.11b). Recall that class II MHC molecules are found only in cells that internalize antigen, such as macrophages and activated B and T cells. Exposure to the presented antigen stimulates helper T cells to secrete cytokines that induce and regulate other immune responses. Importantly, helper T cells do not kill the antigen-presenting cell, which is an immune cell. To assist them, helper T cells have a surface protein called CD4 that binds to the class II MHC molecule and enhances the association between the T cell and macrophage or activated B or T cell. Thus, helper T cells are also called CD4 cells.

Quick Test 23.5

1. List the three major types of T cells, and explain their functions.

2. What are the two classes of MHC molecules? On which types of cells is each class found?

3. Why do MHC molecules pose limits to transplantation between people, except in the case of identical twins?

4. What type of T cell has CD4 on its surface, and with what class of MHC does this cell associate?

5. What type of T cell has CD8 on its surface, and with what class of MHC does this cell associate?

Helper T Cell Activation

Like B lymphocytes, millions of T lymphocytes with different specificities normally circulate in the blood and lymph or reside in peripheral lymphoid tissues. Activation of helper T cells involves two simultaneous events: Helper T cells first bind with class II MHC–foreign antigen complexes on the surfaces of macrophages and B cells, and then helper T cells receive from these cells an inducing signal in the form of IL-1. As a result, the helper T cells proliferate and differentiate. Some of the daughter cells begin to secrete cytokines, while a small proportion of them become long-lived memory T cells. Like humoral responses, cell-mediated responses can take up to 17 days to develop, especially upon first exposure to an antigen. Once memory T cells are present in the body, however, subsequent responses to that same antigen are quicker and much more vigorous.

Activated helper T cells secrete several types of cytokines that help stimulate and regulate the activities of other helper T cells, B cells, cytotoxic T cells, macrophages, mast cells, NK cells, and hematopoietic stem cells (**Table 23.2**). Helper T cells thus appear to be the central coordinators of immune responses—which is why the depletion of helper T cells, a primary characteristic of *acquired immunodeficiency syndrome (AIDS),* is so devastating to the immune system.

Cytotoxic T Cell Activation: The Destruction of Virus-Infected Cells and Tumor Cells

A cytotoxic T cell becomes activated to kill when two events occur simultaneously: when it both binds with a class I MHC–foreign antigen complex on the surface of a virus-infected cell and receives an inducing signal (in the form of IL-2) from a helper T cell. Once activated, the cytotoxic T cell releases perforins, causing the target cell to lyse. The secretion of such lysis-inducing proteins is a common phenomenon in immune reactions; we have already encountered it in the complement cascade and in the way NK cells destroy their target cells. Cytotoxic T cells also release proteins called **fragmentins**, which first gain entry into target cells through the perforin-induced pores and then work inside them to bring about their death through apoptosis. Pathogens released from the dead cell are rapidly eliminated by nearby phagocytic cells or are targeted for destruction by antibodies.

In similar fashion, cytotoxic T cells also defend against tumor cells. These cells sometimes possess distinctive molecules, called *tumor antigens,* that are not present on normal cells. The class I MHC molecules that normally exist on tumor cells present fragments of these antigens to cytotoxic T cells, thereby initiating their killing response. Note that certain cancers and viruses (Epstein-Barr virus, for example) actively inhibit the production of class I MHC molecules on affected cells, which enables such cancers and viruses to escape detection by cytotoxic T cells.

As we have seen, effective immune responses arise from multiple direct and indirect interactions among macrophages, helper T cells, B cells, and cytotoxic T cells, as depicted in **Figure 23.12**.

Quick Test 23.6

1. List four cytokines that are produced by helper T cells, and describe how each helps regulate immune responses.

2. Briefly describe how a cytotoxic T cell kills a virus-infected cell.

3. List the four cell types that are involved in immune responses, and briefly describe the functions of each.

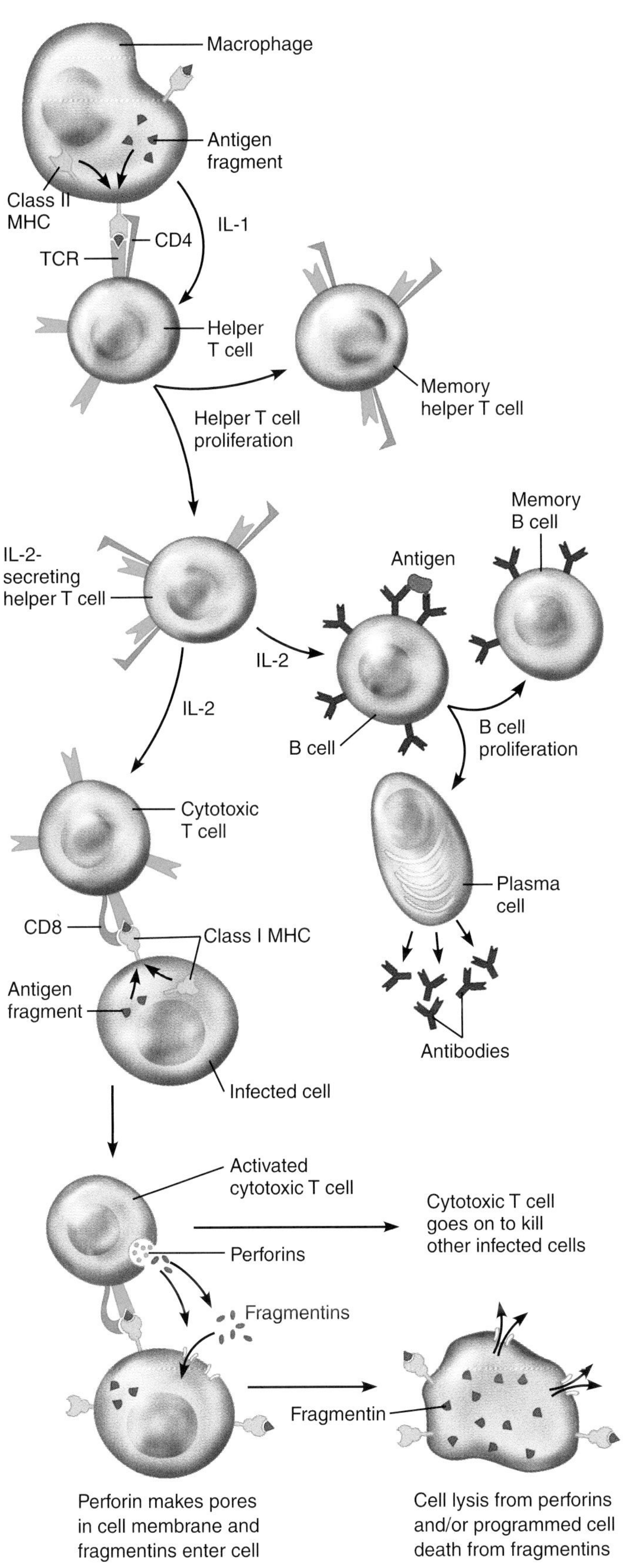

Macrophage

Antigen fragment

Class II MHC

TCR — CD4

IL-1

Helper T cell

Memory helper T cell

Helper T cell proliferation

IL-2-secreting helper T cell

Antigen

Memory B cell

IL-2

IL-2

B cell

B cell proliferation

Cytotoxic T cell

Plasma cell

CD8

Class I MHC

Antigen fragment

Infected cell

Antibodies

Activated cytotoxic T cell

Perforins

Cytotoxic T cell goes on to kill other infected cells

Fragmentins

Fragmentin

Perforin makes pores in cell membrane and fragmentins enter cell

Cell lysis from perforins and/or programmed cell death from fragmentins

Immune Responses in Health and Disease

We have seen that effective immune responses depend on the interplay of many kinds of cells and molecules. In the sections that follow we explore how the immune system generates immunity, and responds to transplantation and transfusion. We'll also see how disease can result when the immune system malfunctions.

Generating Immunity: Immunization

In 1798, a physician named Edward Jenner was investigating the spread of smallpox when he noticed that milkmaids in Gloucestershire, England, who had previously been ill with cowpox (a mild disease contracted from cows) usually escaped smallpox infection, even when this disfiguring, often fatal disease was running rampant in their community. Suspecting that exposure to cowpox conferred some sort of protection against smallpox, Jenner began to deliberately inoculate people with cowpox, boldly predicting that this action would bring an end to the terrible smallpox scourge. This treatment came to be known as *vaccination* after the Latin word *vacca,* meaning "cow." Indeed, thanks to intensive worldwide vaccination programs in the 20th century, smallpox has been eradicated.

In vaccination, also known as *immunization,* a safe form of a microorganism or a collection of its components that are not expected to cause disease is introduced into the body, where the inoculum both stimulates an immune response to that pathogen and—even more significantly—induces immunological memory (**Figure 23.13**). A vaccinated person who subsequently encounters the natural pathogen then mounts a strong immune response similar to a naturally occurring secondary immune response; a successfully vaccinated person is immune to that pathogen. Beginning in the 20th century, routine immunization of infants and children has been extremely effective in preventing many infectious diseases, and such vaccine programs may result in the global eradication of polio by 2005. The term *immunization* is also used to refer to the conferral of immunity that results from a natural infection. Both artificial immunization and natural immunization confer a type

Figure 23.12 Summary of the events in the immune response to a virus. A helper T cell is activated both by specific contact with a macrophage presenting viral antigen via class II MHC and by IL-1, generating memory T cells and IL-2-secreting helper T cells. A cytotoxic T cell is activated by specific contact with an infected cell presenting viral antigen via class I MHC molecules, and by IL-2 secreted by a helper T cell. The activated cytotoxic T cell produces perforins and fragmentins, each of which can induce target-cell death. A B cell is activated by specific contact with free-floating antigen and by IL-2 secreted by a helper T cell, generating memory B cells and plasma cells that secrete antibody specific for the antigen. The immune response to other types of pathogens is similar to that described here.

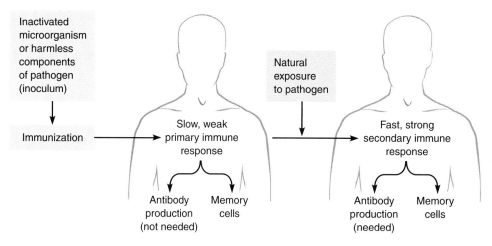

Figure 23.13 Acquisition of long-term immunity through vaccination (immunization).
Introduction into the body of the inoculum stimulates immune responses that both target the introduced materials for destruction and generate memory cells. Upon subsequent natural exposure to the pathogen, memory cells mount a rapid secondary response that prevents or diminishes the symptoms of the disease.

Five-year-old Ivan receives an immunization against chicken pox, and develops antibodies and memory cells specific for the virus. Which of these two components of his immune response is more important for Ivan, and why?

Table 23.2 | Selected Cytokines Secreted by Activated Helper T Cells

Cytokine	Target cells	Effects on target cells
Interleukin-2	Helper T cells and cytotoxic T cells	Stimulates proliferation
	B cells	Stimulates proliferation and plasma cell development
	NK cells	Enhances activity
Interleukin-4	B cells	Stimulates proliferation and plasma cell development; induces plasma cells to secrete IgE and IgG; increases numbers of class II MHC molecules
	Helper T cells	Stimulates proliferation
	Macrophages	Increases numbers of class II MHC molecules; enhances phagocytosis
	Mast cells	Stimulates proliferation
Interleukin-5	B cells	Stimulates proliferation; induces plasma cells to secrete IgA
	Hematopoietic stem cells	Induces proliferation and development of eosinophils
Interleukin-10	Macrophages	Inhibits cytokine production (helps down-regulate the immune response)
Interferon-γ	Multiple cell types	Confers resistance to viruses
	Macrophages	Enhances phagocytosis
	B cells	Enhances antibody production
	Cancer cells	Inhibits proliferation
	Cytotoxic T cells and NK cells	Enhances killing capacity of cytotoxic T cells and NK cells

Memory cells. Antibodies function in the short term; they will help clear the harmless vaccine from Ivan's body. Memory cells, however, can survive a lifetime, and are poised to combat the chicken pox virus whenever he is exposed to it.

of protection referred to as **active immunity** because it depends on the ability of the immunized person's immune system to mount a response. Unfortunately, not all infectious agents (for example, viruses that cause the common cold) are easily managed by vaccination.

Ready-made antibodies to a particular antigen can also be introduced into the body to provide another kind of protection called **passive immunity**. This type of immunization is passive because it does not require a response from the immunized person's immune system. Because no foreign antigens are introduced into the body in this procedure, the person's B cells do not make antibodies, and no memory B cells are made. Although the introduced antibodies help fight an ongoing infection, passive immunization does not induce long-term immunity. The antibodies typically used for passive immunization are first isolated from people who are already immune to a particular disease; these antibodies are then injected into another person's body, conferring in the process a short-lived but immediate protection from that disease. A person who had been bitten by a rabid animal, for example, may be injected with antibodies collected from other people who have been vaccinated against rabies. This measure is important because rabies may progress more rapidly than the time it takes for a person to mount an active immune response. Most individuals exposed to rabies are given both passive and active immunizations; the injected antibodies fight the virus for a few weeks, and then the person's own immune response, which is induced by both the immunization and the infection itself, takes over.

Passive immunity also occurs naturally when IgG antibodies in the blood of a pregnant woman cross the placenta and reach the fetus. IgA antibodies (see Table 23.1) are also passed from mother to nursing infant in breast milk, especially when the milk is in its early form, the colostrum. Passive immunity persists only as long as the transferred antibodies do—a few weeks to a few months—but it provides the infant protection from infections until the baby's own immune system has had a chance to mature.

> ### Quick Test 23.7
>
> 1. How does an immunization such as the one against chicken pox confer lifelong immunity without causing the disease?
>
> 2. What is the difference between passive immunity and active immunity? Do the maternal antibodies a baby acquires through the placenta provide active immunity or passive immunity?

Roles of the Immune System in Transfusion and Transplantation

The immune system's ability to distinguish self from nonself, while essential to a healthy immune system, limits our ability to transfuse blood or to transplant tissue between individuals. For this reason, material from the donor must be matched, as closely as possible, to the recipient in order to minimize immune reactions. In addition, transplant recipients are given medicines that suppress the ability of their immune system to react against the foreign tissue. Note that the body's hostile response to an incompatible transfusion or transplant is not a disorder of the immune system, but instead a normal response to foreign antigens.

Blood-Group Compatibility

Blood is classified into different types (designated by the letters A, B, AB, or O) according to the presence or absence of certain antigens on the surface of a person's red blood cells. An individual who has *type A blood* possesses red blood cells with surface antigens known as A antigens. These antigens are perceived as self by the individual's immune system; they are not antigenic to their "owner." Similarly, B antigens are found on type B red blood cells, and both A antigens and B antigens are found on type AB red blood cells. Type O red blood cells possess neither antigen.

An individual with type B blood does not produce antibodies to B antigen because that antigen is recognized as "self." Surprisingly, however, this person will have antibodies to A antigen (called anti-A antibodies), *even if this individual has never been exposed to type A blood.* The presence of circulating antibodies specific for A antigen seems to suggest that the individual's B lymphocytes have detected and responded against type A blood. But what has in fact occurred is that antibodies to the A antigen have been produced in response to bacteria that are normally present in the body and possess epitopes that are very similar to the blood group antigens A and B. A person with type B blood does not make antibody to the B-like bacterial antigens—those are too much like self—but the individual does make antibodies to A-like bacterial antigens, which are perceived as foreign by his or her immune system. Thus in a type B individual, the anti-A antibodies that are constantly circulating in the blood induce an immediate and devastating *transfusion reaction* in the event that this individual receives a transfusion with type A or type AB blood, both of which contain the A antigen. For these reasons, a person with type O blood is considered to be a *universal donor* because any anti-A or anti-B antibodies that may be present in the recipient will find no target on the type O donor cells. In contrast, a person with type AB blood is considered to be a *universal recipient;* because such an individual lacks antibodies against A antigen or B antigen, transfusion of blood of any blood group into this individual will not induce a transfusion reaction.

One more issue must be considered with blood donations: When whole blood is transfused, it contains antibodies that could, in a mismatched transfusion, attack the *recipient's* red blood cells. This usually is not a problem because *packed* red blood cells, rather than whole blood, are typically used in transfusions.

Tissue Grafts and Organ Transplantation

As previously described, MHC molecules are present in many different forms across the human population. Therefore, it is unlikely that two people would have the same HLA profile, and thus HLA molecules are responsible for stimulating the rejection that occurs in tissue grafts and organ transplants. Note that MHC molecules do *not* play a role in transfusion reactions because red blood cells do not have MHC molecules.

To minimize the chance of rejection, physicians attempt to match the HLA molecules of the donor and recipient as closely as possible, using a procedure called *tissue typing.* When a recipient has no identical twin (who would have exactly the same tissue type), siblings usually provide the closest HLA match. In addition to testing and matching HLA molecules, physicians also attempt to minimize the chance of rejection by prescribing drugs that suppress the recipient's immune responses. The complication with this strategy is that it renders the recipient more susceptible to infections and cancer during the course of treatment. The main effect of drugs such as cyclosporin A and FK506, which have greatly improved the success of organ transplants, is to inhibit production of IL-2, which in turn inhibits B cell and T cell activation and the mounting of immune responses.

Despite the risk of rejection, transplantation of bone marrow is often used successfully to treat leukemia and other cancers, as well as various blood-cell diseases. Before receiving the transplant, the recipient is typically treated with radiation to eliminate his or her own bone marrow cells. Such treatment both eliminates the abnormal cells and effectively inactivates that person's immune system, which minimizes the likelihood that the recipient will reject the graft. The great danger in this procedure is that the donated marrow, which contains lymphocytes, will mount immune responses against the recipient, an example of what is known as a *graft versus host reaction.* The intensity of such a reaction can be minimized if the HLA molecules of the donor and recipient are well matched, which is why bone marrow donor programs continually seek volunteer donors the world over. Because of the diverse array of HLA molecules within the human population, a diverse pool of potential donors is needed.

Quick Test 23.8

1. What cell surface molecules are tested in the process of tissue typing?

2. What is the name of the phenomenon in which a grafted tissue mounts an immune response against a recipient?

Immune Dysfunctions

The complex, highly regulated interplay of foreign substances with lymphocytes and other cells that make up the immune system provides us with extraordinary protection from infections. In addition, a growing body of evidence suggests that immune function is intimately associated with nervous system and endocrine system functions. When this delicately balanced network of interacting cells and molecules is disrupted, the effects on the individual can range from the minor inconvenience of some allergies to the serious and sometimes devastating consequences of autoimmune and immunodeficiency diseases. In the following sections we examine what can happen when immune function goes awry.

Allergy

Allergies (also known as *hypersensitivity reactions*) are exaggerated responses to certain environmental antigens known as *allergens*. The most common allergies involve antibodies of the IgE class (see Table 23.1). Allergies occur in some people who are genetically predisposed to produce more than the usual amount of IgE when they are exposed to allergens. If an individual produces high levels of IgE in response to pollen, the result is an allergic reaction commonly known as *hay fever*. As shown in **Figure 23.14**, some of these IgE antibodies do not bind to the pollen but instead attach by their tails to mast cells. When the person is subsequently exposed to pollen again, the pollen grains and the antigen-binding sites of these IgE antibodies bind, causing adjacent antibody molecules to become cross-linked and inducing the mast cell to degranulate—that is, to release histamine and other inflammatory agents into the surrounding fluid. Following its release, histamine causes dilation and increased permeability of small blood vessels in the immediate vicinity. These inflammatory events lead to the typical symptoms of hay fever: sneezing, runny nose, tearing eyes, and breathing difficulty, which can result from histamine-induced contraction of smooth muscle in respiratory airways. Drugs that act as antihistamines diminish allergy symptoms by blocking receptors located on blood vessel endothelial cells and smooth muscle cells that normally bind to histamine.

The most serious consequence of an acute allergic response is *anaphylactic shock,* a life-threatening reaction to injected or ingested allergens. Anaphylactic shock results from widespread degranulation of mast cells throughout the body, which triggers abrupt dilation of peripheral blood vessels. This dilation is not just confined to a restricted area, as in hay fever, but is widespread and therefore causes a precipitous drop in total peripheral resistance and mean arterial pressure. As a consequence of the drop in pressure, death may occur within a few minutes. Bee venom or penicillin are examples of allergens that can trigger anaphylactic shock in people who

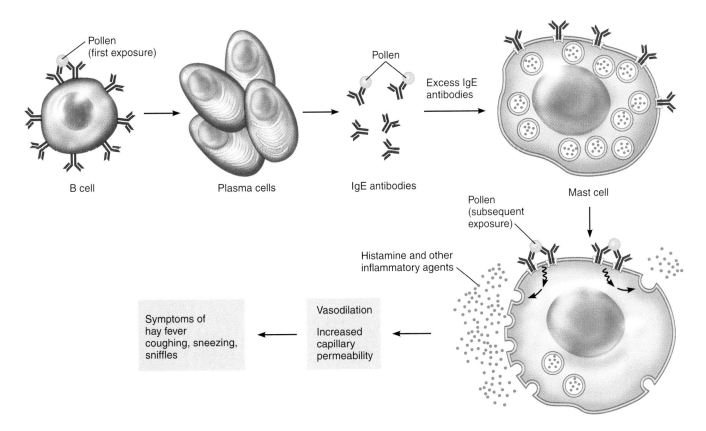

Figure 23.14 Events in hay fever, an allergic response. In response to exposure to allergens such as pollen, an allergic person produces abundant IgE molecules; excess IgE binds to the surface of mast cells. Upon subsequent exposure to the same kind of pollen, the allergens and bound IgE bind, cross-linking the IgE and inducing rapid degranulation (histamine release). Histamine induces increased vascular permeability and the resulting symptoms of hay fever.

In anaphylactic shock there is a drop in blood pressure due, in part, to widespread histamine release. What actions of histamine contribute to this loss of blood pressure?

are extremely allergic to them. Some susceptible individuals carry syringes containing the hormone epinephrine as a prophylactic measure. When injected into the bloodstream, this hormone counteracts the allergic response by stimulating increased cardiac output and constriction of blood vessels, both of which tend to raise blood pressure back toward normal levels.

Autoimmune Diseases

When the immune system loses tolerance to self and begins to react against normal molecules of the body, autoimmune disease can result. In a disease called *systemic lupus erythematosus* (often called *lupus*), for example, the immune system generates antibodies (known as *autoantibodies*) against all sorts of self molecules, resulting in a widespread array of signs including skin rashes, fever,

arthritis, and kidney dysfunction. *Rheumatoid arthritis,* another antibody-mediated autoimmune disease, causes painful inflammation of and eventual damage to the cartilage and bone of joints. In insulin-dependent diabetes mellitus, another autoimmune disease, the insulin-producing beta cells of the pancreas are targeted by cell-mediated immune responses. Another autoimmune disease is *multiple sclerosis (MS),* the most common chronic neurological disease in developed countries. In this disease, T cells that have a propensity to attack normal myelin are thought to infiltrate the central nervous system and cause demyelination of nerve fibers, thereby precipitating a number of serious neurological abnormalities (**Clinical Connections: Multiple Sclerosis**, p. 748).

The causes of autoimmunity are varied and complex. Although much remains to be learned about these diseases, we know that people who inherit particular MHC molecules also are more likely to develop certain auto-immune diseases. For example, individuals who inherit certain class II MHC molecules are at higher risk of developing insulin-dependent diabetes mellitus than are members of the general population.

Histamine induces increased capillary permeability and vasodilation, resulting in the loss of fluid from blood vessels and decreased peripheral resistance, respectively. Both of these events contribute to a drop in mean arterial pressure.

MULTIPLE SCLEROSIS

Autoimmune diseases such as multiple sclerosis (MS) occur when tolerance to self breaks down and the immune system wrongly identifies normal body components as foreign and mounts an attack. In multiple sclerosis, the tissue under attack is myelin, the material that wraps around and insulates the axons of neurons and is abundant in the white matter of the brain and spinal cord. The name of the disease refers to the many lesions or *scleroses* (from the Greek word for scarring or hardening) that result from the destruction of myelin *(demyelination)*.

The symptoms of MS range from mild to severe and may appear in various combinations, depending on where the demyelination occurs. The symptoms may also come and go, or they may persist for the remainder of a person's life. The symptoms of MS typically include blurred vision (or blindness in one eye), muscle weakness, and trouble maintaining balance while walking (ataxia). These symptoms may worsen, leading to a complete inability to walk or stand. Other symptoms include muscle spasticity; tremors; impairment of pain, temperature, and touch sensations; speech disturbances; vertigo; and fatigue. MS is usually diagnosed between the ages of 20 and 40 and occurs more often in women than in men. Approximately 250,000–350,000 people currently live with this disease in the United States. In fact, MS is the most common cause of neurological disability in developed countries.

Because the cerebrospinal fluid of MS patients has been shown to contain activated cytotoxic T cells (which are absent from the CSF of healthy people), it is likely that these cells are primarily responsible for the demyelination that occurs in MS. However, despite intensive research it is still not clear why the body's cytotoxic T cells turn against its own myelin. One possibility is that an infection of some sort either induces inappropriate immune responses to myelin or causes a breakdown in the blood-brain barrier that normally prevents lymphocytes from leaving capillaries within the CNS. Lymphocytes have no opportunity to develop tolerance to self-antigens (including myelin proteins) that are located in and around the brain and spinal cord; when lymphocytes are subsequently exposed to these antigens, they perceive them as foreign and mount defensive responses, causing irreparable CNS damage. Scientists have studied a number of infectious agents (including measles, mumps, and rubella viruses) that may induce the autoimmune responses typical of MS but have been unable to implicate any particular causative agent.

Increasing evidence also suggests that heredity may play a role in determining a person's susceptibility to MS. Some ethnic groups, including Gypsies and Eskimos, never experience MS, and Native Americans, Japanese, and other Asian peoples have a very low incidence of the disease. Among the genes associated with susceptibility to the disease are those encoding HLA or MHC molecules. A higher incidence of MS is associated with people who have inherited a particular kind of class II MHC molecule.

Treatment and perhaps even prevention of this and other devastating autoimmune diseases requires better understanding of the precise mechanisms by which B cells and T cells respond against foreign antigens while remaining tolerant of the body's own tissues.

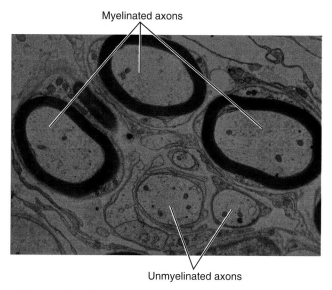

Cross-section of axons.

Immunodeficiency Diseases

In allergies or autoimmune diseases, problems arise due to overactivity in the immune system; *immunodeficiency diseases* arise as a result of underactivity in the immune system. There are almost as many immunodeficiency diseases as there are functions within the immune system. Many congenital immunodeficiencies affect the function of either humoral or cell-mediated immunity, but in *severe combined immunodeficiency disease (SCID),* both branches of the immune system fail to function. For people with this genetic disease, long-term survival may require a transplant of healthy bone marrow that will continue to supply functional B cells and T cells. One type of SCID is caused by deficiency of the enzyme adenosine deaminase (ADA), which plays a role in the breakdown of the DNA building blocks known as *purines*. ADA is particularly active in lymphocytes, and its deficiency results in the accumulation of lethal by-products. ADA has been treated with some success by gene therapy (**Discovery: Gene Therapy for Severe Combined Immunodeficiency Disease**, p. 750); in this procedure an individual's own bone marrow cells are removed, provided with a functional ADA gene, and then returned to the body.

Immunodeficiency is not always a congenital condition; an individual may develop immune dysfunction later in life. For example, certain cancers can cause immunodeficiency by suppressing the immune system. A prime example is *Hodgkin's disease,* which damages the lymphatic system. Another well-known and devastating acquired immunodeficiency is AIDS.

The Role of Stress in the Immune Response

Clinical and personal observations have long suggested a correlation between psychosocial factors and immune function. Indeed, several studies have suggested that a positive outlook is associated with improved health in both cancer and AIDS survivors; conversely, people hospitalized for depression have been found to exhibit diminished immunity. The mind-immunity relationship arises from multiple complex interactions that exist among the immune system, the nervous system, and the endocrine system. The effects of the immune system on these and other systems is depicted in the Systems Integration chart on page 752.

Many of these interactions are well established. Some steroid hormones are known to suppress immune responses; corticosteroids, for example, reduce the number and activity of immune cells and are potent anti-inflammatory agents. Likewise, research on wild animals has demonstrated that chemicals such as PCB and DDT, which mimic naturally occurring endocrine hormones, reduce the effectiveness of immune responses and

increase the incidence of infection. By contrast, both growth hormone and thyroid hormone play important roles in T cell development and function.

The nervous, endocrine, and immune systems interact in a variety of ways. For example, autonomic nerves innervate lymphoid organs such as the bone marrow and the spleen, often terminating at B cell and T cell clusters, and lymphocytes bear receptors for epinephrine and acetylcholine. Additionally, lymphocytes and macrophages secrete cytokines that affect both the endocrine and central nervous systems. For example, IL-1 secreted by macrophages induces fever, suppresses appetite, inhibits thyroid function, and stimulates the release of pituitary hormones.

The correlation between stress and immune function continues to be demonstrated in various ways. One study that tested college students for immune functions just after a vacation and again during final exams found that their immune systems were impaired in various ways (for example, plasma interferon levels were lower) during exam week.

Quick Test 23.9

1. Which class of antibody is induced in allergies such as hay fever? How do these antibodies trigger the symptoms of allergy?

2. Does rheumatoid arthritis result from a humoral autoimmune response or from a cell-mediated autoimmune response?

3. Name two lymphoid organs that are innervated by the autonomic nervous system.

Summary of Factors Influencing Specific and Nonspecific Defenses

Figure 23.15 graphically summarizes many of the concepts discussed in this chapter. It may help you to recall and understand the many components of immune function, and how they lead to the clearance of microorganisms from the body.

▨ *CHAPTER SUMMARY*

The components of the immune system—lymphoid tissues, leukocytes and the molecules they produce—work together to generate and regulate immune responses. The function of the immune response is twofold: to clear foreign materials from the body, and to bring about long-term immunity to infectious diseases.

Anatomy of the Immune System, p. 726

There are five major types of leukocytes. Neutrophils, monocytes, and macrophages (which arise from monocytes) are phagocytic; they engulf and destroy foreign matter and debris. Eosinophils and basophils defend against large parasites, and are

also involved in allergic reactions. The lymphocytes consist of B lymphocytes (B cells), T lymphocytes (T cells), and null cells. Whereas B cells and T cells exhibit specificity, null cells are nonspecific. Most null cells are large granular lymphocytes known as natural killer (NK) cells, which pose an important, early threat against viral

GENE THERAPY FOR SEVERE COMBINED IMMUNODEFICIENCY DISEASE

As its name indicates, severe combined immunodeficiency disease (SCID) is a combination of deficiencies of both the humoral and cell-mediated branches of immunity. It is no surprise, therefore, that people with this disorder exhibit increased susceptibility to all types of infections. Infants with this congenital disease usually begin to develop recurrent infections at 3–6 months of age, and even normally harmless microbes can cause serious or even life-threatening infections. The most common diseases associated with SCID are pneumonia due to the bacterium *Pneumocystis carinii,* and (in infants) diarrhea caused by rotavirus.

About one in every million children are born with adenosine deaminase (ADA) deficiency, the first known cause of SCID. This enzyme plays a role in DNA metabolism, and although it is used by all cells of the body, the disruption of DNA synthesis that its absence causes in B cells and T cells is particularly damaging. Because the antigen-driven proliferation of lymphocytes is the key to immunity, people with ADA deficiency have a markedly low lymphocyte count and therefore a virtually nonexistent immune response.

Although the frequent infections characteristic of SCID are treated by the administration of antiviral medicines, antibiotics, and antibodies, ADA deficiency is fatal—usually by the time the child reaches 2 years of age—unless immune function is restored. Like other blood cell diseases, ADA deficiency can be cured by transplanting bone marrow from a donor with a similar HLA tissue type, but a well-matched donor is hard to find. Fortunately, ADA-deficiency SCID has also been successfully treated by regular injections of the purified enzyme itself.

Now, with the advent of *gene replacement therapy,* or gene therapy, there is new hope for those born with genetic diseases such as ADA deficiency SCID. The concept is simple: If a gene is defective (that is, it produces an aberrant product or inadequate levels of it), then introducing a functional copy of the gene—one that directs normal synthesis of the product—should cure the problem. For gene therapy to be successful in treating ADA deficiency, a normal adenosine deaminase gene must be introduced into T cells and B cells, and the gene must also be replicated normally when the cells divide. Ideally, cells carrying the new gene should persist in the individual, so that repeated treatments are not needed.

Gene therapy is accomplished in four steps: (1) Lymphocytes or stem cells are collected from the blood, bone marrow, or umbilical cord of an infant or child with the disease; (2) the harvested cells are treated with a chemical that causes them to proliferate in order to generate additional target cells for gene insertion; (3) a healthy ADA gene is introduced into the cells, often using a nonpathogenic virus as a carrier; and (4) the cells, now producing normal ADA, are infused back into the bloodstream of the affected child.

ADA deficiency was the first human disorder to be treated by gene therapy. Since 1990, a number of infants and children with the disease have been treated in this manner in the United States, Europe, and Japan. Reports indicate that recipients of this treatment are doing well. However, because most recipients are still receiving enzyme injections regularly, it is not yet possible to attribute their good health to gene therapy alone. Moreover, grave concerns about the viruses used to carry such genes arose from the 1999 death of Jessie Gelsinger, a teenager who received gene therapy (for a different genetic disease) and had a severe inflammatory reaction to the virus used in the treatment. Much has been learned from this tragic event, and scientists continue to pursue this kind of treatment for debilitating and often deadly genetic diseases.

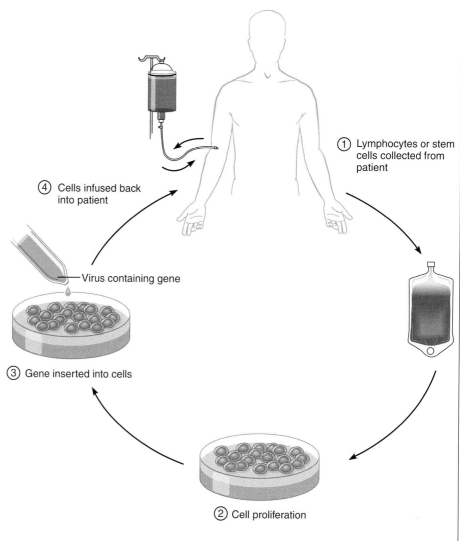

④ Cells infused back into patient

Virus containing gene

③ Gene inserted into cells

① Lymphocytes or stem cells collected from patient

② Cell proliferation

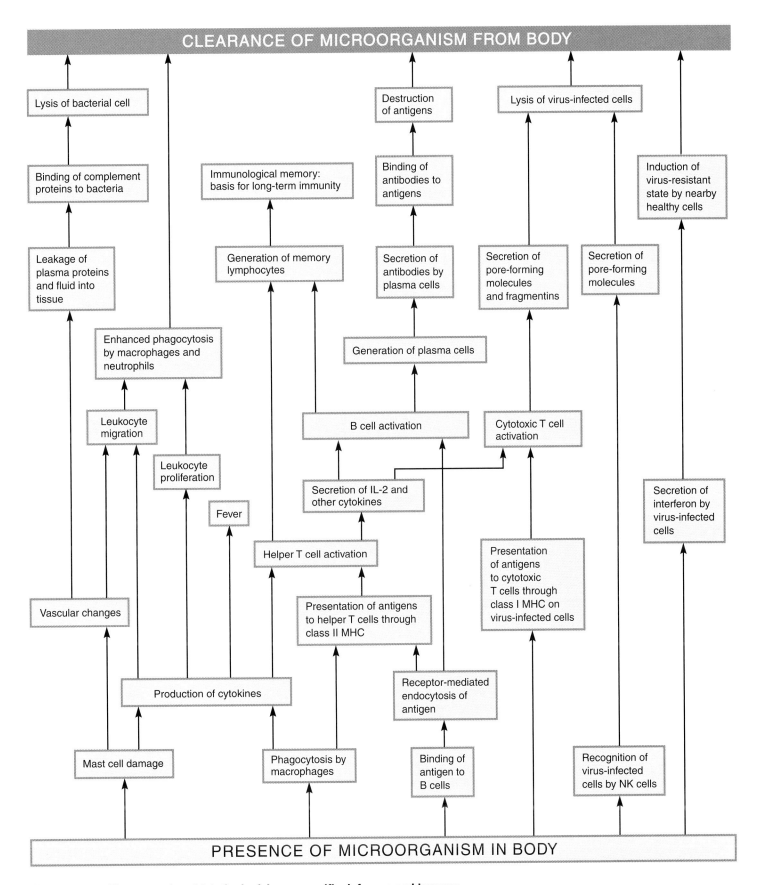

Figure 23.15 **The events by which the body's nonspecific defenses and immune responses clear microorganisms from the body.**

▬ **SYSTEMS INTEGRATION** Immune System

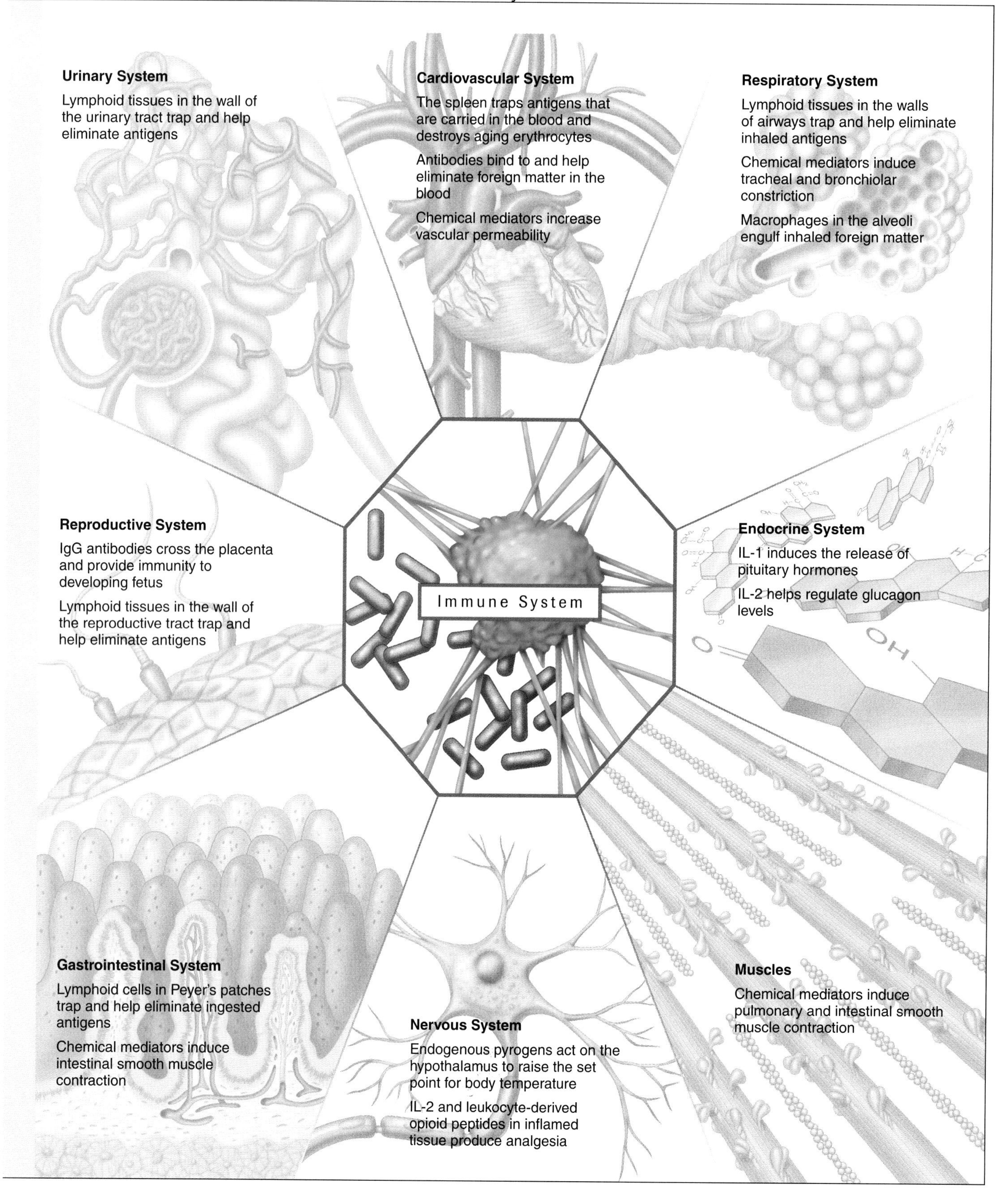

Urinary System

Lymphoid tissues in the wall of the urinary tract trap and help eliminate antigens

Cardiovascular System

The spleen traps antigens that are carried in the blood and destroys aging erythrocytes

Antibodies bind to and help eliminate foreign matter in the blood

Chemical mediators increase vascular permeability

Respiratory System

Lymphoid tissues in the walls of airways trap and help eliminate inhaled antigens

Chemical mediators induce tracheal and bronchiolar constriction

Macrophages in the alveoli engulf inhaled foreign matter

Reproductive System

IgG antibodies cross the placenta and provide immunity to developing fetus

Lymphoid tissues in the wall of the reproductive tract trap and help eliminate antigens

Immune System

Endocrine System

IL-1 induces the release of pituitary hormones

IL-2 helps regulate glucagon levels

Gastrointestinal System

Lymphoid cells in Peyer's patches trap and help eliminate ingested antigens

Chemical mediators induce intestinal smooth muscle contraction

Nervous System

Endogenous pyrogens act on the hypothalamus to raise the set point for body temperature

IL-2 and leukocyte-derived opioid peptides in inflamed tissue produce analgesia

Muscles

Chemical mediators induce pulmonary and intestinal smooth muscle contraction infections. Leukocytes develop to maturity in the central lymphoid tissues (the bone marrow and, in the case of T cells, the thymus). The peripheral lymphoid tissues exhibit a netlike architecture that traps foreign matter present in the blood (spleen), lymphatic fluid (lymph nodes), air (tonsils, adenoids), and in food and water (appendix and Peyer's patches).

Organization of the Body's Defenses, p. 728

When an infectious agent breaks through the first line of defense (the skin and mucous membranes, for example), it faces both nonspecific defenses and immune responses. Nonspecific defenses provide the body's most rapid defenses against infection or injury. In inflammation, oxygen, nutrients, defensive molecules, and phagocytic cells are drawn to the affected region. Interferons, secreted from virus-infected cells, protect the surrounding healthy cells from infection, whereas NK cells can recognize and kill virus-infected cells. The complement system is activated when the first in a series of complement proteins binds to certain types of bacteria, leading to bacterial lysis. The complement cascade can also be activated by antibodies, so it contributes to both nonspecific defenses and immune responses.

B cells and T cells provide for the features of immune responses: specificity, diversity, memory, and self-tolerance. There are two types of immune responses: the humoral response and the cell-mediated response.

Humoral Immunity, p. 737

The humoral response is the result of B cell activation. Upon contact with specific antigen, B cells proliferate and develop into long-lived memory B cells and short-lived plasma cells.

Whereas memory cells provide for long-term immunity to the antigen, plasma cells secrete antibodies that bind to and target the antigen, and then recruit other defenses (such as phagocytic cells), to destroy it.

Cell-Mediated Immunity, p. 740

The cell-mediated response occurs when cytotoxic T cells detect specific antigen presented by a class I MHC molecule (on a virus- or bacteria-infected cell or a tumor cell) and develop into active killers. Cytotoxic T cells destroy their targets in two ways: by releasing perforins that form pores in the stricken cell membrane, and fragmentins that enter the cell and induce apoptosis. Both humoral and cell-mediated responses are supported and regulated by cytokines secreted by activated helper T cells. Helper T cells are activated to proliferate and to secrete cytokines when they contact specific antigen presented by a class II MHC molecule (on a macrophage or a B cell). At the same time, long-lived memory T cells are generated. Thus the responses of helper T cells, cytotoxic T cells, and B cells coordinate the specific disposal of, and long-term immunity to, offending antigens.

Immune Responses in Health and Disease, p. 743

The aim of vaccination, or immunization, is to provide protection from infection. Both immunization and natural infection induce what is known as active immunity because they depend on the response of a person's own immune system and generate memory against the agent. Passive immunity, in contrast, is generated when antibodies are transferred from one person to another. These ready-made antibodies immediately target and mediate the disposal of antigens for which they are specific. Passive immunization is used when a danger-

ous bacteria or virus has entered the body of a person who is not already immune to it.

The immune system's capacity to distinguish self from nonself limits our ability to share tissues through blood transfusion and transplantation. For example, a transfusion reaction results when antibodies induce lysis of mismatched red blood cells. Similarly, a graft rejection can result when HLA, or MHC molecules, which differ from person to person, are not well matched. The survival of a graft requires making the best possible HLA match and using immunosuppressive drugs to diminish the recipient's immune responses. Bone marrow transplantation poses a particular problem when tissues are not well matched: The donated marrow, which contains lymphocytes, can mount immune responses against the recipient, resulting in a type of rejection known as a graft versus host reaction.

Immune dysfunction can result in allergies, autoimmune diseases, or immunodeficiency diseases. An allergy results from an exaggerated response to environmental antigens (allergens). Autoimmune diseases occur when the immune system reacts against self, as in rheumatoid arthritis or multiple sclerosis. An immunodeficiency disease can result when a component of immunity is incapacitated by an inherited or acquired condition. Immunodeficiency diseases may affect humoral or cell-mediated immune function, or both (as in severe combined immunodeficiency, or SCID).

Evidence suggests that the immune system, the nervous system, and the endocrine system are physiologically linked. Neuroendocrine mechanisms have been shown to regulate immune responses, and immune responses in turn can induce changes in both endocrine and neural function.

▮▮▮ *EXERCISES*

Multiple-Choice Questions

1. Which of the following conditions would lead to the most serious immunodeficiency disease?
   a) lack of IgG
   b) lack of neutrophils
   c) lack of B cells
   d) lack of cytotoxic T cells
   e) lack of helper T cells

2. Which of the following molecules can opsonize antigen?
   a) a T cell receptor
   b) interferon
   c) an antibody
   d) a perforin
   e) interleukin-2

3. Lymphocytes contact foreign antigen in all of the following tissues *except*
   a) bone marrow.
   b) spleen.
   c) lymph nodes.
   d) appendix.
   e) Peyer's patches.

4. Which of the following is *not* true about humoral immunity?
   a) It involves B cells.
   b) It involves antibody.
   c) It involves cytotoxic T cells.
   d) It can provide passive immunity when transferred from one person to another.

5. Macrophages
   a) have class I MHC molecules.
   b) have class II MHC molecules.
   c) are phagocytic.
   d) are indirectly involved in specific immunity.
   e) all of the above

6. Activated cytotoxic T cells release pore-forming molecules called
   a) histamines.
   b) complement proteins.
   c) perforins.
   d) immunoglobulins.
   e) ready-porins.

7. An individual with type AB blood
   a) is considered a universal blood donor.
   b) is considered a universal blood recipient.
   c) produces antibodies to the B antigen.
   d) produces antibodies to the A antigen.
   e) is Rh-positive.

8. Which of the following events can result in lifelong immunity?
   a) passage of maternal antibodies to a developing fetus
   b) an inflammatory response to a splinter
   c) phagocytosis of bacteria by a neutrophil
   d) administration of the polio vaccine
   e) administration of antibodies against the rabies virus

9. Foreign antigens phagocytosed by macrophages are presented by
   a) class I MHC molecules to cytotoxic T cells.
   b) class II MHC molecules to helper T cells.
   c) class I MHC molecules to helper T cells.
   d) class II MHC molecules to CD8-bearing cells.
   e) class II MHC molecules to cytotoxic T cells.

10. Of the following events, which occurs earliest in the process of local inflammation?
    a) increased capillary permeability
    b) fever
    c) attack by cytotoxic T cells
    d) release of histamine
    e) lysis of microbes mediated by antibodies and complement

11. Which of the following is *not* true about helper T cells?
    a) They function in both cell-mediated and humoral immune responses.
    b) They secrete antibody.
    c) They bear surface CD4 molecules.
    d) They are subject to infection by HIV.
    e) When activated, they secrete IL-2 and other cytokines.

12. IL-2 is important for the activation of all of the following cell types *except*
    a) B cells.
    b) cytotoxic T cells.
    c) NK cells.
    d) T helper cells.
    e) macrophages.

13. Which of the following is an autoimmune disease in which myelinated neurons become the target of the immune response?
    a) myasthenia gravis
    b) multiple sclerosis
    c) diabetes mellitus
    d) rheumatoid arthritis

Objective Questions

1. Fill in the blank with the abbreviation for the cell type mediating the stated function: helper T cell (T_H), cytotoxic T cell (T_C), B cell (B), or macrophage (M).
   a) _____ Phagocytosis
   b) _____ Secretion of cytokines such as IL-2
   c) _____ Killing of virus-infected cells
   d) _____ Specific binding to free virus
   e) _____ Differentiation into antibody-secreting plasma cells

2. Fill in the blank with the letter that applies to the stated situation: humoral immune response (H), cell-mediated response (CM), both (B), or neither (N).
   a) _____ Occurs in a viral infection
   b) _____ Involves the production of antibodies
   c) _____ Involves the phagocytic activity of neutrophils
   d) _____ Involves killing of virus-infected cells
   e) _____ Involves T cells bearing CD8

3. A person who experiences life-threatening allergic reactions to bee sting venom might be given an experimental drug designed to (block/enhance) the binding of IgE to mast cells.

4. When a macrophage is infected by a virus, viral antigen will be presented by (class I/class II MHC molecules) to a (helper T cell/cytotoxic T cell).

5. Evidence exists of interactions among the immune system, the nervous system, and the endocrine system. (true/false)

6. A young girl who has never been immunized against tetanus cuts her foot on a rusty nail. In the emergency room her wound is cleaned, and she is given an injection of tetanus antitoxin (antibody to tetanus toxin). This is considered (active/passive) immunization.

7. Macrophages internalize foreign antigens by (endocytosis/phagocytosis), whereas B cells do so by (endocytosis/phagocytosis).

8. A person who has had a thymectomy as a treatment for a thymic tumor

will likely experience a diminished (T cell/B cell) count.

Essay Questions

1. You notice that the area surrounding a paper cut on your hand has become red, warm, and swollen. Briefly describe the processes that lead to these symptoms of inflammation.

2. Through what lymphoid organs will a cell destined to be an active, mature T cell travel during its lifetime? Briefly state the function of each organ you mention.

3. Immune responses exhibit four features, all attributed to lymphocyte function. List these four features and discuss how they arise.

4. Draw a diagram showing the central role of helper T cells in both humoral and cell-mediated immune responses.

Critical Thinking

1. Consider a simple artificial antigen containing two distinct epitopes. Describe and diagram the process of clonal selection that results as B cells specific for these epitopes are selected by the antigen and are stimulated to proliferate and to differentiate into memory B cells and plasma cells. Also diagram how secreted antibodies interact with this antigen.

2. An infant who has experienced multiple bacterial infections since birth has been diagnosed with a macrophage deficiency. Describe the effect this deficiency has on the baby's nonspecific defenses and immune responses. Do you think that a neutrophil deficiency would present a more severe or a less severe state of immunodeficiency?

3. A child needs a kidney transplant. Both of his parents and his two older siblings are willing to donate. Tissue typing reveals that his father has the HLA genotye "8,9" (one HLA gene copy is "8," the other is "9"), and his mother has an HLA genotype "2,6." Considering that a child inherits one copy of the HLA gene from each parent, which is the better potential donor—a parent or a sibling? Is it possible to have a 100% donor-recipient match of the HLA type?

Find the answers to these exercises, and additional study tools, at the Physiology Place (www.physiologyplace.com).

Running a marathon.

24

The Whole Body: Integrated Physiological Responses to Exercise

R unning a marathon—26.2 miles—is a major stress to the body. Energy stores become depleted, body fluid volume decreases, muscle tissue tears, and joints and bones are subject to continuous stress. But with training, the body can endure this stress. Training increases the efficiency of energy metabolism, strengthens the heart, increases blood volume and bone density, and enhances the aerobic capacity of muscles and the ability to produce sweat.

Throughout this text, we have followed the experiences of Bill and Jane (first introduced at the end of Chapter 1) as they ran their respective marathons, focusing on what was transpiring in their bodies system by system. In this chapter, we will examine their experiences from an integrated, whole-body perspective. We will study the various physical stresses endured by Bill and Jane, and how their training allowed them to cope with these stresses and complete their challenging runs.

OBJECTIVES

▒ Describe the manner in which different physiological systems interact in the stressful, real-life situation of running a marathon.

▒ Provide an example of how some physiological responses are sensed by subjective experience.

▒ Describe how increased physiological function is accomplished during exercise.

▒ Describe the mechanisms that regulate physiological function during exercise, and explain how regulation of one system influences other systems, thus requiring coordination of multiple systems.

▒ Describe how resources are allocated to various tissues and organs under stressful conditions that cause conflicting physiological demands.

▒ Identify the physiological characteristics that limit exercise intensity and duration and thus influence performance.

Before You Begin

Make sure you have mastered the following topics:

1. *Muscle metabolism, p. 385*

2. *Factors affecting cardiac output, p. 438*

3. *Regulation of energy metabolism by insulin, glucagon, and epinephrine, p. 189*

4. *Factors affecting pulmonary airflow, p. 531*

5. *Mechanisms of thermoregulation, p. 14*

N ow that we've seen how each physiological system operates on its own, let's revisit the marathon experiences of Bill and Jane and see how physiological systems interact with each other. Before reading the rest of this chapter, you may wish to return to the last part of Chapter 1 and read about Bill and Jane's marathons again (the key events are summarized in **Figure 24.1**).

Principles of Physiological Integration

Before we explore the physiological details at each stage of the race, we must cover a few general principles that determine how and why physiological systems interact.

The Metabolic Demands of Exercise

Changing posture, lifting weight, and moving from one place to another are accomplished through the voluntary actions of skeletal muscle. For very brief periods, the energy for muscle contraction can be provided by the energy stored within the muscle itself in molecules such as ATP, creatine phosphate, glucose, and glycogen. More

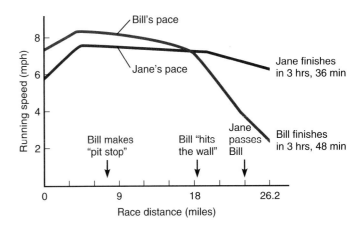

Figure 24.1 Bill and Jane's paces over the course of the marathon. Both started slower than they wished because they were packed in tightly with other runners.

prolonged activity requires increased delivery of oxygen and metabolic fuels (such as glucose and fatty acids) to the muscle from other sources. In addition, contracting muscle cells produce heat and metabolites, including lactic acid and carbon dioxide, that must be carried away in order for the muscle to continue functioning. If these metabolites are not removed, muscle pain and fatigue can result. Moreover, the increased metabolic demand of skeletal muscle must be satisfied without compromising the energy supply to the brain and other vital organs such as the heart.

How Physiological Function Is Increased

Locomotion, heart function, and respiration are all examples of physiological processes that operate in repeating cycles. The speed (or capacity) of these processes can be augmented by increasing the frequency of the cycles and/or the magnitude (or strength) of each cycle. During exercise, an individual can also tap into reserve anatomical capacity to increase function. For example, gas exchange in the lungs can be enhanced by inflating more alveoli, because at rest not all alveoli are inflated with each breath. Also, the contracting skeletal muscles that cause locomotion can aid cardiac and respiratory function.

The ability to achieve and maintain an increased level of physiological function depends on the availability of energy as well as the competing needs of different physiological systems. These needs are profoundly influenced by environmental conditions and the individual's behavior.

The changes leading to increased function within specific physiological systems are regulated by two control mechanisms: (1) intrinsic, or local, controls and (2) extrinsic controls through the endocrine and nervous systems. Neural control may be voluntary or involuntary. Key to increasing function is an increase in blood flow. Under nonstressful conditions, intrinsic control is the dominant mechanism determining blood flow to skeletal muscle and many other individual organs. But when several organs begin competing for increased delivery of blood, regulation requires greater involvement of hormonal and neural control mechanisms.

Quick Test 24.1

1. Name three fuels that can be used for energy by skeletal muscles and that are normally present within the muscle cells themselves.

2. Blood flow to muscles and other organs is regulated by intrinsic and extrinsic control. As the general demand for blood flow increases, which type of control assumes greater importance?

The Start: Transition from Rest to Exercise

We have seen that the body's response to exercise is orchestrated by both intrinsic control mechanisms and extrinsic control mechanisms. Now let's look at how these mechanisms interacted while Bill and Jane ran their marathon.

Autonomic Nervous System Input: Fight or Flight

Marathon running and many other types of exercise are cultural substitutes for primitive activities such as hunting, fighting, and fleeing: activities that require rapid release of stored fuels and the distribution of these fuels (via the blood) to the muscles that need them. The autonomic nervous system activates and regulates this release and distribution of metabolic fuels. The subjective "heart-pounding" excitement Bill and Jane felt just before the start of the race was triggered by reduced parasympathetic nerve activity and increased sympathetic nerve activity and by epinephrine, the circulating hormone of the sympathetic nervous system. In modern society, many stresses still activate this sympathetic response (which can also engender fear and rage), but people tend to participate less and less in the physical activities for which the response evolved. The chronically elevated catecholamine concentrations that can result from such stress have been associated with anxiety, irritability, and depression. Exercise can provide an outlet for pent-up sympathetically mediated excitement, and increase clearance of catecholamines from the circulation. Perhaps this explains why Bill and Jane feel better after their daily exercise routine (although maybe not after a marathon).

The increased sympathetic activity (and decreased parasympathetic activity) that Bill and Jane experienced just before the marathon helped prepare them for the event. Sympathetic input to the SA node and myocardium causes increased cardiac output by increasing both heart rate and stroke volume, and as a result, mean arterial pressure rises in preparation for the increased demand for blood flow to skeletal muscle. Sympathetically mediated increases in venomotor tone promote venous return, which also enhances cardiac output. Sympathetic input to the pancreas stimulates glucagon release and inhibits insulin release, causing a rise in blood glucose in preparation for the increased demand for energy in skeletal muscle.

Energy Sources and Mobilization: Different Fuels for Different Needs

The proportional use of fuels (carbohydrate, fat, and even protein) during exercise depends upon the intensity and duration of the activity. In a marathon, the mix of fuels used at the beginning is different than the mix used later on. As we will see, the rate of carbohydrate expenditure at the beginning of the race has a profound effect on performance at the end. Although Bill and Jane could not directly adjust their fuel usage by force of will (the proportion of each fuel used is dictated by regulation of enzyme activity in metabolic pathways), they could indirectly influence the relative depletion of carbohydrate by adjusting their behavior (exercise intensity).

We have seen that ATP is broken down to ADP during the crossbridge cycling that pulls thin filaments along thick filaments and causes muscle contraction. The instantaneous recharging of ADP back to ATP is accomplished by the donation of high-energy phosphate from creatine phosphate. However, this "energy reservoir" by itself can only supply a few seconds of energy reserve.

Glucose (whether free in the blood or stored as glycogen in muscle and liver) is another energy source that can be called on quickly because it can be broken down to produce ATP through glycolysis under anaerobic conditions (that is, it does not require oxygen). However, recall that although glycolysis liberates energy quickly, it is not a very efficient way of generating ATP. Exercising while this metabolic pathway is dominant (before oxygen delivery can be increased) means that lots of glucose will be broken down to produce rather limited amounts of usable energy. Therefore, this metabolic pathway is appropriate for sprinting, but it cannot support long-duration exercise such as a marathon.

The metabolic pathways involved in fat-burning, aerobic metabolism (that is, metabolic pathways that use oxygen, such as fatty acid oxidation and the Krebs cycle) require more time to respond to the changes in metabolic rate during exercise. Stated in more precise physiological terms, the activities of rate-limiting enzymes in these pathways change with the buildup or depletion of metabolic reaction products. Certain reaction products (including heat, H^+, and CO_2) cause vasodilation, which increases blood flow to exercising muscles. These products also decrease the affinity of hemoglobin for oxygen,

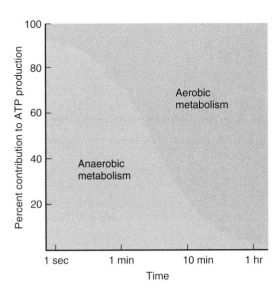

Figure 24.2 The relative contributions of anaerobic and aerobic metabolism to ATP production over the first hour of the marathon. This time-compressed figure indicates the runners' metabolic status at the start, after 1 minute, after a little over 1 mile (10 minutes), and after about 7.5 miles (1 hour), running at approximately 70% of maximal aerobic capacity.

> *Why is glycolysis said to be anaerobic?*

resulting in greater oxygen unloading in the muscle tissue. The time course for the transition from an anaerobic/glycolysis-dominant metabolism to an aerobic/lipolysis-dominant metabolism at the onset of moderate exercise is shown in **Figure 24.2**.

If a person begins exercising relatively gradually (at a low intensity), a greater proportion of ATP is generated through aerobic, lipolytic metabolism. This conserves blood glucose, which is normally the only fuel used by the brain, as well as muscle glycogen, which seems to be an important factor in limiting muscle function in the late stages of a marathon. Thus Jane's slower start conserved both liver and muscle glycogen and contributed to her ability to maintain her pace in the later stages of the marathon. Because Bill started at a higher intensity, he used relatively more glycogen at earlier stages of the race and thus depleted almost all of his muscle glycogen by about 20 miles into the race (when he "hit the wall").

Cardiovascular Adjustments: Anticipating the Needs

Support of aerobic metabolism in exercising muscle requires increased oxygen delivery, which in turn requires increased blood flow to the muscle. Both respiratory and cardiac function increase *immediately* at the onset of exercise. In fact,

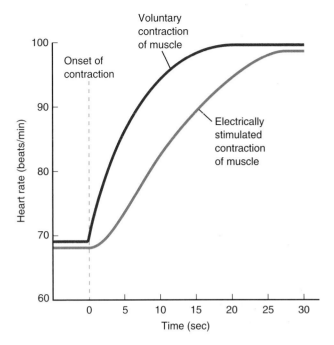

Figure 24.3 Central command. Heart rate increases more rapidly in response to voluntary exercise than to muscle contraction in response to external electrical stimuli.

Source: Redrawn from A. Krogh and J. Lindhard, "A comparison between voluntary and electrically induced muscular work in man." *Journal of Physiology* (London), 51 (1917): 182–201.

> *Is central command an example of intrinsic control or extrinsic control of the heart?*

these functions, especially heart rate, may already be somewhat elevated in the anticipation of exercise, as we just discussed. In addition to the fight-or-flight response, a mechanism that is unique to exercise—a feedforward mechanism called *central command*—is also involved. This feedforward signal originates in the motor cortex or other higher brain centers and is associated with Bill and Jane's conscious decisions to contract their muscles.

An experiment that illustrates the effect of central command is shown in **Figure 24.3**. When muscles are contracted *voluntarily,* heart rate increases with virtually the very next beat, far faster than possible if baroreceptor- or chemoreceptor-mediated autonomic controls based on negative feedback were involved. When muscle contraction is induced by *external electrical stimulation,* increases in heart rate are delayed and take longer to attain a steady state because negative feedback mechanisms activated in response to the involuntary muscle contractions are the only control signals involved. Central command ensures that blood pressure will not decrease at the onset of exercise. This theoretically would occur if a large proportion of blood volume suddenly rushed into exercising skeletal muscle.

> *It does not require oxygen.*

> *Extrinsic control*

Respiratory Responses: Recruiting Reserve Capacities

As exercise begins, active skeletal muscles consume greater amounts of oxygen and produce greater amounts of carbon dioxide, but arterial concentrations of these gases change very little (at least during mild to moderate exercise). This is because feedforward control circuits increase ventilation before the gas concentrations in arterial blood can change. One of these feedforward circuits is the central command associated with the intention to perform exercise (just described); another feedforward signal originates in mechanoreceptors and chemoreceptors in exercising skeletal muscles. Both of these feedforward signals stimulate ventilation.

In Chapter 18 we saw that changes in arterial oxygen and carbon dioxide concentrations regulate ventilation through negative feedback control. During exercise, these feedback mechanisms seem to *inhibit* ventilation in order to prevent hyperventilation. In contrast, the feedforward signals provide the stimulus for *increased* ventilation during exercise.

Ventilation increases during exercise as a consequence of both increased breathing frequency and increased tidal volume. At low to moderate exercise intensities, the dominant adjustment is increased tidal volume. As exercise intensity increases further, tidal volume stabilizes at about 65% of vital capacity, and additional increases in ventilation are due to increased frequency.

Recall that *all* of the cardiac output flows through the lungs. Resting cardiac output (5 liters/min) is easily accommodated by about one-third of pulmonary capillary capacity. Thus only one of every three capillaries is conducting blood at any given time, and more capillaries in the base of the lungs are carrying blood than in the apex. Alveoli near capillaries that are not conducting blood are collectively referred to as the *physiological dead space,* because like the anatomical dead space described in Chapter 17, air in these alveoli does not participate in gas exchange. As cardiac output increases during exercise, more of the capillaries begin carrying blood continuously, and perfusion becomes more uniform throughout the lungs. This constitutes one type of pulmonary reserve capacity.

Once cardiac output exceeds three times the resting value, all the reserve capillaries have been recruited, pulmonary pressure rises modestly, and blood flows faster within individual capillaries. As a consequence, the amount of time each red blood cell spends exchanging gas with an alveolus is reduced. Transit time along the length of an alveolar capillary is normally about 1 second, but a red cell can fully exchange oxygen and carbon dioxide in one-third that time. This constitutes a second type of pulmonary reserve capacity. In combination, the two reserve mechanisms ensure that the pulmonary circuit can accommodate up to a sixfold increase in cardiac output during exercise, with no reduction in red cell oxygenation and carbon dioxide release. Bill and Jane's cardiac outputs never exceeded this limit while they ran.

World-class endurance athletes, by contrast, may attain rates of oxygen consumption and cardiac output that can exceed the reserves of the pulmonary circuit. That both skeletal muscle (the major consumer of oxygen) and the heart (the generator of cardiac output) can hypertrophy with training, but the lungs cannot, suggests to some researchers that the fixed anatomical capacity of the lungs may be the ultimate limitation in aerobic exercise performance. Other researchers, however, have argued that the limitation in aerobic exercise performance is due to decreases in arterial blood pH and increases in body temperature, both of which cause the oxygen-hemoglobin saturation curve to shift to the right (see Figure 18.10), which means that less oxygen is loaded per hemoglobin molecule as it transits a lung capillary.

Temperature Regulation

Before they started running, Bill and Jane's core temperatures were approximately 37°C, and their leg muscle temperatures were about 2°C lower (at this time, body heat was generated primarily by their hearts and livers). When they began exercising, their muscles generated large amounts of heat (only 25% of the energy consumed by exercising muscle generates force; 75% is lost as heat). After 10 minutes of running, Jane and Bill's muscle temperatures reached a steady state (39°C), which were about one-half degree above core temperature. Under the mild environmental conditions at the start of the race (68°F/20°C), these increases in core temperature were proportional to the runners' exercise intensities and independent of ambient temperature. (But in the late stages of the race, when ambient temperature reached 86°F/30°C and the hot sunshine was beating down on them, this would no longer be true.) The increase over resting temperature represents the error signal (actual temperature versus the regulated set-point temperature) in the negative feedback control system for thermoregulation.

The first physiological response implemented to maintain thermal homeostasis is an increase in blood flow to the skin. In this way, heat is carried by the blood from the muscles and the core to the body surface, where it is dissipated to the environment. Skin blood flow can increase from about 0.15 liter/min at rest to more than 7 liters/min during exercise. The heat is transferred to the external environment primarily by convection and evaporation (sweating); conduction to the air is a minor component because air has a low thermal conductance, and conduction to the ground is low because Bill and Jane's feet are insulated by shoes and socks.

At the onset of exercise, sympathetically mediated cutaneous vasoconstriction increases in proportion to exercise intensity (as part of the fight-or-flight response). As body temperature increases, vasoconstriction diminishes (because increased tissue temperatures reduce the sensitivity of venous adrenergic receptors), resulting in a twofold increase in blood flow to the skin. But this increase is minuscule compared to the 30-fold increase caused by a special sympathetically mediated active vasodilation mechanism in the skin. This active vasodilation is the efferent arm of the negative feedback response to an increase in core temperature. Core temperature has a far greater influence on blood flow to the skin than does either skin temperature or even the local metabolic needs of the skin. The neurotransmitter and receptors involved in sympathetic cutaneous vasodilation have not been identified, but this mechanism seems somehow tied to the sweating response.

Quick Test 24.2

1. Prior to the start of exercise, there is often an anticipatory increase in sympathetic nervous activity and decrease in parasympathetic activity. What is the effect of these changes on cardiac output? On blood glucose concentration? Explain.

2. A gradual increase in exercise intensity at the onset of exercise helps conserve a particular fuel that is present in the liver and in muscle. What is this fuel?

3. Define the term *central command* as it relates to the cardiovascular and respiratory adjustments that occur at the start of exercise.

4. What are the two reserve mechanisms that enable the lungs to accommodate up to a sixfold increase in blood flow with little or no change in their ability to deliver oxygen to the blood or remove carbon dioxide?

The Long Haul: Almost Steady State

After the initial transition from primarily anaerobic to primarily aerobic metabolism in the first mile or so, the flow of energy (from glucose and fatty acids to ATP and then to heat and external work by exercising muscles) has settled into a relatively steady-state condition. It is not a perfect steady state, however, because muscle glycogen is slowly but continuously being depleted, and blood glucose is taking up the slack. For the time being, Bill and Jane feel good; let us examine events occurring in their bodies at this time.

Hormonal Control of Energy Metabolism

Mobilization of glucose and fatty acids into the bloodstream from storage sites (liver and adipose tissue, respectively) is regulated by glucagon and epinephrine. A decline in insulin concentration also contributes to this mobilization. The changes in the circulating concentrations of these substrates and regulatory hormones over the course of a 26.2-mile marathon are illustrated

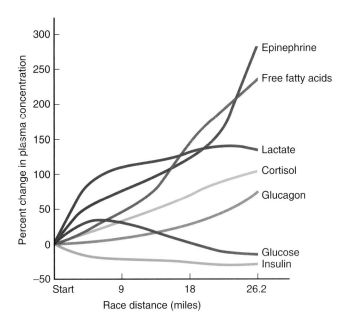

Figure 24.4 Changes in the plasma concentrations of some metabolic fuels and some hormones that regulate them over the course of a marathon.

Compiled from data reported by R. Cade et al., "Marathon running: Physiological and chemical changes accompanying late-race functional deterioration." *European Journal of Applied Physiology*, 65 (1992): 485–491, © Springer-Verlag; M.J. O'Brien et al., "Carbohydrate dependence during marathon running." *Medicine and Science in Sports and Exercise*, 25 (1993): 1009–1017; E.F. Coyle et al., "Muscle glycogen utilization during prolonged strenuous exercise when fed carbohydrate." *Journal of Applied Physiology*, 61 (1986): 165–172.

in **Figure 24.4**. (Lactate is included in the graph because it is involved in gluconeogenesis.)

The slight increase in glucose concentration during the first two-thirds of the race is due to the stimulatory effect of epinephrine on glycogenolysis and gluconeogenesis. Even though an increase in plasma glucose after a meal causes the pancreas to secrete insulin, plasma insulin concentrations decrease during exercise due to inhibition of insulin secretion by sympathetic nervous input to the pancreatic beta cells. Glucagon secretion is stimulated by sympathetic nerve impulses to the pancreatic alpha cells. Lactate (a product of anaerobic metabolism) rises quickly at the beginning of the race but levels off as aerobic metabolism and gluconeogenesis increase. Free fatty acid concentrations rise steadily as a result of epinephrine and a high glucagon-to-insulin ratio. Note that blood glucose levels fall below baseline, and that epinephrine concentrations rise precipitously in the final third of the race; these are signs of physiological distress (as we will see later in the chapter). For the time being (miles 4–18), blood glucose concentrations are slightly elevated, so Bill and Jane have a conscious, subjective feeling of well-being.

Some evidence suggests that the sex of a person influences how various energy substrates are used, but the relative importance of this influence during exercise and the mechanisms involved remain topics of debate. In some

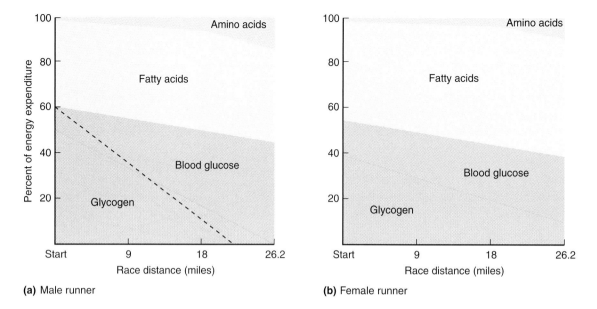

Figure 24.5 Use of nutritional fuels for energy by men and women during a marathon.
(a) The relative dependence of a typical male runner on each fuel source. The contribution from
muscle glycogen declines from 50% at the start of the race to less than 10% at the end; this
decline is offset by increased use of blood glucose (derived from gluconeogenesis) and fatty acid
oxidation. In the late stages of the race, deaminated amino acids contribute significantly to
gluconeogenesis. The dashed line, which represents Bill's rate of muscle glycogen consumption,
indicates that by starting out too quickly, he consumed glycogen at an undesirably high rate that
ultimately resulted in glycogen depletion before the race ended. (b) The relative dependence of a
typical female runner on each fuel source. Compared to men, women rely more heavily on fatty acid
oxidation, which results in less depletion of muscle glycogen and less oxidation of amino acids.

Source: Adapted from Figure 5 in E.F. Coyle et al., "Muscle glycogen utilization during prolonged strenuous exercise when fed
carbohydrate." *Journal of Applied Physiology* (1986): 165–173.

experiments, testosterone (higher in Bill) and estrogen
and progesterone (both higher in Jane) directly alter key
enzymes that control metabolism of glucose and free fatty
acids. These steroid hormones may also indirectly alter
glucose and fat metabolism by influencing the secretion
and action of other hormones such as epinephrine and insulin. Men tend to have higher circulating epinephrine
concentrations during exercise, which would enhance the
rate of glycogenolysis in muscle and liver. Whereas estrogens increase lipolysis in muscle and adipose tissue, they
inhibit gluconeogenesis and glycogenolysis; as a result,
men burn a relatively greater proportion of carbohydrate
for fuel, and women burn a relatively greater proportion
of fat (**Figure 24.5**). Some researchers have estimated
that over the course of a marathon, women may oxidize
as much as 75% more fat and 40% less carbohydrate than
men. However, other researchers have observed that the
differences in substrate use between men and women decrease as the level of training and aerobic fitness increases.

Cardiovascular Control: Factors Influencing Blood Pressure

Although mean arterial pressure is a regulated variable,
Bill and Jane's mean arterial pressures increased approximately 20% during the early and middle stages of the
race. Why should this be? Why isn't peripheral resistance
adjusted by baroreceptor-mediated negative feedback
mechanisms to bring blood pressure back to normal? The
answer is probably related, in part, to blood flow in muscles during exercise. Blood vessels are compressed each
time a muscle contracts; in many muscles, blood flow
stops altogether when muscle tension exceeds 70% of
maximum. Various *metaboreceptors* that are sensitive to the
mechanical stresses of compression, the transient buildup
of metabolites such as potassium, or increases in temperature send afferent signals to integrative centers in the
central nervous system; this information initiates a regulated rise in mean arterial pressure.

The results of a classic experiment that demonstrated
this metabolic sensor system are shown in **Figure 24.6**.
Blood pressure rises gradually during rhythmic contraction
of skeletal muscle and then rapidly returns to resting value
when the contractions cease. Part of this response is due to
increases in metabolite concentrations in the muscle as
blood flow is interrupted during each muscle contraction.
The metaboreceptors send intermittent (but repetitive) signals to increase blood pressure. If the buildup of metabolites is exaggerated by completely blocking blood flow to
the exercising muscle with a tourniquet, blood pressure

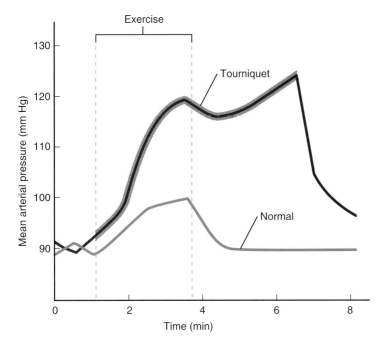

Figure 24.6 Regulation of blood pressure by metabolite buildup in exercising muscle. When greater quantities of metabolites are trapped in an exercising arm by a tourniquet, blood pressure rises more quickly and remains elevated as long as the tourniquet is in place, even if exercise has ceased. The shadowing on the tourniquet curve represents the time during which the tourniquet was applied.

Source: Redrawn from M. Alam and F.H. Smirk, "Observations in man upon a blood pressure raising reflex arising from the voluntary muscles." *Journal of Physiology* (London), 89 (1937): 372–383.

rises faster and higher. Note that blood pressure remains elevated after exercise ceases and does not return to normal until after the tourniquet is removed and the metabolites are "washed out" by the restored flow of blood.

What is the benefit of this metabolite-regulated rise in blood pressure during exercise? After all, it means the heart must work harder. Because blood pressure is the driving force for perfusing the muscle with blood, a higher driving force ensures that flow is maintained through a greater portion of the skeletal muscle contraction cycle, and that flow will be enhanced between contractions to help wash out the metabolites that have built up with each momentary stoppage of flow.

The muscle metaboreceptors are thought to contribute to cardiovascular control in humans in a meaningful way only at high exercise intensities; in contrast, central command signals arising from the motor cortex seem to be active during all exercise intensities (**Figure 24.7**). These signals keep Bill and Jane's blood pressure slightly elevated through at least the first two-thirds of the marathon; in the final stages of the race, blood volume diminishes (via sweating) to the point that adequate perfusion pressure to vital organs (especially the brain and heart) becomes

threatened. At this point, the classical negative feedback system—baroreceptor reflex control—assumes dominance in maintaining mean arterial pressure.

Respiration: The Effort of Breathing

During the race, Jane and Bill feel the effort involved in moving their bodies horizontally along the road and vertically up hills, but not the effort of breathing. However, moving air into and out of their lungs requires a significant amount of effort, and about 15% of cardiac output during exercise flows to respiratory muscles in order to accomplish this work. In this section we see how Bill and Jane have unconsciously learned to breathe with less effort, and how the various movements associated with exercise can provide a boost to respiratory movements.

First, it is necessary to understand how we can measure the work of breathing. Recall that inflation of the lungs starts with contraction of the diaphragm and external intercostal muscles, which causes a decrease in intrapleural pressure, which expands the lung tissue. At rest, expiration is due primarily to the passive recoil of the elastic lung tissue, but during exercise the contraction of internal intercostal and abdominal muscles increases intrapleural pressure, which helps "push" air out of the lungs. Measuring the magnitude of the decreases and increases in intrapleural pressure gives an indication of the effort (work) of breathing. Placing a pressure gauge directly in the pleural space would be difficult and downright dangerous; fortunately, the esophagus runs through the pleural cavity, and (so long as active swallowing is not occurring) the pressures measured there provide a reasonable approximation of intrapleural pressure.

If we were to ask Bill or Jane to hyperventilate under resting conditions in order to mimic their ventilation rates during the marathon, they would have to produce over four times as much pleural pressure during expiration to attain the same volume and flow rate. The explanation for this relates to airway resistance. During exercise, epinephrine relaxes airway smooth muscle (the opposite of what it does to vascular smooth muscle). But unconscious breathing during exercise is also influenced by receptors that monitor either the tension of respiratory muscles, the stretch of the lung tissue, or the pressure in the airways. These receptors send afferent signals to the respiratory control center that coordinate the overall breathing pattern by fine-tuning both the recruitment order of respiratory muscles and the force output of each muscle. This results in an orderly compression of alveoli with minimal compression of the conducting airways, which would increase resistance. Conscious hyperventilation at rest, however, overrides this control mechanism and results in haphazard compression of the lung structure accompanied by increased turbulence, airway resistance, and pockets of trapped air. Exercise training seems to help refine the unconscious neural control mechanism,

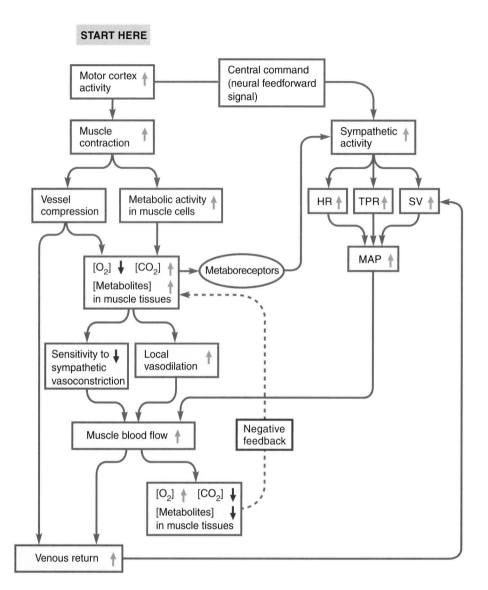

Figure 24.7 The roles of metaboreceptors and central command in control of mean arterial pressure during exercise. The main benefit of increased mean arterial pressure is enhanced blood flow to muscles, which promotes delivery of oxygen and removal of metabolites from the active muscle tissue. Note that muscle contraction has two distinct influences on blood flow: By compressing veins running between muscles, venous return is enhanced (the skeletal muscle pump), but by compressing arterioles and capillaries within the muscle, blood flow and oxygen delivery are briefly interrupted, causing carbon dioxide and metabolite buildup. These metabolites stimulate several mechanisms that promote skeletal muscle blood flow. They cause vascular smooth muscle relaxation, resulting in local vasodilation. They stimulate metaboreceptors that trigger sympathetically mediated increases in mean arterial pressure. Finally, they reduce the sensitivity of the vasculature in the exercising muscle to the sympathetic efferent signals that are causing vasoconstriction elsewhere in the body.

If an increase in muscle metabolism triggers local vasodilation and an increase in muscle blood flow, is it an example of active hyperemia or reactive hyperemia?

Active hyperemia

which may explain why untrained individuals "feel their lungs burning" (irritation due to air turbulence) when they exert themselves.

The muscular actions needed for running are not limited to the legs. The arms swing and the torso moves to counterbalance the leg movements and maintain balance. (This explains Bill's sore biceps muscles after the marathon.) Furthermore, abdominal organs bounce around, pushing on the diaphragm. These muscular contractions and intestinal movements contribute to the expansion and compression of the thoracic cavity and thus have some influence on ventilation. A runner can take advantage of these movements, because when breathing is synchronized with leg movements, 8% less oxygen is consumed compared to desynchronized running. An 8% energy savings compounded over several hours of marathon running is quite significant. People tend to synchronize these movements unconsciously, and trained runners get "in sync" faster than untrained runners.

The Gastrointestinal Tract: The Forgotten System

Because most runners have more than enough energy stores to support their metabolic needs through the completion of the marathon, they do not need to eat during the event. Therefore, the energy needs of gastrointestinal tissues are relatively low, and blood flow through this region is "expendable," at least temporarily. Thus gastrointestinal function during exercise has received relatively little attention. However, as Bill discovered, sometimes the gastrointestinal tract cannot be ignored during exercise.

At rest, ingested food is propelled through the gastrointestinal tract by gastrointestinal motility. During and after a meal, motility and other digestive functions are stimulated by parasympathetic efferent signals and inhibited by sympathetic efferent signals. The regulated transit of digested food through the small intestine and colon allows adequate time for absorption of almost all of the water from the lumen.

Bill's gastrointestinal difficulties during the race are a common experience for runners. Diarrhea results from inadequate absorption of water due to excessively rapid transit of digested food through the gastrointestinal tract. Although intrinsic gastrointestinal motility is thought to be reduced during exercise (because parasympathetic efferent signals are reduced and sympathetic efferent signals are increased), the entire body is bouncing up and down with each stride. Therefore, the contents of the gastrointestinal tract are literally shaken, propelling them by forces that are outside the normal control of gastric and intestinal motility; this might lead to movement through the small intestine and into the colon at a rate faster than appropriate for adequate absorption of water. The movement of radioactive-tracer-labeled food has been studied under controlled exercise conditions. Various studies have reported results ranging from no change to 20% increases in gastric emptying rate during exercise, and 10% increases in transit through the small intestine during exercise. Alternatively, psychological stress (at rest) increases transit through the small intestine by 25%. Runner's diarrhea, therefore, may be caused by a combination of stress and jarring movement. Furthermore, because caffeine stimulates secretion of fluids into the small intestine, Bill's coffee intake may have contributed to his distress at mile 8 in the race.

Some controversy exists regarding the influence of glucose contained in sports drinks on water absorption. As we will discuss shortly, fluid replacement during marathon running is a critical issue. *Hypo*-osmotic carbohydrate-containing beverages are beneficial if runners find them more palatable than water and therefore drink more. However, *hyper*osmotic beverages may inhibit diffusion of water out of the intestinal lumen, leading to discomfort or diarrhea.

Gastrointestinal blood flow may possibly be decreased so drastically during heavy exercise in the heat that the epithelial cells are rendered somewhat ischemic, which can cause increased permeability of the epithelium to relatively large molecules. The consequences of this phenomenon are described in the section on postexercise fever.

Thermoregulation: Intensifying the Competition for Blood

Bill and Jane have been running for 2 hours, and the air temperature has increased to 86°F (30°C). The thermal gradient between their skin and the air is now smaller than it was at the beginning of the race, and thus heat flow due to convection is less effective, and heat loss via evaporation is increased. More blood is diverted to the skin, and increasing amounts of fluid are lost as sweat. The competition among organs for blood is becoming intense, and compromises must be made if both exercise performance and homeostasis are to be maintained.

The changes in the distribution of blood flow that result during heat exposure and exercise are depicted in **Figure 24.8**, in which a triangle represents the body. The vital organs (brain, heart, and lungs) are situated at the center of the triangle; the skin, skeletal muscle, and viscera (gastrointestinal tract) are situated at the corners. The relative amounts of blood flowing through the organs are represented by color: Deeper shades of red represent greater blood flow. The liver and kidneys are not easily categorized as belonging to either the "vital organs" or the viscera in this figure because even though they are vital organs, their blood flow can be reduced somewhat during exercise for the sake of working muscle and temperature regulation.

At rest (Figure 24.8a), about 60–75% of cardiac output flows through the vital organs and viscera, with the balance distributed through the skin and skeletal muscle.

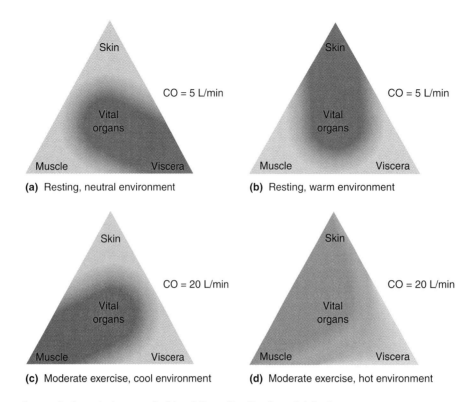

(a) Resting, neutral environment

(b) Resting, warm environment

(c) Moderate exercise, cool environment

(d) Moderate exercise, hot environment

Figure 24.8 Thermoregulatory-induced changes in blood flow distribution. (a) During resting (thermoneutral) conditions, most of the cardiac output is distributed to the vital and visceral organs. (b) When environmental temperatures increase, more blood is directed to the skin at the expense of the viscera and muscles. (c) During exercise in a cool environment, cardiac output can increase by a factor of 4. Although the *percentage* of blood directed to the vital organs has decreased, the *absolute* flow rate to the heart actually increases and flow to the brain remains constant. (d) As environmental temperature increases, more blood is directed away from exercising muscle, which degrades performance, and from visceral organs, which can cause organ damage if flow reduction is of sufficient severity or duration.

Over a modest range of environmental temperatures (called the *thermoneutral zone*), body core temperature can be regulated solely by regulating the amount of warm blood sent to the cooler skin surface. In warm conditions at rest, more blood must be redirected to the skin in order to dissipate heat and maintain constant body temperature (Figure 24.8b). This adjustment can be made simply by changing blood flow distribution (no change in cardiac output is required), and it in no way interferes with the function of the other tissues. Muscle energy needs are low when a person is resting.

During exercise under cool conditions (Figure 24.8c), blood flow to working muscle increases dramatically. This is accomplished not only by redistribution of blood flow away from skin and viscera (including mild reductions in renal and hepatic blood flow), but also by an increase in cardiac output. Blood flow to all organs and tissues is still sufficient to meet their metabolic needs.

As core temperature rises, the cardiovascular system is faced with conflicting demands: providing increased blood flow to the skin for heat dissipation *and* providing increased blood flow to exercising muscle without ever sacri-

ficing flow to the brain and heart. This was Bill and Jane's cardiovascular condition through the first two-thirds of the marathon. In spite of these conflicting needs, they were coping quite well because their premarathon training had expanded their blood volumes. But as was the case with substrate metabolism, their cardiovascular systems were not quite in steady state. Blood volume slowly diminished as the rate of sweating increased. Eventually, blood volume was no longer sufficient to provide all tissues with adequate flow, so vasoconstriction of renal, hepatic, and splanchnic blood vessels (vessels supplying digestive organs) increased. Eventually, blood flow to vital visceral organs (kidney and liver) began to be compromised (Figure 24.8d); this is not a sustainable condition, and if it occurs, exercise performance cannot be maintained in the last stages of the race.

Renal and splanchnic vascular resistance varies in direct proportion to core temperature. A 1.5–2.0°C increase in core temperature causes decreases in blood flow of 25% to the liver, 30% to the kidneys, and over 50% to the stomach and intestines. Moderate, transient reductions in blood flow to the kidney and liver can be tolerated, but the function of these organs must be maintained above certain critical thresholds if they are to perform the necessary filtration and detoxification of the blood.

"Safety valves" exist to prevent blood flow reductions below these critical thresholds. Recall that at rest, renal blood flow is about 1 liter/min and represents 20% of cardiac output. The primary function of the kidneys—to filter waste products from the blood—is clearly a function of importance that cannot be discontinued during exercise. Furthermore, if fluid flow is not maintained through the renal tubules, they can become clogged with protein precipitates (casts). Thus renal blood flow cannot be interrupted and redirected to exercising muscle to the same extent as the flow to the stomach and intestines can be redirected. Although sympathetic activity reduces renal blood flow during exercise, it is counterbalanced by the strong influence of autoregulatory mechanisms within the kidney. In addition, prostaglandins (PGE_2 and PGI_2) produced locally in the kidney stimulate vasodilation that counteracts any excessive vasoconstriction due to sympathetic activity or increased plasma renin/angiotensin concentrations. Because aspirin and several other over-the-counter pain relievers work by blocking an enzyme involved in the synthesis of prostaglandins, taking such medication while exercising in the heat can inhibit an important safety mechanism.

Fluid Balance: The Weak Link in Steady State

Unlike energy stores, which, if managed properly, are more than adequate to complete a marathon, water stores must be replenished. Under warm conditions, rates of sweating can reach 30 mL/min (1.8 liters/hr). It is not unusual for runners to experience a net loss of 2 liters of body water during a marathon, even if conditions are cool and the runners are drinking during the race. Although sweat is a filtrate of plasma, only about 10% of the total fluid loss is evident as a change in plasma volume. In fact, running a marathon under mild conditions may result in very little change in plasma volume, because water (and solutes) redistribute from interstitial and intracellular locations to the plasma compartment.

One way the body counteracts fluid losses during exercise is to reduce urine production. At rest, urine production averages about 60 mL/hr, but this can vary enormously, depending upon fluid intake and ingestion of alcohol or caffeine. During exercise under cool conditions, urine flow rate may not change significantly, but sodium excretion decreases. Under warmer conditions such as those during the later stages of the marathon, both glomerular filtration rate and urine flow rate decrease.

Increased sympathetic nerve activity and reduced glomerular filtration rate cause the juxtaglomerular cells to release more renin. This enzyme is involved in the proteolytic cleavage process that ultimately produces angiotensin II, which in turn stimulates aldosterone

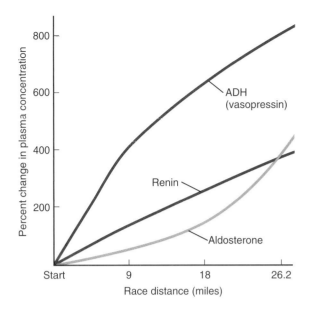

Figure 24.9 Changes in the plasma concentrations of some hormones that regulate fluid balance over the course of the marathon.

Compiled from data reported by J. Pastene et al., "Water balance during and after marathon running." *European Journal of Applied Physiology*, 73 (1996): 49–55; B.J. Freund et al., "Exaggerated ANF response to exercise in middle-aged vs. young runners." *Journal of Applied Physiology*, 69 (1990): 1607–1614; R.A. Irving et al., "The immediate and delayed effects of marathon running on renal function." *Journal of Urology*, 136 (1986):1176–1180.

secretion by the adrenal cortex. Aldosterone promotes sodium reabsorption in the renal tubules (and therefore helps conserve water). The time course for changes in the circulating concentrations of hormones that influence renal function are shown in **Figure 24.9**.

In the later stages of the race, Bill and Jane's plasma became hyperosmotic due to their heavy sweating. Sweat glands are like miniature kidneys in the sense that iso-osmotic fluid from blood plasma is pumped into ducts, and then the sodium and other electrolytes are reabsorbed as the fluid moves through the ducts toward the skin surface. The loss of hypo-osmotic sweat results in hyperosmotic plasma, which is sensed by osmoreceptors in the hypothalamus. This triggers the release from the posterior pituitary gland of antidiuretic hormone (ADH, also known as vasopressin), which promotes water reabsorption by the renal tubules.

Although water conservation occurs via the physiological reflex mechanisms just described, replacing lost water requires a behavioral response: drinking. But the thirst mechanism in humans does not provide a tightly coupled negative feedback loop that keeps pace with fluid loss. After dehydrating exercise, people usually drink only about two-thirds of the water they lost as sweat, a situation called "voluntary dehydration." As a result, people must consciously try to drink more fluid than their thirst indicates. Because Jane drank whether she was thirsty or not, she kept her fluid volume more stable than

Bill, who drank only to satisfy his thirst through most of the race. Voluntary dehydration is the major reason for flavoring and sweetening "sports drinks"—it makes them more palatable than plain water and thus encourages greater fluid intake. Although sports drink manufacturers and some researchers contend that adding sugar to the drink helps maintain blood glucose and therefore improves athletic performance, controlled experiments are equivocal on this matter.

Quick Test 24.3

1. Soon after the start of exercise, insulin secretion decreases despite an increase in plasma glucose levels. What is responsible for the decrease in insulin secretion?

2. What are muscle *metaboreceptors*? How are these involved in the control of cardiovascular function during exercise?

3. As core body temperature rises during exercise, the cardiovascular system is faced with the conflicting demands of providing increased blood flow to muscles and providing increased blood flow to what other organ or structure?

4. Define the term *voluntary dehydration* as it applies to exercise.

The Decline to the End: "The Wall" or the Finish Line?

Exercise diminishes energy stores, decreases blood volume, and places many other physiological stresses on the body that will ultimately cause performance to decline. The degree and speed of this decline are determined by how well the body is able to adapt to these stresses and by behavior, as we will see in this section.

The Prime Directive: Protect the Brain

The organ that directs Bill and Jane's motor control, guides their behavior, processes their objective and subjective sensations—and imposes the strictest, incontrovertible demands for blood flow—is the brain. The energy needs of the brain change very little under varying physiological conditions—"thinking as hard as you can" will not change the rates of total blood flow (about 750 mL/min) or oxygen consumption (about 50 mL/min). But depending on what type of information is being processed, the distribution of blood throughout the various structures of the brain can vary considerably. The energy is used to maintain ion gradients across neural membranes, which in turn allow neural depolarizations—the electrical activity of information processing and thought. Interrupting blood flow to the brain can cause unconsciousness within just a few seconds. Severe interruptions, such as those resulting

from a blood clot lodging in a cerebral blood vessel (stroke) or loss of blood pressure (hemorrhage or cardiac arrest), can permanently affect memory and intellectual processes. By analogy, interrupting the power to a computer results in the loss of all information in memory that was not saved to disk. We do not seem to have any backup system analogous to a disk drive for our memory; thus we cannot ever afford to have our brain power down. As such, all other organs (except the heart which delivers the "power" to the brain) can and will be sacrificed for the sake of the brain.

Fatigue: The Early Warning System

The fatigue that sets in during the later stages of a marathon is caused by a combination of three factors: (1) intrinsic factors within the exercising muscles, (2) systemic factors influenced by metabolic and environmental stresses, and (3) psychological factors that are still poorly understood.

Within the muscle, ATP, creatine phosphate, and glycogen are energy-related factors that could theoretically become depleted and cause muscle fatigue. In reality, ATP concentrations are reduced somewhat during exercise but attain a steady state. Creatine phosphate concentrations may fall progressively during high-intensity exercise, but they do not fall to zero during the moderate exercise intensity of a marathon. Most evidence indicates that the terminal fatigue experienced in an endurance event such as the marathon correlates with the depletion of glycogen. Although more and more of the energy used by muscle originates from beta oxidation of fatty acids as the race progresses, apparently some small energy contribution continues to be supplied by the carbohydrate stored as glycogen. Primarily small-diameter, slow oxidative fibers are recruited in marathon running, but the glycogen stores in these smaller fibers can be depleted, resulting in recruitment of more large-diameter, fast glycolytic fibers. This recruitment allows the muscle as a whole to continue contracting.

Fatigue is also associated with increases in lactic acid, although it is the hydrogen ion and not the lactate anion that is directly responsible for fatigue. In fact, lactate released by contracting muscle is recycled into glucose in the liver by gluconeogenesis. As shown previously (see Figure 24.4), plasma lactate concentrations rise moderately and remain relatively stable during a marathon. (In contrast, plasma lactate can increase over 1000% during exercise requiring maximal oxygen consumption.) The hydrogen ion, by contrast, contributes to the fatigue of maximal intensity exercise in several ways: (1) by inhibiting phosphofructokinase, an enzyme of glycolysis; (2) by displacing calcium ions from troponin, shutting down crossbridge cycling; (3) by stimulating local pain receptors; (4) by acting on the brain to cause pain and nausea; (5) and by interfering with *hormone-sensitive lipase,* an intracellular enzyme that mobilizes free fatty acids.

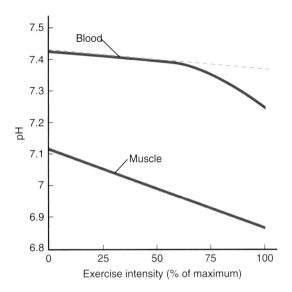

Figure 24.10 Influence of exercise intensity on arterial blood pH and intracellular pH of exercising leg muscle. At moderate exercise intensities (up to about 65% of maximum), the rate of H^+ clearance from the circulation rises in proportion to its rate of release into the circulation, resulting in a linear relationship between arterial blood pH and exercise intensity (dashed line). At higher exercise intensities, H^+ is produced and released into the blood at much greater rates than it is cleared, resulting in the deviation from linearity. For practical reasons, experiments designed to investigate this relationship have maintained each exercise intensity for only a few minutes.

Compiled from data reported by P.A. Mole et al., "Myoglobin desaturation with exercise intensity in human gastrocnemius muscle." *American Journal of Physiology,* 277 (1999): R173–R180; and J.A. Zoladz et al., "Changes in acid-base status of marathon runners during an incremental field test." *European Journal of Applied Physiology,* 67 (1993): 71–76.

> As the kidneys compensate for acidosis, does bicarbonate excretion increase or decrease?

Are these H^+-mediated mechanisms of fatigue important during marathon running? As shown in **Figure 24.10**, the pH of arterial blood falls only modestly (and linearly) as exercise intensity increases up to the level attained during a marathon (about 70% of maximal aerobic capacity); at higher exercise intensities, plasma pH decreases more exponentially. In contrast, noninvasive nuclear magnetic resonance techniques have shown that muscle pH decreases linearly over the entire range of exercise intensities.

Few studies have examined the extent to which plasma pH might change over a long period at a contant exercise intensity. As shown in **Figure 24.11**, arterial pH falls quickly to a level determined by exercise intensity in subjects running at a typical marathon pace, then slowly rises. Why should it rise? Recall that the plasma concentration of any substance reflects the balance between its

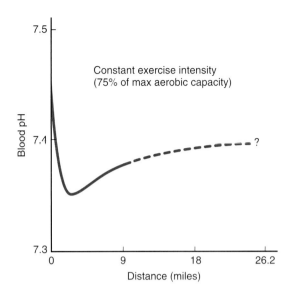

Figure 24.11 Changes in arterial blood pH over a portion of a marathon. The dashed line is an extrapolation; measurements have not been made during later stages of the race, when blood volume and glucose stores are diminished.

Source: Adapted from a study by A. Usaj and V. Starc, "Blood pH and lactate kinetics in the assessment of running endurance." *International Journal of Sports Medicine* 17 (1996): 34–40.

release into, and its removal from, the bloodstream. Many of the hydrogen ions released into the blood by exercising muscle are removed via respiration as they are combined with bicarbonate to yield carbon dioxide, which is subsequently exhaled. In addition, an increase in urinary hydrogen ion excretion during exercise is promoted by the rise in plasma aldosterone (see Figure 24.9), which also increases sodium reabsorption.

The evidence just presented suggests that hydrogen ion accumulation may play a contributing role in *local* muscle fatigue during a marathon but is probably not a major *systemic* factor (although we do not know much about how blood pH changes under the combined stresses of prolonged exercise, hyperthermia, and fluid depletion). Phosphate accumulation in muscle tissue, like accumulation of hydrogen ions, inhibits phosphofructokinase and displaces calcium ions from troponin. Calcium may accumulate in mitochondria, which interferes with oxidative phosphorylation. The significant increases in plasma potassium concentrations that occur during a marathon have been implicated as contributing factors in fatigue.

Even though blood glucose concentrations are usually fairly stable during exercise, even over the course of a marathon (see Figure 24.4), severe hyperthermia and dehydration can reduce renal and hepatic blood flow to the point that gluconeogenesis can no longer sustain the glucose needs of the brain. The resulting hypoglycemia is associated with profound psychological fatigue, disorientation, and possible loss of consciousness.

Tactical and Biological Influences on Fatigue

Based on Bill and Jane's physical characteristics as described in Chapter 1, Bill should have finished first. In this section we look at the reasons why he did not. His 2-minute "pit stop" was not a factor, because it was offset by Jane's eight 15-second stops to fully drink at the aid stations (but these 2 minutes of behavior on her part *were* a factor, as we will see).

When at the start Bill darted through the crowd while Jane remained trapped in the pack, he used up proportionally more of his muscle glycogen reserves than she did. Because he did not spend time at the aid stations, he covered the first 18 miles faster than Jane. Beyond this point, however, Bill paid for his early gains. He was in a much greater state of dehydration, which resulted in reduced venous return, higher heart rate, and reduced ability to dissipate heat. His glycogen stores were depleted from smaller-diameter, slow oxidative fibers (which provide more controlled movement), which necessitated recruitment of large-diameter, fast glycolytic fibers (which provide more explosive movement). The resulting loss of fine motor control led to stiff-legged, inefficient strides that resulted in greater energy expenditure per unit distance traveled—precisely the wrong development when energy stores were dwindling. Thus, in this dehydrated, glycogen-depleted, thermally stressed condition, Bill was unable to maintain his pace, and his final eight miles were run at a much slower pace. He lost whatever time he had gained by speeding past the aid stations. Thus, tactical errors impeded Bill's performance.

But Jane also had some physiological advantages that offset Bill's greater muscle mass and aerobic capacity. First, recall from Figure 24.5 that energy expenditure in a woman seems to differ from that in a man. A greater proportion of the energy Jane expended in the marathon was derived from fat, and thus muscle glycogen was used proportionally less. At the end of the marathon, Jane still had some glycogen remaining in her skeletal muscles. In addition, one research study found that women reported lower perception of leg muscle pain than men exercising at the same relative intensity. If confirmed, this type of pain tolerance might represent another reason Jane was better able to maintain her pace near the end of the marathon.

Conscious Decisions Versus Autonomic Control

Despite the existence of the "safety valve" mechanisms that protect the vital organs under extreme conditions, often the most effective homeostatic response is behavioral: A person ceases the activity or leaves the situation that is causing so much physiological stress. Afferent signals to the brain, including pain, acidosis, hypoglycemia, hypoxia, and hypotension, engender "psychological" fatigue. Therefore, fatigue can be considered as the efferent (and conscious) signal that prompts us to change our be-

havior (that is, stop running). A major element of marathon running requires that a person ignore this signal. However, the autonomic nervous system remains the ultimate arbiter: It can shut down excessively stressful activity no matter how hard runners will themselves to continue.

Quick Test 24.4

1. In an endurance event such as a marathon, terminal fatigue most likely results from depletion of what metabolic fuel?

2. In maximal intensity exercise, fatigue is associated with the accumulation of what chemical substance?

3. In marathon running, what type of skeletal muscle fiber is recruited initially?

The Aftermath

In view of what we have learned about the physiological stresses of exercise, it should come as no surprise that a bout of severe and/or prolonged exercise, such as a marathon race, has an impact on body function that can in some cases last for many hours after the exercise has stopped. In this section, we explore some of these post-exercise consequences.

Syncope: The Autonomic Nervous System's Final Veto Power

Even though Bill was hypovolemic in the last few miles of the race, he did not collapse until after he crossed the finish line. Why? Once Bill had come to a stop, the skeletal muscle pump activities of his legs ceased, and blood began to pool in his leg veins. Venous return to the heart decreased, and therefore ventricular filling pressure, stroke volume, cardiac output, and systemic blood pressure all fell precipitously. At this point, the autonomic nervous system used its final veto power: *syncope* (more commonly known as fainting).

In Bill's condition, syncope provided several homeostatic benefits: First, skeletal muscle activity ceased (except his diaphragm), which ended the competition for blood flow between the brain and muscle. Second, heat production was reduced, which also ended the competition for blood flow between the brain and the skin. Third, because 70% of Bill's blood volume was below the level of his heart when he was standing, venous return was impeded by gravity. Syncope brought Bill's heart, legs, and brain to the same level, allowing improved venous return unimpeded by gravity, better filling pressure, improved stroke volume, and improved cardiac output. However, syncope itself can pose hazards. Fortunately Bill was not hurt by the fall nor was he resting in a dangerous location.

Figure 24.12 Fluid shifts during and after exercise. In each diagram, the entire body (the large triangle) contains the extracellular fluid (ECF) and cells (circles) containing the intracellular fluid (ICF). Of the cells, three are encased in the skull (the small triangle). The darker the shade of blue, the greater the concentration of salts.

Would you expect the decrease in ECF osmolarity that occurs after the race to stimulate ADH secretion by the posterior pituitary or inhibit it?

Water Intoxication

Bill recovered from his fainting spell only to succumb later to a debilitating headache and vomiting. The cause was hypo-osmolarity and the resultant shifts of water between the intracellular fluid and the extracellular fluid (**Figure 24.12**).

Over the course of the marathon, Bill and Jane were gradually losing sodium, chloride, potassium, and other electrolytes ("salts") as well as water in their sweat. Their extracellular fluid volumes decreased steadily throughout the race. They both lost relatively more water than salts because some of the salts were reabsorbed along the sweat gland ducts. As a result, their plasma and interstitial fluid became hyperosmotic (Figure 24.12b), which drew water out of cells by osmosis. Their cell volumes shrank somewhat due to the fluid shifts (Figure 24.12c); this shrinkage may have influenced certain cellular functions, which may be another factor that contributed to fatigue.

After the race, Jane drank a modest amount of water and ate some food, which replenished electrolytes as well as water. In contrast, after the race Bill drank a volume of pure water that was almost equal to the volume he lost through sweating during the race, but because he did not also take in any electrolytes, his plasma and interstitial fluid became hypo-osmotic (Figure 24.12d). Water moved into his cells by osmosis to equilibrate intracellular and extracellular osmolarity. But due to the net loss of electrolytes during the race, more water entered Bill's cells than originally left them, causing the cells to swell (Figure 24.12e). Although this was not a problem for most of his cells, the critical exceptions were the central nervous system cells encased in his skull. The resultant intracranial pressure increase caused Bill's headache, vomiting, and convulsions—a condition termed *water intoxication.* We can now see that voluntary dehydration may be a protective mechanism that limits rehydration and guards against serious osmotic imbalances that can lead to disturbances in the central nervous system. (Voluntary dehydration is diminished if food is ingested with water.) Bill's water intoxication was actually a very serious condition and could have been life threatening. His friends probably should have taken Bill to the hospital for intravenous infusion of electrolytes, instead of letting him sleep.

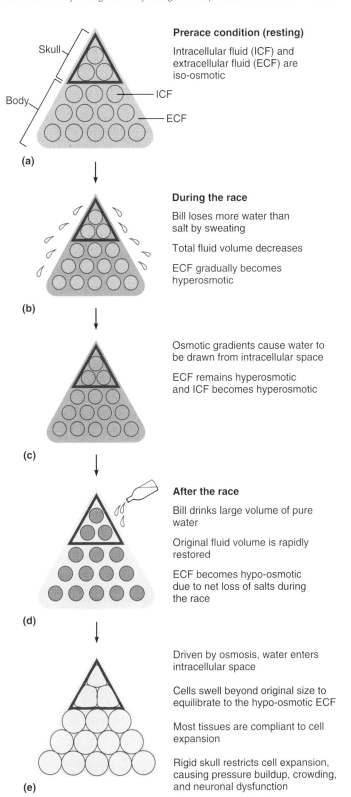

Prerace condition (resting)

Intracellular fluid (ICF) and extracellular fluid (ECF) are iso-osmotic

Skull

Body

ICF

ECF

(a)

During the race

Bill loses more water than salt by sweating

Total fluid volume decreases

ECF gradually becomes hyperosmotic

(b)

Osmotic gradients cause water to be drawn from intracellular space

ECF remains hyperosmotic and ICF becomes hyperosmotic

(c)

After the race

Bill drinks large volume of pure water

Original fluid volume is rapidly restored

ECF becomes hypo-osmotic due to net loss of salts during the race

(d)

Driven by osmosis, water enters intracellular space

Cells swell beyond original size to equilibrate to the hypo-osmotic ECF

Most tissues are compliant to cell expansion

Rigid skull restricts cell expansion, causing pressure buildup, crowding, and neuronal dysfunction

(e)

Postexercise Fever

The elevated body temperature that occurs *during* exercise is characterized as hyperthermia—that is, a body temperature that is above the set-point temperature (an error signal). However, hours *after* long-duration exercise such as a marathon, some people feel chilled and

A decrease in ECF osmolarity tends to inhibit ADH secretion.

wrap up in a blanket (as Bill did), even though their core temperatures are at or above "normal." The initiation of heat conservation mechanisms (vasoconstriction also occurs) at a time of elevated core temperature indicates that the thermoregulatory set point has been shifted upward (the definition of a fever).

The change in set point is due to the action of cytokines on the thermoregulatory center in the hypothalamus. Production of specific cytokines, such as interleukin-1 and interleukin-6, is increased in the period 4–24 hours following exercise such as a marathon. This increased production occurs because fragments of extracellular matrix and damaged cells bind to "scavenger receptors" on macrophages, which initiates production of cytokines. These cytokines, in turn, stimulate the liver to synthesize plasma proteins that serve antioxidative functions and scavenge iron and heme. (After a marathon, the "blood" some runners see in their urine is actually heme from red cells damaged in the foot vasculature by the 42,000 impacts that occurred during the race.)

Another possible trigger for cytokine production after exercise is intestinal-derived endotoxin. We all have bacteria colonized in our lower gastrointestinal tracts. Fragments of the cell walls of some types of bacteria are a complex of carbohydrate and lipid known as endotoxin. If the epithelial barrier of the gastrointestinal tract becomes permeable to endotoxin (not necessarily to whole, viable bacteria), this endotoxin is taken up by the portal circulation and transported to the liver, which contains a high density of specialized macrophages called Kupffer cells. The endotoxin binds to specific receptors on the Kupffer cells, triggering cytokine production.

Exercising under conditions of severe heat stress can result in excessive vasoconstriction in the gastrointestinal tract, leading to ischemia. Ischemic damage to the epithelium allows large amounts of endotoxin to cross it, fully saturate the clearance mechanisms of the liver, and gain entry into the general circulation, activating monocytes and macrophages throughout the body. Overproduction of another cytokine, tumor necrosis factor (TNF), can lead to vascular permeability, peripheral vasodilation, and decreased cardiac contractility. At a time when maintenance of mean arterial pressure is already compromised by the fluid adjustments required to dissipate heat, TNF-induced decrements in cardiovascular function can lead to life-threatening situations. Severe cases of exertional heat stroke usually involve these mechanisms.

Delayed-Onset Muscle Soreness

Bill and Jane had somewhat sore and tired muscles after the race. In Bill's case, these subjective feelings were mild compared to the deep, diffuse aching, heaviness, and inflexibility he experienced upon waking the next morning. Microscopic examination of muscle samples taken after marathons reveals sarcomere damage, especially at the Z

disks. The plasma membranes of muscle fibers become more permeable to large molecules, which is reflected by large increases in plasma concentrations of muscle cell enzymes such as creatine kinase. The changes in membrane permeability are thought to permit intracellular calcium concentration to increase, which disrupts muscle cell function.

Macrophages invade the affected muscle and clear out damaged fibers by phagocytosis. In addition, macrophages appear to deliver various cytokines that promote repair. In fact, evidence suggests that even though blocking the mild inflammatory response that normally follows a marathon by taking aspirin or other drugs alleviates acute symptoms of pain and swelling, taking these drugs can actually delay the long-term healing of contraction-induced skeletal muscle damage.

Jane's symptoms were less severe. Women exhibit much lower plasma creatine kinase concentrations following a marathon than do men. Various experimental approaches have shown that these reduced plasma creatine kinase concentrations are specifically due to circulating estrogens. The exact mechanism involved, and the functional significance in terms of susceptibility to muscle damage or rate of healing, are not presently known.

Effect of Marathon Running on Susceptibility to Sickness

Almost any exercise (even just running up a flight of stairs) leads to a transient increase in circulating white blood cells, primarily because of the mobilization of neutrophils. About half of the neutrophils in the vascular compartment do not circulate freely, but are instead loosely attached (electrostatically) to the inner walls of large-diameter veins, especially veins in the pulmonary circuit. The fluid shear stresses that accompany the increased cardiac output of exercise strip these cells free; epinephrine also triggers this "release." As a result, neutrophil counts can more than double following moderate exercise. The other populations of leukocytes can become activated via stress hormones or fragments of damaged extracellular matrix and cells, as previously mentioned. Once activated, these cells express adhesion molecules that bind to the vascular endothelium and help the leukocytes move out of the bloodstream into the interstitial space. Some subpopulations of lymphocytes may exhibit transient 50% (or greater) decreases in circulating numbers via this mechanism; other leukocyte subpopulations may increase as a result of increased lymphatic flow to the thoracic duct due to skeletal muscle pump activity.

These transient changes in circulating leukocytes and anecdotal reports of chronic illness in elite athletes have been misinterpreted as a change in overall readiness to resist infection. However, little scientific evidence exists to support this view. The changes in circulating leukocyte counts are small in magnitude and transient in nature. As we saw in Chapter 23, the main battlefield during infection is usually in the interstitial space, especially near the barriers to the external environment (skin, respiratory tract), not in the bloodstream. Only a tiny fraction of all leukocytes are actually in the bloodstream at any moment; a twofold change in circulating cell numbers represents a shift in location of only 1% of the cells. Most controlled studies that have actually measured infection rates find no difference in incidence of infection between exercisers and sedentary people.

The resting (chronic) numbers and functioning of leukocytes are generally no different in highly trained versus untrained individuals. However, the spectrum of cytokines produced by the leukocytes of active people may be somewhat different than those of inactive people. Given that some of these cytokines are involved in the pathological mechanisms leading to atherosclerosis

and other "inflammatory" diseases, the small, training-induced changes in the concentrations of these cytokines over long periods of time may contribute mechanistically to the well-established influence of exercise on cardiovascular health.

Quick Test 24.5

1. Explain how syncope can be beneficial in restoring cardiac output following a drop in systemic blood pressure.

2. Explain why it is dangerous for a person to drink a large volume of pure water to replace fluid lost via sweating.

3. The elevation of the thermoregulatory set point that occurs in postexercise fever is triggered by what chemical substances acting on the brain?

CHAPTER SUMMARY

Principles of Physiological Integration, p. 757

Exercise places many demands on the body, which can be met only through the activities of different organ systems working together. One such demand is an increased need for energy to power skeletal muscle contractions, which for very brief periods can be supplied by molecules stored within the muscle cells themselves (ATP, creatine phosphate, glucose, and glycogen). More prolonged activity requires increased delivery of oxygen and fuels (glucose and fatty acids, for example) to the muscle cells from other sources. The body's ability to achieve and maintain increased levels of physiological function depends on the availability of energy and the often competing needs of different physiological systems. Within specific systems, adjustments are regulated by local metabolites (intrinsic control) and neural and hormonal signals (extrinsic control). Under most conditions, blood flow to muscles and other organs is controlled by intrinsic mechanisms. When conditions become more stressful, organs begin to compete with each other for blood delivery, and neural and hormonal signals assume a greater role.

The Start: Transition from Rest to Exercise, p. 758

Often, exercise is preceded by an anticipatory increase in sympathetic nervous activity (and epinephrine secretion) and a decrease in parasympathetic activity. Cardiac ouput increases (due to increased stroke volume, heart rate, and venomotor tone), as does plasma glucose concentration (due to increased glucagon secretion and decreased insulin secretion by the pancreas). At the start of exercise, central command (a feedforward mechanism arising in the cerebral cortex) helps orchestrate the cardiovascular and respiratory adjustments that occur. The proportion of fuels utilized depends on exercise intensity and duration. Early on (before oxygen delivery can be increased), anaerobic glycolysis is the predominant ATP generating mechanism in exercising muscle cells; afterward aerobic metabolism (fatty acid oxidation and the Krebs cycle) becomes predominant.

Pulmonary ventilation increases due to central command and also in response to signals arising in mechanoreceptors and chemoreceptors in exercising muscles. In mild to moderate exercise, arterial oxygen

and carbon dioxide concentrations change very little despite a substantial increase in pulmonary blood flow. A sympathetically mediated increase in blood flow to the skin helps dissipate the excess heat generated by exercising muscles.

The Long Haul: Almost Steady State, p. 761

After the initial shift from anaerobic to aerobic metabolism that occurs in prolonged exercise, fuel utilization settles into a relative steady state. However, glycogen (and body fluids) are continually being depleted. Mobilization of energy stores is regulated by changes in plasma levels of glucagon, epinephrine, and cortisol (which rise), and insulin (which falls). Plasma glucose rises initially due to epinephrine-stimulated glycogenolysis and gluconeogenesis but eventually begins to decline. Lactate rises due to the initial burst of anaerobic metabolism but levels off due to the rise in aerobic metabolism coupled with increased gluconeogenesis. Fatty acid levels rise steadily due to an increased glucagon-to-insulin ratio.

Central command signals appear to be important in maintaining an elevated mean arterial pressure (MAP) during exercise, which helps maintain increased blood flow to muscles. At higher exercise intensities, another regulatory mechanism comes into play that is triggered by signals arising in metaboreceptors that detect changing conditions in exercising muscles. Baroreceptor-mediated negative feedback becomes important in maintaining MAP only after blood volume has diminished (due to sweating) to the point where it threatens the cardiovascular system's ability to sustain adequate perfusion of vital organs. Increased pulmonary ventilation is aided by forced expiration and a decrease in airway resistance (due to epinephrine).

As core body temperature rises, the cardiovascular system becomes faced with the conflicting demands of providing increased blood flow to exercising muscles and to the skin (for heat dissipation). Vasoconstriction in renal, hepatic, and splanchnic vessels helps maintain MAP, especially as blood volume diminishes. Fluid losses

may be counteracted through a decrease in urine production, a consequence of increased sympathetic activity, reduced GFR, and activation of the renin-angiotensin-aldosterone system. Water conservation is also promoted by ADH secretion, which is stimulated as sweating causes the plasma to become hyperosmotic.

The Decline to the End: "The Wall" or the Finish Line?, p. 768

The stresses of exercise ultimately cause a decline in performance; the degree and speed of this decline depend on how well the body is able to adapt to these stresses. Prolonged exercise ultimately leads to fatigue, which arises due to intrinsic factors within the exercising muscles, systemic factors influenced by metabolic and environmental stresses, and psychological factors. Glycogen depletion is most likely the pivotal event that triggers terminal fatigue in endurance exercise; in high-intensity exercise accumulation of hydrogen ions appears to be important.

The Aftermath, p. 770

At the end of exercise, cessation of skeletal muscle pump activity may result in a fall in blood pressure. In extreme situations, syncope may occur, which is actually beneficial in that it ends competition for blood flow between the muscles and the brain, reduces heat production by muscles, and increases venous return and cardiac output. Replacement of lost body fluids with large volumes of pure water after exercise can lead to cell swelling and water intoxication. Following prolonged exercise, a person may experience fever due to increased cytokine levels, which act on hypothalamic thermoregulatory centers to elevate the set point. A person may also experience delayed-onset muscle soreness, a symptom of inflammation triggered by contraction-induced muscle damage. This inflammatory response normally sets the stage for eventual muscle repair.

▓ EXERCISES

Multiple-Choice Questions

1. Although water in sweat is derived from plasma, it is possible that a person could lose a significant volume of water via sweating while experiencing only a small decrease in blood volume. This is because
   a) much of the blood volume is occupied by cells, not water.
   b) water moves into the plasma from other compartments, such as interstitial fluid.
   c) the kidneys divert water from the renal tubules to the peritubular capillaries.
   d) water moves from red blood cells to the plasma.
   e) the gastrointestinal system absorbs more water in response to the loss of blood volume.

2. The skeletal muscle fibers that are first recruited in a marathon are primarily
   a) slow glycolytic fibers.
   b) fast glycolytic fibers.
   c) slow oxidative fibers.
   d) fast oxidative fibers.

3. In the initial stages of a marathon, plasma glucose levels rise and insulin secretion falls. This is because
   a) pancreatic beta cells are sensitive to glucose levels in the fluid surrounding them and normally reduce their secretion of insulin in response to a rise in glucose concentration.
   b) the decrease in insulin secretion triggers the rise in glucose levels.
   c) increased sympathetic input to pancreatic beta cells suppresses insulin secretion despite the rise in plasma glucose.
   d) exercise reduces the responsiveness of many tissues to insulin.
   e) all of the above

4. Which of the following is evidence that negative feedback chemoreceptor control of ventilation is *not* responsible for the increase in pulmonary ventilation that occurs at the onset of moderate exercise?
   a) If negative feedback control were responsible, arterial P_{CO_2} would rise at the beginning of exercise,

triggering the increase in ventilation; these changes in partial pressure are not observed.
   b) If negative feedback control were responsible, arterial P_{CO_2} would fall below normal during exercise; these changes in partial pressure are not observed.
   c) If negative feedback control were responsible, ventilation would not increase at all during exercise; negative feedback normally works to keep alveolar ventilation constant.
   d) If negative feedback control were responsible, ventilation would decrease at the start of exercise. Negative feedback almost always triggers responses that are inappropriate, which is why it is called *negative* feedback.
   e) none of the above

5. When core temperature rises, the resulting increase in blood flow to the skin conflicts with the cardiovascular system's efforts to maintain increased blood flow to exercising

muscles. Why are these two processes said to be in conflict?

a) Blood cools at it flows through the skin but warms as it flows through the muscles.

b) Blood gains oxygen as it flows through the skin but loses oxygen as it flows through the muscles.

c) The increase in blood flow to the skin is due to sympathetically mediated vasodilation, whereas the increase in muscle blood flow is due to parasympathetically mediated vasodilation.

d) The increase in blood flow to the skin is due to sympathetically mediated vasoconstriction, whereas the increase in muscle blood flow is due to local controls.

e) The vasodilation that allows skin blood flow to increase also lowers total peripheral resistance and thus tends to lower mean arterial pressure. Any reduction in mean arterial pressure would tend to decrease muscle blood flow.

6. Assume that the decrease in arterial pH that occurs during exercise is due to an increase in lactic acid production. If so, it is an example of
a) respiratory acidosis.
b) metabolic acidosis.
c) respiratory alkalosis.
d) metabolic alkalosis.

7. Which of the following is an example of negative feedback control?

a) stimulation of increased heart rate by central command signals at the onset of exercise

b) stimulation of increased heart rate by signals arising in muscle metaboreceptors at the onset of exercise

c) stimulation of pulmonary ventilation by signals arising in muscle mechanoreceptors and chemoreceptors at the onset of exercise

d) vasodilation of skin blood vessels in response to a rise in core body temperature

e) all of the above

8. If it were possible for a person to lose 1 liter of *pure water* through sweating, how would it affect extracellular fluid (ECF) osmolarity *prior to any possible movement of water between body fluid compartments*? Assume that all intra- and extracellular solutes are impermeant and that ECF and intracellular

fluid (ICF) osmolarities are initially at the normal value of 300 mOsm.

a) ECF osmolarity will not change.
b) ECF osmolarity will rise above 300 mOsm.
c) ECF osmolarity will fall below 300 mOsm.
d) More information is needed in order to determine how ECF osmolarity will be affected.

9. In the situation described in question 8, assume that any possible water movement between body fluid compartments has occurred and ECF and ICF have reached osmotic equilibrium. At this point, ECF osmolarity will be
a) normal at 300 mOsm.
b) greater than 300 mOsm.
c) less than 300 mOsm.
d) More information is needed in order to determine how ECF osmolarity will be affected.

10. Starting with the situation described in question 9, suppose the person drinks 1 liter of pure water. Assuming that this water has been completely absorbed into the ECF *but no movement of water between body fluid compartments has yet occurred,* how will ECF osmolarity be affected?

a) It will decrease to normal (300 mOsm).
b) It will decrease but remain above normal.
c) It will decrease to below normal.
d) More information is needed in order to determine how ECF osmolarity will be affected.

11. For the situation described in question 10, how will ECF volume compare to normal?
a) ECF volume will be normal.
b) ECF volume will be below normal.
c) ECF volume will be above normal.
d) More information is needed in order to determine how ECF volume will be affected.

12. For the situation described in question 10, how will ECF osmolarity compare to normal *after* any possible water movement between body fluid compartments has occurred and osmotic equilibrium has been attained?
a) ECF osmolarity will be normal at 300 mOsm.
b) ECF osmolarity will be below normal.

c) ECF osmolarity will be above normal.
d) More information is needed in order to determine how ECF osmolarity will be affected.

13. If the person described in question 9 lost 1 liter of sweat containing solutes at a concentration of 50 mOsm and drank 1 liter of pure water, how would ECF osmolarity compare to normal after the water has been absorbed and osmotic equilibrium has been attained?
a) ECF osmolarity will be normal at 300 mOsm.
b) ECF osmolarity will be below normal.
c) ECF osmolarity will be above normal.
d) More information is needed in order to determine how ECF osmolarity will be affected.

Objective Questions

1. Muscle mechanoreceptors send (feedforward/feedback) signals for respiratory control.

2. At the onset of exercise, pulmonary ventilation increases mostly as a result of an increase in (breathing frequency/tidal volume).

3. A gradual initial increase in exercise intensity helps conserve the body's stores of (glycogen/fatty acids).

4. During exercise core temperature normally rises due to elevation of the thermoregulatory set point. (true/false)

5. As the body's glycogen stores become depleted in prolonged exercise, blood glucose is maintained at or near its normal resting level through the conversion of (fatty acids/lactate) to glucose.

6. At the onset of low-intensity exercise, ATP production in exercising muscle cells occurs primarily through (aerobic/anaerobic) metabolism.

7. After exercising, people usually drink less fluid than they lost through sweating. This phenomenon is referred to as _____. (two words)

8. Sweat is (hypo-osmotic/hyperosmotic) relative to plasma.

9. Exercise has been shown to induce a significant increase in the number of leukocytes within the body. (true/false)

10. Under normal resting conditions, the distribution of blood flow to organs is controlled primarily by (intrinsic/extrinsic) regulatory mechanisms.

11. During prolonged exercise, plasma concentrations of renin would be expected to (increase/decrease).

12. As lactate begins to accumulate in the blood during prolonged exercise, some of it is converted to glucose. (true/false)

13. The central command signals that stimulate increased cardiac output at the onset of exercise are triggered in response to afferent signals arising in muscle metaboreceptors. (true/false)

14. As body temperature increases during exercise, skin blood flow increases due to (sympathetically/parasympathetically) mediated vasodilation.

Essay Questions

1. Can heart transplant patients raise their cardiac output during exercise? If so, how?

2. List the different physiological responses that are influenced by body movements or limb movements associated with locomotion.

3. Why is it more efficient to increase tidal volume instead of ventilatory frequency during moderate exercise?

4. As exercise intensity increases further, tidal volume stabilizes at about 65% of vital capacity, and additional increases in ventilation are due to increased frequency. Why?

5. How is an increase in body temperature during exercise different from a fever?

6. If sympathetic input generally promotes vasoconstriction, how is it that blood flow to skeletal muscles increases during exercise?

7. Make a list of all the physiological processes that are enhanced by epinephrine during exercise.

Critical Thinking

1. Many people who do not have use of their legs participate in marathons in wheelchairs. What physiological responses might be different in these individuals? How would someone who lacks motor neuron input to their legs differ from someone with absolutely no neuronal input to their legs?

2. Recall the discussion of the synchronized movement of our runners' leg muscles with their ventilatory muscles. What animals might have tightly coupled ventilatory-motor movements? Can you think of any disadvantages of having these processes too tightly coupled?

3. Aside from blocking pain, how does aspirin influence physiological function?

4. What might happen to free fatty acid mobilization during a marathon if a person ingested sports drinks or fluids with too much sugar or ate orange slices along the way?

5. If you gave a rodent a chemical that inactivated all sympathetic efferent signaling (both neural and hormonal), could it still run? What mechanisms still exist during exercise under these conditions that promote physiological function?

6. If you gave the same drug mentioned in question 5 to human beings, they probably could not run but might be able to swim. Why?

Find the answers to these exercises, and additional study tools, at the Physiology Place (www.physiologyplace.com).

Answers to Odd-Numbered Multiple-Choice and Objective Questions

CHAPTER 1

Multiple-Choice Questions

1. b
3. b
5. e

Objective Questions

1. extracellular fluid
3. true
5. false
7. false
9. connective

CHAPTER 2

Multiple-Choice Questions

1. b
3. a
5. d
7. d
9. c
11. d
13. c

Objective Questions

1. polar
3. cholesterol
5. integral
7. microvilli
9. gene
11. true
13. interphase

CHAPTER 3

Multiple-Choice Questions

1. b
3. c
5. b
7. b
9. a

11. d
13. a
15. e

Objective Questions

1. oxidation
3. negative
5. false
7. gluconeogenesis
9. chemiosmotic coupling
11. lowering
13. false
15. anaerobic

CHAPTER 4

Multiple-Choice Questions

1. a
3. a
5. a
7. a
9. c
11. c
13. e
15. d

Objective Questions

1. hydrophobic
3. true
5. false
7. true
9. true
11. true
13. true
15. false

CHAPTER 5

Multiple-Choice Questions

1. c
3. c
5. e

7. c
9. c

Objective Questions

1. secretory cells
3. hormone / blood
5. adenylate cyclase
7. Lipophilic
9. calmodulin
11. neurotransmitters

CHAPTER 6

Multiple-Choice Questions

1. d
3. b
5. c
7. e

Objective Questions

1. Endocrine
3. medulla
5. exocytosis

CHAPTER 7

Multiple-Choice Questions

1. a
3. b
5. d
7. a
9. b

Objective Questions

1. glucagon
3. true
5. absorptive
7. negative
9. stimulate
11. sex hormones
13. increased
15. inhibitory

CHAPTER 8

Multiple-Choice Questions

1. e
3. e
5. d
7. d
9. a
11. d
13. d
15. e

Objective Questions

1. afferent and efferent
3. glial
5. Schwann cells; oligodendrocytes
7. negative
9. ganglia
11. false
13. saltatory
15. out of

CHAPTER 9

Multiple-Choice Questions

1. a
3. b
5. d
7. b
9. a

Objective Questions

1. gap junctions
3. chemical
5. false
7. larger
9. monoamine oxidase; catechol-*O*-methyltransferase

CHAPTER 10

Multiple-Choice Questions

1. b
3. c
5. c
7. d
9. b
11. c

Objective Questions

1. choroid plexus
3. white
5. cerebellum
7. right
9. thalamus
11. limbic
13. false

CHAPTER 11

Multiple-Choice Questions

1. d
3. e
5. a
7. a
9. b
11. b
13. b

Objective Questions

1. warm; cold
3. quickly
5. left
7. ganglion cells
9. false
11. opsin
13. false
15. false

CHAPTER 12

Multiple-Choice Questions

1. c
3. c
5. a
7. a
9. e
11. c

Objective Questions

1. parasympathetic
3. false
5. cranial nerves III, VII, IX, X
7. all autonomic nervous system pre-ganglionic neurons, parasympathetic postganglionic neurons, and motor neurons
9. phospholipase C
11. excitation
13. nicotinic
15. acetylcholinesterase

CHAPTER 13

Multiple-Choice Questions

1. c
3. d
5. c
7. d
9. b

Objective Questions

1. troponin
3. true
5. sarcolemma
7. isometric

9. longer
11. false

CHAPTER 14

Multiple-Choice Questions

1. b
3. c
5. a
7. b
9. c
11. a

Objective Questions

1. SA
3. false
5. semilunar
7. systolic
9. heart rate
11. increases
13. diastole
15. pulse

CHAPTER 15

Multiple-Choice Questions

1. c
3. b
5. c
7. e
9. e
11. c
13. b
15. b
17. b

Objective Questions

1. false
3. decreases
5. increases
7. pressure gradient
9. false
11. an increase
13. true

CHAPTER 16

Multiple-Choice Questions

1. d
3. d
5. d
7. b
9. d

Objective Questions

1. serum
3. thrombin

5. bilirubin
7. thromboxane A2

CHAPTER 17

Multiple-Choice Questions

1. e
3. a
5. d
7. b
9. e

Objective Questions

1. false
3. more
5. decreases
7. conducting zone
9. goblet cells

CHAPTER 18

Multiple-Choice Questions

1. c
3. d
5. a
7. e
9. c
11. d
13. c

Objective Questions

1. the same as
3. false
5. carbonic anhydrase
7. increases
9. carbaminohemoglobin
11. peripheral
13. higher
15. vasodilation

CHAPTER 19

Multiple-Choice Questions

1. b
3. b
5. a
7. c
9. b
11. c

Objective Questions

1. urethra
3. afferent
5. higher

7. internal
9. glomerular filtration rate
11. PAH
13. decrease

CHAPTER 20

Multiple-Choice Questions

1. c
3. b
5. d
7. e
9. d
11. e
13. c
15. c

Objective Questions

1. increases
3. false
5. atrial natriuretic peptide
7. aldosterone
9. false
11. decreases
13. increase
15. false

CHAPTER 21

Multiple-Choice Questions

1. a
3. e
5. d
7. e
9. a

Objective Questions

1. true
3. cephalic-phase
5. serosa
7. lipases
9. stimulates
11. migrating motility complexes
13. amplitude
15. brush border enzymes

CHAPTER 22

Multiple-Choice Questions

1. c
3. b
5. d
7. b
9. a

Objective Questions

1. false
3. persist
5. true
7. FSH
9. decrease
11. uterine tube
13. granulosa
14. LH
17. LH
19. estrogens
21. blastocyst
23. placenta
25. prolactin

CHAPTER 23

Multiple-Choice Questions

1. e
3. a
5. e
7. b
9. b
11. b
13. b

Objective Questions

1. a) M   b) T_H c) T_C d) B   e) B
3. block
5. true
7. phagocytosis; endocytosis

CHAPTER 24

Multiple-Choice Questions

1. b
3. c
5. e
7. d
9. b
11. c
13. b

Objective Questions

1. feedforward
3. glycogen
5. lactate
7. voluntary dehydration
9. false
11. increase
13. false

Credits

Illustrations

**All illustrations by Precision Graphics unless noted below.
The following illustrations by Imagineering STA Media Services:**

Chapter 1: 1.3, 1.8

Chapter 3: Clinical Connections, Discovery, Toolbox

Chapter 6: Clinical Connections, Figure c

Chapter 7: 7.1

Chapter 8: 8.11

Chapter 10: 10.1

Chapter 11: 11.19

Chapter 13: 13.10, 13.11

Chapter 14: 14.13, 14.23, 14.24, 14.25, 14.26, Discovery, Critical Thinking
Exercise 4

Chapter 15: 15.7, 15.10, 15.13, 15.15, 15.18, 15.26

Chapter 16: 16.4, 16.5

Chapter 17: 17.6, Critical Thinking Exercise 1

Chapter 21: Discovery

Chapter 23: Clinical Connections, Discovery

Photos

Chapter 1: Opener: Hughes Martin/Corbis. Clinical Connections: RNT
Productions/Corbis.

Chapter 2: Opener: Dennis Kunkel. Exercise Link (throughout book):
PhotoDisc/Getty Images. 2.16: Don W. Fawcett/Visuals Unlimited. 2.17b:
Visuals Unlimited. 2.20b: Omnikron/Photo Researchers. 2.21: Biophoto
Associates/Photo Researchers. 2.24b: Don W. Fawcett/Visuals Unlimited.
2.25c: David M. Phillips/Visuals Unlimited. Clinical Connections:
Moredun Scientific/Photo Researchers.

Chapter 3: Opener: Dennis Kunkel/Phototake.

Chapter 4: Opener: Kathleen Olson.

Chapter 5: Opener: Alfred Pasieka/Photo Researchers. Discovery: Bart's
Medical Library/Phototake.

Chapter 6: Opener: Educational Images/Custom Medical Stock Photo. 6.9b:
John D. Cunningham/Visuals Unlimited. 6.10c: Don W. Fawcett/Visuals
Unlimited. Clinical Connections, a: John Croye/Custom Medical Stock
Photo. Clinical Connections, b: B. Bates/Custom Medical Stock Photo.
Discovery: Powerstock/SuperStock.

Chapter 7: Opener: CNRI/Phototake. Clinical Connections: Rudi Von Briel/
Index Stock.

Chapter 8: Opener: Science Photo Library/Photo Researchers. Clinical Connections: Cousteau Society/The ImageBank/Getty Images. Discovery:
John Olschowka/Phototake.

Chapter 9: Opener: Dennis Kunkel. Clinical Connections: Corbis.

Chapter 10: Opener: Science Photo Library/Photo Researchers. Clinical
Connections: Zepher/Photo Researchers. Discovery: from Damasio,

H., Grabowski, T., Frank, R., Galaburba, A. M., Damasio, A. R. "The
Return of Phineas Gage: Clues about the Brain from a Famous Patient."
Science, 264: 1102–1105, 1994. Department of Neurology and Image
Analysis Facility, University of Iowa.

Chapter 11: Opener: Science Photo Library/Photo Researchers. 11.18b:
Custom Medical Stock Photo. 11.21: Richard Megna/Fundamental Pho-
tographs. 11.39g: P. Motta, Dept. of Anatomy, University "La Sapienza,"
Rome/Science Photo Library/Photo Researchers. Clinical Connections:
Bojan Brecelj/Corbis.

Chapter 12: Opener: Ed Reschke/Peter Arnold. Clinical Connections:
M. Kalab/Custom Medical Stock Photo. Discovery: Corbis.

Chapter 13: Opener: Educational Images/Custom Medical Stock Photo. 13.3:
top: H. Ris. 13.3: bottom: James E. Dennis/Phototake. 13.5e: John
Heuser. 13.6b: James E. Dennis/Phototake. 13.25: Biophoto Associates/
Photo Researchers. 13.27a: Eric Graves/Photo Researchers. 13.27b: Pearson
Education/PH College. 13.27c: Marian Rice. Clinical Connections: Custom
Medical Stock Photo. Discovery: AFP/Corbis.

Chapter 14: Opener: Cindy Stanfield. Discovery: Dennis Kunkel/Phototake.

Chapter 15: Opener: John Henley/Corbis. Clinical Connections 2: BSIP/
Custom Medical Stock Photo.

Chapter 16: Opener: Dennis Kunkel. 16.1: Oliver Meckes/Eye of Science/
Photo Researchers. 16.7: O. Meckes and N. Ottawa/Photo Researchers.
Clinical Connections: Stan Flegler/Visuals Unlimited. Discovery: Science
Photo Library/Photo Researchers.

Chapter 17: Opener: SIU/Visuals Unlimited. 17.4a: Biophoto Associates/
Science Source/Photo Researchers. 17.4b: Carolina Biological Supply/
Phototake. 17.5: CNRI/Science Photo Library/Photo Researchers. 17.6a:
Martin Dohrn, Royal College of Surgeons/Photo Researchers. 17.6b:
Richard Kessel/Visuals Unlimited. Clinical Connections: Corbis.

Chapter 18: Opener: Jeffrey Rotman/Corbis. Clinical Connections 1: David
Mechlin/Phototake. Clinical Connections 2: B. Bates/Custom Medical
Stock Photo. Discovery: Karl Kummels/SuperStock.

Chapter 19: Opener: CNRI/Photo Researchers.

Chapter 20: Opener: CNRI/Phototake. Clinical Connections: Alan Boyde/
Visuals Unlimited. Discovery: Volvox/Index Stock.

Chapter 21: Opener: Ed Reschke/Peter Arnold. 21.30: Science Photo Library/
Photo Researchers. Clinical Connections 1: S. Benjamin/Custom Medical
Stock Photo.

Chapter 22: Opener: David Phillips/Visuals Unlimited. 22.5d: CNRI/Phototake.
22.13c: Carolina Biological Supply/Phototake. 22.19: D. Phillips/Photo
Researchers. Clinical Connections: Steve Rosen/Custom Medical Stock
Photo. Discovery: M. Long/Visuals Unlimited.

Chapter 23: Opener: Dennis Kunkel/Phototake. 23.4b: O. Meckes and
N. Ottawa/Eye of Science/Photo Researchers. Clinical Connections:
Donald Fawcett/Visuals Unlimited.

Chapter 24: Opener: PhotoDisc/Getty Images.

Glossary

absolute refractory period period during and immediately following an action potential during which a second action potential cannot be generated in response to a second stimulus, regardless of the strength of that stimulus

absorption movement of substance from the external environment to the internal environment by transport across an epithelium; in cardiovascular system, bulk flow of fluid from interstitial space into capillary

absorptive state period following a meal during which nutrients are absorbed

accommodation reflex adjustments of the eye to view near objects; includes rounding of the lens, constriction of the pupil, and posterior movement of the lens

acetylcholine (ACh) (ass-ih-teel-koh'-leen) a neurotransmitter widely employed in both the central and peripheral nervous systems; the most abundant neurotransmitter in the peripheral nervous system, found in efferent neurons of both the somatic and autonomic nervous systems

acetylcholinesterase (AChE) (assih-teel-koh-lin-es'-ter-ase) enzyme that degrades acetylcholine

acidosis condition in which arterial blood pH is 7.35 or lower

acini small sacs of secretory cells leading to a duct; singular, acinus

acrosome large vesicle within the head of the sperm; contains enzymes and other proteins that allow the sperm to fuse with the egg during fertilization

acrosome reaction process whereby acrosome releases its enzymes

actin most common microfilament; found in thin filaments in muscle fibers; provides structural support for microvilli

action potentials large changes in the membrane potential of excitable cells in which the inside of the cell becomes positive relative to the outside; function in transmitting information over long distances in axons

activation energy additional energy that molecules must acquire in order to react; the difference between the energy of the transition state and the energy of the reactants or products

activation gate one of two gates that regulate the opening and closing of voltage-gated sodium channels, represents the portion that opens with depolarization

active hyperemia (hy-per-ee'-me-ah) the increase in blood flow occurring in response to an increase in metabolic activity

active immunity a type of immune protection that depends on the ability of the immune system to mount a response

active transport any method of protein-mediated transport of molecules across a membrane that requires the use of energy

acuity in sensory systems, a measure of the precision of perception

adaptation decreased responsiveness of a sensory receptor to a continued stimulus

additive in the endocrine system, indicates that the effect of two hormones is simply the sum of the effects of each hormone

adenosine triphosphate (ATP) (ah-den'-oh-seen) compound that serves as primary direct energy source for cell activities; synthesized from adenosine diphosphate (or ADP) and inorganic phosphate (P_i)

adenylate cyclase (ad-den'-ah-late sy'-klase) an intracellular enzyme that catalyzes the conversion of ATP to cAMP

adequate stimulus the energy form or stimulus type to which a sensory receptor responds best

adipocytes fat cells that store triglycerides

adrenal cortex outer portion of the adrenal gland; secretes adrenocorticoids

adrenal glands primary endocrine gland located above the kidneys; divided into an outer cortex and an inner medulla

adrenal medulla inner portion of the adrenal gland; secretes catecholamines

adrenergic pertaining to epinephrine (adrenaline) or norepinephrine (noradrenaline)

adrenocorticoids steroid hormones secreted from the adrenal cortex

adrenocorticotropic hormone (ACTH) (ad-ren-oh-kor-tih-koh-troh'-pik) tropic hormone secreted from the anterior pituitary that stimulates secretion of glucocorticoids from the adrenal cortex

afferent neurons neurons that transmit either sensory information or visceral information to the central nervous system for further processing

affinity a measure of how tightly ligand molecules bind to proteins

after-hyperpolarization third phase of an action potential during which the membrane potential is more negative than at rest due to high permeability to potassium

afterload the pressure that the ventricles have to work against as they pump blood

agglutination the linking together of many antigens through the specific binding of many antibodies

agonists molecules that bind and activate a receptor to produce the same effect as the endogenous messenger

aldosterone (al-dos'-stir-own) a steroid hormone released from the adrenal cortex that regulates the reabsorption of sodium and secretion of potassium

alkalosis condition in which arterial blood pH is 7.45 or greater

all-or-none principle states that action potentials either occur fully or not at all

allosteric regulation regulatory mechanism in which a modulator binds reversibly to the regulatory site on an enzyme, inducing a change in its conformation and activity

alpha receptors class of adrenergic receptors

alveolar sacs clusters of alveoli at the end of an alveolar duct

alveolar ventilation a measure of the volume of fresh air reaching the alveoli each minute, which is minute ventilation corrected for dead space volume; also called minute alveolar ventilation

alveoli (al-vee-oh'-lie) terminal sacs of the respiratory tract, where most gas exchange occurs; usually grouped in clusters

amines chemical messengers derived from amino acids

amino acid (ah-meen'-oh) biomolecule containing amine group, carboxyl group, hydrogen, and an R or residual group attached to a central carbon; found in proteins

amnion (am'-nee-on) a membrane that forms a fluid-filled sac around the embryo; also amniotic sac

amphipathic (am-fuh-path'-ick) molecule having both polar and nonpolar regions, as in phospholipid molecules

ampulla enlargement at the base of each semicircular canal; contains the hair cells for detecting angular acceleration

amylase enzyme that digests starch and glycogen; includes salivary amylase and pancreatic amylase

anabolism (an-nab'-oh-lizm) synthesis of large molecules from smaller molecules, generally requiring an input of energy

analgesia absence of pain perception

anatomical dead space conducting zone of the respiratory tract; air in this region does not participate in gas exchange

androgens a class of sex hormones secreted by the testes

anemia decrease in the oxygen-carrying capacity of blood

angiotensin II (an'-gee-oh-ten-sin) a protein derived from a precursor called angiotensin I that stimulates aldosterone secretion by the adrenal cortex and other physiological responses that serve to maintain or increase blood volume and blood pressure

angiotensin converting enzyme (ACE) enzyme that converts angiotensin I to angiotensin II

antagonism in endocrine system, when one hormone opposes the actions of another hormone

antagonist molecule that binds to a receptor but does not activate it, preventing the endogenous messenger from binding

anterior cavity chamber of eye in front of lens and ciliary muscle

anterior pituitary anterior lobe of pituitary gland; primarily secretes tropic hormones

antibodies proteins present in the blood and interstitial fluids that target particular antigens for destruction; also known as immunoglobulins

anticoagulant a substance that prevents the clotting of blood

antidiuretic hormone (ADH) hormone secreted by the posterior pituitary that regulates water reabsorption by the kidneys; also called vasopressin

antigen presentation the movement of an antigen from inside the infected cell to the cell surface by MHC molecules

antigens protein or polysaccharide components of viruses, bacteria, fungi, protozoa, parasitic worms, pollen, transplanted tissue, and tumor cells that can be identified by immune cells

antrum fluid-filled cavity in the developing follicle

aorta a major artery whose branches carry blood to all organs and tissues in the systemic circuit

aortic semilunar valve a valve located between the left ventricle and the aorta

apical membrane membrane on the side of an epithelial cell that faces the lumen of a body cavity; the lumen-facing membrane

aquaporins water channels

aqueous humor fluid in the anterior cavity of the eye that nourishes the lens and cornea

arachidonic acid phospholipid found in the plasma membranes of cells and platelets; it is the precursor molecule for synthesis of eicosanoids

arachnoid mater (ah-rak'-noyd) one of the three meninges, located between the dura mater and pia mater

arterial blood pressure pressure exerted by the force of blood acting on the walls of arteries

arteries large vessels that carry blood away from the heart

arteriolar tone partial vasoconstriction of arterioles due to inherent contractile activity of single-unit smooth muscle in arteriole walls

arterioles small blood vessels that carry blood to the capillaries; walls contain smooth muscle that contracts and relaxes to regulate blood flow

ascending tracts neural pathways that transmit information from spinal cord or brainstem toward the brain

aspartate (ah-spar'-tate) amino acid that also functions as a neurotransmitter

association areas areas of the cerebral cortex involved in complex processing that requires integrating different types of information

association fibers neural pathways that connect one area of cerebral cortex to another on the same side of the brain

associative learning learning by making connections between two or more stimuli

astrocytes (ass'-trow-sites) glial cells in the CNS that provide support to neurons and are critical to the formation of the blood-brain barrier

atmospheric pressure (P_{atm}) pressure of outside air; at sea level, 760 mm Hg or 1 atmosphere

ATP synthase the enzyme that synthesizes ATP during oxidative phosphorylation; located in the inner mitochondrial membrane

atria (ay'-tree-ah) the heart's two upper chambers, which receive blood carried to the heart in veins; singular, atrium

atrial natriuretic peptide (ANP) hormone secreted from the atrium that regulates plasma sodium levels

atrioventricular (AV) node (ay-tree-oh-ven-trik'-you-lar) part of the conduction system of the heart located near the tricuspid valve

atrioventricular valve (AV valve) (ay-tree-oh-ven-trik'-you-lar) valves that separate the atrium and ventricle on either side of the heart

auditory cortex portion of the temporal lobe of the brain that processes auditory information

autocrine (au'-toh-krin) type of chemical messenger for which the secretory cell and target cell are the same

autonomic ganglia clusters of synapses between preganglionic and postganglionic neurons of the autonomic nervous system; singular, autonomic ganglion

autonomic nervous system the division of the nervous system that encompasses efferent neurons that synapse with and regulate the function of internal organs and other structures not under voluntary control

autoreceptors receptors on a target cell that bind the same messenger that is released from the secretory cell

autorhythmicity the ability of the heart to generate signals that trigger its contractions on a periodic basis; the heart's ability to generate its own rhythm

autosome any chromosome other than the sex chromosomes

axon branch that extends from the cell body of a neuron and sends information to other neurons or effector cells via action potentials and release of neurotransmitter

axon hillock the site where the axon originates from the cell body of a neuron and the point of initiation of action potentials; trigger zone

axon terminal the end of the axon that forms a synapse with another neuron or an effector cell

B cells B lymphocytes

B lymphocytes cells that produce antibodies that mark selected foreign substances for destruction; also called B cells

baroreceptor a type of sensory receptor found in blood vessels and the heart; responds to changes in pressure within the cardiovascular system

baroreceptor reflex negative feedback loop for regulating blood pressure

basal cells in olfactory system, precursor cells for development of new olfactory receptor cells

basal metabolic rate (BMR) the slowest metabolic rate of the body, indicative of the energy necessary to sustain vital functions

basal nuclei a particular group of nuclei located deep in the cerebrum that are important in regulating voluntary movements

basement membrane a layer of noncellular material that is relatively permeable to most substances; anchors the basolateral membrane and provides physical support for the epithelial layer

basic electrical rhythm (BER) a pattern of electrical activity in gastrointestinal smooth muscle, in which waves of depolarization occur at regular intervals

basilar membrane membrane in the cochlea of the inner ear that separates the scala tympani from the scala media

basolateral membrane membrane on the side of an epithelial cell that faces the internal environment and is in contact with interstitial fluid; blood-facing membrane

basophils (bay'-so-fils) leukocytes that defend against larger parasites; release histamine, heparin, and other chemicals

beta receptors class of adrenergic receptors

bicuspid valve the AV valve on the left side of the heart, which possesses two cusps; also called the mitral valve

bile fluid secreted by the liver that contains bicarbonate and bile salts

bile salts cholesterol derivatives manufactured by the liver and found in bile; function is to emulsify fats in the small intestine

bilirubin product of hemoglobin catabolism

biogenic amines a class of neurotransmitters derived from amino acids

biomolecules molecules synthesized by living organisms that contain carbon

blastocyst early stage of embryo development recognizable by presence of inner cell mass

blind spot anatomically, the optic disk of the retina where there are no photoreceptors; physiologically, the visual field where light strikes the optic disk and therefore cannot be detected

blood vessels conduits through which blood flows

blood-brain barrier physical barrier that exists between the blood and the interstitial fluid in the CNS, formed by tight junctions between endothelial cells of the cerebral capillaries

Bohr effect effect of hydrogen ions on the ability of hemoglobin to bind oxygen

Bowman's capsule in each nephron, a cup-shaped structure that surrounds the glomerulus and conducts filtrate into the renal tubule at the inflow end of the renal tubules; site where filtrate enters the renal tubules

Boyle's law law showing the inverse relationship between pressure and volume

brain portion of the central nervous system that resides in the cranium

brainstem bottom-most part of the brain that connects the forebrain and cerebellum to the spinal cord; consists of three main regions: midbrain, pons, and medulla oblongata

Broca's area area of association cortex devoted to language expression; located in the frontal lobe

bronchi branched tubes of the respiratory tract, located between the trachea and bronchioles of the lungs

bronchioles small tubules leading from the bronchi to the alveoli; less than 1 mm in diameter

brush border the collection of microvilli that are located on the apical membranes of epithelial cells lining the small intestine or renal tubule

bulbourethral glands (bul-bo-you-reeth'-ral) accessory glands of the male reproductive system

bundle of His part of the conduction system of the heart located in the interventricular septum

calcification formation of bone

calcitonin peptide hormone released from C cells of the thyroid gland that regulates plasma calcium levels

calcitriol steroid hormone derived from vitamin D that regulates plasma calcium levels

calmodulin (kal-mod'-you-lin) cytosolic calcium-binding protein; modulates the activity of intracellular proteins

calorie amount of energy that must be put into one gram (or one mL) of water to raise its temperature by one degree centigrade (°C) under a standard set of conditions

capillaries the smallest blood vessels in the body; possess thin walls that permit material exchange between blood and tissues

carbamino effect decrease in affinity of hemoglobin for oxygen when carbon dioxide binds to hemoglobin

carbaminohemoglobin (kar-bah-meen'-oh) hemoglobin with carbon dioxide bound to it

carbohydrates biomolecules composed of carbon, hydrogen, and oxygen in a ratio of 1:2:1

carbonic anhydrase enzyme that catalyzes the reversible reaction converting carbon dioxide and water to carbonic acid

cardiac cycle a series of mechanical and electrical events occurring within the heart during a single beat

cardiac output (CO) the volume of blood ejected from each ventricle per minute

cardiovascular system organ system consisting of the heart, blood vessels, and blood

carrier a transmembrane protein that binds molecules on one side of a membrane and transports them to the other side by means of a conformational change, or a change in shape

cartilage a connective tissue secreted by chondrocytes that is similar to uncalcified bone

cascade a series of sequential steps that progressively increase in magnitude

catabolism breakdown of large molecules into smaller molecules; generally releases energy

catalytic rate for enzyme-catalyzed reactions, a measure of the number of reactants converted to product per unit time

catechol-*O*-methyltransferase (COMT) enzyme that degrades catecholamines

catecholamines amine compounds that contain a catechol group and are derived from the amino acid tyrosine

cell body portion of neuron where nucleus is located; also called soma

cell-mediated immunity reaction of certain types of T cells to kill abnormal or infected body cells

cells smallest living units; separated from environment by a membrane

cellulose polysaccharide of glucose found in plants that humans are unable to digest or absorb

central canal long thin cylindrical canal that runs the length of the spinal cord and is continuous with the cerebral ventricles; contains cerebrospinal fluid

central chemoreceptors chemoreceptors located in the medulla oblongata that respond directly to changes in hydrogen ion concentration in the cerebrospinal fluid and indirectly to arterial P_{CO_2}; function in regulating ventilation

central lymphoid tissues tissues where leukocytes develop to maturity, including the bone marrow and thymus

central nervous system (CNS) the division of the nervous system that includes the brain and spinal cord; consolidates information received from the organs and develops commands to be sent to the organs; and is also the site of learning, memory, emotions, and cognition

central pattern generator in the respiratory system, network of neurons responsible for establishing the breathing rhythm

central venous pressure (CVP) the pressure in the large veins in the thoracic cavity that lead to the heart

cephalic-phase control in gastrointestinal system, the regulation of GI function by stimuli originating in the head

cerebellum (ser-ah-bel'-um) a bilaterally symmetrical brain structure, with an outer cortex and inner nuclei; located below the forebrain and posterior to the brainstem

cerebral cortex thin layer of gray matter that covers the cerebrum

cerebral hemispheres the two halves of the cerebrum

cerebrospinal fluid (CSF) clear, watery fluid that surrounds and protects the CNS and is similar in composition to plasma

cerebrum (seh-ree'-brum) largest structure of the brain, which contains both gray and white matter; gray matter areas include the cerebral cortex at the surface and deep subcortical nuclei

cervical nerves spinal nerves originating from the cervical spinal cord

cervix the lower, narrower portion of the uterus, containing a central canal that leads directly to the vagina

channel a transmembrane protein that transports molecules by way of a passageway or pore that extends from one side of the membrane to the other

channel-linked receptor protein that functions as both an ion channel and a receptor; binding of a messenger to the receptor portion of the protein opens the channels

chemical driving force difference in energy due to a concentration gradient that causes a molecule to move from high concentration to low

chemical synapse synapse where a neuron secretes a neurotransmitter into the extracellular fluid in response to an action potential arriving at the axon terminal

chemiosmotic coupling entire process that couples electron transport to ATP synthesis; the utilization of energy released during electron transport to transport hydrogen ions across the inner mitochondrial membrane up their concentration gradient

chemoreceptors receptors that monitor the concentrations of certain chemicals in various locations in the body

chest wall structures that protect the lungs and form an airtight compartment around them; includes the rib cage, sternum, thoracic vertebrae, muscles, and connective tissue

chief cells specialized cells in the gastric glands that secrete pepsinogen

chloride shift the movement of chloride ions into erythrocytes in exchange for the movement of bicarbonate into plasma

cholecystokinin (CCK) a hormone secreted by the duodenum and jejunum that inhibits gastric secretion and motility, and stimulates pancreatic enzyme secretion, gallbladder contraction, and relaxation of the sphincter of Oddi

choline acetyl transferase (CAT) enzyme that catalyzes the synthesis of acetylcholine

cholinergic (koh-lin-er'-jik) pertaining to acetylcholine

chondrocytes (kon'-droh-sites) cells that produce cartilage

chorion (kor'-ee-on) the outermost membrane that encapsulates the embryo, forming a tough envelope that isolates it from its surroundings

choroid middle layer of the posterior two-thirds of the eye; contains pigment that absorbs light

choroid plexus (kor'-oid) vascularized tissue lining the cerebral ventricles, synthesizes cerebrospinal fluid

chromatid one of two strands of a chromosome pair that is undergoing cell division

chromatin the form in which DNA, along with its associated proteins, exists throughout most of the cell cycle; loosely coiled DNA and proteins scattered throughout the nucleus

chylomicrons lipoprotein particles formed during the process of lipid absorption in the small intestine

chyme a mixture of food particles and gastrointestinal secretions; found in the stomach and intestines

cilia hairlike processes found on certain epithelial cells in the respiratory tract and oviduct

ciliary muscles smooth muscles in the eye attached to the lens by zonular fibers; regulate the curvature of the lens for focusing light

circadian rhythm (sir-kay'-dee-an) endogenous fluctuations in body functions that occur on a 24-hour cycle

circular muscle inner layer of smooth muscle of the iris; also called constrictor muscle

class I MHC molecules found on the surfaces of all nucleated cells; identify cells as self

class II MHC molecules found on specialized cell types, including macrophages, activated B and T cells, and the cells of the interior of the thymus

clearance virtual measure of the volume of plasma from which a substance is completely removed or "cleared" by the kidneys per unit time

clonal selection antigen-driven proliferation of lymphocytes necessary for specific immune responses

coagulation cascade sequence of reactions leading to the formation of a blood clot

coagulation factors plasma proteins that take part in the coagulation cascade

coated pit an indentation of the plasma membrane that eventually forms an endocytotic vesicle; coated on its inner surface by specific proteins

coccygeal nerve spinal nerve originating from the coccyx

cochlea a spiral-shaped structure in the inner ear that contains the receptor cells for hearing

codon three-base sequence in mRNA that codes for a specific amino acid

coenzymes (koh-en'-zimes) molecules that do not themselves have catalytic activity but that are necessary for proper enzyme function and participate directly in reactions catalyzed by enzymes; often serve to transfer certain chemical groups from one reactant to another

cofactors nonprotein components of some enzymes necessary for them to hold normal conformation during metabolic reactions

cold receptors thermoreceptors that increase in responsiveness as the temperature decreases

collateral branch of an axon

collateral ganglion sympathetic ganglion independent of the sympathetic chain

collecting ducts ducts that collect fluid from several different renal tubules and carry it to the renal pelvis for eventual elimination

colon an organ of the GI tract that absorbs water and ions from the chyme, and stores feces; comprises the ascending colon, descending colon, transverse colon, and sigmoid colon

colony-stimulating factors substances that stimulate the production of a specific type of blood cell

commissural fibers fibers that connect one region of cerebral cortex to corresponding region on the other side of the brain

complement system 30 proteins that act to destroy invading microorganisms, especially bacteria

compliance a measure of the ability of blood vessels or other hollow structures to stretch as the pressure inside them rises

concentration gradient difference in the concentration of a substance between regions

conductance ease of movement; in electricity, the ease of ion movement

conducting zone the upper part of the respiratory tract; conducts air from the larynx to the lungs; no gas exchange occurs

conduction the transfer of thermal energy from one object to another that occurs when the objects are in direct contact with each other; in the nervous system, the propagation of action potentials along an axon

conduction fibers specialized muscle cells that rapidly conduct action potentials through the heart

conduction system a set of specialized heart muscle cells that initiate and conduct action potentials

cones photoreceptors that enable visibility during relatively bright light and are responsible for color vision

connective tissue tissue whose primary function is to provide physical support for other structures, to anchor them in place, or to link them together

contractile component thick and thin filaments of muscle; compared to elastic components

contractility capacity of a muscle to generate force

contralateral referring to ascending and descending pathways that are on the side opposite their origin

convergence the transmission of information from two or more neurons to one cell

convex lens type of lens that causes parallel light waves to converge at a single point

copulation act of mating

core temperature temperature deep within the body

cornea transparent structure at the front of the eye that allows light waves to enter

corpus callosum (kor'-pus kal-loh'-sum) the primary band of nervous tissue that connects the two cerebral hemispheres

corpus luteum (kor'-pus loo'-tee-um) a gland that develops from the ruptured follicle following ovulation and secretes progesterone and estrogen

cortex outer portion of an organ such as cerebral cortex, adrenal cortex, and renal cortex

evaporation conversion of a liquid to a gas; one of the mechanisms of thermoregulation

excitable cells cells capable of producing action potentials

excitation-contraction coupling in a muscle cell, the sequence of events that links the action potential to the contraction

excitatory postsynaptic potential (EPSP) a graded depolarization caused by neurotransmitter binding to receptors on the postsynaptic neuron

excitatory synapse site of communication at which activity in the presynaptic neuron causes depolarization of the post-synaptic neuron

excretion elimination from the body through the kidneys (urine) or GI tract (feces)

exocrine glands glands that are specialized for the transport of materials from the body's internal environment to the external environment

exocytosis (ex-oh-sy-toh'-sis) transport of material out of a cell via vesicles that fuse with the plasma membrane; involved in the cellular secretion of hydrophilic molecules

exopeptidase enzyme that digests proteins by cleaving off amino acids from the end of the polypeptide

expiratory neurons neurons in the central nervous system that increase their firing rate during expiration

expiratory reserve volume (ERV) the maximum volume of air that can be expired from the end of a normal expiration

external anal sphincter ring of skeletal muscle that regulates the excretion of feces

external auditory meatus ear canal

external intercostals inspiratory muscles of the chest wall

external respiration the exchange of oxygen and carbon dioxide between the atmosphere and the tissues of the body; involves both the respiratory and cardiovascular systems

external urethral sphincter ring of skeletal muscle that regulates the excretion of urine

extracellular fluid (ECF) fluid located outside cells, account-ing for one-third of total body water; synonymous with internal environment

extrapyramidal tract all motor control pathways outside the pyramidal system; controls supportive movements

extrinsic control regulation of an organ or tissue by neural input, circulating hormones, or any other factor originating from outside the organ

facilitated diffusion the passive movement of molecules across a membrane by way of a transport protein

fast glycolytic fibers type of skeletal muscle with high myosin ATPase activity and high glycolytic capacity

fast pain a sharp, pricking sensation that can be easily local-ized and is produced by activation of nociceptors; is transmitted by Aδ fibers

fatigue a decline in a muscle's ability to maintain a constant force of contraction in the face of long-term, repetitive stimulation

fatty acids long hydrocarbon chains with a carboxyl group (—COOH) at one end

feedback inhibition regulatory mechanism in which an enzyme in a metabolic pathway is inhibited by an intermediate appearing downstream

feedforward activation regulatory mechanism in which an enzyme in a metabolic pathway is stimulated by an intermediate appearing upstream

fertilization the process by which two gametes (one from each parent) fuse together to produce a new cell; known as conception in humans

fetus the developing human after two months following conception

fibrin the last active clotting factor in the coagulation cascade, which forms polymers that make up the actual clot

fight-or-flight response the group of physiological changes coordinated by the sympathetic nervous system that prepares the body to cope with threatening situations

filtered load the quantity of a certain solute that is filtered at the glomerulus per unit time; equals' the product of the GFR and the plasma concentration of the solute

filtration the movement of fluid across capillary walls from plasma to the interstitium

flux the number of molecules crossing a membrane in a given length of time; a measure of the rate at which a substance is being transported

follicle spherical structure in the ovary containing a single developing ovum

follicle stimulating hormone (FSH) a gonadotropic hormone that stimulates gametogenesis and regulates other gonadal functions in either sex

follicular phase (fuh-lik'-you-lar) the initial phase of the ovarian cycle, during which follicular maturation occurs; terminates with ovulation

forced expiratory volume (FEV) a measure of the percent-age of the forced vital capacity that can be exhaled within a certain time frame

forced vital capacity (FVC) the maximum amount of air a person can forcefully expire following a maximum inspiration

forebrain largest and uppermost part of the brain, divided into left and right halves, or hemispheres; consists of the cerebrum and diencephalon

fovea (foh'-vee-ah) central point on the retina of the eye, in which visual acuity is greatest

fragmentin protein released from cytotoxic T cells that enters target cells and induces apoptosis

free nerve endings somatosensory receptors in the skin that lack identifiable specialized sensory structures; includes some mechanoreceptors, thermoreceptors, and nociceptors

frequency coding the coding of stimulus intensity by the frequency of action potentials in a neuron, in which a stronger depolarizing stimulus above threshold causes the action potential frequency to increase

frontal lobe one of four lobes of the cerebrum, located in the anterior portion of the cerebrum; important in voluntary motor control, behavior, and personality traits

functional residual capacity (FRC) the volume of air in the lungs at the end of a resting expiration

G cells endocrine cells of the stomach that secrete gastrin

G proteins membrane proteins with the ability to bind guano-sine nucleotides; function in coupling an extracellular messen-ger to a response inside the target cell

G protein–linked receptor plasma membrane receptor that is coupled to a G protein

GABA gamma-aminobutyric acid, an amino acid neurotransmitter

gallbladder small muscular sac located immediately adjacent to the liver; stores bile in between meals

gametes (gam-eets') cells of reproduction produced in meiosis that contain half of a person's genetic material; examples are sperm cells in males and ova in females

gametogenesis (gah-mee-toh-jen'-ih-sis) the production of gametes from a pool of germ cells

ganglion (gang'-glee-on) cluster of neuron cell bodies in the peripheral nervous system; plural ganglia

gap junctions areas where two adjacent cells are connected together by membrane proteins called connexons that form small channels between the cells, enabling ions and small molecules to move freely between them

gastric glands glands in the lining of the stomach that secrete gastric juice

gastric juice watery secretion of the stomach containing hydrochloric acid, pepsinogen, and mucus

gastric mucosal barrier mucus layer that protects the lining of the stomach from the effects of acid and pepsin

gastric pits openings in the lining of the stomach that lead to gastric glands

gastric-phase control regulation of GI function by stimuli originating in the stomach

gastrin a hormone secreted by the stomach that regulates many gastrointestinal functions, including gastric acid secretion

gastrointestinal tract (GI tract) a number of organs joined in series to form a passageway through which food and digestion products are conducted

gene section of DNA that codes for a particular protein or proteins

generator potential a change in membrane potential, also known as receptor potential

genetic code the correspondence between DNA triplets and specific amino acids that governs the expression of all genetic information

gestation pregnancy

glial cells (glee'-al) cells in the nervous system that provide various types of support to the neurons, including structural and metabolic support

glomerular filtration the bulk flow of protein-free plasma from the glomerular capillaries into Bowman's capsule

glomerular filtration pressure sum of the Starling forces acting to move fluid across the capillary walls of the glomerulus

glomerular filtration rate (GFR) the volume of the plasma filtered per unit time from all renal glomeruli combined

glomerulus (gloh-mer'-you-lus) in each nephron, a ball-like cluster of capillaries in the renal corpuscle; site of filtration

glottis opening to the larynx

glucagon a peptide hormone secreted by alpha cells of the pancreas; promotes metabolic processes of the postabsorptive state

glucocorticoids steroid hormones secreted from the adrenal cortex that regulate the body's response to stress, regulate protein, carbohydrate, and lipid metabolism in a variety of tissues, and regulate blood glucose levels; primary glucocorticoid is cortisol

gluconeogenesis (gloo-koh-nee-oh-jen'-ih-sis) process during which new glucose molecules can be synthesized from noncarbohydrate precursors by the liver

glucose most common monosaccharide; provides important source of cellular energy

glucose sparing process by which non-nervous tissues convert to using fatty acids for energy rather than glucose, which is spared for use by the central nervous system

glucose-dependent insulinotropic peptide (GIP) a hormone secreted by the duodenum and jejunum that stimulates insulin secretion by pancreas

glutamate amino acid neurotransmitter

glycerol (gliss'-er-ol) 3-carbon alcohol that functions as the "backbone" of a triglyceride or phospholipid

glycine (gly'-seen) amino acid neurotransmitter

glycogen (gly'-coh-jen) a glucose polymer found in animal cells; functions as an energy store

glycogenesis (gly-koh-jen'-eh-sis) synthesis of glycogen from glucose monomers

glycogenolysis (gly-koh-jen-nol'-ih-sis) breakdown of glycogen to glucose monomers

glycolysis the first stage of glucose oxidation, occurring in the cytosol, in which each glucose molecule is broken down to two molecules of pyruvate

goblet cells epithelial cells in the respiratory tract and GI tract that secrete mucus

Golgi apparatus (goal'-jee) an organelle consisting of membrane-bound flattened sacs called cisternae that process molecules synthesized in the endoplasmic reticulum and prepares them for transport

gonadotropin releasing hormone (GnRH) a hypothalamic tropic hormone that stimulates the secretion of gonadotropins by the anterior pituitary

gonadotropins two hormones, follicle stimulating hormone (FSH) and luteinizing hormone (LH), that are secreted by the anterior pituitary and regulate gonadal function in either sex

gonads the primary reproductive organs; serve as sites of gametogenesis and sex hormone secretion

Graafian follicle fully matured follicle, just prior to ovulation

graded potentials small changes in membrane potential that occur when ion channels open or close in response to a stimulus, such as the binding of neurotransmitters to receptors; magnitude of the potential change varies with the strength of the stimulus

granular cells specialized cells in the wall of the afferent and efferent arterioles that secrete renin; also called juxtaglomerular cells

granulosa cells specialized cells that surround a developing ovum and support its development

gray matter areas of the CNS consisting primarily of cell bodies, dendrites, and axon terminals of neurons; where synaptic transmission and neural integration occur

growth hormone inhibiting hormone (GHIH) tropic hormone released by the hypothalamus that inhibits growth hormone secretion from the anterior pituitary; also known as somatostatin

growth hormone (GH) peptide hormone secreted by the anterior pituitary, essential for normal growth

growth hormone releasing hormone (GHRH) tropic hormone released by the hypothalamus that stimulates growth hormone secretion from the anterior pituitary

gyri (jy'-ri) ridges in the highly convoluted gray matter of the cerebral cortex; singular: gyrus

habituation decrease in response to a repeated stimulus; a form of learning

hair cells cells with stereocilia; receptor cells for hearing and equilibrium

Haldane effect effect of oxygen on the binding of carbon dioxide to hemoglobin

half-life time it takes for half the amount of a molecule to be degraded

haploid in reference to a cell's chromosome number, possessing half the number of chromosomes found in most other cells of the body

haustration mixing movements of the colon

helper T cells T cells that secrete many different cytokines and thereby influence the action of other lymphocytes

hematocrit (heh-mat'-oh-krit) the fraction of the blood volume that is occupied by red blood cells

hemoglobin (hee'-moh-gloh-bin) a protein in red blood cells that carries oxygen and carbon dioxide

hemostasis process of stopping bleeding; includes vascular spasms, platelet plug formation, and blood clot formation

hepatocytes liver cells

homeostasis (hom-ee-oh-stay'-sis) maintenance of relatively constant conditions within the body's internal environment

hormones chemical messengers released from endocrine cells or glands into the interstitial fluid, where they then diffuse into the blood and travel to target cells

human chorionic gonadotropin (hCG) a hormone secreted by the chorion that maintains the corpus luteum during the first three months of pregnancy

humoral immunity a specific immune response generated by B lymphocytes and conferred by antibodies that circulate in the blood and lymph

hydrolysis use of water to split another molecule

hyperopia (hi-per-oh'-pee-ah) common visual defect of the eye causing far-sightedness

hyperosmotic a solution having an osmolarity that is higher than another

hyperplasia increase in cell number

hyperpnea (hy-perp-nee'-ah) an increase in alveolar ventilation to match increased metabolic demands

hyperpolarization any change in membrane potential in which the inside of the cell becomes more negative than it is at rest

hypertension higher than normal arterial blood pressure

hyperthermia a condition of higher than normal body temperature

hypertonic a solution that draws water out of a cell causing the cell to shrink

hypertrophy increase in cell size

hyperventilation an increase in alveolar ventilation such that metabolic demands of the tissue are exceeded

hypo-osmotic a solution with a lower osmolarity than another

hypothalamus a region at the base of the brain that regulates autonomic functions and secretes several hormones, most of which regulate secretory activity of the pituitary gland

hypothermia a condition of lower than normal body temperature

hypotonic a solution that causes water to enter a cell causing the cell to swell

hypoventilation a decrease in alveolar ventilation such that metabolic demands of the tissue are not met

ileocecal sphincter ring of smooth muscle that regulates movement of chyme from the ileum to the cecum

ileum terminal portion of the small intestine

immune response series of events that destroy and eliminate foreign and abnormal substances from the body

immunity capacity of the immune system to defend against disease by recognizing and eliminating disease-causing agents from the body

immunoglobulins (Ig) (im-mun-oh-glob'-you-lins) proteins present in the blood and interstitial fluids that target particular antigens for destruction; also known as antibodies.

implantation process by which an embryo adheres to and embeds in the wall of the uterus

inactivation gates one of two types of gates in voltage-gated sodium channels; close slowly in response to depolarization

inclusions cytosolic particles composed of triglycerides or glycogen; serve as energy stores for cellular metabolism

inflammation a complex series of events that culminate in the accumulation of proteins, fluid, and phagocytic cells in an area where tissue has been injured or invaded by microorganisms

inhibin a protein hormone secreted by the gonads that suppresses the release of FSH from the anterior pituitary

inhibitory postsynaptic potential (IPSP) a graded hyperpolarization caused by neurotransmitter binding to receptors on the postsynaptic neuron

inhibitory synapse site of communication whereby activity in the presynaptic neuron decreases the likelihood of an action potential occurring in the postsynaptic neuron

innervate to make a synaptic connection with an effector organ

inositol triphosphate (IP₃) a second messenger released by the phosphatidylinositol system that stimulates the release of calcium

inspiratory capacity (IC) the maximum volume of air that can be inspired at the end of a resting expiration

inspiratory neurons CNS neurons that have an increased firing rate during inspiration

inspiratory reserve volume (IRV) the maximum volume of air that can be inspired from the end of a normal inspiration

insulin a peptide hormone secreted by beta cells of the pancreas; promotes metabolic processes of the absorptive state

insulin-like growth factor (IGF) peptide hormone secreted by the liver in response to growth hormone; promotes protein synthesis and growth; also known as somatomedin

integral membrane proteins proteins that are tightly embedded within the lipid bilayer

integrating center group of cells, usually in the CNS or an endocrine gland, that use sensory information to determine a response and communicate commands to effectors

intercalated disk specialized junction between cardiac muscle cells that contain both desmosomes and gap junctions

intercostal nerves spinal nerves that innervate the intercostal muscles of respiration

interferons a group of related proteins that interfere with virus replication

interleukins cytokines that communicate messages between leukocytes

intermediate filaments fibrous proteins with a diameter between that of microfilaments and microtubules; stronger and more stable than microfilaments

internal anal sphincter ring of smooth muscle that regulates the excretion of feces

internal environment fluid that surrounds the cells inside the body, including fluid in the bloodstream that surrounds blood cells; synonymous with extracellular fluid

internal intercostals muscles of expiration that are located between the ribs

internal respiration cellular respiration that occurs in the mitochondria

internal urethral sphincter ring of smooth muscle that regulates the excretion of urine

interneurons neurons located entirely in the central nervous system; account for 99% of all neurons in the body

internodal pathways systems of conducting fibers that run from the SA node to the AV node through the walls of the atria

interphase period in the life cycle of the cell during which it is carrying on its normal physiological functions

interstitial fluid (ISF) extracellular fluid that is present outside the blood, and that bathes most of the cells of the body

intestinal-phase control regulation of GI function by stimuli originating in the small intestine

intra-alveolar pressure (P_{alv}) the pressure exerted by the air within the alveoli

intracellular fluid (ICF) fluid located inside cells, accounting for two-thirds of total body water

intrapleural pressure (P_{ip}) the pressure of the fluid inside the pleural space

intrapleural space a fluid-filled compartment located between the lungs and chest wall; is bounded by the visceral and parietal pleura

intrinsic control regulation of an organ or tissue by factors originating from within the organ or tissue itself; also known as autoregulation or local regulation

intrinsic factor a protein secreted by parietal cells of the stomach that binds to vitamin B_{12} and that is necessary for the absorption of this vitamin in the small intestine

ionotropic receptor (eye-oh-no-troh'-pik) a receptor protein that also functions as an ion channel that opens or closes in response to the binding of a chemical messenger

ipsilateral (ip-sih-lat'-er-al) referring to ascending and descending pathways that are on the same side as their origin

iris pigmented smooth muscle in the eye that sits in front of the lens and regulates the diameter of the pupil to control the amount of light entering the eye

ischemia (iss-key'-me-ah) a condition in which blood flow in a tissue is insufficient to keep up with metabolic demand

isometric twitch (eye-soh-met'-rick) a twitch during which a muscle generates force but does not shorten

iso-osmotic two solutions having the same osmolarity

isotonic a solution that will not alter cell volume

isotonic twitch (eye-soh-tah'-nik) a twitch during which a muscle shortens and lifts a constant load

isovolumetric contraction contraction of the ventricles with all heart valves closed, such that the volume of blood contained within the ventricles is constant; occurs early in systole

isovolumetric relaxation relaxation of the ventricles with all heart valves closed, such that the volume of blood contained within the ventricles is constant; occurs early in diastole

jejunum middle portion of the small intestine

juxtaglomerular apparatus (jux-tah-gloh-mer'-you-lar) a collection of specialized cells in the distal tubules and the afferent and efferent arterioles near where the three structures come together in the kidney; regulates glomerular filtration and renin secretion

juxtaglomerular cells specialized cells in the wall of the afferent and efferent arterioles that secrete renin; also called granular cells

kinetic energy energy associated with motion

kinocilium large stereocilia projecting from the receptor cells for equilibrium; direction of bending in response to acceleration of the body determines direction of receptor potentials in receptor cells

Krebs cycle a cyclical metabolic pathway occurring in the mitochondrial matrix, in which acetyl coenzyme A is a primary reactant and carbon dioxide and reduced coenzymes are produced; also called the citric acid cycle, tricarboxylic acid cycle, or TCA cycle

lactation milk production that takes place in the mammary glands

lacteal blind-ended lymphatic vessel found within each villus in the lining of the small intestine

lamina propria in the wall of the GI tract, a layer of connective tissue within the mucosa

larynx (lar'-inks) the initial passageway of the respiratory tract, which contains the vocal cords

latent period the time between a stimulus and the beginning of a response; in muscle physiology, the lag of a few milliseconds that occurs between the action potential in a muscle and when the muscle first begins to generate force

lateral geniculate body nucleus of the thalamus that transmits auditory information to the auditory cortex

lateral horn a region of the gray matter of the spinal cord where certain autonomic preganglionic neurons originate; synonymous with intermediolateral cell column

lateral inhibition process during which a stimulus that strongly excites receptors in a certain location inhibits activity in the afferent pathways of other receptors located nearby

peripheral lymphoid tissues tissues that trap foreign matter present in the blood, including the spleen, lymph nodes, tonsils, adenoids, appendix, and Peyer's patches

peripheral membrane proteins proteins that are loosely bound to the lipid bilayer by associations with integral membrane proteins or phospholipids

peripheral nervous system the division of the nervous system that contains nerves that provide communication between the central nervous system and organs of the body; includes afferent and efferent branches

peristalsis a wave of momentary contraction that travels along the length of a hollow vessel or organ, such as the GI tract

peritoneum a membrane lining the inside of the abdominal cavity

peritubular capillaries capillary bed that branches off the efferent arterioles of cortical nephrons and is located close to the renal tubules; functions in exchange with renal tubules during reabsorption and secretion

permeability measure of the ease with which molecules are able to move through the cell membrane

permissiveness phenomenon in which one hormone is needed for another hormone to exert its actions

peroxisomes spherical membrane-bounded organelles that function in the degradation of molecules such as amino acids, fatty acids, and toxic foreign matter

phagocytosis (fag-uh-sy-toh'-sis) process by which a cell engulfs microorganisms, abnormal cells, and foreign particles present in blood and tissues

pharynx (fair'-inks) a passageway leading from the mouth to the esophagus or larynx that serves as a common passageway for food and air

phasic receptors receptors that decrease in responsiveness to a continuing stimulus; also called rapidly adapting receptors

phosphatase an enzyme that catalyzes dephosphorylation

phosphatidylinositol system (fos-fah-tide'-il-in-os'-ih-tol) a signal transduction system that produces two second messengers, diacylglycerol and inositol triphosphate

phospholipid amphipathic lipid molecule consisting of a glycerol backbone to which two fatty acids and a phosphate-containing chemical group are attached

phosphorylation addition of a phosphate group to a molecule

photopigment the molecule in the photoreceptors that absorbs light, the first step of phototransduction

photoreceptors cells located in the retina of the eye that detect light waves; includes rods and cones

phrenic nerve (fren'-ik) nerve that innervates the diaphragm

physiology the study of the phenomena of living things

pia mater innermost of the three meninges, adjacent to the nervous tissue

pineal gland primary endocrine gland located in the brain; secretes the hormone melatonin

pinocytosis a form of endocytosis in which a cell takes up fluid and dissolved molecules via endocytotic vesicles that pinch off the plasma membrane

pituitary gland primary endocrine gland connected to the hypothalamus at the base of the brain; divided into the anterior pituitary and posterior pituitary

placenta a structure consisting of maternal and fetal tissues that allows exchange of gases, nutrients, and wastes between the mother's circulatory system and the circulatory system of the fetus

placental lactogen a hormone secreted by the placenta that promotes breast growth and the mobilization of energy stores in the mother

plasma liquid in the blood made up of water and dissolved solutes, including proteins; represents about 20% of the total volume of extracellular fluid

plasma cells mature form of B cells that secrete antibodies

plasma membrane barrier separating a cell from the extracellular fluid; consists of phospholipids, proteins, and cholesterol

plasmin plasma protein that dissolves a blood clot

plasticity ability to change

platelets cell fragments that play an important role in blood clotting

pleura (plur'-ah) the membrane that lines the chest wall and lung, forming a pleural sac around each lung

pleural sac membrane surrounding each lung

pneumothorax condition in which air enters the pleural space, causing the lungs to collapse and the chest wall to expand

polypeptide polymer containing amino acids joined together by peptide bonds

polysaccharide polymer composed of many monosaccharides joined together by covalent bonds

pons middle portion of the brainstem; connects to the cerebellum

pontine respiratory group (PRG) respiratory center of the pons containing both inspiratory and expiratory neurons

positive feedback a type of feedback in which the response of a system goes in the same direction as the change that set it in motion

postabsorptive state period between meals during which stored nutrients are mobilized

posterior pituitary posterior lobe of pituitary gland; secretes antidiuretic hormone (ADH) and oxytocin

postganglionic neuron the neurons of the autonomic nervous system that travel from autonomic ganglia to the effector organs

postsynaptic cell cell that receives communication from a neurotransmitter released at a synapse

postsynaptic neuron at a synapse, the neuron that receives signals from another neuron

postsynaptic potential (PSP) a change in the membrane potential in a postsynaptic neuron that occurs in response to the binding of neurotransmitters to receptors

potential energy stored energy

potentiation in endocrinology, the amplification of one hormone's effects by another hormone

precapillary sphincters smooth muscle surrounding capillaries; regulate blood flow through capillary beds

preganglionic neuron the neurons of the autonomic nervous system that travel from the central nervous system to autonomic ganglia, where they communicate with postganglionic neurons

preload ventricular end-diastolic pressure

presynaptic facilitation a phenomenon occurring at an axo-axonic synapse such that activity in the presynaptic neuron enhances the release of neurotransmitter from the postsynaptic neuron

presynaptic inhibition a phenomenon occurring at an axo-axonic synapse such that activity in the presynaptic neuron decreases the release of neurotransmitter from the postsynaptic neuron

presynaptic neuron at a synapse, a neuron that transmits signals to a second neuron or an effector cell

primary active transport active transport of molecules utilizing a protein (pump) that uses ATP as the energy source

primary endocrine organs organs whose primary function is the secretion of hormones

primary immune response antigen-induced lymphocyte response that occurs when a person is exposed to an antigen for the first time

primary motor cortex region of the frontal lobe where voluntary movement is initiated

primary somatosensory cortex region of the parietal lobe of the cerebral cortex specialized for the processing of somatic sensory information

primordial follicle follicle at the earliest stage of development

procedural memory memory of learned motor skills and behaviors

progesterone a sex hormone secreted by the ovaries, primarily during the luteal phase of the menstrual cycle

projection fibers axons in the CNS that connect the cerebral cortex with lower levels of the brain or the spinal cord

proliferative phase phase of the uterine cycle during which the uterus renews itself after the menstrual phase

promoter sequence a specific base sequence in DNA to which the enzyme RNA polymerase can bind, thereby initiating transcription

proprioception (pro-pree-oh-cep'-shun) the perception of the position of limbs and the body

prostacyclin eicosanoid released from healthy endothelial cells that inhibits platelet aggregation

prostate gland an accessory gland of the male reproductive system that secretes fluid into the urethra

protein a polymer containing amino acids joined together by peptide bonds; usually refers to chains containing more than 50 amino acids

protein C plasma protein that inhibits blood coagulation

protein kinase a type of enzyme that catalyzes phosphorylation of a target protein

protein phosphatase enzyme that catalyzes the dephosphorylation of a protein

proteolysis (proh-tee-oh-ly'-sis) process by which proteins are broken down to amino acids

proteolytic activation rendering an enzyme active by cleaving off certain amino acids

proximal tubule portion of the renal tubule nearest the renal corpuscle; includes the proximal convoluted tubule and the proximal straight tubule

puberty period of sexual maturation

pulmonary arteries arteries that carry blood to the lungs from the heart

pulmonary circuit the portion of the vasculature that encompasses all the blood vessels within the lungs and those connecting the lungs with the heart

pulmonary semilunar valve valve located between the right ventricle and pulmonary trunk

pulmonary surfactant detergent-like substance secreted by type II alveolar cells; decreases the surface tension in the lungs

pulmonary veins veins that carry blood from the lungs to the heart

pulmonary ventilation the movement of air into and out of the lungs by bulk flow

pumps proteins that actively transport molecules across a membrane

pupil hole through which light can enter the eye

Purkinje fibers (purr-kin'-gee) an extensive network of conducting fibers that spread through the ventricular myocardium

pyloric sphincter ring of smooth muscle that regulates gastric emptying

pylorus a narrow passage between the stomach and the duodenum

pyramidal tract a direct pathway from the primary motor cortex to the spinal cord; controls fine voluntary movements

QRS complex in ECGs, waveform representing depolarization of the ventricles

radial muscle outer layer of smooth muscle in the iris; also called dilator muscle

radiation transfer of thermal energy in the form of electromagnetic waves

rapidly adapting receptors receptors that decrease in responsiveness to a continuing stimulus; also called phasic receptors

reabsorption transport of a substance into blood

reactants molecules in a chemical reaction that are converted to product

reactive hyperemia a local increase in blood flow that occurs following termination of an occlusion of blood flow to the same area

receptive field the area over which an adequate stimulus can produce a response, either excitatory or inhibitory, in an afferent neuron or higher-order neurons

receptor potential graded potential caused by the opening or closing of ion channels on sensory receptors, and triggered by sensory stimuli

receptor-mediated endocytosis a form of endocytosis in which the plasma membrane contains receptors that recognize and bind specific molecules in the extracellular fluid; enables cells to selectively take up certain molecules

receptors proteins on a target cell that recognize and bind a chemical messenger; in sensory systems, detectors of stimuli

recruitment an increase in the number of active neurons

reduction addition of electrons to a molecule

referred pain the perception of a painful stimulus as originating at a site on the body distinct from the location of the stimulus

reflex automatic patterned response to a stimulus

reflex arc pathway by which a stimulus reflexively induces a response

refraction bending of light waves as they travel through media of different densities at an angle other than perpendicular

refractory period period of reduced membrane excitability; during and immediately after an action potential when the membrane is less excitable than it is at rest

regenerative a signal that feeds into itself by positive feedback

regulatory site binding site on an enzyme molecule that is specific for molecules known as modulators

relative refractory period period immediately following the absolute refractory period during which it is possible to generate a second action potential, but only with a stimulus stronger than that needed to reach threshold under resting conditions

REM sleep sleep characterized by high-frequency waves in the EEG and periodic rapid movements of the eyes; REM = rapid eye movement

renal arteries arteries that branch off the aorta and provide the kidneys with their blood supply

renal corpuscle site of glomerular filtration; consists of glomerulus and Bowman's capsule

renal pelvis funnel-shaped passage forming the initial portion of the ureter

renal threshold the plasma concentration of solute at which the transport maximum is exceeded and excess solute appears in the urine

renal tubule a portion of a nephron, consisting of a long, coiled tube

renal veins transport blood from the kidneys back into general circulation

renin enzyme released by the kidney; converts angiotensinogen to angiotensin I

replication duplication of DNA prior to cell division

repolarization return of the membrane potential of a cell to the resting potential following a depolarization

reproductive tract system of passageways through which gametes are transported

residual volume (RV) the volume of air remaining in the lungs after a maximum expiration

resistance hindrance to movement

resorption the breakdown of bone tissue

respiration the process of gas exchange within the body; includes internal respiration and external respiration

respiration rate frequency of breaths

respiratory acidosis decrease in blood pH caused by increases in plasma carbon dioxide levels

respiratory alkalosis increase in blood pH caused by decreases in plasma carbon dioxide levels

respiratory bronchioles small tubules of the respiratory tract located between terminal bronchioles and alveolar ducts

respiratory membrane the structure across which gas exchange occurs in the lungs; a barrier between blood and air consisting of capillary endothelial cells and their basement membranes and alveolar epithelial cells and their basement membranes

respiratory quotient the ratio of carbon dioxide produced by the body to the amount of oxygen consumed

respiratory tract air passages leading from pharynx to lungs

respiratory zone the lower part of the respiratory tract; the site of gas exchange within the lungs

resting membrane potential the voltage that exists across a cell membrane when the cell is not transmitting electrical signals; polarity is such that the inside of the cell is negative with respect to the outside

reticular formation a diffuse network of nuclei located in the brainstem that is important in sleep-wake cycles, arousal of the cerebral cortex, and consciousness

retina innermost layer of the eye; consists of neural tissue and contains photoreceptors

reuptake active transport of neurotransmitter back into the presynaptic neuron that released it

ribonucleic acid (RNA) polynucleotide molecules found in a cell's nucleus and cytoplasm that are necessary in the expression of genetic information; includes mRNA, tRNA, and rRNA

ribosomes complexes of rRNA and proteins that function in protein synthesis; found in the cytosol and rough endoplasmic reticulum

rods photoreceptors that enable visibility during relatively low light and are responsible for black-and-white vision

round window membrane between the middle and inner ear; dissipates sound waves in the inner ear

saccule (sak'-yool) structure of the inner ear; detects up and down linear acceleration

sacral nerves spinal nerves originating from the sacral spinal cord

salivary glands glands that secrete saliva, located in the mouth and pharynx

saltatory conduction type of action potential conduction that occurs in myelinated axons; action potentials "jump" from node to node

sarcolemma (sar-co-lem'-ah) a muscle fiber's plasma membrane

sarcomeres (sar'-kuh-meers) the fundamental repeating units that make up myofibrils

sarcoplasmic reticulum (SR) modified smooth endoplasmic reticulum in muscle cells that surrounds the myofibrils and stores calcium

scala media fluid-filled duct in the cochlea; also called cochlear duct

scala tympani fluid-filled duct in the cochlea; also called tympanic duct

scala vestibuli fluid-filled duct in the cochlea; also called vestibular duct

Schwann cells glial cells that form myelin around axons in the peripheral nervous system; one Schwann cell provides a segment of myelin for one axon

sclera (sklee'-rah) tough connective tissue that makes up the white of the eye

scrotum a sac that houses the testes

second messenger an intracellular messenger molecule that is produced in response to the binding of an extracellular messenger (the first messenger) to a receptor

secondary active transport active transport of molecules utilizing a carrier that uses a concentration gradient or electrochemical gradient as a source of energy

secondary bronchi branches off the bronchi leading to the lungs

secondary endocrine organs organs whose secretion of hormones is secondary to another function

secondary immune response antigen-induced lymphocyte response caused by subsequent exposure to an antigen

secrete release from a cell, gland, or epithelium

secretin a hormone secreted by the duodenum and jejunum that inhibits gastric secretion and motility and stimulates pancreatic bicarbonate secretion

secretion movement of substance from the internal environment to the external environment by transport across an epithelium; movement of substance from inside a cell to outside the cell by movement across the plasma membrane

secretory phase phase of the uterine cycle during which the uterine endometrium prepares for the possible arrival and subsequent implantation of the fertilized egg

secretory vesicles intracellular vesicles containing molecules destined for secretion from the cell

segmentation mixing motility pattern of the small intestine

semicircular canals structures in the inner ear that contain the receptor cells for rotational acceleration

semilunar valves valves located between the ventricles and arteries on either side of the heart and that prevent blood from flowing back into the ventricles when the ventricles are relaxed

seminal vesicles accessory glands of the male reproductive system; the first to secrete fluid into the reproductive tract during sperm transport

seminiferous tubules (sem-ih-nif'-er-ous) thin, highly coiled hollow tubes in the testes where sperm are produced

sensitization increased responsiveness to repetitive stimuli; a form of learning

sensory homunculus map that indicates which areas of the primary somatosensory cortex are devoted to particular regions of the body

sensory receptors specialized neuronal structures that detect a specific form of energy in either the internal or external environment

sensory unit a single afferent neuron and all sensory receptors associated with it

series elastic component connective tissue that is in series with the sarcomeres of skeletal muscle

serosa the outermost of the four major layers of the wall of the digestive tract; composed mostly of connective tissue

Sertoli cells epithelial cells lining the wall of a seminiferous tubule whose primary function is to support sperm development

serum plasma with coagulation factors removed

set point the normal or desired value of the regulated variable in a homeostatic regulatory system

sex chromosomes X and Y chromosomes that determine a person's sex

sex determination the role of genes in deciding a person's sex

sex differentiation the role of testosterone and Müllerian inhibiting substance in determining sexual characteristics of a person

sex hormones steroid hormones including estrogens, progesterone, and androgens secreted from the adrenal cortex and gonads that regulate reproductive function, promote

gametogenesis, growth and maintenance of reproductive organs, and the development of secondary sex characteristics

signal transduction the process by which the binding of a chemical messenger to receptors brings about a response in a target cell

simple diffusion passive movement of molecules in response to chemical or electrical forces

single-unit smooth muscle smooth muscle with gap junctions linking the cells together so they function as a unit

sinoatrial (SA) node (sy-noh-ay'-tree-al) a region in the wall of the upper right atrium where pacemaker cells are concentrated; normally determines the heart rate

size principle the correspondence between the size of motor units and the order of recruitment

skeletal muscle the type of muscle that is generally connected to bone

sliding filament mechanism process of muscle contraction whereby thick and thin filaments slide past each other

slow oxidative fibers muscle cells with slow ATPase activity and high oxidative capacity

slow pain a poorly localized, dull, aching sensation produced by activation of nociceptors; is transmitted by C fibers

slow waves cyclical fluctuations in the resting membrane potential

slowly adapting receptors receptors that maintain responsiveness to a continuing stimulus; also called tonic receptors

slow-wave sleep (SWS) sleep characterized by multiple stages of low frequency waves in the EEG

small intestine an organ of the GI tract, consisting of a coiled tube 8–10 feet long; the primary site for the digestion and absorption of all the nutrients in food

smooth muscle the type of muscle found in internal organs, blood vessels, and other structures that are not under voluntary control; lacks striations

sodium-potassium pump a protein that utilizes ATP to actively transport sodium ions out of the cell and potassium ions into the cell against their electrochemical gradients

somatic nervous system the division of the nervous system that encompasses nerve cells that regulate skeletal muscle contractions

somatomedin (soh-mat-oh-me'-din) peptide hormone secreted by the liver in response to growth hormone; promotes protein synthesis and growth; also known as insulin-like growth factor

somatosensory system (suh-mat-uh-sen'-suh-ree) branch of the nervous system associated with perception of somatic sensations; associated with receptors in the skin and proprioception

somatostatin tropic hormone released by the hypothalamus that inhibits growth hormone secretion from the anterior pituitary; also known as growth hormone inhibiting hormone (GHIH)

somesthetic sensations sensations that arise from receptors in the skin

spatial summation the addition of graded potentials generated at different locations that occurs when they are stimulated more or less simultaneously

special senses senses of vision, olfaction, taste, hearing, and equilibrium

specific immune response mechanism that targets and eliminates specific substances

spermatogenesis (sper-mah-toh-jen'-ih-sis) sperm production

spermatogonia undifferentiated germ cells that are precursors for sperm

spermatozoa (spur-ma-toh-zoh'-ah) newly formed sperm cells with characteristic head, midpiece, and tail

sphincter a ring of muscle that surrounds an orifice and regulates the passage of material through it by altering its diameter

sphincter of Oddi a ring of smooth muscle that regulates the flow of bile and pancreatic juice into the duodenum

spinal cord portion of the central nervous system that travels in the vertebral column

spinal nerves a total of 31 pairs of nerves originating in the spinal cord and traveling to the periphery

spinothalamic tract somatosensory pathway that transmits information from thermoreceptors and nociceptors to the thalamus

spirometer device for measuring lung volumes

srY gene gene located on the Y chromosome that codes for testis-determining factor

starch polysaccharide found in plants

Starling forces hydrostatic and osmotic pressures that exist across capillary walls and determine the direction of fluid movement

Starling's law of the heart law stating that when there is a change in the rate at which blood flows into the heart from the veins, the heart automatically adjusts its output to match the inflow

stereocilia (ster-ee-oh-sil'-ee-ah) hairlike projections on the upper surfaces of hair cells in the inner ear that move in response to sound vibrations or acceleration of the head

steroids lipids derived from cholesterol consisting of three 6-carbon rings and one 5-carbon ring, many of which function as hormones

stomach an organ of the GI tract that stores food and releases it into the small intestine; major anatomical regions are the fundus, body, and antrum

striated muscle (stry'-ay-ted) muscle in which the cells have a striped appearance due to the presence of sarcomeres; includes skeletal and cardiac muscle

stroke volume (SV) the volume of blood ejected from each ventricle during a single heartbeat

subarachnoid space the space between the pia mater and arachnoid mater, filled with cerebrospinal fluid

submucosa one of the four major layers of the wall of the GI tract, located between the mucosa and muscularis externa; composed mostly of connective tissue

submucosal plexus one of the two neural networks that make up the enteric nervous system; located in the submucosa

substance P hormone that decreases gastrointestinal motility; functions as a neurotransmitter in some neurons

substrate a reactant in an enzyme-catalyzed reaction; binds to the active site of an enzyme molecule

substrate-level phosphorylation a mechanism of ATP synthesis, in which an enzyme transfers a phosphate group from a substrate to ADP; occurs in steps 7 and 10 of glycolysis and step 5 of the Krebs cycle

subthreshold in nervous system, a stimulus too weak to generate an action potential

sulci (sul'-sigh) grooves in the highly convoluted gray matter of the cerebral cortex; singular, sulcus

summation in neurophysiology, the adding together of graded potentials that occurs within a neuron; in muscle physiology, the adding together of twitches that occurs when a muscle is stimulated at high frequency

suprathreshold in nervous system, a stimulus that exceeds the threshold to generate an action potential

sympathetic chains structures parallel to the spinal column on either side in which the sympathetic ganglia are linked together in rows; also called sympathetic trunks

synaptic cleft the extracellular space between the axon terminal of the presynaptic cell and the postsynaptic cell at a synapse

synaptic delay time between the arrival of an action potential at the axon terminal of the presynaptic cell and when a response occurs in the postsynaptic cell

synaptic vesicles vesicles in the axon terminal that contain neurotransmitter molecules

synergistic pertaining to a process in which the net effect is greater than the sum of the individual effects

systemic circuit the portion of the vasculature that encompasses all of the body's blood vessels, except those belonging to the pulmonary circuit

systole (sis'-toh-lee) the period of ventricular contraction during a cardiac cycle

systolic pressure (SP) the maximum aortic pressure attained during the cardiac cycle; occurs during systole

T cell receptor (TCR) receptors that detect foreign antigens on body cells

T cells leukocytes that kill particular abnormal or infected body cells; also called T lymphocytes

T lymphocytes cells that kill particular abnormal or infected body cells; also called T cells

T wave in ECG recordings, the waveform representing ventricular depolarization

T₃ one of the thyroid hormones

T₄ one of the thyroid hormones

tastants chemical substances that give foods their flavors

taste buds structures on the tongue, roof of the mouth, and pharynx that contain taste receptor cells

tectorial membrane (tek-tor'-ee-al) membrane in the organ of Corti in which the tips of stereocilia are embedded

temporal lobe one of four lobes of the cerebrum located at the side and separated from the frontal lobe by a deep groove; important in processing auditory information and language

temporal summation the addition of graded potentials generated at a particular site that occurs when stimulated at a high frequency

tendons cords of elastic connective tissue that transmit force from skeletal muscles to bones

terminal bouton the axon terminal of a motor neuron, which stores and releases acetylcholine

terminal bronchioles bronchioles that lead directly to the airways of the respiratory zone of the respiratory tract; the last component of the conducting zone

testes male gonads

tetanus (tet'-ah-nus) in a muscle being stimulated at high frequency, the plateau phase of the contraction, during which the tension is relatively constant

theca cells (thee'-ca) cells that surround the granulosa cells in a follicle

thermoreceptors sensory receptors that detect temperature; include peripheral thermoreceptors that detect skin temperature and central thermoreceptors that detect core body temperature

thermoregulation utilization of feedback mechanisms to maintain normal body temperature

thick filaments filaments composed of myosin that form part of the contractile machinery of a muscle cell

thin filaments filaments composed of actin that form part of the contractile machinery of a muscle cell; also contain troponin and tropomyosin in striated muscle cells

thoracic nerves spinal nerves originating from the thoracic spinal cord

threshold in an excitable cell, the critical value of the membrane potential to which the cell must be depolarized in order to trigger an action potential

thrombin one of the active clotting factors in the coagulation cascade; reacts with fibrinogen to convert it to fibrin and activates factor XIII

thromboxane A_2 (TXA$_2$) eicosanoid released from aggregated platelets that promotes hemostasis

thrombus blood clot

thymus primary endocrine gland located near the heart; secretes the hormone thymosin; also is site of T lymphocyte maturation

thyroid gland butterfly-shaped primary endocrine gland located on the ventral surface of the trachea; secretes tetraiodothyronine (T_4), triiodothyronine (T_3), and calcitonin

thyrotropin releasing hormone (TRH) tropic hormone secreted from the hypothalamus that stimulates secretion of TSH from the anterior pituitary

tidal volume (V_T) the volume of air that moves into and out of the lungs during a normal, unforced breath

tight junction a junction that connects two adjacent cells together, forming a nearly impenetrable barrier that limits the passage of molecules between them

tissue group of similar cells that carry out a specific function

tissue factor a protein found in subendothelial tissues that initiates the extrinsic clotting pathway when exposed to blood

tissue factor plasma inhibitor substance secreted from healthy endothelial cells to inhibit the extrinsic coagulation pathway

tonic receptors receptors that maintain responsiveness to a continuing stimulus; also called slowly adapting receptors

total body water (TBW) the volume of water that is contained in all the body's compartments

total lung capacity (TLC) the volume of air in the lungs at the end of a maximum inspiration

total peripheral resistance (TPR) in the systemic circuit, the combined resistance of all the blood vessels

trachea (tray'-key-ah) cartilaginous tube of the respiratory tract, located between the larynx and the bronchi

transcytosis the transport of macromolecules across epithelial cells; involves endocytosis at one membrane, followed by exocytosis at the opposite membrane

transduction conversion of the energy form of a stimulus into an electrical signal in the form of changes in membrane potential

transmembrane proteins integral membrane proteins that span the lipid bilayer, with surfaces exposed to both the cytosol and interstitial fluid

transport maximum (Tm) rate of transport by carrier proteins when carriers are 100% saturated

transpulmonary pressure the difference between the intrapleural pressure and the intra-alveolar pressure, which represents the distending pressure acting on the lungs; $P_{alv} - P_{ip}$

transverse tubules (T tubules) structures that transmit action potentials from the sarcolemma into the cell's interior, triggering the release of calcium from the sarcoplasmic reticulum

treppe increase in muscle tension with repeated twitch contractions

tricuspid valve the AV valve on the right side of the heart, which has three cusps

triglyceride a lipid consisting of three fatty acids linked to a glycerol backbone; commonly called fat

triplet three-base sequence in DNA that codes for a specific amino acid

tropic hormones hormones that regulate the secretion of hormone; also called trophic hormones

tropomyosin (troh-poh-my'-oh-sin) one of the two regulatory proteins in striated muscle; a long fibrous molecule that acts to block myosin-binding sites on thin filaments when a muscle is not contracting

troponin (troh-poh'-nin) one of the two regulatory proteins in striated muscle; binds calcium reversibly and is responsible for starting the crossbridge cycle by moving tropomyosin out of its blocking position

tubuloglomerular feedback autoregulatory mechanism in which a change in glomerular filtration rate is regulated by paracrines secreted from the macula densa, located downstream from the glomerulus

turbulent flow a type of blood flow in which the blood moves in different directions in different places within a vessel; increases the resistance to blood flow

twitch the mechanical response of a muscle to a single action potential

tympanic membrane eardrum

type I cells epithelial cells lining alveoli

type II alveolar cells cells that line alveoli and secrete surfactant

umbilical cord a rope-like structure that extends from the placenta to the fetus; contains vessels that provide blood to the fetus

upper airways air passages in the head and neck; include the nasal cavity, oral cavity, and pharynx

upper esophageal sphincter ring of skeletal muscle that controls movement of food from pharynx to esophagus

upper motor neurons neurons of the pyramidal tract that originate in the primary motor cortex and terminate in the spinal cord ventral horn

urinary system organ system that consists of two kidneys, two ureters, the urinary bladder, and the urethra

urine a fluid produced by the kidneys and eliminated from the body

uterine cycle cyclic changes in uterine structure and function that occur during the menstrual cycle

uterine tubes tubes protruding from the upper part of the uterus on either side that function in egg transport and fertilization; also called fallopian tubes or oviducts

uterus a hollow pear-shaped organ located in the center of the pelvic cavity that functions to house and nourish the developing human

utricle structure of the inner ear; detects forward and backward linear acceleration

vagina a canal leading from the cervix to the outside of the body, receives the penis during intercourse; the female organ of copulation

vagus nerve (vay'-gus) major parasympathetic nerve that originates in the medulla oblongata and innervates much of the viscera; cranial nerve X (CN X)

varicosities axon swellings of autonomic postganglionic neurons; store and release neurotransmitter

vas deferens (vas def'-er-enz) a duct leading from each testis to the ejaculatory duct; carries sperm and fluid

vasa recta capillary bed that branches off the efferent arteriole of a juxtamedullary nephron and surrounds the loop of Henle; functions in maintaining medullary osmotic gradient

vasculature all the blood vessels in the body

vasoconstriction decrease in the radius of a blood vessel

vasodilation increase in the radius of a blood vessel

vasopressin hormone secreted by the posterior pituitary that regulates urine output by the kidney; stimulates water reabsorption by the kidneys; functions as a neurotransmitter in some neurons; also known as antidiuretic hormone (ADH)

vaults barrel-shaped cell organelles that may be involved in the transport of molecules across the nuclear envelope and may have other functions

veins large blood vessels that carry blood toward the heart

vena cava (vee'-nah kay'-vah) one of the two large veins that carry blood into the right atrium; plural, venae cavae

venomotor tone the degree of tension exerted by smooth muscle in the walls of veins; venous compliance decreases as venomotor tone increases

venous pooling the accumulation of blood in the veins

venous return blood flow into the heart

ventilation-perfusion ratio relationship of ventilation to perfusion in alveoli, or $\dot{V}_A/\dot{Q}$

ventral horn anterior half of the gray matter on either side of the spinal cord

ventral respiratory group (VRG) respiratory control center in the medulla; contains inspiratory and expiratory neurons

ventral root point at which a spinal nerve bifurcates before joining the spinal cord, the ventralmost of the two resulting branches; contains efferent nerve fibers

ventricles in the heart, the two lower chambers, which pump blood into the arteries; in the brain, chambers that contain cerebrospinal fluid

ventricular ejection the exit of blood from the ventricles

venules blood vessels that carry blood from capillaries to veins

vertebral column bony structure that surrounds and protects the spinal cord

very low density lipoproteins (VLDLs) one of a group of lipoprotein particle types consisting of lipids and proteins in various ratios; VLDLs contain a high ratio of lipids to proteins, thus their low density

vestibular apparatus structures of the inner ear that contain the receptor cells for equilibrium, including the semicircular canals, utricle, and saccule

vestibular membrane membrane in the cochlea of the inner ear that separates the scala vestibuli from the scala media

vestibulocochlear nerve nerve that contains the afferents for hearing and equilibrium; cranial nerve VIII

villi folds in the mucosal surface of the small intestine that facilitate the transport of materials by increasing the surface area of the epithelium; singular, villus

visceral reflexes automatic changes in the functions of organs that occur in response to changing conditions inside the body

vital capacity (VC) the maximum volume of air that can be expired following a maximum inspiration

vitamins organic molecules that are required in trace amounts in the body; often act as coenzymes

vitreous chamber cavity of the eye posterior to the lens and ciliary muscle

vitreous humor jelly-like material found in the vitreous chamber of the eye; maintains the spherical structure of the eye

voltage-gated channels channels that open or close in response to a change in membrane potential

von Willebrand factor (vWF) protein that anchors platelets to collagen fibers during the platelet adhesion phase of hemostasis

vulva external genitalia of the female; includes the mons pubis, labia majora, labia minora, the vestibule, and vestibular glands

warm receptors thermoreceptors that increase responsiveness with an increase in temperature

Wernicke's area area of association cortex devoted to language comprehension; located in the posterior and superior portion of the temporal lobe and the inferior parietal lobe

white matter areas of the CNS consisting primarily of myelinated axons; specialized for the rapid transmission of information over relatively long distances in the form of action potentials

withdrawal reflex reflex removal of a limb from a noxious stimulus

Wolffian ducts rudimentary precursor to the male reproductive system

zona pellucida a thick layer of noncellular material that forms between an oocyte and surrounding granulosa cells

zonular fibers strands of connective tissue in the eye that connect the ciliary muscles to the lens and are involved in adjusting the shape of the lens to focus light

zygote (zy'-goat) fertilized ovum before cleavage begins

zymogens inactive precursor forms of pancreatic digestive enzymes that are stored in secretory cells and released by exocytosis

Index

Dear Student:

As the authors of *Principles of Human Physiology,* **Second Edition** we wanted to make the complex topics in physiology easy to understand and integrate. We would appreciate hearing about your experiences with this textbook and its multimedia support, and we invite your suggestions for improvements!

Many thanks,

William J Germann   *Cindy Lou Stanfield*

William J. Germann   Cindy Lou Stanfield

1. What did you like most about *Principles of Human Physiology,* **Second Edition**? Please provide examples in order of priority. _____

_____

_____

_____

_____

2. Which study tool(s)—e.g. Before You Begin, Quick Tests, Chapter Summaries, End-of-Chapter Exercises — did you find most useful? _____

_____

_____

_____

_____

3. Do you have any ideas about how this book could be improved? Please write your specific suggestions, citing page numbers if appropriate. _____

_____

_____

_____

_____

4. Did you use the **InterActive Physiology**® **CD-ROM** that was included with the book?
☐ Yes ☐ No
If so, which topic(s) did you find most useful? _____

_____

_____

_____

5. Are there topics that should be added to **InterActive Physiology**®? How can we improve **InterActive Physiology**®? _____

_____

_____

_____

continued on back ➡

6. Did you use **The Physiology Place** web site? ☐ Yes ☐ No

If so, what did you like or use most? Was it easy to use, or did you experience problems? How can we improve **The Physiology Place**? ____

____

____

____

NO POSTAGE
NECESSARY
IF MAILED
IN THE
UNITED STATES

BUSINESS REPLY MAIL

FIRST-CLASS MAIL PERMIT NO. 275 SAN FRANCISCO CA

POSTAGE WILL BE PAID BY ADDRESSEE

BENJAMIN CUMMINGS
PEARSON EDUCATION
1301 SANSOME STREET
SAN FRANCISCO CA 94111-9328

School: ______________________

Instructor's Name: ______________________

Optional

Your name: ______________________

Email: ______________________

Date: ______________________

May Benjamin Cummings have permission to quote your comments in promotions for ***Principles of Human Physiology,*** **Second Edition?** ☐ Yes ☐ No

Frequently Used Abbreviations and Symbols

ACh acetylcholine

ACTH adrenocorticotropic hormone

ADH antidiuretic hormone

ADP adenosine diphosphate

ATP adenosine triphosphate

AV atrioventricular

CA carbonic anhydrase

cAMP cyclic AMP

CCK cholecystokinin

cGMP cyclic GMP

CNS central nervous system

CO cardiac output

CRH corticotropin releasing hormone

CSF cerebrospinal fluid

CVP central venous pressure

C_X clearance of X

DAG diacylglycerol

DNA deoxyribonucleic acid

ECF extracellular fluid

EDV end-diastolic volume

ECG electrocardiogram

EPSP excitatory postsynaptic potential

ESV end-systolic volume

E_X equilibrium potential of X

FAD flavin adenine dinucleotide

FRC functional residual capacity

FSH follicle stimulating hormone

GDP guanosine diphosphate

GFR glomerular filtration rate

GH growth hormone

GnRH gonadotropin releasing hormone

GTP guanosine triphosphate

HR heart rate

ICF intracellular fluid

IP_3 inositol triphosphate

IPSP inhibitory postsynaptic potential

ISF interstitial fluid

LH luteinizing hormone

MAP mean arterial pressure

MHC major histocompatibility complex

mm Hg millimeters of mercury

mV millivolts

NAD^+ nicotinamide adenine dinucleotide

π osmotic pressure

P_{alv} intra-alveolar pressure

P_{atm} atmospheric pressure

P_{ip} intrapleural pressure

PIP_2 phosphatidylinositol-4,5-biphosphate

PSP postsynaptic potential

RNA ribonucleic acid

SA sinoatrial

SR sarcoplasmic reticulum

SV stroke volume

TRH thyrotropin releasing hormone

TSH thyroid stimulating hormone

TPR total peripheral resistance

V_m membrane potential

[X] concentration of X

The Metric System

Prefixes

Prefix (abbreviation)	Standard Notation	Scientific Notation
mega- (M)	multiply by 1,000,000	multiply by 1×10^{6}
kilo- (k)	multiply by 1000	multiply by 1×10^{3}
deci- (d)	multiply by 0.1	multiply by 1×10^{-1}
centi- (c)	multiply by 0.01	multiply by 1×10^{-2}
milli- (m)	multiply by 0.001	multiply by 1×10^{-3}
micro- (μ)	multiply by 0.000001	multiply by 1×10^{-6}
nano- (n)	multiply by 0.000000001	multiply by 1×10^{-9}
pico- (p)	multiply by 0.000000000001	multiply by 1×10^{-12}

Conversion Factors

Metric Unit (abbreviation)	English Unit	Conversion: Metric to English	Conversion: English to Metric
Lengths			
millimeters (mm)	inches	mm × 0.03937 = inches	inches × 25.4 = mm
centimeters (cm)	inches	cm × 0.3937 = inches	inches × 2.54 = cm
meters (m)	feet	m × 3.2808 = feet	feet × 0.3048 = m
meters (m)	yards	m × 1.0936 = yards	yards × 0.9144 = m
Weights			
grams (g)	ounces	g × 0.0353 = ounces	ounces × 28.3495 = g
grams (g)	pounds	g × 0.0022 = pounds	pounds × 453.6 = g
kilograms (kg)	pounds	kg × 2.2046 = pounds	pounds × 0.4536 = kg
Volumes			
milliliters (mL)	ounces	mL × 0.0338 = ounces	ounces × 29.574 = mL
milliliters (mL)	quarts	mL × 0.00106 = quarts	quarts × 946.3 = mL
liters (L)	gallons	L × 0.2643 = gallons	gallons × 3.784 = L
Temperature			
Celsius (°C)	Fahrenheit (°F)	(°C × 9/5) + 32 = °F	(°F − 32) × 5/9 = °C